Simple Views on

Condensed Matter

THIRD EDITION

SERIES IN MODERN CONDENSED MATTER PHYSICS

Editors-in-charge: I. Dzyaloshinski and Yu Lu

Published

Vol. 1: Exotic Properties of Superfluid Helium 3
by G. E. Volovik

Vol. 2: Quantum Dissipative Systems
by U. Weiss

Vol. 4: Simple Views on Condensed Matter
by P.-G. de Gennes

Vol. 5: Lectures on Electron Correlation & Magnetism
by P. Fazekas

Vol. 6: Low-Dimensional Quantum Field Theories for Condensed Matter Physicists
eds. L. Yu, S. Lundqvist and G. Morandi

Vol. 7: The Physics of Quantum Well Infrared Photodetectors
by K. K. Choi

Vol. 8: Simple Views on Condensed Matter (2nd Edition)
by P.-G. de Gennes

Vol. 9: Theory of Critical Phenomena in Finite-Size System: Scaling and Quantum Effects
by J. G. Brankov, D. M. Danchev and N. T. Tonchev

Vol. 10: Quantum Dissipative Systems (2nd Edition)
by U. Weiss

Vol. 11: Magnetism in Heavy Fermions
by H. B. Radousky

Forthcoming

Vol. 3: Modulated Structures: Commensurate, Incommensurate, Stochastic Structures and Quasicrystals
by V. Pokrovsky and G. Uimin

Series in Modern Condensed Matter Physics – Vol. 12

Simple Views on

Condensed Matter

THIRD EDITION

Pierre-Gilles de Gennes
Collège de France

World Scientific
New Jersey • London • Singapore • Hong Kong

Published by

World Scientific Publishing Co. Pte. Ltd.
5 Toh Tuck Link, Singapore 596224
USA office: Suite 202, 1060 Main Street, River Edge, NJ 07661
UK office: 57 Shelton Street, Covent Garden, London WC2H 9HE

The author and publisher would like to thank the following publishers of the various journals and books for their assistance and permission to include the selected reprints found in this volume:

Academic des Sciences, Institut de France (*C. R. Acad. Sci. (Paris)*); American Chemical Society (*J. Phys. Chem., Macromolecules, Langmuir*); American Institute of Physics (*J. Chem. Phys.*); American Physical Society (*Phys. Rev., Rev. Mod. Phys.*); Cambridge University Press (*Introduction to Polymer Dynamics*); CNRS Publications (*De la Physique Theorique à la Biologie*); EDP Sciences (*The Eur. Phys. J.*); Elsevier Science Publishers (*Advances in Colloid and Interface Science, Phys. Lett., Physica A, Colloids and Surfaces A*); IDSET, Paris (*Symmetries and Broken Symmetries in Condensed Matter Physics*); Indian Academy of Sciences (*Pramana*); Institute of Physics (*J. Phys.: Condens. Matter*); IUPAC (*Proc. 4th Int. Symp. Molecular Order and Mobility in Polymer Systems*); Les Editions de Physique (*Journal de Physique, J. Phys. Rad.*); National Academy of Sciences (*PNAS*); Pergamon Press (*Solid State Comm., Chem. Eng. Sci.*); Physics Publishing Co. (*Physics*); Royal Society of Chemistry (*Faraday Discussions of the Royal Society of Chemisty*); Societé Française de Chimie (*J. Chim. Phys.*); Società Italiana di Fisica (*Rivista del Nuovo Cimento*); Springer (*Physics of Amphiphilic Layers*)

British Library Cataloguing-in-Publication Data
A catalogue record for this book is available from the British Library.

SIMPLE VIEWS ON CONDENSED MATTER
THIRD EDITION

ISBN 981-238-278-X
ISBN 981-238-282-8 (pbk)

Printed in Singapore by World Scientific Printers (S) Pte Ltd

Foreword to the First Edition

Books of "collected papers" are frequent in our days. Are they useful? I am hesitant about this. Looking at my own articles, I found a number of ideas which ultimately lead to nowhere... . Also a certain fraction was wrong.

However, on closer scrutiny, I listed two lots which could still, possibly, be useful. The articles of the first lot are relatively general, but simple, and can be used as an elementary starting point for a beginner. The second lot is formed by papers which never really reached an international readership, because they appeared in relatively obscure journals. I was tempted to give them a second chance — but finally we took only a few (because of a language problem).

It rapidly became obvious that the articles in the list — being rather ancient — would give a distorted view of their own subfield, since science moves fast. It was necessary then to add some comments, or "afterthoughts". These afterthoughts are highly subjective! They sometimes focus on one later discovery, by younger colleagues, based on the same general ideas. Some other comments concern a flaw in my original views.

On the whole, I found it amusing to observe how short-sighted I have been — unable to imagine certain natural consequences of one idea. Some of these surprises are also described in the "afterthoughts". On the other hand, I *never* give a detailed list of references to more recent work. Only a few key references show up: I owe a lot here to Y. Heffer and M. Monnerie, who traced them. I am also thankful to the various publishers who allowed us to reprint the papers — with one exception (Gordon and Breach), where the permission fees for one article were at such a high cost, that we were forced to take this paper out of our list.

The whole construction is fragile. But I still hope that it can give, to a young reader, a certain feeling about the enthusiasm (and novelty) of our condensed matter research during these post war decades... .

Pierre Gilles de G.
Paris, March 1992

Foreword to the Expanded Edition

The first version of this book came out 6 years ago; it was apparently useful for a (small) community of physical chemists. Thus, the temptation of adding new material came naturally:

a) a few general papers (e.g. on hysteresis in wetting) had been stupidly omitted from the first set,

b) new topics emerged in polymer science: the slippage of melts against solid walls; the behavior of branched chains in nanopores, ...

c) three years ago we started a (naive but enthusiastic) reflection on the mechanics of granular materials largely based on ideas by J. P. Bouchaud, M. Cates and others. This field is still opaque: even the most clumsy attempts, like ours, may be slightly helpful.

The truth is that, during the last 6 years, I spent a significant amount of time trying to explain condensed matter physics to children in high schools: a very different exercise, which crystallised into another book and a CD ROM. But I am equally fond of all levels of communication. Explaining the motions of a gyroscope without equations, or explaining the late stages of an avalanche with many equations, these exercises are equally rewarding. Let me wish that this heavier form of "Simple Views" will not look like a stone falling in a sleepy pond ...

Pierre Gilles de G.
Paris, June 1998

Contents

Part III. Polymers

Part IV. Interfaces

Part V. Wetting and Adhesion

Part VI. Chirality

Part VII. Granular Matter

Part I. Solid State

LE JOURNAL DE PHYSIQUE ET LE RADIUM TOME 20, JUIN 1959, PAGE 624.

SUR UN EXEMPLE DE PROPAGATION DANS UN MILIEU DÉSORDONNÉ

Par P. G. DE GENNES, P. LAFORE et J. P. MILLOT,

Centres d'Études Nucléaires de Saclay et de Fontenay-aux-Roses.

Résumé. — On envisage une solution solide où des atomes « actifs » A sont substitués, au hasard, aux nœuds d'une matrice inactive B. On étudie un phénomène de propagation qui fait seulement intervenir les atomes A. Ce cas se rencontre notamment pour les bandes d'impuretés des semi-conducteurs, et pour les ondes de spin d'un alliage à un constituant ferromagnétique.

Nous montrons ici comment certaines propriétés géométriques de la répartition des diffuseurs, qui ont été étudiées précédemment, jouent un rôle essentiel sur : *a*) le spectre de valeurs propres de l'équation d'ondes ; *b*) les phénomènes de transport.

Abstract. — This paper deals with propagation phenomena in binary solid solutions AB, the waves being carried by the A atoms only. Such a situation is met in impurity bands en semi-conductors, and also for spin waves in alloys with one ferromagnetic component. We here investigate the influence of accidental clusters formed by A atoms on the eigenvalue spectrum and on the transport properties.

I. Introduction. — Soit une solution solide AB, où des atomes « actifs » A sont substitués au hasard, aux nœuds (i) du réseau périodique de la matrice B. Deux exemples nous serviront à préciser ce que nous entendons par atomes « actifs ».

1er *exemple :* Les atomes A sont porteurs de spins S, et couplés ferromagnétiquement entre eux par une énergie d'échange. La matrice B est, elle, magnétiquement inactive. La question de savoir dans quelles conditions un tel alliage admet un point de Curie bien défini a été soulevée par J. Friedel. On peut y apporter une réponse au moins partielle, en étudiant, aux basses températures, la stabilité de l'ordre ferromagnétique à l'égard des premières excitations de spin. Il faut donc déterminer les niveaux d'onde de spin de l'alliage désordonné. Ce sont les valeurs propres E de l'équation de propagation :

$$Ea_i = 2 \sum_j' a_j J_{ij} \tag{1}$$

où a représente l'amplitude de la déviation du i^e spin, et où la somme Σ_j' est limitée aux sites qui sont également occupés par des atomes A.

2e *exemple :* Les atomes A forment une « bande d'impuretés » dans un semi-conducteur. A chaque atome A, situé au site (i), est associé une orbitale $\varphi_i(R)$, et on cherche, dans l'approximation des liaisons fortes, des fonctions d'onde à un électron de la forme :

$$\psi(\boldsymbol{R}) = \sum_i a_i \varphi_i(\boldsymbol{R}). \tag{2}$$

Si les φ_i relatives à des sites différents sont orthogonales, on a encore pour les a_i une équation du type (1).

Dans la suite, nous nous limiterons au cas où seuls les premiers voisins sont couplés entre eux : cette simplification correspond mal aux situations physiques usuelles, mais elle sera utile pour mettre en relief les résultats essentiels (ceux-ci ne sont pas modifiés qualitativement tant que la portée des couplages reste finie). Par un choix convenable de l'échelle des énergies, (1) peut alors être écrite :

$$Ea_i = \sum_j'' a_j \tag{3}$$

où la somme Σ_j'' est limitée aux premiers voisins, occupés par des atomes A, de l'atome A. C'est à cette forme réduite que nous nous intéresserons dans la suite.

Lorsque tous les nœuds du réseau sont occupés par des atomes A, les solutions de (3) sont des ondes planes $a_i = e^{ik.R_i}$ (où R_i est la coordonnée du Ier Site) et les niveaux correspondants sont :

$$E(\boldsymbol{k}) = \sum_j^{(1)} e^{i\mathbf{k}.(\mathbf{R}_j - \mathbf{R}_i)}. \tag{4}$$

Dans (4) la somme $\Sigma^{(1)}$ est étendue à tous les premiers voisins d'un site quelconque (i) [1]. Le fait essentiel est que ces niveaux forment un spectre *continu*. Introduisons maintenant des atomes B. Les premiers termes du développement de la densité de niveaux en série de puissances de la concentration en atomes B, sont convergents (section II) ; ils représentent une simple déformation du spectre continu.

Au contraire, aux faibles concentrations en atomes A, ceux-ci sont pour la plupart isolés. Il leur est associé un seul niveaux $E = 0$ fortement dégénéré. Quelques atomes A forment des paires et donnent deux niveaux $E = \pm 1$. Quelques-uns

[1] Nous nous limitons à des réseaux de Bravais.

encore forment des triplets et donnent trois niveaux, etc... La série ainsi engendrée converge rapidement dans ce domaine (voir section III) et le spectral global est *discret*.

Que se passe-t-il dans le domaine intermédiaire ? Chaque atome A appartient à un « amas » de n atomes A. L'équation de propagation ne couple que les atomes d'un même amas. Le rang n de l'amas peut être fini ou infini. Les amas infinis jouent un rôle à part : ils contribuent seuls au spectre continu et aux phénomènes de transport sur des distances macroscopiques. Il est donc essentiel de déterminer le pourcentage d'atomes A engagés dans des amas infinis : deux méthodes d'approximation sont appliquées à ce problème dans la section III. Enfin, dans la section IV, on indique comment appliquer ces résultats à l'étude du spectre ou de la conductivité aux hautes températures, grâce à une méthode de moments.

II. Fortes concentrations en atomes actifs. — Soit N le nombre de sites du cristal, Nx le nombre d'atomes A répartis au hasard), $Ny = N(1 - x)$ le nombre d'atomes B. Nous nous intéressons ici aux faibles valeurs de y : nous utiliserons donc une technique de perturbation, l'hamiltonien non perturbé étant celui du cristal A pur :

$$< i|\varkappa_0|j > \; = 1 \text{ pour } i \text{ et } j \text{ voisins} \qquad (5)$$
$$= 0 \text{ dans tous les autres cas.}$$

Ses niveaux sont donnés par (4) et ses vecteurs propres $|\boldsymbol{k} >$ sont définis par :

$$< i|\boldsymbol{k} > \; = N^{-1/2}\, e^{i\mathbf{k}.\mathbf{R}_i}. \qquad (6)$$

La perturbation introduite par les atomes B peut être décrite comme un potentiel infiniment répulsif agissant sur les sites B

$$< i|\varkappa - \varkappa_0|j > \; = V\, \delta_{ij} \qquad (7)$$

pour (i) occupé par un atome B, et nul dans le cas contraire. V est un nombre positif que nous ferons tendre vers l'infini à la fin du calcul, ce qui assurera la nullité de l'amplitude au site correspondant ([2]).

La densité de niveaux dans la solution solide est donnée par :

$$N\, \rho(E) = \overline{\text{Trace} \left\{ \delta(E - \varkappa) \right\}} \qquad (8)$$

([2]) Le lecteur attentif sera peut être surpris par notre définition (7) de la perturbation. On est tenté de prendre en effet $< i|\varkappa - \varkappa_0|j > \; = -1$ pour (i) et (j) voisins, (i) ou (j) étant un atome B, et nul dans tous les autres cas. Avec ce potentiel, on trouve, au lieu de (13),

$$< \boldsymbol{k}|T_i(E)|\boldsymbol{l} >$$
$$= N^{-1}\, e^{i(\mathbf{l}-\mathbf{k})\mathbf{R}_i} \left[T(E) + \frac{(E - E(\boldsymbol{k}))\,(E - E(\boldsymbol{l}))}{E} \right].$$

Cette modification est due à l'existence d'un niveau parasite ($E = 0$) sur l'atome B, qui est automatiquement éliminé par (7).

la moyenne étant prise sur toutes les configurations possibles des atomes A et B. Introduisons l'opérateur $\tilde{T}(E)$ défini par :

$$\tilde{T}(E) = [\varkappa - \varkappa_0]\,[1 + (E - \varkappa_0 + i\,\varepsilon)^{-1}\,\tilde{T}(E)] \qquad (9)$$

on a :

$$\rho(E) = \frac{i}{2\pi N} \left[\text{Trace} \left\{ \frac{1}{E + i\,\varepsilon - \varkappa} \right\} - CC \right]$$
$$= \rho_0(E) + \frac{i}{2\pi N} \sum_{\mathbf{k}} (E + i\,\varepsilon - E(\boldsymbol{k}))^{-2} \,.$$
$$[< \boldsymbol{k}|\tilde{T}(E)|\boldsymbol{k} > - CC]. \qquad (10)$$

Soit $\tilde{T}_i(E)$ l'opérateur auquel se réduit $\tilde{T}(E)$ lorsqu'il existe un seul atome B dans le réseau, au site (i). $\tilde{T}(E)$ peut être exprimé en fonction des $\tilde{T}_i(E)$ par l'équation de Watson [1].

$$\tilde{T} = \sum_L \tilde{T}_j + \sum_{i \neq j} \tilde{T}_L\, D^{-1}\, \tilde{T}_j + \sum_{i \neq j \neq k} \tilde{T}_i\, D^{-1}\, \tilde{T}_j\, D^{-1}\, \tilde{T}_k \qquad (11)$$

où *a*) les sommations sont étendues aux sites occupés par les atomes B, *b*) deux indices successifs sont toujours différents, *c*) on a posé

$$D = (E + i\,\varepsilon - \varkappa_0).$$

D'autre part, on calcule facilement $\tilde{T}_i(E)$ grâce à la forme spéciale du potentiel perturbateur (7)

$$< \boldsymbol{k}|\tilde{T}_i(E)|\boldsymbol{l} >$$
$$= N^{-1}\, e^{i(\mathbf{l}-\mathbf{k}).\mathbf{R}_i} \frac{V}{1 - VN^{-1} \sum_{\mathbf{m}} < \boldsymbol{m}|D^{-1}|\boldsymbol{m} >}. \qquad (12)$$

Nous prenons comme convenus la limite $V \to \infty$ et obtenons

$$< \boldsymbol{k}|\tilde{T}_i(E)|\boldsymbol{l} > \; = N^{-1}\, e^{i(\mathbf{l}-\mathbf{k}).\mathbf{R}_i}\, T(E) \qquad (13)$$

avec

$$[T(E)]^{-1} = -\, V_0 (2\pi)^{-3} \int_{\text{zone}} d_3\, m \frac{1}{E + i\,\varepsilon - E(m)} \qquad (14)$$

V_0 désigne le volume atomique et l'intégrale est étendue à la première zone de Brillouin du réseau réciproque.

Nous pouvons maintenant effectuer la moyenne de (11) sur toutes les configurations possibles des atomes A et B. Au premier ordre en y, seule la première somme $\sum_i \tilde{T}_i$ contribue. En reportant dans la densité de niveaux (10) on trouve :

$$\rho(E) - \rho_0(E) = \frac{iy}{2\pi} \left\{ T(E)\, V_0 (2\pi)^{-3} \right.$$
$$\left. \int_{\text{zone}} d_3\, m \frac{1}{[E + i\,\varepsilon - E(m)]^2} - CC \right\} + \text{O}.y$$
$$= \frac{y}{\pi}\, Im. \left\{ \frac{1}{T(E)} \frac{dT(E)}{dE} \right\} + \text{O}.y. \qquad (15)$$

Ce résultat peut être transformé en examinant la diffusion par un atome B unique (pris à l'origine). La solution en ondes sortantes s'écrit :

$$a_i = e^{i\mathbf{k}.\mathbf{R}_i} - \Gamma_E^{+}(\mathbf{R}_i)$$

$$\Gamma_E^{+}(\mathbf{R}_i) = - T(E)\, V_0 (2\pi)^{-3} \int_{\text{zone}} d_3 m \frac{e^{i\mathbf{m}.\mathbf{R}_i}}{E + i\varepsilon - E(\mathbf{m})}. \qquad (16)$$

L'onde sortante a toutes les symétries du groupe ponctuel : c'est l'analogue réticulaire d'une onde s, Le déphasage δ correspondant est donné par [2]

$$\pi\, T(E)\, \rho_0(E) = e^{i\delta} \sin \delta. \qquad (17)$$

En reportant cette valeur dans (15) on obtient

$$\rho(E) - \rho_0(E) = \frac{y}{\pi}\frac{d\delta}{dE}. \qquad (18)$$

C'est un cas particulier d'une formule due à Friedel pour les milieux continus [3] et généralisé par de Witt [4]. Il est également possible de montrer que, aux grandes longueurs d'ondes (18) est équivalent à la relation classique entre l'indice de réfraction n et la longueur de diffusion par un atome B, soit a,

$$n^2 - 1 = \frac{y}{V_0}\frac{a\,\lambda^2}{\pi}. \qquad (19)$$

On en déduit également que le niveau le plus haut, dans cette approximation est

$$E_{\max} = E(0) - yK(0) \qquad (20)$$

$$[K(E)]^{-1} = V_0 (2\pi)^{-3} \int_{\text{zone}} d_3 m \frac{\mathfrak{P}}{E - E(m)} \qquad (21)$$

ou le symbole $\mathfrak{P}$ représente une partie principale au sens de Cauchy. La relation (20) est liée à certaines propriétés combinatoires remarquables des réseaux périodiques (voir appendice 1).

Passons maintenant à l'étude des termes en y^2 dans la densité de niveaux. Pour les sommer exactement, il faut regrouper dans (11) tous les termes faisant intervenir deux atomes B situés en (i) et (j) : soit $\tilde{T}_{ij}(E)$ la contribution correspondante. On peut sommer l'élément diagonal de cet opérateur en utilisant (11) et (13) :

$$< \mathbf{k}|T_{ij}(E)|\mathbf{k} >$$
$$= N^{-1}\, T(E) \left\{ \frac{1 - \Gamma_E^{+}(\mathbf{R}_j - \mathbf{R}_i)\, e^{i\mathbf{k}(\mathbf{R}_L - \mathbf{R}_j)}}{1 - \Gamma_E^{+}(\mathbf{R}_j - \mathbf{R}_i)\, \Gamma_E^{+}(\mathbf{R}_i - \mathbf{R}_j)} + \text{sym.} \right\}. \qquad (22)$$

Le terme symétrique étant déduit du précédent par l'échange de i et j. (Sur la couche d'énergie $E = E(\mathbf{k})$, on aurait pu obtenir (22) par un calcul direct de l'onde diffusée.) On obtient ainsi :

$$< \mathbf{k}|\tilde{T}(E)|\mathbf{k} > = yT(E)$$
$$+ \frac{y^2}{2} \sum_{i\neq0} \left\{ \frac{1 - \Gamma_E^{+}(\mathbf{R}_i)\, e^{-i\mathbf{k}.\mathbf{R}_i}}{1 - \Gamma_E^{+}(\mathbf{R}_i)\, \Gamma_E^{+}(-\mathbf{R}_i)} + \text{sym.} - 2 \right\}. \qquad (23)$$

Le fait important est que la somme qui figure dans (23) converge : pour $|\mathbf{R}| \to \infty$, à trois dimensions, $\Gamma_E^{+}(\mathbf{R})$ décroît comme $1/\mathbf{R}$. Les termes qui devront être examinés séparément sont :

$$- \sum_{i\neq0} \Gamma_E^{+}(\mathbf{R}_i)\, e^{-i\mathbf{k}.\mathbf{R}_i} = 1 + \frac{T(E)}{E - E(\mathbf{k}) + i\varepsilon} \qquad (24)$$

$$\sum_{i\neq0} \Gamma_E^{+}(\mathbf{R}_i)\, \Gamma_E^{+}(-\mathbf{R}_i) = -1 - \frac{dT}{dE} \text{ etc...} \qquad (25)$$

On voit qu'ils sont finis, et, en reportant dans (10), que le terme en y^2 de la densité de niveaux est convergent. Jusqu'à l'ordre y^2, on obtient donc un spectre déformé, mais qui reste essentiellement continu. Cet état de choses ne sera en fait modifié que par l'apparition *d'états liés*, dus aux atomes A qui sont complètement entourés par des atomes B. La première contribution de ces amas finis correspond à un atome A isolé par Z voisins B ; elle est donc d'ordre y^Z. Pour éliminer la contribution de ces amas finis, nous allons en premier lieu, dans la section suivante, estimer le pourcentage d'atomes A engagés dans de tels amas.

Il n'est pas inutile de souligner enfin les différences qui existent entre le cas tridimensionnel envisagé ci-dessus et celui d'une chaîne linéaire : pour une chaîne linéaire $T(E)$ est imaginaire pur (il n'y a pas de propagation cohérente) et le spectre est discret sur tout l'intervalle de concentration $0 \leqslant x < 1$.

III. **Étude géométrique des amas.**

— Faisons choix d'un site origine 0 (quelconque) et soit $Pn(x)$ la probabilité pour que 0 soit occupé par un atome A qui appartiennent à un amas de rang n. Nous poserons par convention

$$P_0(x) = y = 1 - x. \qquad (26)$$

Les probabilités $Pn(x)$ pour n fini se calculent de la façon suivante : pour chaque forme possible d'amas contenant 0, on détermine : *a*) le rang n de l'amas, *b*) le nombre m d'atomes B qui sont voisins d'au moins un atome A de l'amas. (Ces sites m constituent l'adhérence de l'amas.) Soit A_{nm} le nombre total d'amas qu'il est possible de réaliser, contenant 0, de rang n et d'adhérence formée de m atomes B. On a

$$P_n(x) = \sum_m A_{nm}\, x^n\, y^m. \qquad (27)$$

Les facteurs y spécifient que tous les points de l'adhérence sont bien des atomes B. Nous donnons

TABLE 1

1. Réseau cubique simple.

$P_1 = xy^6$
$P_2 = 6\, x^2 y^{10}$
$P_3 = 36\, x^3 y^{13} + 9\, x^3 y^{14}$
$P_4 = x^4(32\, y^{15} + 204\, y^{16} + 96\, y^{17} + 12\, y^{18})$
$P_5 = x^5(60\, y^{17} + 495\, y^{18} + 1\,140\, y^{19} + 780\, y^{20} + 180\, y^{21} + 15\, y^{22})$

2. Réseau cubique centré.

$P_1 = xy^8$

$P_2 = 8\,x^2\,y^{14}$

$P_3 = x^3(12\,y^{20} + 36\,y^{19} + 36\,y^{17})$

3. Réseau cubique a faces centrées.

$P_1 = xy^{12}$

$P_2 = 12\,x^2\,y^{18}$

$P_3 = x^3(24\,y^{23} + 126\,y^{24})$

dans la table I les valeurs des $Pn(x)$ (pour les premiers ordres) relatives à quelques types de réseaux.

La probabilité pour que 0 appartienne à un amas infini d'atomes A est obtenue par différence

$$P_\infty(x) \equiv S(x) = 1 - \sum_0^\infty P_n(x). \qquad (28)$$

En limitant dans (28) la somme à ses premiers termes, on obtient une valeur approchée par excès de $S(x)$. L'aspect des courbes ainsi obtenues (*fig.* 1)

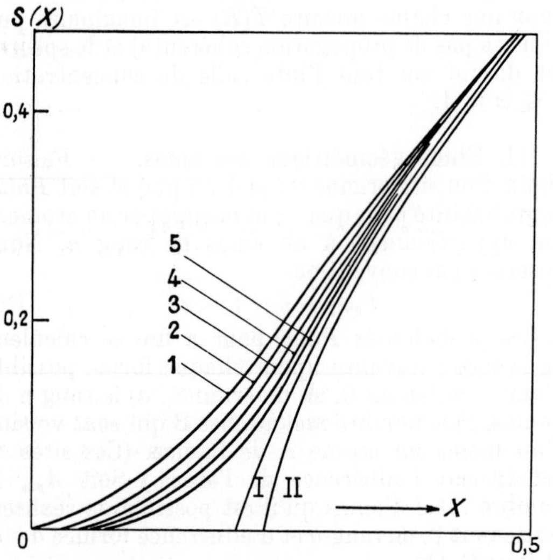

Fig. 1.

suggère l'existence de deux domaines de concentration dans lesquels le comportement de $S(x)$ diffère. Aux fortes concentrations (pratiquement pour $x > 0.5$) $S(x)$ est très voisin de x d'après la figure 1. Presque tous les atomes A sont engagés dans un même amas infini et la série (28) converge rapidement. (Toutefois, nous ne sommes pas parvenus à démontrer la convergence uniforme de (28) dans ce domaine).

Aux faibles concentrations $S(x)$ est *identiquement nulle.* En effet :

a) $S(x)$ et ses dérivées convergent uniformément sur un segment fini (Ox_c). Pour un réseau cubique simple par exemple, l'adhérence m est inférieure à $4n + 2$ (cette valeur étant obtenue par un amas en forme de chaîne linéaire)

$$P_n(x) > \left(\sum_m A_{nm}\right) x^n\, y^{4n+2}$$

la série $\sum_n P_n$ converge (puisque sa somme est bornée par I) et la valeur maxima de xy^4 est $4^4/5^5$. Donc $\left(\sum_m A^{mn}\right)$ est majoré par le terme de rang n d'une série géométrique de raison $5^5/4^4$, ce qui suffit à montrer la propriété.

b) Les dérivées $\left(\frac{d^n}{dx^n} S\right)_{x=0}$ sont toutes nulles ; Pour la dérivée d'ordre n, seuls interviennent les amas de rang $q \leqslant n$: subtituons donc au cristal infini un cristal fini de f_n atomes, assez grand pour contenir tous les amas de rang $q \leqslant n$ et leurs adhérences. Soient P'_q les probabilités qui se substituent aux P_q dans ces conditions

$$\sum_0^f P'_q = 1$$

$$P'_q = P_q \quad \text{pour} \quad q \geqslant n$$

donc

$$\left(\frac{d^n}{dx^n} S\right)_{x=0} = \left[\frac{d^n}{dx^n} \sum_0^n P_q(x)\right]_{x=0} = \left[\frac{d^n}{dx^n} \sum_0^n P'_q(x)\right]_{x=0}$$

$$= \left[\frac{d^n}{dx^n} \sum_0^f P'_q(x)\right]_{x=0} = 0. \qquad (29)$$

Donc sur un segment fini (Ox_c), $S(x)$ est égale à $S(0)$, donc est nulle : *il n'y a pas d'amas infinis dans ce domaine de concentration.*

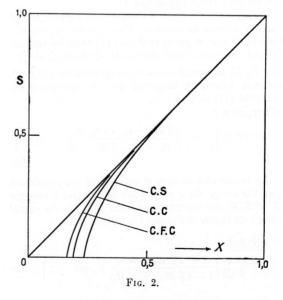

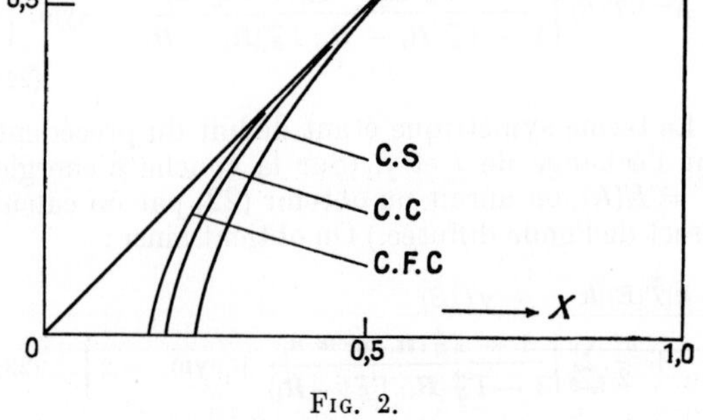

Fig. 2.

Nous allons maintenant substituer à (28) un développement approché qui lui est identique aux premiers termes, que nous pouvons sommer, et qui fait apparaître directement un « point critique » x_c. La méthode consiste à dénombrer les premiers amas construits à partir de 0, puis à itérer la construction de ces amas pour en obtenir de plus grands. L'itération est approchée, car elle n'exclut pas certains recouvrements entre les parties successives de l'amas ou de l'adhérence. Nous en exposerons le principe sur le réseau cubique simple, et nous citerons simplement les résultats pour d'autres réseaux.

Posons

$$F(x) = \sum_0^\infty P_n(x) = 1 - S(x) \tag{30}$$

$$= y + x[G(xy)]^6$$

$$(y = 1 - x). \tag{31}$$

Dans (31) nous avons séparé le cas où 0 est un atome B, et la fonction $G(xy)$ doit engendrer des amas à partir d'un premier voisin P_1 de 0, sachant que 0 et P_1 sont des atomes A.

On écrit donc dans la première approximation

$$G = y + xG^5. \tag{32}$$

(L'exposant 5 au lieu de 6, tient compte du fait que 0 est déjà un atome A.) Dans la deuxième approximation, on remplace (32) par un système d'équations, destiné à construire au mieux les amas de trois atomes A. Ici, ces amas sont de deux types, selon que les trois points sont alignés ou forment un angle droit. Pour engendrer les amas à partir de P_2 (*fig.* 3) sachant que 0 et P_1 sont des

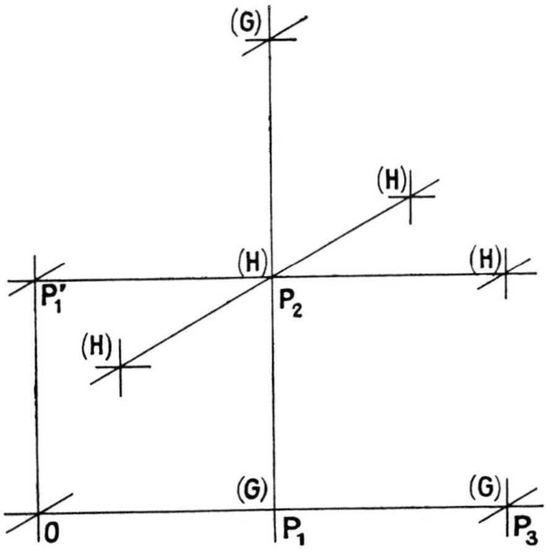

Fig. 3.

atomes A, nous introduirons une nouvelle fonction $H(xy)$. Par contre, nous engendrons ceux qui sont issus de P_3, nous utilisons encore G. (En effet, la présence d'un atome A en O ne limite la construction des amas à partir de P_3 que pour des rangs élevés.)

On pose donc

$$G = y + xGH^4. \tag{33}$$

Il faut maintenant une équation pour H. Lorsque 0, P_1, P_2 sont des atomes A, il reste seulement 4 sites voisins de P_2 à explorer (P_1' a été exploré en même temps que P_1 grâce au facteur G^6 dans (31)). Pour ces 4 sites on néglige, comme pour P_3, le fait que 0 est occupé, ce qui donne

$$H = y + xGH^3 \tag{34}$$

(31) et (32) ou bien (31), (33) et (34) déterminent F en fonction de x et $y = 1 - x$ dans la 1re et dans la 2e approximation respectivement. (Les systèmes correspondants pour différents réseaux sont regroupés dans la table II.)

TABLE 2

	1re approximation	2e approximation
Cubique simple :	$G = y + xG^5$	$G = y + xH^4 G$
		$H = y + xH^3 G$
Cubique face centrée :	$G = y + xG^7$	$G = y + xG^5 H^2$
		$H = y + xG^3 H$
Cubique centré :	$G = y + xG^7$	$G = y + xGH_1^3 H_2^2$
		$H_1 = y + xGH_1^2 H_2$
		$H_2 = y + xGH_1^3 H_2^2$

Si l'on voulait réaliser des approximations d'ordre supérieur, il ne faudrait pas seulement modifier le système qui donne G, mais aussi l'équation (31). Que peut-on déduire en pratique des système de la table II ? On vérifie d'abord, en développant la solution en série de x et y considérés comme variables indépendantes, que les premiers termes du développement (30) sont correctement reproduits, et que les termes suivants sont peu différents numériquement de la valeur exacte.

D'autre part, quelque soit le réseau (tridimensionnel) et l'ordre d'approximation (1 ou 2) lorsque l'on fait $y = 1 - x$ chaque système admet pour $F(x)$ deux solutions réelles sur l'intervalle (1), dont l'une est $F = 1$, et dont l'autre est telle que $F(1) = 0$.

Les deux déterminations se coupent pour une valeur x_c de l'intervalle 01. Partant de $x = 0$ on doit d'abord suivre la première, puis, à partir de x, la seconde détermination. Les résultats sont représentés sur les figures 1 et 2. On suit clairement, grâce à cette approximation, l'apparition des amas infinis. Les points critiques x_c estimés de cette façon sont

$$x_c = 0{,}224$$
$$0{,}192$$
$$0{,}162$$

pour les réseaux cubiques simple, centré et face centrée respectivement.

Mentionnons ici encore les propriétés particulières de la chaîne linéaire : Pour tout $x \neq 1$ il n'y a que des amas finis d'atomes A. $S(x)$ est nul pour $x < 1$ et discontinu pour $x = 1$. Ceci est dû au fait qu'un nombre même faible, d'atomes B répartis sur la chaîne, suffit à la couper en segments finis, et se relie directement à l'absence de propagation cohérente signalée dans la section II.

IV. **Principe de la séparation du spectre continu.**

— Nous avons vu comment, dans le domaine des fortes concentrations en atomes A, une généralisation de l'approximation « optique » appliquée à la diffusion par les atomes B permet de déterminer le spectre perturbé. La conductivité dans la bande correspondante pourrait être calculée par les mêmes méthodes. Dans le domaine des faibles concentrations, le spectre se réduit à quelques raies discrètes calculables directement et la conductivité statique, elle, est nulle. Il est plus particulièrement intéressant d'interpoler ces résultats aux concentrations intermédiaires. Nous indiquons ici le principe d'une méthode de moments qui permet d'y parvenir, et les résultats numériques des toutes premières approximations correspondantes.

L'hamiltonien $\varkappa$ associé à (3) n'a d'élément de matrice qu'entre atomes A voisins (et ils sont alors égaux à 1). Il est facile de calculer ses moments

$$\overline{\overline{\varkappa^n}} = \frac{\overline{\mathrm{Tr}\ \varkappa^n}}{\overline{\mathrm{Tr}\ 1}} = \frac{1}{x}\int E^n\ \rho(E)\ \mathrm{d}E \tag{35}$$

$$= \sum_p N_{np}(0)\ x^p. \tag{36}$$

$N_{np}(\boldsymbol{R}_i)$ est le nombre de chemins reliant les sites (0) et (i) en n sauts de 1er en 1er voisin, et qui utilisent en tout p points distincts du réseau (autres que 0). On peut se proposer de calculer à la main les premiers moments, puis de construire $\rho(E)$ à partir de ces moments en lui imposant une forme analytique plausible. Cette opération n'est relativement permise que lorsque $\rho(E)$ est une fonction continue et régulière de E ; pour l'appliquer ici, il faut d'abord soustraire la contribution des amas finis aux moments.

D'une façon générale, étant donnée une quantité Q supposée additive pour des amas disjoints, nous désignons par Q_∞ la partie de Q qui provient des amas infinis.

En premier lieu nous connaissons

$$\int \rho_\infty(E)\ \mathrm{d}E = S(x).$$

Pour calculer $\overline{\overline{\varkappa^2_\infty}}$ il faut déterminer la probabilité $S_2(x)$ pour que deux voisins soient simultanément occupés par des atomes A et qu'ils appartiennent à un même amas infini. $S_2(x)$ s'obtient comme $S(x)$, par différence :

$$S_2(x) - x^2 - x^2\,y^{13} - x^3(8y^{13} + 2y^{14}) - x^4(8y^{15} + 52y^{16} + 24y^{17} + 3y^{18}) \tag{37}$$

(pour un cubique simple). $S_2(x)$ comme $S(x)$, est une série de polynômes qui représente deux fonctions différentes selon que x est inférieur ou supérieur à une certaine valeur critique, d'ailleurs égale à x_c comme le montrent les inégalités :

$$S(x) \geqslant S_2(x) \geqslant x\ S(x). \tag{38}$$

On peut aussi faire une sommation approchée de $S_2(x)$. Dans l'exemple ci-dessus, on trouve, par la deuxième approximation

$$S_2(x) = x^2(1 - G^2\,H^8). \tag{39}$$

On en tire ensuite le deuxième moment

$$\overline{\overline{\varkappa^2_\infty}} = \frac{\int E^2\ \rho_\infty(E)\ \mathrm{d}E}{\int \rho_\infty\ \mathrm{d}E} = z\,\frac{S_2(x)}{S(x)} \tag{40}$$

où $Z = E(0)$ est le nombre de voisins d'un site donné.

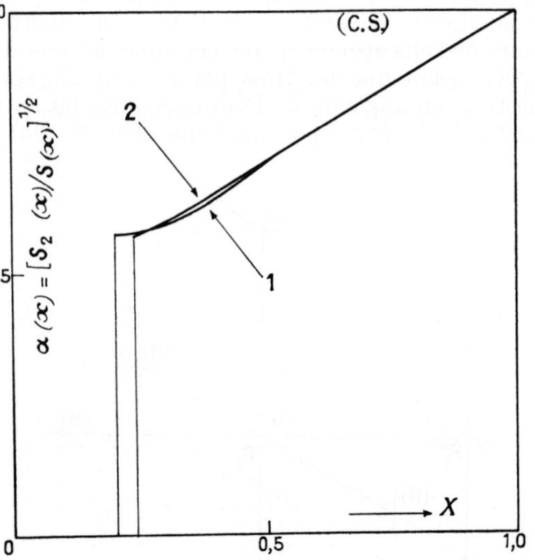

FIG. 4.

On a porté sur la figure 4, la quantité $[S_2(x)/S(x)]^{1/2}$ qui donne une information sur la largeur de la bande continue. Cette quantité est finie pour $x = x_c$: dès qu'il apparaît des amas infinis, ils engendrent une bande de largeur finie. Les moments d'ordre supérieur se calculeraient de la même facon.

Nous passons maintenant à l'étude de la conductivité de notre bande d'impuretés : les amplitudes a_i décrivent ici des porteurs de courant qui ne peuvent se déplacer qu'en sautant d'atome A en atome A ($2N \times f$ électron occupant $2Nx$ niveaux, le facteur 2 provenant du spin). Soit $\sigma(\omega)$ la conductivité (scalaire pour un cristal cubique) à la fréquence angulaire. Nous savons déjà que pour $x < x_c$ la conductivité statique $\sigma(0)$ est nulle, et qu'elle est non nulle pour $x > x_c$ (si l'on imagine une tranche de cristal placée entre deux électrodes planes finies, un amas infini a une probabilité 1 de rencontrer les deux électrodes). Nous allons préciser ces notions par un calcul de moments.

Nous supposons toujours nos porteurs indépendants. En outre, le problème étant plus difficile que celui de la densité de niveaux, nous le simplifions encore en nous plaçant dans la limite des hautes températures ($k_B T \gg 1$ dans nos unités). On peut alors en appliquant la méthode de Kubo [5], exprimer la conductivité en fonction des fluctuations de courant par une formule très compacte :

$$\sigma(\omega) = \frac{f(1-f)}{3k_B T} \int_{-\infty}^{\infty} dt\, e^{-i\omega t}\, \mathrm{Tr}\left\{ \boldsymbol{I}(0) . \boldsymbol{I}(t) \right\} \tag{41}$$

$\boldsymbol{I}$ est l'opérateur courant à un électron

$$\boldsymbol{I} = \frac{ie}{\hbar} [\varkappa, \boldsymbol{R}] \tag{42}$$

$$< i|\boldsymbol{I}|j > = \frac{ie}{h} (\boldsymbol{R}_i - \boldsymbol{R}_i) \quad \text{(pour } (i) \text{ et } (j) \text{ voisins et occupés par des atomes A.} \tag{43}$$

$$= 0 \qquad \text{(dans tous les autres cas)} \tag{43}$$

$$\boldsymbol{I}(t) = e^{i\varkappa t} \boldsymbol{I} e^{-i\varkappa t}. \tag{44}$$

Les moments d'ordre pair de $\sigma(\omega)$ (seuls sont nuls) sont donnés par

$$\int d\omega\, \omega^{2n} \sigma(\omega) = \frac{2n}{3} \frac{f(1-f)}{k_B T} \mathrm{Tr} \left\{ \boldsymbol{I} . \frac{d^{2n}}{dt^{2n}} \boldsymbol{I} \right\}_{t=0} . \tag{45}$$

Les traces se calculent dans la représentation $|i>$ en utilisant de façon répétée l'équation du mouvement déduite de (44)

$$\frac{d}{dt} \boldsymbol{I} = \frac{i}{h} [\varkappa, \boldsymbol{I}]. \tag{46}$$

Au voisinage de $x = 1$ la contribution des amas finis étant négligeable, on peut postuler avec sécurité que $\sigma(\omega)$ est uen courbe continue. Si plus spécifiquement on prend une forme de Lorentz tronquée ($\sigma \cong \tau(1 + \omega^2 \tau^2)^{-1}$ pour $|\omega| < \omega_c$ et nulle au delà), on trouve par l'étude des moments d'ordre 2 et 4, pour un cubique simple

$$k\, \omega_c = 6{,}48$$

$$\tau = 0{,}34\, \hbar \frac{1}{y}. \tag{47}$$

(Rappelons encore que notre unité d'énergie est l'élément de matrice de $\varkappa$ entre deux sites voisins.) L'intérêt de τ vient de ce qu'il représente une

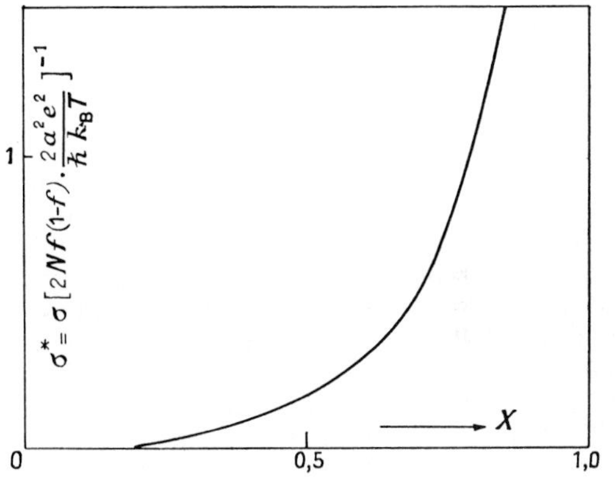

Fig. 7.

moyenne des temps de relaxation sur l'ensemble des niveaux. Aux concentrations intermédiaires, on doit, ici encore, éliminer le spectre discret par les mêmes méthodes. Le résultat d'un tel calcul (fait à la première approximation) est représenté sur la figure 5. Les valeurs numériques sont imprécises, mais l'aspect général doit être retenu, et en particulier l'annulation pour $x = x_c$.

VI. **Conclusions.** — La propriété la plus remarquable exhibée par notre modèle est l'existence d'une concentration critique x_c qui résulte d'un effet purement géométrique, et qui sépare un domaine « macromoléculaire ». On retrouverait le même phénomène dans tous les cas tridimensionnels pourvu que la portée des couplages entre atomes actifs soit finie (x_c étant d'autant plus faible que cette portée est grande).

Ces résultats sont à rapprocher de ceux d'Anderson [6] relatifs à une situation beaucoup plus générale. Le modèle d'Anderson diffère du nôtre en particulier par le fait que les niveaux des atomes A isolés sont supposés eux-mêmes aléatoires. L'exemple (schématisé à l'extrême) que nous avons choisi, permet de se faire une image correcte de l'existence d'une concentration critique et du comportement en deçà et au delà de cette concentration. L'existence d'un seuil critique pour les amas infinis a aussi été envisagée par Broadbent et Hammersley [7], dans une situation voisine où ce sont les couplages entre atomes qui sont aléatoires et non leur répartition.

Transposés à notre situation, leurs théorèmes permettent d'écrire

$$\lim_{n\to\infty} \sum_{L} N_{nn}(R_i) = \mathrm{e}^{Kn+0(n)} \qquad (48)$$

$$x_{\mathrm{c}} \geqslant \mathrm{e}^{-K}.$$

Du point de vue de l'analyse combinatoire, la confrontation de la méthode « optique » de la section II avec la méthode de trace de la section IV, conduit à des résultats remarquables, que nous résumons dans l'appendice.

Enfin, pour en revenir aux problèmes physiques qui nous ont servi de point de départ, nous pouvons formuler les remarques suivantes :

1) Lorsque les atomes A sont des porteurs de moments magnétiques couplés ferromagnétiquement, le spectre des ondes de spin se déforme au fur et à mesure que l'on ajoute des atomes B, puis disparaît pour $x = x_c$. Pour $x < x_c$, le système se comporte comme une collection d'amas finis d'atomes A et sa fonction de partition est donc une fonction régulière de la température : il n'y a pas de point de Curie bien défini dans ce domaine.

2) Lorsque les atomes A forment une bande d'impuretés dans un semi-conducteur comme *Ge* ou *Si*, il est tentant de relier x_c à la concentration critique qui apparaît aux basses températures pour leur conductivité et pour le temps de relaxation spin réseau [8], [9]. (L'interprétation de cette dernière propriété repose sur le fait que des électrons itinérants subissent plus fortement l'action perturbatrice du réseau que des électrons liés à des atomes A isolés [10].)

Toutefois, nous ne pensons pas que ce problème puisse être résolu sans tenir compte des corrélations de Coulomb entre porteurs, qui ont été négligées ici sans justification, et qui pourraient bloquer la conductivité statique dans un domaine où celle-ci serait « géométriquement permise » [11].

Nous voudrions enfin remercier les P^rs J. Friedel et A. Herpin pour plusieurs discussions sur ces questions, et le Dr P. W. Anderson qui a eu l'amabilité de nous transmettre son manuscrit avant publication. Nous remercions d'autre part, Mme Pillard et M. Grandidier qui ont effectué les calculs numériques.

Appendice 1. — Nous montrons ici comment l'approximation « optique » de la section II détermine certaines propriétés purement combinatoires des chemins tracés sur le réseau. Dans le domaine de validité de cette approximation, et pour les ordres inférieurs à Y^z, on peut associer à chaque vecteur d'onde $\boldsymbol{k}$ (d'énergie $E(\boldsymbol{k})$ dans le cristal A pur) une énergie perturbée $E_1(\boldsymbol{k})$, et l'on a d'après (36)

$$\overline{\overline{x}}{}_1^n = \sum_p N_{np}(0)\, x^p = V_0(2\pi)^{-3} \int_{\text{zone}} \mathrm{d}_3\, m E_1^n(\boldsymbol{m}). \quad (48)$$

Pour $x = 1$, (48) restitue une formule classique de marche aléatoire. Si l'on prend la dérivée logarithmique de (48) on trouve :

$$\bar{p}_n \equiv \frac{\sum_p N_{np}(0)\, p}{\sum_p N_{np}(0)} = n \frac{\int_{\text{zone}} \mathrm{d}_3\, m\, E^n(\boldsymbol{m}) \left[\frac{\partial}{\partial x} \alpha n\, E_1(\boldsymbol{m})\right]_{x=1}}{\int_{\text{zone}} \mathrm{d}_3\, m\, E^n(\boldsymbol{m})}. \quad (49)$$

Pour $n \to \infty$ seules interviennent dans (49) les valeurs extrêmales de $|E(\boldsymbol{m})|$ qui sont

$$|E(\boldsymbol{m})| = E(0) = Z.$$

Nous nous limitons aux réseaux où ces valeurs ne sont obtenues qu'en un point de la zone (ex. : cfc) ou en des points de propriétés équivalentes à un changement de signe pour E (ex. : c. s.). On a alors

$$\lim_{n\to\infty} \frac{1}{n} \bar{p}_n = \frac{1}{E(0)} \left[\frac{\partial}{\partial x} E_1(0)\right]_{x=1} = -\frac{K(Z)}{Z} \qquad (50)$$

où nous avons fait usage de (20). On peut calculer numériquement l'intégrale (21) qui conduit à $K(E)$. On trouve ainsi que (50) prend la valeur 0.702 pour un cubique simple.

Appendice 2. — On peut se demander si x_c n'est pas toujours égal à 1, c'est-à-dire si, dès qu'il existe un faible pourcentage d'atome B, les atomes A ne forment plus que des amas finis (quoique très grands).

En fait ceci n'a lieu que pour la chaîne linéaire. Soient en effet deux atomes A situés aux nœuds (i) et (j) d'un réseau à 2 dimensions, ou plus. Dans un domaine fini de concentrations $x < 1$, la probabilité ϖ pour que ces deux atomes appartiennent à un même amas n'est pas nulle, même lorsque (i) et (j) sont arbitrairement éloignés. Nous résumerons ici la démonstration de cette propriété sur le réseau carré plan. Soit $(i_1)\,(i_2)\,\ldots\,(i_N)$ le (ou l'un des) plus court chemin reliant (i) et (j). Pour que (i) et (j) soient isolés, il faut que (i) soit entouré d'une « enveloppe » d'atomes B n'enfermant pas (j) ou réciproquement, ou qu'ils soient séparés par une même « enveloppe » d'atomes B de rang infini. De telles enveloppes forment elle-mêmes des amas d'atomes B, au sens généralisé (couplages entre premiers ou seconds voisins dans notre cas). Chacune d'elles coupe le chemin $(i_1)\,\ldots\,(i_N)$ en au moins un point (i_p). Soit θ_p la probabilité pour que (i_p) fasse partie d'une telle enveloppe. On a

$$1 - w < \sum_{1}^{N} \theta_p.$$

Toutes les enveloppes passant pas (i_p) ont un rang supérieur à $r(p)$, où r est le plus petit des nombres p et $N - p$. Donc θ_p est inférieur à la probabilité pour que (i_p) appartienne à un amas généralisé quelconque d'atomes B, de rang supérieur ou égal à $r(p)$. Par extension des raisonnements de la section III, on montre que

$$\theta_p < \frac{(\alpha y)^r}{1 - \alpha y}$$

où α est un nombre caractéristique du réseau. Par conséquent

$$1 - W < \frac{2}{(1 - \alpha y)^2} [\alpha y - (\alpha y)^{N/2}]$$

$$W > 1 - \frac{2\alpha y}{(1 - \alpha y)^2}$$

quel que soit N. Par exemple pour $y < \frac{1}{4\alpha}$, w est supérieur à $1/q$. On pourrait évidemment, tout au long du calcul, chercher des majorantes plus serrées.

Manuscrit reçu le 7 février 1959.

BIBLIOGRAPHIE

[1] Watson (K. M.), *Phys. Rev.*, 1953, **89**, 575.
[2] Lippman (B.) et Sehwinger (J.), *Phys. Rev.*, 1950, **79**, 669.
[3] Friedel (J.), *Phil. Mag.*, 1952, **43**, 153.
[4] De Witt (B. S.), *Phys. Rev.*, 1956, **103**, 1565.
[5] Kubo (R.), *Canad. J. Phys.*, 1956, **34**, 1274.
[6] Anderson (P. W.), *Phys. Rev.*, 1958, **109**, 1492.
[7] Broadbent (S. R.) et Hammersley (J. M.), *Proc. Camb. Phil. Soc.*, 1957, **53**, 629.
[8] Conwell (E. M.), *Phys. Rev.*, 1956, **103**, 51.
[9] Feher (G.), Flecher (R. C.) et Gore (E. A.), *Phys. Rev.*, 1955, **100**, 784.
[10] Bardeen (J.), Pines (D.) et Slichter (C.), *Phys. Rev.*, 1957, **106**, 489.
[11] Mott (N. F.), *Canad. J. Phys.*, 1956, **34**, 1356.

Afterthoughts: Sur un exemple de propagation dans un milieu désordonné

1) When this work was performed, we had not heard about Hammersley's work — the historical starting point of percolation theory (Ref. 7)! We heard about it with some dismay through P.W. Anderson's preprint on localisation (Ref. 6), and inserted the reference in the final version.

2) At the time, I was naively convinced that all the electron states built on the infinite cluster were delocalised. But in fact (as first observed by S. Kirkpatrick many years later), some of these states are localised on small "pockets" in the infinite cluster.

PHYSICAL REVIEW VOLUME 118, NUMBER 1 APRIL 1, 1960

Effects of Double Exchange in Magnetic Crystals*

P.-G. DE GENNES†
Department of Physics, University of California, Berkeley, California

(Received October 9, 1959)

This paper discusses some effects of mobile electrons in some antiferromagnetic lattices. It is shown that these electrons (or holes) always give rise to a distortion of the ground state spin arrangement, since electron transfer lowers the energy by a term of first order in the distortion angles. In the most typical cases this results in: (a) a nonzero spontaneous moment in low fields; (b) a lack of saturation in high fields; (c) simultaneous occurrence of "ferromagnetic" and "antiferromagnetic" lines in neutron diffraction patterns; (d) both ferromagnetic and antiferromagnetic branches in the spin wave spectra. Some of these properties have indeed been observed in compounds of mixed valency such as the manganites with low Mn^{4+} content. Similar considerations apply at finite temperatures, at least for the (most widespread) case where only the bottom of the carrier band is occupied at all temperatures of interest. The free energy is computed by a variational procedure, using simple carrier wave functions and an extension of the molecular field approximation. It is found that the canted arrangements are stable up to a well-defined temperature T_1. Above T_1 the system is either antiferromagnetic or ferromagnetic, depending upon the relative amount of mobile electrons. This behavior is not qualitatively modified when the carriers which are responsible for double exchange fall into bound states around impurity ions of opposite charge. Such bound states, however, will give rise to local inhomogeneities in the spin distortion, and to diffuse magnetic peaks in the neutron diffraction pattern. The possibility of observing these peaks and of eliminating the spurious spin-wave scattering is discussed in an Appendix.

I. INTRODUCTION

WE are concerned here with magnetic compounds of mixed valency, of which the best known example is the series $(La_{1-x}Ca_x)(Mn_{1-x}^{3+}Mn_x^{4+})O_3$. At both ends of the composition diagram, these manganites behave like antiferromagnetic insulators.[1,2] However, to take a definite example, if we substitute 10% of calcium in pure $LaMnO_3$ the room temperature conductivity is increased by two orders of magnitude.[1] This shows that the 10% extra holes which have been added are comparatively free to move from one manganese ion to another, and are able to carry a current. These carriers also have a strong effect on the magnetic properties of the material: at low temperatures there is a nonzero spontaneous magnetization (approximately 0.4 of what is expected for complete lining up of the spins on the above example), indicating that some sort of ferromagnetic coupling is present. This was first explained by Zener[3] in the following way: (1) intra-atomic exchange is strong so that the only important configurations are those where the spin of each carrier is parallel to the local ionic spin; (2) the carriers do not change their spin orientation when moving; accordingly they can hop from one ion to the next only if the two ionic spins are not antiparallel; (3) when hopping is allowed the ground state energy is lowered (because the carriers are then able to participate in the binding). This results in a lower energy for ferromagnetic configurations. This "double exchange" is completely different from the usual (direct or indirect) exchange couplings, as pointed out by Anderson and Hasegawa.[4] The coupling energy is shared between the carriers, and cannot be written as a sum of terms relating the ionic spins by pairs. Also, the dependence of the carrier energy on the angle between different ionic spins is quite remarkable. This brings in special effects which do not seem to have been considered up to now. For instance, if the pure material is antiferromagnetic, it will turn out that the carrier energy in the mixed material is lowered if the sublattices become canted. This gain in energy is of first order with respect to the angle of canting, while the loss of antiferromagnetic exchange energy is only of second order; as a result, the canted arrangement is indeed more stable. It is the

* Supported by the National Science Foundation.
† On leave from the Centre d'Etudes Nucleaires de Saclay, Gif-sur-Yvette, France.
[1] G. H. Jonker and J. H. Van Santen, Physica 16, 337 (1950); 19, 120 (1953).
[2] E. O. Wollan and W. C. Koehler, Phys. Rev. 100, 545 (1955).
[3] C. Zener, Phys. Rev. 82, 403 (1951).
[4] P. W. Anderson and H. Hasegawa, Phys. Rev. 100, 675 (1955).

purpose of this paper to discuss these distorted arrangements, especially as regards their stability and their effects on magnetic properties.

As a first step, we now outline the very simplified physical model which will be used in the calculations of the later sections. We shall describe the carrier wave functions, in a tight binding approximation, as a linear combination of some orthogonal functions φ_i localized on each magnetic site (i):

$$\psi=\sum \alpha_i \varphi_i. \tag{1}$$

The φ's are such that off-diagonal elements of the one-electron Hamiltonian between them and the anion orbitals are zero. The eigenvalue equation satisfied by the amplitudes α_i is then of the form

$$E\alpha_i=\sum_j t_{ij}\alpha_j \tag{2}$$

where $t_{ij}=(\varphi_i|\mathcal{H}|\varphi_j)$ is a matrix element of the one-carrier Hamiltonian $\mathcal{H}$, commonly referred to as the transfer integral between ions i and j. In practice t_{ij} connects only neighboring magnetic sites. When the ionic spins $\mathbf{S}_i$ and $\mathbf{S}_j$ are parallel, t_{ij} is maximum and equal to some constant b_{ij}. When $\mathbf{S}_i$ and $\mathbf{S}_j$ are antiparallel, $t_{ij}=0$. More generally, as shown in reference 4, when $\mathbf{S}_i$ makes an angle θ_{ij} with $\mathbf{S}_j$ the transfer integral for carriers of spin $\frac{1}{2}$ is

$$t_{ij}=b_{ij}\cos(\theta_{ij}/2). \tag{3}$$

Equations (1), (2), and (3) rely on the following simplifications: (a) the fivefold degeneracy of the d band is neglected; (b) the intra-atomic exchange integral is assumed larger than b_{ij} (so that transfer for $\theta_{ij}=\pi$ is indeed negligible); (c) the ionic spins $\mathbf{S}_i$ are described as classical vectors; (d) the ions are held rigidly at their equilibrium positions; (e) Coulomb interactions between carriers are not considered; (f) interactions between the carriers and the compensating charges of opposite sign (e.g., Ca^{++} substituted for La^{+++}) are averaged out. Removal of (a) would not qualitatively modify the considerations to be developed in the following but would only complicate matters by introducing more unknown parameters; (b) and (c) are good starting approximations in all cases; (d) neglects the strong coupling between carriers and lattice vibrations, the importance of which has been stressed by Zener.[5] However, we shall be interested only in the lowest energy levels of the carriers and only in the part of this energy which depends on the orientations of the ionic spins. It is then reasonable to treat the polaron self-energy as an additive constant, so that (d) is an acceptable assumption. On the other hand a study of effective masses and mobilities would require a more detailed treatment of the carrier-phonon interaction, so that we do *not* expect the b_{ij}'s to be simply related to the electrical conductivity. Assumption (e) restricts us to dilute carrier systems. Fortunately this is a mild requirement because double exchange effects are often strong even in this limit. Assumption (f) is a drastic simplification. It amounts to neglecting all possible bound states of the carriers around the impurities which have been used to create a state of mixed valency. One might argue that as soon as the conductivity of the mixed specimens is much larger than the conductivity of the pure material, a band picture is appropriate. However, this conductivity, although high, often shows a temperature dependence corresponding to an activation energy; we believe that for low concentrations only a small fraction of the Zener electrons is involved in the conduction process, while all of them (bound or not bound) participate in double exchange. The essential observation here is that the over-all effects of bound carriers on the ionic spins is in fact very similar to the effect of the free carriers described by Eq. (2), as will be shown in Sec. IV. (In both cases the carrier energy is lowered by a distortion of the ionic spin arrangement.) The magnetic properties of our assembly are not strongly affected by assumption (f), and our simple model is indeed applicable.

To study effects of thermal excitation it is very important to recognize that in many cases the over-all band width of the carriers is expected to be large when compared with the temperatures of interest. (From the transition temperatures in the manganite series we infer that the band width is at least $\sim$0.1 ev and probably larger.) This shows that the anomalies in paramagnetic behavior predicted by Anderson and Hasegawa,[4] which are due to a uniform filling of the band, cannot be observed in general. The opposite situation, where the carriers fill only the bottom of the band, is much closer to the actual state of affairs, and we shall deal uniquely with this limiting case. The behavior of the system at finite temperatures still remains an extremely complicated problem, and the difficulties are twofold: first, we have to know the ground state energy of the one-carrier Hamiltonian for all arrangements of the ionic spins. This is a question of wave propagation in a three-dimensional disordered medium, and can be solved only by means of very rough approximations. Second, there is the problem of the statistical behavior of the ionic spins submitted to the double exchange coupling. As we shall see in Sec. III, even the molecular field approach involves some labor in this instance, but, apart from these computational difficulties, we are able to get a clear picture of the successive transitions that occur. The major difficulty with which we are left is then due to the rather large number of interaction constants which have to be derived from experiment; for instance, when dealing with a "layer" antiferromagnet, we need both the intra-layer and inter-layer exchange couplings of the pure material, and the two corresponding transfer integrals, that is to say four constants. Possible means of deriving these constants are discussed in Sec. V.

[5] C. Zener, J. Phys. Chem. Solids 8, 26 (1959).

II. CANTED SPIN ARRANGEMENTS AT LOW TEMPERATURES

We now consider a Bravais lattice of magnetic ions, and further assume that the spin ordering of the unperturbed system is of the *"antiferromagnetic layer"* type. Each ionic spin of length S is coupled ferromagnetically to z' neighboring spins in the same layer, and antiferromagnetically to z spins in the adjacent layers. The exchange integrals are called $J'(>0)$ and $J(<0)$. The Zener carriers are allowed to hop both in the layer (with transfer integrals b') and also from one layer to the other (with transfer integrals b). The number of magnetic ions per unit volume is called N, and the number of Zener carriers Nx. We shall consider configurations in which all spins within each layer remain parallel, but where the angle between magnetizations of successive layers takes a prescribed value θ. We determine Θ_0 by minimizing the sum of exchange and double exchange energies. The exchange contribution is

$$E_{ex}=-Nz'J'S^2+Nz|J|S^2\cos\Theta_0. \quad (4)$$

To obtain the double exchange contribution we first compute the energy E_k of a Zener carrier of wave vector k. The amplitude of the carrier wave function is $\alpha_i=e^{i\mathbf{k}\cdot\mathbf{R}_i}$ and Eqs. (1) and (3) give us

$$E_k=\sum t_{ij}e^{i\mathbf{k}\cdot(\mathbf{R}_j-\mathbf{R}_i)} \quad (5)$$

$$=-b'\gamma_k'-b\gamma_k\cos(\Theta_0/2) \quad (6)$$

where $\gamma_k'=\sum_j' e^{i\mathbf{k}\cdot(\mathbf{R}_j-\mathbf{R}_i)}$ and $\gamma_k=\sum_j e^{i\mathbf{k}\cdot(\mathbf{R}_j-\mathbf{R}_i)}$ are sums extended to the nearest neighbors of site (i) which are respectively in the same layer or in different ones (note that $\gamma_0'=z'$, $\gamma_0=z$). As explained in the introduction, we are only interested in the bottom of the energy band defined by (6) because we restrict ourselves to a small carrier concentration ($x\ll 1$). The minimum of (6) depends on the geometrical configuration of the magnetic lattice, and on the sign of b and b'. We shall simplify the discussion by assuming (arbitrarily) that b and b' are positive. Then

$$E_m=-\gamma_0'b'-\gamma_0 b\cos(\Theta_0/2). \quad (7)$$

Other signs for b and b' would lead to the same physical results but sometimes require a more complicated notation (when the band is not symmetrical). For simple layer configurations like the one shown on Fig. (2a), Eq. (7) is always valid provided we replace b and b' by their absolute value. The carriers, being few in number, occupy only energy levels close to E_m and the total double exchange energy is

$$E_d=NxE_m. \quad (8)$$

Minimizing the sum of (4) and (8) with respect to Θ_0 we get

$$\cos(\Theta_0/2)=bx/4|J|S^2 \quad (9)$$

$$E=E_{ex}+E_d=N[-z'J'S^2-xz'b' -z|J|S^2-(z/8)b^2x^2/|J|S^2]. \quad (10)$$

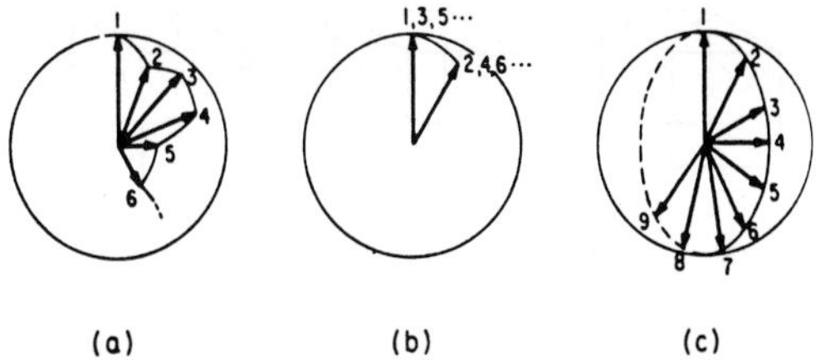

FIG. 1. Allowed configurations for the magnetizations $\mathbf{I}_1$, $\mathbf{I}_2$, $\mathbf{I}_3$, $\cdots$, $\mathbf{I}_n$, $\cdots$ corresponding to successive layers. The angle between $\mathbf{I}_n$ and $\mathbf{I}_{n+1}$ is equal to Θ_0. (a) disordered; (b) two sublattice system; (c) helical arrangement. (a), (b), and (c) are degenerate from the standpoint of nearest neighbor exchange and double exchange. External fields, anisotropy energies, or small ferromagnetic couplings between next nearest layers favor configuration (b). Antiferromagnetic coupling between next nearest layers favors (c).

For $x<4|J|S^2/b$ Eq. (9) defines an angle Θ_0 between 0 and π, and the magnetizations in the successive layers point in different directions.

All arrangements which satisfy Eq. (9) are degenerate. There is a large number of them, because, when we go from layer n to layer $n+1$ the magnetic moment I_{n+1} of layer $(n+1)$ is allowed to take any orientation on a cone (of some angle θ) around I_n. A typical arrangement is shown on Fig. 1(a). The degeneracy is removed by energy terms not included in our model, of which the most important are probably magnetocrystalline energies. For general values of θ, the simple two-sublattice ordering of Fig. 1(b) is then stabilized. For some special values of θ (e.g., $\theta=2\pi/n$, where n is a small integer) a helical arrangement as shown on Fig. 1(c) might still be degenerate with the two-sublattice arrangement. However, this is an exceptional situation, and even then a small magnetic field is enough to restore the two-sublattice ordering, where a nonzero spontaneous moment is present. It is probably worthwhile at this stage to point out the difference between the ordering of Fig. 3(c) and the helical spin arrangements considered by Villain,[6] Yoshimori and Kaplan[7] in materials with pure exchange forces. The latter arrangements are nondegenerate (apart from trivial degeneracies due to crystal symmetries) and can be destroyed only by magnetic fields of the order of the exchange field. In our case, on the other hand, all the orderings shown in Fig. 1 are degenerate, and only a few of the, such as 1(b), are not destroyed by application of an external field. In the following we shall restrict our attention to two-sublattice systems of type 1(b). The general case will be considered briefly in Sec. V. We now derive a few important properties of our spin assembly at $T=0$.

(1) The spontaneous magnetization of the two sublattice arrangement is

$$M=I\cos(\Theta_0/2) \quad (11)$$

[6] J. Villain, J. Phys. Chem. Solids 11, 303 (1959).

[7] A. Yoshimori, J. Phys. Soc. (Japan) 14, 807 (1959); T. H. Kaplan, Phys. Rev. 116, 888 (1959).

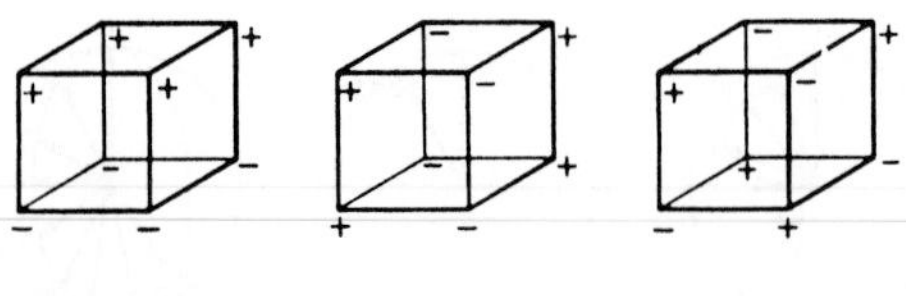

FIG. 2. Examples of "layer" antiferromagnet (2a), "chain" antiferromagnet (2b), and "alternating" antiferromagnet (2c) in a simple cubic lattice. The (2a) structure is observed in $LaMnO_3$, the (2c) structure in $CaMnO_3$.[2]

where $I/2$ is the magnetization of one sublattice. By making use of Eq. (9) this becomes

$$M/I = bx/4|J|S^2. \qquad (12)$$

The ferromagnetic moment is accordingly proportional to x (at low x).

(2) There is a nonzero susceptibility in high fields, due to a field induced change in angle between sublattices. This susceptibility is isotropic, because in high fields all magnetocrystalline terms are expected to be negligible. The spin system rotates to set its moment parallel to the field. The Zeeman energy is

$$E_Z = -HI\cos(\Theta/2), \qquad (13)$$

and the moment is always given by (11), but with an angle Θ between sublattices different from Θ_0. We derive Θ by taking the minimum of the sum (4), (7), and (13):

$$\cos(\Theta/2) = (bx/4|J|S^2) + HI/4|J|NS^2\gamma_0. \qquad (14)$$

The susceptibility is then

$$\chi = I^2/4|J|\gamma_0 S^2 N. \qquad (15)$$

Apart from possible small corrections in I, it is identical to the transverse susceptibility of the pure material. It could be measured by the same experimental techniques which have been applied to triangular spin arrangements in spinels.[8] In the manganite series such a lack of saturation in high fields has been qualitatively observed.[2]

(3) The magnetic scattering of neutrons shows an unusual pattern. Let us call $\boldsymbol{\mu}_1$ and $\boldsymbol{\mu}_2$ respectively the moments carried by one spin on each sublattice, and put $\mu = \mu_1 = \mu_2$. The scattering intensity if proportional to the square of: $\boldsymbol{\mu} = (1/k)\mathbf{k}\times(\boldsymbol{\mu}_1 \pm \boldsymbol{\mu}_2)$ where $\mathbf{k}$ is the scattering vector,[9] and the + or − sign correspond respectively to "lattice" reflections (L) (where both sublattices are in phase) and to "superlattice" reflections (S) (where they are opposite in phase). *Both types of reflections are simultaneously observed* in general. For instance, if $\mathbf{k}$ is perpendicular to the plane ($\boldsymbol{\mu}_1\boldsymbol{\mu}_2$) of the spins the ($L$) and ($S$) intensities are respectively proportional to $\cos^2(\Theta_0/2)$ and $\sin^2(\Theta_0/2)$ where Θ_0 is defined by (9). We see that neutron diffraction measures Θ_0 directly. Using the notation of Wollan and Koehler[2] we may write

$$\mu_{\text{ferro}}^2 = \mu^2\cos^2(\Theta_0/2), \qquad (16a)$$

$$\mu_{\text{antif}}^2 = \mu^2\sin^2(\Theta_0/2). \qquad (16b)$$

We now apply the above considerations to the experimental data on the manganites[2] for low Mn^{4+} content. Pure $LaMnO_3$ is indeed a layer antiferromagnet, as shown on Fig. 2. For the mixed compounds, we derive from (16a) and (16b) two sets of values for $\cos(\Theta_0/2)$ as a function of x. These values are plotted on Fig. 3. It may be seen that they coincide rather well, and that the linear relation (9), with all effects of higher order in x neglected, accounts reasonably for the data. From the slope of this plot we infer that $b/|J|S^2 \sim 16$. (Unfortunately we have no information on the interlayer exchange constant $|J|$ in $LaMnO_3$.) This figure shows incidentally that the carrier band width is indeed large when compared with the exchange energies in the pure material, as mentioned in the introduction.

We shall now discuss briefly the effect of an applied magnetic field on the neutron lines. If this field is applied parallel to the scattering vector $\mathbf{k}$, and if it is strong enough to overcome all anisotropy forces, the ferromagnetic reflections are extinguished. On the other hand, if the anisotropy fields are comparable in strength to the applied field, a very complicated situation is obtained, and the (L) lines are not completely extinguished. This appears to be the case in the mixed manganites with low Mn^{4+} content.

We observe incidentally that if helical arrangements were stabilized by some auxiliary coupling (such as a small antiferromagnetic exchange between next nearest layers) they would give rise to another class of neutron lines, which cannot in general be indexed in any multiple of the unit cell.[7]

(4) Another interesting question is related to possible nuclear resonance experiments on nonmagnetic ions belonging to the structure. Assume for instance that the magnetic atoms within different layers are separated

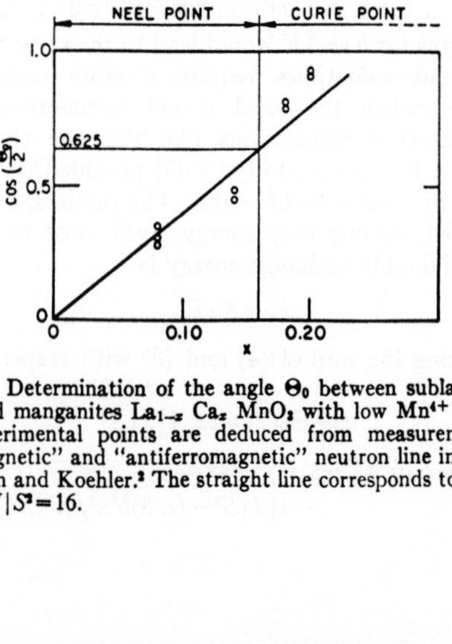

FIG. 3. Determination of the angle Θ_0 between sublattices in the mixed manganites $La_{1-x}Ca_xMnO_3$ with low Mn^{4+} content. The experimental points are deduced from measurements of "ferromagnetic" and "antiferromagnetic" neutron line intensities by Wollan and Koehler.[2] The straight line corresponds to Eq. (9) with $b/|J|S^2 = 16$.

[8] I. S. Jacobs, J. Phys. Chem. Solids 11, 1 (1959).
[9] O. Halpern and M. H. Johnson, Phys. Rev. 55, 898 (1939).

by fluorine ions, and that the environment of each fluorine contains an equal number of spins from sublattices 1 and 2. Then, in the pure material, we expect that the hyperfine fields acting on the F^{19} nucleus due to both sublattices will cancel exactly. In the mixed compounds however, they will not, and there will be a shift in the resonance, proportional to $\cos(\theta/2)$. One expects two types of nuclear relaxation due to coupling with (a) the carriers and (b) the spin waves. From the spin lattice times measured both in antiferromagnetic insulators[10] and in ferromagnetic metals,[11] we infer that relaxation effects should not prevent observation of the line. Unfortunately, there are inhomogeneities in the hyperfine field due to local spin distortions not included in the present model (see Sec. IV). This inhomogeneous broadening is expected to be large, and the experiment does not seem feasible.

This section has been restricted to calculations on "layer" antiferromagnets. We would like to mention that there is another class of antiferromagnetic materials where carrier motion is allowed in the unperturbed structure, namely the "chain" structure, an example of which is shown on Fig. 2(b). Considerations very similar to the above may be applied there. The number of allowed canted arrangements which are degenerate with respect to exchange and double exchange is increased: the only requirement is again that the angle between the magnetizations carried by neighboring chains is equal to some fixed value Θ_0. This can be accomplished in many ways and configurations with more than two sublattices may have to be considered here even when magnetocrystalline forces are present.

III. MAGNETIC BEHAVIOR AT FINITE TEMPERATURES

The statistical properties of our spin assembly at finite temperatures may be approached by a study of elementary excitations (spin waves) or by some extension of the molecular field concept. Both techniques are much increased in complexity when double exchange carriers are present. The most interesting property finally to be displayed is the existence of a first transition point when the canted arrangement collapses, the system becoming ferro- or antiferromagnetic at higher temperatures. This property is beyond the scope of spin-wave analysis, and we accordingly concentrate here on the more useful molecular field description. (The low-frequency spin-wave spectrum is considered in Appendix 1.) A first step is to derive the energy of carriers at the bottom of the band where the ionic spins are not completely ordered. We get an approximate value for this energy by taking as a variational wave function in Eq. (2) the very simplest one where all α_i's are equal. (This is the exact eigenfunction for all ordered states with our choice of sign for the transfer integrals.) We obtain

$$E_m=-\sum_j b_{ij}\langle\cos(\theta_{ij}/2)\rangle$$

where the $\langle\ \rangle$ symbol represents a thermal average on the possible states of the ionic spins. We may rewrite the preceding equation for our layer antiferromagnet in the form

$$E_m=-\gamma_0 b\langle\cos(\theta/2)\rangle-\gamma_0' b'\langle\cos(\theta'/2)\rangle \qquad (17)$$

where θ' and θ are the angles between any ionic spin and its neighbors respectively in and out of the layer. We now make use of the molecular field approximation in the following way: we neglect all correlations between different ionic spins, and assume for each of them a statistical distribution corresponding to a molecular field.

$$w_n(\mathbf{S})=(1/\nu)\exp(-\boldsymbol{\lambda}_n\cdot\mathbf{S}/S). \qquad (18)$$

The index n specifies the sublattice to which $\mathbf{S}$ belongs ($n=1$ or 2); $\boldsymbol{\lambda}_n$ is proportional to the molecular field acting on sublattice (n). The two molecular fields are equal in length ($\lambda_1=\lambda_2=\lambda$) and the angle between them is Θ. The normalization constant ν is:

$$\nu=\int_{-1}^{1} du e^{-\lambda u}$$
$$=2(\sinh\lambda)/\lambda. \qquad (19)$$

The relative amount of saturation of each sublattice is:

$$m=\frac{1}{\nu}\int_{-1}^{1} du u e^{-\lambda u}$$
$$=-(1/\lambda)+\mathrm{ctanh}\lambda. \qquad (20)$$

Our aim is to insert the assumed distribution function (18) into a variational principle for the free energy. We first derive the entropy term:

$$-TS=Nk_BT\int_{-1}^{1} w(x)\ln w(x)dx$$
$$=Nk_BT(\lambda m-\ln\nu). \qquad (21)$$

The calculation of the energy term is somewhat more complicated. Consider for instance the first term of (17), which corresponds to double exchange between sublattices. Let us write

$$\cos(\theta/2)=\sum_{l=0}^{l=\infty} A_l P_l(\cos\theta). \qquad (22)$$

The coefficients A_l can be obtained from the generating function of the P_l's (see Appendix 2). They are given by

$$A_l=(-)^{l+1}\frac{2}{(2l-1)(2l+3)}. \qquad (23)$$

[10] R. G. Shulman and V. Jaccarino, Phys. Rev. 108, 1219 (1957).
[11] A. C. Gossard and A. M. Portis, Phys. Rev. Letters 3, 164 (1959).

We then express $P_l(\cos\theta)$ in terms of the polar angles of the two spins S_1, S_2 which serve to define the angle θ. The average orientations of S_1 and S_2 are not parallel; since they belong to different sublattices, the angle between them is Θ. It is convenient to refer the polar angles $(\theta_1\varphi_1)$ of $\mathbf{S}_1$ to a polar axis parallel to $\boldsymbol{\lambda}_1$ and $(\theta_2\varphi_2)$ to an axis parallel to $\boldsymbol{\lambda}_2$. With the usual notation for spherical harmonics $Y_{lm}(\theta\varphi)$ and Wigner rotation coefficients D^l, we get:

$$P_l(\cos\theta)=N_l\sum_{mm'} Y_{lm}(\theta_1\varphi_1)Y_{lm'}^*(\theta_2\varphi_2)D_{mm'}{}^l(0,\Theta,0). \tag{24}$$

The N_l are normalization factors. Let us take the thermal average of (24) with independent statistical distributions w_1, w_2 as defined in (18). The only nonzero term corresponds to $m=m'=0$. $D_{00}{}^l(0,\Theta,0)$ is simply a Legendre polynomial, and the final result is

$$\langle P_l(\cos\theta)\rangle=\langle P_l(\cos\theta_1)\rangle\langle P_l(\cos\theta_2)\rangle P_l(\cos\Theta). \tag{25}$$

Furthermore

$$\langle P_l(\cos\theta_1)\rangle=\frac{1}{\nu}\int_{-1}^{1} e^{-\lambda u}P_l(u)du$$
$$=(2/\nu)i^l j_l(-i\lambda) \tag{26}$$

where j_l is the usual spherical Bessel function, taken here for an imaginary argument. Collecting the results of Eq. (22) to Eq. (26) we obtain

$$\langle\cos(\theta/2)\rangle=-\frac{2}{j_0{}^2(-i\lambda)}\sum_{l=0}^{\infty}\frac{j_l{}^2(-i\lambda)P_l(\cos\Theta)}{(2l-1)(2l+3)}. \tag{27}$$

In a similar way, if we deal with the angle θ' of two neighboring spins belonging to the same sublattice, we obtain

$$\langle\cos(\theta'/2)\rangle=-\frac{2}{j_0{}^2(-i\lambda)}\sum_{l=0}^{\infty}\frac{j_l{}^2(-i\lambda)}{(2l-1)(2l+3)}. \tag{28}$$

We insert (27) and (28) in (17), and multiply by the number Nx of carriers to obtain the total double exchange energy

$$E_D=\frac{2Nxz}{j_0{}^2(-i\lambda)}\sum_{l=0}^{\infty}\frac{j_l{}^2(-i\lambda)}{(2l-1)(2l+3)}\times[\gamma_0 bP_l(\cos\Theta)+\gamma_0'b']. \tag{29}$$

Since λ is related to the sublattice magnetization m by Eq. (20), we may consider (29) as expressing the double exchange energy in terms of the macroscopic parameters m and Θ. We will have to add the usual exchange energy

$$E_{\text{ex}}=-Nm^2(z'J'+zJ\cos\Theta). \tag{30}$$

The free energy F is the sum

$$F=-TS+E_D+E_{\text{ex}}. \tag{31}$$

This has to be a minimum with respect to λ (or m) and with respect to Θ. Let us first carry out the variation with respect to $\cos\Theta$. We obtain

$$m^2+\frac{2\xi}{j_0{}^2(-i\lambda)}\sum_l\frac{j_l{}^2(-i\lambda)}{(2l-1)(2l+3)}\frac{dP_l(v)}{dv}=0 \tag{32}$$

where $\xi=bx/|J|S^2$ and $v=\cos\Theta$. For each λ (31) is an implicit equation for v. For large λ (low temperatures) the solution v_m is such that $-1<v_m<1$: the spin arrangement is canted. For small λ (higher temperatures) v_m is outside the interval $(-1, 1)$: the equilibrium configuration is ferromagnetic ($v_m>1$) or antiferromagnetic ($v_m<-1$). The critical value of λ (or m) is obtained by setting $v=\pm1$ in Eq. (32), and observing that

$$\left(\frac{dP_l(v)}{dv}\right)_{v=1}=\tfrac{1}{2}l(l+1)$$
$$\left(\frac{dP_l(v)}{dv}\right)_{v=-1}=(-)^{l+1}\tfrac{1}{2}l(l+1). \tag{33}$$

The resulting equations for the critical λ's may be written in the form

$$\frac{1}{\xi}=\frac{1}{j_1{}^2(-i\lambda)}\sum_{l=1}^{\infty}\frac{l(l+1)}{(2l-1)(2l+3)}j_l{}^2(-i\lambda)$$

(canted to ferro) (34a)

$$\frac{1}{\xi}=-\frac{1}{j_1{}^2(-i\lambda)}\sum_{l=1}^{\infty}(-)^l\frac{l(l+1)}{(2l-1)(2l+3)}j_l{}^2(-i\lambda)$$

(canted to antiferro). (34b)

For finite λ the convergence of these series is good. λ and m are related by Eq. (20), and the best graphical representation of Eq. (34) is in fact obtained by plotting m^2 as a function of ξ (Fig. 4). We see immediately from this plot that for small m the arrangement is never canted (except for the special value $\xi=2.5$.) The upper transition point always corresponds to a colinear configuration for the magnetizations of both sublattices. By lowering the temperature, however, we increase m and finally intersect the critical curve at a well-defined temperature T_1 which we refer to as the second critical point. It may be shown that m and Θ are both continuous functions of temperature at $T=T_1$. A complete study of the thermal behavior is complicated by the large number of independent parameters which come into play (J,J',b,b'). This is why Eq. (34), which involves only one parameter (ξ), is of particular interest.

We now consider the dependence of the free energy (31) on m in the "colinear" range (above T_1), and more specifically the limit of small m, which corresponds to the upper Curie point. Equation (31) may be expanded in the following form:

$$F=F_0+F_2m^2+F_4m^4+\cdots \tag{35}$$

where

$$(1/N)F_2=\tfrac{3}{2}k_BT-S^2(z'J'+zJv)-(2x/5)(z'b'+vzb). \tag{36}$$

We know from the preceding argument that only the parallel ($v=1$) and antiparallel ($v=-1$) cases have to be considered, in which case F_4 takes the simple form

$$(1/N)F_4=(6x/7\times 25)(z'b'+zb)+(9/20)k_BT. \tag{37}$$

We now write that (35) is a minimum and obtain

$$m^2=-(F_2/2F_4). \tag{38}$$

The upper transition point corresponds to $F_2=0$. It is given by the greater of the two quantities quoted in the following:

$$T_C=(2/3k_B)[(zJ+z'J')S^2+(2x/5)(z'b'+zb)] \quad \text{(Curie point)} \tag{39a}$$

$$T_N=(2/3k_B)[(-zJ+z'J')S^2+(2x/5)(z'b'-zb)] \quad \text{(Néel point)}. \tag{39b}$$

(Remember that for the cases in which we are interested J is negative and J' positive.) We emphasize the fact that only two transition points are observed in all cases (T_1 and T_C or T_1 and T_N). Another interesting quantity is the paramagnetic Curie point T_P defined through the asymptotic form of the paramagnetic susceptibility $\chi=C/(T-T_P)$. When the carrier band width is small compared to k_BT there is no contribution to T_P from the double exchange effect, as emphasized by Anderson and Hasegawa.[4] On the other hand, when k_BT remains small compared with the band width, as it probably

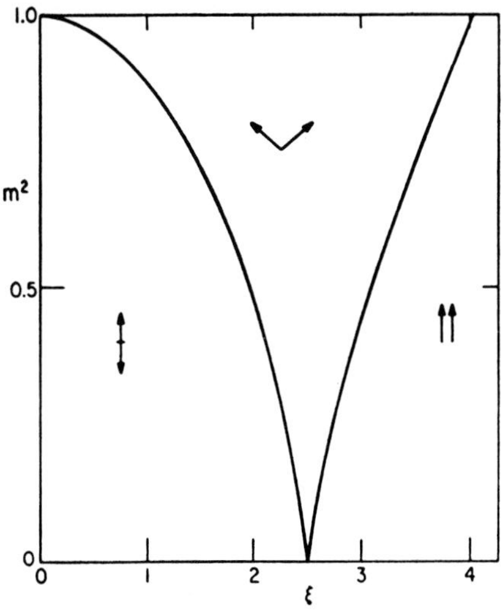

FIG. 4. Square of the relative saturation $m_{T_1}=M_{T_1}/M_0$ of each sublattice at the lower transition point T_1, as a function of $\xi=bx/|J|S^2$.

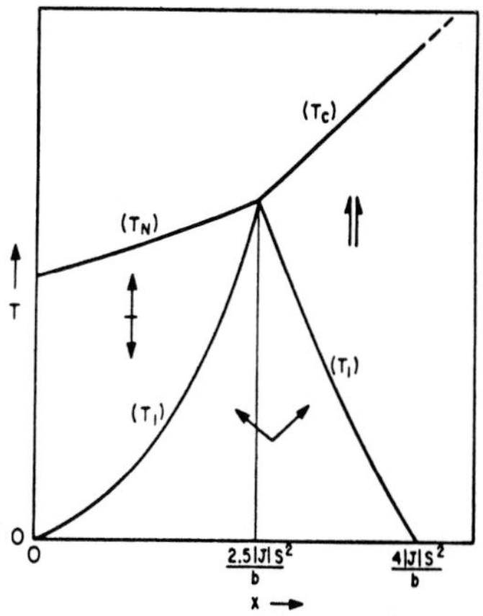

FIG. 5. Typical "magnetic phase diagram" for a layer antiferromagnet. Theoretical values of T_1, T_c, and T_N are given in the text.

does in practice, our approximation scheme remains valid. If we retain the same simple form of variational wave function to obtain the carrier ground-state energy, we can show by the Van Vleck trace method that

$$k_BT_P=\tfrac{2}{3}S^2\sum_j J_{0j}+(4x/15)\sum_j b_{0j} \tag{40}$$

so that T_P and T_C coincide within the molecular approximation, as in the more familiar case of pure exchange. It is not possible to derive simple expressions for the lower transition point T_1, except in the limiting case where ξ is not very different from 2.5. It is then easy to see from Fig. 4 that the transition (T_1) occurs for small values of m, so that the expansion (35) for the free energy may be used on the whole temperature interval between the two transition points. In this range of ξ values Eqs. (34a) and (34b) for $[m]_{T_1}$ take the simple form

$$[m^2]_{T_1}=\frac{175}{18}\left|\frac{1}{\xi}-\frac{2}{5}\right|. \tag{41}$$

By eliminating m^2 between (41) and (38) we get

$$T_1=T_0-\frac{2}{3k_B}\left|\frac{1}{\xi}-\frac{2}{5}\right|\left[\frac{2x}{3}(zb+z'b')+\frac{35}{4}k_BT_0\right]\times\left|\frac{1}{\xi}-\frac{2}{5}\right|\ll 1 \tag{42}$$

where T_0 is the upper transition point as defined by (39). Figure 5 shows qualitatively the stability domains of the different spin configurations when both the temperature T and the parameter ξ (proportional to x) are raised. The angle Θ_0 between sublattices in the ground state is given by (9). When Θ_0 is larger than 103° ($\xi<2.5$) the canted arrangement becomes antiferromagnetic by raising the temperature above T_1. When Θ_0 is smaller than 103° ($4>\xi>2.5$) it becomes ferromagnetic.

We close this section by a brief discussion of the

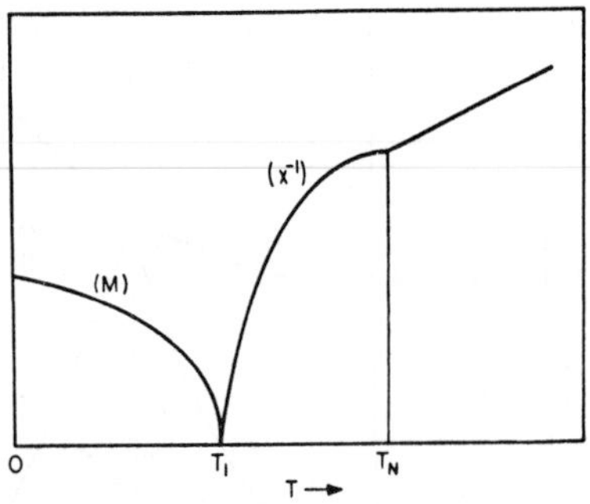

FIG. 6. Typical plot of inverse susceptibility and spontaneous magnetization when $\xi<2.5$ (magnetocrystalline energies are neglected).

physical anomalies which are expected of the lower transition point T_1.

(1) The behavior of the susceptibility χ is remarkable when the upper transition point is antiferromagnetic. By decreasing T we first observe the usual discontinuity in slope at T_N. Then χ increases and becomes infinite at T_1. The qualitative behavior of $1/\chi$ is shown on Fig. 6. Experimental results very similar in their general appearance to the graph of Fig. 6 have been obtained in antiferromagnetic CrSb doped with a small amount of MnSb.[12] The pure compound CrSb is known to be of the antiferromagnetic layer type.[13] It may be that electron transfer is allowed between chromium and manganese atoms, in which case our model could be applied.

(2) Below T_1 both ferromagnetic and antiferromagnetic neutron lines occur simultaneously. Above T_1 only one type is observed. A phenomenon of this type has been indeed quoted by Wollan and Koehler for some mixed manganite samples.[2]

(3) The specific heat shows a (small) discontinuity at $T=T_1$.

(4) The electrical conductivity is favored by interlayer transfer. We accordingly expect a slight discontinuity in slope at $T=T_1$. The sign of this discontinuity will depend on the type of order above T_1. However, our oversimplified model is not applicable to this problem, and we are not able to make more detailed predictions as regards this point.

IV. LOCAL SPIN DISTORTIONS

1. Bound States

The simple model of Secs. II and III did not take into account any possible bound states of the carriers. Such bound states may in fact occur, especially when the amount of impurities is very small. For instance, each Ca^{2+} substituted for La^{3+} in $LaMnO_3$ acts as an effective charge $-e$ and is able to accept one carrier (a hole) in a localized orbit. We intend to show that the over-all effects of an assembly of such centers on the ionic spin system are similar to those of the free carriers considered in the preceding Section. We consider in particular the extreme case where the wave functions relative to different impurity centers do not overlap (very small x). In analogy with the known properties of color centers in alkali halides, we expect the wave functions to be of rather small extension; in the above example, the hole will be shared by the eight manganese atoms surrounding the impurity, as shown on Fig. 7.

We first consider a single impurity center in an unperturbed antiferromagnetic matrix, and restrict ourselves to the following simple example: the pure material is an "alternating" ferromagnet of one simple cubic [see Fig. 2(c)] or bcc structure, with exchange integral J coupling each spin to its z neighbors. The Zener electron (or hole) is strongly bound and can only occupy *two* neighboring sites (1) and (2). The transfer integral is $t_{12}=b\cos(\theta_{12}/2)$ as before, and the propagation equations which replace (2) are

$$\begin{aligned}(E-U)\alpha_1&=b\cos(\theta_{12}/2)\alpha_2\\(E-U)\alpha_2&=b\cos(\theta_{12}/2)\alpha_1\end{aligned}\tag{43}$$

where U is the binding energy. (Note that the value of b may be modified by the attractive potential.) The ground state corresponds to

$$E_D=U-b\cos(\theta_{12}/2).\tag{44}$$

This is the analog, for a localized state, of Eq. (7) which applied to an extended state. In both cases, a small departure of θ_{12} from π decreases the double exchange energy in first order and increases the exchange energies only in second order, so that a canted arrangement will be stable. Here, however, the final configuration corresponds to a local distortion of the spin system, and is accordingly more difficult to compute. We shall first derive the amount of canting by a simple "rigid field approximation" (RFA) where all ionic spins other than $\mathbf{S}_1$ and $\mathbf{S}_2$ are assumed to retain their original orienta-

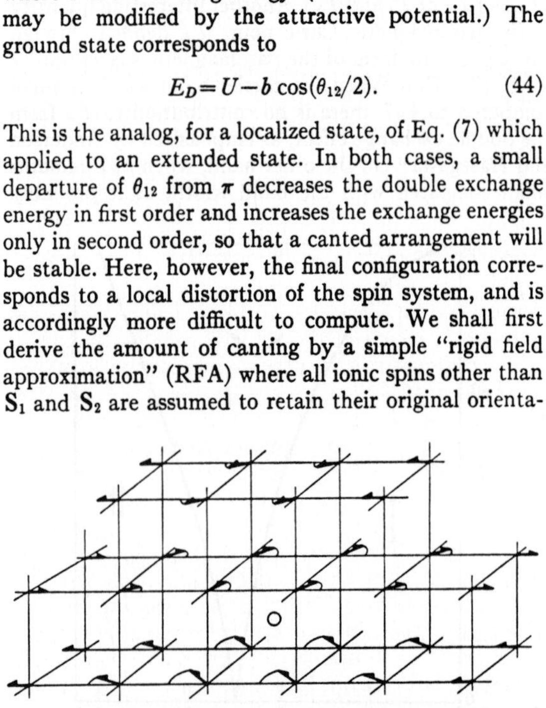

FIG. 7. Local spin distortion in $LaMnO_3$. The bound hole is localized on eight manganese atoms (black circles) around the impurity center Ca^{2+} (open circle). All ionic spins remain in one plane [here taken to be (001)]. Deflections are maximum close to the impurity center, but they decrease only slowly with distance.

[12] T. Hirone, S. Maeda, and I. Tsubokawa, J. Phys. Soc. (Japan) 11, 1083 (1956); E. W. Gorter and F. K. Lotgering, J. Phys. Chem. Solids 3, 238 (1957).

[13] A. I. Snow, Revs. Modern Phys. 25, 127 (1953).

tion. S_1 and S_2 have opposite deflections ϵ and $-\epsilon$ and $\theta_{12}=\pi-2\epsilon$. The increase in exchange energy is

$$E_{ex}=2|J|S^2[2(z-1)(1-\cos\epsilon)+1-\cos2\epsilon]. \quad (45)$$

The minimum of the sum of (44) and (45) corresponds to

$$-b\cos\epsilon+2|J|S^2[2(z-1)+4\cos\epsilon]\sin\epsilon=0. \quad (46)$$

This always gives a solution ϵ, with the limiting forms

$$\epsilon=\frac{b}{4|J|S^2(z+1)}; \quad (b/z|J|S^2\ll1)$$
$$\epsilon=\frac{\pi}{2}-\frac{4|J|S^2(z-1)}{b}; \quad (b/z|J|S^2\gg1). \quad (47)$$

A bound Zener electron *always* gives rise to a local distortion of the spin system. This is to be contrasted with the effects which are obtained with pure exchange, when some adequate substitution simply changes the sign of one exchange integral, between (1) and (2). In the latter case there is also a strongly inhomogeneous constraint applied to the spin system. However, this results in a local spin distortion only if the new (ferromagnetic) exchange integral exceeds a well-defined threshold value ($J_{new}>\frac{1}{2}(z-1)|J|$ in RFA, and $J_{new}>\frac{1}{2}(z-2)|J|$ by an exact calculation).

We have proved that in the vicinity of each impurity center there is an unbalanced magnetic moment due to the special effects of double exchange. There is however one feature which is deliberately neglected in the RFA: a local deflection of spins (1) and (2) is always accompanied by smaller distortions on the neighboring sites, the amplitude of which decreases only slowly with distance. These "wings" may be studied with good accuracy by making use of Green's function techniques, an example of which is given in Appendix 3. If we now consider not only one impurity but a dilute assembly of these, the "wings" result in a coupling of the unbalanced moments carried by the different impurity centers. At low temperatures it is energetically favorable for these moments to line up, thus reducing the spin distortions in the matrix. As a result we expect to observe a nonzero spontaneous moment, increasing linearly with x, and the over-all magnetic behavior is very similar to what we found in Sec. II by discussing free carriers. We also expect the "interlocking" between different centers to preserve this ordered state up to some critical temperature T_1, above which the unbalanced moments exhibit a paramagnetic behavior. Here again, the conclusions of the "free carrier model" and of the "color center model" are not very different.

We now mention briefly some phenomena for which both models do *not* lead to the same predictions for very dilute impurity systems. These are: (1) optical absorptions; (2) electrical conductivities; (3) nuclear resonance at anion sites; (4) neutron diffraction. The nuclear resonance lines are widely broadened, because the long-range part of the spin distortion due to each individual impurity center is responsible for inhomogeneous hyperfine fields. Consequently this type of experiment seems difficult to carry out. Neutron diffraction studies, on the other hand, could give some very interesting information on local spin distortions. This is discussed in Appendix 3, the main conclusions of which may be summarized as follows. (a) The distortions lead to diffuse peaks around the superlattice line. Superlattice lines for which the scattering vector is parallel to the spin direction in the pure material are extinguished and consequently most favorable. (b) There is always a parasitic spin wave scattering even at low temperatures (because of the strong zero point motion effects in antiferromagnets). (c) However, spin wave emission is an inelastic process while scattering by static distortions is strictly elastic. As a result, by a suitable geometrical arrangement one can make the spin wave scattering very diffuse so that it will be automatically subtracted with the background. The final requirement is that the neutron spectrometer should be able to analyze an angular distribution corresponding to a diffuse peak whose integrated intensity is roughly x times smaller than the intensity of a typical magnetic line of the unperturbed structure. A typical value of x (small enough to reduce overlap between separate distortions) is 0.1, so that this figure is not prohibitive.

2. Self-Trapped Carriers

Sections II and III applied the "free carrier" model to layer and chain antiferromagnets, where the carriers are always allowed to move within each chain or layer. The "alternating" antiferromagnetic structures (where all neighbors of a + spin are − spins) are somewhat different; in the unperturbed structure the carriers are not allowed to move; it is then more favorable for each individual carrier to build up a *local* distortion of the spin lattice in which it becomes "self-trapped." The resulting centers are able to move only slowly, and their physical properties are modified. In practice, we are not very much concerned in effects such as a change in effective mass, because the slow carriers will always fall into bound states. We are mainly interested in the shape of the local distortions and in interactions between them.

We first show that self-trapping will indeed occur. The argument will be written down for a simple cubic- or body-centered cubic "alternating" antiferromagnet, each ionic spin having z equivalent neighbors, with exchange couplings $2JS^2\cos\theta_{ij}$, transfer integrals $b\cos(\theta_{ij}/2)$, and carrier concentration x defined as before. Let us first compute the gain in energy for a uniform canting of both sublattices in the free carrier model, as in Sec. II. We obtain

$$E_{free}=-\tfrac{1}{8}Nzb^2x^2/|J|S^2. \quad (48)$$

It is convenient to introduce the paramagnetic Néel point $T_N=[(2|J|S^2z)/3k_B]$ of the pure material, and the dimensionless ratio $\eta=b/k_BT_N$, so that Eq. (48) becomes:

$$E_{\text{free}}/k_BT_N=-(N/12)z^2\eta^2x^2. \tag{49}$$

We now consider another possible configuration, with noninteracting localized states, and compute the corresponding energy by making use of a variational principle. We assume that the carrier to be considered is trapped on some magnetic site (which we call 0 for instance) with an amplitude α_0, but we also allow it to have a nonzero amplitude α_1 on the z neighboring spins. As far as the ionic spins are concerned, we assume that they all retain the same orientation which they had in the pure material, except for $\mathbf{S}_0$, which is allowed to make an arbitrary angle θ with the common direction of the z neighboring spins. The exchange energy, counted from the initial configuration ($\theta=\pi$) is

$$E_{\text{ex}}=2|J|S^2z(1+\cos\theta). \tag{50}$$

The wave equation for the trapped carrier takes the restricted form

$$\begin{aligned} E\alpha_0&=zb\cos(\theta/2)\alpha_1\\ E\alpha_1&=b\cos(\theta/2)\alpha_0 \end{aligned} \tag{51}$$

and the ground-state carrier energy is

$$E_D=-b\sqrt{z}\cos(\theta/2). \tag{52}$$

By taking the minimum of $E_{\text{ex}}+E_D$ we obtain an overestimate of the energy per trapped carrier E_1

$$E_1/k_BT_N=-(z/24)\eta^2 \quad \text{if} \quad \eta<12/\sqrt{z} \tag{53a}$$

$$=6-\sqrt{z}\eta \qquad \eta>12/\sqrt{z}. \tag{53b}$$

Actually this estimate is not accurate for large couplings ($\eta\gg1$) where the "radius" of the trapped carrier exceeds one interatomic distance. Equation (53) is sufficient for our purposes, however, and shows that E_1 is negative. The total energy of the trapped carriers NxE_1 is proportional to x, while the energy (49) corresponding to uniform canting goes like x^2. At low x the self-trapped configuration is always more stable. The following remarks should be made.

(1) Equation (49) represents the average effect of an attractive interaction between carriers via the ionic spins; each carrier tends to cant the antiferromagnetic lattice and decreases the energy of all other carriers at the bottom of the band. This explains the x^2 dependence of (48).

(2) In layer antiferromagnets, the energy of the free carrier model contains a negative term, due to intralayer motion, and proportional to the number of carriers x; self-trapped states are much less favored in such structures. (Of course, in practice, we shall find bound states as explained earlier.) Another viewpoint, leading to the same result, is the following: in a layer or chain antiferromagnet spin distortions contribute only to a part of the band energy, and the coupling between carriers and ionic spins is moderately strong when compared with the unperturbed energy of the carriers. On the other hand, the coupling is very strong in an alternating antiferromagnet, since *all* the carrier energy is due to distortions of the spin arrangement.

The over-all effect of self-trapped carriers on magnetostatic properties is complex, and probably unobservable because of the existence of bound states. We shall restrict ourselves here to a few qualitative remarks, related to the behavior of the mixed manganites with *high* Mn^{4+} content. Pure $CaMnO_3$ is an alternating antiferromagnet (simple cubic). The mixtures of neighboring compositions are not conducting and do not show any ferromagnetic behavior. This has been interpreted by Goodenough[14] by means of a qualitative model which is somewhat related to self-trapping. He assumes that the extra electron reverses the spin $\mathbf{S}_0$ of the central site (θ going from π to 0) but leaves all other spins unaltered. This then results in a simple reduction of the sublattice magnetization, and does not bring in any noncompensated moment, since there is an equal number of trapped carriers on both sublattices. We are not quite satisfied with this explanation, for the following reason: from Eqs. (50) and (52) we may show that the actual configuration of $\mathbf{S}_0$ is canted ($0<\theta<\pi$) when $\eta<12z^{-\frac{1}{2}}$. For larger η's, our simple variational wave function yields a ground state which agrees with the Goodenough picture, but it is not a good approximation any more. In fact it is easy to see that for such values of η the whole configuration built up by $\mathbf{S}_0$ and its six nearest neighbors becomes canted with respect to the more distant spins. In all cases there is a noncompensated magnetic moment directed perpendicularly to the spin direction of the unperturbed structure. Furthermore, this moment is always accompanied by a long-range distortion of the spin lattice. Exactly like in the case of bound states, we expect these long-range distortions to couple ferromagnetically the moments due to different carriers, in contradiction with experiment. We do not believe that the discrepancy can be explained within our simple model.

V. CONCLUSIONS

The special form of the double exchange coupling is such that all antiferromagnetic (and also all ferrimagnetic) spin arrangements are distorted as soon as some Zener carriers are present. This is due to the fact that electron transfer lowers the energy by a term of first order in the distortion, while the initial exchange energy is increased only in second order. If the Zener carriers are free to move in the structure, the distortion will usually correspond to a uniform canting of the sublattices. If they are bound or nearly bound, the distortion is nonhomogeneous, but thc average effects are similar to the above. In practice the second alternative

[14] J. B. Goodenough, Phys. Rev. 100, 564 (1955).

is closer to the actual state of affairs. However the "free carrier model," corresponding to the first alternative, is expected to provide a good starting point in all cases, when one is only interested in the macroscopic magnetic properties of the system.

We now write down a short list of the physical quantities which may be used to determine the relevant exchange and transfer integrals in a given "layer" antiferromagnet. These are: (1) the paramagnetic Curie point, the Néel point, and the low-temperature susceptibility of the pure material (from which one can extract J and J'). (2) The spontaneous moment at $T=0$ of the mixed materials (from which we get the ratio $b/|J|$ as shown by Eq. (11). This moment may be obtained both by neutron diffraction or by magnetostatic measurements (the latter being clear cut only if the moment is obtained by an extrapolation to low fields of the high field B-H curve, as has been done by Jacobs[8] for triangular spin arrangements in spinels). (3) The paramagnetic Curie point and the two transition points of the mixed compounds, as given by (39) and (40). Location of the transition points is achieved with the best accuracy by using specific heats and electrical conductivities. The numbers derived from low-temperature data are of course the most reliable, because they do not make use of the very naive approximations of Sec. III as regards the carrier energy levels and the statistical behavior. The assumption that the carrier bandwidth is large when compared to k_BT should also be checked, and could eventually be improved.

In Sec. II we emphasized the fact that many degenerate canted configurations are always allowed in the presence of double exchange as illustrated by Fig. 1. Only the simplest (two sublattice) types were considered and the question may be raised whether this is a serious restriction or not. We now present some remarks related with the more general situation. (a) In spite of the degeneracy one and only one arrangement is stable at each temperature. This may be seen in the following way: if we go to a partially disordered state as shown in Fig. 1(a) we lower the free energy by an entropy term proportional to the number of layers ($\sim N^{1/3}$) or of chains ($\sim N^{2/3}$). On the other hand, the small magnetocrystalline or next nearest neighbor exchange terms are modified by an amount proportional to N; their contribution always dominates, and the only observable arrangements are those for which it is a minimum. (b) Of course, even these small energy terms do not remove the degeneracy between arrangements which can be deduced from one another by a symmetry operation of the lattice group. The crystal may accordingly break up into domains, and the domain wall energy may be extremely small in some cases. As an example think of two helical arrangements corresponding to Fig. 1(c), one right-handed, the other left-handed, with a sharp boundary parallel to the plane of the layers; the wall energy has no contribution from nearest neighbor exchange or double exchange. (c) Whatever the arrangement is, the angle Θ between neighboring units and the relative saturation m of each of them is still given by the formulas of Secs. II and III and the lower transition point T_1 is always observable.

Finally, we would like to mention the possible extension of all the above considerations to more complicated systems of mixed valency such as those derived from MnSe (rocksalt structure) or CrSb (NiAs structure). The case of $Cr_{1-x}Mn_xSb$ has already been mentioned in Sec. III. The compound $Mn_{0.9}Li_{0.1}Se$[15] is known to show two transition points T_1 and T_C. It is ferromagnetic between T_1 and T_C, and antiferromagnetic (with a strong parasitic ferromagnetism) at lower temperatures. Conductivity data have not appeared, and from the few experimental results presently available it is not yet possible to decide whether the low T arrangement is canted or not. We stress the fact that for the simpler case considered here a material which is ferromagnetic at high temperatures always displays a strong moment even below T_1, the angle between sublattice magnetizations being smaller than 103°, as explained in Sec. III.

ACKNOWLEDGMENTS

It is a pleasure to thank Professor C. Kittel and Professor A. M. Portis for discussions on these and related matters, and P. Pincus for checking the calculations.

APPENDIX 1. LOW-FREQUENCY SPIN WAVES

We consider the antiferromagnet layer structure of Sec. II, and study small amplitude motions of the ionic spin systems around the two sublattice equilibrium arrangement. The z axis is taken parallel to the spin direction in the pure material. The y axis is parallel to the spontaneous moment. The components of a spin $\mathbf{S}_i$ located on the first sublattice will be written as

$$S_i^x = S a_{ix}$$

$$S_i^y = S[\cos(\Theta_0/2) + a_{iy}]$$

$$S_i^z = S\sin(\Theta_0/2)\left(1 - \frac{a_{ix}^2 + a_{iy}^2}{2\sin^2(\Theta_0/2)} - \frac{a_{iy}\cos(\Theta_0/2)}{\sin^2(\Theta_0/2)} - \frac{1}{8}\frac{a_{iy}^2\cos^2(\Theta_0/2)}{\sin^4(\Theta_0/2)}\right) \quad (54a)$$

where terms up to second order in a_x, a_y have been retained. For a spin $\mathbf{S}_j$ located on the second sublattice we shall put

[15] S. J. Pickart, R. Nathans, and G. Shirane, Bull. Am. Phys. Soc. 4, 52 (1959); R. R. Heikes, T. R. McGuire, and R. J. Happel, Jr., Bull. Am. Phys. Soc. 4, 52 (1959).

$$S_j^x = Sb_{jx}$$

$$S_j^y = S[\cos(\Theta_0/2)+b_{jy}]$$

$$S_j^z = -S\sin(\Theta_0/2)\left(1-\frac{b_{jx}^2+b_{jy}^2}{2\sin^2(\Theta_0/2)} - \frac{b_{jy}\cos(\Theta_0/2)}{\sin^2(\Theta_0/2)} - \frac{1}{8}\frac{b_{jy}^2\cos^2(\Theta_0/2)}{\sin^4(\Theta_0/2)}\right). \quad (54b)$$

From these formulas we compute the exchange energy $-\sum J_{ij}S^2$ as θ_{ij} up to second order in a and b. To get the double exchange energy, we make use of three approximations: (a) low-frequency approximation; the Zener carriers are at every instant in the ground state corresponding to the distortion (54). (b) Long wavelength approximation; the distortion in (54) is nearly homogeneous. At every lattice point we may think of the carriers as occupying the bottom of a band whose width corresponds to the local value of the spin distortion. (c) Electrical neutrality approximation; we assume that the density of carriers is not modulated by the spin wave. (a), (b), and (c) may be shown to be entirely correct for the low-frequency part of the spin wave spectrum. We may then write the total energy as a sum of two terms, related with intralayer and interlayer couplings respectively:

$$E = E_{\text{intra}} + E_{\text{inter}} \quad (55)$$

and get:

$$E_{\text{inter}} = |J|S^2 \sum_{ij} (a_{iy}+b_{jy})^2 \quad (56)$$

$$E_{\text{intra}} = J'' \sum_{i>j} \{(a_{ix}-a_{jx})^2 + [1/\sin^2(\Theta_0/2)](a_{iy}-a_{jy})^2\} \quad (57)$$

where

$$J'' = J' + \tfrac{1}{4}xb'. \quad (58)$$

Equations (56) and (57) show that the energy is of second order in a and b, as expected. The fact that E_{inter} depends only on the y components of the distortion is simply a consequence of the degeneracy illustrated on Fig. 1. The dependence of E_{inter} on b has been eliminated by making use of Eq. (9) for Θ_0. From the energy formulas we can derive an effective field $\mathbf{H}_i$ acting on each spin $\mathbf{S}_i$ (our special choice of independent variables does not change the vector product $H_i \times S_i$). Putting $a_{ix} = A_x e^{i(\mathbf{k}\cdot\mathbf{R}_i+\omega t)}$, etc., we get finally:

$$\begin{aligned}
i\hbar\omega A_x &= [2J'/\sin(\Theta_0/2)](\gamma_0'-\gamma_k')A_y + 2J\sin(\Theta_0/2)(\gamma_0 A_y+\gamma_k B_y)\\
i\hbar\omega A_y &= -2J'\sin(\Theta_0/2)(\gamma_0'-\gamma_k')A_x\\
i\hbar\omega B_x &= -[2J'/\sin(\Theta_0/2)](\gamma_0'-\gamma_k')B_y - 2J\sin(\Theta_0/2)(\gamma_0 B_y+\gamma_k A_y)\\
i\hbar\omega B_y &= 2J'\sin(\Theta_0/2)(\gamma_0'-\gamma_k')B_x.
\end{aligned} \quad (59)$$

The secular equation derived from (59) is

$$\begin{aligned}
&(\hbar\omega)^4 - 2(\hbar\omega)^2[2J''(\gamma_0'-\gamma_k') + 2|J|\sin^2(\Theta_0/2)\gamma_0]2J''(\gamma_0'-\gamma_k')\\
&+(2J'')^2(\gamma_0'-\gamma_k')^2[2J''(\gamma_0'-\gamma_k') + 2|J|\sin^2(\Theta_0/2)(\gamma_0-\gamma_k)]\\
&\times 2|J|\sin^2(\Theta_0/2)(2\gamma_0)=0.
\end{aligned} \quad (60)$$

For small k (which is the only case where our approximations are meaningful) the solutions are

$$\begin{aligned}
(\hbar\omega_1)^2 &= 8\gamma_0|J|J''\sin^2(\Theta_0/2)(\gamma_0'-\gamma_k')\\
(\hbar\omega_2)^2 &= 2J''(\gamma_0'-\gamma_k')[2J''(\gamma_0'-\gamma_k') + 2|J|\sin^2(\Theta_0/2)(\gamma_0-\gamma_k)].
\end{aligned} \quad (61)$$

In the first branch ω is proportional to k (as it is in antiferromagnets), and in the second it is proportional to k^2 (as it is in ferromagnets). It is of interest to observe that both branches collapse when the intralayer coupling J'' vanishes.

APPENDIX 2. EXPANSION OF $\cos\theta/2$ IN TERMS OF LEGENDRE POLYNOMIALS $P_l(\cos\theta)$

We are interested in the coefficients A_l of Eq. (22). They are given by

$$A_l = \frac{2l+1}{2}\int_0^\pi P_l(\cos\theta)(\cos\theta/2)\sin\theta d\theta = \frac{2l+1}{2\sqrt{2}}\int_1^1 P_l(u)(1+u)^{\frac{1}{2}}du. \quad (62)$$

These integrals may be derived from the generating function

$$\frac{1}{(1-2uh+h^2)^{\frac{1}{2}}} = \sum_l h^l P_l(u) \quad (63)$$

by writing:

$$\int_{-1}^1 du\, y = 2\sqrt{2}\sum_l h^l \frac{A_l}{2l+1} \quad (64)$$

where $y^2 = [(1+u)/(1-2uh+h^2)]$. Equation (64) can be integrated by parts and yields

$$\frac{1}{2h}\left[h-1+(1+h)^2\frac{\tan^{-1}(h^{\frac{1}{2}})}{h^{\frac{1}{2}}}\right] = 2\sum_l \frac{A_l}{2l+1}h^l. \quad (65)$$

By expanding the left-hand side and identifying coefficients of h^l on both sides we get Eq. (23).

APPENDIX 3. LONG-RANGE PART OF A LOCAL SPIN DISTORTION

We choose to deal here with the simple example already used in the first part of Sec. IV (simple cubic alternating antiferromagnet with only one bond perturbed, between atoms 1 and 2). It was shown there that the spins S_1 and S_2 rotated by angles ϵ and $-\epsilon$

from their initial (antiparallel) orientations. We now consider the smaller deflections ϵ_i (assumed all in the same plane) which are found on the other magnetic sites ($i\neq 1, 2$). All these sites are submitted only to exchange forces, and for small ϵ_i then equilibrium conditions take the simple form

$$\gamma_0\epsilon_i=\sum_j{}' \epsilon_j \tag{66}$$

where the sum $\sum_j{}'$ is extended to all 6 nearest neighbors of site i, and

$$\gamma_k=\sum_j{}' e^{i\mathbf{k}\cdot\mathbf{R}_{ij}}. \tag{67}$$

[Note that γ_k, defined here by (67), for an alternating antiferromagnet, is different from the γ_k defined in Sec. II, for a layer antiferromagnet.] The solution of Eq. (66) which exhibits the required values on the perturbed sites is easy to express in terms of the Green's function

$$\Gamma_{lm}=A\sum_k \frac{\gamma_0}{(\gamma_0-\gamma_k)} e^{i\mathbf{k}\cdot(\mathbf{R}_l-\mathbf{R}_m)} \tag{68}$$

where the sum $\sum_k$ is extended over the first Brillouin zone and where A is a normalization factor chosen to give $\Gamma_{ll}=1$.

$$A^{-1}=\sum_k \frac{\gamma_0}{\gamma_0-\gamma_k}. \tag{69}$$

The numerical value of A is 0.65947.[16] The quantity Γ_{lm} considered as a function of l (or m) satisfies Eq. (66) except when $l=m$. It describes the deflection of spin $\mathbf{S}_m$ when one spin $\mathbf{S}_l$ has been submitted to a prescribed deflection. Γ_{lm} is a slowly decreasing function of $|\mathbf{R}_l-\mathbf{R}_m|$. It has the following asymptotic form

$$\Gamma_{lm}=(3A/2\pi)a/|\mathbf{R}_l-\mathbf{R}_m| \tag{70}$$

where a is the length of the cube edge. Let us first consider the case where all deflections are small, even on the perturbed sites (1) and (2). Then the complete solution of Eq. (66) satisfying the boundary condition is

$$\epsilon_i=\epsilon(\Gamma_{1i}-\Gamma_{2i})/(1-\Gamma_{12}). \tag{71}$$

This may still be simplified by observing that $1-\Gamma_{12}=A$.

We now discuss a few important features of this result. (1) The angles ϵ_i for (i) different from (1) and (2) are substantially smaller than ϵ. Consider for instance the sites (i') which are nearest neighbors of (1), (2) being excluded. The average deflection of these sites is $\bar{\epsilon}_{(i')}=\frac{1}{5}\sum_{i'}\epsilon_{i'}$. By manipulating the Γ functions, one easily shows from (71) that

$$|\bar{\epsilon}_{i'}|=\tfrac{1}{5}\epsilon. \tag{72}$$

This explains *a posteriori* why the rigid field approximation used in Sec. IV is a good starting point. (2) The deflection ϵ_i is proportional to $\Gamma_{1i}-\Gamma_{2i}$. At long distances from the impurity center, Γ_{1i} is similar to the potential of a point charge, as shown by Eq. (70) and $\Gamma_{1i}-\Gamma_{2i}$ behaves like the potential of a dipole. This gives rise to a very specific neutron diffraction pattern, which we now compute.

For small ϵ_i the scattering amplitude for a scattering vector $\mathbf{q}$ is, apart from polarization factors and normalization constants:

$$a(\mathbf{q})=\sum_i \epsilon_i\sigma_i e^{i\mathbf{q}\cdot\mathbf{R}_i}, \tag{73}$$

where $\sigma_i=\pm 1$ depending on the sublattice to which (i) belongs. Equation (73) takes into account only the scattering due to the spin distortion. (The amplitude due to the ordered structure vanishes except on Bragg peaks, which we discard.) Equation (73) may be transformed by making use of the solution (71) and of the defining equation (68) for the Γ_{lm}. Finally we compute $|a(\mathbf{q})|^2$, take an average over the three possible orientations of the pair (12) and multiply by the number Nx of modified bonds in the crystal. The result is

$$\langle|a(\mathbf{q})|^2\rangle=\frac{2Nx\epsilon^2\gamma_0}{\gamma_0-\gamma(\mathbf{q}-\boldsymbol{\tau}_s)} \tag{74}$$

where $\boldsymbol{\tau}_s$ is the scattering vector corresponding to a superlattice line [they are all equivalent here; for instance we may take $\boldsymbol{\tau}_s=(\pi/a, \pi/a, \pi/a)$]. We see that the distortion gives rise to a diffuse scattering around the magnetic peaks of the unperturbed matrix. When $\mathbf{q}-\boldsymbol{\tau}_s$ is small (74) takes the limiting form

$$|a(\mathbf{q})|^2=12Nx\epsilon^2/(|\mathbf{q}-\boldsymbol{\tau}_s|^2a^2). \tag{75}$$

We now discuss the parasitic scattering, due to thermal distortions (spin waves) which may prevent observation of this effect. It often happens that this scattering is also concentrated around the magnetic peaks.[17] However, the spin-wave effects may be made diffuse by a suitable geometrical arrangement, as we shall now show. The parasitic scattering is due to neutrons, of initial wave vector $\mathbf{k}_0$, emitting or absorbing a spin wave ($\mathbf{q}$). Energy and "momentum" conservation require that

$$\hbar\omega_{\mathbf{q}}=\pm(\hbar^2/2M)[(\mathbf{q}+\mathbf{k}_0)^2-k_0^2] \tag{76}$$

where M is the neutron mass and $\omega_{\mathbf{q}}$ the spin wave frequency, which is given by

$$\omega_{\mathbf{q}}=c|\mathbf{q}-\boldsymbol{\tau}_s| \tag{77}$$

when $\mathbf{q}$ is close to $\boldsymbol{\tau}_s$. The constant $c=[(12|J|Sa)/\sqrt{3}\hbar]$ is the velocity of the spin wave. If the velocity of the ingoing neutron $\hbar k_0/M$ is smaller than c, Eqs. (76) and (77) have no solution in the vicinity of $\mathbf{q}=\boldsymbol{\tau}_s$. The inelastic scattering is then distributed on a very wide surface in $\mathbf{q}$ space, and shows no anomaly near the magnetic peaks; it does not prevent the observation of the scattering due to static spin distortions [the latter

[16] M. Tickson, J. Research Natl. Bur. Standards **50**, 177 (1953).

[17] R. J. Elliott and R. D. Lowde, Proc. Roy. Soc. (London) **A230**, 46 (1955).

being indeed singular, as shown by Eq. (75)]. On the other hand k_0 must not be too small. We get superlattice reflections only if $2k_0 > \tau_s$. Both requirements may be satisfied if $\tau_s < 2Mc/\hbar$. Expressing J in terms of the Néel temperature of the pure material by the approximate relation $4|J|S(S+1) = k_BT_N$ we find that this condition may be written

$$\rho = \frac{k_BT_NMa^2}{(S+1)\hbar^2} > \frac{\pi}{2}. \tag{78}$$

Taking $a = 3$ A, $T_N = 80°$K and $S = \frac{3}{2}$ we get $\rho = 7$, so that the inequality (78) seems easy to satisfy. In our example, the experiment is feasible with neutron wavelengths $2\pi/k_0$ between 1.5 and 3 A.

Our formula (71) was restricted to cases where the deflection $\epsilon = \epsilon_1 = -\epsilon_2$ on the perturbed sites was small. In practice this is not the case, and (71) does not apply. However, it is always a good approximation to assume that all deflections other than ϵ_1 and ϵ_2 are indeed small. It may then be shown that the simple solution (71) is still acceptable provided that we replace ϵ by $\sin\epsilon$. With this slight modification all the later formulas also retain their validity.

Afterthoughts: Effects of double exchange in magnetic crystals

25 years after this work, when the cuprate supraconductors were discovered, I was tempted to think that double exchange might be useful concept for their understanding, P.G. de Gennes, *Comptes Rendus Acad. Sci. (Paris)* **305**, 345 (1987).

Indeed one can generate an interesting attraction electron-electron coupling via magnetic excitations. But the double exchange picture predicts a ferromagnetic state at strong doping, which has never been seen in the cuprates — hence this idea failed.

What is the actual source of superfluidity in the cuprates? A large fraction of the community believes that they are standard BCS materials, but with strong couplings due to motions (harmonic or anharmonic) of the oxygens. Another fraction claims that magnetic features are essential for instance in the form of "spin bags". I am hesitant about this view, since the magnetic features in strongly superconducting YBaCuO are so weak. A third group, led by P. Anderson, believes that the electrons in the cuprate form a "Luttinger liquid" which is not a simple Fermi liquid. I like the aesthetics of this approach, but I do not have a deep understanding of it.

Reprinted from THE PHYSICAL REVIEW, Vol. 129, No. 3, 1105–1115, 1 February, 1963

Nuclear Magnetic Resonance Modes in Magnetic Material. I. Theory

P. G. DE GENNES
Faculte des Sciences d'Orsay, Seine-et-Oise, France

P. A. PINCUS*
Centre d'Etudes Nucléaires de Saclay, Seine-et-Oise, France
and
Department of Physics, University of California, Los Angeles, California

AND

F. HARTMANN-BOUTRON AND J. M. WINTER
Centre d'Etudes Nucléaires de Saclay, Seine-et-Oise, France

(Received 30 July 1962; revised manuscript received 22 October 1962)

In a magnetic medium the nuclear spins are coupled by the Suhl-Nakamura indirect interaction. This interaction is weak but has a very long range b. It is known to contribute to the nuclear spin relaxation. It also gives rise to a displacement of the nuclear magnetic resonance frequency which is important at temperatures in the helium range for materials with a large concentration of nuclei and large b. More generally the indirect interaction gives rise to a spectrum of nuclear spin waves. This spectrum may be shown to be meaningful even when the nuclei are far from order because of the long range of the interaction. Different methods to observe this spectrum are discussed.

I. INTRODUCTION

NUCLEAR magnetic resonance has been observed in a large number of ferromagnetic and antiferromagnetic compounds.[1] The resonance frequency ω_n is usually derived by the following arguments: If the hyperfine interaction between the nuclear spin **I** and the electron spin **S** is $A\,\mathbf{I}\cdot\mathbf{S}$,there is an average hyperfine field acting on the nucleus, of magnitude

$$H_n = -A\langle S\rangle/\hbar\gamma_n, \qquad (1.1)$$

where γ_n is the nuclear gyromagnetic ratio, and $\langle\ \rangle$ denotes a thermal average. Equation (1.1) for the resonance field is in fact correct only to first order in the coupling constant A. It neglects all correlations between the motions of **I** and **S**. This is obviously a good approximation in many cases, since the hyperfine interaction is very small compared to the exchange coupling J between electron spins. In fact, if one looks at the exact eigenmodes of a single nuclear spin in a ferromagnetic matrix,[2] one finds that the deviations from Eq. (1.1) are of relative order A/J and thus completely negligible at all temperatures.

The situation is changed if we have a large density of nuclei; then if one of them, $\mathbf{I}_i$, is tilted by an angle θ with respect to the electron magnetization, it distorts the electron spin arrangement. The distortion is of order $(A/J)\langle I_z\rangle\theta$, and thus very small, but it has a long range in space and thus reacts on the motion of many other nuclear spins. The number of active neighbors is of order $J/\hbar\omega_e$ (where ω_e is the electron resonance frequency) and the relative correction to Eq. (1.1) is then of order $(A/J)\langle I_z\rangle(J/\hbar\omega_e) = A\langle I_z\rangle/\hbar\omega_e$. This "frequency pulling" will thus be important at low temperatures ($\langle I_z\rangle$ not too small) provided that (a) the concentration of nuclear spins is large (this will be realized with Co^{59} and Mn^{55}) and (b) the electron spin resonance frequencies are low (small anisotropy fields). When these conditions are realized the nuclear resonance behavior may be drastically different from what it is in the usual case. This possibility was mentioned in a previous note.[3] The purpose of the present paper is to give a more thorough discussion of the effect and of its consequences on nuclear induction signals. In Sec. II we derive the macroscopic equations for the coupled nuclear-electron spin system for various cases of interest (ferromagnets, antiferromagnets, Bloch walls···). In Sec. III we come back to the microscopic description, and derive the spectrum of the nuclear spin waves. The most remarkable result is that these spin waves correspond to well-defined excitations of the nuclear spin system even at comparatively high temperatures (~1°K), where the nuclear spins are strongly disordered. Section IV is concerned with a discussion of some effects of demagnetizing fields on the nuclear spin wave spectrum. In Sec. V we discuss the effects of the frequency pulling on nuclear resonance signals for both steady state and transient behavior.

II. MACROSCOPIC EQUATIONS OF MOTION

Ferromagnetic Case

Let us first consider a saturated ferromagnetic material, of electronic magnetization **M** and nuclear magnetization **m**. The field acting on the nuclei is the sum of the external field $\mathbf{H}_0$ (along the z direction) plus the

* National Science Foundation Post Doctoral Fellow during the part of this work performed at CENS.
[1] J. M. Winter, J. Phys. Radium (to be published).
[2] A discussion of a similar problem can be found in P. G. DeGennes and F. Hartmann-Boutron, Compt. Rend. **253**, 2922 (1961).
[3] P. G. DeGennes, F. Hartmann-Boutron, and P. A. Pincus, Compt. Rend. **254**, 1264 (1962).

hyperfine field $H_n(\mathbf{M}/M_0)=\alpha\mathbf{M}$. The equation of motion for the nuclear spins is thus

$$d\mathbf{m}/dt=\gamma_n\mathbf{m}\times(\mathbf{H}_0+\alpha\mathbf{M}). \quad (2.1)$$

Similarly, the field acting on the electron spin is $\mathbf{H}_0+\mathbf{H}_A+\alpha\mathbf{m}$ (where $\mathbf{H}_A$ is the anisotropy field); at the comparatively low nuclear frequencies in which we are interested the electrons follow this field adiabatically, i.e., $\mathbf{M}$ is parallel to $\mathbf{H}_0+\mathbf{H}_A+\alpha\mathbf{m}$. For small transverse motions of $\mathbf{M}$ we may write:

$$M_z=M_0;\quad M^+=M_x+iM_y=M_0\alpha m^+/(H_0+H_A). \quad (2.2)$$

We shall now assume that Eq. (2.1) may be linearized in the usual way to describe small amplitude motions of $\mathbf{m}$ around its equilibrium value $\mathbf{m}_0$. This assumption is not entirely trivial since the nuclear spins are usually far from saturation. It will be justified by microscopic considerations in Sec. III. We then obtain from (2.1)

$$dm^+/dt=-i\omega m^+ =-i\gamma_n[(H_0+\alpha M_0)m^+-m_0\alpha M^+], \quad (2.3)$$

and inserting (2.2) we arrive at the resonance condition:

$$\omega=\gamma_nH_n[1-\eta(m_0/M_0)]+\gamma_nH_0; \quad (2.4)$$

$\eta=\alpha M_0/(H_0+H_A)=H_n/(H_0+H_A)$ is the enhancement factor relating the effective rf field to the applied rf field in the nuclear resonance experiment.[4] Equation (2.4) predicts a frequency pulling in zero external field of relative value

$$-\delta\omega/\omega=\eta m_0/M_0 \quad (2.5)$$

$$=(\omega_n/\omega_e)(\langle I_z\rangle/\langle S_z\rangle) \cong[I(I+1)/3S](\omega_n/\omega_e)(\hbar\omega_n/k_BT), \quad (2.6)$$

where $\omega_n=\gamma_nH_n$ and $\omega_e=\gamma_e(H_0+H_A)$. Equation (2.5) applies for 100% isotopic abundance. Taking $\omega_n=3\times10^9$; $H_A=10^3$ Oe, $I=S=5/2$ (these being plausible values for cubic compounds of Mn^{2+}) we would have $\omega_e\sim3\times10^{10}$ and $\delta\omega/\omega\sim3\times10^3/T$. The shift in frequency would be important for such a case at temperatures below 2°K. (A more complete comparison between the shift and the width of the resonance line will be given in Sec. III.)

Equation (2.5) shows that we may expect a large frequency pulling in all situations where the enhancement factor is large. This suggests that the effect might be important in Bloch walls where η values of the order 10^3 and more are met. In fact it may be shown that a formula very similar to (2.5) applies in a Bloch wall. This is discussed in detail in Appendix A.

Equation (2.4) also implies that the relation $\omega(H_0)$ between resonance frequency and external field is nonlinear since η depends on H_0. The apparent gyromagnetic ratio defined as $\gamma_{app}=\partial\omega/\partial H_0$ is given by

$$\gamma_{app}=\gamma_n[1+\eta^2(m_0/M_0)]. \quad (2.7)$$

The effect on γ_{app} is more spectacular since an extra factor of η (~100) comes into play. Of course this is somewhat fictitious since to measure γ_{app} one needs rather large magnetic fields for which η is strongly reduced.

Finally, if the electron motion is damped, the nuclear frequency acquires a small imaginary part $(1/T_1)_n$. This relaxation effect however is very small: it vanishes in the adiabatic approximation of Eq. (2.2). If one replaces (2.2) by the complete equation of motion for the electron spins, including a relaxation of $\mathbf{M}$ towards the *instantaneous* value of the field $\mathbf{H}_0+\mathbf{H}_A+\alpha\mathbf{m}$, one obtains:

$$(T_{1n}\delta\omega_n)^{-1}=\eta(\gamma_n/\gamma_e)(T_{2e}\omega_e)^{-1}. \quad (2.8)$$

For $\omega\ll\omega_e$ and $\omega_eT_{2e}\gg1$ and taking $(T_{2e}\omega_e)^{-1}=10^{-1}$, $(\gamma_n/\gamma_e)=10^{-3}$, $\eta=100$ we obtain $(T_{1n}\delta\omega_n)^{-1}\sim10^{-2}$. Thus the relaxation rate T_{1n}^{-1} given by Eq. (2.8) is very small; and there are other more important effects contributing to the linewidth (some of them will be dis cussed later).

Antiferromagnetic Case

We shall now consider the frequency pulling in an antiferromagnet where we must distinguish two sublattices of electronic spins $\mathbf{M}_1$ and $\mathbf{M}_2$ and the corresponding two nuclear spin sublattices $\mathbf{m}_1$ and $\mathbf{m}_2$.

Neglecting damping, the equations of motion corresponding to (2.1) and (2.2) are

$$\begin{aligned} d\mathbf{M}_1/dt&=\gamma_e\mathbf{M}_1\times(\mathbf{H}_0+\mathbf{H}_A-\lambda\mathbf{M}_2+\alpha\mathbf{m}_1),\\ d\mathbf{M}_2/dt&=\gamma_e\mathbf{M}_2\times(\mathbf{H}_0-\mathbf{H}_A-\lambda\mathbf{M}_1+\alpha\mathbf{m}_2),\\ d\mathbf{m}_1/dt&=\gamma_n\mathbf{m}_1\times(\mathbf{H}_0+\alpha\mathbf{M}_1),\\ d\mathbf{m}_2/dt&=\gamma_n\mathbf{m}_2\times(\mathbf{H}_0+\alpha\mathbf{M}_2), \end{aligned} \quad (2.9)$$

where λ is the molecular field constant and is related to the effective exchange field by $\lambda M_0=H_{ex}$. In the small oscillation approximation (2.9) becomes

$$\begin{aligned} -\omega M_1^+&=\gamma_e[-H_{ex}M_2^++\alpha M_0m_1^+ -M_1^+(H_0+H_A+H_{ex}+\alpha m_0)],\\ -\omega M_2^+&=\gamma_e[H_{ex}M_1^+-\alpha M_0m_2^+ -M_2^+(H_0-H_A-H_{ex}-\alpha m_0)],\\ -\omega m_1^+&=\gamma_n[\alpha m_0M_1^+-m_1^+(H_0+\alpha M_0)],\\ -\omega m_2^+&=\gamma_n[-\alpha m_0M_2^+-m_2^+(H_0-\alpha M_0)], \end{aligned} \quad (2.10)$$

where $M_0=M_{10}=-M_{20}$ and $m_0=m_{10}=-m_{20}$. On solving the secular equation (2.10), one finds the nuclear resonance frequencies are given by:

$$\omega/\gamma_n\cong H_n[1-(2\omega_{ex}\omega_n\gamma_em_0/\omega_1\omega_2\gamma_nM_0)]^{1/2} \pm H_0[1+(\omega_n{}^2\gamma_e{}^2m_0/\omega_1\omega_2\gamma_n{}^2M_0)], \quad (2.11)$$

where $\omega_{ex}=\gamma_eH_{ex}$ and $\omega_{1,2}$ are the two antiferromagnetic resonance frequencies, i.e., $\omega_{1,2}/\gamma_e=(2H_{ex}H_A)^{1/2}\pm H_0$. For many antiferromagnetics where $(2H_{ex}H_A)^{1/2}\gg H_0$, the fractional frequency pulling reduces to $\frac{1}{2}(m_0/M_0)$

[4] A. M. Portis and A. C. Gossard, J. Appl. Phys. **31**, 205S (1960). A. C. Gossard, thesis, University of California, Berkeley, 1960 (unpublished).

$\times(H_n/H_A)$, i.e., one half of the value (2.5) obtained for the ferromagnetic case in zero field. The nuclear g shift given by the second term in (2.11) is of the order of $\gamma_e\omega_n/\gamma_n\omega_{ex}$ smaller than the frequency pulling and is thus negligible. Notice that there exist two degenerate nuclear resonance frequencies which are split by an external field. The two frequencies clearly correspond to the two nuclear sublattices one of which is magnetized parallel to H_0 and the other antiparallel.

We conclude that, here again, the effect will be important only if we have a very low anisotropy field H_A. This is actually the case in $KMnF_3$,[5] and might also happen in $MnCO_3$.[6] However, both these compounds are canted antiferromagnets for which Eqs. (2.9) and (2.10) must be modified. This is done in the second paper of this series[7] and the result is

$$\omega/\gamma_n = H_n[1-(2\omega_{ex}\omega_n\gamma_e m_0/\omega_{1,2}^2\gamma_n M_0)]^{1/2}. \quad (2.12)$$

Notice that the two nuclear resonance frequencies are split even in the absence of an external field. This arises because the electronic resonance modes are linearly polarized in a canted antiferromagnet, but circularly polarized in a usual Néel antiferromagnet. In a case such as $MnCO_3$,[6] this splitting may be considerable because one electronic resonance mode may be at an infrared frequency of the order of 10^{12} cps, while the other branch is at microwave frequencies ($\sim 10^9$ cps). Of course, the pulling caused by the infrared mode will always be negligible, but, using the previous estimates for Mn^{55} nucleus, the pulling arising from the microwave electronic mode may be of the order of several percent. For example, at an electronic resonance frequency of 9 kMc/sec,

$$|\delta\omega_n/\omega_n| \sim |\gamma_e\omega_{ex}\omega_n m_0/\gamma_n\omega^2 M_0| \sim 0.4, \quad (2.13)$$

for $\omega_n \simeq 4\times10^9$ cps and an exchange field of 10^6 Oe.

Such a large effect may occur in both $KMnF_3$ and $MnCO_3$. Heeger *et al.*[5] have shown that the effective anisotropy field acting on the electronic spins and arising from the nuclear spins through the hyperfine interaction $A\langle I_z\rangle/\hbar\gamma_e$ gives a significant contribution to the microwave resonance frequency. In fact, the frequency pulling in a canted system is just the square of the fraction of the electronic resonance frequency arising from the hyperfine interaction. The reason for the large effect is just that the anisotropy field arising from the nuclei (of the order of a few oersteds at liquid helium temperature) is just of the same order of magnitude as the ordinary anisotropy field, which is enhanced by the exchange field, and is enhanced in the same manner. This frequency pulling seems to be the basis for the nonlinear effects observed by Heeger *et al.*[7,8] in $KMnF_3$.

[5] A. J. Heeger, A. M. Portis, D. T. Teaney, and G. Witt, Phys. Rev. Letters **7**, 307 (1961).
[6] M. Date, J. Phys. Soc. Japan **15**, 2251 (1960).
[7] A. M. Portis, A. J. Heeger, and G. Witt (to be published).
[8] A. J. Heeger, A. M. Portis, and G. Witt, presented at the International Conference on Magnetic and Electric Resonance and Relaxation, Eindhoven, July, 1962 (unpublished).

III. MICROSCOPIC INTERPRETATION

Connection with the Suhl-Nakamura Interaction

We now proceed to show that the frequency shift $\delta\omega$ of Eq. (2.5) is due to the indirect interaction between nuclei, as derived by Suhl[9] and Nakamura.[10] Let us consider for instance a ferromagnetic Bravais lattice, each electron spin $\mathbf{S}_n$ being coupled only to one nuclear spin $\mathbf{I}_n$. The linearized equations of motion for both systems are

$$\begin{aligned}\omega S_n^+ &= \textstyle\sum_{n,m}\omega_{nm}(S_n^+ - S_m^+) + \omega_e S_n^+ \\ &\quad -(A/\hbar)[\langle I_z\rangle S_n^+ - SI_n^+]; \quad (3.1)\\ \omega I_n^+ &= -(A/\hbar)(SI_n^+ - \langle I_z\rangle S_n^+) + \gamma_n H_0 I_n^+,\end{aligned}$$

where $S^{-1}\hbar\omega_{nm}$ is the exchange coupling between electron spins located at site n and m and we have assumed that the electron system is completely saturated ($\langle S_z\rangle = S$). The eigenmodes of Eq. (3.1) are the traveling waves

$$S_n^+ = ue^{i\mathbf{q}\cdot\mathbf{R}_n}; \quad I_n^+ = ve^{i\mathbf{q}\cdot\mathbf{R}_n}.$$

In terms of u and v Eq. (3.1) becomes

$$\begin{aligned}[\omega-\omega_{0q}+(A/\hbar)\langle I_z\rangle]u-(A/\hbar)Sv&=0;\\ -(A/\hbar)\langle I_z\rangle u+[\omega+(A/\hbar)S-\gamma_n H_0]v&=0,\end{aligned} \quad (3.2)$$

where $\omega_{0q} = \omega_e + \sum_m \omega_{nm}[1-e^{i\mathbf{q}\cdot(\mathbf{R}_m-\mathbf{R}_n)}]$ is the usual electronic spin wave frequency. From (3.2) we derive the secular equation

$$[\omega-\omega_{0q}+(A/\hbar)\langle I_z\rangle][\omega+(A/\hbar)S-\gamma_n H_0] - (A/\hbar)^2 S\langle I_z\rangle = 0. \quad (3.3)$$

When $\hbar\omega_e \gg AS$ this has a high frequency solution ($\omega\sim\omega_{0q}$) representing the electron spin waves and a low frequency solution

$$\omega_q \cong -(AS/\hbar)[1+(A\langle I_z\rangle/\hbar\omega_{0q})]+\gamma_n H_0, \quad (3.4)$$

representing nuclear spin waves. The constant term $(AS/\hbar)$ stems from the average hyperfine field and the q dependent term comes from the indirect interaction between nuclei through the electron spins. Equation (3.4) can also be obtained by writing the linearized equations of motion for the nuclear spins alone when coupled by the Suhl-Nakamura interaction

$$\begin{aligned}\mathcal{H}_1 &= -\tfrac{1}{2}A^2 S\sum_{m,n} I_m^+ I_n^- N^{-1}\sum_{\mathbf{q}}\frac{e^{i\mathbf{q}\cdot(\mathbf{R}_m-\mathbf{R}_n)}}{\hbar\omega_{0q}}\\ &= \sum_{m,n} u_{mn} I_m^+ I_n^-. \quad (3.5)\end{aligned}$$

The dispersion relation (3.4) is represented in Fig. 1. For $q=0$, Eq. (3.4) leads again to the shifted frequency Eq. (2.6). For large q's, ω_{0q} being of the order of the exchange frequencies, the interaction term becomes negligible, and ω is equal to the unshifted frequency

[9] H. Suhl, Phys. Rev. **109**, 606 (1958)
[10] T. Nakamura, Progr. Theoret. Phys. (Kyoto) **20**, 542 (1958).

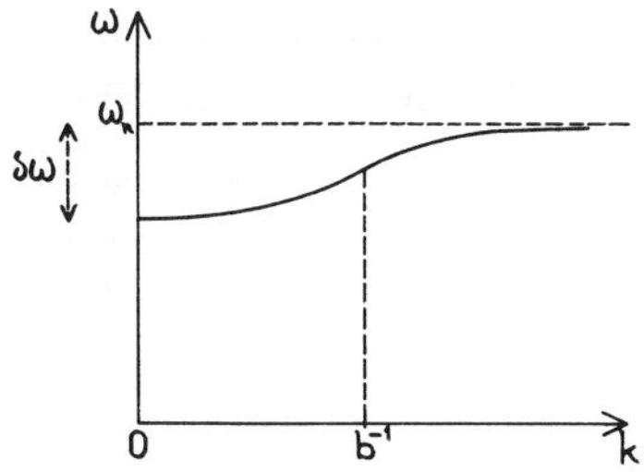

FIG. 1. The nuclear spin wave spectrum.

$\gamma_n(H_n+H_0)$. The transition between these two limiting behaviors takes place for q values such that $\omega_{0q}\sim 2\omega_e$. Since $\omega_{0q}\cong\omega_e+\omega_{ex}a^2q^2$, where a is an interatomic distance and ω_{ex} an exchange frequency, this occurs when $q^{-1}\sim a(\omega_{ex}/\omega_e)^{1/2}\gg a$. The physical meaning of this result is simply that the Suhl interaction has a range $b=a(\omega_{ex}/\omega_e)^{1/2}$ and thus gives a significant contribution to the nuclear precession frequency only at wavelength larger than b. Since $b\gg a$ in the cases of interest, it is only for a very small fraction of the modes that the frequency is shifted from the conventional value.

These remarks also give a clearer meaning to the change in gyromagnetic ratio Eq. (2.7). When we apply an external magnetic field H_0 we change the Zeeman energy of the nuclei, but we also change the range b of the interaction between them; the whole nuclear spin wave spectrum is then modified.

It is also of interest to estimate the velocity $\partial\omega/\partial q$ of typical nuclear spin waves from Eq. (3.4). $\partial\omega/\partial q$ is maximum when $ab\sim 1$ and is then

$$(\partial\omega/\partial q)_{\max}\sim b\delta\omega\sim a\omega_n\eta(m_0/M_0)(\omega_{ex}/\omega_e)^{1/2}.$$

Using our previous values $\omega_n=3\times10^9$, $\eta(m_0/M_0)=3\times10^{-3}T^{-1}$, $(\omega_{ex}/\omega_e)^{1/2}=30$ and taking $a=3$ Å we get $(\partial\omega/\partial q)_{\max}=10^2T^{-1}$ cm/sec, a nonnegligible velocity. It must be realized however, that the mean free path l is small. $l\sim b\delta\omega/\Delta\omega$ for the high velocity spin waves (where $\Delta\omega$ is the line width of the corresponding mode). As we shall see later, $(\delta\omega/\Delta\omega)$ is of the order 50 in favorable cases such as $KMnF_3$ at 4°K, and then $l\sim 15\ \mu$. Of course, the mean free path could be increased by working at lower temperatures.

Validity of the Linearization Procedure

We now proceed to show that the linearization procedure used to derive Eq. (3.3) is indeed justified even at rather high temperatures. For this purpose we consider the power spectrum $P_q(\omega)$ of the quantity

$$A_q^+=\sum_i e^{i\mathbf{q}\cdot\mathbf{R}_i}I_i^+; \tag{3.6}$$

$$P_q(\Omega)=\int dt\,\langle A_q^-(0)A_q^+(t)\rangle e^{i\Omega t}. \tag{3.7}$$

When the spin wave excitations are meaningful, $P_q(\omega)$ shows a sharp peak at the spin wave frequency ω_q. At very low temperatures ($k_BT\sim A$), the nuclear spins being all lined up, this situation will clearly be realized. We are interested, however, in higher temperatures (in the helium range) for which the average polarization of the nuclei ($\langle I_z\rangle/I$) is small.

For such cases of strong disorder $P_q(\omega)$ is best studied by a method of moments. Let us consider the first moment

$$\bar{\Omega}_q=\left(\int P_q(\Omega)\Omega d\Omega\right)\Big/\left(\int P_q(\Omega)d\Omega\right) \tag{3.8}$$

$$=i\langle A_q^-dA_q^+/dt\rangle/\langle A_q^-A_q^+\rangle. \tag{3.9}$$

We compute $dA_q^+/dt=(i/\hbar)[\mathcal{H},A_q^+]$ with the Hamiltonian $\mathcal{H}=\mathcal{H}_0+\mathcal{H}_1$, where $\mathcal{H}_0$ is the first order Hamiltonian $\mathcal{H}_0=\sum_n ASI_n^z$ and $\mathcal{H}_1$ is the second-order Suhl interaction (3.5). The thermal averages are also to be taken with respect to this Hamiltonian:

$$\langle O\rangle=\mathrm{Tr}[Oe^{-\beta\mathcal{H}}]/\mathrm{Tr}[e^{-\beta\mathcal{H}}].$$

Then

$$\hbar\Omega_q=-AS+\frac{2\sum_{n,m,p}\langle I_n^-I_m^+I_p^z\rangle u_{mp}e^{i\mathbf{q}\cdot(\mathbf{R}_p-\mathbf{R}_m)}}{\sum_{n,m}\langle I_n^-I_m^+\rangle}. \tag{3.10}$$

At the temperatures of interest we may, to a very good approximation, make the substitution

$$\langle I_n^-I_m^+I_p^z\rangle\to\langle I_n^-I_m^+\rangle\langle I^z\rangle\to (\tfrac{2}{3})I(I+1)\langle I^z\rangle\delta_{nm} \tag{3.11}$$

for $p\neq m$, n. This neglects only some small correlations brought in by the Suhl interaction of order A^2/Jk_BT while the term retained is of order $\langle I^z\rangle\sim A/k_BT$. For the same reason the terms $p=n\neq m$ and $p=m\neq n$ can be neglected. Finally the term $n=m=p$ which corresponds to the self energy of nucleus I_n is always negligible as explained in the introduction (being only one particular term of a long range interaction), and we may as well make approximation (3.11) on this term too. The result is then simply

$$\hbar\bar{\Omega}_q=-AS+2\langle I^z\rangle\sum_p u_{mp}e^{i\mathbf{q}\cdot(\mathbf{R}_m-\mathbf{R}_p)} \tag{3.12}$$

and thus the first moment $\bar{\Omega}_q$ coincides with the spin wave frequency (3.4).

Our next problem is then to determine the width of the power spectrum $P_q(\omega)$. If this width comes out to be small (when compared with $|\bar{\Omega}_q+(AS/\hbar)|$), then we may say that the frequency $\bar{\Omega}_q$ corresponds to a well defined elementary excitation of wave vector $\mathbf{q}$. We derive the width from the reduced second moment $\langle\Omega_q^2\rangle_{av}-\bar{\Omega}_q^2=(\Delta\Omega_q^2)$ and also simplify the calculation by going to the limit of complete disorder in the nuclear spins ($AS\ll k_BT$). This clearly will give us an overestimate of the width. It is found that $(\Delta\Omega_q)^2$ is independent of q in this limit and thus identical to the value $(\Delta\Omega_0)^2$ computed by Suhl[9] for the uniform mode:

$$\frac{(\Delta\Omega_q)^2}{4\pi^2}=\frac{(\Delta\Omega_0)^2}{4\pi^2}=\frac{I(I+1)}{24\pi S^2}\frac{\omega_n^4}{\omega_{ex}^{3/2}\omega_e^{1/2}}. \tag{3.13}$$

As pointed out above we are mainly interested in the ratio $[\bar{\Omega}_q+(AS/\hbar)]/\Delta\Omega_q \leq (\delta\omega_0/\Delta\Omega_0)$ between the total bandwidth of the spin wave spectrum and the width of one individual spin wave mode. This is given by

$$(\delta\omega_0/\Delta\Omega_0) = (2\pi)^{-1}[24\pi/I(I+1)]^{1/2}\langle I_z\rangle(\omega_{ex}/\omega_e)^{3/4}. \quad (3.14)$$

For nuclei such as Mn^{55} and Co^{59}, assuming a low anisotropy ($\omega_e/\gamma_e \sim 10^3$ Oe) and a reasonable value of exchange ($\omega_{ex}/\gamma_e \sim 10^6$ Oe), we find that $(\delta\omega_0/\Delta\Omega_0)$ is larger than 1 when T is below $\sim 3°$K. In such a case the nuclear spin waves are well defined at temperatures $\sim 1°$K where the nuclear polarization level is only of order 1%.

Apart from all theoretical consideration (3.14) is of course very important since it gives us the temperature region in which the frequency shift $\delta\omega_0$ of the uniform mode is large when compared to the Suhl width $\Delta\Omega_0$: all frequency pulling effects can be observed only when $|\delta\omega_0/\Delta\Omega_0|>1$.

In fact, (3.14) can be obtained by the following simple argument. Let us consider the field F_i acting on the ith nuclear spin arising from all the other nuclear spins. Because the nuclei are far from saturation F_i is a random field with its associated distribution function $P(F_i)$. The average field $\bar{F}_i$ is proportional to the polarization:

$$\bar{F}_i = C\langle I_z\rangle/I.$$

One can estimate the width of the distribution by noticing that the number, N, of spins contributing to the local field is very large: $N\sim(b/a)^3\sim[H_{ex}/(H_0+H_A)]^{3/2}$. In the temperature range of interest, the field F_i is therefore the sum of a large number of random, nearly independent contributions. The width of the distribution is therefore of the order of $CN^{-1/2}$. There will exist well defined collective modes if this width is small relative to the mean displacement $\bar{F}_i$.

The condition is, therefore,

$$N^{1/2}\langle I_z\rangle/I>1,$$

which is within a numerical factor of the order of unity identical to (3.14).

Specific Heat of the Nuclear Spin System

The standard expression for the nuclear specific heat in the "high temperature" limit ($k_BT>\hbar\omega_n$) is[11]

$$C=Nk_B[I(I+1)/3](\hbar\omega_n/k_BT)^2. \quad (3.15)$$

At first sight one might wonder whether the frequency shift should not be added to ω_n in (3.15). Actually this is not so, and (3.15) is the correct formula, even in the presence of frequency pulling in the range $\hbar\omega_n \ll k_BT$. This may be seen in a formal way by a trace method carried to 4th order in A. More physically we may argue as follows: the frequency pulling affects only the frequency of the long wavelength nuclear spin waves ($kb<1$). As pointed out before the relative number of shifted modes is very small of order $(a/b)^3 \cong (\omega_e/\omega_{ex})^{3/2} \sim 10^{-4}$. When $\hbar\omega_n<k_BT$ all modes are excited, and the contribution of the shifted ones to the specific heat is negligible. On the other hand when $\hbar\omega_n>k_BT$ (i.e., at temperatures below 10^{-2}°K), only the bottom of the nuclear spin wave spectrum is excited: This is precisely the region of the shifted modes and their effect on the specific heat will then become important.

[11] W. Marshall, Phys. Rev. **110**, 1280 (1958), Appendix A.

IV. DEMAGNETIZATION FIELD EFFECTS

Until now, we have neglected the demagnetizing fields. It is known that they modify the electronic spin wave spectrum for low value of the wave vector **q**[12] in a ferromagnet. Here we calculate the modifications of the nuclear spin wave spectrum arising from these effects. It will be shown that the nuclear spin wave spectrum is also anisotropic and therefore, effects such as parallel pumping[13] and Suhl instability[14] should be investigated.

Consider the equations of motion for the **q** component of the nuclear magnetization:

$$\begin{aligned} dm_{qx}/dt &= \gamma_n m_{qy}\alpha M_z - \gamma_n m_z \alpha M_{yq} + \gamma_n(H_0-N_zM)m_{qy}; \\ dm_{qy}/dt &= -\gamma_n m_{qx}\alpha M_z + \gamma_n m_z\alpha M_{qx} - \gamma_n(H_0-N_zM)m_{qx}. \end{aligned} \quad (4.1)$$

The corresponding equations for M_{qx} and M_{qy} are[12]

$$\begin{aligned} \frac{dM_{qx}}{dt} &= M_{qy}\left[\gamma_eH_A+\gamma_e(H_0-N_zM) + \frac{4\pi}{q^2}q_y^2\gamma_eM + \omega_{ex}(aq)^2\right] + \left[\frac{4\pi\gamma_eM}{q^2}q_xq_y\right]M_{qx} - \alpha\gamma_eMm_{qy}; \\ \frac{dM_{qy}}{dt} &= -M_{qx}\left[\gamma_eH_A+\gamma_e(H_0-N_zM) + \frac{4\pi}{q^2}q_x^2\gamma_eM + \omega_{ex}(aq)^2\right] - \left[\frac{4\pi\gamma_eM}{q^2}q_xq_y\right]M_{qy} + \alpha\gamma_eMm_{qx}. \end{aligned} \quad (4.2)$$

Without any loss of generality q_y may be assumed to be zero. The nuclear frequency being low the time derivatives of M_{qx} and M_{qy} are neglected.

[12] C. Herring and C. Kittel, Phys. Rev. **81**, 869 (1951).
[13] E. Schlomann, J. J. Green, and U. Milano, J. Appl. Phys. **31**, 386S (1960).
[14] H. Suhl, J. Phys. Chem. Solids **1**, 209 (1957).

M_{qy} and M_{qx} may then be expressed in terms of m_{qy} and m_{qx}; the nuclear equations of motion then become:

$$dm_{qx}/dt=\gamma_n m_{qy}[\alpha M+H_0-N_zM] -\gamma_n\alpha m(\alpha M/C_q)m_{qy};$$
$$dm_{qy}/dt=-\gamma_n m_{qx}[\alpha M+H_0-N_zM] +\gamma_n\alpha m(\alpha M/D_q)m_{qx}, \tag{4.3}$$

with

$$C_q=H_0-N_zM+H_A+(\omega_{\text{ex}}/\gamma_e)(aq)^2;$$
$$D_q=H_0-N_zM+H_A+(\omega_{\text{ex}}/\gamma_e)(aq)^2+4\pi M\sin^2\theta_q,$$

where θ_q is the angle between the wave vector $\mathbf{q}$ and the z axis. The nuclear spin wave frequency is easily obtained as

$$\omega_{nq}^2=\gamma_n^2[H_n+H_0-N_zM-\alpha m(\alpha M/C_q)] \times[H_n+H_0-N_zM-\alpha m(\alpha M/D_q)]. \tag{4.4}$$

The nuclear frequency depends on the angle θ_q. The equations (4.3) show that a small ellipticity is induced in the nuclear motion. This ellipticity is usually very small because it affects only the second term in (4.3). As for a ferromagnet, Eq. (4.4) is not valid when the wave vector q goes to zero. For $q=0$ we have to consider the coupled motion of the two uniform modes. The resonance frequency is

$$\omega_{n0}^2=\gamma_n^2[H_n+H_0-N_zM-\alpha m(\alpha M/C_0)] \times[H_n+H_0-N_zM-\alpha m(\alpha M/D_0)],$$

where

$$C_0=H_A+H_0+(N_x-N_z)M;$$
$$D_0=H_A+H_0+(N_y-N_z)M.$$

When $0<q<1/L$, L being the dimension of the sample, the electronic eigenmodes are magnetostatic modes and the nuclear eigenmodes have the spatial behavior of these electronic modes.

We notice that the perturbation of the nuclear spin wave spectrum due to demagnetizing effects occurs for wave vectors less than $(1/c)=(1/a)(4\pi\gamma_eM/\omega_{\text{ex}})^{1/2}$, and c is usually of the same order as b, the range of the Suhl-Nakamura interaction.

The occurrence of an ellipticity in the nuclear motion suggests that a parallel pumping experiment is possible within the nuclear system. We add an oscillating field $h\cos2\omega_0t$ along the z axis and also include a transverse damping term. Then the equations of motion become

$$dm_{qx}/dt=\gamma_n m_{qy}(\alpha M+H_0-N_zM)-\gamma_n m\alpha M_{qy} +\gamma_n m_{qy}h\cos2\omega_0t-(m_{qx}/T_2);$$
$$dm_{qy}/dt=-\gamma_n m_{qx}(\alpha M+H_0-N_zM)+\gamma_n m\alpha M_{qx} -\gamma_n m_{qx}h\cos2\omega_0t-(m_{qy}/T_2). \tag{4.5}$$

By the same technique M_{qy} and M_{qx} are expressed in terms of m_{qx} and m_{qy}, but now there is in C_q and D_q an oscillating term $h\cos2\omega_0t$. This term may be neglected, its effect is to change h into $h[1-\eta^2(m/M)]$.

Equations (4.5) are written now as

$$dm_{qx}/dt=\gamma_n m_{qy}c_q-(m_{qx}/T_2);$$
$$dm_{qy}/dt=-\gamma_n m_{qx}d_q-(m_{qy}/T_2),$$

with

$$c_q=H_n+H_0-N_zM-(\alpha^2mM/C_q);$$
$$d_q=H_n+H_0-N_zM-(\alpha^2mM/D_q).$$

The problem is now formally identical to the problem of parallel pumping in a ferromagnetic system. The critical field h_c which produces instability is given by

$$(\gamma_nh_c/4)(c_q-d_q)(c_qd_q)^{-1/2}>1/T_2. \tag{4.6}$$

It is quite clear that this condition is very difficult to satisfy for three reasons:

(a) there is no enhancement factor for a longitudinal field h;
(b) the transverse damping is very large;
(c) the ellipticity of the nuclear motion, which is measured by $e=(c_q-d_q)(c_qd_q)^{-1/2}$ is very small. This term is approximately given by

$$e=\alpha^2mM/(C_q-D_q)/H_nC_qD_q \simeq\alpha^2mM/H_nH_0\simeq H_nm/H_0M.$$

Thus, even at very low temperature, this term is very small.

V. NUCLEAR RESONANCE IN THE PRESENCE OF PULLING

Steady State Behavior

In this section we would like to show in what way an appreciable frequency pulling may change the nuclear resonance absorption. In particular we shall show that nonlinear absorption effects may arise from the nonlinear nature of the frequency pulling. The nonlinear effect of the frequency pulling arises because the resonance frequency shift is proportional to the nuclear polarization itself. Thus the nuclear absorption line at low power levels will be centered near the pulled frequency, but at high power levels when the nuclear polarization may be small, the absorption line may then be centered near the unshifted frequency ω_n. In other words, if one were to apply a rf field at ω_n, at low powers one would find very little absorption since the applied signal would be well off resonance. As one increases the power, at a certain power level the line would be able to snap over to ω_n and thus a nonlinear absorption versus power curve might be observed. Such is the observed type of behavior in $KMnF_3$.[7,8]

In order to determine the response of the nuclear system to an effective rf field h^*, where $h^*=\eta h$, we may consider a rate equation for the z polarization of the nuclear spin system as is, for example, given in Abragam,[15] by

$$dm_z/dt=-2Wm_z+(m_0-m_z)/T_1, \tag{5.1}$$

[15] A. Abragam, *The Principles of Nuclear Magnetism* (Oxford University Press, New York), Chaps. II and IV.

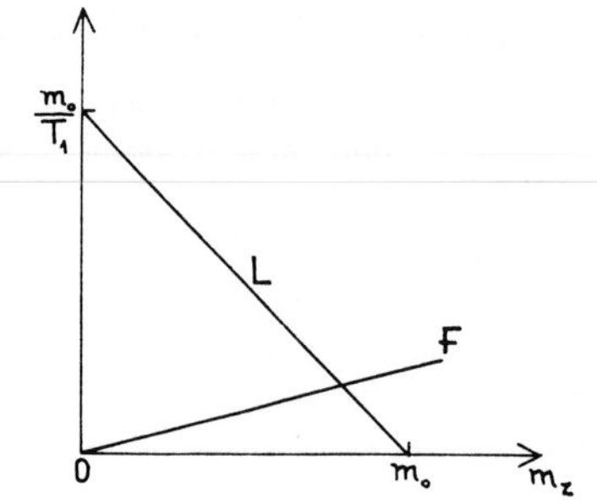

FIG. 2. A parametric plot of Eq. (4.5) in the absence of frequency pulling. The line L represents the left-hand side of (4.5) and F represents the right-hand side. The stable solution is given by the intersection of F and L.

where m_0 is the thermal equilibrium value of the magnetization and where the transition probability per unit time, induced by the rf field is given by

$$W=(\gamma_n h^*)^2(\pi/2)f(\omega). \tag{5.2}$$

The function $f(\omega)$ is the normalized shape function of the resonance line. For example for a Gaussian line $f(\omega)$ is given by

$$f(\omega)=(2\pi)^{-1/2}\Delta^{-1}e^{-(\omega-\omega_0)^2/2\Delta^2}, \tag{5.3}$$

where Δ is the linewidth in frequency units and ω_0 is the resonance frequency. Under steady state conditions, (5.1) and (5.2) give

$$(m_0-m_z)/T_1=\pi f(\omega)(\gamma_n h^*)^2 m_z. \tag{5.4}$$

The characteristic transverse time T_2 is defined as $\pi f(\omega_0)$ and is given by

$$T_2^{-1}=(\pi/2)^{1/2}\Delta \tag{5.5}$$

for a Gaussian line. The right-hand side (rhs) of (5.4) represents the rate at which energy is absorbed from the radio frequency field, and the left-hand side (lhs), the rate at which energy is transmitted to the lattice; of course, in equilibrium these two quantities must be equal. The solution of (5.4) gives the power and frequency dependence of the nuclear absorption signal. For the usual case, where there is negligible frequency pulling, $f(\omega)$ is independent of m_z and (5.4) can easily be solved to give the standard results. However, in the presence of appreciable frequency pulling ω_0 is a function of the nuclear magnetization m_z and thus one must resort to graphical solutions of (5.4). In Fig. 2 we give the trivial graphical solution of (5.4) in the absence of frequency pulling. The ordinate of the line of negative slope, L, is proportional to the rate at which energy is delivered to the lattice [lhs of (5.4)], while the line of positive slope, F, is given by the rhs of (5.4). The slope of F is proportional to the power and for slopes greater than unity the line is effectively saturated. On resonance ($\omega=\omega_0$) this just gives the usual saturation requirement that $(\gamma_n h^*)^2 T_1 T_2 \sim 1$. At a fixed power level, the frequency dependence of the ordinate of intersection will trace out the nuclear absorption line.

In the presence of appreciable frequency pulling the curve F will be nonlinear and there exists the possibility of multiple intersections or multiple roots of (5.4). We shall graphically investigate several possibilities under the assumption of a Gaussian line (5.3). We shall take the frequency pulling shift in the form

$$\delta\omega=-\beta\omega_n m_z, \tag{5.6}$$

where β is assumed to be independent of the nuclear magnetization. This assumption is not strictly valid since the electronic resonance frequency will depend appreciably on m_z when the frequency pulling is large. However, this will usually be a correction to the large m_z dependence given in (5.6). For a Gaussian line with frequency pulling, (5.4) becomes

$$f(\omega)=(2\pi)^{-1/2}\Delta^{-1}\exp[-(\Delta\omega+\beta\omega_n m_z)^2/2\Delta^2], \tag{5.7}$$

where $\Delta\omega=\omega-\omega_n$. Let us suppose that an rf field is applied between ω_n and $\omega_n-\delta\omega$. In Fig. 3 we show qualitatively the behavior of (5.4). At low power levels, F_1, there is only one solution corresponding to very little absorption. However at sufficiently high powers, F_2, new solutions appear corresponding to much smaller values of m_z and correspondingly higher energy absorption. For the special case $\omega_n-\omega\gg\Delta$ the first point of a contact for the high absorption solution may easily be seen to occur at

$$(\pi/2)(\gamma_n h^*)^2 T_1 T_2 \approx [\omega-(\omega_n-\delta\omega)](\omega_n-\omega)^{-1}. \tag{5.8}$$

This is generally of the order of the usual saturation condition. The two extreme solutions for high and low absorption may be seen to be stable with respect to small fluctuations in m_z. The central solution is unstable. For example, for the central solution, if there is a small fluctuation with $\delta m_z>0$ the relaxation rate will exceed the absorption and thus the system will be driven to the low absorption solution. Similarly for $\delta m_z<0$ the system will be driven to the high absorption solution. In fact, the condition for stability is clearly that the slope of the rf field absorption curve, F, be greater than the slope of the relaxation time L. At frequencies $\omega<\omega_n$ such that $\Delta\omega<0$, the curves F_1 and F_2 will shift to the right, with decreasing absorption for a fixed power. For $\omega>\omega_n$, the curves F_1 and F_2 will shift to the left, making it rapidly very difficult to obtain a high absorption solution. For a given power level, the maximum absorption will occur at that frequency such that the high absorption root first appears. Consequently, the frequency for peak

FIG. 3. A parametric plot of Eq. (4.5) in the presence of frequency pulling. F_1 and F_2 represent two different power levels. The two extreme solutions (for F_2) are stable. This sketch is not drawn to scale in order to give a better demonstration of the qualitative behavior.

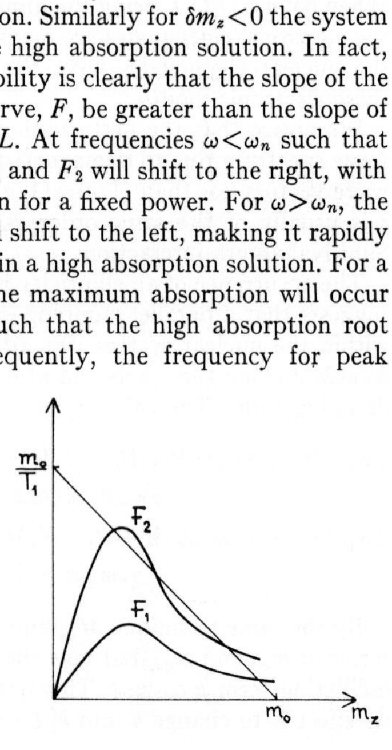

absorption will be a function, both of power level and temperature, and will usually lie between the shifted and unshifted nuclear resonance frequencies. As the power increases, the peak frequency will approach ω_n, and as the power decreases, the peak frequency will approach $\omega_n-\delta\omega$. As the temperature increases, the pulling effect will decrease and the peak frequency will tend toward ω_n. The line shape in the presence of strong frequency pulling will probably be very asymmetric. The absorption on the high frequency side of the maximum should fall off much more steeply than on the low-frequency side and the width of the observed line will be of the order of a fraction of the frequency pulling, and may be much greater than the natural width. Of course, the precise critical power level will depend strongly on T_1 and T_2. This nuclear saturation at the critical power level is just the type of effect which has been observed in $KMnF_3$.

There remains the question of how the spin system can be pulled to small values of m_z from the other stable solution $m_z\approx m_0$. There exists a critical rf field, h_c, above which the high absorption solution is the only equilibrium solution to (4.5). For a Gaussian line this critical field is given by

$$(\pi/2)(\gamma_n h_c^*)^2T_1T_2 \approx\Delta^2\exp\{[\omega-(\omega_n-\delta\omega)]^2/2\Delta^2\}/\delta\omega(\omega_n-\omega), \quad (5.9)$$

for $\omega_n-\omega\gg\Delta$. For sufficiently broad lines and/or frequencies sufficiently near the pulled frequency, $\omega_n-\delta\omega$, the high absorption solution will be reached in a time of the order of T_1. As the power is decreased from h_c, the low-absorption solution would not appear until the first contact point (given by 5.8) is reached. Thus, one could observe a region of hysteresis in the nuclear saturation for power levels between these two contact points. Of course, the precise value of h_c depends very strongly on the exact nuclear resonance line shape. An experiment in which one sweeps from low to high frequency should also give rise to hysteresis in the same way. In fact, the condition for the second contact (5.9) is strictly not valid when $(\gamma_n h^*)^2T_1T_2>1$. Then it is known[16] that there exist large changes in the line shape especially in the wings which may change the condition for h_c drastically. However, the excitation of the pulled solution will probably be nucleated by other phenomena[17] such as spin pinning.

At high power levels, one might wonder whether or not Suhl instabilities[14] might occur, as in ferromagnets, via the coupling, through the dipolar fields, of the $q=0$ nuclear spin wave mode with degenerate $q\neq0$ modes. In Appendix B, it is shown that such an effect would occur at much higher power levels than, for example, given by (5.8).

[16] A. Redfield, Phys. Rev. **98**, 1787 (1955).

[17] A. M. Portis, G. Witt, and A. J. Heeger, to be presented at the Eighth Annual Conference on Magnetism and Magnetic Materials, Pittsburgh, November, 1962 (unpublished).

Spin Echoes

We now investigate the behavior of a coupled electron nuclear system when a strong rf field pulse H_1 is applied at a frequency ω in the vicinity of the nuclear resonance frequency.

At the beginning of the pulse, the magnetization **m** is tilted, m_z is reduced, and the instantaneous precession frequency $\omega_n[1-\eta(m_z/M_0)]$ increases; the nuclear system is then submitted to a rf field which is not any more tuned to the correct frequency and the magnetization becomes insensitive to H_1. The main conclusion is that it is difficult to make 90 pulses when the frequency shift $\delta\omega$ is large. Numerically, for a ferromagnet the effect is computed as follows: in a frame rotating at frequency ω, the equations of motion for the nuclear magnetization are:

$$dm_x/dt=(\Delta\omega+\mu\delta\omega)m_y; \quad (5.8)$$

$$dm_y/dt=-(\Delta\omega+\mu\delta\omega)m_x+\gamma_nH_1\eta m_0\mu; \quad (5.9)$$

$$d\mu/dt=-\gamma_nH_1\eta(m_y/m_0), \quad (5.10)$$

where $\Delta\omega=\omega-\omega_n$ and $\mu=(m_z/m_0)$. Combining (5.8) and (5.10) we obtain

$$(\Delta\omega+\mu\delta\omega)(d\mu/dt)=-\gamma_nH_1\eta m_0^{-1}(dm_x/dt). \quad (5.11)$$

This may be integrated (with the boundary condition $\mu=1$ for $m_x=0$) and gives m_x explicitly as a function of μ. Eliminating then m_y between (5.9) and (5.10), one obtains a second-order equation for μ, formally identical to the equation of motion of a classical point in a static one-dimensional field:

$$d^2\mu/dt^2=dF(\mu)/d\mu; \quad (5.12)$$

$$F(\mu)=8^{-1}(\delta\omega)^2(1-\mu)P(\mu); \quad (5.13)$$

$$P(\mu)=\mu^3+(1+4x)\mu^2 +(4y^2+4x^2-1)\mu+4y^2-4x^2-4x-1; \quad (5.14)$$

$$x=\Delta\omega/\delta\omega; \quad y=\gamma_n\eta H_1/\delta\omega. \quad (5.15)$$

The "energy integral" for this problem yields

$$(d\mu/dt)^2=4^{-1}(\delta\omega)^2(1-\mu)P(\mu),$$

where the integration constant is chosen to ensure $d\mu/dt=0$ when $\mu=1$. This system starts from this value μ, then decreases, and $d\mu/dt$ remains negative until $P(\mu)$ vanishes. At this point, μ starts to increase and goes back to 1. Thus, if we want to make a 90° pulse (μ going from 1 to 0) we need that $P(\mu)$ does not vanish in the interval $0<\mu\leq1$. A detailed study of $P(\mu)$ shows that the minimum value of y (or H_1) for which this is realized is $y=4^{-1}(2^{1/2}-1)\sim0.1$. The pulse frequency for which this is obtained corresponds to $x=-\frac{1}{2}$.

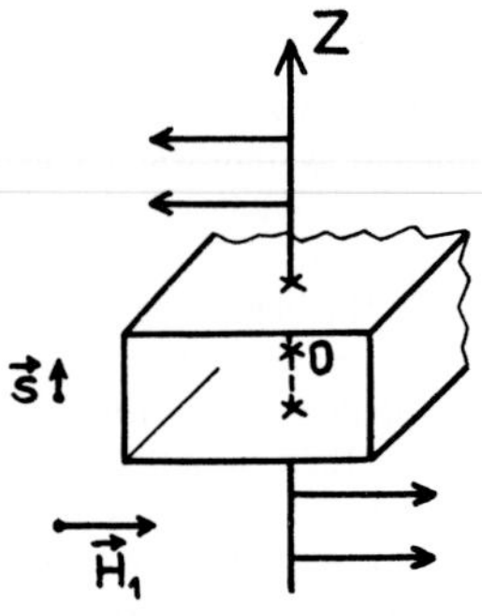

FIG. 4. Bloch wall. In the interior of the Bloch wall the magnetization rotates remaining parallel to the xy plane perpendicular to O_3.

Thus, to make 90° pulses, in the most favorable case we need an rf field H_1 such that

$$(\delta\omega)^{-1}(\eta\gamma_n H_1)=0.1 \quad \text{or} \quad H_1=0.1\alpha m_0. \tag{5.16}$$

In the liquid helium range this leads to fields of order 0.1 to 1 G (for 100% isotopic abundance). At lower temperatures it would be considerably more difficult to apply echo techniques.

Up to now, we have discussed only the motion when the rf field was applied. The last question is: What happens after the rf pulse? The answer is given simply by Eqs. (5.8), (5.9), and (5.10) with $H_1=0$ (for times shorter than the spin lattice relaxation time T_1). Then m_z retains the constant value which has been prescribed by the pulse conditions and the transverse components of the magnetization process at the frequency, $\omega_n[1-\eta(m_z/M_0)]$.

ACKNOWLEDGMENTS

The authors are especially indebted to Professor A. Portis and Professor A. Heeger for their discussions and correspondence concerning the experimental situation in $KMnF_3$ and to Professor N. Bloembergen for his suggestion of the occurrence of a degeneracy in the nuclear spectrum. We have also benefited from discussions with C. Robert. One of us (P. P.) would like to express his thanks to Professor A. Abragam and the entire Magnetic Resonance Group at Saclay for their kind hospitality dubing the tenure of a National Science Foundation Postdoctoral fellowship.

APPENDIX A. FREQUENCY SHIFT IN BLOCH WALLS

We derive the frequency shift for a 180° wall in a uniaxial crystal. The general notation and structure of the wall at rest is represented on Fig. 4. In a state of motion there are small, z dependent deflections M_z, M_θ, and m_z, m_θ, of the electron and nuclear magnetization. The component M_z (along the axis perpendicular to the wall) is always negligible. The equations of motion for the nuclear spins are

$$\begin{aligned} i\omega m_\theta &= dm_\theta/dt=\gamma_n\alpha M_0 m_z, \\ i\omega m_z &= dm_z/dt=\gamma_n\alpha(-M_0 m_\theta+m_0 M_\theta). \end{aligned} \tag{A1}$$

The variation $M_\theta(z)$ of the electronic magnetization is related to the overall displacement of the wall S by

$$M_\theta=-M_0(\partial\theta/\partial z)S, \tag{A2}$$

where θ is the turn angle in the undisturbed Bloch wall. For the simple case at hand

$$\partial\theta/\partial z=e^{-1}\sin\theta, \tag{A3}$$

where e is the wall thickness.

The displacement S can be related to the fields acting on the wall; the external rf field H_1 and the fields from the nuclei $\alpha\mathbf{m}$:

$$m_w\frac{d^2S}{dt^2}+\beta\frac{dS}{dt}+CS=2M_0H_1-\alpha M\int m_\theta d\theta. \tag{A4}$$

In this equation, m_w is the wall mass (per cm^2), β a damping coefficient, and C is related to the static permeability of the material. The right-hand side dE/dS gives the change in magnetic and hyperfine energy per cm^2, dE, when the wall is displaced by an amount dS; it may correctly be called the pressure acting on the wall.

It is convenient to introduce at this stage the maximum enhancement factor $\eta_{\max}$ for the rf field acting on the nuclei. $\eta_{\max}$ is obtained for nuclei at the center of the wall; it may be calculated by neglecting the small hyperfine contribution to the pressure in (A4).

$$\begin{aligned} \eta_{\max} &= -(\alpha M_\theta/H_1)_{\theta=\pi/2} \\ &= 2\alpha M_0{}^2e^{-1}(-m_w\omega^2+i\beta\omega+C)^{-1}. \end{aligned} \tag{A5}$$

$\eta_{\max}$ as defined here is generally complex. We know, however, from the results of Portis and Gossard,[5] that at room temperature in cobalt $\eta_{\max}$ is essentially real and of order 1500. Let us now come back to the calculation of the resonance frequency for which we can put $H_1=0$. We obtain from (A2), (A4), and (A5)

$$M_\theta=(\tfrac{1}{2})\eta_{\max}e(\partial\theta/\partial Z)\int m_\theta d\theta. \tag{A6}$$

Thus, the dependence of all amplitudes M_θ, m_θ, m_z is given by $\partial\theta/\partial z$. Putting $m_\theta=R(\partial\theta/\partial z)=Re^{-1}\sin\theta$, we

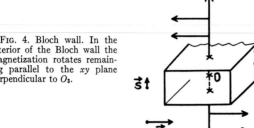

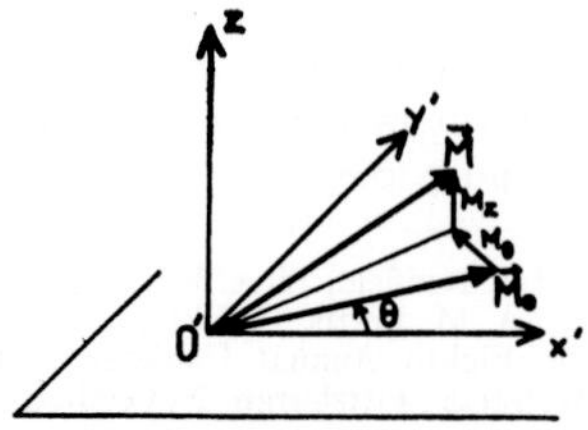

FIG. 5. Vectors relating to Bloch wall. M_0 is the equilibrium position of the magnetization in the $x'y'$ plane. M is the deflected magnetization.

get $\int m_\theta d\theta = 2R/e$ and from (A6)

$$M_\theta = \eta_{\max} m_\theta. \quad \text{(A7)}$$

If we now insert (A7) in (A1) we obtain the secular equation:

$$\omega^2 = \omega_n^2[1 - \eta_{\max}(m_0/M_0)];$$
$$\omega \cong \omega_n[1 - \tfrac{1}{2}\eta_{\max}(m_0/M_0).] \quad \text{(A8)}$$

Equation (A8) is very similar to Eq. (2.4) for the frequency shift in the bulk material. The factor $\frac{1}{2}$ obviously comes from the fact that, in the Bloch walls, the effective rf field acting on the nuclei is linearly polarized. Equation (A8) was derived for a simple case but remains qualitatively correct for all situations.

In fcc cobalt, with $\eta_{\max} = 1500$, $M = 1400$ emu, $m_0 = 3.3 \times 10^{-2} T$, we obtain from (A8), $(\omega - \omega_n)/\gamma_n \sim 10^3$ Oe at $T = 4°$K. This estimate, however, is correct only if $\eta_{\max}$ does not change significantly from room temperature to 4°K. In fact, in the case of a metal like cobalt, the damping coefficient β is probably due to eddy current effects, and is proportional to the conductivity,[12] σ. From the data of Portis and Gossard[5] we estimate $\beta\omega/C \sim 0.1$ at room temperature. At low temperatures, β depends critically on the residual resistivity of the cobalt particles and we cannot make a very definite prediction. Assuming, for instance, $\sigma_{T=0}/\sigma_{T=300} = 10$, we would have, at very low temperatures, $\beta\omega \sim C$. The real and imaginary parts of $\eta_{\max}$ are then equal, the shift and the intrinsic width of the resonance are then comparable and of order 500 Oe. It would thus be of interest to study simultaneously $\eta_{\max}$ and $\delta\omega$ on cobalt particles in the helium range. Of course, as in the case of the bulk material, the resonance frequency is power dependent through m_0 and the shift $\omega - \omega_n$ can be observed only at very low power levels.

APPENDIX B. SUHL INSTABILITY WITHIN THE NUCLEAR SYSTEM

In this Appendix we calculate the critical field for the Suhl instability. We start from the equations of motion (4.1) for the nuclear magnetization

$$dm_{qx}/dt = \gamma_n[\alpha M_z + H_0 - N_z M_z]m_{qy} - \gamma_n \alpha m_z M_{qy}.$$

The electronic term M_{qy} is now expressed in terms of m_{qy} and we obtain

$$dm_{qx}/dt = \gamma_n[\alpha M_z + H_0 - N_z M_z - (\alpha^2 m_z M_z / C_q)]m_{qy}, \quad \text{(B1)}$$

and a similar equation for m_{qy}.

If a strong rf field is applied at a frequency corresponding to the uniform nuclear mode, m_z and M_z are strongly affected. The effects related to the reduction of m_z are discussed in Sec. V. If the electronic and nuclear motions are elliptical, terms oscillating at a frequency $2\omega_0$ will appear in m_z and M_z. These terms produce an instability for the nuclear spin waves degenerate with the uniform mode.

Let us calculate now the amplitudes δM_z and δm_z of the oscillating parts of M_z and m_z.

We write the equations for the motion of M_z and m_z disregarding damping terms and nonuniform components:

$$dM_{0z}/dt = \gamma_e[M_{0y}(\alpha m_{0x} + h_{0x}) - M_{0x}(\alpha m_{0y} + h_{0y})],$$
$$dm_{0z}/dt = \gamma_n \alpha(m_{0y} M_{0x} - m_{0x} M_{0y}). \quad \text{(B2)}$$

We neglect the direct coupling between the rf field and the nuclear magnetization.

M_{0x} and M_{0y} are expressed in terms of h_x, h_y, m_{0x}, and m_{0y}. We obtain

$$M_{0x} = M_0(\alpha m_{0x} + h_x)/D_0,$$
$$M_{0y} = M_0(\alpha m_{0y} + h_y)/C_0. \quad \text{(B3)}$$

C_0 and D_0 are defined in Sec. IV.

It is interesting to compare the magnitude of the field due to the nuclear motion αm_{0x} to the magnitude of the external applied field h_x. m_{0x} at resonance is given by

$$m_{0x} \sim \gamma_n \eta h_x T_2 m_0.$$

The ratio p of the two quantities is

$$p = \alpha m_{0x}/h_x = \alpha m_0 \eta \gamma_n T_2 = (\omega_0 T_2)\eta(m_0/M_0). \quad \text{(B4)}$$

At low temperature, p may be larger than unity. If $\eta = 10^2$, $(m_0/M_0) = 3 \times 10^{-4}$, $\omega_0 T_2 = 10^3$, then $p = 30$.

The calculations will be done assuming $p \gg 1$ and, therefore, h_x and h_y will be neglected in the Eqs. (B3). If this condition is not valid, the critical fields will be larger than our estimation. Equations (B2) become

$$dM_{0z}/dt = \gamma_e M_0 p^2 h_x h_y (D_0 - C_0)/D_0 C_0,$$
$$dm_{0z}/dt = \gamma_n M_0 p^2 h_x h_y (D_0 - C_0)/D_0 C_0. \quad \text{(B5)}$$

$(D_0 - C_0)/(D_0 C_0)^{1/2}$ is a measure of the ellipticity of the electronic uniform mode; for a thin plate perpendicular to the z axis, we have

$$(D_0 - C_0)/D_0 C_0 = 4\pi M/(H_0 + H_A)(H_0 + H_A + 4\pi M),$$

and if $4\pi M \gg H_0$, Eqs. (B5) become

$$dM_{0z}/dt = \gamma_e M_0 p^2 h_x h_y/(H_0 + H_A),$$
$$dm_{0z}/dt = \gamma_n M_0 p^2 h_x h_y/(H_0 + H_A).$$

The amplitudes δM_z and δm_z are easily obtained:

$$\delta M_z = p^2(\gamma_e M_0/\gamma_n H_n)h^2/4(H_0 + H_A),$$
$$\delta m_z = p^2(M_0/H_n)h^2/4(H_0/H_A),$$

h being the amplitude of the rotating applied field.

The $\alpha\delta M_z$ term behaves exactly as an external oscillating field applied along the z axis, and the problem is identical to the parallel pumping problem. The critical field is given by

$$\gamma_n \alpha \delta M_z \eta(m/M) \geq 1/T_2$$

which may be written as

$$(\gamma_n \eta h T_2)^2 \geq (\omega_n T_2)^{-1}[\eta(m/M)]^{-2}(M/m)(\gamma_n/\gamma_e).$$

With the usual assumptions, the equation is

$$(\gamma_n \eta h T_2)^2 \geq 3.$$

The critical field we obtain is much larger than the field given by the usual saturation condition. With such a field all the approximations made in this Appendix are not valid.

The effect of the δm_z term is examined. The calculations are similar and lend to the condition

$$\alpha^2 M_z \delta m_z / C_q \geq 1/T_{2q}.$$

We obtain

$$(\gamma_n \eta h T_2)^2 \geq 8(\omega_n T_2)^{-1}(\eta m/M)^{-2}.$$

Or with the assumptions, $(\gamma_n \eta h T_2)^2 \geq 10$.

The conclusion is that it is impossible to reach a critical field giving rise to Suhl instability.

Volume 7, number 5 PHYSICS LETTERS 15 December 1963

ONSET OF SUPERCONDUCTIVITY IN DECREASING FIELDS

D. SAINT-JAMES
Service de Physique du Solide et de Résonance Magnétique, C.E.N. de Saclay

and

P. G. GENNES
Faculté des Sciences d'Orsay (S & O), France

Received 16 November 1963

We show that in ideal samples the nucleation of superconducting regions in decreasing fields should always occur near the surface of the sample. As a result the nucleation field is not equal to the Landau value [1] $H_{c2} = \varkappa\sqrt{2}\, H_c$ (where $\varkappa$ is the Landau-Ginzburg parameter) but is given by $H_{c3} = 2.392\, \varkappa H_c$. For a superconductor of the first kind, this implies that the values of $\varkappa$ derived from supercooling experiments must be corrected. For a superconductor of the second kind, the conclusion is that in fields H between H_{c2} and H_{c3} there remains a superconducting sheath on some parts of the sample.

We derive the nucleation field in the Landau-Ginzburg region. The linearised equation to be solved is (in the conventional notation [1])

$$\frac{1}{2m}[-i\hbar\nabla - 2e\frac{A}{c}]^2 \psi + \alpha\psi = 0 \;, \qquad (1)$$

where $H = \mathrm{curl}\, A$ is simply the applied field. In connection with eq. (1) it is often convenient to introduce the characteristic length $\xi(T)$ defined by

$$\xi^2(T) = -\frac{\hbar^2}{2m\alpha} \;. \qquad (2)$$

We assume that a) $\xi(T)$ is small compared to the bulk dimensions of the sample [2], b) the surface polish is such that the local radii of curvature of the boundary are large compared with $\xi(T)$. This enables us to consider only the problem of a plane boundary. It will turn out that the favourable situation for nucleation occurs when the field H is parallel to the surface. Thus we take the boundary plane as yOz, the field being in the z direction and the metal occupying the half space $X > 0$. The half-space $X < 0$ is assumed to be a vacuum or an insulator. In this case the boundary condition to be applied to eq. (1) is, to a good approximation

$$[(-i\hbar\frac{\partial}{\partial x} - \frac{2eA_x}{c})\psi]_{x=0} = 0 \;. \qquad (3)$$

We choose the gauge $A_X = A_Z = 0$, $A_Y = HX$ and look for solution of the form *

$$\psi = f(x)\, e^{iky} \;. \qquad (4)$$

Eqs. (1) and (3) become:

$$-\frac{\hbar^2}{2m}\frac{d^2 f}{dx^2} + \frac{1}{2m}[\hbar k - \frac{2e}{c}HX]^2 f = -\alpha f \;, \qquad (5)$$

* Taking $\psi = e^{ik'z} f(x,y)$ give an extra k'^2 contribution to the value of $-\alpha$. As we are interested in the lowest value of $-\alpha$ we take $k' = 0$.

Volume 7, number 5 PHYSICS LETTERS 15 December 1963

$$df/dx = 0, \quad x = 0, \quad x = \infty. \tag{6}$$

Eq. (5) is of the form of the Schrödinger equation for a harmonic oscillator of frequency $\omega = 2eH/mc$, the minimum of the potential being located at:

$$X_0 = \frac{\hbar ck}{2eH}. \tag{7}$$

The wave function is concentrated in region of dimension $\sim\xi(T)$ around X_0. When $X_0 \gg \xi(T)$ the boundary condition (6) is unimportant and we find the usual harmonic oscillator solution:

$$f = \exp\left[-\tfrac{1}{2}\left(\frac{X-X_0}{\xi(T)}\right)^2\right], \tag{8}$$

$$-\alpha = \tfrac{1}{2}\hbar\omega = \frac{e\hbar H}{mc}. \tag{9}$$

In the opposite limit $X_0 = 0$, the solution (8) again applies, since it satisfies (6) exactly; thus for $X_0 = 0$ we still find the eigenvalue (9).

We shall now show that for intermediate values of X_0 the eigenvalue is lowered below the value (9). We replace eq. (5) (applying for $x > 0$) and the boundary condition (6) by another Schrödinger equation, applying for $-\infty < x < \infty$, where the potential $V(X)$ is symmetrised:

$$V(X) = \frac{2eH^2}{mc^2}(X-X_0)^2 \qquad (X > 0);$$
$$V(X) = V(-X) \qquad (X < 0). \tag{10}$$

The lowest eigenvalue for the potential V has no nodes, is even, and thus satisfies automatically to condition (6).

For $X_0 > 0$, $V(X) < (2e^2H^2/mc^2)(X-X_0)^2$ in all region $X < 0$. Thus the lowest eigenvalue associated with $V(X)$ is smaller than the eigenvalue (9) associated with the potential $(2e^2H^2/mc^2)(X-X_0)^2$. This means that *nucleation will be easier in the vicinity of the surface.*

The next step is to find the value of X_0 for which the eigenvalue is a minimum. Using the variational principle on the free energy, this minimum condition can be written as:

$$\int f^2\left(\hbar k - \frac{2eH}{c}x\right)dx = 0. \tag{11}$$

Eq. (11) shows that in the optimum state the over-all surface current vanishes. A detailed solution of (5) and (6) in terms of Weber functions shows that the optimum value of X_0 is $X_0 = \mu^2\xi(T)$ the corresponding eigenvalue being:

$$-\alpha = \mu^2\frac{e\hbar H}{mc} \tag{12}$$

with $\mu^2 = 0.59010$. The coefficient μ is defined exactly by the implicit equation

$$\int_0^\infty dt(2t-\mu)\, t^{-\frac{1}{2}+\frac{1}{2}\mu^2}\, e^{-(t-\mu)^2} = 0. \tag{13}$$

Eq. (12) corresponds to a field $H_{c3} = (1/\mu^2)\sqrt{2}\varkappa H_c$. This completes our discussion when the field is in the plane of the surface.

On the other hand, when H is normal to the surface, the Landau result (9) remains valid. We have not yet performed the calculation for intermediate angles, but according to all indications the field will vary smoothly with angle between H_{c2} and H_{c3}. The physical conclusions are:

1. Nucleation in an ideal specimen will always occur at the field H_{c3} (higher than the Landau value $H_{c2} = \varkappa\sqrt{2}H_c$) except if some very special steps are taken to avoid surface effects (e.g., temperature gradients or field inhomogeneities).

2. For superconductors of the first kind, starting from the experimental values of Faber [3] (reviewed by Lynton [1]) of the supercooling field we arrive at the following revised value of $\varkappa$: 0.0153(Al), 0.066(In), 0.0968(Sn). The theoretical values derived from $\varkappa = 0.96\,(\lambda_L(0)/\xi_0)$ (where $\lambda_L(0)$ and ξ_0 are taken from specific heat and anomalous skin effect measurements) are [1,4]: 0.01(Al), 0.051(In), 0.149(Sn).

3. For superconductors of the second kind, in fields H such that $H_{c2} < H < H_{c3}$ there will be a superconducting sheath near the surface of the sample. If the sample is a long cylinder with H along the axis, the sheath will cover all the surface of the cylinder. If it is a sphere, the sheath will be restricted to a small band near the equatorial plane when $H \sim H_{c3}$, but if H is decreased toward H_{c2} the sheath will progressively extend up to the poles.

These effects may explain some apparent discrepancies which occur between magnetic flux measurement of the transition field (determining H_{c2}) and resistivity measurements (determining more or less H_{c3} in simple geometries such as the cylinder described above [5]).

Of course, in non-ideal samples the volume defects can also participate in the nucleation process.

We end up with two remarks

a) The above calculation is valid only when the superconducting material is surrounded by an insulator. If the surface was coated by a normal metal, the boundary condition to be imposed on the Landau-Ginzburg wave function is strongly different from (3) and the nucleation field is modified. Calculations are in progress to investigate this situation.

b) A very different situation, where $\xi(T)$ is comparable to the sample dimensions, has been achieved in a recent experiment by Parks [7] with a thin film in a perpendicular field. The threshold

Volume 7, number 5 PHYSICS LETTERS 15 December 1963

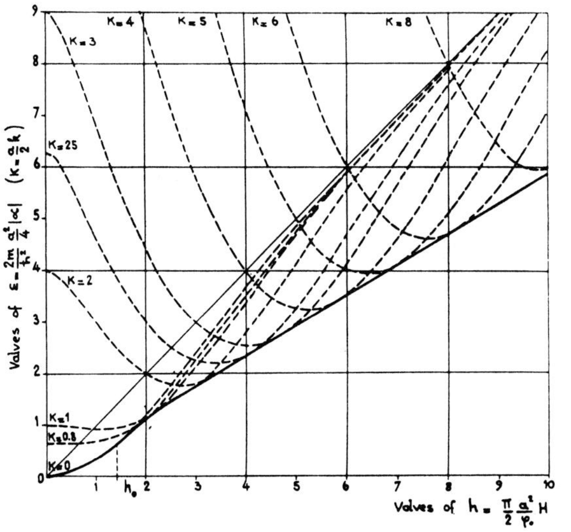

Fig. 1. Variation of the nucleation field in a slab.

field can again be derived from eq. (5), the boundary conditions (6) being modified to:

$$(\mathrm{d}f/\mathrm{d}x) = 0 \text{ for } x = \pm \tfrac{1}{2}a$$

(a is the sample thickness, i.e., in Park's experiment the cylinder diameter). We have also calculated this case. For $H < H_0$ ($H_0 \sim 2.75\ \varphi_0/\pi a^2$) (where $\varphi_0 = ch/2e$) we find that the optimum X_0 is 0: nucleation starts symmetrically. For $H > H_0$, X_0 moves toward the boundaries and progressively we recover the one boundary situation described above. The theoretical field is given on fig. 1. In practice Parks observes some anomalies at $H \sim 6.4\varphi_0/\pi a^2$ but a detailed interpretation of his results should take into account the (unknown) dissipative mechanisms which are responsible for the finite width of the resistive transition.

The present work was initiated during a visit of one of us (P. G. de Gennes) at the University of Rochester and he wishes to acknowledge in this connection some very stimulating conversations with Professor R. D. Parks. We are also indebted to Drs P. Nozières, G. Sarma and R. Pick for fruitful discussions on these and related topics.

References

1) V. L. Ginzburg and L. D. Landau, J. Exptl. Theoret. Phys. USSR 20 (1950) 1064.
V. L. Ginzburg, J. Exptl. Theoret. Phys. USSR 30 (1956) 593; translation Sov. Phys. JETP 3, 621; J. Exptl. Theoret. Phys. USSR 34 (1956) 113; translation Sov. Phys. JETP 7 (1958) 78.
E. A. Lynton, Superconductivity (J. Wiley, New York; Methuen and Co., London, 1961) p. 46.
2) C. Caroli, P. G. de Gennes and J. Matricon, J. phys. 23 (1962) 707.
3) T. E. Faber, Proc. Roy. Soc. A231 (1957) 353.
J. F. Cochran, D. E. Mapother and R. E. Mould, Phys. Rev. 103 (1958) 1657.
4) J. Bardeen and J. R. Schrieffer, Progress in Low Temperature Physics, vol. 3, ed. C. J. Gorter (North-Holland Publ. Comp., Amsterdam) p. 170.
5) W. F. Druyvesteyn and D. J. Van Ooijen, Communication at the Int. Conf. on the Science of Superconductivity, Colgate University, August 1963, to be published in Revs. Modern Phys.
6) P. G. de Gennes, Communication at the Int. Conf. on the Science of Superconductivity, Colgate University, August 1963, to be published in Revs. Modern Phys.
7) R. Parks and J. M. Mochel, Post deadline communication at the Int. Conf. on the Science of Superconductivity Colgate University August 1963, Phys. Rev. Letters 11 (8) 354.

* * * * *

Afterthought: Onset of superconductivity in decreasing fields

Here, with Cooper pairs, we see an example of "edge states" — quantum states of a charged particle performing cyclotron orbits near a reflecting surface, and bouncing many times on the surface. For the classical superconductors, with large coherence lengths, the roughness of the surface is not relevant. But for other systems, with smaller lengths, the role of roughness (in its various guises) could be interesting: I have not seen a theoretical study of this problem — but I may have missed it.

Reprinted from
REVIEWS OF MODERN PHYSICS Volume 36, No. 1 (Part 1 of Two Parts) January 1964

Boundary Effects in Superconductors

P. G. DE GENNES
Faculté des Sciences, Orsay (S & O) France

I. INTRODUCTION

The present paper is concerned with the properties of layered structures of superconducting (and non-superconducting) materials, and with the related boundary problems. The basic experimental facts in this field are the following:

(a) for small superconducting samples surrounded by a nonmetallic substrate (e.g., films evaporated on glass), the transition temperature is very close to the bulk value. Tunneling experiments also show that the energy gap is close to the bulk value (derived for instance by ultrasonic attenuation).

(b) for "NS sandwiches" (thin superconducting film S deposited on a normal metal substrate N) the transition temperature may be significantly lower than in the bulk S material—or even vanish completely. Such effects were observed in the **Pb** Ag and **Sn** Ag systems.[1,2] However, as pointed out by Rose-Innes and Serin,[3] some of these experiments cannot be trusted entirely because of spurious atomic migration effects. To circumvent this difficulty one must deposit the films and keep them constantly at

[1] D. Smith, S. Shapiro, J. L. Miles, and J. Nicol, Phys. Rev. Letters 6, 686 (1961).
[2] W. A. Simmons and D. H. Douglass, Phys. Rev. Letters 9, 153 (1962).
[3] A. C. Rose-Innes and B. Serin, Phys. Rev. Letters 7, 278 (1961).

low temperatures, as done by Hilsch.[4] The resulting samples have a very short electron mean free path l in the normal state (all the defects being quenched) and thus their coherence length is modified. This effect can be taken into account accurately if l is determined from resistivity measurements in the normal state.[5] Paradoxically, we expect more accurate results from such "dirty superconductor" sandwiches than from a "clean" NS system: In the former case the detailed atomic structure at the interface is less important and the large-scale motions of the superconducting electrons are ruled by a simple diffusion equation. In the latter case, on the other hand, the reflection and transmission properties of the transition region play an important role, and unfortunately they cannot be controlled at the present time. Thus in the following we shall restrict our attention to "dirty" systems.

(c) a thin (~1000 Å) normal slab N separating two superconductors S and S′ is able to carry a finite supercurrent from S to S′. These SNS′ junctions have been studied first in the pioneer work of Meissner.[6] Their interest is twofold: (1) from the dependence of the critical current on the thickness of the N slab one may obtain an estimate of V_N, the electron–electron interaction in N. (2) the SNS junctions have a wide range of critical currents and critical fields: this could be useful for some low-temperature devices. These properties are discussed in Sec. V.

II. ORDER PARAMETER AND EXCITATION SPECTRUM IN A NONHOMOGENEOUS SYSTEM

Let us first recall briefly how the excitation spectrum of a superconducting system is constructed by the self-consistent field method. We assume that the electrons are coupled by a point interaction $-V(\mathbf{r}_i)\delta(\mathbf{r}_i - \mathbf{r}_j)$. This is a good approximation, since the range of the exact interaction is of the order of a Fermi wavelength,[7] while the effects which we shall discuss take place on a much larger scale. Note that $V(\mathbf{r})$ will not be the same in the N and S regions. In the S regions, $V = V_S$ is positive (attractive interaction). In the N regions, $V = V_N$ may be of either sign, depending on a delicate balance between Coulomb repulsion and phonon-induced attraction. (If $V_N > 0$ the N material is also a superconductor at low enough temperatures $T < T_N$. However, as soon as $V_N \lesssim \frac{1}{2}V_S$, the corresponding transition temperature becomes so small that we never observe it.) One of the major interests of the thin-film experiments stems from the fact that, when done under suitable conditions to be discussed below, they may give us measurements of V_N.

To derive the Fermi-type excitations, we write down the equation[8] for the one-electron operator $\psi_\alpha^+(\mathbf{r})$

$$i\frac{\partial\psi_\alpha^+}{\partial t}(r) = \left[\frac{p^2}{2m} + U(r)\right]\psi_\alpha(r) - V(r) \times \sum_\beta \psi_\alpha^+(r)\psi_\beta^+(r)\psi_\beta(r) \quad (2.1)$$

(α and β are spin indices; the origin of energies is at the Fermi level) and U is the one-electron potential, with different values in the two metals; it also includes the effects of impurities and boundaries. We linearize the last term of (2.1) according to the rule

$$V_n\psi_\alpha^+(r)\psi_\beta^+(r)\psi_\beta(r) \rightarrow V_n\langle\psi_\alpha^+(r)\psi_\beta^+(r)\rangle\psi_\beta(r)\,. \quad (2.2)$$

The bracket denotes a thermal average. (The Hartree and exchange contribution which are essentially T-independent are incorporated in V). The only nonvanishing terms in (2.2) come from

$$V(\mathbf{r})\langle\psi_\downarrow^+(\mathbf{r})\psi_\uparrow^+(\mathbf{r})\rangle = -V(\mathbf{r})\langle\psi_\uparrow^+(\mathbf{r})\psi_\downarrow^+(\mathbf{r})\rangle = \Delta^+(\mathbf{r})\,. \quad (2.3)$$

We call $\Delta(\mathbf{r})$ the pair potential. We now look for eigenmodes of the linearized equation, of the form

$$\psi_\uparrow(\mathbf{r}t) = \sum_n (u_n(r)e^{-iE_nt}\gamma_{n\uparrow} + v_n^+(r)e^{iE_nt}\gamma_{n\downarrow}^+)\,,$$
$$\psi_\downarrow(\mathbf{r}t) = \sum_n (u_n(r)e^{-iE_nt}\gamma_{n\downarrow} - v_n^+(r)e^{iE_nt}\gamma_{n\uparrow}^+)\,, \quad (2.4)$$

where $\gamma_{n\mu}^+\gamma_{n\mu}$ are new fermion operators $[\gamma_{n\mu}^+\gamma_{m\sigma}] = \delta_{nm}\delta_{\mu\sigma}$, and the excitation energy E_n is restricted to positive values. u and v are the eigenfunctions of the following system of equations:

$$Eu = [(1/2m)p^2 + U(r)]u + \Delta v\,,$$
$$Ev = -[(1/2m)p^2 + U(r)]v + \Delta^+u\,. \quad (2.5)$$

Once we have solved for the u's and v's, we must write down the self-consistency requirement obtained from Eqs. (2.3) and (2.4). The prescription for calculating the average in (2.3) is that the new fermion states γ_n^+ of energy E_n have their thermal equilibrium population given by the Fermi function $f(E_n) = 1/1 + \exp E_n/T$. This gives

$$\Delta(\mathbf{r}) \equiv V(\mathbf{r})\langle\psi_\uparrow(\mathbf{r})\psi_\downarrow(\mathbf{r})\rangle = V(r) \times \sum_n v_n^+(r)u_n(\mathbf{r})[1 - 2f(E_n)]\,. \quad (2.6)$$

[4] P. Hilsch, Z. Physik **167**, 511 (1962).
[5] Unfortunately we do not have such data for the films of Refs. 1 and 2.
[6] H. Meissner, Phys. Rev. **117**, 672 (1960).
[7] P. W. Anderson and P. Morel, Phys. Rev. **125**, 1263 (1962).
[8] We shall use a system where $\hbar = k_B = 1$.

To ensure the convergence of this equation, we cut off the interaction V when the excitation energy ϵ_n is higher than ω_D, the local Debye frequency, in agreement with the original BCS procedure. This prescription may be justified by a detailed calculation with the original retarded interaction, even for the nonhomogeneous systems which we consider here.

The pair potential $\Delta(r)$ will be space-dependent for these systems: typical variations of $\Delta(r)$ in an NS sandwich and in an SNS junction are represented on Fig. 1. This space dependence has a first important

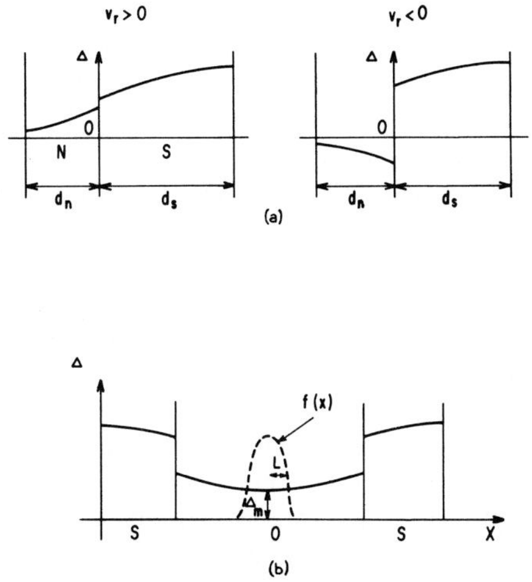

FIG. 1. Spatial dependence of the pair potential $\Delta(X)$ at temperatures close to the transition point in an NS sandwich (a), and in an SNS junction (b). For the sandwich, the two cases $V_N > 0$ and $V_N < 0$ have been represented. (b) also shows the function $f(X)$ corresponding to the wavefunction of a low-energy excitation.

consequence: the eigenfunctions $u_n(\mathbf{r})$ $v_n(\mathbf{r})$ of Eq. (2.5) are *not* simply proportional to the one electron wave functions in the normal state $w_n(r)$ defined by

$$[p^2/2m + U(\mathbf{r})]w_n = \epsilon_n w_n\,. \qquad (2.7)$$

In more physical terms we may say that the optimum electron pairing is *not* obtained by pairing one electron in a state w_n, and another in the time-reversed state w_n. This procedure works only when $\Delta(\mathbf{r})$ may be taken as independent of r: in an infinite, pure metal or homogeneous alloy.[9]

The pair potential $\Delta(\mathbf{r})$ is a natural "order parameter" for our inhomogeneous systems. This is not the only possible choice, however: we could also use for instance the "condensation amplitude" $F(\mathbf{r})$ defined by

$$F(\mathbf{r}) = \langle\psi_\uparrow(\mathbf{r})\psi_\downarrow(\mathbf{r})\rangle\,. \qquad (2.8)$$

[9] $\Delta(\mathbf{r})$ is only approximately constant for an alloy. (For a detailed discussion of this point see Ref. 12.)

F is the probability amplitude of finding two electrons in the condensed state at point $\mathbf{r}$. There are two important properties which we shall now emphasize in connection with Δ and F.

(*a*) *boundary conditions*: On an atomic scale, $F(\mathbf{r})$ and $\Delta(\mathbf{r})$ are continuous functions of $\mathbf{r}$. But if, as usual, we are interested in a larger scale, and describe the interface between two metals as a sharp boundary, *neither F nor Δ are continuous on this surface.* Actually, as we shall see, for our dirty systems, the quantity which is continuous is $F(\mathbf{r})/N(\mathbf{r}) = \Delta(\mathbf{r})/N(\mathbf{r})V(\mathbf{r})$, where $N(\mathbf{r})$ is the local density of states (per energy unit and per volume unit) at the Fermi level.

(*b*) *relation between the pair potential and the energy gap E_0*: We define E_0 (a quantity independent of $(\mathbf{r})$ as the lowest excitation energy of the fermions in the self-consistent field Δ: i.e., the minimum positive eigenvalue of Eq. (2.5). We shall now prove the following theorem: when $\Delta(\mathbf{r})$ depends on only one space coordinate X, the energy gap E_0 is equal to the minimum value of $|\Delta(X)|$ in the sample.

This is a rather surprising result, since we might expect at first sight that E_0 is increased beyond $|\Delta|_{\min}$ by some sort of 0-point energy. In fact, this 0-point energy exists, but it is very small in the usual limit when the coherence length ξ is much larger than the Fermi wavelength.

To prove this statement, we first rewrite Eq. (2.5) in spinor form

$$\hat{\psi} = \begin{pmatrix} u \\ v \end{pmatrix}, \qquad (2.9)$$

$$E\psi = [T\sigma_z + \Delta(X)\sigma_x]\hat{\psi} = \mathcal{H}\hat{\psi}\,, \qquad (2.10)$$

$$T = p^2/2m + U\,. \qquad (2.11)$$

We choose Δ as real (this is allowed for a static, non magnetic problem). The general aspect of $\Delta(X)$ is represented on Fig. 1(b). The "potential well" corresponds to a minimum value Δ_m, and a spatial range δ (this in practice will be of the order of the coherence length ξ, or at least of the order of the thickness d of the film in which Δ_n is obtained). We want to use a variational principle to show that the minimum E is close to $|\Delta|_{\min}$. But the operator $\mathcal{H}$ is not positive definite (E and $-E$ are simultaneous eigenvalues). Thus we consider the operator $\mathcal{H}^2$, which, thanks to the properties of the Pauli matrices σ, can be written as

$$\mathcal{H}^2 = T^2 + \Delta^2 + i[T\Delta]\sigma_y\,. \qquad (2.12)$$

Now, as a trial function, we choose the spinor

$$\hat{\psi}(r) = e^{ik_F Z} f(X)\hat{\varphi}_y\,, \qquad (2.13)$$

where $\hat{\varphi}_y$ is (either) one of the eigenfunctions of $\sigma_y(\sigma_y\hat{\varphi}_y = \pm\hat{\varphi}_y)$. The $0Z$ axis is perpendicular to the $0X$ direction (and otherwise arbitrary). Since our energies are counted from the Fermi level $E_F = k_F^2/2m$. We have simply

$$T\hat{\psi} = e^{ik_F Z}\left(-\frac{1}{2m}\frac{\partial^2}{\partial X^2}\right)f(X)\hat{\varphi}_y\,. \qquad (2.14)$$

The particular choice of wave vector in (2.13) will lead to a function $f(X)$ with no nodes in the lowest state. We now compute the expectation value

$$E^2 = (\hat{\psi}|\mathcal{H}^2|\psi) = \int dx f^+(X)\left[\left(-\frac{1}{2m}\frac{\partial^2}{\partial X^2}\right)^2 + |\Delta(X)|^2\right]f(X)$$
$$\pm i\int f^+(X)\left[-\frac{1}{2m}\frac{\partial^2}{\partial X^2}\,,\ \Delta(X)\right]f(X)dx\,. \qquad (2.15)$$

If we choose for f a real function the third term vanishes; we shall now estimate the first and second terms, by inserting for $f(X)$ a smooth function as shown on Fig. 1(b), with a range L. Then the kinetic energy contribution will be of order $(1/2mL^2)^2$. For the potential energy, if $|\Delta|$ increases according to the law

$$|\Delta(X)|^2 = |\Delta|^2_{\min}(1 + X^2/\delta^2)\,,$$

then the second term of (2.15) is of order $|\Delta|^2_{\min}(1 + L^2/\delta^2)$ and

$$E^2 = |\Delta|^2_{\min}(1 + L^2/\delta^2) + (1/2mL^2)^2\,.$$

Taking the minimum of this expression with respect to the range L of our trial function we get

$$E^2 = |\Delta|^2_{\min}(1 + \mu^{\frac{2}{3}})\,,$$

where μ = const $\times$ $(1/2m\delta^2|\Delta|_{\min})$.

We now observe that μ is extremely small in practice: for instance when the variations of Δ take place in one coherence length

$$\delta \backsim v_F/|\Delta|_{\min}\,,$$

then $\mu \sim \lambda_F/\delta \sim 10^{-2}$ to 10^{-3}. Thus the lowest eigenvalue $|E|$ is very close to $\Delta_{\min}$ and the theorem is proven.

Let us now see what consequences this has for tunneling experiments performed on the outer faces of an (NS) sandwich: the minimum of $|\Delta(X)|$ is obtained on the edge of the N side. $(X = -d_n)$ If we perform a tunneling experiment on this side we measure directly as a gap the pair potential at the surface. Some preliminary experiments in this field have been reported.[1] Unfortunately they measure only the absolute value of $|\Delta(-d_n)|$ and thus they do not provide a direct information on the sign of the electron electron interaction V_N in the normal metal. (However, as we shall see later on some examples we might derive the sign of V_N if the dependence of $|\Delta(-d_n)|$ on d_n was measured.) A particularly simple case is one where V_N is negligible: then $\Delta(-d_N) \to 0$ and there is no gap. The density of states to be expected in an extreme case of this sort has been computed (in the highly idealized limit of specular reflection of the electrons on the boundaries).[10] It may be that some negative results of tunneling experiments on superconductors with short coherence lengths ξ are due to this effect: if, below the oxide surface, there is a slab of thickness $\sim\xi$ where the BCS parameter NV is small (because of partial oxidation, say), then there will be a strong density of states at low energies due to excitations in the slab, and no gap will be observed.

Let us now consider another type of tunneling experiment on an NS sandwich: this time we perform the experiment on the S side. Here again the energy gap is $E_0 = |\Delta(-d_n)|$ much smaller in this case than the local pair potential $\Delta(d_S)$ (as is clear from Fig. 1). This identity of the gap on the two sides of a thin sandwich has apparently not been verified up to now. Of course, to get a significant tunnel current, at voltages just beyond E_0, on the S side, the thickness of the S slab must not exceed one coherence length, since the one-fermion excited states which may contribute in this range of voltages are rather localized in the N region. (A discussion of the spread out of the states on a simple example can be found in Ref. 10.)

Finally we turn to the case of an SnS junction: here the minimum of $|\Delta(X)|$ is obtained in the central plane $E_0 = \Delta_m = \Delta(0)$ [Fig. 1(b)]. Again, if we perform a tunneling experiment on the outer face of the S slabs, and if they are not much thicker than a coherence length, we should measure E_0 as a gap. The general conclusion is that the energy gap measured by a tunneling experiment on the outer surface of a layered system is not, in general, simply related to the value of the pair potential on this surface. (The only case where the two coincide, in the above example, is for tunneling on the N side of an NS sandwich).

III. SELF-CONSISTENT EQUATION FOR THE PAIR POTENTIAL

We now discuss the numerical determination of the pair potential $\Delta(X)$ for a layered structure. The naive method is to guess a shape of $\Delta(X)$, solve the eigenfunction problem (2.5) for this potential, insert

[10] P. G. De Gennes and D. Saint-James, Phys. Letters 4, 151 (1963).

the resulting u's and v's in the self-consistency requirement (2.6), obtain a new value for $\Delta(X)$ and iterate the process. The technique may be improved by the use of a variational principle, but even so it remains rather complicated. To reach a simpler situation we shall restrict our attention to the vicinity of the transition point T of the layered system, and we shall *assume* that the superconducting transition is always of second order: when this is true, the pair potential Δ is small at all points in space when the temperature is close to the transition point. Then it can be treated as a perturbation in the eigenvalue equations (2.5) and the self-consistency condition (2.6) becomes a linear integral equation for Δ, which is not too difficult to solve in practice.

This equation, due to Gor'kov[11] may be written as

$$\Delta(\mathbf{r}) = V(\mathbf{r}) \sum_{\omega} \int d_3\mathbf{r}' \Delta(\mathbf{r}') H_\omega(\mathbf{r}\mathbf{r}') , \quad (3.1)$$

$$H_\omega(\mathbf{r}\mathbf{r}') = T \sum_{\omega} \sum_{nm} \frac{1}{\epsilon_n - i\omega} \frac{1}{\epsilon_m + i\omega} \times w_n^+(\mathbf{r}) w_m^+(\mathbf{r}) w_n(\mathbf{r}') w_m(\mathbf{r}') , \quad (3.2)$$

where $\omega = 2\pi T(\nu + \frac{1}{2})$ and the sum $\sum_\omega$ represents a sum over all (positive or negative or 0) integers ν. The functions w_n are the one-electron wave functions in the normal state, defined by Eq. (2.7). They include the effects of impurity and boundary scattering. The one-electron Hamiltonian $\hat{T}$ being real we can choose the w's to be real. This choice of standing, rather than running waves is convenient as usual when impurity scattering is important; thus from now on we drop the stars in Eq. (3.2).

The main properties of the symmetric kernel $H_\omega(\mathbf{r}\mathbf{r}')$ are the following.

A. Sum Rule

From the orthogonality of the (real) functions w's we get

$$\int H_\omega(\mathbf{r}\mathbf{r}') d_3\mathbf{r}' = \sum_n \frac{1}{\epsilon_n^2 + \omega^2} |w_n(\mathbf{r})|^2$$

$$= N(\mathbf{r}) \int \frac{d\epsilon}{\epsilon^2 + \omega^2} = \frac{\pi}{|\omega|} N(\mathbf{r}) , \quad (3.3)$$

where $N(\mathbf{r})$ is the local density of states.[12]

B. Relation with a One-Electron Correlation Function[13]

Consider the sum

$$\mathcal{G}_\Omega(\mathbf{r}\mathbf{r}') = \overline{\textstyle\sum_m w_n(\mathbf{r}) w_m(\mathbf{r}) w_n(\mathbf{r}') w_m(\mathbf{r}')} \, \delta(\epsilon_n - \epsilon_m + \Omega) , \quad (3.4)$$

where the average is taken over all states n with a fixed energy ϵ_n. When $\mathcal{G}$ is known, $H_\omega(\mathbf{r}\mathbf{r}')$ can be derived simply according to (3.2). Now we observe that the Fourier transform

$$\mathcal{G}(\mathbf{r}\mathbf{r}'t) = \int \frac{dt}{2\pi} e^{i\Omega t} \mathcal{G}_\Omega(\mathbf{r}\mathbf{r}') \quad (3.5)$$

is a one-electron correlation function, namely

$$\mathcal{G}(\mathbf{r}\mathbf{r}'t) = \overline{\delta(\mathbf{r}(0) - \mathbf{r})[\mathbf{r}(t) - \mathbf{r}']} , \quad (3.6)$$

where the average is over all one-electron states of energy ϵ_n (in practice at the Fermi energy) $\mathbf{r}(t) = e^{i\hat{T}t}\mathbf{r}e^{-i\hat{T}t}$ is the electron position operator in the Heisenberg representation; it describes the motion of an electron in the normal metal, with the Hamiltonian $\hat{T} = p^2/2m + U(\mathbf{r})$. Equation (3.6) is easily verified in the w_n representation. It is useful because the form of the correlation function or the right-hand side is immediately known in many cases. We shall now discuss this on some examples.

C. Value of $H_\omega(\mathbf{r}\mathbf{r}')$ for an Infinite "Dirty" Medium

When the electron mean free path l is small, the correlation function (3.6) spreads according to a diffusion equation

$$(\partial/\partial|t|)\mathcal{G}(\mathbf{r}\mathbf{r}'t) - D\nabla^2\mathcal{G}(\mathbf{r}\mathbf{r}'t) = \text{const} \times \delta(\mathbf{r}\mathbf{r}')\delta(t) . \quad (3.7)$$

Here $D = \frac{1}{3} v_F l$ is the diffusion coefficient, v_F is the Fermi velocity. Equation (3.7) is valid when (1) uncertainty relations do not come into play (distances $|r - r'| \gg \lambda_F$) (2) the diffusion approximation applies ($|\mathbf{r} - \mathbf{r}'| \gg l$ and $t \gg l/v_F$).

Taking Fourier transforms we get

$$\mathcal{G}_\Omega(\mathbf{q}) = \int \mathcal{G}_\Omega(\mathbf{r}\mathbf{r}') e^{i\mathbf{q}\cdot(\mathbf{r}-\mathbf{r}')} d_3r'$$

$$= \text{const} \left\{ \frac{1}{i\Omega + Dq^2} + \text{c.c.} \right\} . \quad (3.8)$$

When $\mathcal{G}$ is known we can derive H_ω by Eqs. (3.2) and (3.4),

[11] L. P. Gor'kov, Zh. Eksperim. i Teor. Fiz. **37**, 1407 (1959) [English transl.: Soviet Phys.—JETP **10**, 998 (1960)]. See also C. Caroli, P. G. De Gennes, and J. Matricon, Phys. Cond. Matter **1**, 176 (1963).

[12] C. Caroli, P. G. De Gennes, and J. Matricon, J. Phys. Rad. **23**, 707 (1962).

[13] P. G. De Gennes and E. Guyon, Phys. Letters **3**, 168 (1963).

$$H_\omega(\mathbf{q}) = \int H_\omega(\mathbf{r}\mathbf{r}')e^{i\mathbf{q}\cdot(\mathbf{r}-\mathbf{r}')}d_3r'$$
$$= \text{const} \times \int d\epsilon d\Omega \frac{1}{\epsilon - i\omega} \frac{1}{\epsilon + \Omega + i\omega}$$
$$\times \left\{\frac{1}{i\Omega + Dq^2} + \text{c.c.}\right\} = 2\pi N \frac{1}{2|\omega| + Dq^2}. \quad (3.9)$$

The constant in (3.9) has been obtained from the sum rule (3.3). Eq. (3.9) can also be obtained by the Green's function method—but the latter derivation is much more tedious and slightly less general.[14]

Since in the following we shall be concerned with situations where the pair potential Δ depends only on one space coordinate ($\Delta(X)$), it is of interest to write down the one-dimensional Fourier transform of $H_\omega(q)$,

$$H_\omega(XX') = \frac{1}{2\pi}\int H_\omega(q)e^{iq(X-X')}dq$$
$$= (\pi N/2|\omega|\xi_\omega)e^{-|X-X'|/\xi_\omega}. \quad (3.10)$$

$\xi_\omega = (D/2|\omega|)^{\frac{1}{2}}$ gives us the range of $H_\omega(XX')$. The largest range corresponds to $\omega = \omega_0 = \pi T$,

$$\xi_{\omega_0} = \xi = (D/2\pi T)^{\frac{1}{2}}. \quad (3.11)$$

ξ plays the role of a coherence length for our alloy (The dirty superconductor approximation requires $l \ll \xi$.)

D. Boundary Effects

The first, and most simple problem here is that of a *free surface* (the metal occupying, for instance, the half space $X > 0$); Eq. (3.7) for the correlation function remains valid. Since there is no electron flow out from the surface, we must have $D(d\mathcal{G}/dX)_{X=0} = 0$. This condition can be achieved by a method of images, and we get finally

$$H_\omega(XX') = \frac{\pi N}{2|\omega|\xi_\omega}[e^{-|X-X'|/\xi_\omega} + e^{-(X+X')/\xi_\omega}]. \quad (3.12)$$

Since

$[(\partial H_\omega/\partial X)(XX')]_{X=0} = 0$, we see in Eq. (3.1) that

$$(d\Delta/dX)_{X=0} = 0. \quad (3.13)$$

The pair potential has 0 slope at the surface.[15]

It is slightly more complicated to discuss the boundary problem for two metals A and B (separated by the plane $X = 0$) with different densities of state N_A, N_B and diffusion coefficients D_A, D_B. By the correlation function method or with Green's functions, we may show that in each of the metals $H_\omega(XX')$ is still ruled by the equation

$$2|\omega|H_\omega(XX') - D(X')(d^2/dX'^2)H_\omega(XX') = 2\pi N(X)\delta(X - X'). \quad (3.14)$$

What are the boundary conditions on the surface? One of them can be obtained by a sum rule argument: we integrate (3.14) on X', obtaining

$$2|\omega|\int_{-\infty}^{\infty} H_\omega(XX')dX' - \left[D\frac{d}{dX'}H_\omega(XX')\right]_{X'=0+}^{X'=\infty}$$
$$- \left[D\frac{d}{dX'}H_\omega(XX')\right]_{X'=-\infty}^{X'=0-} = 2\pi N(X). \quad (3.15)$$

The contributions for $X' = \infty$ vanish, and by comparison with (3.3) we get

$$\left[D(X)\frac{d}{dX'}H_\omega(XX')\right]_{X'=0-}^{X'=0+} = 0. \quad (3.16)$$

The other boundary condition will have the general form

$$[H_\omega(X,X')]_{X'=0+} = [\alpha H_\omega(X,X') + \beta(d/dX')H_\omega(X,X')]_{X'=0-}. \quad (3.17)$$

β/α is an effective length, which could be derived from the microscopic transport equation satisfied by H_ω, and which is familiar from similar neutron transport problems at the surface of a moderator. If there is no insulating barrier between A and B,[16] we know from the neutron case that β/α is comparable to the mean free path l. Since $(1/H_\omega)(d/dX')H_\omega \sim (1/\xi)$, the second term of (3.17) is of order l/ξ, and is negligible in our limit. Thus the second boundary condition will be simply of the form

$$H_\omega(X,0_+) = \alpha H_\omega(X,0_-). \quad (3.18)$$

We now proceed to show that α is determined by the symmetry properties of $H_\omega(XX')$. From Eq. (3.14) we have

$$\left.\begin{aligned}
H_\omega(XX') &= \frac{N_A\pi}{2|\omega|\xi_A}[e^{-|X-X'|/\xi_A} + \lambda e^{-(X+X')/\xi_A}], \\
&\qquad X > 0,\ X' > 0 \quad (\text{side A}), \\
H_\omega(XX') &= (N_B\pi/2|\omega|\xi_B)\mu e^{-X/\xi_A + X'/\xi_B}, \\
&\qquad X > 0 \quad (\text{side A}), \\
&\qquad X' < 0 \quad (\text{side B}).
\end{aligned}\right\} \quad (3.19)$$

[14] Note that the diffusion concept applies even for concentrated alloys.

[15] When the finite thickness of the metal–vacuum transition layer is taken into account, the condition is not exactly one of 0 slope; see Reference 12. On all physical effects, however, this leads only to corrections of order λ_F/ξ.

[16] More accurately, if the transmission coefficient of the barrier is much larger than l/ξ.

Here λ and μ are unknown parameters, to be determined from the boundary conditions (3.16), (3.18), which yield

$$(1-\lambda)N_A = \mu N_B\,, \quad (1+\lambda)(N_A/\xi_A) = \alpha\mu(N_B/\xi_B)\,,$$
$$\mu = [2\xi_B/(\xi_B + \alpha\xi_A)](N_A/N_B)\,,$$
$$H_\omega(XX') = (\pi/|\omega|)\,[N_A/(\xi_B+\alpha\xi_A)]e^{-X/\xi_A + X'/\xi_B} \quad (X>0\,,\ X'<0)\,.$$

We interchange X and X', repeat the argument and write that $H_\omega(XX')$ is symmetric; we get

$$[N_A/\xi_B + \alpha\xi_A)] = [N_B/(\xi_A + \alpha^{-1}\xi_B)]\,,$$
$$\alpha = N_A/N_B\,. \qquad (3.20)$$

Thus we conclude that $H_\omega(XX')/N(X')$ is continuous when X' crosses the boundary. If we now return to Eq. (3.1) for the pair potential, we observe that $\Delta(X)/V(X)$ obeys the same boundary conditions than $H_\omega(XX')$. Thus the boundary conditions for dirty superconductors are, from Eqs. (3.16) and (3.20),

$$\left.\begin{array}{l}\Delta/NV \quad \text{continuous} \\ (D/V)(d/dX)\,\Delta \ \text{continuous}\end{array}\right\}\text{at a metallic interface,}$$
$$d\Delta/dX = 0 \qquad \text{at a free surface.} \qquad (3.21)$$

These conditions will play a crucial role in the following. Finally, it is sometimes of interest to translate these results in terms of the "Landau–Ginsburg wavefunction" ψ. For a dirty superconductor, the relation between ψ and Δ is[11]

$$\psi = [(\tfrac{1}{8}\pi)(n\tau/T)]^{\frac{1}{2}}\Delta\,, \qquad (3.22)$$

where n is the number of electrons per cm^3, and $\tau = l/v_F$ is the transport relaxation time. Comparing (3.22) and (3.21) we see that ψ is *not* continuous at a metallic interface.

IV. NS SANDWICHES: TRANSITION TEMPERATURE

We now consider a two-layer system of the type represented on Fig. 1. Our aim is to solve the integral equation (3.1) for the pair potential and see at what temperature T it has a nontrivial solution: this will give us the transition point (provided that the transition is of second order). Of course, this is a rather formidable task, and we shall perform it only in some limiting cases.

A. The Cooper Limit

We assume that the thicknesses d_N, d_S of the slabs are much smaller than the respective coherence lengths $\xi_N(T)\xi_S(T)$ defined by Eq. (3.11). (Note that the ξ'_S are temperature dependent: in particular, if the transition point of the sandwich is low, the ξ'_S will be large and the requirement is mild.) Then the kernel $H_\omega(XX')$ is essentially constant when X or X' is varied in one of the slabs. There are three values to be derived:

$$H_\omega(NN)(\text{for both } X,X' \text{ in } N),\quad H_\omega(SN) = H_\omega(NS)\,,$$
$$\text{and } H_\omega(SS)\,.$$

They are derived from the equations

$$d_n H_\omega(NN) + d_S H_\omega(NS) = N_N(\pi/|\omega|)\,,$$
$$d_n H_\omega(SN) + d_S H_\omega(SS) = N_S(\pi/|\omega|)\,, \qquad (4.1)$$
$$H_{NN}/N_N = H_{NS}/N_S\,, \qquad H_{SN}/N_n = H_{SS}/N_S\,. \qquad (4.2)$$

The group (4.1) is derived from the sum rule (3.3). The group (4.2) is derived from the boundary condition (3.20).

From (4.1) and (4.2) we get

$$\frac{H_\omega(\mathrm{NS})}{N_n N_S} = \frac{H_\omega(\mathrm{NN})}{N_n^2} = \frac{H_\omega(\mathrm{SS})}{N_S^2} \quad \frac{\pi}{|\omega|}\frac{1}{N_n d_n + N_S d_S}\,. \qquad (4.3)$$

We can then write down Eq. (3.1) for the pair potential Δ. Since $\Delta(X)$ will be constant in each slab in our limit, this becomes simply

$$\Delta_n = \sum_\omega V_n T\frac{\pi}{|\omega|}\frac{1}{N_n d_n + N_S d_S}(N_n^2 d_n\Delta_n + N_n N_S d_S \Delta_S)$$
$$\Delta_S = \sum_\omega V_S T\frac{\pi}{|\omega|}\frac{1}{N_n d_n + N_S d_S}(N_n N_S d_n\Delta_n + N_S^2 d_S\Delta_S)\,. \qquad (4.4)$$

We now require that (4.4) has a nontrivial solution. We shall write down the result only for the case where the frequency cutoff ω_0 of the interaction V is the same in N and S. (This is often close to the experimental situation, and greatly simplifies the algebra). When this cutoff is taken into account, we must effect the standard replacement

$$T\sum_\omega \frac{\pi}{|\omega|} \to \log\frac{1.14\,\omega_0}{T} = \frac{1}{\rho}\,, \qquad (4.5)$$

and the nontrivial solution is obtained for

$$\rho = \frac{V_n N_n^2 d_n + V_S N_S^2 d_S}{N_n d_n + N_S d_S}\,. \qquad (4.6)$$

ρ plays the role of a "effective NV" in the BCS formula for T [Eq. (4.5)]. A formula of this type

(4.6) has been first derived by Cooper[17] by a very simple argument. [There is a difference however in the weighting factors which, according to (4.6) are $N_n d_n$ and $N_S d_S$, while the Cooper argument leads to d_n and d_S]. The main consequences of (4.6) are the following:

(1) when both V_S and V_n are attractive (positive), there is always a finite transition temperature T,

(2) when V_n is repulsive and $|V_n|d_n > V_S d_S \times (N_S/N_n)^2$, the system does never become superconducting. There are some experimental data on Ag Pb[1] and Ag Sn[2] sandwiches which might be interpreted in terms of Eq. (4.6). The thicknesses are in the 300-Å range. In these experiments the transition temperature seems to vanish rather sharply when d_n exceeds some critical value. This might suggest that V_{Ag} is repulsive. However, we do not think that any definite conclusion can be reached in the present state of affairs, since the intrinsic mean free path l (and thus the coherence length ξ) was not controlled. It is thus not clear whether the limit $\xi > d$ applies, and if not, as we shall see later, the conclusions might be very different.

Finally, we should mention the possible effects of a thin oxide layer separating the two films.[17] When the transmission of the oxide layer becomes small, the boundary condition (3.17) cannot be simplified, and we must compute the effective length β/α. This "oxide effect" is difficult to control, and is one of the main limitations for the determination of V_N in the Cooper limit.

B. Thick Films: the One-Frequency Approximation

We now consider NS sandwiches where the thicknesses d_S d_N are somewhat larger than the corresponding coherence lengths ξ_S ξ_N. It turns out that in this situation we can obtain the shape of the pair potential $\Delta(X)$ in closed form if we make a slight simplification on the integral equation (3.1). The argument is as follows: the kernel $H_\omega(XX')$ has a range $\xi_\omega = (D/2|\omega|)^{\frac{1}{2}}$. For the lowest frequency $\omega = \pm\omega_0 = \pm\pi T$, the range $\xi_{\omega_0} = \xi$ is maximum. All the other frequencies' components have shorter ranges $\xi/\sqrt{3}$, $\xi/\sqrt{5}$, etc. The approximation amounts to retaining the lowest-frequency component H_{ω_0}, and replacing all the other ones by a δ function suitably normalized,

$$T\sum_\omega H_\omega(XX') \to 2TH_{\omega 0}(XX') + CN(X)\delta(X-X'). \tag{4.7}$$

From the sum rule (3.3) and the cutoff prescription $T\sum_\omega (\pi/|\omega|) \to \log(1.14\,\omega_0/T)$, we get

$$C = \log(1.14\,\omega_0/T) - 2\,.$$

Equation (3.1) for the pair potential becomes

$$\Delta(X)[1 - CN(X)V(X)] = 2V(X)T \times \int H_{\omega 0}(XX')\Delta(X')dX'\,. \tag{4.8}$$

The main interest of this form is the following: in each slab we know that $H_{\omega 0}(XX')$ obeys the "diffusion equation" (3.14) with respect to X or X'. Applying the operator $2|\omega| - D(d^2/dX^2)$ to both sides of (4.8) we obtain

$$(1 - CVN)2\pi T\Delta - D(d^2\Delta/dX^2) = 2NV\Delta\,. \tag{4.9}$$

Thus in this approximation, the pair potential in each slab is ruled by an elementary differential equation and the only problem left is to match the boundary conditions [to be derived from (4.8)]. How accurate is the approximation? We can get an estimate of the accuracy by returning to the case of an infinite dirty metal, where the solutions of (3.1) are plane waves $\Delta \sim e^{iqX}$. Making use of the Fourier transform (3.9), and performing the sum over ω, we get for this case the exact relation[13]

$$v \equiv (1/NV) - \log(1.14\,\omega_0/T) = \psi(\tfrac{1}{2}) - \psi(\tfrac{1}{2} + \tfrac{1}{2}y)\,,$$
$$y = Dq^2/2\pi T\,, \quad \psi(X) = \Gamma'(X)/\Gamma(X)\,. \tag{4.10}$$

This has to be compared to the approximate relation derived from (4.9),

$$v = [-2y/(1+y)]\,. \tag{4.11}$$

The two curves $v(y)$ are plotted on Fig. 2, both for $y > 0$ and for $y < 0$ (the latter corresponding to imaginary values of q, i.e., exponential decays which are of interest for our slab problems). Near $y = 0$, the exact slope dv/dy is equal to -2.44, while the approximate one is -2. (This means that the coefficients of the Landau–Ginsburg equation as deduced from the approximate form are correct within 20%). Near $y = -1$, the exact relation is

$$v = -1.39 + 2/(1+y)\,, \tag{4.12a}$$

while the approximate one is

$$v = -2 + 2/(1+y)\,. \tag{4.12b}$$

---

[17] L. Cooper, Phys. Rev. Letters **6**, 698 (1961). Earlier theoretical discussions include: A. D. Misener and J. O. Wilhelm, Trans. Roy. Soc. Can. Sec. III **29**, 5 (1935); R. H. Parmenter, Phys. Rev. **118**, 1173 (1960). A recent phenomenological theory has been proposed by D. H. Douglass, Phys. Rev. Letters **9**, 155 (1962).

The singular term of v is exactly reproduced (Note that large $|v|$ corresponds to $N\,|V|$ small, a situation often realized in the normal slab.) Thus we conclude that the one-frequency approximation is satisfactory, except in the range $y > 1$. In practice, the positive y's will be encountered in the S slab, and the condition $y < 1$ means that the thickness of the S slab must be larger than one coherence length.

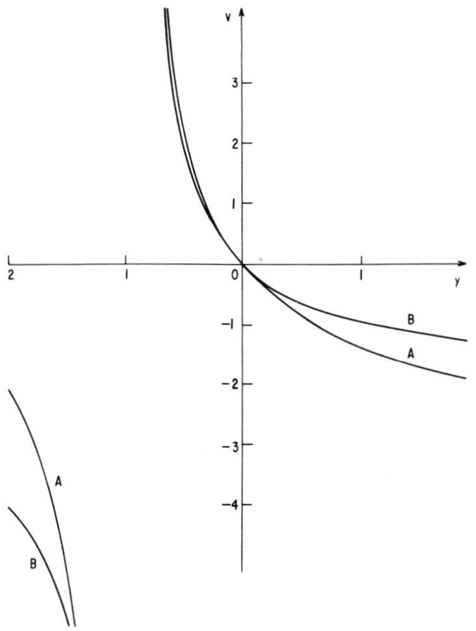

FIG. 2. Fourier transform of the kernel of the integral equation (3.1) for the pair potential in an infinite dirty material. Abscissa $y = Dq^2/2\pi T$ (q is the wave vector). Ordinate $v = (1/NV) - \log(1.14\,\omega_0/T)$. Curve A: exact (Eq. 4.10); Curve B: one-frequency approximation (Eq. 4.11). The region $y < -1$ corresponds to negative V (repulsive interactions).

We now discuss the boundary conditions to be applied to $\Delta(X)$ at a metallic interface when the approximate Eq. (4.8) is used. From the boundary conditions satisfied by $H_\omega(XX')$ [Eqs. (3.16) and (3.20)], we see that in the one-frequency approximation

$$\Delta(1 - CNV)/NV \quad \text{is continuous}\,,$$
$$(D/V)(1 - CNV)(d\Delta/dX) \quad \text{is continuous}\,. \tag{4.13}$$

These boundary conditions are slightly different from the exact ones (Eq. 3.21). In the exact solution, for instance, Δ/NV is continuous, but shows a rapid variation near the boundary because of the high-frequency terms in the kernel (3.1). This rapid variation is here approximated by an extra discontinuity. At a free surface we still have the condition $d\Delta/dX = 0$.

We are now able to write down the shape of $\Delta(X)$ for an (NS) sandwich (corresponding to Fig. 1) as the solution of Eq. (4.9) with the above boundary conditions,

$$\Delta(X)\,\frac{1 - CNV}{NV} = A\,\frac{\cos q(X - d_S)}{\cos qd_S}\,, \quad X > 0 \quad \text{(side S)}\,,$$

$$\Delta(X)\,\frac{1 - CNV}{NV} = A\,\frac{\cosh K(X + d_n)}{\cosh Kd_n}\,, \quad X < 0 \quad \text{(side N)}\,, \tag{4.14}$$

where A is an arbitrary constant, and

$$q^2 = \frac{2\pi T}{D_S}\left(-1 + \frac{2N_SV_S}{1 - CN_SV_S}\right) > 0\,, \tag{4.15a}$$

$$K^2 = \frac{2\pi T}{D_n}\left(+1 - \frac{2N_nV_n}{1 - CN_nV_n}\right) > 0\,. \tag{4.15b}$$

The solution (4.14) has 0 slope on the free surfaces. On the NS boundary there remains to satisfy the second equation (4.12). This yields

$$qtgqd_S = \eta KthKd_n\,, \tag{4.16}$$

where $\eta = D_nN_n/D_SN_S$.

Equation (4.16) is an implicit equation for the transition temperature T.[18] To discuss experimental data, it is sometimes preferable to invert the procedure: when N_S, V_S, N_n, D_n, D_S are known, and the transition temperature is measured, we know q from (4.15a). Then (4.16) gives K, from which we may derive by Eq. (4.15b) the unknown electron–electron interaction V_n in the normal metal. Finally we must check that we are in the range of validity of the one-frequency approximation ($d_S > \xi_S$); in practice this implies that T is not much lower than the transition point T_{S0} of the bulk S metal.

Two characteristic lengths are of interest in this boundary problem: (1) the depth of penetration of the pairs on the N side is K^{-1}. In the (usual) situation where V_N is small (or arbitrary sign), K^{-1} is close to $(D_n/2\pi T)^{\frac{1}{2}}$ (slightly larger if $V_N > 0$, slightly smaller if $V_N < 0$). The only case where K^{-1} may become anomalously large is when the metal N is also a superconductor ($V_N > 0$) and when T is just above T_{N0} (the transition temperature of the bulk N metal). (2) The "extrapolation length"

$$b = \frac{\Delta(0+)}{(d\Delta/dX)_{0+}} = \left(\frac{D_SN_S}{D_nN_n}\right) K^{-1}\coth Kd_n \tag{4.17}$$

defines the macroscopic boundary condition satisfied

[18] More accurately, the largest root T of (4.16) is the transition temperature.

by the pair potential on the S side. Consider for instance the case $d_n \to \infty$ $(b \to b_\infty)$. Then, when N_S and N_n are comparable, we see from (4.8) that $b_\infty \sim K^{-1}$. On the other hand, if N is a semimetal or a heavily doped semiconductor, N_S/N_n may be of order 100, and λ_∞ is much larger than K^{-1}. Finally, when $N_n \to 0$ (insulator on the N side), b_∞ becomes infinite and we come back to the requirement $(d\Delta/dX)_{0+} = 0$.

In the limit where $d_S \gg \xi_S$ we may simplify (4.16) and write

$$q(d_S + b) = \tfrac{1}{2}\pi .$$

In this limit, q is small, and from (4.15a) we see that T is close to T_{SO}. Writing down the explicit value of C we obtain, in the one-frequency approximation,

$$T = T_{SO} - (1/\pi)D_S q^2 = T_{SO} - \tfrac{1}{4}\pi[D_S/(d_S + b)^2] . \tag{4.18}$$

[Note that in the right-hand side of (4.18) we may use the value of b corresponding to $T = T_{SO}$.] When $d_n \to \infty$ the transition temperature (4.18) reaches a lower limit

$$T_{\text{lim}} = T_{SO} - \tfrac{1}{4}\pi[D_S/(d_S + b_\infty)^2] . \tag{4.19}$$

When d_n is finite, but still larger than K^{-1}, we may expand $b \cong b_\infty(1 + 2e^{-Kd_n})$ and we obtain

$$T = T_{\text{lim}} + \pi D_S[b_\infty/(d_S + b_\infty)^3]e^{-2Kd_n} . \tag{4.20}$$

This exponential law of approach to T_{lim} has been first proposed, on experimental grounds, by Hilsch.[4] In the experiments by Hilsch (on Cu Pb sandwiches) the electron mean free path l_{Cu} was measured and could be varied (from 40 to 800 Å) by changing the evaporation temperature. The transition temperature could be described by a law of the form $T = T_{\text{lim}} + Ce^{-2Kd_n}$ and the decay constant $2K$ was shown to be inversely proportional to $(l_{Cu})^{\frac{1}{2}}$. This in agreement with Eq. (4.15b) since $D = \frac{1}{3}v_F l$. For a Pb Cu sandwich with $l_{Cu} = 40$ Å, Hilsch measures $K^{-1} = 200$ Å. Taking $v_{F(Cu)} = 1.58\ 10^8$ cm/sec and $T = 7°$K, we have $(2\pi T/D_n)^{\frac{1}{2}} = 190$ Å, smaller than K^{-1}. Thus we expect the interaction v_{Cu} to be attractive, and in fact we compute from (4.15b) $(NV)_{Cu} \cong 0.05$. Of course this conclusion is very provisory: if a 10% uncertainty is allowed on K, $(NV)_{Cu}$ may range between -0.06 and 0.10. It is clear, however, that from such experiments we may derive important informations on the electron–electron interaction in "normal" metals.[19] There are two favorable points: (1) the significance of K is independent of the physical state of the NS boundary; for instance, if there was a thin oxide layer between N and S, the amplitude of the temperature shifts $T - T_{\text{lim}}$ would be reduced, but the exponential factor in (4.20) would still hold with the same value of K.[20] (2) When the BCS parameter $(NV)_n$ is small, as will often be the case, the one-frequency approximation is very accurate on the N side [since (4.12a) and (4.12b) are very similar in the limit $y \to -1$]. In fact, in the discussion of reference 13, the complete Eq. (4.10) was used to relate K and $(NV)_{Cu}$ and the results were essentially identical to those derived from (4.12b) or (4.15b).

C. Thin Superconducting Layer on a Massive Normal Substrate

The only case which is not covered by the above discussions (A and B) corresponds to $d_S \ll \xi_S(T)$, $d_n > \xi_n(T)$. We shall discuss the limiting case $d_n \to \infty$, which is comparatively simple, since when $d_S \ll \xi_S$, $\Delta(X)$ and $H_\omega(XX')$ are nearly constant in the S region $(0 < X < d_S)$. It is then easy to determine $H_\omega(XX')$ from (3.14), (3.16), (3.20), and the equation for Δ finally reads

$$\Delta_S\left(1 - N_S V_S T \sum_\omega \frac{\pi}{|\omega|}\frac{N_S d_S}{D(\omega)}\right) = N_S V_S T \sum_\omega \frac{\pi}{|\omega|}\frac{N_n}{D(\omega)} \times \int_{-\infty}^{0} dX \Delta(X) e^{X/\xi_\omega} , \tag{4.21a}$$

$$\Delta(X) = N_n V_n T \sum_\omega \frac{\pi}{|\omega|}\left\{\Delta_S \frac{N_S d_S}{D(\omega)} e^{X/\xi_\omega} + \int_{-\infty}^{0} dX' \Delta(X') \frac{1}{2\xi_\omega}\left[e^{-|X-X'|/\xi_\omega} + \left(1 - \frac{N_S d_S}{2D(\omega)}\right) e^{(X+X')/\xi_\omega}\right]\right\} \quad (X < 0) , \tag{4.21b}$$

$$\Delta_S = \Delta(0 < X < d_S) ,$$

$$D(\omega) = N_n \xi_\omega + N_S d_S ,$$

$$\xi_\omega = (D_n/2|\omega|)^{\frac{1}{2}} .$$

Consider first the case $V_N = 0$. Then $\Delta(X) = 0$ for $X < 0$ and we get the equation for the transition

[19] A preliminary discussion of these effects was given in reference 10. It was not realized, however, in this reference, that Eq. (4.10) applied even for repulsive interaction $V_n < 0$.

[20] In fact, for a typical Hilsch experiment ($d_{Pb} = 400$ Å), the theoretical value of $T_{SO} - T_{\text{lim}}$ as derived from (4.16) assuming no oxide layer is $\sim 2°$K, the experimental value is closer to 1°K.

point directly from (4.21a). After rearrangement, this becomes

$$\log \frac{T_{SO}}{T} = \sum_{n\geq 0} \frac{1}{n+\frac{1}{2}} \times \left[1 - \frac{1}{1+Z(2n+1)^{-\frac{1}{2}}}\right] = m(Z)\,, \qquad (4.22)$$

$$Z = \frac{N_n \xi_n}{N_S d_S} = \frac{N_n}{N_S d_S}\left(\frac{D_n}{2\pi T}\right)^{\frac{1}{2}}. \qquad (4.23)$$

we shall be mainly interested in the limit of low T, large Z, for which

$$m(Z) \rightarrow 2\log Z + \log 2\gamma + \beta/Z + \cdots\,, \qquad (4.24)$$

where γ is Euler's constant ($\gamma = 1.78$) and $\beta = (\sqrt{2}-2)\zeta(\frac{1}{2}) = 0.86$. Then (4.22) may be cast into the explicit form

$$d_S = \frac{d_1}{1-\beta[\frac{1}{2}\,\gamma(T/T_{SO})]^{\frac{1}{2}}} \quad \text{for } d_n = \infty\,, \quad d_S \ll \left(\frac{D_n}{2\pi T}\right)^{\frac{1}{2}}, \qquad (4.25)$$

$$d_1 = \frac{N_n}{N_S}\left(\frac{\gamma}{\pi}\,\frac{D_n}{T_{SO}}\right)^{\frac{1}{2}}. \qquad (4.26)$$

The diagram (d_S,T) is represented on Fig. 3. Note

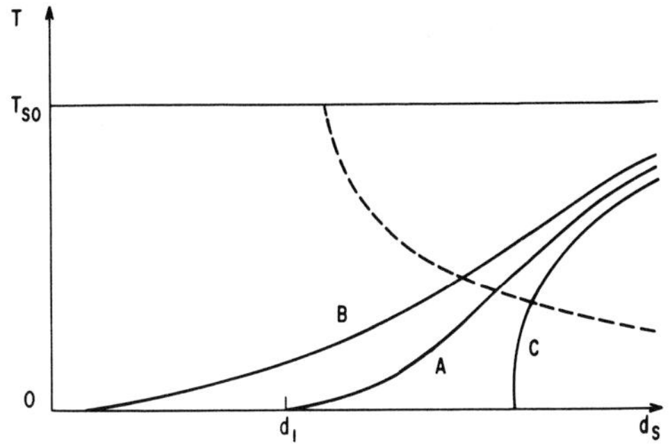

FIG. 3. Transition temperature of a thin film of superconducting material (thickness d_S) deposited on a massive normal substrate ($d_n = \infty$). Curve A corresponds to $V_N = 0$ (no electron–electron interaction in N). Curve B cooresponds to a weak attractive interaction, and curve C to a weak repulsive interaction. The region of validity of the calculation of Sec. IV.C corresponds to $d_S < \xi_S(T)$, and is bounded by the dotted curve.

that there is no second-order transition point when $d_S < d_1$; thus the absence of all superconductivity points in a finite range of d_S does not necessarily show that the interaction V_n is repulsive. The physical origin of this surprising result may be understood qualitatively by an extension of the Cooper argument: the interaction V_S is reduced by a factor $\sim d_S/(d_S + \xi_n)$, where $\xi_n = (D_n/2\pi T)^{\frac{1}{2}}$ is the range of penetration of the pairs on the N side. The important point is that ξ_n becomes large when T is low: this tends to accelerate the decrease of T when d_S decreases, and leads to a critical thickness d_1. A similar argument can be applied to the nonlinear self-consistency equation for Δ_S at $T = 0$.

These considerations can be extended to the case of a small, nonvanishing V_N (or arbitrary sign) by applying the one-frequency approximation to (4.21b). The results are represented qualitatively on Fig. 3.

V. JUNCTIONS

Consider a thin normal layer N (of thickness $2a$) embedded in a superconducting metal S. We ask what is the maximum supercurrent density J_m which may cross this SNS junction. When $2a \gg \xi_N$, we expect this current to be very small, while when $2a < \xi_N$, J_m is certainly very large; the interest of these junctions is precisely that the critical current may be adjusted to (nearly) any desired value by a proper choice of the thickness $2a$.

We shall compute this critical current for temperatures T which are only slightly lower than T_{SO}. Then, in the S regions, the pair potential is ruled by the local Landau–Ginsburg equation[21]

$$\frac{T_{SO}-T}{T_{SO}}\Delta + L_1\frac{d^2\Delta}{dX^2} - R_1|\Delta|^2\Delta = 0 \qquad (5.1)$$

$$(|X| > a)\,,$$

where, for a dirty superconductor,[11]

$$L_1 = \tfrac{1}{8}\,\pi(D/T_{SO})\,,$$
$$R_1 = \tfrac{7}{8}\,[\zeta(3)/(\pi T_{SO})^2]\,. \qquad (5.2)$$

The superfluid current is given by

$$J = -\frac{i}{2}C\left(\Delta^+\frac{d\Delta}{dX} - \Delta\frac{d\Delta^+}{dX}\right), \qquad (5.3)$$

where $C = \frac{1}{2}\,\pi e(N_S\,D_S/T_{SO})$. (5.4)

The variations of Δ predicted by (5.1) take place within a characteristic distance $p \sim \xi_S(T_{SO}/T_{SO} - T)^{\frac{1}{2}} \gg \xi_S$. Since p is large, the effect of the N slab may be included by imposing appropriate boundary conditions to Eq. (5.1) on the planes $X = \pm a$. We shall now derive these conditions.

The major point is the following: in the region of the junction we may derive the shape of Δ from the linearized self-consistency equation (3.1), and even set $T = T_{SO}$; all the corrections neglected by this

[21] L. D. Landau and V. L. Ginsburg, Zh. Eksperim. i Teor. Fiz. 20, 1064 (1950).

procedure are of order $\xi_S/p = (1 - T/T_{SO})^{\frac12} \ll 1$. Thus we return to (3.1), use the one-frequency approximation, and write down immediately the solutions for $|X| < a$:

$$\Delta(X) = [N_nV_n/(1 - CN_nV_n)](\cosh KX/\cosh Ka) \quad \text{(even)},$$

$$\Delta(X) = [N_nV_n/(1 - CN_nV_n)][\sinh KX/\sinh ka) \quad \text{(odd)}, \quad (5.6)$$

where K is given by Eq. (4.15b) with $T = T_{SO}$. The solutions for $T = T_S$ for $|X| > a$ are linear in X,

$$\Delta(X) = [N_SV_S/(1 - CN_SV_S)](\alpha(X/a) + 1 - \alpha), \quad X > a. \quad (5.7)$$

Equations (5.6) and (5.7) satisfy the first boundary condition (4.13a). The second one, (4.13b), yields finally

$$Ka \tanh Ka = \eta\alpha \quad \text{(even)},$$
$$Ka/\tanh Ka = \eta\alpha \quad \text{(odd)}. \quad (5.8)$$

These results may be rewritten in a form applicable to a solution of no given parity:

$$(d\Delta/dX)_a - (d\Delta/dX)_{-a} = \eta K \tanh Ka(\Delta_a + \Delta_{-a}),$$
$$(d\Delta/dX)_a + (d\Delta/dX)_{-a} = \eta K \coth Ka(\Delta_a - \Delta_{-a}). \quad (5.9)$$

The conditions (5.9) are to be imposed on the solutions of the Landau–Ginsburg equation (5.1). They can also be used to transform the expression of the current crossing the junction. Solving for $(d\Delta/dX)_a$ in (5.9) we get

$$J_{(X=a)} = -\frac{i}{2}C\left[\left(\frac{d\Delta}{dX}\right)_a \Delta_a^+ - \text{c.c.}\right] = \frac{-i\eta CK}{2\sinh 2Ka}(\Delta_a^+\Delta_{-a} - \Delta_{-a}\Delta_a^+). \quad (5.10)$$

Equation (5.10) is somewhat similar to the Josephson formula for the current through an insulating junction.[22] In most actual situations we will have $|\Delta_a| = |\Delta_{-a}|$, the two potentials differing only by a phase $\Delta_{-a} = \Delta_a e^{i\varphi}$. Then $J = J_m \sin\varphi$ with the maximum value J_m given by

$$J_m = \eta CK/\sinh 2Ka\,|\Delta_a|^2. \quad (5.11)$$

At this point, however, there is an important difference from the Josephson case. For an insulating junction, $|\Delta_a|$ is very close to $|\Delta_\infty|$, the value of the pair potential in the bulk S material. This is shown explicitly elsewhere.[23] For a metallic junction, on the other hand, $|\Delta_a|$ is much smaller than $|\Delta_\infty|$. As an example, we shall now compute $|\Delta_a|$ explicitly in the limit of a thick junction $Ka \gg 1$. In this case, the currents are small and we can neglect the phase of Δ in Eq. (5.1). Then multiplying (5.1) by $(d\Delta/dX)$ and integrating we obtain

$$L_1(d\Delta^2/dX) + \Delta^2(\epsilon - \tfrac12 R_1\Delta^2) = \tfrac12\Delta_\infty^2\epsilon, \quad (5.12)$$
$$\epsilon = T_{SO} - T/T_{SO}, \quad \Delta_\infty^2 = \epsilon/R_1.$$

For a thick junction we may write the boundary condition (5.9) in the form (4.17),

$$\frac{1}{\Delta_a}\left(\frac{d\Delta}{dX}\right)_a = \frac{1}{b_\infty} = \frac{D_nN_n}{D_SN_S}K. \quad (5.13)$$

Inserting (5.13) in (5.12) we get a second-order equation for $|\Delta_a|^2$,

$$\tfrac12 R_1\Delta_a^4 - (L_1/b_\infty^2 + \epsilon)\Delta_a^2 + \tfrac12\Delta_\infty^2\epsilon = 0. \quad (5.14)$$

When the N material is a metal b_∞, and $L_1^{\frac12}$ are of the same order of magnitude, the term of order Δ_a^4 is negligible and the correct root corresponds to

$$\left|\frac{\Delta_a}{\Delta_\infty}\right|^2 = \frac{\epsilon}{2}\frac{b_\infty^2}{L_i}, \quad (5.15)$$

$$\frac{J_m}{e} \cong 4\eta^{-1}e^{-2Ka}\frac{N}{KR_1}\left(\frac{T_{SO} - T}{T_{SO}}\right)^2 \quad (Ka \gg 1). \quad (5.16)$$

We see that J_m is proportional to $(T_{SO} - T)^2$, while for an insulating junction, J_m is proportional only to $T_{SO} - T$.[23] As already pointed out, the great interest of the present metallic case is that Ka is a very flexible parameter, from the experimental point of view. Unfortunately, at the present time, the only experiments in this field are the early ones by Meissner,[6] where the mean free paths were not controlled. Qualitatively, for various nonmagnetic metals, Meissner found values of $K^{-1} \sim 10^3$ Å, a very reasonable figure as can be seen from Eq. (4.15b). It is very much to be hoped that more accurate data will soon be taken.

Once J_m is known, it is of course possible to derive an effective penetration depth δ for the junction currents, using the Ferrell Prange procedure.[24] δ is proportional to e^{Ka}. Consider for example a Pb Cu Pb junction with $l_{Cu} = 40$ Å and $K \sim 200$ Å^{-1} at 7°K. If the copper thickness is $2a \sim 4000$ Å, the penetration depth δ will be of order 1 mm, the critical current will be in the milliampere range. To force a normal current of the same magnitude through the junction, we

[22] It is, in fact, possible to derive the Josephson formula in a completely self-consistent way by similar methods.[23]

[23] P. G. De Gennes, Phys. Letters **5**, 22 (1963).

[24] R. A. Ferrell and R. E. Prange, Phys. Rev. Letters **10**, 479 (1963).

would need an applied voltage of order 1 mV. Thus in such a case we expect interesting signals even for rather large thicknesses, for which the crystallographic state of the N slab can be accurately controlled.

We might finally mention the magnetic junctions such as Pb Fe Pb. They do not seem very promising for the following reason: if Γ is the energy difference between a state ($k\uparrow$) with spin up and the time reversed state ($-k\downarrow$) in the magnetic metal, the penetration range of the pairs in (N) is roughly $\hbar v_F/\Gamma$, and thus very small (of order 100 Å or less[25] in agreement with the experimental results of Meissner)[6]. Thus for the thicknesses of interest $2a \sim \hbar v_F/\Gamma$, it would be very difficult to prepare an N slab free from holes, and of well defined properties.

VI. CONCLUDING REMARKS

We have seen that the transition temperature T of "dirty" NS sandwiches can be related to the electron–electron interaction V_N in the normal metal. The best procedure is to deduce K from a plot of T vs normal slab thickness and then to relate K to V_N, since this method does not imply any specific assumption about the "transition" layer between the two metals (provided that this layer is smaller than K^{-1}). The experiments could be carried out with metals or semimetals on the N side (but for semimetals, the shift of T is comparatively smaller).

The strong dependence of T on the mean free paths l forbids the discussion of experiments where l was not measured. On the other hand, it probably explains the spectacular "aging effects" observed on the transition temperature of Sn Au sandwiches by Rose-Innes and Serin,[3] since a slow atomic diffusion at room temperature will react markedly on l in the vicinity of the interface, and on the transmission coefficient. It is remarkable to observe that, in a homogeneous alloy, the effect of impurity concentration on the transition temperature is weak, while it is strong in our inhomogeneous NS systems, where impurity scattering controls the leakage of superconducting pairs towards an unfavorable region.

Finally, we notice that V_N could also be derived from measurements of critical current vs thickness in the SNS junctions.

[25] P. G. De Gennes and G. Sarma, J. Appl. Phys. **34**, 1380 (1963).

ACKNOWLEDGMENTS

We would like to thank J. Friedel, E. Guyon, D. Saint-James, and G. Sarma for various discussions on superconducting contacts.

Afterthoughts: Boundary effects in superconductors

1) An interesting different approach on proximity effects was built up soon after by the late W. McMillan [*Phys. Rev.* **175**, 537,559 (1968)] using one electron Green's functions as the basic object. The main advantage is that this can easily incorporate the effect of surface roughness on the spectra.

2) More recently, some very careful studies have been carried out at very low T, giving long range proximity effects in normal metals N: see A.C. Mota in *Josephson Effect: Achievements and Trends*, ed. A. Barone (World Scientific, 1986), p. 248.

3) The effects of magnetic fields on the weak superconductivity induced in N are extremely spectacular: see the review by the Orsay group on superconductivity in *Quantum Fluids,* ed. D. Brewer (North-Holland, 1966).

Part II. Liquid Crystals

THE JOURNAL OF CHEMICAL PHYSICS VOLUME 48, NUMBER 5 1 MARCH 1968

Soluble Model for Fibrous Structures with Steric Constraints

P.-G. DE GENNES

Laboratoire de Physique des Solides, Faculte des Sciences, 91-Orsay, France*

(Received 8 September 1967)

We consider a set of thin, long, flexible chains in two dimensions. The chains are in thermal equilibrium under a strong unidirectional stretching force. Configurations where different chains intersect each other are excluded. The chain configurations $y_n(x)$ are interpreted as the trajectories of an assembly of quantum mechanical particles, x playing the role of the time. The model is thus brought into correspondence with a one-dimensional gas of free fermions, and can be solved exactly. The system of chains has no long-range order, but a remarkably strong short-range order: the x-ray form factor, for a wave vector $\mathbf{q}$ normal to the stretching direction, shows a sharp peak at $\mathbf{q}=2k_F$, where k_F is the Fermi wave vector of the associated set of particles.

I. INTRODUCTION

Rather sharp x-ray-diffraction lines are observed in a certain number of dilute solutions made of rodlike, or lamellar, building units. For instance in certain lipid-water systems we can find lamellar structures where the water sheets are remarkably thick[1] (100 Å or more). It is of great interest to ascertain what are the forces between lipidic units which are responsible for the stability of such open systems.

In the present paper we have a much more modest objective: we consider a two-dimensional system of thin flexible chains, with no forces between them, except that the chains are not allowed to intersect each other. A preferred orientation is given to the chains by stretching, as shown in Fig. 1. This situation differs significantly from the physical conditions occurring in a lamellar or rodlike phase of the lipid-water type, where (a) the rigidity of the individual units may be important, and (b) a preferred orientation in the liquid phase would result from flow properties and not from tensions. On the other hand, this model leads to rather simple results, and some of its qualitative conclusions may thus be of some use both in fiber physics and in the physics of lipids.

The analysis, to be described in Sec. II, relies heavily on the analogy between a configuration of a flexible molecule and the path of a quantum mechanical particle. This analogy was first exploited by Edwards.[2] Here, however, our definition of the "time" variable will differ from Ref. (2). A configuration $y_n(x)$ of the nth chain will be interpreted as a trajectory of a particle (the nth particle) moving along the y axis, as a function of the time x.

For our highly stretched system, the chains do not fold back, i.e., $y_n(x)$ is a single-valued function of x. In our quantum mechanical picture this means that all particles travel forward in time (no antiparticles).

This technique reduces the statistics of Fig. 1 to a study of a one-dimensional many-body system. Now, if we want the chains not to intersect each other, we must impose that the wavefunctions vanish when two particles coincide. This is satisfied with antisymmetric wavefunctions; thus we can take into account the constraints by imposing Fermi statistics to our particles.

The main drawback of this approach is that it is expressed naturally in the slightly esoteric language of path integrals. A few helpful references in this connection are listed under Ref. (3).

II. THEORY

Consider first *one chain* $y(x)$ submitted to a tension A. The chain energy is of the form

$$A\int_0^{L_x} dx\left[1+\left(\frac{dy}{dx}\right)^2\right]^{1/2}=\text{const}+\tfrac{1}{2}A\int_0^{L_x} dx\left(\frac{dy}{dx}\right)^2. \quad (1)$$

(We shall discuss later the approximations involved in the expansion of the square root.)

Let us call $G(x'y' \mid x''y'')$ the total weight of all configurations of our chain linking $(x'y')$ to $(x''y'')$; $(x''>x')$.

G can be written as a sum $\int\mathfrak{D}y$ [over all paths $y(x)$ linking the two points] of Boltzmann factors

$$G(x'y' \mid x''y'')=\int\mathfrak{D}y\exp\left[-\frac{A}{2T}\int_{x'}^{x''}dx\left(\frac{dy}{dx}\right)^2\right]. \quad (2)$$

Writing Eq. (2) for times x'' and $x''+\epsilon$ ($\epsilon\to0$) we can derive a differential form of Eq. (2)[3]:

$$-\frac{\partial G}{\partial x''}(x'y' \mid x''y'')=-\frac{T}{2A}\frac{\partial^2 G}{\partial y''^2} \qquad (x''>x'). \quad (3)$$

This has the form of a Schrödinger equation for a nonrelativistic particle (of mass A/T) moving along y, and x'' being an imaginary time [since there is no i on the left-hand side of Eq. (3)]. G is the quantum

* Laboratoire associe au Centre National de la Recherche Scientifique.

[1] A recent review is by V. Luzzati in *Biological Membranes*, D. Chapman, Ed. (Academic Press Inc., New York, to be published).

[2] S. F. Edwards, Proc. Phys. Soc. (London) **85**, 613 (1965); **88**, 265 (1966).

[3] R. P. Feynman, Rev. Mod. Phys. **20**, 347 (1948). I. M. Gelifand and A. M. Yaglom, J. Math. Phys. **1**, 48 (1960). R. P. Feynman and A. R. Hibbs, *Quantum Mechanics and Path Integrals* (McGraw-Hill Book Co., New York, 1965).

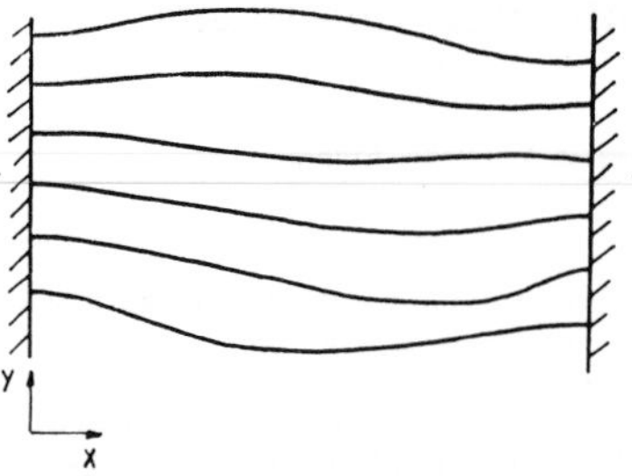

FIG. 1. Model for a two-dimensional fiber structure. The component chains are assumed to be attached to two plates I and F and placed under tension. The chains are bent by thermal fluctuations. Different chains cannot intersect each other.

mechanical amplitude for a process where the particle leaves y' at time x' and reaches y'' at time x''.

Let us now turn to a system of chains $y_1(x)$, $y_2(x)$, $\cdots$, $y_N(x)$. We can again define an amplitude (or "propagator") $G(x', y_1'\cdots y_N', x'', y_1''\cdots y_N'')$ for a process where the N particles (each with mass A/T) start from given positions $y_1'\cdots y_N'$ at time x', and reach the positions $y_1''\cdots y_N''$ at time t. [To shorten the notation we will often write $Y=(y_1\cdots y_N)$ and G as $G(x', Y'|x''Y'')$.] The propagator G can be expanded in terms of the eigenfunctions $\psi_i(y_1\cdots y_N)$ of the N-body problem

$$G(x'Y'|x''Y'')=\sum_i \exp[-E_i(x''-x')]\psi_i^*(Y')\psi_i(Y'') \quad (x''>x'). \quad (4)$$

If there were no further restriction on the chain configurations, each eigenfunction ψ_i would be a simple product of one-particle wavefunctions. But we want G to vanish whenever two arguments y_n' and y_m' (or y_n'' and y_m'') coincide. This will be achieved by antisymmetrization. More precisely, we can define a one-to-one correspondence between the wavefunctions ψ_i of our problem and the wavefunctions ϕ_i for N free fermions moving along the y axis,

$$\psi_i(y_1\cdots y_N)=(N!)^{1/2}\phi_i(y_1\cdots y_N), \quad \text{for } Y\in R_1$$
$$=0, \quad \text{elsewhere,}$$

where the region R_1 is defined by the inequalities

$$0\leq y_1\leq y_2\leq\cdots\leq y_N\leq L_y$$

and is the physical region allowed in our chain problem. The $(N!)^{1/2}$ factor is required for normalization. The energy levels E_i of ψ_i and ϕ_i are the same. Also, if $f(y_1\cdots y_N)$ is an arbitrary, symmetric, function of $(y_1\cdots y_N)$ the matrix element of f between two functions ψ_i, ψ_j is identical with the matrix element computed for the corresponding states in the fermion system

$$\int dy_1\cdots dy_N\psi_i^* f\psi_j=\int dy_1\cdots dy_N\phi_i^* f\phi_j =(i|f|j). \quad (5)$$

As a first, simple, application, let us compute the average of one Fourier component ρ_q of the density, for a wave vector q normal to the stretching direction,

$$\rho_q(Y)=\sum_n \exp(iqy_n).$$

The thermal average of ρ_q at a given x is

$$\langle\rho_q[Y(x)]\rangle=\int dY\,\frac{G(0Y_i|xY)\rho_q(Y)G(xY|L_xY_f)}{G(0Y_i|L_xY_f)},$$

where $Y_i\equiv(y_{1i}, y_{2i}\cdots y_{Ni})$ and $Y_f\equiv(y_{1f}\cdots y_{Nf})$ represent the points of binding of the chains on both end plates. We now insert the eigenfunction expansion (4) in this equation and make use of the following limiting properties (valid when L_x is large and x is far from both ends):

$$G(0Y_i|L_xY_f)\to\exp(-E_gL_x)\psi_g^*(Y_i)\psi_g(Y_f),$$
$$G(0Y_i|xY)\to\exp(-E_gx)\psi_g^*(Y_i)\psi_g(Y),$$
$$G(xY|L_xY_f)\to\exp[-E_g(L_x-x)]\psi_g^*(Y)\psi_g(Y_f),$$

where ψ_g is the ground state, of energy E_g. The thermal average then takes a form independent of the detailed structure of the boundary conditions described by Y_i and Y_f,

$$\langle\rho_q[Y(x)]\rangle=(g|\rho_q|g).$$

The free fermion system does not display any crystalline order in y space; thus the only nonvanishing component of $\langle\rho_q\rangle$ corresponds to $q=0$: there are no Bragg peaks in the x-ray pattern of our fiber. However, as we shall now see, there is a sharp accident in this pattern!

For our highly stretched chains, we can approximate the x-ray form factor as follows:

$$S(q)=(L_xL_y)^{-1}\int_0^{L_x}dx'\int_0^{L_x}dx''\langle\rho_q[Y(x')]\rho_{-q}[Y(x'')]\rangle. \quad (6)$$

Again we have considered only the case where the

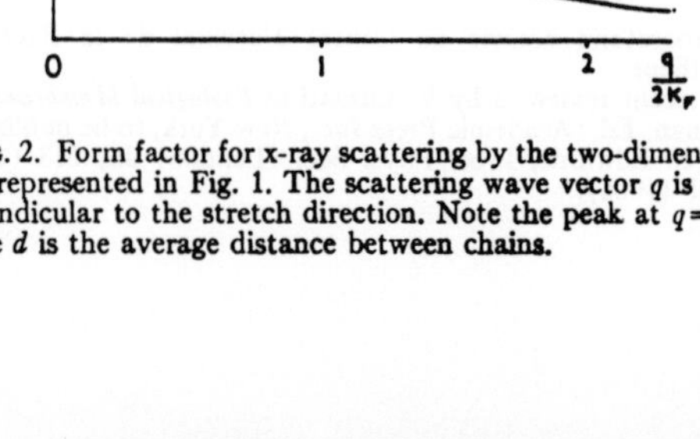

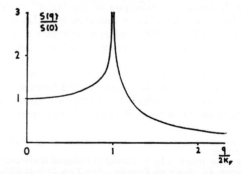

FIG. 2. Form factor for x-ray scattering by the two-dimensional fiber represented in Fig. 1. The scattering wave vector q is taken perpendicular to the stretch direction. Note the peak at $q=2\pi/d$ where d is the average distance between chains.

scattering wave vector **q** is normal to the direction of stretch. The factors L_x and L_y, corresponding to the dimensions of the fiber, have been extracted in (6) to make $S(q)$ dimensionless and independent of sample size. The bracket can be written explicitly as:

$$\langle\rho_q[Y(x')]\rho_{-q}[Y(x'')]\rangle$$
$$=\left[\int dY'dY''G(0Y_i\,|\,x'Y')\rho_q(Y')G(x'Y'\,|\,x''Y'')\rho_{-q}(Y'')G(x''Y''\,|\,L_xY_f)\Big/G(0Y_i\,|\,L_xY_f)\right].\quad(7)$$

Again we insert in each G factor the eigenfunction expansion (4) and make use of the limiting forms valid when x' is far from 0, x'' far from L_x (x' and x', on the other hand, can be close to each other). This leads to

$$\langle\rho_q[Y(x')]\rho_{-q}[Y(x'')]\rangle=\sum_i\exp[-(E_i-E_g)\,|\,x''-x'\,|\,]\,|\,(g\,|\,\rho_{-q}\,|\,i)\,|^2.\quad(8)$$

Integrating Eq. (8) on x' and x'' we arrive at

$$S(q)=(2/L_y)\sum_i(E_i-E_g)^{-1}\,|\,(g\,|\,\rho_{-q}\,|\,i)\,|^2.\quad(9)$$

In the state ϕ_g the fermions occupy plane wave states $\exp(iky)$ (of energy $k^2T/2A$) where k ranges from $-k_F$ to k_F. k_F is the Fermi wave vector, related to the average number of chains per unit length along y, $1/d$, by the relation $k_F=\pi/d$. The excited states ϕ_i coupled to ϕ_g in Eq. (8) are obtained from ϕ_g by transferring one fermion from an occupied state (k) to an empty state ($k+q$). Thus

$$S(q)=(2/L_y)\sum_k(2A/T)[(q+k)^2-k^2]^{-1}$$
$$(|\,k\,|<k_F,\ |\,k+q\,|>k_F).\quad(10)$$

Replacing $\sum_k$ by $L_y/2\pi\int dk$ and performing the integration, we obtain

$$S(q)=(A/\pi Tq)\log|\,(q+2k_F)/(q-2k_F)\,|.\quad(11)$$

The q dependence of this form factor is shown in Fig. 2. As announced, there is a peak at $q=2k_F=2\pi/d$. This peak corresponds to the well-known 'Kohn anomaly"[4] of the associate fermion gas.

It may be appropriate at this point to discuss briefly the validity of the high tension approximation [Eq. (1)]. The expansion of $[(dy/dx)^2+1]^{1/2}$ assumes $|\,dy/dx\,|\ll1$. The "velocities" must be small, or the momenta $A/T\ dy/dx$ must be much smaller than A/T. The order of magnitude of the momenta is k_F. Thus we require $k_F\ll A/T$ or $Ad\gg T$.

[4] W. Kohn, Phys. Rev. Letters **2**, 393 (1959).

III. CONCLUDING REMARKS

A purely steric hindrance between our chains is enough to create very strong correlations between them. In fact, an x-ray experimentalist, recording a curve of intensity vs scattering vector, and finding a plot similar to Fig. 2, might be tempted to identify the peak as a Bragg line, and to conclude that some long-range order exists in the system—while in reality there is none.

These conclusions might retain some validity for more complex and more realistic systems, e.g., nonintersecting chains in three dimensions, or nonintersecting sheets in three dimensions. But, unfortunately, these cases cannot be solved by a simple generalization of the methods of Sec. II. On the other hand, it is possible to superimpose on the present (two-dimensional) model, a set of weak, long-range forces reminiscent of the couplings which may occur in the lipid–water system. However (as already pointed out in the introduction) a crucial feature of our model is the existence of an external tension A, giving a preferred orientation to the chains; this has no direct counterpart in the lipid–water phases.

ACKNOWLEDGMENTS

The author is greatly indebted to Dr. V. Luzzati for a general introduction to the field of lamellar and rodlike structures, and to Dr. A. Parsegian for a correspondence on related matters.

JOURNAL DE PHYSIQUE *Colloque C 4, supplément au n° 11-12, Tome 30, Nov.-Déc. 1969, page C 4 - 65*

CONJECTURES SUR L'ÉTAT SMECTIQUE

P. G. de GENNES

Faculté des Sciences, Orsay, et C. E. N., Saclay

Résumé. — On présente une description phénoménologique des déformations dans un smectique impliquant *deux* paramètres scalaires : le déplacement u des couches perpendiculairement à leur plan et la dilatation en volume θ. L'intensité de la lumière diffusée par les fluctuations de u et de θ est faible sauf lorsque le vecteur de diffusion est presque parallèle au plan des couches. On analyse aussi le spectre des vibrations de grande longueur d'onde : pour un vecteur d'onde $\mathbf{q}$ oblique par rapport au plan des couches, on attend *deux* branches acoustiques observables en effet Brillouin.

Abstract. — A continuum theory for smectic liquid crystals is constructed in terms of *two* scalar functions : the displacement u of the layers normal to their plane and the volume dilatation θ. The intensity of the light scattering due to thermal fluctuations of u and of θ is expected to be weak — the only exception corresponding to a scattering wave vector parallel to the plane of the layers. The frequency spectrum of the vibrations (for a wave-vector $\mathbf{q}$ neither parallel nor normal to the optical axis) shows *two* acoustic branches which should be observable by Brillouin scattering.

I. Introduction et description de Landau-Peierls. — Les smectiques sont des matériaux stratifiés, avec, dans la configuration au repos, des couches successives équidistantes (intervalle d) et perpendiculaires à une même direction OZ (Fig. 1). La structure détaillée

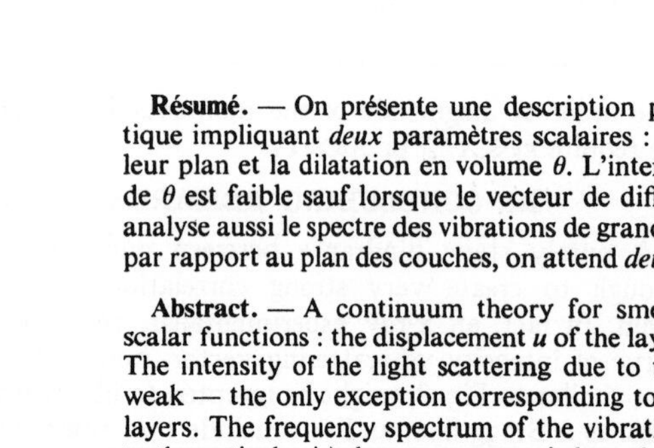

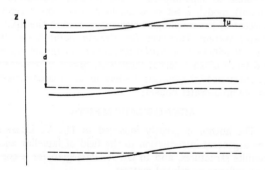

FIG. 1. — Représentation schématique de l'état smectique.. Lignes pointillées : plan des couches dans la configuration au repos. Lignes pleines : aspect de la déformation dans le mode d'ondulation ($q_z = 0$).

de ces couches est en général mal connue [1]. Nous nous contentons ici d'admettre que de tels systèmes existent, avec les symétries suivantes :

1) pas d'ordre à longue distance dans le plan de chaque couche (différence avec un cristal tridimensionnel),

2) aucun axe privilégié dans le plan de la couche (système optiquement uniaxe d'axe z),

3) les directions + Oz et − Oz sont équivalentes (smectique non ferroélectrique).

Partant d'une telle configuration au repos, nous voulons faire ici une analyse phénoménologique des déformations de grande longueur d'onde. Ceci a déjà été entrepris dans le passé, en partant de deux points de vue un peu différents :

a) Le point de vue d'Oseen [2], remanié par Frank [3], peut être présenté de la façon suivante : repérons en chaque point $\mathbf{r}$ l'axe optique local par un vecteur unitaire $\mathbf{n}(\mathbf{r})$ ou « *directeur* ». La forme la plus générale pour l'énergie libre de déformation est alors (par cm^3)

$$F = F_0 + 1/2\, K_1(\text{div}\,\mathbf{n})^2 + 1/2\, K_2(\mathbf{n}.\text{rot}\,\mathbf{n} + q_0)^2 + 1/2\, K_3(\mathbf{n}_\wedge \text{rot}\,\mathbf{n})^2. \quad (\text{I}.1)$$

Telle quelle, cette forme s'applique aux cholestériques ($q_0 \neq 0$) ou aux nématiques ($q_0 = 0$), qui sont aussi des matériaux localement uniaxes, mais sans structure en couches. Dans les smectiques, il y a une exigence supplémentaire : pour un système de couches incompressibles, l'intégrale

$$\frac{1}{d}\int_A^B \mathbf{n}.\mathrm{d}\mathbf{r} \qquad (\mathrm{I}.2)$$

représente le nombre de couches traversées le long d'un chemin reliant A à B. Ce nombre doit être indépendant du chemin choisi, donc

$$\mathrm{rot}\,\mathbf{n} \equiv 0\,. \qquad (\mathrm{I}.3)$$

On est alors amené à une énergie libre qui contient uniquement la divergence de $\mathbf{n}$

$$F = F_0 + 1/2\,K_1(\mathrm{div}\,\mathbf{n})^2 - 1/2\,\chi_a(\mathbf{n}.\mathbf{H})^2\,. \qquad (\mathrm{I}.4)$$

On a ajouté dans cette dernière expression l'effet d'un champ magnétique $\mathbf{H}.\chi_a = \chi_\parallel - \chi_\perp$ est l'anisotropie de la susceptibilité magnétique (on se limitera ici au cas $\chi_a > 0$).

b) La description de Peierls et Landau [4] adopte comme variable fondamentale le déplacement unidimensionnel $u(\mathbf{r})$ des couches perpendiculairement à leur plan. Pour de faibles déplacements, l'énergie libre a alors la forme

$$F = F_0 + \frac{\bar{B}}{2}\left(\frac{\partial u}{\partial z}\right)^2 + \frac{\chi_a H^2}{2}\left[\left(\frac{\partial u}{\partial x}\right)^2 + \left(\frac{\partial u}{\partial y}\right)^2\right] + \frac{1}{2}K_1\left(\frac{\partial^2 u}{\partial x^2} + \frac{\partial^2 u}{\partial y^2}\right)^2 + \frac{1}{2}K'\left(\frac{\partial^2 u}{\partial z^2}\right)^2 + + K''\frac{\partial^2 u}{\partial z^2}\left(\frac{\partial^2 u}{\partial x^2} + \frac{\partial^2 u}{\partial y^2}\right). \qquad (\mathrm{I}.5)$$

Le champ H est ici supposé parallèle à la direction de l'axe optique au repos Oz. Pour $H = 0$ il n'y a pas de terme en $\left(\frac{\partial u}{\partial x}\right)^2$ par exemple, car un tel terme impliquerait une valeur de $F - F_0$ non nulle pour une rotation d'ensemble du système par rapport à Oy. L'expression (I.5) ne contient pas le terme en

$$\frac{\partial u}{\partial z}\left(\frac{\partial^2 u}{\partial x'} + \frac{\partial^2 u}{\partial y'}\right)$$

de la réf. [4] : ce terme disparaît en effet lorsque les directions + Oz et − Oz sont équivalentes. En pratique les termes K' et K'' sont négligeables par rapport au terme B, et peuvent être omis. Par contre le terme K_1 doit être conservé, car il figure seul (en champ H nul) lorsque les déformations sont indépendantes de la coordonnée $z\left(\frac{\partial u}{\partial z} = 0\right)$.

Il est facile de vérifier que les définitions (I.4) et (I.5) du coefficient K_1 coïncident.

Au total, la description de Peierls-Landau recouvre celle de Oseen (pour les déformations faibles) mais tient compte en plus (par le terme $\bar{B}$) d'une modification éventuelle de la distance entre couches. On peut exploiter la formule (I.5) en passant aux composantes de Fourier :

$$u_q = \int \mathrm{d}\mathbf{r}\,u(\mathbf{r})\,\mathrm{e}^{i\mathbf{q}.\mathbf{r}}$$

$$F - F_0 = = \frac{1}{2}\sum_q |u_q|^2\{\bar{B}q_z^2 + K_1(q_\perp^2 + \xi^{-2})q_\perp^2\} \qquad (\mathrm{I}.6)$$

où l'on a posé $q_\perp^2 = q_x^2 + q_y^2$ et où l'on a introduit la longueur de cohérence magnétique ξ, définie par [5]

$$\xi = \sqrt{\frac{K_1}{\chi_a}}\,\frac{1}{H}\,. \qquad (\mathrm{I}.7)$$

En appliquant le théorème d'équipartition à (I.6) on obtient finalement les moyennes thermodynamiques

$$\langle |u_q|^2 \rangle = \frac{k_B T}{\bar{B}q_z^2 + K_1(q_\perp^2 + \xi^{-2})\,q_\perp^2} \qquad (\mathrm{I}.8)$$

$$\langle u^2(\mathbf{r}) \rangle = \frac{k_B T}{4\,\pi(\bar{B}K_1)^{1/2}}\log(\xi/d)\,. \qquad (\mathrm{I}.9)$$

Pour $H \to 0$, $\langle u^2 \rangle$ a une divergence logarithmique, qui est discutée dans la réf. [4]. A cause de cette divergence, la notion d'ordre smectique n'est pas rigoureusement self consistente en champ nul. Nous avons calculé ailleurs les corrélations qui subsistent alors. sur un exemple soluble à deux dimensions [6]. Mais, pour les systèmes qui nous intéressent ici, il importe de réaliser que la divergence logarithmique est très faible : considérons par exemple un grand monocristal smectique ayant atteint son équilibre dans le champ terrestre. Avec $E \sim 10^{11}$ ergs/cm^3 $K_1 \sim 10^{-6}$ dyne, $\chi_a \sim 10^{-7}$ on, attend $\sqrt{\langle u^2 \rangle} \sim 1$ Å. Donc en pratique $\langle u^2 \rangle$ n'est pas beaucoup plus grand que dans un solide (soit à cause d'un champ résiduel, soit en raison des dimensions finies du spécimen).

Si nous revenons maintenant sur le principe général de la description de Landau-Peierls, nous constatons que son hypothèse fondamentale est d'utiliser comme seule variable d'état le déplacement unidimensionnel des couches $u(\mathbf{r})$. Or, en réalité, pour spécifier à l'échelle macroscopique l'état d'un petit élément smectique, il faut non seulement connaître la configuration des couches, décrite par $u(\mathbf{r})$, mais aussi la densité ν d'atomes présente dans chaque couche. On peut par exemple définir une dilatation transversale ψ en posant

$$\nu = \nu_0(1 - \psi)$$

où ν_0 est la valeur au repos. La dilatation en volume θ est reliée à ψ et u par la formule

$$\theta = \psi + \frac{\partial u}{\partial z}. \qquad \text{(I.10)}$$

Dans ce qui suit, nous décrirons l'état local de déformation du smectique par les variables $u(\mathbf{r})$ et $\theta(\mathbf{r})$. Nous étudierons d'abord les propriétés statiques (énergie libre et intensité de la lumière diffusée par les fluctuations thermiques) puis les propriétés dynamiques (énergie et spectres de vibration).

Nous vérifierons que l'introduction de la variable θ était effectivement nécessaire, et ce pour deux raisons :

a) les fluctuations thermiques de u et θ donnent des contributions comparables en diffusion de la lumière,

b) les mouvements de u et de θ s'effectuent à des fréquences comparables et sont fortement couplés.

Ces propriétés (*a*) et (*b*) sont radicalement opposées à ce que l'on rencontre dans les *nématiques*, où (*a*) la diffusion de la lumière est dominée par les fluctuations d'orientation (les fluctuations de θ ne représentant qu'une faible correction [5]), et (*b*) les relaxations d'orientation sont lentes par rapport à celles de θ [7].

II. **Statique des fluctuations** — En l'absence d'effets piézoélectriques, avec les deux variables d'état $u(\mathbf{r})$ et $\theta(\mathbf{r})$ la forme de l'énergie libre de déformation (par cm^3) est :

$$F = \frac{1}{2}A\theta^2 + \frac{1}{2}B\gamma^2 + C\theta\gamma + \frac{1}{2}\chi_a H^2\left[\left(\frac{\partial u}{\partial x}\right)^2 + \left(\frac{\partial u}{\partial y}\right)^2\right] + \frac{1}{2}K_1\left(\frac{\partial^2 u}{\partial x^2} + \frac{\partial^2 u}{\partial y^2}\right)^2 \qquad \text{(II.1)}$$

où $\gamma \equiv \frac{\partial u}{\partial z}$, et les coefficients A, B, C représentent des rigidités isothermes (la stabilité exige $AB > C^2$). Pour une configuration statique de u imposée la valeur d'équilibre de la dilatation θ est

$$\theta = -\frac{C}{A}\frac{\partial u}{\partial z} \qquad \text{(II.2)}$$

En réinsérant cette valeur dans (*II.1*), on retrouve la forme de Landau-Peierls avec :

$$\bar{B} = B - \frac{C^2}{A} \quad (> 0) \qquad \text{(II.3)}$$

Mais la relation (II.2) n'est pas applicable aux fluctuations thermiques qui nous intéressent en vue de la diffusion de la lumière. Nous conservons donc la forme (II.1), qui, après transformation de Fourier, devient :

$$\left.\begin{aligned} F &= \sum_q F_q \\ F_q &= \frac{1}{2}A\,|\,\theta_q\,|^2 + \frac{1}{2}B'\,|\,\gamma_q\,|^2 + C\theta_q\,\gamma_{-q} \\ B' &= B + K_1\,\frac{q_\perp^2(q_\perp^2 + \xi^{-2})}{q_z^2} \end{aligned}\right\} \qquad \text{(II.4)}$$

On en tire les moyennes thermodynamiques :

$$<|\,\theta_q\,|^2> = \frac{\int |\,\theta_q\,|^2\,\mathrm{e}^{-\beta F_q}\,\mathrm{d}\theta_q}{\int \mathrm{e}^{-\beta F_q}\,\mathrm{d}\theta_q} \quad \text{etc.}$$

$$\left.\begin{aligned} <|\,\theta_q\,|^2> &= \frac{B'\,k_B\,T}{AB' - C^2} \\ <|\,\gamma_q\,|^2> &= \frac{Ak_B\,T}{AB' - C^2} \\ <\theta_q\,\gamma_{-q}> &= \frac{-\,Ck_B\,T}{AB' - C^2}. \end{aligned}\right\} \qquad \text{(II.5)}$$

Examinons maintenant l'effet de ces fluctuations sur la diffusion de la lumière. Dans la configuration au repos, le tenseur constante diélectrique a pour éléments non nuls $\varepsilon_{zz} = \varepsilon_\parallel$ et $\varepsilon_{xx} = \varepsilon_{yy} = \varepsilon_\perp$ ($\varepsilon_\parallel$ et $\varepsilon_\perp$ sont les valeurs relatives à la fréquence optique considérée). Au premier ordre en déformations, les modifications du tenseur diélectrique ont la forme :

$$\left.\begin{aligned} \delta\varepsilon_{zz} &= a_z\,\theta + b_z\,\gamma \\ \delta\varepsilon_{xx} &= \delta\varepsilon_{yy} = a_\perp\,\theta + b_\perp\,\gamma \\ \delta\varepsilon_{xy} &= 0 \\ \delta\varepsilon_{zx} &= (\varepsilon_\perp - \varepsilon_\parallel)\frac{\partial u}{\partial x} \\ \delta\varepsilon_{zy} &= (\varepsilon_\perp - \varepsilon_\parallel)\frac{\partial u}{\partial y}. \end{aligned}\right\} \qquad \text{(II.6)}$$

Dans ces formules, $a_z, a_\perp, b_z, b_\perp$ sont quatre coefficients phénoménologiques sans dimensions, et de l'ordre de l'unité. La forme simple pour $\delta\varepsilon_{zx}$(ou $\delta\varepsilon_{zy}$) vient de ce que u/x = constante correspond à une rotation pure (au 1er ordre en u) autour de Oy.

Lorsque la différence $|\varepsilon_\perp - \varepsilon_\parallel|$ est faible, l'intensité de la lumière diffusée, pour un vecteur de diffusion $\mathbf{q}$, est proportionnelle à [8]

$$I(\mathbf{q}) = \langle |\mathbf{i}.\widehat{\delta\varepsilon}(\mathbf{q}).\mathbf{f}|^2 \rangle \qquad (II.7)$$

où $\mathbf{i}$ et $\mathbf{f}$ sont des vecteurs unitaires repérant les polarisations initiale et finale, et $\widehat{\delta\varepsilon}(\mathbf{q})$ la transformée de Fourier du tenseur (II.6). On peut calculer $I(\mathbf{q})$ à partir de (II.5) et (II.6).

Les conclusions générales sont les suivantes :

a) si $\mathbf{i}$ et $\mathbf{j}$ sont parallèles entre eux, et de surcroît soit parallèles, soit perpendiculaires à l'axe optique Oz, l'intensité diffusée est *faible*. Par exemple :

$$\langle |\delta\varepsilon_{zz}(\mathbf{q})|^2 \rangle = \frac{k_B T}{AB' - C^2}\{a_z^2 B' - 2 a_z b_z C + b_z^2 A\} \qquad (II.8)$$

En ordre de grandeur

$$\langle |\delta\varepsilon_{zz}(\mathbf{q})|^2 \rangle \sim \frac{k_B T}{A}$$

comme pour un fluide isotrope de compressibilité A^{-1}.

b) Le cas le plus intéressant correspond à une polarisation incidente $\mathbf{i}$ parallèle à l'axe optique, et une polarisation sortante $\mathbf{f}$ perpendiculaire à cet axe (ou l'inverse).

Par exemple, pour $\mathbf{f}$ parallèle à Ox, on mesure

$$\langle |\delta\varepsilon_{zx}(\mathbf{q})|^2 \rangle = \frac{(\varepsilon_\perp - \varepsilon_\parallel)^2 k_B T A q_x^2}{B' A - C^2}. \qquad (II.9)$$

En général, pour $q_z \neq 0$, cette intensité est faible elle aussi. Mais pour $q_z = 0$ (vecteur de diffusion normal à l'axe optique) elle devient grande. Par exemple, avec $\mathbf{q}$ parallèle à Ox, on obtient :

$$\langle |\delta\varepsilon_{zx}(q)|^2 \rangle = \frac{(\varepsilon_\perp - \varepsilon_\parallel)^2 k_B T}{K_1(q^2 + \xi^{-2})}. \qquad (II.10)$$

Cette expression est très analogue, en forme et en ordre de grandeur, à ce que l'on a dans un nématique [9] [8]. Mais la forte diffusion décrite par (II.10) est restreinte à un très petit domaine en $\mathbf{q}$. Plus précisément il faut réaliser la condition

$$B q_z^2 < K_1 q^4$$

soit pour l'angle Φ entre $\mathbf{q}$ et le plan xOy :

$$\Phi \sim \frac{q_z}{q} < \sqrt{\frac{K_1}{B}}\, q \qquad (II.11)$$

$\sqrt{\frac{K_1}{B}}$ sera de l'ordre de la distance intermoléculaire d et cette condition peut être écrite en gros

$$\Phi < qd(\sim 10^{-2})$$

c'est-à-dire que l'angle entre $\mathbf{q}$ et le plan xOy doit être inférieur à ~ ½ degré. Il n'est pas sûr que cette condition puisse être réalisée sur l'ensemble du spécimen, au moins pour les cristaux liquides smectiques assez imparfaits dont on dispose couramment. Mais la recherche de ce pic de diffusion pour $\Phi = 0$ mériterait en tout cas d'être tentée.

III. **Dynamique.** — Dans un *solide* nous trouvons pour chaque vecteur d'onde $\mathbf{q}$ trois modes de propagation acoustique (Fig. 2*a*). Dans un *liquide isotrope* nous avons un mode acoustique (oscillations de densité) et deux modes transverses complètement amortis, contrôlés par la viscosité $\mathrm{Im}\,\omega = \eta q^2/\rho$ (où η est la viscosité, ρ la densité) (Fig. 2*b*). Qu'advient-il dans un smectique ?

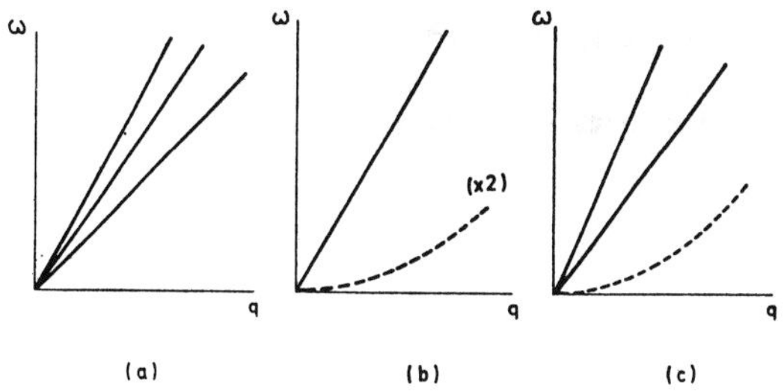

FIG. 2. — Spectre des excitations basse fréquence pour un solide (a), un liquide isotrope (b), et un smectique (c). Les branches en trait plein représentent des fréquences réelles, les branches pointillées des fréquences imaginaires (régions purement amorties).

Comme nous allons le voir, il y a deux types de comportement possibles selon l'orientation de $\mathbf{q}$ par rapport à l'axe optique :

a) *pour* $q_z \neq 0$ on trouve deux branches acoustiques (vibrations couplées de θ et de $\gamma = \partial u/\partial z$) plus une branche amortie transversale analogue à celle d'un liquide (Fig. 2*c*).

b) *pour* $q_z = 0$ on attend un mode de compression acoustique (vibrations de θ), un mode d'ondulation (Fig. 1) fortement amorti, et un mode hydrodynamique (glissement plan sur plan avec frottement visqueux).

Etudions d'abord le cas $q_z \neq 0$. Ici nous allons avoir des vibrations acoustiques de relativement haute fréquence, pour lesquelles le comportement est 1) adiabatique, 2) peu amorti. Pour engendrer ces vibrations il faut partir de *l'énergie* (et non plus de l'énergie libre) qui a la forme (par cm^3)

$$E = E_0 + 1/2\, A_0\, \theta^2 + 1/2\, B_0\, \gamma^2 + C_0\, \theta\gamma \quad \text{(III.1)}$$

Dans cette expression A_0, B_0, C_0 sont des coefficients de rigidité adiabatique. Les termes impliquant des dérivées d'ordre supérieur (comme le terme en K_1 de (II.1)) et les termes associés au champ magnétique ne sont pas importants pour $q_z \neq 0$ et ont été omis dans (III.1). Le système de tensions mécaniques $\sigma_{\alpha\beta}$ résultant de (III.1) peut être calculé par le travail fourni à la surface d'un spécimen déformé, et il se réduit aux composantes suivantes :

$$\left.\begin{aligned} \sigma_{zz} &= (B_0 + C_0)\,\gamma + (A_0 + C_0)\,\theta \\ \sigma_{xx} &= \sigma_{yy} = C_0\,\gamma + A_0\,\theta \end{aligned}\right\} \quad \text{(III.2)}$$

Le champ de vitesses dans le smectique est un vecteur de composantes (v_x, v_y, $\partial u/\partial t$). En négligeant l'amortissement, les équations d'accélération sont :

$$\rho \frac{\partial^2 u}{\partial t^2} = \frac{\partial}{\partial z}\sigma_{zz} = = (B_0 + C_0)\frac{\partial \gamma}{\partial z} + (A_0 + C_0)\frac{\partial \theta}{\partial z} \quad \text{(III.3)}$$

$$\left.\begin{aligned} \rho \frac{\partial v_x}{\partial t} &= \frac{\partial}{\partial x}\sigma_{xx} = C_0 \frac{\partial \gamma}{\partial x} + A_0 \frac{\partial \theta}{\partial x} \\ \rho \frac{\partial v_y}{\partial t} &= C_0 \frac{\partial \gamma}{\partial y} + A_0 \frac{\partial \theta}{\partial y} \end{aligned}\right\} \quad \text{(III.4)}$$

Il faut y adjoindre l'équation de conservation

$$\frac{\partial v_x}{\partial x} + \frac{\partial v_y}{\partial y} + \frac{\partial}{\partial t}(\gamma - \theta) = 0\,. \quad \text{(III.5)}$$

En cherchant des modes propres de la forme

$$\exp\{\, i\mathbf{q}.\mathbf{r} - i\omega t \,\}$$

on aboutit à l'équation séculaire

$$\omega \{ [\rho\omega^2 - (B_0 + C_0)\, q_z^2]\, [\rho\omega^2 - A_0\, q_\perp^2] - - (A_0 + C_0)\, q_z^2 [\rho\omega^2 + C_0\, q_\perp^2] \} = 0\,. \quad \text{(III.6)}$$

Le mode $\omega = 0$ correspond à des vitesses hydrodynamiques perpendiculaires à $\mathbf{q}$ et à l'axe optique Oz. (Dans une approximation plus raffinée, tenant compte de la friction, il prendrait la forme

$$\rho\omega = i(q_z^2\, \eta_z + q_\perp^2\, \eta_\perp)$$

où η_z et $\eta\perp$ sont deux coefficients de viscosité.)

Les deux autres racines de (III.6) sont les plus intéressantes, car elles décrivent les modes acoustiques qui sont seuls couplés en pratique à la diffusion de la lumière. On peut les mettre sous la forme :

$$\omega_1 = s_1(\Phi)\, q$$
$$\omega_2 = s_2(\Phi)\, q$$

où Φ est l'angle entre Oz et x. En posant $\rho s^2 = \mathbf{q}$, on obtient ρs_1^2 et ρs_2^1 comme les racines (positives) de l'équation :

$$x^2 - x\, \{ A_0 \cos^2 \Phi + (A_0 + B_0 + 2\, C_0) \sin^2 \Phi \} + + \sin^2 \Phi \cos^2 \Phi\, \{ B_0\, A_0 - C_0^2 \} = 0\,. \quad \text{(III.7)}$$

En général les vitesses s_1 et s_2 seront comparables à la vitesse du son dans la phase liquide isotrope. Il faut noter toutefois que l'une des vitesses tend vers 0 quand $\Phi \to \pi/2$ (**q** paralèlle à l'axe optique) et aussi quand $\Phi \to 0$ (**q** perpendiculaire à l'axe optique). Nous allons maintenant discuter brièvement le cas $\Phi = 0$, qui est particulièrement intéressant puisqu'il correspond aux fortes diffusions de la lumière.

Pour $q_z = 0$ nous avons un mode hydrodynamique banal ($\rho\omega = i\eta q^2$) et un mode acoustique de vitesse $\sqrt{A_0/\rho}$, qui tous deux contribuent peu à la diffusion. Le mode véritablement intéressant est le mode d'ondulation (Fig. 1). Dans la discussion de ce mode, il faut inclure le terme K_1 de l'équation (II.1) qui décrit la seule force de rappel présente. Il faut aussi tenir compte des amortissements, qui font intervenir ici un troisième coefficient de viscosité η_u. Au total, on obtient alors pour u l'équation de mouvement :

$$- \rho\omega^2\, u = - K_1\, q^2 (q^2 + \xi^{-2})\, u - i\omega\eta_u\, q^2\, u \quad \text{(III.8)}$$

Considérons d'abord le cas sans champ magnétique ($\xi = \omega$) et limitons-nous plus particulièrement au domaine

$$\frac{K_1\, \rho}{\eta_u^2} \ll 1\,.$$

Dans ces conditions on trouve à partir de (III.8) un mode de relaxation purement hydrodynamique

$$\omega_1 = + i\eta_u q^2$$

et une racine décrivant les fluctuations d'orientation :

$$\omega_2 = + i \frac{K_1 q^2}{\eta_u} . \qquad \text{(III.9)}$$

C'est cette dernière racine qui définit l'élargissement de la raie Rayleigh observable pour $q_z = 0$. Sa dépendance en q est très analogue à ce que l'on attend dans un nématique [7].

Terminons cette discussion du mode d'ondulation par une remarque concernant le cas de champs magnétiques H très forts : plus précisément, supposons que les inégalités suivantes sont réalisées :

$$\left.\begin{aligned} q\xi &< 1 \\ q\eta_u &< H\sqrt{\chi_a \rho} \end{aligned}\right\} \qquad \text{(III.10)}$$

Alors les termes réactifs dominent sur les termes dissipatifs, et on attend d'après (III.8) une onde « magnétoacoustique » lente, de vitesse

$$S_{//} = H \sqrt{\frac{\chi_a}{\rho}} \sim \text{qq cm/s} . \qquad \text{(III.11)}$$

Mais la deuxième inégalité (III.10) implique des valeurs de q extrêmement faibles qui seraient très difficiles à réaliser en pratique.

IV. **Remarques finales.** — Notre description phénoménologique des smectiques prévoit :

a) une diffusion lumineuse remarquable lorsque le vecteur de diffusion **q** est perpendiculaire à l'axe optique. Dans ce cas, et si en outre l'inégalité $K_1 \rho \ll \eta_u^2$ est satisfaite, le spectre en fréquence de la lumière diffusée devrait correspondre à un pic central étroit, dont la largeur est définie par l'équation (III.9),

b) dans les autres directions de **q**, une diffusion Brillouin d'intensité comparable à ce qu'elle est dans un fluide isotrope, mais comportant deux branches acoustiques.

Il faut insister ici sur certaines limitations de notre traitement :

1) La divergence logarithmique des fluctuations de position des couches (en champ magnétique nul) a été ignorée complètement. Comme nous l'avons vu dans l'introduction, cette divergence est faible, et probablement observable non pas dans le domaine visible ($qd \ll 1$) mais seulement dans le domaine des rayons X ($qd \sim 1$).

2) Nous nous sommes limités à deux variables d'état, le déplacement des couches u et la dilatation en volume θ. On peut envisager aussi d'inclure la température T, mais les effets qui en résultent (diffusion de la lumière par les fluctuations de T, amortissement par conduction thermique des vibrations) sont en général faibles et peu intéressants. Dans un cas toutefois les effets thermiques sont peut-être non triviaux : celui du mode d'ondulation ($q_z = 0$) ; il faudrait une étude plus détaillée pour savoir si ce mode est adiabatique ou isotherme [10].

3) Une autre variable éventuellement importante est le champ électrique **E**, puisqu'une déformation de la structure smectique fait en général apparaître des polarisations et des charges [11]. Il est possible de montrer toutefois que dans un smectique la forme de Landau-Peierls pour l'énergie libre (éq. I.5) n'est pas sérieusement affectée par ces effets. L'argument est le suivant : la polarisation **P** associée aux déplacements u a la forme :

$$\left.\begin{aligned} P_z &= - e_1\left(\frac{\partial^2 u}{\partial x^2} + \frac{\partial^2 u}{\partial y^2}\right) - e_2 \frac{\partial^2 u}{\partial z^2} \\ P_x &- e_3 \frac{\partial^2 u}{\partial z\, \partial x} \\ P_y &= - e_3 \frac{\partial^2 u}{\partial z\, \partial y} \end{aligned}\right\} \qquad \text{(IV.1)}$$

La densité de charges correspondante est :

$$- \operatorname{div} \mathbf{P} = (e_1 + e_3)\left(\frac{\partial^2 \gamma}{\partial x^2} + \frac{\partial^2 \gamma}{\partial y^2}\right) + e_2 \frac{\partial^2 \gamma}{\partial z^2} \qquad \text{(IV.2)}$$

L'équation de Poisson pour le potentiel électrique V est :

$$\operatorname{div}(\hat{\varepsilon} \nabla V) = 4\pi \operatorname{div} \mathbf{P} \qquad \text{(IV.3)}$$

(où ε est ici le tenseur constante diélectrique à fréquence nulle).

Soit pour une composante de Fourier de vecteur d'onde **q** :

$$\begin{aligned} &- (\varepsilon_{\|} q_z^2 + \varepsilon_\perp q_\perp^2) V_q = \\ &= 4\pi \{ (e_1 + e_3) q_\perp^2 + e_2 q_z^2 \} \gamma_q . \qquad \text{(IV.4)} \end{aligned}$$

La correction à l'énergie libre (par cm^3) est :

$$\Delta F = \frac{\mathbf{E}\hat{\varepsilon}\mathbf{E}}{8\pi} \qquad \text{(IV.5)}$$

où $\mathbf{E} = -\nabla V$ est le champ électrique. Pour une composante de Fourier, ceci donne :

$$\Delta F_q = \frac{2\pi\{(e_1 + e_3) q_\perp^2 + e_2 q_z^2\}^2}{\varepsilon_\parallel q_z^2 + \varepsilon_\perp q_\perp^2} \mid \gamma_q \mid^2. \quad \text{(IV.6)}$$

Dans la forme non perturbée (éq. I.6) on avait

$$\frac{1}{2}\bar{B}q_z^2 \mid u_q \mid^2 \equiv \frac{1}{2}\bar{B} \mid \gamma_q \mid^2 .$$

En ordre de grandeur, ΔF_q est plus petit que ce terme par un facteur $q^2 d^2 \ll 1$. Donc les effets piézoélectriques sont négligeables [12].

Bibliographie

[1] CHISTYAKOV (G.), *Soviet Physics Uspekhi*, 1967, **9**, 551. Le cas des phases smectiques de lipides doit être disjoint, ces matériaux ayant fait l'objet d'études intensives : voir la revue par V. LUZZATI à cette conférence.

[2] OSEEN (C. W.), *Trans. Faraday Soc.*, 1933, **29**, 883.

[3] FRANK (F. C.), *Disc. Faraday Soc.*, 1958, **25**, 1.

[4] LANDAU (L. D.), LIFSHITZ (I. M.), *Statistical Physics*, (Pergamon Press, Londres, 1958), p. 411.

[5] DE GENNES (P. G.). Proceedings of the 2nd Kent Conference on liquid Crystals (à paraître dans *Molecular Crystals*).

[6] DE GENNES (P. G.), *J. Chem. Phys.*, 1968, **48**, 2257.

[7] GROUPE D'ORSAY, *J. Chem. Phys.*, 1969, **51**, 816.

[8] DE GENNES (P. G.), *C. R. Acad. Sci. Paris*, 1968, **266**, 15.

[9] CHATELAIN (P.), *Acta crystallographica*, 1948, **1**, 315.

[10] Pour le mode « péristaltique » des films de savon, qui est un peu comparable à l'ondulation décrite ici, le régime est probablement isotherme : voir : DE GENNES (P. G.), *C.R. Acad. Sci. Paris*, 1969, **268 B**, 1207.

[11] MEYER (R. B.), *Phys. Rev. Lett.*, 1969, **22**, 918.

[12] La situation est tout à fait différente dans un nématique, où l'intensité de la lumière diffusée peut être fortement affectée par les effets piézoélectriques. GROUPE D'ORSAY, à paraître.

Afterthought: Conjectures sur l'état smectique

This paper gives a simple description of elasticity and friction in the simplest form of smectics (the smectics A, where the director is normal to the layers).

The main advance since these early days has been the discovery that fluctuations in the layered structure can deeply alter the elastic coefficient B [G. Grinstein and R. Pelcovitz, *Phys. Rev. Lett.* **47**, 856 (1981)]. They also lead to a divergence (at low frequencies) of certain friction coefficients [G. Mazenko, S. Ramaswamy and J. Toner, *Phys. Rev. Lett.* **49**, 51 (1982)]. I have never succeeded in providing a simple, intuitive, description of these effects.

Reprinted from THE JOURNAL OF CHEMICAL PHYSICS, Vol. 51, No. 2, 816–822, 15 July 1969

Dynamics of Fluctuations in Nematic Liquid Crystals

GROUPE D'ETUDE DES CRISTAUX LIQUIDES (ORSAY)
Laboratoire de Physique des Solides, Faculté des Sciences, 91—Orsay, France*

(Received 12 November 1968)

In a nematic system, the molecular axes tend to be aligned parallel to each other, but there are some fluctuations in the alignment. They give rise to a strong scattering of light. By a frequency analysis of the outgoing light, one can measure the power spectrum of these fluctuations, i.e., their dynamical behavior. The small motions of an incompressible nematic liquid crystal with coupled flows and molecular rotations are analyzed here in terms of the Leslie equations. The structure of the modes depends essentially on the parameter $\mu = K\rho\eta^{-2}$, where K is an average elastic constant, ρ is the density, and η is an average viscosity. In the most usual case of small μ values the modes are purely dissipative. For a given wave vector $\mathbf{q}$ and polarization α there is one slow mode describing the relaxation of a twisted nematic configuration and a fast mode where the molecules exert no torque on the hydrodynamic motion. The light-scattering data should measure the relaxation frequency of the slow mode: from such data taken at different q values the six friction coefficients introduced by Leslie may in principle be determined.

I. INTRODUCTION

In a nematic liquid crystal[1] the molecules tend to be aligned along a constant direction, labeled by a unit vector (or "director") $\mathbf{n}_0$. However, there are fluctuations from this average configuration: the local value of the director $\mathbf{n}(\mathbf{r})$ at point $\mathbf{r}$ differs slightly from $\mathbf{n}_0$. These fluctuations are very large for long wavelengths and give rise to a strong scattering of light. This has been studied experimentally by Châtelain[2]; a theoretical analysis which reproduces the main features of his results has been given recently.[3] At the time of the Châtelain experiments, it was not feasible to measure the *frequency* distribution of the outgoing light. But with the present laser sources, such experiments become possible and are currently under way.[4] For this reason, we attempt, in the present paper, to analyse the dynamical behavior of the orientation fluctuations.

We first write down the fundamental formulas for a (slightly) inelastic scattering of light, using the language of response functions and the fluctuation–dissipation theorem (Sec. II). Then we introduce the fundamental hydrodynamic equations, coupling internal and overall motion, as derived by Ericksen[5] and Leslie[6] (Sec. III). These equations involve six phenomenological coefficients describing the coupling and friction. We show in Sec. IV how these coefficients can all be determined from the light scattering data.

* Associé au C.N.R.S.

[1] A useful general review on liquid crystals is I. G. Chistyakov, Usp. Fiz. Nauk. **89**, 563 (1966) [Sov. Phys.—Uspekhi 9, 551].

[2] P. Chatelain, Acta Cryst. **4**, 453 (1951).

[3] P. G. de Gennes, Compt. Rend. **266**, 15 (1968). See also: "Proceedings of the Kent Conference on Liquid Crystals, 1968," Mol. Cryst. (to be published).

[4] Orsay Liquid Crystal Group (to be published).

[5] J. L. Ericksen, Arch. Ratl. Mech. Anal. **4**, 231 (1960); **9**, 371 (1962).

[6] F. M. Leslie, Quart. J. Mech. Appl. Math. **19**, 357 (1966).

II. FLUCTUATIONS, CORRELATIONS, AND LIGHT SCATTERING

We assume that at each point $\mathbf{r}$ the medium is optically uniaxial, and described by a dielectric tensor of the form

$$\hat{\epsilon} = \bar{\epsilon} + \epsilon_a(\mathbf{n}:\mathbf{n} - 1/3) \qquad \text{(II.1)}$$

The main fluctuations of $\hat{\epsilon}$ will then be due to the fluctuations of $\mathbf{n}(\mathbf{r})$. We consider an ingoing light beam of frequency ω_0, wave vector $\mathbf{k}_0$, polarization unit vector $\mathbf{i}$, and an outgoing beam $(\omega_1, \mathbf{k}_1, \mathbf{f})$. We also restrict our attention to the case $\epsilon_a \ll \bar{\epsilon}$: then both light beams propagate in a nearly isotropic medium, and this simplifies the algebra considerably. We define a frequency transfer

$$\omega = \omega_0 - \omega_1 \qquad (\omega \ll \omega_0) \qquad \text{(II.2)}$$

and a wave vector transfer

$$\mathbf{q} = \mathbf{k}_0 - \mathbf{k}_1. \qquad \text{(II.3)}$$

In terms of these quantities, the differential scattering cross section, per unit scattering volume, per unit solid angle of the outgoing beam $(d\Omega)$ and per unit angular frequency $(d\omega)$ is given by a formula of the Van Hove type[7]:

$$d\sigma/d\Omega d\omega = \pi^2\lambda^{-4}(2\pi)^{-1} \int d\mathbf{r} \int_{-\infty}^{\infty} dt$$
$$\times \exp[i(\mathbf{q}\cdot\mathbf{r} - \omega t)]\langle \delta\epsilon_{fi}(00)\delta\epsilon_{if}(\mathbf{r}t)\rangle. \qquad \text{(II.4)}$$

The bracket $\langle\,\rangle$ denotes a thermal average, $\delta\epsilon_{if}(\mathbf{r}t) = \mathbf{i}\cdot\delta\hat{\epsilon}\cdot\mathbf{f}$ is the fluctuating part of the dielectric tensor at a given point $\mathbf{r}$ and time t ($\lambda = 2\pi c/\omega_0$ is the vacuum wavelength). We consider only the fluctuations $\delta\hat{\epsilon}$

[7] L. Van Hove, Phys. Rev. **95**, 249 (1954).

which are due to the fluctuations $\delta\mathbf{n}(\mathbf{r})$ of the director

$$\mathbf{n}(r)=\mathbf{n}_0+\delta\mathbf{n}$$

($\delta\mathbf{n}\cdot\mathbf{n}_0=0$ to first order in δn). Then from Eq. (II.1)

$$\delta\epsilon_{if}=\epsilon_a[\delta\mathbf{n}\cdot\mathbf{i}f_0+\delta\mathbf{n}\cdot\mathbf{f}i_0]$$

($f_0=\mathbf{f}\cdot\mathbf{n}_0$, $i_0=\mathbf{i}\cdot\mathbf{n}_0$). For a given spatial Fourier component of wave vector $\mathbf{q}$ it is convenient to analyse $\delta\mathbf{n}$ into two components, one in the $(\mathbf{q}, \mathbf{n}_0)$ plane, and the other normal to it (Fig. 1): from the symmetry of the problem the two components will be uncoupled. Thus we introduce two unit vectors $\mathbf{e}_\alpha$ ($\alpha=1, 2$):

$$\mathbf{e}_2=(\mathbf{n}_0\times\mathbf{q})(q\sin\phi)^{-1},$$

$$\mathbf{e}_1=\mathbf{e}_2\times\mathbf{n}_0 \tag{II.5}$$

and define $i_\alpha=\mathbf{e}_\alpha\cdot\mathbf{i}$, $f_\alpha=\mathbf{e}_\alpha\cdot\mathbf{f}$. The mode where $\delta\mathbf{n}$ is parallel to $\mathbf{e}_1$ is a combination of bending and splay while the mode where $\delta\mathbf{n}$ is parallel to $\mathbf{e}_2$ is a combination of bending and torsion, as shown on Fig. 1. In terms of these modes we may write for the cross section:

$$d\sigma/d\Omega d\omega=\pi^2\lambda^{-4}\sum_{\alpha=1,2} I_\alpha(\mathbf{q},\omega)(i_\alpha f_0+i_0 f_\alpha)^2, \tag{II.6}$$

$$I_\alpha(\mathbf{q}\omega)=(2\pi)^{-1}\int_{-\infty}^{\infty} dt\,\exp(-i\omega t)\langle\delta n_\alpha(-\mathbf{q},0)\delta n_\alpha(\mathbf{q},t)\rangle, \tag{II.7}$$

$$\delta n_\alpha(\mathbf{q}t)=\int d\mathbf{r}\,\exp(i\mathbf{q}\cdot\mathbf{r})\delta n_\alpha(\mathbf{r}t). \tag{II.8}$$

The merit of the cross-section formulas (II.6) is to separate clearly a purely geometric factor $(i_\alpha f_0+i_0 f_\alpha)^2$, and a fundamental correlation function I_α.

How do we compute I_α? Consider a "gedanken experiment," where we add to the Hamiltonian of our liquid crystal a term of the form

$$-\int\mathbf{n}(\mathbf{r})\cdot\mathbf{h}^e(\mathbf{r},t)\,d\mathbf{r},$$

where $\mathbf{h}^e$ is some external source. From the equations of motion we will find that the director is modified at all points: $\mathbf{n}_0\rightarrow\mathbf{n}_0+\delta\mathbf{n}$. The most general form, to first order in $\mathbf{h}^e$, is

$$\delta n_i(\mathbf{r}t)=\int d\mathbf{r}'\int_{-\infty}^{t}\chi_{ij}(\mathbf{r}-\mathbf{r}',t-t')h_j^e(\mathbf{r}',t')\,dt' \tag{II.9}$$

or for a given Fourier component

$$\delta n_\alpha(\mathbf{q},t)=\int_{-\infty}^{t}\chi_\alpha(\mathbf{q},t-t')h_\alpha^e(\mathbf{q}t')\,dt'. \tag{II.10}$$

[In Eq. (II.10) we have again used the fact that the modes $\alpha=1$ and $\alpha=2$ are uncoupled.]

If h_α^e is proportional to $\exp(i\omega t)$, then δn_α is also proportional to $\exp(i\omega t)$, and we have

$$\delta n_\alpha=\chi_\alpha(q\omega)h_\alpha^e,$$

$$\chi_\alpha(\mathbf{q}\omega)=\int_0^\infty dt\chi_\alpha(\mathbf{q}t)\exp(-i\omega t).$$

When the "response function" $\chi_\alpha(\mathbf{q}\omega)$ is known, we can derive from it the function $I_\alpha(\mathbf{q}\omega)$ which describes the light scattering [Eq. (II.7)]. The relation between the two is given by the fluctuation dissipation theorem.[8] In the limit of interest here ($\hbar\omega\ll k_BT$) this reduces to

$$I_\alpha(\mathbf{q}\omega)=-(k_BT/\pi\omega)\,\mathrm{Im}[\chi(\mathbf{q}\omega)]. \tag{II.11}$$

Thus our program is the following: we write the equations of motion in the presence of an external perturbation h^e modulated at a frequency ω. We find the resulting $\delta\mathbf{n}$ and thus obtain $\chi(\mathbf{q}\omega)$. Finally by Eq. (II.11) we shall arrive at $I_\alpha(\mathbf{q}\omega)$.

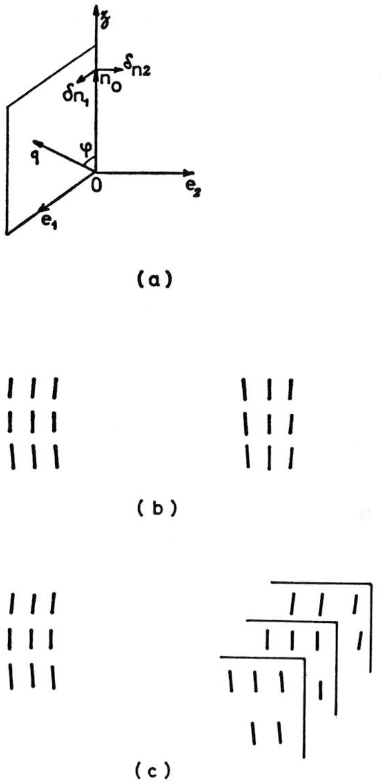

FIG. 1. The two uncoupled modes δn_1 and δn_2 (a); components of distorsion in the δn_1 mode: bending and splay (b); and in the δn_2 mode: bending and torsion (c).

III. HYDRODYNAMIC AND ROTATIONAL MOTIONS

A. Molecular Field and Resulting Torque

The static free energy in a nematic crystal undergoing long wavelength distortions and (eventually)

[8] See for instance R. Kubo, Rept. Progr. Phys. 29, 255 (1966).

submitted to a constant external magnetic field H, has the form

$$\mathfrak{F}=\frac{1}{2}\int d\mathbf{r}[K_{11}(\operatorname{div}\mathbf{n})^2+K_{22}(\mathbf{n}\cdot\operatorname{curl}\mathbf{n})^2$$
$$+K_{33}(\mathbf{n}\times\operatorname{curl}\mathbf{n})^2-\chi_a(\mathbf{n}\cdot\mathbf{H})^2]+\text{const} \quad (\text{III.1})$$

with three independent elastic constants K_{ii}. (A fourth elastic term introduced by Frank[9] does not in fact contribute to volume energies[10] and may be omitted). Each of the K's is of order k_BT_c/a, where k_BT_c is a typical molecular interaction energy and a is a molecular length. $\chi_a=\chi_{\parallel}-\chi_{\perp}$ is the anisotropy of the magnetic susceptibility. We assume $\chi_a>0$ in all that follows: this case is often met in practice; it is also the most favorable if one wishes to prepare a single crystal by field alignment.

A useful notion in connection with $\mathfrak{F}$ is that of a *molecular field* $\mathbf{h}$, defined by a functional derivative:

$$h_i(\mathbf{r})=-\delta\mathfrak{F}/\delta n_i(\mathbf{r}). \quad (\text{III.2})$$

The various contributions to $\mathbf{h}$, as obtained from Eq. (III.1) are the following:

Splay: $\mathbf{h}=K_{11}\operatorname{grad}\operatorname{div}\mathbf{n}$,

Torsion: $\mathbf{h}=-K_{22}[A\operatorname{curl}\mathbf{n}+\operatorname{curl}(A\mathbf{n})]$,

Bending: $h=K_{33}[\mathbf{B}\times\operatorname{curl}\mathbf{n}+\operatorname{curl}(\mathbf{n}\times\mathbf{B})]$,

External field: $\mathbf{h}=\chi_a(\mathbf{H}\cdot\mathbf{n})\mathbf{H}$. (III.3)

In these equations $A=\mathbf{n}\cdot\operatorname{curl}\mathbf{n}$ and $\mathbf{B}=\mathbf{n}\times\operatorname{curl}\mathbf{n}$.

The case where all three elastic constants are equal leads to a very simple form of $\mathbf{h}$:

$$\mathbf{h}=K\nabla^2\mathbf{n}+\chi_a\mathbf{H}(\mathbf{n}\cdot\mathbf{H}).$$

Note that it is always possible to add a term proportional to $\mathbf{n}^2$ in the free-energy density: since $|\mathbf{n}|$ is a constant, this does not change the physical situation, but implies the change $\mathbf{h}\rightarrow\mathbf{h}+\lambda\mathbf{n}$. It is only the components of $\mathbf{h}$ normal to $\mathbf{n}$ which have a physical meaning.

These notions are useful in particular when one writes down the equation of motion for the *internal rotation*: defining

$$\mathbf{\Omega}=\mathbf{n}\times(d\mathbf{n}/dt)$$

as the rotational velocity, we have the equation

$$\mathfrak{I}(d\mathbf{\Omega}/dt)=\mathbf{n}\times\mathbf{h}-\mathbf{\Gamma};$$

$\mathfrak{I}$ is the moment of inertia per unit volume; $\mathbf{n}\times\mathbf{h}$ is the torque on one molecule due to its coupling with the neighbors; $\mathbf{\Gamma}$ is the frictional torque which the molecules exert on the over-all hydrodynamic motion. In practice, for frequencies and wavelengths compatible with the continuum model, the inertial term is completely negligible. (We shall discuss this point in more detail in Sec. V). Then one may write simply:

$$\mathbf{n}\times\mathbf{h}=\mathbf{\Gamma}. \quad (\text{III.4})$$

The most general form for the frictional torque $\mathbf{\Gamma}$ in an incompressible liquid crystal was written by Leslie[6]

$$\mathbf{\Gamma}=\mathbf{n}\times\{\gamma_1[(d\mathbf{n}/dt)-\boldsymbol{\omega}\times\mathbf{n}]+\gamma_2\hat{A}\mathbf{n}\}, \quad (\text{III.5})$$

where $\boldsymbol{\omega}=\frac{1}{2}\operatorname{curl}\mathbf{v}$ is the local rotational velocity of the fluid, while $\hat{A}$ is the shear rate tensor

$$A_{ij}=\tfrac{1}{2}[(\partial v_j/\partial x_i)+(\partial v_i/\partial x_j)],$$

and γ_1 and γ_2 are two constants with the dimension (and the order of magnitude) of a viscosity.

B. Acceleration Equations for an Incompressible Fluid

The stress tensor σ_{ij} may be written as a sum of three parts

$$\sigma_{ij}=-p\delta_{ij}+\sigma_{ij}^0+\sigma_{ij}'; \quad (\text{III.6})$$

p is the pressure (and is an unknown to be determined from the equations, since we are restricting our attention to an incompressible fluid), σ_{ij}^0 is the part of the stress which is related to the free energy $\mathfrak{F}$, and has been studied in detail by Ericksen[5]. For the small amplitude motions to be discussed here σ^0 may be neglected completely, since it is quadratic in the amplitudes. The dissipative part σ_{ij}' is a linear function of the velocity gradients A_{ij}, and of the velocity $\mathbf{N}$ of the internal motion relative to the background fluid

$$\mathbf{N}\equiv(d\mathbf{n}/dt)-\boldsymbol{\omega}\times\mathbf{n}. \quad (\text{III.7})$$

We take the functional form of σ_{ij}' in an incompressible nematic fluid from the work of Leslie[6]: in the notation of Ref. 6 we have

$$\sigma_{ij}'=\alpha_1n_kn_pA_{kp}n_in_j+\alpha_2n_iN_j+\alpha_3n_jN_i+\alpha_4A_{ij}$$
$$+\alpha_5n_in_kA_{kj}+\alpha_6n_jn_kA_{ki}. \quad (\text{III.8})$$

The coefficients α have the dimension of a viscosity. The tensor σ_{ij}' is not symmetric, since the internal degrees of freedom create a torque $\mathbf{\Gamma}$ per unit volume. Writing the standard condition

$$\Gamma_z=-\sigma_{xy}+\sigma_{yx}$$

and comparing with Eq. (III.5) one arrives at the

[9] F. C. Frank, Discussions Faraday Soc. 25, 19 (1958).
[10] J. L. Ericksen, Arch. Ratl. Mech. Anal. 10, 189 (1962).

conditions

$$\gamma_1=\alpha_3-\alpha_2,$$

$$\gamma_2=\alpha_6-\alpha_5. \qquad \text{(III.9)}$$

Thus the hydrodynamic properties of our nematic fluid are described by six independent coefficients.

For small amplitude motions we may replace $\mathbf{n}$ by $\mathbf{n}_0$ in the Eq. (III.8) for σ_{ij}'. Let us choose a frame of reference such that $\mathbf{n}_0$ is parallel to Oz. Then we have the following formulas:

$$\sigma_{zz}'=(\alpha_1+\alpha_4+\alpha_5+\alpha_6)A_{zz},$$

$$\sigma_{xx}'=\alpha_4A_{xx},$$

$$\sigma_{yy}'=\alpha_4A_{yy},$$

$$\sigma_{xy}'=\sigma_{yx}=\alpha_4A_{xy},$$

$$\sigma_{xz}'=\alpha_3N_x+(\alpha_4+\alpha_6)A_{xz},$$

$$\sigma_{zx}'=\alpha_2N_x+(\alpha_4+\alpha_5)A_{xz},$$

$$\sigma_{yz}'=\alpha_3N_y+(\alpha_4+\alpha_6)A_{yz},$$

$$\sigma_{zy}'=\alpha_2N_y+(\alpha_4+\alpha_5)A_{yz}. \qquad \text{(III.10)}$$

Let us now write down the equation of motion of the hydrodynamic velocity $\mathbf{v}$. With the above definition of the stress tensor it is[6]

$$\rho(dv_i/dt)=(\partial/\partial x_j)\sigma_{ji}, \qquad \text{(III.11)}$$

where ρ is the density. For our small amplitude motions, we may put

$$d/dt\to\partial/\partial t=i\omega,$$

$$\partial/\partial x_j=iq_j$$

and obtain

$$i\omega\rho\mathbf{v}=-i\mathbf{q}p+\mathbf{F}, \qquad \text{(III.12)}$$

where

$$+F_i=+iq_j\sigma_{ji}' \qquad \text{(III.13)}$$

is the total friction force. The incompressibility condition implies

$$\mathbf{q}\cdot\mathbf{v}=0$$

and this allows to eliminate the scalar pressure p. The acceleration equation then becomes

$$i\omega\rho\mathbf{v}=\mathbf{F}-\mathbf{q}[(\mathbf{q}\cdot\mathbf{F})/q^2]. \qquad \text{(III.14)}$$

IV. EIGENMODES AND RESPONSE FUNCTIONS

We shall now solve explicitly the coupled hydrodynamic equations describing the internal motion [Eqs. (III.4) and (III.5)] and its reaction on the hydrodynamic velocity field [Eq. (III.14)]. The internal variables are the small components n_x, n_y, of the director. For a given wave vector $\mathbf{q}$ we may, without loss of generality, take $Ox\parallel\mathbf{e}_1$, and $Oy\parallel\mathbf{e}_2$: thus $n_x\equiv n_1$, $n_y\equiv n_2$.

Let us first construct the molecular field $\mathbf{h}$ and the resulting torques. For a Fourier component of given $\mathbf{q}$ the free energy (per cubic centimeter), as given by Eq. (III.1), takes the form

$$F=F_0+\tfrac{1}{2}\sum_{\alpha=1,2}K_\alpha(\mathbf{q})\mid n_\alpha\mid^2, \qquad \text{(IV.1)}$$

where

$$K_\alpha(\mathbf{q})=K_{33}q_z^2+K_{\alpha\alpha}q_\perp^2+\chi_aH^2 \qquad \text{(IV.2)}$$

(we call $q_\perp$ the component of $\mathbf{q}$ normal to $\mathbf{n}_0$). The corresponding molecular field has only two nonzero components:

$$h_\alpha=-K_\alpha(\mathbf{q})n_\alpha+h_\alpha^e \qquad (\alpha=1,2) \qquad \text{(IV.3)}$$

where h_1^e and h_2^e are the external "sources" introduced to define a response function [see Eq. (II.10)].

Combining Eqs. (III.4) and (III.5) we have

$$\mathbf{n}\times\mathbf{h}=\mathbf{n}\times(\gamma_1\mathbf{N}+\gamma_2\hat{A}\mathbf{n}), \qquad \text{(IV.4)}$$

where we use Leslie's notation[6]:

$$\mathbf{N}=(d\mathbf{n}/dt)-\boldsymbol{\omega}\times\mathbf{n}. \qquad \text{(IV.5)}$$

This gives two equations:

$$\gamma_1N_\alpha+\gamma_2A_{\alpha z}=h_\alpha \qquad (\alpha=1,2). \qquad \text{(IV.6)}$$

Let us write explicitly the nonzero components of $\hat{A}$ and $\boldsymbol{\omega}$ which come into play:

$$A_{zz}=iq_zv_z,$$

$$A_{11}=iq_\perp v_1=-iq_zv_z,$$

$$A_{z1}=(i/2)(q_\perp v_z+q_zv_1)=(i/2q_\perp)(q_\perp^2-q_z^2)v_z,$$

$$A_{z2}=(i/2)q_zv_2,$$

$$A_{12}=(i/2)q_\perp v_2,$$

$$\omega_1=-(i/2)q_zv_2,$$

$$\omega_2=(i/2)(q_zv_1-q_\perp v_z)=-(i/2q_\perp)q^2v_z. \qquad \text{(IV.7)}$$

Inserting Eqs. (IV.7) and (IV.3) into (IV.6) we arrive at

$$iC_1(\mathbf{q})v_z+n_1[i\omega\gamma_1+K_1(\mathbf{q})]=h_1^e,$$

$$iC_2(\mathbf{q})v_2+n_2[i\omega\gamma_2+K_2(\mathbf{q})]=h_2^e \qquad \text{(IV.8)}$$

with the following definitions for $C_\alpha(\mathbf{q})$:

$$C_1(\mathbf{q}) = (2q_\perp)^{-1}[q_\perp^2(\gamma_1+\gamma_2)+q_z^2(\gamma_1-\gamma_2)],$$

$$C_2(\mathbf{q}) = \tfrac{1}{2}q_z(\gamma_2-\gamma_1). \quad \text{(IV.9)}$$

We now turn to the acceleration Eqs. (III.14), inserting for the force $\mathbf{F}$ the detailed form derived from Eqs. (III.13) and (III.10), and obtain two equations for the two independent components v_z, v_2 of the form

$$v_z[\rho\omega - iP_1(\mathbf{q})] = i\omega Q_1(\mathbf{q})n_1,$$

$$v_2[\rho\omega - iP_2(\mathbf{q})] = i\omega Q_2(\mathbf{q})n_2 \quad \text{(IV.10)}$$

where

$$P_1(\mathbf{q}) = (2q^2)^{-1}[\alpha_s q_\perp^4 + \alpha_v q_z^4 + q_\perp^2 q_z^2 \alpha_m],$$

$$P_2(\mathbf{q}) = \tfrac{1}{2}(q_\perp^2\alpha_4 + q_z^2\alpha_v),$$

$$Q_1(\mathbf{q}) = (q_\perp/q^2)(\alpha_3 q_\perp^2 - \alpha_2 q_z^2),$$

$$Q_2(\mathbf{q}) = \alpha_2 q_z, \quad \text{(IV.11)}$$

and we have introduced the symbols

$$\alpha_s = \alpha_3 + \alpha_4 + \alpha_6,$$

$$\alpha_v = -\alpha_2 + \alpha_4 + \alpha_5,$$

$$\alpha_m = 2(\alpha_1+\alpha_4) + \alpha_5 + \alpha_6 + \alpha_3 - \alpha_2. \quad \text{(IV.12)}$$

We can now solve explicitly for the small displacements n_α in terms of the external sources h_α^e: eliminating the velocities between Eqs. (IV.8) and (IV.10), we arrive at

$$n_\alpha = \chi_\alpha(\mathbf{q}, \omega)h_\alpha^e,$$

$$\chi_\alpha(\mathbf{q}\omega) = \frac{\rho\omega - iP_\alpha(\mathbf{q})}{[\rho\omega - iP_\alpha(\mathbf{q})][i\omega\gamma_1 + K_\alpha(\mathbf{q})] - C_\alpha(\mathbf{q})Q_\alpha(\mathbf{q})\omega}. \quad \text{(IV.13)}$$

Equation (IV.13) is our central result, and we shall now discuss some of its consequences.

A. Fluctuation Modes

For a given wave vector $\mathbf{q}$ and polarization α, we find in general two complex eigenfrequencies which correspond to the poles of χ_α, and are the roots of the equation

$$i\gamma_1\rho\omega^2 + \omega[\rho K_\alpha(\mathbf{q}) + \gamma_1 P_\alpha(\mathbf{q}) - C_\alpha(\mathbf{q})Q_\alpha(\mathbf{q})] - iP_\alpha(\mathbf{q})K_\alpha(\mathbf{q}) = 0. \quad \text{(IV.14)}$$

Let us first analyze these modes for the case of 0 *magnetic field*, and for a general direction of $\mathbf{q}(q_z \sim q_\perp \sim q)$. We shall assume that the six constants α have the same order of magnitude and are of order η, where η has the significance of an average viscosity. Similarly, we assume that the three elastic constants are comparable, and of order of magnitude K. The functions which are involved in Eq. (IV.14) may be estimated as follows

$$P_\alpha(\mathbf{q}) \sim \eta q^2,$$

$$C_\alpha(\mathbf{q}) \sim Q_\alpha(\mathbf{q}) \sim \eta q,$$

$$K_\alpha(\mathbf{q}) \sim Kq^2. \quad \text{(IV.15)}$$

The structure of Eq. (IV.14) is then critically dependent on the value of the following dimensionless parameter

$$\mu = (K\rho/\eta^2) \quad \text{(IV.16)}$$

Typically $K = 10^{-6}$ dyn, $\rho = 1$ g/cm^3, and $\eta = 0.1$ P leading to $\mu = 10^{-4}$. Thus in most cases we expect that $\mu \ll 1$. This observation leads to a considerable simplification on Eq. (IV.14); in the limit $\mu \ll 1$ the two eigenfrequencies become

$$\omega_{s\alpha} = i\frac{K_\alpha(\mathbf{q})P_\alpha(\mathbf{q})}{\gamma_1 P_\alpha(\mathbf{q}) - C_\alpha(\mathbf{q})Q_\alpha(\mathbf{q})} \equiv iu_{s\alpha},$$

$$\omega_{F\alpha} = (i/\rho\gamma_1)[\gamma_1 P_\alpha(\mathbf{q}) - C_\alpha(\mathbf{q})Q_\alpha(\mathbf{q})] \equiv iu_{F\alpha}. \quad \text{(IV.17)}$$

The subscripts (s) and (F) stand for "slow" and "fast", since

$$u_s \sim Kq^2/\eta, \qquad u_F \sim (\eta/\rho)q^2, \qquad u_s/u_F \sim \mu \ll 1.$$

The (s) and (F) modes are both admixtures of hydrodynamic motion and internal rotation. The mode (s) corresponds to a slow relaxation of a twisted nematic structure, the torques due to the twist being balanced by friction terms. All inertial effects in the background fluid are negligible. On the other hand, in the mode (F) these inertial effects are important, but there is no torque on the molecules. For both modes the rotation velocity of the molecules ωn_α is comparable to the velocity gradients qv.

Note that in the present limit $(\mu \ll 1)$ the frequencies $\omega_{s\alpha}$, $\omega_{F\alpha}$ are pure imaginary: a disturbed nematic system returns to equilibrium without any oscillations. For both modes, with a fixed direction for $\mathbf{q}$, the relaxation frequencies are proportional to q^2. The particular case where $\mathbf{q}$ is parallel to the nematic axis Oz deserves a special mention: here both polarizations give the same frequencies:

$$\omega_s = iK_{33}q^2[\gamma_1 + (\alpha_2/\alpha_v)(\gamma_1-\gamma_2)]^{-1}$$

$$\omega_F = (i/2\rho)q^2[\alpha_v + (\alpha_2/\gamma_1)(\gamma_1-\gamma_2)] \qquad (q_\perp = 0; \mu \ll 1). \quad \text{(IV.18)}$$

Up to now we have analyzed the eigenfrequencies only in the limit of 0 magnetic fields. Let us now investigate the opposite limit, namely high magnetic fields: inspection of Eq. (IV.13) shows that this domain is defined by the condition

$$\chi_a H^2 \gg (q^2\eta^2)/\rho. \qquad \text{(IV.19)}$$

The inequality (IV.19) is not easy to fulfill: with $H=10$ kOe, $\eta=10^{-1}$, $\chi_a=10^{-5}$, $\rho=1$, it requires $q<100\ \text{cm}^{-1}$. Such small q values correspond to very small angles ($\sim10^{-3}$ rad) between ingoing and outgoing beam in a light scattering experiment, and are thus difficult to achieve. However the high-field limit is of some theoretical interest: here we find from Eq. (IV.14) one slow hydrodynamic mode

$$\omega_{s\alpha}=(i/\rho)P_\alpha(\mathbf{q}) \qquad \text{(IV.20a)}$$

and one fast orientational mode:

$$\omega_{F\alpha}=i\chi_a H^2/\gamma_1. \qquad \text{(IV.20b)}$$

B. Frequency Spectrum of the Fluctuations

We shall now write down some explicit formulas for the power spectrum $I_\alpha(\mathbf{q}\omega)$ of the fluctuations, defined as in Eq. (II.7). The value I_α is related to the response function χ_α by Eq. (II.11) and χ_α itself is given in Eq. (IV.13). The resulting formula for I_α is rather complicated and not very enlightening physically. In the present discussion we shall restrict our attention to the 0 field case, with $\mu\ll1$. Then the formulas become comparatively simple and more instructive. The eigenfrequencies ($iu_{s\alpha}$, $iu_{F\alpha}$) are given by Eq. (IV.17) and the response function χ_α may be cast in the form

$$\chi_\alpha(\mathbf{q}\omega)=\frac{\rho\omega-iP_\alpha(\mathbf{q})}{i\gamma_1\rho(\omega-iu_{s\alpha})(\omega-iu_{F\alpha})}$$

$$=\frac{\rho\omega-iP_\alpha}{(u_{F\alpha}-u_{s\alpha})\gamma_1\rho}[(\omega-iu_{s\alpha})^{-1}-(\omega-iu_{F\alpha})^{-1}]. \qquad \text{(IV.21)}$$

In the limit $\mu\ll1$ we may replace $u_{F\alpha}-u_{s\alpha}$ by $u_{F\alpha}$. Finally, taking the imaginary part of χ, we obtain the power spectrum

$$I_\alpha(\mathbf{q}\omega)\cong\frac{k_BT}{\pi\gamma_1\rho u_{F\alpha}}\left[\frac{P_\alpha}{\omega^2+u_{s\alpha}^2}-\frac{C_\alpha Q_\alpha}{(\omega^2+u_{F\alpha}^2)\gamma_1}\right]. \qquad \text{(IV.22)}$$

Thus, for fixed q, I_α is a superposition of two Lorentz curves, a narrow one (of width $u_{s\alpha}$) and a broad one (of width $u_{F\alpha}$). The coefficients P_α and $C_\alpha Q_\alpha/\gamma_1$ are of comparable magnitude (ηq^2). The integrated area of the first Lorentzian in the bracket of Eq. (IV.22) is of order $\eta q^2/u_s$ while the area of the second Lorentzian is of order $\eta q^2/u_F$, i.e., smaller by a factor μ. Thus, in the present limit of small μ's, we may drop completely the second Lorentzian, and we have finally

$$I_\alpha(\mathbf{q}\omega)=(k_BT/\pi\gamma_1\rho u_{F\alpha})\,[P_\alpha(\mathbf{q}\omega)/(\omega^2+u_{s\alpha}^2)] \qquad \text{(IV.23)}$$

Note that I_α is *entirely controlled by the slow mode.* A useful check on Eq. (IV.23) is obtained if we investigate the integrated intensity:

$$\int I_\alpha(\mathbf{q}\omega)\,d\omega=k_BTP_\alpha/\gamma_1\rho u_{F\alpha}u_{s\alpha}$$

$$=k_BT/K_\alpha(\mathbf{q}). \qquad \text{(IV.24)}$$

This does agree with the simple equipartition result of Ref. 3. It is then possible to rewrite (IV.23) in the more compact form

$$I_\alpha(\mathbf{q}\omega)=[k_BT/\pi K_\alpha(\mathbf{q})][u_{s\alpha}/(\omega^2+u_{s\alpha}^2)]. \qquad \text{(IV.23}'\text{)}$$

The only dynamical information given by the light scattering data (in the small μ limit) is the relaxation frequency of the slow mode $u_{s\alpha}$. How can we use this in practice? First, as explained in Ref. 3, from a study of the integrated intensities under magnetic fields, one should measure the elastic constants. Then a useful way of plotting the data is to consider $K_\alpha(\mathbf{q})/u_{s\alpha}(\mathbf{q})$, where $u_{s\alpha}$ is the experimental width. Collecting Eqs. (IV.17) and (IV.9), (IV.11), (IV.12) we arrive at the following theoretical forms:

$$\frac{K_1(\mathbf{q})}{u_{s1}(\mathbf{q})}=\gamma_1-\frac{q_\perp^4(\gamma_1+\gamma_2)\alpha_3+q_\perp^2q_z^2[\alpha_3(\gamma_1-\gamma_2)-\alpha_2(\gamma_1+\gamma_2)]+q_z^4\alpha_2(\gamma_2-\gamma_1)}{q_\perp^4\alpha_s+q_\perp^2q_z^2\alpha_m+q_z^4\alpha_v}, \qquad \text{(IV.25a)}$$

$$K_2(\mathbf{q})/u_{s2}(\mathbf{q})=\gamma_1+[\alpha_2(\gamma_1-\gamma_2)q_z^2/(q_\perp^2\alpha_4+q_z^2\alpha_v)]. \qquad \text{(IV.25b)}$$

If the data do have these analytical forms in q_z, $q_\perp$, we can extract six relations between coefficients from (IV.25a) and three relations from Eq. (IV.25b). Thus the six unknown Leslie coefficients may be determined, and some internal checks are even possible.

It may be useful to conclude this section by a word of caution: our final, simple formulas for the fluctuation power spectrum (IV.23) and for the modes (IV.17) apply only in the limit $\mu\ll1$ and when all the friction coefficients α_i are of comparable magnitude (η). If some of the α's turn out to be in fact much larger than the average viscosity η, it may become necessary to return to the complete equation (IV.13). In such cases, propagating modes might be found (as opposed to the purely aperiodic modes which were discussed here); it will be interesting, therefore, to test this point in particular, experimentally.

V. CONCLUSIONS

We have found two types of modes for the small amplitude motions of a nematic liquid. In the most usual case of small μ values and weak or medium

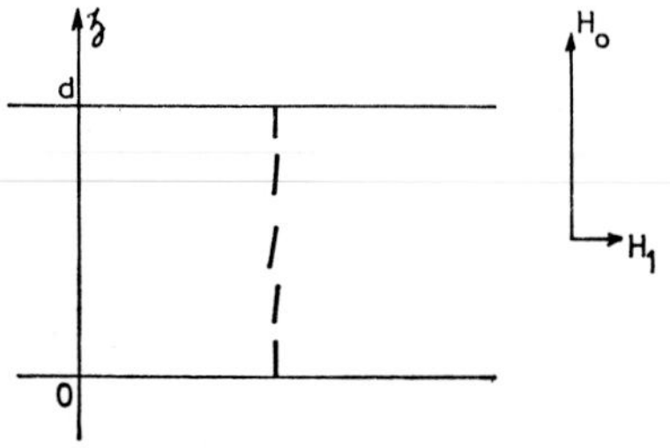

FIG. 2. Instantaneous distorsion under the action of a rf magnetic field $H_1 \exp(i\omega t)$.

magnetic fields, these modes may be visualized as follows:

(a) In the slow mode, a distorted molecular configuration relaxes exponentially with a rate $u_s \sim (Kq^2/\eta)$ ($\sim 10^3$ sec^{-1} for $q = 10^4$ cm^{-1}).

(b) The fast mode describes a diffusion of vorticity with a rate $u_F \sim \eta q^2/\rho$ (comparable to what it is in a conventional fluid). Typically $u_F = 10^7$ sec^{-1} for $q = 10^4$ cm^{-1}.

For both modes the angular deflexions δn of the molecular axis are of order vq/u where v is the amplitude of the velocity oscillations. For a given v, δn is much larger for the slow mode (small μ): for this reason the scattering of light, which measures the fluctuations δn, is entirely controlled by the slow mode.

Our calculation was based on a complete neglect of the inertial term $\mathfrak{I}(d\Omega/dt)$ in the equation of motion (III.4) for the internal rotation. It is of some importance to comment on this point, since, in the opposite limit where inertial terms are dominant, very different modes occur, as has been shown by Ericksen[11]. We have to compare $\mathfrak{I}(d\Omega/dt) \sim \mathfrak{I}\omega^2 \delta n$ to the nematic restoring torque $\mathbf{\Gamma} = \mathbf{n} \times \mathbf{h} \sim Kq^2 \delta n$. The moment of inertia per unit volume $\mathfrak{I}$ is of order ρa^2 where a is a molecular dimension.

Thus:

$$\Gamma^{-1}\mathfrak{I}(d\Omega/dt) \sim (\rho\omega^2 a^2/Kq^2).$$

Let us consider now more specifically the small μ limit:

For the fast mode ($\omega \sim \eta q^2/\rho$)

$$\Gamma^{-1}\mathfrak{I}(d\Omega/dt) \sim (\eta^2/K\rho)(qa)^2 \sim [(qa)^2/\mu].$$

For the slow mode ($\omega \sim Kq^2/\eta$)

$$(1/\Gamma)\mathfrak{I}(d\Omega/dt) \sim \mu (qa)^2.$$

We see that for the slow mode the inertial terms are always negligible. For the fast mode, they are negligible only if

$$qa < \mu^{1/2}.$$

Apart from the light scattering problem, the response functions χ which we have derived may be useful to analyze certain experiments where a nematic crystal is submitted to a small rf magnetic field H_1. To take a definite example, consider a slab of thickness d, and assume that the boundary conditions imposed by both walls (at $z=0$ and $z=d$) correspond to a director normal to the walls (parallel to Oz; see Fig. 2). The static field H_0 is also along z, and the equilibrium configuration corresponds to $\mathbf{n}_0$ parallel to the z axis. We add a small rf field $H_1 \exp(i\omega t)$ in the x direction. At the frequencies of interest the field H_1 may be taken as uniform in space. To H_1 we may associate a source

$$h_x^e = \chi_a(\mathbf{n}\cdot\mathbf{H})H_1 \cong \chi_a H_0 H_1. \qquad \text{(V.1)}$$

The resulting amplitudes n_x and v_x must vanish at the boundaries. It is then convenient to analyze H_1 in a Fourier series[12]

$$H_1 = (4/\pi)\sum_p [H_1/(2p+1)] \sin[(2p+1)\pi z/d]. \qquad \text{(V.2)}$$

Each Fourier component involves a response function

$$R_p(\omega) \equiv \chi_1\{q_\perp = 0, q_z = [(2p+1)\pi/d], \omega\}$$
$$\cong (K_{33}q_z^2 + \chi_a H_0^2)^{-1}[u_s/(u_s + i\omega)], \qquad \text{(V.3)}$$

where we have again taken the small μ limit. In terms of these functions we have

$$n_x = \chi_a H_0 H_1 \frac{4}{\pi} \sum_p \frac{R_p}{2p+1} \sin\left(\frac{(2p+1)\pi z}{d}\right). \qquad \text{(V.4)}$$

This shows that the amplitude n_x is comparable to its static value up to frequencies of order

$$u_s(\pi/d) \sim (1/\eta)[(K\pi^2/d^2) + \chi_a H_0^2]$$
$$(\text{for } H_0 < [(\pi\eta)/d](\chi_a \rho)^{-1/2})$$

This result is of a rather broad practical interest, since it tells us what are the frequencies up to which a nematic sample of dimensions d is able to follow an external perturbation. Unfortunately, these frequencies seem to be rather low: for $d = 10\ \mu$, $K = 10^{-6}$, $\eta = 10^{-1}$, H_0 small, we expect $u_s(\pi/d) \sim 100$ sec^{-1}.

Finally, we should mention briefly the limitations inherent to our approximation of an incompressible fluid: in actual fact, there will exist compressional waves, which cause a high frequency modulation of the dielectric constant: this gives rise to a Brillouin doublet in the scattered light, with frequencies $\omega = sq$, where the sound velocity s is in the range 10^4–10^5 cm/sec. Thus for $q = 10^4$ cm^{-1}, $\omega \gtrsim 10^8$ sec^{-1} and the doublet is well separated from the slow modes which we have discussed here.

ACKNOWLEDGMENTS

We wish to thank Dr. J. L. Ericksen and Dr. F. M. Leslie who kindly sent us a number of reprints and preprints concerning the fundamental hydrodynamic equations in liquid crystals.

[11] (a) J. L. Ericksen, J. Acoust. Soc. Am. (to be published); (b) Quart. J. Mech. Appl. Math. (to be published).

[12] H_1 is constant in space, but it is convenient (and permissible) to analyze it in a series of trigonometric functions, each of which vanishes at the boundaries.

Afterthought: Dynamics of fluctuations in nematic-liquid crystals

This mode analysis is useful, but one simplifying feature was not observed at this stage, and was discovered only later by O. Parodi [*J. Physique* **31**, 581 (1970)]. This is an Onsager relation between two off-diagonal friction coefficients, giving a relation between the Leslie coefficients:

$$\alpha_2 + \alpha_3 = \alpha_6 - \alpha_5 \ .$$

I asked Eriksen once why he (together with Leslie) did not automatically apply the Onsager relations in their original papers. The answer is, apparently, that Truesdell had some deep doubts concerning their applicability for unbounded media.

Note on the dynamics of prenematic fluids

J. Prost, P.G. de Gennes (unpublished)

Abstract : The order parameter of a nematic is a symmetric traceless tensor $Q_{\alpha\beta}$ defining the anisotropy of the magnetic susceptibility. In the isotropic phase, the average $< Q_{\alpha\beta} >$ vanishes, but anisotropy can be restored by flow. One of us constructed the relaxation equations coupling orientation rates, shear rates and the corresponding driving forces (1). A different, deeper presentation was given by P. C. Martin and coworkers (2). The purpose of this note is to show how one can switch from one presentation to the other.

The starting point is a free energy F ($Q_{\alpha\beta}$) depending on the orientation level $Q_{\alpha\beta}$: in the isotropic phase this has the structure :

$$F = \frac{1}{2} A(T) Q_{\alpha\beta} Q_{\alpha\beta} - H_\alpha H_\beta Q_{\alpha\beta} + \ldots \tag{1}$$

where H_α is the magnetic field. (Similar terms can be written for electric effects). The factor A (T) is positive but small. Thus (when $\underset{\sim}{H} = 0$), the average $< Q_{\alpha\beta} >$ vanishes, but the fluctuations $< Q^2 >$ are large.

In the original presentation (1) , the entropy source was written as :

$$T\dot{S} = \phi_{\alpha\beta} \dot{Q}_{\alpha\beta} + \frac{1}{2} \sigma_{\alpha\beta} e_{\alpha\beta} \tag{2}$$

where

$$\phi_{\alpha\beta} = - \frac{\partial F}{\partial Q_{\alpha\beta}} \tag{3}$$

is the force conjugate to Q, and $R_{\alpha\beta} \equiv \dot{Q}_{\alpha\beta}$ is the time derivative of Q.

$\sigma_{\alpha\beta}$ is the mechanical stress, and :

$$e_{\alpha\beta} = \partial_\alpha v_\beta + \partial_\beta v_\alpha \tag{4}$$

is the shear rate (v_α being the velocity) ; we assume incompressible flow ($e_{\alpha\alpha} = 0$).

The Onsager relations between flux and force then take the form :

$$\left.\begin{aligned} \frac{1}{2}\sigma_{\alpha\beta} &= \frac{1}{2}\eta\, e_{\alpha\beta} + \mu R_{\alpha\beta} \\ \phi_{\alpha\beta} &= \mu\, e_{\alpha\beta} + \gamma R_{\alpha\beta} \end{aligned}\right\} \tag{5}$$

where η is the viscosity in zero alignment ($Q \equiv 0$).

Eq. (5) gives a framework for the interpretation of many dynamical experiments in the isotropic phase :

- flow birefringence
- inelastic scattering of light
- shear wave attenuation
- transient Kerr effects

However, another decomposition of fluxes and forces has been introduced by Paul Martin and coworkers, and it gives more insight in the dissipation terms (2). The basic idea is that a deformation of the fluid pattern $\varepsilon_{\alpha\beta}$ (ie : of the molecular positions) causes automatically an alignment $Q^R_{\alpha\beta}$ where the index R stands for "reactive" :

$$Q^R_{\alpha\beta} = \psi\, e_{\alpha\beta} \tag{6}$$

where the constant ψ will be ascertained later.

Taking time derivatives, this introduces a reactive component in the alignment rate, proportional to the deformation rate :

$$R^R_{\alpha\beta} = \psi\, e_{\alpha\beta} \tag{7}$$

Comparing with eq (5) and imposing that $Q = Q^R$ when the alignment is equilibrated ($\phi = 0$), we see that :

$$\psi = -\frac{\mu}{\gamma} \tag{8}$$

For general situations we now write :

$$R_{\alpha\beta} = R^R_{\alpha\beta} + R^D_{\alpha\beta} \tag{9}$$

where R^D is the dissipative part. Then the entropy source (4.6) becomes :

$$T\dot{S} = \phi_{\alpha\beta} R^D_{\alpha\beta} + \frac{1}{2} \sigma^d_{\alpha\beta} e_{\alpha\beta} \tag{10}$$

where

$$\sigma^d_{\alpha\beta} = \sigma_{\alpha\beta} - \frac{2\mu}{\gamma} \phi_{\alpha\beta} \tag{11}$$

is the dissipative component of the stress.

With this choice of flux and forces, the constitutive relations become diagonal :

$$\left.\begin{aligned} R^D_{\alpha\beta} &= \gamma^{-1} \phi_{\alpha\beta} \\ \sigma^D_{\alpha\beta} &= \left(\eta - \frac{2\mu^2}{\gamma}\right) e_{\alpha\beta} \end{aligned}\right\} \tag{12}$$

References

(1) P.G. de Gennes, "Short range order in the isotropic phase of nematic", Mol. Cryst. 12, 193 (1971).

(2) P.C. Martin, O. Parodi, P.S. Pershan, Phys. Rev. A 6, 2401 (1972).

Solid State Communications, Vol. 10, pp. 753–756, 1972. Pergamon Press.

AN ANALOGY BETWEEN SUPERCONDUCTORS AND SMECTICS A

P.G. de Gennes

Laboratoire de Physique des Solides, Bât. 510, Faculté des Sciences Orsay 91

(*Received* 27 *January* 1972 *by P.G. de Gennes*)

The conformation of a smectic A can be described by a phase function $\phi(R)$, the n-th layer corresponding to $\phi(R) = 2\pi n$. The role of the phase in smectics A and in superfluids is similar. This analogy leads to the following predictions for a *second order* smectic A↔nematic transition: (1) the transition temperature is lowered if twist, or bend distortions are imposed: these distortions correspond to a magnetic field in superconductors. (2) the Frank coefficients K_2 and K_3 of the nematic phase must show pretransitional anomalies.

1. PRINCIPLES

SMECTICS A are layered systems:[1] the layers may be planar, or deformed into focal conic textures. But the interlayer distance d is essentially fixed. In terms of the director $\underset{\sim}{n}$ (a unit vector parallel to the local optical axis) this implies that the contour integral

$$\frac{1}{d} \oint \underset{\sim}{n}.\underset{\sim}{dl}$$

on a closed circuit, must be equal to 0 in a dislocation-free case, and to an integer ν in more general situations. This is the analog of flux quantization in a superfluid,[2] $\underset{\sim}{n}$ playing the role of the magnetic vector potential.

This remark becomes particularly useful if we have a second order smectic A↔nematic transition point T_c. (That such a transition may be of second order was noticed first by McMillan,[3] using a specific interaction model). Let us consider such a transition and start from a nematic phase aligned along the z axis ($\underset{\sim}{n} = \underset{\sim}{n}_0$). In terms of the Fourier components ρ_k of the density, the free energy may be expanded as:

$$F = F_0 + \frac{1}{2} \sum_k A(\underset{\sim}{k},T)\,|\rho_k|^2 \qquad (1)$$

The X-ray intensity for a scattering wave vector $\underset{\sim}{k}$ is then proportional to $T/A(\underset{\sim}{k},T)$. The function $A(\underset{\sim}{k})$ is a minimum at $\underset{\sim}{k} = \pm\, 2\pi/d.\underset{\sim}{n}_0$.

Near one of these points:

$$A\left(\underset{\sim}{k} \to \frac{2\pi}{d}\underset{\sim}{n}_0\right) = A(T) + \frac{1}{2M_v}(k_z - q_0)^2 + \frac{1}{2M_T}(k_x^2 + k_y^2) \qquad (2)$$

At the transition point T_c, $A(T_c) = 0$. In the vicinity of T_c, we may write the density $\rho(\underset{\sim}{R})$ as:

$$\rho(\underset{\sim}{R}) = 2^{-1/2} e^{2\pi i z/d}\, \psi(\underset{\sim}{R}) + \text{cc} \qquad (3)$$

where $\psi(\underset{\sim}{R})$ is a slowly varying function of $\underset{\sim}{R}$, and is complex. The phase of ψ indicates the position of the layers.[4] In terms of ψ, including terms up to order 4, the free energy[1] becomes:

$$F - F_0 = A(T)\,|\psi|^2 + \frac{1}{2} B|\psi|^4 + \nabla\psi : \frac{1}{2M} : \nabla\psi \qquad (1')$$

where $1/2M$ is the tensor defined in equation (2). Let us now include the fluctuations of the director $\underset{\sim}{n} = \underset{\sim}{n}_0 + \delta\underset{\sim}{n}$. F must be invariant if we rotate simultaneously $\underset{\sim}{n}$ and the layers. This imposes:

$$F - F_0 = A|\psi|^2 + \frac{1}{2} B|\psi|^4 + \left(\nabla + i\frac{2\pi}{d}\delta\underset{\sim}{n}\right)\psi^* : \frac{1}{2M} : \left(\nabla - i\frac{2\pi}{d}\delta\underset{\sim}{n}\right)\psi. \quad (4)$$

This has the Landau–Ginsburg form,[2] familiar for a charged superfluid: many effects which occur in these systems should have their counterpart in smectics A. We list a few of them below. Note that, for practical applications, equation (4) must be supplemented by the Frank elastic terms of the unperturbed nematic:[5]

$$F_{el} = \frac{1}{2} K_1(\text{div } \underset{\sim}{n})^2 + \frac{1}{2} K_2(\underset{\sim}{n} . \text{ curl } \underset{\sim}{n})^2 + \frac{1}{2} K_3(\underset{\sim}{n} . \text{ curl } \underset{\sim}{n})^2 \quad (5)$$

2. APPLICATIONS BELOW T_c

Neglecting for the moment the difference between M_v and M_T, we find two characteristic lengths:

(a) *a coherence length* $\xi(T) = (2MA)^{-1/2}$ giving the distance over which a local perturbation affects the amplitude of ψ: for instance, $\xi(T)$ is the core radius for dislocations in the smectic phase.

(b) *a penetration depth* $\lambda(T) = d/2\pi\ (MKB/A)^{1/2}$ describing the following gedanke experiment: at the free surface of the smectic, a *weak* twist or bend deformation is imposed (Fig. 1): the deformation penetrates only in a thickness λ (K is the appropriate elastic constant, K_2 for twist, K_3 for bend).

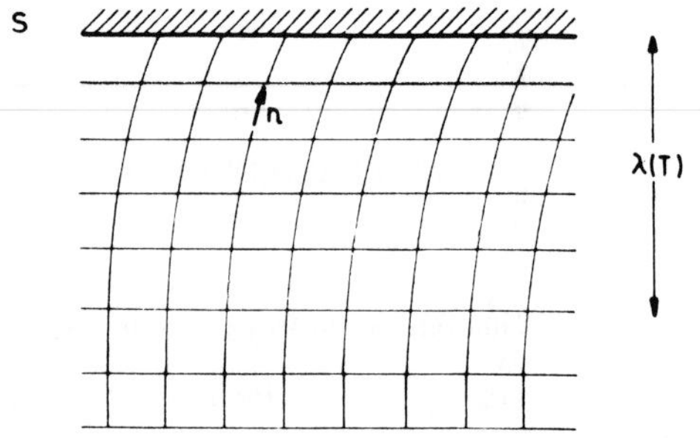

FIG. 1. Analog of the penetration depth for a smectic A. The layers are parallel to the free surface S. A weak bend deformation is imposed from S. The director $\underset{\sim}{n}$ (initially normal to S) is perturbed in a thickness $\lambda(T)$. A similar penetration depth may also be defined for twist deformations.

The ratio $\kappa = \lambda/\xi$ is temperature independent. κ controls the phase diagram under imposed deformations (Fig. 2)

(1) a splay deformation does not change T_c

(2) when $\underset{\sim}{h} = 2\pi/d$ curl $\delta\underset{\sim}{n}$ is non zero (i.e. under twist or bend), T_c is decreased. If $\kappa > 2^{-1/2}$, we expect a 'Shubnikov phase' (see reference 2) where the deformation is relaxed by a regular array of dislocations (Fig. 3).

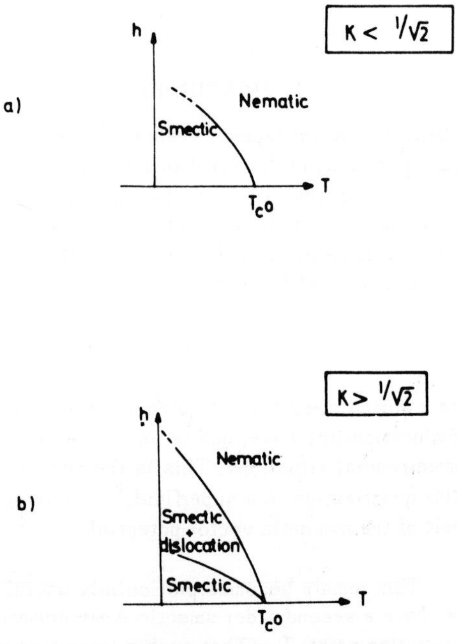

FIG. 2. Phase diagram near a second order smectic A ⟷ nematic transition point T_{c0}. h measures the amplitude of an imposed twist (or bend) deformation. Depending on the numerical value of the Landau–Ginsburg parameter κ one may have a 'type I' material [Fig. 2(a)] or a 'type II' material [Fig. 2(b)].

The (h,T) limiting curve is defined by:

$$A(T) \cong A'(T_c - T) = h/2M \quad (6)$$

where $M = M_T$ for twist ($\underset{\sim}{h}$ parallel to $\underset{\sim}{n}_0$) and $M = (M_T M_v)^{1/2}$ for bend ($\underset{\sim}{h}$ normal to $\underset{\sim}{n}_0$). Qualitatively, if the distortion described by $\underset{\sim}{h}$

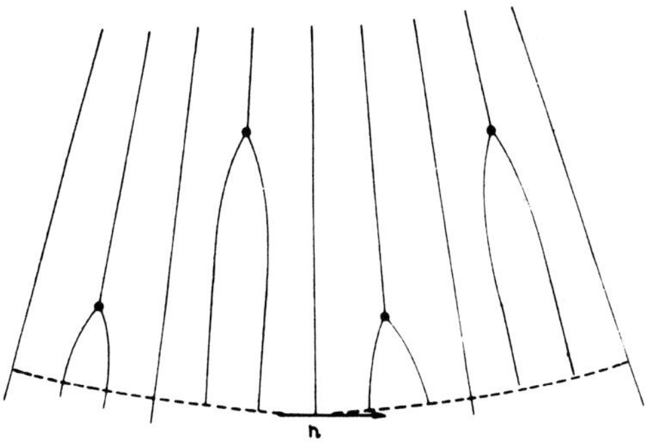

FIG. 3. The 'Shubnikov phase' of a smectic A with $K > 1/\sqrt{2}$ under an imposed bending deformation. The planes of the layers tend to remain equidistant. This is achieved by a regular network of edge dislocations. (The specific model chosen for the cores need not be correct). A similar network (with screw dislocations) is expected under twist.

takes place over a length L ($h \sim 1/dL$), equation (6) gives $T_c - T/T_c \sim d/L$. This may be easily measurable; instead of using a mechanical agent, twist may be imposed by starting from a cholectric phase.

If $\kappa < 2^{-1/2}$, we expect a first order transition under a finite h; the threshold curve corresponding to:

$$\frac{A^2}{2B} = \frac{1}{2} K_i (\text{curl } \underset{\sim}{n})^2 \qquad (7)$$

(where $K_i = K_2$ for twist, and K_3 for bend). For distortions weaker than the threshold,[7] and in suitable geometric conditions, it may be possible to obtain an 'intermediate state'[2] with coexisting nematic and smectic regions.

3. PRETRANSITIONAL EFFECTS IN THE NEMATIC PHASE

Above T_c, the order parameter ψ vanishes on the average, but it has significant fluctuations, leading to the 'cybotactic clusters' of de Vries:[6] These clusters have a size $\xi(T)$. They do not accept twist and bend, and thus give rise to an increase δK_i in the twist and bend elastic constants. A rise δK_3 has indeed been observed by Gruler[7] in certain nematics just above the smectic domain. The fluctuations of ψ, when incorporated in equation (4) give extra terms proportional to h^2 in the free energy: this is the analog of the fluctuation diamagnetism in superconductors above T_c, which has been calculated most simply by Schmid[8] in the limit of weak perturbations ($h\xi^2 \ll 1$). For the present problem, the new h^2 contributions represent a correction to the elastic constants. A transposition of the results of reference 7 gives:

$$K_2 = \frac{\pi k_B T}{6 M_T d^2} \left(\frac{M_v}{A(T)}\right)^{\frac{1}{2}} \qquad (8)$$

$$K_3 = \frac{\pi k_B T}{6 d^2} (M_v A(T))^{-1/2}$$

Thus, in the mean field approximation, δK_i should be proportional to $(T - T_c)^{-1/2}$. However, the mean field results are probably not very accurate for the smectic–nematic transition (while they are quite sufficient for superconductors). In fact, using the Wilson calculation of critical exponents,[9] applied to a complex order parameter ψ, one expects $A(T) \sim (T - T_c)^{\gamma}$ with $\gamma = 1.30$, and $\xi(T) \sim (T - T_c)^{-\gamma/2-\eta}$ with $\eta \sim 0.04$. Neglecting η and using simple scaling arguments, one is led to $\delta K \sim \xi(T) \sim (T - T_c)^{-0.66}$

Similarly, below T_c, the critical 'distortion fields' h defined in equations (6) and (7) will probably not be exactly linear in $T_c - T$, and will contain weak logarithmic operation.

Acknowledgements – This work was done at P.U.C (Rio de Janeiro) during the 4th Brazilian Symposium on theoretical physics; it is a pleasure to thank Professors E. Ferreira and R. Lobo for their hospitality on this occasion. The author is also indebted to P. Martin for various related discussions.

REFERENCES

1. General references on smectics A: FRIEDEL G., *Ann. Phys.* **18**, 273 (1922); DE GENNES P. G., *J. Phys.* **30**, Colloque C 4 (suppt to No. 11–12) P. C4–65 (1969).
2. General references on superconductors: LYNTON E. A., *Superconductivity*, 2nd ed., Methuen, London (1964); *Superconductivity*, ed. R. D. Parks, M. Dekker, N.Y. (1969). In particular Chap. 6 and 8.
3. McMILLAN W. L., *Phys. Rev.* **A4**, 1238 (1971).
4. The phase of ψ differs from the phase defined in the abstract by a term $2\pi z/d$.
5. FRANK F. C., *Disc. Faraday Soc.* **25**, 19 (1958).
6. DE VRIES A., *Molecular Crystals and Liquid Crystals*, **10**, 31 (1970); **10**, 219 (1970); **11**, 361 (1970).
7. GRULER H., Oral presentation at the Pont-à-Mousson meeting, June (1971).
8. SCHMID A., *Phys. Rev.* **180**, 527 (1969).
9. WILSON K. G., *Phys. Rev. Lett.* (to be published).

La conformation d'un spécimen de smectique A peut être décrite par une 'fonction phase' $\phi(\underset{\sim}{R})$, telle que la n^{eme} couche à $\phi(\underset{\sim}{R}) = 2n\pi$. Le rôle de la phase dans les smectiques A et dans les superfluides chargés est essentiellement le même. Cette analogie conduit aux prédictions suivantes, pour le voisinage d'une transition smectique A $\longleftrightarrow$ nématique du 2eme ordre: (1) la température de transition biasse si une distorsion de torsion ou de flexion est imposée: ces distortions sont l'analogue d'un champ magnétique pour un supraconducteur. (2) les coefficients de Frank K_2 et K_3 dans la phase nématique doivent montrer des anomalies prétransitionnelles.

Afterthought: An analogy between superconductors and smectics A

This analogy was found independently by the late W.L. McMillan — but he concentrated mostly on the phase transition properties — which were analysed in detail later by T. Lubensky and his coworkers.

Another interesting byproduct of the analogy is the invention of the *smectic A* phase:* a smectic A with chiral molecules. S. Renn and T. Lubensky [*Phys. Rev.* **A38**, 2132 (1988)] noticed that in type-II superconductors a magnetic field H can penetrate (as an array of quantized vortices). Here, twist is the analog of H; thus a twisted smectic, with an array of screw dislocations, could exist: it was indeed found soon after [J.W. Goodby *et al., J. Am. Chem. Soc.* **III**, 8119 (1989)].

Pramāṇa, Suppl. No. 1, 1975, pp. 1–21

Hydrodynamic properties of fluid lamellar phases of lipid/water

F BROCHARD* and P G DE GENNES**
* Laboratoire de Physique des Solides, Université de Paris–Sud, 91405 Orsay, France
** College de France, 11, Place Marcelin–Berthelot, 75231 Paris Cedex 05, France.

(Translated by G. Clairon)

Abstract. We investigate here the hydrodynamic equations and the low frequency collective modes of a fluid lamellar phase (lipid+water) of the L_α type. We find (1) the modes expected for smectics A, and in particular a second sound propagating with low velocities, (2) a mode of *slip* for the lipid layers. The latter should be observable by Rayleigh scattering, and might give information on the elasticity of the lipid layer. We also make an attempt to correlate the macroscopic parameters governing the modes to the microscopic interactions, and especially to the long range interactions between layers.

1. Introduction

A lipid molecule is composed of a hydrophilic group and one or several long chains of paraffinic hydrophobes. In the presence of water, such molecules display a variety of phases[1]. In all cases, the aliphatic chains are gathered in regions where water is excluded. We are concerned here with the '*lamellar*' phases and specifically with the phase L_α (according to the classification of ref. 2). In this phase, shown in figure 1, the paraffin chains are approximately liquid and there is no positional order within each lipid layer. Besides, we are assuming that there is no privileged direction in the plane of the layers (no cooperative tilt of the chains). Under these conditions, the symmetry of the phase L_α is that of a *smectic A*[3].

From a physicochemical point of view the (water/lipid) L_α phases can be classified into three groups, with slightly different properties[4].

(a) *Un-charged lipids:* Esters (monoglycerids) or molecules with compensated charge ('zwitter ionic') (lecithins). In this type of materials, the thickness h_E of the water layer is always small (from 5 to 25 Å).

(b) *Charged lipids with only one aliphatic chain* (*Common soaps*)*:* Here on account of the electrostatic repulsions (mainly proportional to

$1/h_E{}^5$) between loaded layers, one could have greater thickness of water h_E. However, in practice, $L\alpha$ phases with a large h_E are not observable in this group because they are in competition with other arrangements (rod phases, etc.) which are energetically preferable when the percentage of water is high.

(c) *Lipids charged with several chains : (Mitochondrial lipids and chloroplast lipids) :* Here, it appears that (by steric hindrance) the rod phases (etc.) are not favoured. Consequently, the phases $L\alpha$ remain observable even for strong percentages of water (h_E reaches up to 250 Å).

For a theoretical understanding, the case of a thick water layer is more attractive, since the forces and movements can be described by quasi-microscopic arguments. Unfortunately, the materials of group (c) are often poorly defined, heterogenous ; and not accessible to quantitative experimentation. Besides, when the water thickness is large the forces between the layers are very small and the ideal lamellar distribution of figure 1 is almost always disturbed by numerous structural defects. One is thus forced towards group (a) and (b) for which the microscopic analysis is more uncertain. This leads us to study the collective modes of the lamellar phase, based mainly on a *phenomenological analysis* independent of the detailed nature of the forces – (sections 2 and 3). However, when the thickness h_E is large, it is possible to express the phenomenological parameters in terms of simple microscopic parameters, such as the viscosity of the water. This will be analysed in section 4.

The elasticity and large wavelength oscillations of a classical smectic A have been investigated from a microscopic point of view in ref.[6, 7]. Let us recall here the fundamental aspects of this analysis.

(i) The *static* elasticity of a smectic A can be reduced to a Landau–Peierls description where only the unidimensional displacement u of layers (perpendicular to their plane) comes into play. The elastic

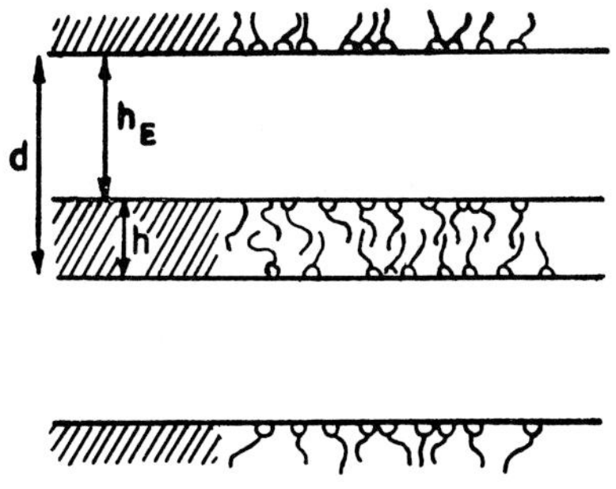

Figure 1 Relative distribution of molecules of lipid and water in the L_α phase. (The polar head of each lipid molecule is shown as a circle) (According to V. Luzzati ref.[1]).

energy has a 'mixed' structure where both the derivatives of order 1 and 2 appear

$$F = \tfrac{1}{2}\bar{B}\left(\frac{\partial u}{\partial z}\right)^2 + \tfrac{1}{2}K\left(\frac{\partial^2 u}{\partial x^2} + \frac{\partial^2 u}{\partial y^2}\right)^2 \tag{1}$$

(in which z is the normal to the layers).

This mixed elasticity has remarkable consequences : long range effects of distortion on the surface[8] ; abnormal structure of defects[9] ; instability under mechanical tension[10], etc. It will remain valid for the lamellar lipids.

(ii) The *dynamics*, takes into consideration not only the displacement u (with a dilatation of layers $\gamma = \partial u/\partial z$) but also the total change of the density (total dilatation θ). The variables γ and θ are coupled : they give rise to two distinct modes of acoustical propagation. One (first sound) is mainly a longitudinal density wave. The other (second sound) is an oscillation of layers with an almost fixed density. (The name 'second sound' is justified by a close analogy with superfluid helium[11]). The second sound has been recently observed in Brillouin scattering experiments[12], and also by sound transmission[13].

In the lamellar lipid phases, one also expects a second sound. A very rough estimation of the parameters (section 3) leads us to anticipate velocities S_2 on the order 10 m/sec. But the most original aspect of the system L_α is different : it is a system with two chemically independent components. To use the language of ref. 7 the number of 'conserved quantities' is increased by one. Comparing with ref. 6, one can state that the variables γ and θ are no longer sufficient to describe the local state of the system : it is necessary to know also the *local concentration of lipid c* or its shift from the value at equilibrium c_0. We shall write

$$\varepsilon_3 = \frac{c_0 - c}{c_0} \tag{2}$$

In order to build up the formal hydrodynamics of the phase L_α, it is the variable ε_3 which is the most convenient. On the other hand, in order to connect the hydrodynamical parameters to the concrete properties of the layer, it is often much easier to introduce, instead of ε_3, the relative variation of the surface per polar head A, that is

$$\delta = \frac{A - A_{eq}}{A_{eq}} \tag{3}$$

The quantities ε_3 are δ are linked by the relation

$$\varepsilon_3 = \gamma + \delta \tag{4}$$

We now analyze the coupled hydrodynamic modes of ε_3, θ and γ : we see that including ε_3 (or δ) leads to a new collective mode with low frequency in which the lipid layers are sliding with respect to the water (figure 2).

2. Principles of the dynamics

2.1. *Equations without friction*

Let us start by writing the form of the elastic energy E in a lamellar system with 2 chemical components, described by the variables θ (total dilatation), u (unidimensional displacement of layers) and δ (surface dilatation per polar head). The energy E (defined here by cm^3) comprises at first the quadratic terms with respect to 3 deformations

$$\begin{aligned} \theta &= \varepsilon_1 \\ \gamma &= \frac{\partial u}{\partial z} = \varepsilon_2 \\ \gamma + \delta &= \varepsilon_3 \end{aligned} \tag{5}$$

This contribution can be expressed by

$$E' = \tfrac{1}{2} \sum_{ij} C^{(0)}_{ij} \, \varepsilon_i \, \varepsilon_j \tag{6}$$

(As in ref. 6 we are utilizing the symbol $C^{(0)}$ for the adiabatic coefficients, and the symbol C for the isothermal coefficients). To the energy E', as in any smectic A, further terms associated with the curvature of the layers, should be added[6] :

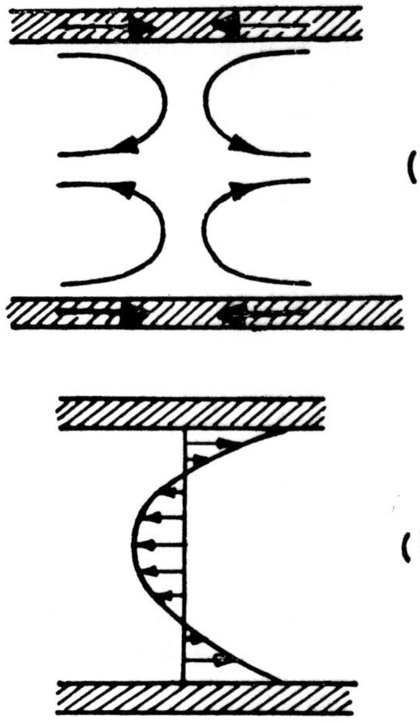

Figure 2 The mode of pure slip in the L_α phase (lipid regions hatched) (a) lines of flow (b) profile of the horizontal velocities.

$$E'' = \tfrac{1}{2} K \left(\frac{\partial^2 u}{\partial x^2} + \frac{\partial^2 u}{\partial y^2} \right)^2 \tag{7}$$

(Let us stress that for the coefficients K there is no difference between adiabatic and isothermal.) From the energy E one can define the main mechanical agents :

(a) the pressure

$$p = -\frac{\partial E}{\partial \theta} = -\sum_j C_{1j}^{(0)} \varepsilon_j \tag{8}$$

(b) the 'vertical' force (parallel to Oz) on the layers, defined as in a classical smectic A[7]

$$g = -\frac{\delta E}{\delta u} = \frac{\partial}{\partial z} \sum_j C_{2j}^{(0)} \varepsilon_j \tag{9}$$

(where the $\delta / \delta u$ is a functional derivative).

(c) the force on the lipid fraction, f

$$f = \nabla \left[\frac{\partial E}{\partial E_3} \right] = \nabla \sum_j C_{3j}^{(0)} \varepsilon_j \tag{10}$$

We can then write the equation of acceleration of the fluid under the form

$$\rho \dot{\mathbf{v}} = -\nabla p + f + g + \text{viscous terms} \tag{11}$$

where ρ represents the overall density, $\rho \dot{\mathbf{v}}$ is the mechanical momentum per 1 cm^3 of material and $\dot{\mathbf{v}} = \frac{\partial \mathbf{v}}{\partial t}$. (We are limiting ourselves to an approximation of small motions and we are neglecting therefore the difference between $\frac{\partial}{\partial t}$ and $\frac{d}{dt}$). To this equation of acceleration, one must add the equations of conservation for the total mass :

$$\dot{\theta} = \text{div } \mathbf{v} \tag{12}$$

and for the lipid mass :

$$\dot{\varepsilon}_3 = \dot{\gamma} + \dot{\delta} = \text{div } \mathbf{v}_L \tag{13}$$

where $\mathbf{v}_L$ is the velocity of the lipid fraction. For the case which is of interest to us here, $\mathbf{v}_L$ will differ from $\mathbf{v}$ mainly in the slip mode of figure 2.

2.2. *Scattering effects*

The structure of the viscous terms of equation 11 as well as the dissipation associated with *relative* flows ($\dot{u} - v_z$ or $\mathbf{v}_L - \mathbf{v}$), remain to be defined.

In the first place, it is convenient to rewrite the terms of friction in eq. 11 by introducing a system of viscous tensions:

$$\sigma'_{\alpha\beta} \quad (\alpha, \beta = x, y, z)$$
$$\rho \dot{v} \,|_{\text{viscous}} = \partial_\beta \; \sigma'_{\beta\alpha} \tag{14}$$

To analyze the scattering effects, one then calculates the entropy source $T\dot{S}$ by writing the evolution of the total energy from the acceleration equation. It is found that

$$T\dot{S} = \boldsymbol{\sigma}' : \nabla \mathrm{v} + g(\dot{u} - v_z), + \mathbf{f}.(\mathbf{v}_L - \mathbf{v}) + \mathbf{E}.\mathbf{J} \tag{15}$$

In the form (15) we have included all possible type of currents **J** and the conjugate fields **E**. For instance if **J** is a heat current, $E = -\nabla T/T$. However, in the major part of this work, we shall ignore thermal phenomena for simplicity.

Writing that the total dissipation remains unchanged, if one superposes to the flow studied a simple rotation (modifying in the same manner the three velocities $\mathbf{v}$, u and $\mathbf{v}_L$) one can show that the tensor σ' is *symmetrical.* Therefore eq. (15) involves only the symmetrical part of the velocity gradient tensor

$$\sigma' : \nabla v \rightarrow \sigma'_{\alpha\beta} A_{\alpha\beta} \tag{16}$$

with
$$A_{\alpha\beta} = \tfrac{1}{2}(\partial_\alpha v_\beta + \partial_\beta v_\alpha) \tag{17}$$

$$\partial_\alpha = \frac{\partial}{\partial x_\alpha}$$

Finally, from the entropy source (15), we can define fluxes $(A_{\alpha\beta}, \dot{u} - v_z, \mathbf{v}_L - \mathbf{v}, \mathbf{J})$ and forces $(\sigma_{\alpha\beta}, g, \mathbf{f}, \mathbf{E})$ in the thermodynamics of irreversible processes. In the domain of small fluxes, one can assume linear relations between fluxes and forces, with the following properties:

(a) The symmetry of the smectic phase A imposes that the pair (A, σ') is not coupled to the other variables. The relation between σ' and A is of the form *

$$\sigma'_{\alpha\beta} = A_{\gamma\delta} \, \eta_{\alpha\beta\gamma\delta} \tag{18}$$

where $\eta_{\alpha\beta\gamma\delta}$ is a viscosity matrix, explicitly stated in ref. 7 and which involves five independent parameters. Later on we will write this in a simplified form.

(b) The forces and fluxes normal to the plane of the layers are themselves coupled. We shall write these couplings without including the pair (E, J). Then one has

* Summation over repeated indices is assumed throughout this paper.

$$g = L_{22} (\dot{u} - v_z) + L_{23} (v_{Lz} - v_z) \quad (19)$$

$$f_z = L_{32} (\dot{u} - v_z) + L_{33} (v_{Lz} - v_z) \quad (20)$$

with $L_{23} = L_{32}$.

The coefficient L_{22} describes an effect of *permeation,* present even in a system with only one chemical component, and considered first by Helfrich[14]. The coefficient L_{33} would describe an effect of diffusion of the lipid molecules from one layer to the other. The coefficient L_{23} connects these two mechanisms. But, in fact, for the lipids which interest us, all these processes are extremely weak, on account of the small solubility of water in the fat regions and vice versa. For this reason, in the rest of this work, *we will completely neglect the permeation and diffusion normal to the layers.* This amounts to making the coefficients L tend to infinity and therefore taking

$$\dot{u} = v_z$$

$$v_{Lz} = v_z \quad (21)$$

(c) For the component $\mathbf{f}_\perp$ of force $\mathbf{f}$, projected on the plane of the layers, we have another simple phenomenological relation : *

$$\mu \mathbf{f}_\perp = \mathbf{v}_{L\perp} - \mathbf{v}_\perp \quad (22)$$

The relative migration of the lipid with respect to water can be done by slipping, without overcoming any potential barrier : $\mathbf{v}_L$ can easily be different from $\mathbf{v}_\perp$. We shall call *the coefficient of slippage* μ.

2.3. *The incompressible limit*

In the thermotropic smectics A, (with only one chemical component) it is known by direct measurements[12] that the second sound is *very slow* compared to the first one : these two modes are then nearly uncoupled, and it becomes possible to calculate the properties of the second sound by imposing from the beginning $\theta = 0$, *i.e.*, by considering the substance as incompressible[15]. The lipid–water systems have interactions between layers which are looser than the thermotropic smectics A : therefore this approximation is surely excellent. It also simplifies the calculation of the modes. The consequences are the following :

(a) the elastic energy (2) is reduced to a form with two variables. Here we find it more convenient to select as independent variables γ and δ (rather than γ and ε_3) and we write :

$$E' \rightarrow \tfrac{1}{2} D^0_{22} \gamma^2 + D^0_{23} \gamma\delta + \tfrac{1}{2} D^0_{33} \delta^2 \quad (23)$$

The corresponding free energy F' is obtained by replacing the coefficient D^0 by a similar set D. The coefficient $\overline{B}$ which comes in the static

*For any vector, $\mathbf{v}$, the symbol $\mathbf{v}_\perp$ represents the component in the plane of the layers.

description (eq. 1) is obtained by minimising F' as regard to δ (for γ fixed. It is equal to

$$\bar{B} = D_{22} - (D_{23}^2/D_{33}) \tag{24}$$

The stability of the system imposes the inequalities :

$$\bar{B} > 0 , \quad D_{22} > 0 , \quad D_{33} > 0 \tag{25}$$

(b) From a dynamic point of view the incompressibility imposes :

$$\operatorname{div} \mathbf{v} = 0 \tag{26}$$

The pressure p is no longer defined by (8), but becomes an unknown function, to be determined in the process of calculating, by condition (26). In the limit (21) of no permeation, the equation of conservation of the lipid system can be rewritten in the form :

$$\operatorname{div} \mathbf{v}_{L\perp} = \dot{\delta} \tag{27}$$

In the incompressible regime, the viscous tensions σ' are simplified : eq. (18) takes the explicit form

$$\left.\begin{aligned} \sigma'_{zz} &= 2\eta_V \frac{\partial v_z}{\partial z} \\ \sigma'_{\alpha z} &= \eta_M (\partial_\alpha v_z + \partial_z v_\alpha) \\ \sigma'_{\alpha\beta} &= \eta_T (\partial_\alpha v_\beta + \partial_\beta v_\alpha) \end{aligned}\right\} \; (\alpha, \beta \text{ restricted to } x \text{ or } y) \tag{28}$$

A priori there is still the possibility of adding to $\sigma'_{\alpha\beta}$ a diagonal term

$$\sigma'_{xx} = \sigma'_{yy} = \sigma'_{zz} = \text{const} \frac{\partial v_z}{\partial z} \tag{29}$$

But such a diagonal term can always be incorporated into the pressure p, and must therefore be omitted.

Let us finish this section by a remark on the change of variables carried out in the equation (23) : *i.e.*, in passing from the *volume* dilatation ε_3 of the lipid fraction to the *bidimensional* dilatation (in a layer) δ.

This change of variable modifies the definition of forces. With ε_3 and γ one had the forces

$$\begin{aligned} \mathbf{f} &= \nabla \left(\frac{\partial E}{\partial \varepsilon_3} \right)_\gamma \\ g &= \frac{\partial}{\partial z} \left(\frac{\partial E}{\partial \gamma} \right)_{\varepsilon_3} \end{aligned} \tag{30}$$

with δ and γ as variables, one has new forces $\mathbf{f}'$, g'

$$f'_z = 0$$

$$\mathbf{f}'_\perp = \nabla_\perp \left(\frac{\partial E}{\partial \delta}\right)_\gamma$$

$$g' = \frac{\partial}{\partial z}\left(\frac{\partial E}{\partial \gamma}\right)_\delta \qquad (31)$$

Writing $\varepsilon_3 = \gamma + \delta$, one can show from (31) that

$$f'_\perp = f_\perp$$

$$g' = g + f_z \qquad (32)$$

That is to say that all the physical observables are unchanged, as they must be : the two descriptions are indeed equivalent.

3. Collective modes of low frequency

3.1. *Simplified calculation of oblique modes*

In order to simplify further the discussion of the modes, we are going to make an additional approximation : we will completely neglect the effect of viscous tensions σ'. However, we will still consider the slip coefficient μ. This procedure may seem unorthodox since we are thus isolating a friction mechanism among others. It is in fact justified for the following reasons :

(a) The *second sound* modes are but weakly attenuated in the limit of very small vectors q : our approximation bears only on these small attenuation effects.

(b) As for the mode of *slip* shown in figure 2 the shearing stress described by the parameter μ takes place within *one* layer, and is much more important than the shearing stresses on the scale of the wavelength $2\pi/q$ which are described by the viscous tensions σ'.

This being admitted, let us find the modes for which the amplitudes vary like exp ($i\mathbf{q}.\mathbf{r} + i\omega t$) and let us assume that $q_x \neq 0$, $q_z \neq 0$ but $q_y = 0$. Let us first study the following modes

$$v_z \neq 0 \qquad v_x \neq 0 \qquad v_y = 0$$

which are more interesting to us. The equation of acceleration (11), (free from the viscous tensions) taken in the form of equation (31) gives now the following :

$$\left.\begin{aligned} i\omega\rho v_z &= -iq_z p + g' \\ i\omega\rho v_x &= -iq_x p + f_x \end{aligned}\right\} \qquad (33)$$

with

$$\left.\begin{aligned} g' &= iq_z\,[D^0_{22}\gamma + D^0_{23}\delta] \\ f_x &= iq_x\,[D^0_{23}\gamma + D^0_{33}\delta] \end{aligned}\right\} \qquad (34)$$

(Let us stress that the curvature terms obtained from equation (7) are negligible in g' except for a particular case with which we shall deal later on).

While writing

$$\text{div } \mathbf{v} = i(q_x v_x + q_z v_z) = 0$$

we obtain

$$\nabla^2 p = -q^2 p = i(q_z g' + q_x f_x) \tag{35}$$

We can eliminate p from the equation (33) and get

$$i\omega \frac{\partial v_z}{\partial z} = -\omega^2 \gamma = -\frac{q_z^2 q_x^2}{q^2 \rho} [\gamma (D_{22}^0 - D_{23}^0) + \delta (D_{23}^0 - D_{33}^0)] \tag{36}$$

Now using the slip equation (22) and the conservation laws (27) and (26), we are led to

$$\begin{aligned}\dot{\delta} &= \frac{\partial v_{Lx}}{\partial x} = \frac{\partial v_x}{\partial x} + \mu \frac{\partial f_x}{\partial x} \\ &= -\frac{\partial v_z}{\partial z} + \mu \frac{\partial f_x}{\partial x} \\ &= -\dot{\gamma} + \mu \frac{\partial f_x}{\partial x}\end{aligned} \tag{37}$$

or

$$(i\omega + \mu D_{33}^0 q_x^2)\,\delta + (i\omega + \mu D_{23}^0 q_x^2)\,\gamma = 0 \tag{38}$$

Equations (36) and (38) form a linear homogenous system for δ and γ. The solution is non-vanishing when the determinant Δ is equal to 0:

$$\Delta = (\omega^2 - \omega_0^2)(i\omega + \mu D_{33}^0 q_x^2) - \omega_1^2 (i\omega + \mu D_{23}^0 q_x^2) = 0 \tag{39}$$

We have defined the following auxiliary frequencies

$$\left.\begin{aligned}\omega_0 &= \frac{q_z q_x}{q}\left[\frac{D_{22}^0 - D_{23}^0}{\rho}\right]^{\frac{1}{2}} \\ \omega_1 &= \frac{q_z q_x}{q}\left[\frac{D_{33}^0 - D_{23}^0}{\rho}\right]^{\frac{1}{2}}\end{aligned}\right\} \tag{40}$$

We assume for the moment that q_x and q_z are both not equal to 0 and are comparable (oblique modes) so that $\omega_0 \neq 0$ and $\omega_1 \neq 0$. The roots of equation (39) are separated into two well defined groups:

(a) *Modes of second sound* (ω relatively large)

$$\omega_2^2 \simeq \omega_0^2 + \omega_1^2 = \frac{q_x^2 q_z^2}{q^2} \frac{D_{22}^0 + D_{33}^0 - 2D_{23}^0}{\rho} \tag{41}$$

The form of the dispersion relation (41) is exactly the same as for a classical smectic A. (We will discuss the orders of magnitude for the velocities in section 4). In the mode (41) one has practically $\delta = -\gamma$, that is to say $\varepsilon_3 = 0$ (the lipid concentration remains unchanged).

(b) *Modes of slip* (ω small)

$$-i\omega_{g1} = \mu q_x^2 \frac{D_{22}^0 D_{23}^0 - D_{23}^{0\,2}}{D_{22}^0 + D_{33}^0 - 2 D_{23}^0} \tag{42}$$

This new mode is completely damped (ω is purely imaginary). In this mode the two velocities $v_\perp$ and v_x are different. The pressure p balances the horizontal tensions

$$p = \frac{\partial E}{\partial \delta}$$

The pressure p balances also the vertical tensions

$$p = \frac{\partial E}{\partial \gamma}$$

The tensions are totally balanced and the terms of acceleration $\rho\dot{\mathbf{v}}$ in the equation of motion are negligible. The equality

$$\frac{\partial E}{\partial \delta} = \frac{\partial E}{\partial \gamma}$$

gives the ratio of the amplitudes :

$$\frac{\gamma}{\delta} = \frac{D_{33}^0 - D_{23}^0}{D_{22}^0 - D_{23}^0} \tag{43}$$

It is interesting to note that $\gamma / \delta \neq 0$ even if the coefficient of coupling D_{23}^0 is equal to 0. A slip (associated with δ) and a swelling (associated with γ) are always coupled by means of the pressure p. Let us point out finally that the mode of slip is probably isothermal : therefore the coefficient D^0 in (42) and (43) should be replaced by the coefficients D.

Besides these modes in which v_x and v_z are not zero which we have just classified, we *must* mention another mode of different polarisation

$$v_y \neq 0$$
$$v_x = v_z = 0$$

With our approximation of zero viscous tensions these modes appear static ($\omega = 0$). A more detailed calculation including viscous tensions leads to a hydrodynamic transversal mode which is expected in all the smectics A and which has the following dispersion relation :

$$-i\omega = \frac{\eta_M q_z^2 + \eta_T q_x^2}{\rho} \tag{44}$$

This mode gives practically no contribution to the scattering of light and is therefore of small interest.

3.2. *Inclusion of damping effects*

Let us come back now to the slip modes with v_x and $v_z \neq 0$ and let us include the viscous tensions $\sigma_{\alpha\beta}$ which had been omitted in the previous paragraph. They are defined in the equation (28). Writing once more the complete equation of acceleration (11) and eliminating the pressure by div $\mathbf{v} = 0$, one is led to the following simple relation

$$[i\omega\,(i\omega + \tilde{\nu} q^2) + \omega_0^2]\,\gamma - \omega_1^2\,\delta = 0 \tag{45}$$

in which ω_0 and ω_1 remain defined by equation (39). The quantity $\tilde{\nu}$ is an effective kinematic viscosity depending on the orientation of the vector $\mathbf{q}$, and defined explicitly by :

$$\rho\tilde{\nu} = q^{-4}\,[\eta_M\,(q_z^4 + q_x^4) + 2\,(\eta_V - \eta_M + \eta_T)\,q_x^2\,q_z^2] \tag{46}$$

(Naturally if by chance the three viscosities η_V, η_M and η_T become equal, $\rho\tilde{\nu}$ coincides with their common value).

The equation of slip (38) remains unchanged : the determinant obtained from (38) and (45) is of the following type

$$\begin{aligned}\Delta' &= [\omega_0^2 - \omega^2 + i\omega\tilde{\nu}\,q^2]\,[i\omega + \mu\,D_{33}^0\,q_x^2] \\ &\quad + \omega_1^2\,(i\omega + \mu\,D_{23}^0\,q_x^2) = 0\end{aligned} \tag{47}$$

The discussion of the roots of Δ' is not significantly modified when $\mathbf{q}$ is *small* and *oblique*

(a) For the *second sound* one has now a slightly damped frequency

$$\omega = \omega_2 + \tfrac{1}{2}i\,[\tilde{\nu} q^2 + \frac{\omega_1^2}{\omega_2^2}\,(D_{33}^0 - D_{23}^0)\,\mu\,q_x^2]$$

$$\omega = \omega_2 + \tfrac{1}{2}\,[\tilde{\nu} q^2 + \frac{(D_{33}^0 - D_{23}^0)^2}{D_{33}^0 + D_{22}^0 - 2D_{23}^0}\,\mu\,q_x^2] \tag{48}$$

The damping contains two contributions. One is due to the macroscopic shearing stress of the material and is described by $\tilde{\nu}$. The other results from the slip and depends only on q_x^2. Detailed studies of the damping as a function of the orientation of $\mathbf{q}$ should in principle give all the coefficients of friction which we have introduced (η_V, η_M, η_T and μ).

(b) For the mode of *slip* which takes place at a lower frequency one easily verifies that the damping caused by the viscosities η does *not modify* the law of relaxation (42) as one considers only the lowest order in q.

3.3. *Case where the wavefronts are perpendicular to the layers*

Let us examine now the particular case $q_z = 0$. Then the frequencies ω_0 and ω_1 defined by equation (40) vanish, and (in the approximation without viscous tensions) one is led to equation (39) with a double root $\omega = 0$ plus a simple root $\omega \neq 0$. We are going to discuss separately the meaning of these three roots :

(a) One of the roots $\omega \simeq 0$ corresponds to a mode of *pure undulation*[6, 7] as shown in figure 3. In order to determine a more accurate value of ω we must include the terms of curvature occuring in the elastic energy (eq. 6) and also the viscous tensions. We obtain then

$$-i\omega_{\rm und} = \frac{Kq_x^2}{\eta_{\rm M}} \qquad (49)$$

In this mode $\varepsilon_3 \simeq 0$ (the lipid concentration is not modulated). The mode of undulation must make an important contribution to the scattering of light[6]. In practice it is often masked by *static* undulations existing in the material and mainly due to the irregularities of their limiting surface[8, 23].

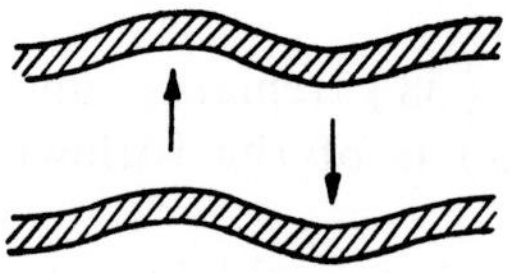

Figure 3 Pure undulation mode : the distance between layers and concentration in lipid are not modulated.

(b) The second root $\omega \simeq 0$ is associated with *simple shearing* (v_z varying with x) analogous to the transversal mode of isotropic fluids. Here also to calculate correctly the relation of dispersion we must take into consideration the viscous tensions : the result being the following :

$$-i\omega_{\rm shear} = \frac{\eta_{\rm M}\, q_x^2}{\rho} \qquad (50)$$

This mode occurs at a higher frequency than the previous one and is probably difficult to observe in *light scattering*.

(c) The third root which appears directly while making $\omega_0 = \omega_1 = 0$ in equation (39) describes the pure mode of *slip* in figure 2. Here the total velocity $\mathbf{v}$ vanishes, but the relative velocity of the lipid $\mathbf{v}_L$ is not equal to zero. The dispersion relation is

$$-i\omega_{\rm slip} = \mu\, D_{33}^0\, q_x^2 \qquad (51)$$

Let us note that (51) differs from (42): this special behaviour of the modes $q_z = 0$ is a constant property of lamellar structures. The frequencies of slip and of undulation are probably comparable.

From an experimental point of view the mode (51) ought to have an intensity comparable to that of Brillouin doublet in oil — that is to say, much weaker than the intensity due to a mode of a pure undulation. It is in principle possible to eliminate the latter by a suitable selection of polarisation[6–8] but it is not certain that this precaution will be sufficient. It is, therefore, preferable to study the mode of slip by utilizing *oblique* wave vectors **q**: the slip must then give a Rayleigh scattering well separated in frequency from the second sound (figure 4).

After the case $q_z = 0$ we ought to discuss the other special case $q_x = 0$. But the behaviour for $q_x = 0$ is controlled by the *permeation* process[7]. This seems difficult to control in a phase L_α: the migration of molecules through the layers will always be facilitated by defects in the structure (dislocations, focal conics, ...). For this reason the modes $q_x = 0$ do not deserve special attention at the moment.

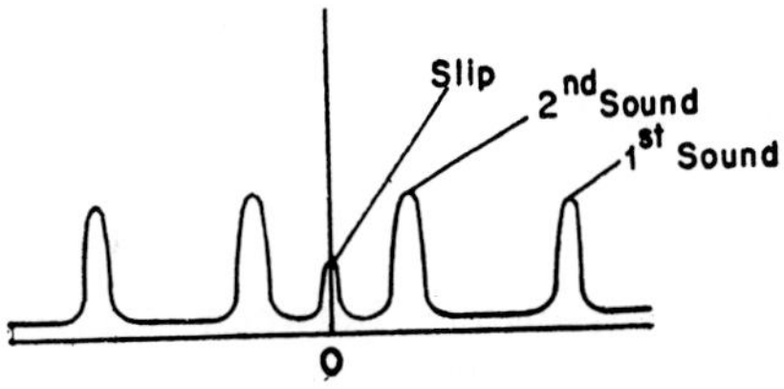

Figure 4 Spectrum of inelastic scattering of light from the L_α phase with the scattering wave vector q oblique with respect to the layers. The central Rayleigh component would be caused mainly by slip.

4. The water-rich phases

As explained in the introduction, the case of lamellar systems L_α with a large water thickness h_E is difficult to study experimentally, but interesting theoretically:

(a) the lipid layers are only weakly coupled, hence the elastic parameters will be much smaller than those obtained for thermotropic smectics.

(b) the dominant interactions are more or less independent of the details of the molecular structure.

(c) the mechanical behaviour of water is (probably) well described by the viscosity η_E of bulk water.

Therefore, we are going now to concentrate on the limit of large h_E, and try to connect the macroscopic coefficients introduced in section 2 to some microscopic parameters.

4.1. *Energies of interaction and static elasticity*

We are going to limit ourselves to the case of charged lipids to which *pure* water has been added (no salt dissolved); this is the only case where h_E is large. The energy of the electrostatic repulsion between layers has been analysed long ago[5]: for large h_E it takes a form totally independent from the charge density (due to the polar heads on one interface). For two opposite lipid layers one gets (per polar head)

$$F_{el} = \frac{QA}{h_E} \sim \frac{QA}{d} \tag{52}$$

$$Q = \frac{\pi D}{4}\left(\frac{k_B T}{e}\right)^2 \tag{53}$$

where A is the surface per polar head, D the dielectric constant of pure water and e the charge of the counter–ions. Instead of the thickness d of a layer ($d = h_E + h_L \sim h_E$) it is sometimes convenient to take into consideration the total concentration in lipid c (molecules / cm^3). Since each lipid layer has two surfaces one can write

$$c\,A\,d = 2 \tag{54}$$

and for the electrostatic energy

$$F_{el} = \tfrac{1}{2}\,Q\,c\,A^2 \tag{55}$$

The Van der Waals attractions between layers give a contribution (per head) of the following form[16]:

$$F_{vw} = \frac{-MA}{12\pi d^2} = -\frac{M}{48\pi}\frac{n^2}{A} \tag{56}$$

The quantity M has the dimensions of energy. When d is very small compared to the ultraviolet wavelengths of absorption of the material, M is independent of d. For greater values of d, M decreases according to a complex law described in ref.[16]. A typical value of M is 10^{-14} ergs for $d = 100$ Å.

Finally, as Parsegian[5] has noted very early, one must include the interfacial energy lipid–water. In order to discuss it let us first define a surface A_0 per polar head which is the equilibrium surface for a double lipid layer immersed in a very great thickness of water ($d \rightarrow \infty$). This surface is realized in a Mueller membrane[17]. For the values of d which are of interest to us the surface A will be slightly different from A_0. It is then correct to write the *interfacial* energy under the simplified form (per polar head)

$$F_{int} = \text{constant} + \tfrac{1}{2}\,\chi\,(A - A_0)^2 \tag{57}$$

Here χ is a coefficient of compressibility (of the order of 10^{16} CGS). The form (57) ensures that $A \rightarrow A_0$ in equilibrium whereas the other energies (F_{el} and F_{VW}) are negligible ($d \rightarrow \infty$).

While minimising $F_{el} + F_{VW} + F_{int}$ for a concentration c well defined and fixed, one can calculate the surface of the equilibrium A_{eq}(b). The resulting law is usually complicated, but it is simplified within the limit for small c (d large) because when d increases the Van der Waals interactions are decreasing much quicker, than the electrostatic interactions. Therefore, one gets

$$A_{eq} = \frac{A_0}{1 + QC/\chi} \tag{58}$$

This simple law gives a reasonable fit to the classical measurements by Skoulios[1] on soaps (for which however the thickness h_E is not very great.)

We are now going to study the fluctuations around the equilibrium value (58) and around the average thickness d, to calculate the *isothermal* rigidities D_{22}, D_{23}, D_{33} defined in analogy with the equation (23). For that, we have to relate the variables A and d to the microscopic dilatations δ and γ

$$\left.\begin{aligned} \delta &= \frac{A - A_{eq}}{A_{eq}} \\ \gamma &= \frac{d - d_{eq}}{d_{eq}} \end{aligned}\right\} \tag{59}$$

After two differentiations of the form $F_{el} + F_{VW} + F_{int}$, we obtain the following equations :

(a) For the coefficient of ' vertical rigidity ' of the layers

$$D_{22} = \frac{12Q}{d^2} - \frac{4Q}{d^2}\,\frac{A_{eq} - A_0}{A_{eq}} - \frac{M}{\pi d^3} \tag{60}$$

$$D_{22} \cong \frac{12Q}{d^2} \quad (d \text{ large}) \tag{61}$$

This corresponds to values $D_{22} \sim 10^5$ CGS for $d = 100$ Å .

(b) For the coefficient of ' horizontal elasticity '

$$D_{33} = \chi\, c\, A_0^2 \tag{62}$$

Typically $D_{33} \sim 10^7$ CGS.

(c) For the cross coefficient

$$D_{23} = \frac{2Q}{d^2}\left(1 + \frac{A_0}{A_{eq}}\right) \sim \frac{4Q}{d_2} \tag{63}$$

$$D_{23} \sim \tfrac{1}{3} D_{22} \tag{64}$$

Finally, we have to estimate the coefficient of the *energy* of *curvature* defined by (7). This term is due mainly to the lipid regions.

One can therefore, write

$$K = k\frac{1}{d}$$

in which k is a characteristic coefficient for the energy of distortion in an isolated double layer. This coefficient has been discussed recently by Helfrich[18] in connection with the elastic properties of vesicules. He estimates $k \sim 5 \cdot 10^{-13}$ ergs.

Finally, from these order of values we can try to draw a few conclusions regarding parameter B (defined by eq. 24) and the characteristic length $\lambda = (\bar{B}/K)^{\frac{1}{2}}$. These two parameters are those which control the whole static elasticity (refer equation 1). They could be measured by experiments of the type described in reference[8]. Our estimate (very approximate), for $d = 100$ Å, is $\bar{B} \sim 10^5$ CGS and $\lambda \sim 100$ Å. If we ignore the difference between adiabatic coefficients and isothermal coefficients this leads us to second sound velocities (eq. 41) on the order of 10 m/sec.

4.2. *Coefficients of friction*

The most interesting factor here is the coefficient of *slip* μ. One can obtain it simply by examining a case of uniform slip where water flows parallel to the lipid layers (figure 2b). It is important to note that there can be no velocity within the lipid region : 'tails' cannot have a velocity different from the 'heads' and the 'heads' on the two sides of a lipid layer have the same velocity for the mode under consideration.

Let us use a reference frame where the lipid layers are at rest ($v_L = 0$). The profile of the velocity in the aqueous region has the Poiseuille form :

$$v(z) = -a\left(z^2 - \tfrac{1}{4}h_E^2\right) \tag{65}$$

for which we have taken the origin of all the z's at the middle plane of the aqueous region. The constant in equation (65) ensures that $v = v_L = 0$ at the lipid–water interface ($z = \pm h_E/2$). The average water velocity associated with the profile (65) is

$$v_E = \frac{1}{h_E}\int_{-h_E/2}^{h_E/2} dz\, v(z) = \frac{a\, h_E^2}{6} \tag{66}$$

The hydrodynamic velocity v is the weighted average

$$v = \frac{h_E\, \rho_E\, v_E + h_L\, \rho_L\, v_L}{h_E\, \rho_E + h_L\, \rho_L} \tag{67}$$

where ρ_E and ρ_L are the density in the two phases. Here, since $v_L = 0$, we have

$$v = \frac{1}{6}\,\frac{a\, \rho_E\, h_E^3}{\rho_E\, h_E + \rho_L\, h_L} \tag{68}$$

and the equation which defines a as a function of v. Besides the force per cm^2 on an lipid–water interface is

$$\eta_E \left. \frac{\partial v}{\partial z} \right|_{z = h_E/2} = - \eta_E \, a \, h_E \tag{69}$$

The force on a lipid layer is twice as large (because there are two sides) and the force per cm^3 is :

$$f = \frac{-2a \, \eta_E \, h_E}{d} \tag{70}$$

We introduce in equation (70) the value of a, which has been inferred from (68) and we compare the result with the definition of μ (eq. 22) :

$$\mu f = v_L - v = - v \tag{71}$$

The result is

$$\mu = \frac{1}{12} \frac{\rho_E \, h_E}{\rho_E \, h_E + \rho_L \, h_L} \frac{h_E . d}{\eta_E} \tag{72}$$

In the limit $h_E \gg h_L$ this reduces to $\mu = \dfrac{d^2}{12 \, \eta_E}$

In view of the difficulty of the problem we have not attempted a systematic study of the macroscopic viscosities η_V, η_T, η_M as functions the microscopic parameters. However, one can think that the coefficient η_T must contain an important contribution due to the lipid fraction : η_T is associated to distortions for which v_x, for instance, varies along y ; these distortions are also involved in the surface viscosities of a monomolecular film[19]. One expects, therefore, a coefficient η_T of the following type

$$\eta_T = \frac{h_L \, \eta_L + h_E \, \eta_E}{d} \tag{73}$$

in which η_L is a high viscosity. As for the coefficient η_M, which is related to a relative slip of successive layers, one is tempted to infer it from an equation of the following form :

$$\frac{d}{\eta_M} = \eta_E \frac{h_E}{\eta_E} + \frac{h_L}{\tilde{\eta}_L} \tag{74}$$

in which $\tilde{\eta}_L$ is another viscosity of the lipid medium associated to a relative slip of the two halves of the hydrophobic layer and is probably, rather large. One would get then

$$\eta_M = \eta_E \frac{d}{h_E} \simeq \eta_E \tag{75}$$

Finally the viscosity η_V probably involves complex processes (connected perhaps to the change of thickness of the lipid double layer when its surface per polar head changes) and we are not able to make predictions on η_V.

It is important to note finally that the components of the viscosities which are associated with the lipid (for instance $\tilde{\eta}_L$ and η_L in the above discussion probably show *a strong dependence on frequencies :* the lipid has long relaxation times. Some of these relaxation times (going up to 10^{-5} sec) are already known through nuclear relaxation[20] : it would be very interesting to compare this information with mechanical measurements (second sound and first sound) determining $\eta(\omega)$.

5. Conclusions

The observations of the collective modes in a lamellar phase L_α could give us some information regarding the interactions between layers and also on specific dissipative processes of the lipid region.

The second sound may be difficult to observe because of the viscosities which are introduced by the lipidic component : In order to have narrow lines, one should, in principle, utilise very small values of q, that is to say, develop a special technique of Brillouin scattering with small angles and low frequencies. However it is also possible that the classical Brillouin measurements (high $\omega \sim 10^{10}$ sec^{-1}) may give an acceptable spectrum if, for these high frequencies, the lipid behaves just like an elastic solid (the viscosity $\eta_L(\omega)$ becoming purely imaginary).

The mode of slip could be detected in Rayleigh scattering with oblique wave vectors. (This is advantageous since one does not need a very well oriented specimen). Incidentally this mode must exist not only in a lamellar phase, but also in hexagonal phases.

One should insist on the fact that our discussion of the orders of magnitude in Section 4 is very brief. For instance, we have estimated only the isothermal elastic coefficients and not the adiabatic coefficients which are taken into consideration for the second sound. We have poor knowledge of the friction coefficients except (fortunately) for the parameter μ. We have also ignored all the phenomena which appear outside the macroscopic limit $q \to 0$, $\omega \to 0$. In fact, at least for large water thickness, it is possible to extend the analysis to $q \sim 1/d$. If one assumes that the predominant friction is due to the water (rather than to the lipid) one finds[21] the so-called 'peristaltic' modes already discussed by several authors in connection with thick soap films[22].

References

1 SKOULIOS A *Adv. Coll. Int. Sci.* **1** 79 (1967); LUZZATI V, MUSTACCHI M, SKOULIOS A and HUSSON F *Acta Cryst.* **13** 660 (1960)
For a general review including biological lipids, *see*, LUZZATI V, Biological Membranes, E. and D. Chapman, Acad. Press (1968)
2 TARDIEU A *These d'Orsay* (1972); TARDIEU A *J. Mol. Biol.* **75** 711–733 (1973)
3 SACKMANN H and DEMUS D *Mol. Cryst.* **2** 81 (1966)
4 GULIK-KRZYWICKI T, TARDIEU A and LUZZATI V *Mol. Cryst. Liquid Cryst.* **8** 285 (1969)
5 PARSEGIAN V A *Sci.* **156** 939 (1969); PARSEGIAN V A *Trans. Far. Soc.* **62** 848 (1966); LANGMUIR I *J. Chem. Phys.* **6** 873 (1938); VERWEY E J *J. Theor. Overbook* The stability of lyophobic colloids (Elsevier, Amsterdam 1948); LANDAU L Collected papers, Pergammon Press)
6 DE GENNES P G *J. Phys. (Paris)* **30 C4** 65 (1969)
7 MARTIN P C, PARODI O and PERSHAN P S *Phys. Rev.* **A-6** 2401 (1972)
8 DURAND G *C.R.A.S.* **275B** 629 (1972); RIBOTTA R, DURAND G and LITSTER J D *Solid State Commun.* **12** 27 (1973); CLARK N A, PERSHAN P S *Phys. Rev. Lett.* **30** 3 (1973)
9 DE GENNES P G *C.R.A.S.* **275B** 939 (1972); BIDAUX R, BOCCARA N and SARMA G *J. de Phys. (Paris)* **34** 661 (1973)
10 DELAYE M, RIBOTTA R and DURAND G *Phys. Lett.* **44A** 139 (1973)
11 DE GENNES P G *Solid State Commun.* **10** 753 (1972)
12 LIAO Y, CLARK N and PERSHAN P S *Phys. Rev. Lett.* **30** 639 (1973)
13 CANDAU S *Private Commun.*
14 HELFRICH W *Phys. Rev. Lett.* **23** 372 (1969)
15 DE GENNES P G (Liquid Crystals), *Oxford* (to be published)
16 NINHAM B W and PARSEGIAN V A *J. Chem. Phys.* **53** 9 (1970)
17 MUELLER P *J. Theor. Biol.* **4** 268 (1963); *Nature, Lond.* **194** 979 (1962a)
18 HELFRICH W *Z. Naturforsch* **28c** 693 (1973)
19 GAINES G L *Interscience, New York* (1966); JOLY M *J. Chem. Phys.* **44** (1947)
20 CHARVOLIN J and RIGNY P *J. Chem. Phys.* **58** 3999 (1973)
21 BROCHARD F *These d'Orsay* (1974)
22 VRIJ A *J. Col. Sci.* **19** 1 (1964)
23 A temporal analysis of the scattered light has enabled recently to separate the thermal fluctuations from the strong static signal, and to study the dynamic mode of undulation (G. Durand – Proceedings of this Conference).

DISCUSSION

(The oral presentation by Dr de Gennes included a more general coverage of the smectic state, to which the following discussion refers).

Janik: Don't you think that the Apollonius effect could be studied by the positron annihilation method?

de Gennes: I know very little about positron annihilation in insulators. For smectics A or for nematics, an interesting situation would occur if the annihilation took place preferentially on *defects* of the structure (disclinations, dislocations, focal conics).

Schnur: Could precise specific heat or volume measurements be an appropriate technique to determine the reality of the model of the texture in smectic A you have described?

de Gennes : I fear that the volume fraction which is strongly distorted is too small to give an observable effect.

Schnur : Is ΔV always a scalar when doing dilatometric experiments on smectic phases?

de Gennes : ΔV probably scales like the energy in most transitions; the dilation coefficient and the specific heat should have similar singularities.

Rustichelli : I think that x-ray diffraction, although it is not able to confirm the Apollonius solution, could be used to exclude the grain boundary solution. Indeed this solution should produce two peaks differing by $\pi/2$ angle when the sample is rocked in a monochromatic beam.

de Gennes : We do not know the density of grain boundaries. Scattering of x-rays by dislocations in smectics seems to be a much more complicated problem than scattering of x-rays by dislocations in solids. We must first have the optical analogue of x-ray topography. By light beams one should be able, if you have a few dislocations, to see them as curved wedges. It comes close in spirit to your Borrmann effect.*

I am worried about all small angle scattering studies. There are too many effects to be considered.

---

* Nityananda R, Kini U D, Chandrasekhar S and Suresh K A, this conference.

Afterthought: Hydrodynamic properties of fluid lamellar phases of lipid/water

These modes are practically important, because they are the starting point to understand: a) the rheological behavior; b) the dynamics of swelling, of lipid water systems such as neat soaps. In our days we have a much better understanding of the forces between lamellae, and this domain has become more controlled: see for instance F. Nallet, D. Roux and J. Prost, *J. Physique* **50**, 3147 (1990).

Part III. Polymers

Physics Vol. 3, No. 1, pp. 37-45, 1967. Physics Publishing Co.

QUASI-ELASTIC SCATTERING OF NEUTRONS BY DILUTE POLYMER SOLUTIONS: I. FREE-DRAINING LIMIT

P.-G. de GENNES

Laboratoire de Physique des Solides associé au C.N.R.S.
Faculté des Sciences d'ORSAY - 91 - France

(*Received* 20 *June* 1966)

Abstract

In the so-called "free-draining limit", successive units of a long molecule equalise their average orientation by a diffusion process along the chain. For low values of the momentum transfer $\hbar\mathbf{q}$ *and of the energy transfer* $\hbar\omega$ *the dynamical form factor* $S(\mathbf{q}\,\omega)$ *for neutron scattering is controlled by this effect, and is independent of the vibrational spectrum. In this regime, we show that the frequency width* $\Delta\omega_q$ *of* $S(\mathbf{q}\,\omega)$ *is small and proportional to* q^4. *The unusual* q^4 *law is related to the fact that, in a time t, a signal travels a distance* $d \sim \sqrt{t}$ *along the chain, but the corresponding distance in space is only of order* $d^{1/2}$ *for a coiled polymer. On the other hand, if the chain is stretched this argument breaks down and the width* $\Delta\omega_q$ *for coherent scattering is predicted to increase.*

The existing experiments on neutron inelastic scattering by polymers [1] have been concerned mainly with the high frequency part of the spectrum ($\omega \sim \omega_{max} \sim 10^{14}$ sec^{-1}). The corresponding short wavelength vibrations and viscous motions are then characteristic of very small sections of the chain [2]. In order to achieve situations where the length of the molecule has important effects on the motion, one should concentrate on experiments where the momentum transfer $\hbar\mathbf{q}$ of the neutron and the energy transfer $\hbar\omega$ are small [3]. In this limit the energy spectrum of the neutrons scattered by the molecule (at fixed q) will consist of a quasi-elastic peak (width $\Delta\omega_q \ll \omega_{max}$) plus a weak background extending up to $\omega \cong \omega_{max}$. Our aim is to discuss the structure of this quasi-elastic peak. The slow motions of a long molecule in a solvent have been discussed by Rouse (4a) and by Zimm (4b). Rouse considered only the so-called "free-draining limit", where all hydrodynamic effects coupling distant segments are neglected. Zimm has included these effects. In the present paper we restrict our attention to the Rouse model: this corresponds to an extreme case, which is probably not often realised in practice, but it displays some rather interesting geometrical properties. The consequences of the long-range hydrodynamic interactions (leading to a very different q dependence of $\Delta\omega_q$) will be taken up in a second paper.

1. Fundamental Equations for the Free-draining Limit

The long chain is divided into N subunits, marked by the points $\mathbf{r}_0, \mathbf{r}_1, \ldots, \mathbf{r}_N$. The intervals are $\mathbf{a}_n = \mathbf{r}_{n+1} - \mathbf{r}_n$. In the absence of external forces the orientations of successive units measured at the same time are assumed independent [5].

$$\langle \mathbf{a}_n(t) \cdot \mathbf{a}_m(t) \rangle = \sigma^2 \delta_{nm} \tag{1}$$

In a solvent at rest, the equation of irreversible motion is of the form

$$\frac{\partial \mathbf{r}_n}{\partial t} = -\sum_m B_{nm} \frac{\partial G}{\partial \mathbf{r}_n} \tag{2}$$

where B_{nm} are coefficients and G is the thermodynamic potential for fixed external forces. The crucial assumption of Rouse (4a) is to assume that B_{mn} is a rapidly decreasing function of $|m - n|$. Then, by a suitable redefinition of the submolecules, one can always reduce the problem to the case $B_{mn} = B\delta_{mn}$. In particular, for O applied forces, G contains only an entropy contribution. For small distortions

$$G = \frac{3k_BT}{2\sigma^2} \sum_n \mathbf{a}_n^2 + \text{const.} \tag{3}$$

This leads to [6]:

$$\frac{\partial \mathbf{a}_n}{\partial t} = W(\mathbf{a}_{n+1} - 2\mathbf{a}_n + \mathbf{a}_{n-1})$$

$$W = \frac{3\ k_BT\ B}{\sigma^2} \tag{4}$$

From equation (4) we derive the basic time dependent correlation functions

$$< \mathbf{a}_n(o) \cdot \mathbf{a}_m(t) > = \sigma^2 G_{mn}(t) \tag{5}$$

where $G_{mn}(t)$ is the Green's function

$$G_{nm}(t) = \frac{1}{2\pi} \int_{-\pi}^{\pi} dp\ e^{ip(n-m)} e^{-2W(1-\cos p)|t|} \tag{6}$$

We consider only the limit of very long chains and the above G_{mn} applies for this case. Note that $G_{nm}(o) = \delta_{nm}$ as required by equation (1).

2. Incoherent Scattering

This is of particular interest, since most polymers contain hydrogen and the incoherent cross section of H is very large. The dynamical form-factor for this case is [3]

$$S(\mathbf{q},\ \omega) = \frac{1}{2\pi} \int dt\ S(\mathbf{q},\ t)\ e^{i\omega t}$$

$$S(\mathbf{q},\ t) = < e^{-i\mathbf{q}\cdot\mathbf{r}_n(o)} e^{i\mathbf{q}\cdot\mathbf{r}_n(t)} > \tag{7}$$

For $t >> W^{-1}$ (W^{-1} being a typical correlation time), $\mathbf{r}_n(t) - \mathbf{r}_n(o)$ is a sum of a large number of independent contributions and has a gaussian distribution. Then

$$S(\mathbf{q},\ t) = e^{-\frac{1}{2} q^2 <[x_n(o) - x_n(t)]^2>} \tag{8}$$

where x_n is the projection of r_n along $\mathbf{q}$. For an infinite chain we can write

$$\mathbf{r}_n = \sum_{m=-\infty}^{m=n-1} \mathbf{a}_m$$

$$\frac{1}{2} < \left[x_n(o) - x_n(t)\right]^2 > = \frac{1}{3} < \mathbf{r}(o) \cdot \left[\mathbf{r}_n(o) - \mathbf{r}_n(t)\right] >$$

$$= \frac{1}{3} \sum_{m<n} \sum_{m'<n} < \mathbf{a}_m(o) \left[\mathbf{a}_m{}'(o) - \mathbf{a}_m{}'(t)\right] > \qquad (9)$$

$$= \frac{\sigma^2}{3} \int_{-\pi}^{\pi} \frac{dp}{2\pi} \sum_{m<n} \sum_{m'<n} e^{ip(m-m')}[1 - e^{-2W(1-\cos p)t}]$$

The sum $\sum_{mm'}$ is easily carried out:

$$\sum_{m<n} \sum_{m'<n} e^{ip(m-m')} = \frac{1}{2(1 - \cos p)}$$

We are only interested in the region $Wt >> 1$ and the integrand in $\int dp$ is non-vanishing only for small p. Then

$$\frac{1}{2} < \left[x_n(o) - x_n(t)\right]^2 > \to \frac{\sigma^2}{6\pi} \int_{-\infty}^{\infty} \frac{dp}{p^2} \left(1 - e^{-Wp^2|t|}\right)$$

$$= \frac{\sigma^2}{3} \sqrt{\frac{W|t|}{\pi}} \qquad (10)$$

The unusual square root dependence of equation (10) (can be understood as follows: in time t, a signal starting from one given submolecule will reach $Q \sim \sqrt{Wt}$ neighboring submolecules. But since the polymer is coiled, the distance in space corresponding to this length is $\Delta x \sim \sigma \sqrt{Q}$. Thus $\Delta x^2 \sim \sigma^2 \sqrt{Wt}$. It is of interest to verify that equation (10) is independent of the size chosen to define the submolecules. Introducing the diffusion coefficient for the center of gravity of the total molecule $D = B\, k_B T/N = (1/3\, N) W\sigma^2$ and the total mean square extension $R^2 = N\sigma^2$, we may write in fact [7] $\sigma^2 \sqrt{Wt} = R\sqrt{3Dt}$.

The quantity which is directly measured is $S(\mathbf{q}\ \omega)$ as defined by (7). It is expressible in terms of the Fresnel integrals $C(x)$, $S(x)$ [8].

$$S(\mathbf{q}, \omega) = \left(\frac{3}{q^2\sigma^2}\right)^2 \frac{1}{W} Z(\rho)$$

$$Z(\rho) = \sqrt{\frac{\pi}{2}}\, \rho^{-3/2} \left\{ \left[\frac{1}{2} - S(\frac{1}{4\rho})\right] \sin\frac{1}{4\rho} + \left[\frac{1}{2} - C(\frac{1}{4\rho})\right] \cos\frac{1}{4\rho} \right\} \qquad (11)$$

$$\rho = \frac{\pi\,\omega}{W} \left(\frac{3}{q^2\sigma^2}\right)^2$$

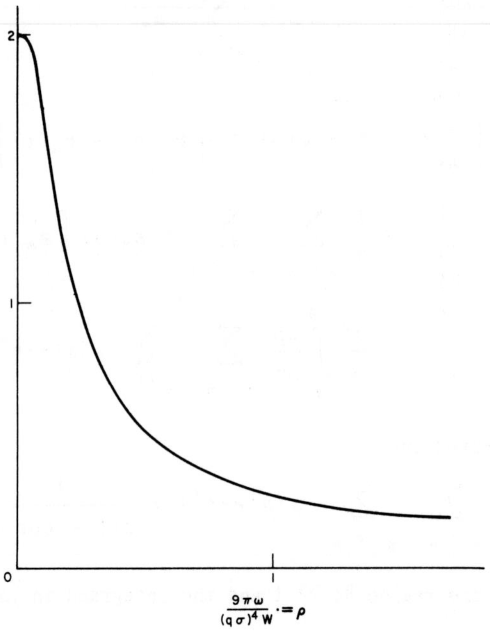

FIGURE 1

The dynamical form factor for incoherent scattering at fixed (small) momentum transfer $\hbar\mathbf{q}$, as a function of the energy transfer $\hbar\omega$.

A plot of $Z(\rho)$ is given on Fig. 1. The half-width at half-maximum $\Delta\rho$ of $Z(\rho)$ corresponds roughly to $\Delta\rho = 1/4$ or

$$\Delta\omega q \sim 10^{-2}\, W\, \sigma^4\, q^4 \cong 3 \times 10^{-2}\, D\, R^2\, q^4 \tag{12}$$

The range of validity of this formula is defined by

$$qR >> 1 >> qb$$

where b is an atomic distance (more precisely, for a simple chain with fixed valence angle θ and interatomic distance a, $b = a(1 + \cos\theta/1 - \cos\theta)$).

There are at least two difficulties to overcome if one wishes to study the quasi-elastic peak:

(a) the energy width $\hbar\Delta\omega q$ predicted by (12) is very small: taking $\sigma q = 1/2$ and $W = 10^{14}$ $\sec^{-1}$ (a rather high value corresponding to a solvent of low viscosity), we get $\hbar\Delta\omega q \sim 1°K$;

(b) the experiment should be done on a relatively dilute solution, but the "noise" coming from the solvent molecules must be low: in particular, the solvent should not contain hydrogen atoms.

3. Coherent Scattering

This situation may be of interest if the coherent scattering amplitude (per unit volume) for the polymer is much larger than the amplitude for the solvent. The correlation function to be studied is:

$$S_{coh}(\mathbf{q}\ t) = \sum_{ij} a_i a_j < e^{i\mathbf{q}\cdot[\mathbf{r}_i(t) - \mathbf{r}_j(o)]} > \tag{13}$$

where the sum Σ_{ij} runs on all atoms in the long molecule, and a_i, a_j are characteristic amplitudes. For an atom i belonging to the nth submolecule we separate

$$\mathbf{r}_i = \mathbf{r}_n + \rho_i$$

At the time of interest ($Wt >> 1$) the vectors $\rho_i(t)$ and $\rho_j(o)$ are uncorrelated between themselves and uncorrelated to the $\mathbf{r}_n$'s. Then, for two atoms i and j belonging respectively to the submolecules n and m (n and m being different or equal)

$$e^{i\mathbf{q}\cdot[\mathbf{r}_i(t) - \mathbf{r}_j(o)]} = < e^{i\mathbf{q}\cdot\rho_i} > < e^{-i\mathbf{q}\cdot\rho_j} > < e^{i\mathbf{q}\cdot[\mathbf{r}_n(t) - \mathbf{r}_m(o)]} > \tag{14}$$

Introduce a form factor

$$F(\mathbf{q}) = \sum_i a_i < e^{i\mathbf{q}\cdot\rho_i} > \tag{15}$$

where the sum is carried over the atoms in one submolecule. Then

$$S_{coh}(\mathbf{q}\ t) = |F(q)|^2 \sum_{nm} < e^{i\mathbf{q}\cdot[\mathbf{r}_n(t) - \mathbf{r}_m(o)]} > \tag{16}$$

For $Wt >> 1$ (and for the distance scale specified by $qb << 1$) we can again assume that the distribution of $\mathbf{r}_n(t) - \mathbf{r}_m(o)$ is gaussian and write

$$S_{coh}(\mathbf{q}\ t) = |F(q)|^2 \sum_{nm} e^{-\frac{1}{2} q^2 <[x_n(t) - x_m(o)]^2>} \tag{17}$$

Expressing the $\mathbf{r}_n$'s in terms of $\mathbf{a}$'s and using equation (5), we arrive at

$$\begin{aligned} \frac{1}{2} < \left[x_n(t) - x_{n+s}(o)\right]^2 > &= \frac{\sigma^2}{6\pi} \int_{-\infty}^{\infty} \frac{dp}{p^2} (1 - e^{-W p^2|t|} \cos p\ s) \\ &= \frac{\sigma^2}{6} \left[|s| + 2\sqrt{\frac{Wt}{\pi}}\ g\left(\frac{s^2}{4W|t|}\right)\right] \end{aligned} \tag{18}$$

where

$$g(u) = \int_1^{\infty} \frac{d\tau}{\tau^2} e^{-u\tau^2} \tag{19}$$

The double sum Σ_{nm} in equation (17) then reduces to $N\ \Sigma_s$ and, for the long distance limit of interest ($q\sigma << 1$) Σ_s may be replaced by $\int ds$. Setting

$$s = 2\sqrt{W|t|}\,\lambda$$

$$\theta = \frac{q^2 s^2}{3} (W|t|)^{1/2} \tag{20}$$

we obtain:

$$\frac{S_{coh}(\mathbf{q}\ t)}{S_{coh}(\mathbf{q},\ t = 0)} = \theta \int_0^\infty d\lambda\ e^{-\theta[\lambda + (1/\sqrt{\pi}) g(\lambda^2)]} \tag{21}$$

$$S_{coh}(\mathbf{q},\ t = 0) = N|F(q)|^2 \frac{12}{\sigma^2 q^2} = \int_{\infty}^{\infty} d\omega\ S_{coh}(\mathbf{q}\ \omega) \tag{22}$$

Equations (21) and (22) deserve a few comments: (a) Equation (22) displays the well-known $1/q^2$ singularity of the total scattering power at small angles. This large intensity may be useful to separate background effects and represents a definite advantage of coherent versus incoherent scattering. (b) Equation (21) shows that $S_{coh}(\mathbf{q}\ t)$ depends on only one particular combination of q and t, namely the variable θ defined in equation (20). Just as in the incoherent case we can verify that θ is independent of the size chosen to define the submolecules.

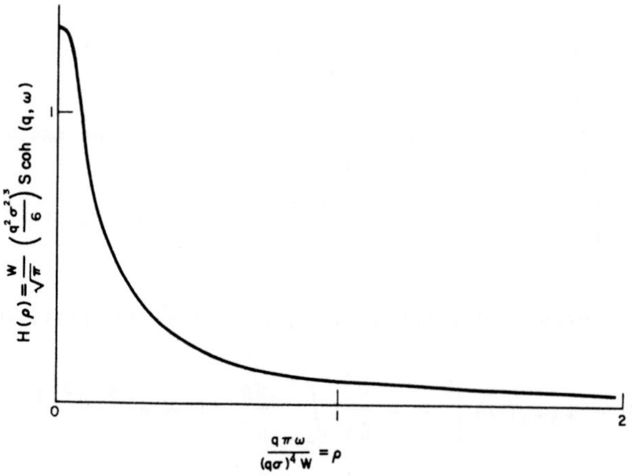

FIGURE 2.

The dynamical form factor for coherent scattering for a long molecule (under 0 applied forces) at fixed (small) momentum transfer $\hbar\mathbf{q}$, as a function of the energy transfer $\hbar\omega$.

Finally, the calculation of the energy distribution $S_{coh}(\mathbf{q}\ \omega)$ (i.e. the Fourier transform of equation (21)) requires some numerical work. The results are shown on Fig. 2. The general aspect is not very different from what was obtained for incoherent scattering. The half-width at half-maximum is

$$\Delta\omega_q \cong 2 \times 10^{-2}\, D\, R^2\, q^4$$

4. Effect of Stretching the Molecule

We now show that if the molecule is stretched, the frequency width $\Delta\omega_q$ expected for coherent scattering (with $\mathbf{q}$ along the direction of elongation) can increase significantly, $\Delta\omega_q$ being now proportional to q^2 instead of q^4. In practice, the elongation might be achieved either on a connected (but still rather dilute) polymer network (swollen rubber) or by electric fields on polar molecules.

At equilibrium, the average orientation of each submolecule is now nonzero:

$$< \mathbf{a}_n > = \mathbf{l} \neq 0$$

In the following we shall consider only weak elongations ($l < \sigma$). Then the Rouse equation (4) still applies [9] and the basic correlation functions are

$$< a_{n\alpha}(o)\; a_{m\beta}(t) > = \left(\frac{1}{3}\,\sigma^2\,\delta_{\alpha\beta} - l_\alpha\, l_\beta\right) G_{nm}(t) + l_\alpha l_\beta \tag{23}$$

where α, $\beta = x, y, z$ and G_{nm} is always defined by (6).

For incoherent scattering this modification has only minor effects. But for coherent scattering, we obtain, instead of equations (17,18):

$$S(\mathbf{q}\; t) = \sum_{s=-\infty}^{s=\infty} e^{is\mathbf{q}\cdot\mathbf{l}}\; e^{-(q^2\sigma^2/6)\left[|s| + 2\sqrt{Wt/\pi}\; g(S^2/4Wt)\right]} \tag{24}$$

If $\mathbf{q}\cdot\mathbf{l} \ll q^2\,\sigma^2$ equation (24) goes back into equation (18). But if $q \cdot l \gg q^2\,\sigma^2$ (namely if $\mathbf{q}$ is parallel to $\mathbf{l}$, and $q < l/\sigma^2$), the behavior is profoundly different. The times of interest are then $t \sim (W\,q^2\,l^2)^{-1}$. The last exponential in equation (24) can then be expanded

$$e^{-q^2\sigma^2/3\sqrt{Wt/\pi}\; g} = 1 - \frac{q^2\,\sigma^2}{3}\sqrt{Wt/\pi}\; g$$

and after some manipulation this leads to

$$S(\mathbf{q}\; t) = \frac{q^2\,\sigma^2}{3(\mathbf{q}\cdot\mathbf{l})^2}\; e^{-Wt(\mathbf{q}\cdot\mathbf{l})^2} \tag{25}$$

The Fourier transform $S_{coh}(\mathbf{q}\;\omega)$ is then a Lorentz curve of width

$$(\Delta\omega_q)_{stretched} = W(\mathbf{q}\cdot\mathbf{l})^2 \qquad (\mathbf{q}\cdot\mathbf{l} \gg q^2\,\sigma^2) \tag{26}$$

Note that in this case $\Delta\omega_q$ is larger than the value at rest ($\sim W\,q^4\,\sigma^4$). Thus in certain conditions an elongation should cause a broadening of the quasi-elastic peak.

5. Concluding Remarks

Our discussion was based on the Rouse assumption: the drift velocity $\frac{d}{dt}\mathbf{r}_n$ is proportional to the local curvature $\mathbf{a}_n - \mathbf{a}_{n+1}$. This assumption is interesting because of it's simplicity, and it has led us to results which have a simple geometrical interpretation. But it is not realistic: in actual fact $\frac{d}{dt}\mathbf{r}_n$ depends on the configuration of submolecules (m) very remote from (n) along the chain - for (at least) two reasons:

(a) The motion of (n) relative to the solvent creates a backflow which reacts on (n). This coupling leads to very different results ($\Delta\omega_q \sim q^3$) and will be discussed separately.

(b) There may be some direct friction between (n) and (m).

Thus (and in spite of obvious experimental difficulties), neutron scattering experiments could reveal some interesting motional properties of long molecules.

Acknowledgments

The author is greatly indebted to J.P. Hurault for assistance in the calculations of section 3, and to S. Alexander for a discussion of long range interactions.

References

1. W.M. MYERS, G.C. SUMMERFIELD and J.S. KING, *Symposium on Inelastic Scattering of Neutrons*, edited by Brookhaven National Laboratory, p. 126 (1965). W.L. WHITTEMORE, *ibid.*, p. 131.

2. Theoretical attempts to describe the vibration spectrum include the following:

   D.T. GOLDMAN and F.T. FEDERIGHI, *Nucl. Sci. Eng.* **16**, 165 (1963).

   J.U. KOPPEL and J.A. YOUNG, *ibid.* **21**, 268 (1965).

   R.A. HARRIS and J.E. HEARST, *J. Chem. Phys.* **44**, 2595 (1966).

3. For the definition of $h\omega$, hq, and of the dynamical form factors, see L. VAN HOVE, *Phys. Rev.* **95**, 249 (1954).

   For a review of experiments and theory in liquids see for instance P.G. DE GENNES, in *Inelastic Scattering of Neutrons by Solids and Liquids*. International Atomic Energy Agency, Vienna (1961).

4a. P.E. ROUSE, *J. Chem. Phys.* **21**, 1272 (1953).

4b. B.H. ZIMM, *J. Chem. Phys.* **24**, 269 (1956).

5. Excluded volume effects are neglected.

6. There is a striking analogy between the problem of a freely orienting chain, of units

$\vec{a}_1$, $\vec{a}_2$, ..., $\vec{a}_N$, and a one-dimensional assembly of spins $\vec{S}_1$, ..., $\vec{S}_N$ coupled by exchange forces. In both cases the fundamental processes contributing to the entropy source are local motions which do not change the total vector length $\vec{S}_1 + \ldots + \vec{S}_N$. From this conversation property the diffusion form of equation (4) follows immediately.

7. Note that, although R and D depend on the molecular weight M of the polymer, the product $R\sqrt{D}$ is independent of M. All the properties discussed assume only that the chain is long ($qR \gg 1$).

8. The functions C and S are defined as in Jahnke-Emde's *Tables of Higher Functions* (Teubner ed.), p. 28, Stuttgart (1960).

9. Except of course for the end points, where the contribution of the applied forces must be included.

Physics Vol. 3, No. 4, pp. 181-198, 1967. Physics Publishing Co.

QUASI-ELASTIC SCATTERING BY DILUTE, IDEAL, POLYMER SOLUTIONS: II. EFFECTS OF HYDRODYNAMIC INTERACTIONS

E. DUBOIS-VIOLETTE and P.-G. de GENNES

*Laboratoire de Physique des Solides**
Faculté des Sciences d'Orsay, 91 - ORSAY, France

(*Received* 21 *February* 1967)

Abstract

Each moving unit of a long, flexible, molecule induces in the surrounding solvent a velocity field which reacts on the motion of other segments. This long-range hydrodynamic interaction modifies strongly the dynamical form factor $S(\mathbf{q}\,\omega)$ at low frequencies ω and small scattering vectors $\mathbf{q}$. For neutron scattering ($qR_G \gg 1$, where R_G is the radius of gyration of the polymer), the frequency width $\Delta\omega_q$ of $S(q\omega)$ at fixed q becomes proportional to q^3(for an ideal coil). Also the effect of stretching the molecule becomes more dramatic, since stretching greatly reduces the hydrodynamic interactions.

1. Introduction

IN A preceding paper [1] we have discussed the slow, quasi macroscopic motions of a flexible polymer chain in the so-called "Rouse limit": namely when the velocity of each monomer depends only on the forces applied on it. This led to comparatively simple laws for the inelastic scattering of neutrons by a long chain. In particular the frequency width $\Delta\omega_q$ of the scattered beam (for a monochromatic incident beam and a fixed momentum transfer $\hbar q$) was found to be proportional to q^4.

We now turn to the more realistic case where the motion of the solvent is taken into account. Then a number of new effects come into play:

(a) the motion of the solvent can be of interest in itself: typically one could measure the inelastic incoherent scattering of neutrons by hydrogen - containing solvent molecules moving in a macromolecular mesh of low scattering power: this experiment gives a diffusion coefficient, and should supplement in an interesting way the nuclear resonance data. However, in the present paper, we shall not be concerned with this class of problems : if we discuss incoherent scattering, for instance, we assume that the only scattering centers of importance are nuclei of the polymer chain. For coherent scattering, we assume that our solution can be treated as essentially incompressible [2]. Then we can again consider that the solute alone participated in the scattering, the specific scattering amplitude α (per cm^3) being the difference $\alpha_{solute} - \alpha_{solvent}$.

* Laboratoire associe au C.N.R.S.

(b) the presence of the solvent leads to long-range interactions between monomers. Let us for instance idealise each monomer as a small sphere of radius b. Then if a force φ_m acts on the m-th monomer, it takes a velocity $\mathbf{w}_m$ relative to the solvent

$$\mathbf{w}_m = \frac{1}{6\pi\eta b}\varphi_m = B\,\varphi_m \tag{1}$$

where η is the viscosity of the solvent. This induces a velocity field around the moving sphere, of the form*

$$\delta\,\mathbf{v}_m(\mathbf{r}) = \frac{3b}{4|\mathbf{r}-\mathbf{r}_m|}[\mathbf{w}_m + (\mathbf{u}\cdot\mathbf{w}_m)\mathbf{u}] + \frac{b^3}{4|\mathbf{r}-\mathbf{r}_m|^3}[\mathbf{w}_m - 3(\mathbf{u}\cdot\mathbf{w}_m)\mathbf{u}] \tag{2}$$

where $\mathbf{u}$ is a unit vector in the direction $\mathbf{r}-\mathbf{r}_m$.

Each monomer drifts in the velocity field $\delta\mathbf{v}$. The total velocity of the n-th unit is thus

$$\frac{d\mathbf{r}_n}{dt} = B\,\varphi_n + \sum_m \delta\mathbf{v}_m(\mathbf{r}_n) + \mathbf{v}^{\circ}(\mathbf{r}_n) \tag{3}$$

where $\mathbf{v}^0(\mathbf{r}_n)$ is the velocity field of the solvent in the absence of any polymer molecule (for our scattering problems we always have $\mathbf{v}^0 \equiv 0$).

Equations (1-3) show that $d\mathbf{r}_n/dt$ depends not only on the force φ_n applied on the n-th unit, but also on all other forces φ_m. We assume that the φ's are small, and compute effects only to first order in φ : then we can average equation (2) over all polymer configurations in a state of 0 forces. The terms in $|\mathbf{r}_n - \mathbf{r}_m|^{-3}$ disappear when we perform the angular average, and assuming an ideal coil we are left with :

$$\delta\mathbf{v}_m[\mathbf{r}_n] = \mathbf{w}_m b \left\langle\frac{1}{|\mathbf{r}_n - \mathbf{r}_m|}\right\rangle = \xi\,\mathbf{w}_m|n-m|^{-1/2} \tag{4}$$

where

$$\xi = \left[\frac{6}{\pi}\right]^{1/2}\frac{b}{\sigma} \tag{5}$$

and σ^2 is the mean square end-to-end dimension of one sub-unit, as defined in (I). The consequence of this long-range hydrodynamic interaction for the viscosity and other macroscopic properties of the solution have been worked out by Kirkwood and Risemann [3] and more accurately by Zimm [4]. The present paper represents essentially an extension of ref. [4] to cover some problems of inelastic scattering.

We shall discuss first the limit of an infinitely long polymer chain (Section 2), since the calculations are comparatively simple for this case. In practice, this limit will be realised if $qR_G >> 1$ where R_G is the radius of gyration of the molecule and q is the scattering vector

* See for instance Landau-Lifshitz, *Fluid mechanics*, Chap. 2, Pergamon Press, Oxford.

$$q \cong \frac{4\pi}{\lambda} \sin \theta/2 \tag{6}$$

(θ : scattering angle; λ : incident wavelength). For thermal or subthermal neutrons q^{-1} will usually be smaller than 150 Å (corresponding to λ = 6 Å and θ = 2°) and the condition $qR_G >> 1$ is satisfied in general. For light scattering, on the other hand, we usually have $qR_G \gtrsim 1$ and a more complicated discussion allowing for the finite size of the molecule must be carried out.

The physical principles of this discussion are given in Section 3. The calculational aspects are described in an appendix.

2. Inelastic Scattering by a Very Long Chain ($qR_G >> 1$)

(1) Equation of motion and relaxation modes of an ideal coil

The force φ_n acting on the n-th sub-unit has been rederived in (I) and is given by

$$\varphi_n = \mathbf{F}_n + \frac{3k_BT}{\sigma^2}[\mathbf{a}_n - \mathbf{a}_{n+1}] \tag{7}$$

where $\mathbf{a}_n = \mathbf{r}_{n+1} - \mathbf{r}_n \cong \partial\mathbf{r}/\partial n$. The first term $\mathbf{F}_n$ is the external force. The second term is a force proportional to $\partial^2\mathbf{r}/\partial n^2$, i.e. to the curvature of the chain. We now restrict our attention to a free chain ($\mathbf{F}_n \equiv 0$) in a solvent at rest ($\mathbf{v}_0 \equiv 0$). Using equations (3) and (4) we arrive at an irreversible equation of motion of the form

$$\frac{d\mathbf{r}_n}{dt} = W\left[\mathbf{a}_n - \mathbf{a}_{n-1} + \xi \sum_{m\neq n} |m-n|^{-1/2}[\mathbf{a}_m - \mathbf{a}_{m-1}]\right] \tag{8}$$

where

$$W = \frac{3k_BT}{\sigma^2} B = \frac{1}{2\pi} \frac{k_BT}{\eta_0\, b\, \sigma^2} \tag{9}$$

η_0 being the viscosity of the solvent.

For our infinite chain the eigenmodes of equation (8) are still very simple, namely

$$\mathbf{r}_n = \text{const.} \quad e^{ipn}e^{-t/\tau_p} \tag{10}$$

with relaxation frequencies $1/\tau_p$ given by

$$\frac{1}{W\tau_p} = 2(1-\cos p) + 4\xi \sum_{s=1}^{\infty} s^{-1/2}\cos(ps)(1-\cos p). \tag{11}$$

Again we are only interested in the modes of low relaxation frequency ($p << 1$) for which equation (11) can be reduced to:

$$\frac{1}{W\ \tau_p} = p^2 \left[1 + 2\xi \int_0^\infty ds\ s^{-1/2} \cos ps \right]$$
$$= p^2 + \sqrt{2\pi}\ \xi\ p^{3/2}\ . \tag{12}$$

We conclude that, as soon as the parameter ξ is non zero, the hydrodynamic interactions dominate the relaxation behavior at low frequencies ($p \to 0$): in this limit we can drop the p^2 term and write

$$\frac{1}{\tau_p} = \widetilde{W}\ p^{3/2} \tag{13}$$

with

$$\widetilde{W} = \sqrt{2\pi}\ \xi\ W$$
$$= \frac{\sqrt{3}}{\pi} \frac{k_B T}{\eta_0 \sigma^3} \tag{14}$$

We can think of $\widetilde{W}$ as of the microscopic jump frequency of a single monomer moving in the solvent, and expect for $\widetilde{W}$ values in the range $10^{10} - 10^{13}$ $\sec^{-1}$.

An important question arises at this stage : the Stokes formula for the velocity field (equation (2)), on which our analysis is based, is valid only for motions at very low frequencies and not too long distances $|\mathbf{r}_n - \mathbf{r}_m|$. Is it in fact correct for our purposes? The answer is yes in the limit $p \to 0$, and the proof proceeds as follows: If the frequency scale in which we are interested is $\Omega \sim 1/\tau_p$, equation (2) applies for distances $|\mathbf{r}_n - \mathbf{r}_m| < L$ where $L^2 = \eta_0/\rho_0\Omega$ and ρ_0 is the solvent density. This corresponds to $s = |n - m| < L^2/\sigma^2$, and leads to a cut-off in the summation $\sum_s$ of equation (11) at the value

$$s_{max} \cong \frac{L^2}{\sigma^2} \cong \frac{\eta_0\ \tau_p}{\rho_0\ \sigma^2} \cong \frac{\eta_0{}^2 \sigma}{\rho_0 k_B T}\ p^{-3/2} = p_0{}^{1/2}\ p^{-3/2} \tag{15}$$

where

$$p_0 = \eta_0^4 \left[\frac{\sigma}{\rho_0 k_B T} \right]^2 \tag{16}$$

If p is much smaller than p_0, s_{max} is much larger than $1/p$: then the convergence of the sum $\sum_s$ in equation (11) is controlled by the factor cos ps, and is independent of the existence of the cut-off s_{max}. For all typical solvents at room temperature p_0 is large (of the order 10^3). The values of p in which we are interested are much smaller than 1 : thus $p \ll p_0$ and equation (2) is correct.

(2) Self correlation function and incoherent scattering

The basic correlation function derived from equation (8) is

$$\begin{aligned} \langle \mathbf{a}_n(o) \cdot \mathbf{a}_m(t) \rangle &= \sigma^2 G_{mn}(t) \\ &= \sigma^2 \frac{1}{2\pi} \int_{-\pi}^{\pi} dp \; e^{ip(n-m)} \; e^{-|t|/\tau_p} . \end{aligned} \tag{17}$$

This is identical with equation (6) of I : the only new feature is that $1/\tau_p$ is now proportional to $p^{3/2}$ as shown by equation (13) while in the Rouse limit of I ($\xi \to 0$), $1/\tau_p$ was proportional to p^2.

We now apply these results to the self correlation function

$$\begin{aligned} S(q, t) &= \langle e^{-i\mathbf{q}\cdot\mathbf{r}_n(o)} \; e^{-i\mathbf{q}\cdot\mathbf{r}_n(t)} \rangle \\ &= \exp\left[-\frac{1}{2} q^2 \langle [x_n(o) - x_n(t)]^2 \rangle \right] \end{aligned} \tag{18}$$

(the latter form being valid in the small q, large t, limit - as explained in I - and x_n being the projection of $\mathbf{r}_n$ along $\mathbf{q}$). Writing the $\mathbf{r}_n$'s in terms of the $\mathbf{a}_m$'s and using equation (17), we arrive at the formula

$$\begin{aligned} \frac{1}{2} \langle [x_n(o) - x_n(t)] \rangle &= \frac{\sigma^2}{6\pi} \int_{-\alpha}^{\infty} \frac{dp}{p^2} \left[1 - \exp\left[- \tilde{W} |p|^{3/2} |t| \right] \right] \\ &= \frac{\sigma^2}{6\pi} \Gamma(\tfrac{1}{3}) \; |\tilde{W} t|^{2/3} . \end{aligned} \tag{19}$$

Finally we construct the dynamic form factor for incoherent scattering

$$\begin{aligned} S(q, \omega) &= \frac{1}{2\pi} \int_{-\infty}^{\infty} S(qt) \; e^{i\omega t} \, dt \\ &= \frac{1}{\pi \tilde{W} \varepsilon^{3/2}} \int_{0}^{\infty} d\theta \; \exp\left[- \sigma\theta^{2/3} \right] \cos(\omega\theta) \end{aligned} \tag{20}$$

where we have introduced the reduced variables

$$\begin{aligned} \varepsilon &= \frac{q^2 \sigma^2}{6} \\ \theta &= \varepsilon^{3/2} \tilde{W} |t| \\ \tilde{\omega} &= \frac{\omega}{\tilde{W} \varepsilon^{3/2}} \end{aligned} \tag{21}$$

and the constant $\alpha = (2/\pi)\Gamma(1/3)$.

It is useful to compute the dynamic form factor at 0 frequency

$$S(q,\ \omega = 0) = \frac{1}{\pi\ W\ \varepsilon^{3/2}} \int_0^\infty d\theta\ \exp\left[-\ \alpha\theta^{2/3}\right] = \frac{3}{4\sqrt{\pi}\ \alpha^{3/2}\ \tilde{W}\ \varepsilon^{3/2}} \tag{22}$$

and to plot the results in terms of the dimensionless function

$$\frac{S(q,\ \omega)}{S(q,\ 0)} = \frac{4\alpha^{3/2}}{3\sqrt{\pi}} \int_0^\infty d\theta\ \exp\left[-\ \alpha\theta^{2/3}\right] \cos(\tilde{\omega}\theta) = g_i(\tilde{\omega}). \tag{23}$$

A graph of this function $g_i(\tilde{\omega})$ is given on Fig. 1. The half width at half maximum is $\Delta\tilde{\omega} = 1.1$ corresponding to

$$\Delta\omega_q = 0.075\ \tilde{W}\ q^3\ \sigma^3 \tag{24}$$

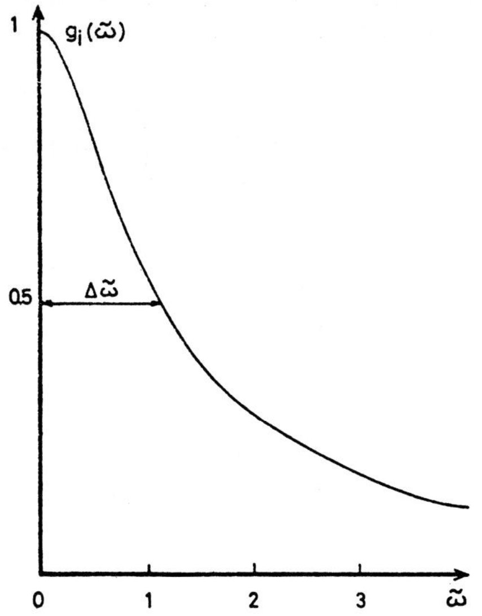

FIGURE 1

$\Delta\omega$ is now *proportional to* q^3 : this is somewhat more favorable, from an experimental point of view, than the q^4 dependence which we had obtained in the Rouse limit : $\Delta\omega$ at a fixed, small q

is now larger.

What are the respective domains of validity of the Rouse limit and of the present approximation? In all cases we have $\Delta\omega_q \sim 1/\tau_p$ and $p \sim q\sigma$. The relaxation frequency $1/\tau_p$ is given by equation (2.6). The contribution of the hydrodynamic effects dominates when $\xi p^{3/2} >> p^2$ or $p << \xi^2$. Thus we arrive at the following set of rules:

$$q\sigma << \xi^2 \rightarrow \text{Apply equations (23, (24)}$$

$$1 >> q\sigma >> \xi^2 \rightarrow \text{Apply equations (11) and (12) of I}$$

In practice, if ξ is in a suitable range and if the transition between the two domains can be observed, one may derive ξ from the experimental data.

(3) Coherent scattering

Our starting point here is equation (17) of I : the time-dependent correlation function to be studied is

$$S_{coh}(q, t) = |F(q)|^2 \sum_{nm} e^{-\frac{1}{2}q^2 < [x_n(0) - x_m(t)]^2 >} \tag{25}$$

where $F(q)$ is a form factor for the monomer (and can be replaced by $F(o)$ for most cases of interest). Expressing the $\mathbf{r}_n$'s in terms of the $\mathbf{a}_m$'s and making use of equation (17) we arrive at

$$\begin{aligned} \frac{q^2}{2} < [x_n(o) - x_{n+s}(t)]^2 > &= \frac{q^2\sigma^2}{6\pi} \int_{-\infty}^{\infty} \frac{dp}{p^2} (1 - e^{-t/\tau_p} \cos ps) \\ &= \varepsilon|s| \left[1 + h(u)\right] \end{aligned} \tag{26}$$

where

$$\begin{aligned} \varepsilon &= \frac{q^2\sigma^2}{6} \\ |s|h &= \frac{2}{\pi} \int_0^{\infty} \frac{dp}{p^2} (1 - e^{-t/\tau_p}) \cos ps \\ h(u) &= \frac{4}{\pi} \int_0^{\infty} \frac{dy}{y^3} [1 - \exp\left[- y^3 u^{-3/2}\right]] \cos y^2 \\ u &= |s| \, |\tilde{W}t|^{-2/3} \end{aligned} \tag{27}$$

In the small q limit we can replace the sum $\sum_m$ in equation (25) by an integral and write

$$S_{coh}(q,\ t) = 2N|F(q)|^2 \int_0^\infty ds\ e^{-\varepsilon s(1+h)}$$

$$= \frac{2}{\varepsilon} N|F(q)|^2\ f(\theta) \tag{28}$$

where θ is still defined by equation (21) and

$$f(\theta) = \theta^{2/3} \int_0^\infty du\ \exp\left[-\ \theta^{2/3}\ u[1 + h(u)]\right]$$

$$f(o) = 1. \tag{29}$$

Finally, the dynamic form factor per monomer unit is

$$\frac{1}{N}\ S_{coh}(q\omega) = \frac{1}{2\pi} \int_{-\infty}^{\infty} dt\ S_{coh}(qt)\ e^{i\omega t}$$

$$= \frac{2}{\pi\ \tilde{W}\ \varepsilon^{5/2}} |F(q)|^2 \int_0^\infty d\theta\ f(\theta)\ \cos(\omega\theta) \tag{30}$$

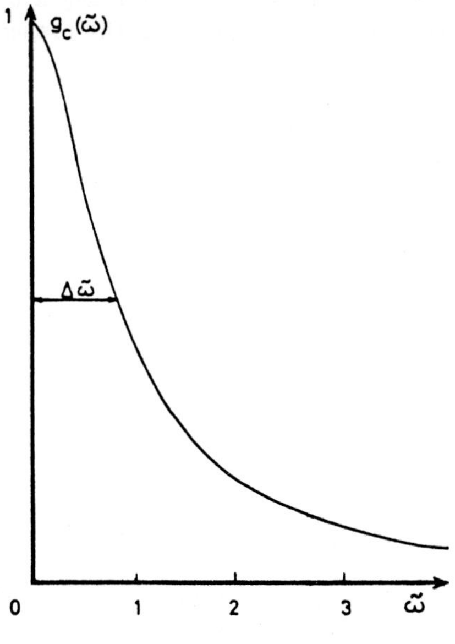

FIGURE 2

The function $f(\theta)$ and the Fourier transform $g_c(\tilde{\omega}) = \int_0^\infty f \cos \tilde{\omega}\,\theta\, d\theta$ have been computed numerically, and the results for g_c are shown on Fig. 2. The half-width at half maximum of g_c is $\Delta\tilde{\omega} = 0.8$, and this leads to

$$\Delta\omega_{q(coh)} = 0.055\ \tilde{W}\ q^3 s^3 \ . \tag{31}$$

Again the width is proportional to q^3 and all the qualitative remarks of the preceding paragraph remain valid.

3. Light Scattering by a Chain of Finite length

(1) Weakness of internal motion effects

As explained in the introduction, the experiments on the inelastic scattering of light correspond to situations where qR_G is of order unity, or smaller: the finite size of the chain must be taken into account.

We shall first consider the simplest type of scattering, where depolarisation effects can be neglected : we assume that the induced dipole $\mathbf{P}_n$ induced on the n-th monomer by the electric field $\mathbf{E}(\mathbf{r}_n)$ is of the form

$$\mathbf{P}_n = \alpha\ E(\mathbf{r}_n)$$

α being a scalar. We also assume that the frequency dependence of α on the frequency band of interest (width $\sim \Delta\omega_q$) can be neglected : this is quite correct in general. Then the frequency distribution of the scattered light at a fixed scattering angle θ is still given by the coherent dynamic form factor $S_{coh}(q\omega)$ with q defined by (6). For the present problem we can again reduce $S_{coh}(qt)$ to the gaussian form of equation (25).

(α) *when the molecule is comparatively small* ($qR_G \ll 1$), it is clear that we cannot study its internal motions by light scattering : we can replace without error the coordinates x_n of each monomer by the coordinate g of the center of gravity, and we have

$$S_{coh}(qt) = |NF|^2\ e^{-q^2/2 < [g(o) - g(t)]^2 >} \tag{32}$$

If the molecule diffuses in the solvent with a diffusion coefficient D, then in the large time limit of interest we can write:

$$< [g(o) - g(t)]^2 > = 2Dt$$

$$\tag{33}$$

$$S_{coh}(qt) = |NF|^2\ e^{-Dq^2 t}$$

and the Fourier transform $S_{coh}(q\omega)$ is a Lorentz curve of width Dq^2. Thus in the small q limit the only parameter which can be derived from this type of experiment is the diffusion coefficient D.

(β) *when* $qR_G \sim 1$, can we extract from the inelastic scattering data a significant information on the *internal* motions of the molecule? We shall now show that the answer to this question is *no*, unless qR_G reaches really high values (of the order of 2.5) which are not usually realisable in practice.

The discussion is particularly simple in one case : namely when the internal motions are completely decoupled from the motion of the center of gravity : let us write the coordinate x_n of the n-th monomer in the form

$$x_n = g + \rho_n \tag{34}$$

where $\sum_n \rho_n \equiv 0$. Then :

$$\begin{aligned} < [x_n(0) - x_m(t)]^2 > &= < [g(0) - g(t)]^2 > + < [\rho_n(0) - \rho_m(t)]^2 > \\ &+ 2 < [g(0) - g(t)][\rho_n(0) - \rho_m(t)] > \end{aligned} \tag{35}$$

We say that the internal motions are decoupled from g when the third term in equation (35) vanishes. Physically, this decoupling property is related to the following question: if we act on each monomer with a constant force $F_n = F$, the center of gravity of the molecule takes a uniform motion : is the internal shape of the molecule distorted? In the Rouse limit there is no distortion; but, with hydrodynamic interactions, there is a (small) distortion : the two ends of the molecule tend to lag backwards*. However, even in this case, the deviations from the uncoupled behavior lead to corrections of order 2 per cent only (see Appendix). Thus, in the following, we shall drop the last term in equation (35). Then, returning to equation (25), we can write:

$$\frac{1}{|NF|^2} S_{coh}(qt) = e^{-Dq^2 t} S_{red}(qt) \tag{36}$$

$$S_{red}(qt) = \frac{1}{N^2} \sum_{nm} \exp\left[-\frac{q^2}{2} < [\rho_n(0) - \rho_m(t)]^2 > \right] \tag{37}$$

For $t = 0$, $S_{red}(q, 0)$ is the familiar Debye function for coherent scattering by a freely orienting chain

$$\begin{aligned} S_{red}(q, t = 0) &= \frac{2}{y}\left[1 - \frac{1}{y}[1 - e^{-y}] \right] \\ y = q^2 R_G{}^2 &= \frac{N q^2 \sigma^2}{6} . \end{aligned} \tag{38}$$

---

*This coupling between overall and internal motions is reflected mathematically in the Zimm analysis by the non-orthogonality of the various relaxation modes. It is amusing to note that for a ring-shaped chain (no arms to be left behind) the orthogonality is restored and equation (35) is decoupled.

For $t \to \infty$, $S_{red}(q, t)$ tends towards a finite limit $\bar{S}(q)$: We might call $\bar{S}$ the Debye-Waller factor for the long chain. At large times t, $\rho_n(0)$ and $\rho_m(t)$ are uncorrelated. Thus we may write

$$\bar{S} = \frac{1}{N^2} \sum_{nm} \exp\left[-\frac{q^2}{2} < \rho_n^2 + \rho_m^2 > \right] \tag{39}$$

$$\bar{S}^{1/2} = \frac{1}{N} \sum_{n} \exp\left[-\frac{q^2}{2} < \rho_n^2 > \right] .$$

Making use of classic results for a freely orienting chain, we have

$$< \rho_n^2 > = \frac{N\sigma^2}{9} \left| \left[\frac{n}{N}\right]^3 + \left[1 - \frac{n}{N}\right]^3 \right|$$

and replacing the sum $\sum_n$ by an integral we arrive at

$$\bar{S} = e^{-y/6} \left[\int_{-1/2}^{1/2} du \; e^{-yu^2} \right]^2 . \tag{40}$$

The two curves giving $S_{red}(q, 0)$ and $S_{red}(q, \infty) = \bar{S}(q)$ as a function of $y = q^2 R_G{}^2$ are shown

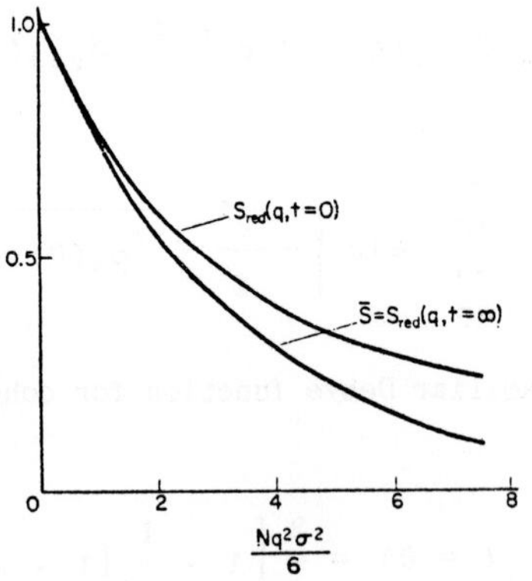

FIGURE 3

on Fig. 3. The contribution to the scattering function of the internal motions is measured by the difference $S_{red}(q, 0) - S_{red}(q, \infty)$. To observe clearly the internal motions we require

$$\frac{S_{red}(q, 0) - S_{red}(q, \infty)}{S_{red}(q, 0)} \gtrsim \frac{1}{2} .$$

This will be realised only for $qR_G \gtrsim 2.5$. Typical values for R_G in flexible chains are of order

500 Å. The maximum q for a given wavelength λ is $4\pi/\lambda$ (corresponding to $\theta = \pi$). Thus, to get an interesting signal, one would probably have to work with $\lambda < 3000$ Å. Such experiments with ultraviolet laser sources may become feasible in the future. General formulae, applicable for all values of qR_G, are given in the appendix.

4. Conclusions

(1) Experiments using neutrons or light

The existence of long-range hydrodynamic interactions between monomers modifies profoundly the frequency spectrum of the internal motions in an ideal coil. This change is rather favorable from the point of view of inelastic neutron scattering experiments, since the frequency width $\Delta\omega_q$ vanishes now less rapidly (like q^3) in the small q limit of interest : thus $\Delta\omega_q$ becomes slightly less difficult to detect.

As regards the inelastic scattering of photons, we have seen that the experiments with unpolarised light can bring in information on the internal motions only for very short wavelengths or very large molecules ($R_G \gtrsim 1300$ Å for $\lambda = 6000$ Å). Such large values of R_G occur only in semi-rigid molecules such as DNA : in such a case the subunits are large, the frequency scale is reduced, and also the analysis must be modified. An important parameter for semi-rigid molecules is the ratio of the persistence length L [6] to q^{-1}. For $qL \gg 1$, we are essentially dealing with a rigid rod, and expect a rather well defined phonon spectrum : $S(q\omega)$ should give some information on this spectrum (rather blurred, however, because of the orientational disorder). For $qL \ll 1$, we expect a central peak of the type described here. What happens for $qL \sim 1$ is not known.

Can the frequency distribution of the *depolarised* light scattered by the polymer give us any interesting information? What is measured here is a quadrupole correlation function. It is certainly sensitive to internal motions even when $qR_G \ll 1$. But it is much more difficult to compute than the vector-vector correlation functions (17). The calculation can be done within the model used by Zimm [4] in his study of flow birefringence. But the validity of this model is dubious : each monomer is described as a deformable object with a gaussian probability distribution. Of course, we can always define larger subunits which are gaussian; but then the average quadrupole of the subunit is not the quantity which dominates the depolarisation effect. Thus, at the present stage, we cannot make any clear-cut prediction on the depolarised light.

Another type of problems is related to what happens when the chain is extended - in practice by associating the chains in a loose, but connected, network on which mechanical stresses can be applied. In the Rouse limit, as explained in (I), this reacts on $\Delta\omega_q$ only for coherent scattering. With hydrodynamic interactions on the other hand, the effects are more drastic: on extension, the coupling terms of the equation of motion (8) go progressively from an $|n - m|^{-½}$ dependence to a $|n - m|^{-1}$ dependence*. Thus the relaxation times $1/\tau_p$ are modified : as a result we expect $\Delta\omega_q$ (for fixed q) to be changed by the extension, even in the case of incoherent scattering.

* This is one of the main causes for the non - Newtonian viscous behvaior observed in dilute polymer solutions. Another complication brought in by stretching, and to be mentioned, is that, in this case, the static Stokes formula (2) cannot be used.

(2) Excluded volume effects

All the analysis of the present paper was restricted to the case of an ideal coil : the power laws derived here (such as $\Delta\omega_q \sim q^3$ in Section 2) can at best apply only in poor solvents where an attraction between neighboring monomers compensates the large repulsion for overlapping monomers : that is to say, very close to the compensation temperature θ, where the first virial coefficient vanishes.

But, in general, excluded volume effects may be important, and they will react on the relaxation time spectrum in two different ways:

(1) The distribution of relative distances $\mathbf{r}_n - \mathbf{r}_m$ is strongly modified: according to the self-consistent calculation of Edwards [7] the average in equation (4) now becomes:

$$\left\langle \frac{1}{|\mathbf{r}_n - \mathbf{r}_m|} \right\rangle = \text{const. } s^{-3/5}, \qquad s = |n - m| .$$

(2) During the motion, each chain portion cannot be crossed by another chain portion (this occurs even at the θ temperature) : no serious predictions concerning this effect are available at the present time.

If we boldly assumed that (1) is the only important effect, we would again be led to rather simple power laws, namely

$$\frac{1}{\tau_p} = \hat{W}\, p^{8/5}$$

$$\Delta\omega_q \sim \hat{W}\, q^{8/3}$$

where $\hat{W}$ is a characteristic rotational frequency of the monomer. It is very much to be hoped that the future high-resolution, high-flux neutron spectrometers will allow for a detailed comparison between these proposed power laws and the experimental data.

Acknowledgements

We have greatly benefited from discussions with S. Lifson, L. Monnerie, A. Silberberg. Parts of the present paper were written down while one of us (P.G.G.) was attending the Batsheva seminar on Solid State Physics, and he wishes to thank Dr. S. Alexander for his hospitality on this occasion.

APPENDIX

Correlation Functions for Finite Chains

(1) Introduction of response functions

Our aim is to compute averages such as

$$< [x_n(0) - x_m(t)]^2 >$$

in an ideal coil of N segments. In the gaussian approximation of equation (18) and equation (25), such averages suffice to compute all dynamic form factors.

Our starting-point is to introduce a set of response functions $\beta_{nm}(t)$ defined as follows: assume that a system of small external forces $F_m(t)$ (all parallel to the x axis) is applied on the chain. The average velocity of the n-th unit is then of the form

$$\overline{\mathring{x}_n(t)} = \sum_m \int_{-\infty}^{t} dt' \ \beta_{nm}(t - t') \ F_m(t') \ .$$

The β_{nm}'s will be obtained from the equations of motion (3). When they are known, we can derive correlation functions from them by the Kubo formula [8]: for a classical system, this takes the form

$$< \mathring{x}_n(t_1) \ \mathring{x}_m(t_2) > = k_B T \ \beta_{nm}(|t_1 - t_2|) \ . \qquad \text{(A.1)}$$

We can write

$$< [x_n(t_1) - x_m(t_2)]^2 > = g_{nm}(t_2 - t_1)$$

$$2 < \mathring{x}_n(t_1)[x_n(t_1) - x_m(t_2)] > = + \frac{\partial}{\partial t_1} g_{nm}(t_2 - t_1)$$

$$2 < \mathring{x}_n(t_1) \ \mathring{x}_m(t_2) > = \frac{-\partial^2}{\partial t_1 \ \partial t_2} g_{nm}(t_2 - t_1) = + \frac{\partial^2}{\partial t_2^2} g_{nm}(t_2 - t_1) \ .$$

Thus we can go from the response functions (A.1) to the correlation functions g by two successive integrations. As regards the integration constants, we note that for a classical system (with 0 magnetic field) $g_{nm}(t)$ is real and even in t; thus $\partial g_{nm}/\partial t|_{t=0} = 0$. Finally we obtain

$$g_{nm}(t) = |n - m| \frac{\sigma^2}{3} + 2k_B T \int_0^t dt_1 \int_0^{t_1} dt_2 \ \beta_{nm}(t_1 - t_2) \ . \qquad \text{(A.2)}$$

(2) Response functions in the Zimm model

The equation of motion, derived from (3) for a solvent at rest ($v_0 \equiv 0$), is:

$$\frac{dx_n}{dt} = \sum_m B_{nm} \ \varphi_m \qquad \text{(A.3)}$$

$$B_{nn} = W/\rho$$

$$B_{nm} = W/\rho \, \xi |n - m|^{-1/2} \qquad (n \neq m) \tag{A.4}$$

$$\rho = \frac{3 k_B T}{\sigma^2}$$

The force φ_m has been written down in equation (7)

$$\varphi_m = F_m - \rho \sum_{m'} A_{mm'} \, x_{m'} \tag{A.5}$$

where $\hat{A}$ is the symmetric matrix introduced by Zimm [4]

$$A_{mm} = 2 \qquad A_{m+1,m} = -1 \qquad \text{for } 1 < m < N \qquad \text{etc.}$$

In an operator notation we rewrite equations (A.3) and (A.5) as :

$$\frac{d|x)}{dt} = \hat{B}|\varphi) = \hat{B}\left[|F) - \hat{A}|x)\right] \ . \tag{A.6}$$

We then introduce the eigenvectors $|p)$ and eigenvalues $1/\rho\,\tau_p$ of the matrix $\hat{B}\hat{A}$

$$\hat{B}\hat{A}\ |p) = \frac{1}{\rho\tau_p} |p) \ .$$

As explained in ref. [4], the matrix $\hat{B}\hat{A}$ is not hermitian, and the vectors $|p)$ do not form an orthogonal set (except for the Rouse limit). But we have the property

$$(p|\hat{A}|q) = 0 \quad \text{for} \quad p \neq q \ .$$

We also introduce a normalised vector $|0)$, for which all components are equal to $N^{-1/2}$, and a vector $|\beta_0) = \hat{B}^{-1}|0)\ [(0|\hat{B}^{-1}|0)]^{-1}$. This is such that

$$(\beta_0|0) = 1$$

$$(\beta_0|p) = 0 \qquad (p \neq 0) \ .$$

We can analyse $|x)$ in the eigenmodes $|p)$ now :

$$|x) = \sum_p c_p(t)\ |p) \ .$$

Taking the scalar product with $(p|\hat{A}$ (for $p \neq 0$), we get

$$\frac{dc_p}{dt} + \frac{c_p}{\tau_p} = \frac{[p|F(t)]}{\rho\ \tau_p(p|A|p)}$$

(A.7a)

$$\frac{dc_p}{dt} = \frac{1}{\rho\ \tau_p(p|\hat{A}|p} \left[[p|F(t)] - \frac{1}{\tau_p} \int_{-\infty}^{t} dt'\ e^{(t'-t)/\tau_p} [p|F(t')] \right]$$

For $p = 0$ we take the scalar product with $(\beta_0|$ and get

$$\frac{dc_0}{dt} = \frac{[0|F(t)]}{(0|\hat{B}^{-1}|0)}, \tag{A.7b}$$

Writing $F_n = (n|F)$, etc., we obtain the response functions from equation (A.7)

$$\beta_{nm}(t) = \frac{(n|0)(0|m)}{(0|\hat{B}^{-1}|0)}\delta_+(t) + \sum_{p\neq 0} \frac{1}{\rho\tau_p} \frac{(n|p)(p|m)}{(p|\hat{A}|p)} \left[\delta_+(t) - \frac{1}{\tau_p} e^{-t/\tau_p} \right] \tag{A.8}$$

The first term is independent of n and m since $(n|0) = (0|m) = N^{-1/2}$. In both terms the symbol δ_+ represents a δ function with a peak immediately after $t = 0$ (such that $\int_0^\infty \delta_+(t)dt = 1$).

(3) Correlation functions

Inserting (A.8) and (A.2) and performing the time integrals, we obtain:

$$< [x_n(0) - x_m(t)]^2 > = |n - m| \frac{\sigma^2}{3} + \frac{2k_BT\ t}{N(0|\hat{B}^{-1}|0)} + \frac{2\sigma^2}{3} \sum_{p\neq 0} \frac{(n|p)(p|m)}{(p|\hat{A}|p)}(1 - e^{-t/\tau_p}). \tag{A.9}$$

Also a similar calculation gives for the motion of the center of gravity (along x)

$$< [g(0) - g(t)]^2 > = 2\ Dt + \frac{2\sigma^2}{3N} \sum_{p\neq 0} \cdot \frac{|(0|p)|^2}{(p|\hat{A}|p)} (1 - e^{-t/\tau_p}) \tag{A.10}$$

where $D = k_BT\ N^{-1}[(0|\hat{B}^{-1}|0)]^{-1}$ is the diffusion coefficient. The last term in equation (A.10) shows the coupling between the motion of g and the internal deformations. At large times equation (A.10) is dominated by the term $2Dt$: this corresponds to the approximation of equation (33), and is valid when $t \gg \tau_1$, where τ_1^{-1} is the lowest relaxation frequency of the chain ($\tau_1^{-1} \sim D\ R_G^{-2}$). In the opposite limit ($t \ll \tau_1$), equation (A.10) becomes:

$$< [g(0) - g(t)]^2 > = \frac{2 k_B T\ t}{N} (0|\hat{B}|0), \quad (t << \tau_1).$$

(4) Expansion of $S_{coh}(q\omega)$ to order $q^2 R_G{}^2$

Equations (A.9) and (18) can be used to compute numerically the dynamical form factors at all values of qR_G. Here we consider only the corrections of order $q^2 R_G{}^2$ to the coherent form factor : the result is :

$$S_{coh}(q, \omega) = L_0(\omega) \left[1 - \frac{q^2}{3} (R_G^2 + \delta R^2) \right] + \frac{q^2 \sigma^2}{3N} \sum_{p \neq 0} \frac{|(0|p)|^2}{(p|\hat{A}|p)} L_p(\omega) \quad (A.11)$$

with

$$L_p(\omega) = \frac{1}{\pi} \frac{Dq + \tau_p{}^{1}}{\omega^2 + (Dq^2 + \tau_p^{-1})^2}$$

$$\delta R^2 = \frac{\sigma^2}{N} \sum_{p \neq 0} \frac{|(0|p|^2}{(p|\hat{A}|p)}$$

(a) *in the Rouse limit* the modes $|p)$ are orthogonal to $|0)$ and we are left with

$$S_{coh}(q, \omega) = L_0(\omega) \left[1 - \frac{q^2 R_G{}^2}{3} \right]$$

Thus, to order $q^2 R_G{}^2$, there is no effect of the internal motions on the dynamical form factor.

(b) *in the hydrodynamic limit* there is an effect, the relative importance of which is measured by $\delta R^2 / R_G{}^2$. Making use of the numerical calculations of ref. [4] we find that the dominant contribution comes from $p = 2$ (first non-trivial even mode). The corresponding relaxation frequency is

$$\frac{1}{\tau_2} = 12.79\ \xi W (2/N)^{3/2}.$$

Keeping only this mode we find $\delta R^2 / R_G{}^2 \sim 1.7\ 10^{-2}$: thus there is now an effect of the internal motions, but it represents less than 2 per cent of the signal.

References

1. P.G. de GENNES, *Physics* 3, 37 (1967) - hereafter referred to as (I).

2. The Brillouin doublet occurring at $\omega = \pm sq$ (where s is the sound velocity of the solution) can be extracted rather easily from the interesting signal.

3. J.G. KIRKWOOD and J. RISEMAN, *J. Chem. Phys.* **16**, 565 (1948).

4. B.H. ZIMM, *J. Chem. Phys.* **24**, 269 (1956).

B.H. ZIMM, G.M. ROE and L.F. EPSTEIN, *J. Chem Phys.* **24**, 279 (1956).

5. See for instance C. TANFORD, *Physical Chemistry of Macromolecules*. Wiley, New York (1961).

6. G. POROD, *Z. Naturf.* **4a**, 401 (1949)

7. S.F. EDWARDS, *Proc. Phys. Soc.* **85**, 613 (1965).

8. R. KUBO, *Rep. Prog. Phys.* **29**, 255 (1966).

2

Minimum number of aminoacids required to build up a specific receptor with a folded polypeptide chain

2.1 Introduction

From the point of view of polymer physics, globular proteins are reminiscent of certain polysoaps, where the hydrophobic part of the chain clusters in a central core, while the hydrophilic residues tend to lie on the outer surface (see Stryer, 1968). This conformation provides both *stability* and *solubility*. The crucial difference with polysoaps lies of course in the presence of *specific receptors* on the protein. A schematic representation for such a receptor is shown in Fig. 2.1. The active site directly involves a number p of aminoacids. These are linked together by comparatively long loops of the peptidic chain. As has been emphasized by Monod (1969), it is of some interest to estimate the *minimum size* required for each of these loops, when the conformation of the active site itself is prescribed. This should lead in particular to one lower bound for the molecular mass of a globular protein carrying one active site.

One possible approach to this problem would be to use Monte Carlo methods on a computer. This, however, (a) is expensive and (b) gives very little insight. Here we shall restrict ourselves to a much more modest, but explicit, calculation, based on rough statistical arguments. We deliberately neglect all the effects related to the hydrophilic/hydrophobic affinities of the aminoacids, although these effects are certainly very important. In the present chapter, we consider first a single loop,

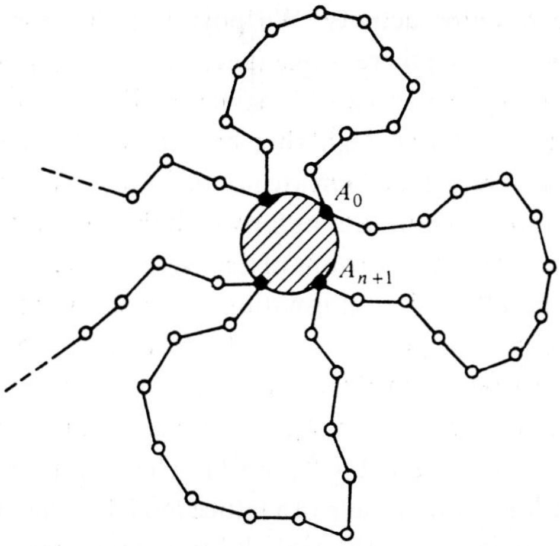

Fig. 2.1. A receptor site (hatched area) involving four aminoacids (black dots) directly. The connecting chain (white dots) comprises three loops and two open ends.

inserted in a dense proteic medium. We show that there does exist a well-defined (and rather large) minimum size for such a loop. Then we extend the argument to a set of $p-1$ loops converging towards the same active site, and show that the minimum size per loop remains essentially unaltered in this case.

2.2 The one-loop problem

2.2.1 Statement of the problem and number of allowed conformations

Consider a sequence $A_0, A_1, \ldots, A_n, A_{n+1}$ of aminoacids, following each other in the primary structure of the protein. We assume that A_0 and A_{n+1} belong to the active site, as shown in Fig. 2.1. Their relative positions and orientations are fixed. The other aminoacids ($A_1 \ldots A_n$) are assumed not to play a direct rôle in the receptor activity. We postulate that for a given sequence $(A_1 \ldots A_n)$ there is one (and only one) conformation of lowest energy which can be achieved. Depending on the choice of $(A_1 \ldots A_n)$ we might then have 20^n conformations for our loop. However, all of them are not realizable. First, we must retain only those conformations for which the loop under study does *not* intersect other portions of the overall protein chain: we call these the allowed conformations. Counting the allowed conformations corresponds to an excluded volume problem in a dense homogeneous medium, familiar from the theory of concentrated polymer systems (for a discussion of this concept applied to proteins, see Flory, 1961). The result (originally derived by Flory with the help of a lattice model) is that, for each unit A added to the chain, the probability of not intersecting any other part of the chain is a constant, and roughly of order e^{-1} (where e is the base of Neperian logarithms). With this estimate, the number of allowed conformations is reduced to $(20/e)^n$. Our problem is to find how many of these are compatible with the geometrical requirements imposed by the active site.

2.2.2 *Position and orientation of the end group*

To each of the $(20/e)^n$ conformations would correspond a well-defined position and orientation of the end group A_{n+1} with respect to A_0. The positions may, for instance, be labeled by the coordinates (x, y, z) of the α carbon in A_{n+1} with respect to the c carbon in A_0. Similarly the angular properties may be specified in terms of a rotation vector $(\Omega_x, \Omega_y, \Omega_z)$ relating the actual orientation of, say, the amide group in A_{n+1}, to a chosen reference state. The set $(x, y, z, \Omega_x, \Omega_y, \Omega_z)$ defines one point M in a six-dimensional space. There are $(20/e)^n$ such points M to consider. How are they distributed?

As soon as $n \geqslant 4$ the angular variables are spread more or less at random (each of them over an interval 2π). The spatial variables, on the other hand, have a nearly gaussian distribution, even in the presence of strong excluded volume effects, provided that the medium is densely filled and homogeneous. Homopeptide chains are known to become gaussian in solution only at very large $n(\sim 30)$. But here we are interested in heteropeptide chains with a statistical distribution of sequence: such systems tend to be gaussian even for rather small $n(\gtrsim 10)$. This leads us to a distribution function $\rho(M)$ for the points M of the form:

$$\rho(M) = \left(\frac{20}{\mathrm{e}}\right)^n (2\pi)^{-3} \left(\frac{2\pi}{3} nb^2\right)^{-3/2} \exp\left(-\frac{3x^2+y^2+z^2}{nb^2}\right) \tag{1}$$

where nb^2 is the mean square elongation of a chain containing n arbitrary aminoacids. The general aspect of this distribution is shown in Fig. 2.2.

In all that follows, we shall be interested only in *loops*, i.e. in conformations for which the end point is rather close to the starting point, or $|x|, y|, |z| < n^{1/2}b$. In this limit $\rho(M)$ takes the simple form:

$$\rho(M) \rightarrow \rho_0 = C\left(\frac{20}{\mathrm{e}}\right)^n n^{-3/2} b^3 \tag{2}$$

$$C = 3^{3/2}(2\pi)^{-9/2}$$

2.2.3 *Number of 'successful' conformations*

We call successful the conformation for which A_{n+1} is indeed located as desired to build up the active site. This implies that x, for instance, be in a certain interval:

$$x_0 \leqslant x \leqslant x_0 + x$$

and that the rotations Ω_x also belong to a certain small interval:

$$\Omega_{2x} \leqslant \Omega_x \leqslant \Omega_{2x} + \Delta\Omega_x$$

(with similar inequalities for the other components). The amplitudes Δx, $\Delta\Omega_x$, etc. are imposed by the thermal fluctu-

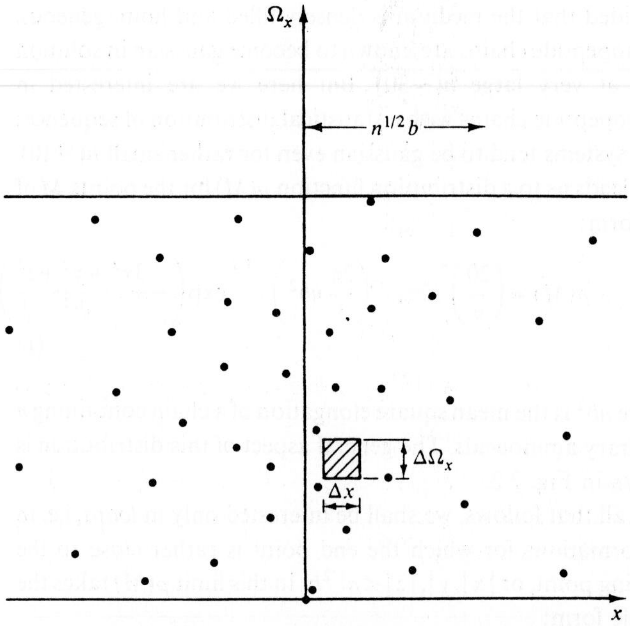

Fig. 2.2. Distribution of the positions (x) and rotation angles (Ω_x) of the terminal aminoacid A_{n+1} with respect to the first A_0, for various primary sequences ($A_1 \ldots A_n$) of the interconnecting loop. The region which satisfies the requirements on A_{n+1} to build up the active site is hatched.

ations of the overall structure, and depend essentially on the rigidity of the protein. In what follows we shall take typically:

$$\Delta x = \Delta y = \Delta z = 0.2\,\text{Å}$$
$$\Delta\Omega_x = \Delta\Omega_y = \Delta\Omega_z = 1/10\ \text{rad} \cong 6°$$

The points M associated with successful conformations of the loop are thus all contained in a small six-dimensional volume w:

$$w = (0.2)R^3 \times 10^{-3}.$$

The probability of failure p_f is the probability that *no* point M falls into the volume w. Let us assume that, in the region of interest, the points M are distributed at random, with an

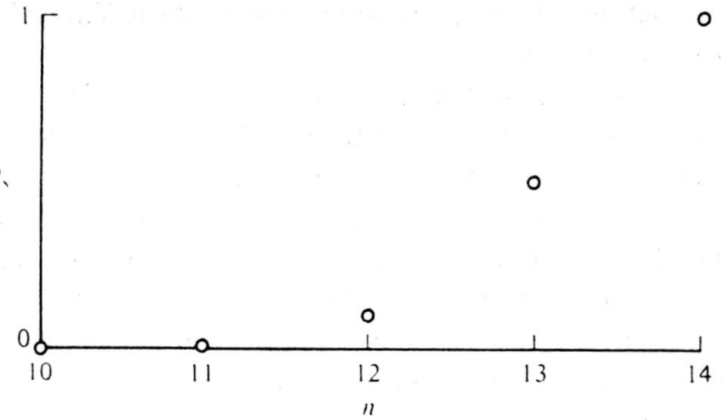

Fig. 2.3. Probability of success versus loop length. Note the abrupt variation near $n=13$.

average density ρ_c. (We shall discuss this assumption in more detail in the following section.) Then p_f is given by a standard Poisson formula for random events:

$$p_f = \exp(-\rho_c w) = \exp\left[-\left(\frac{20}{e}\right)^n \frac{Cw}{n^{3/2}b^3}\right]$$

The probability of success is $p_s = 1 - p_f$. The dependence of p_s on n, as deduced from this equation, is shown in Fig. 2.3. Because of the 'doubly exponential' character of equation (3), p_s is essentially 0 for all values of n below a certain threshold n_c, and $p_s \cong 1$ for $n > n_c$. Taking $b = 4$ Å, and the aforementioned value for w, we are led to $n_c \cong 13$. It turns out in fact that n_c is remarkably insensitive to the exact values chosen for b and w (the dependence being only logarithmic).

2.3 Discussion

2.3.1 Could the same conformation be obtained with more than one sequence?

One basic assumption underlying equation (3) is that, in the region around w, the points M are spread out at random, without any strong correlation between them. This might be incorrect in some cases.

Consider, for instance, the 'stereochemical code' proposed for aminoacids residues by Liquori (1968). In this model, the aminoacids may have only five different conformations (i.e. different sets of angles ϕ and ψ in the conventional notation: see Edsall *et al.* (1966). If this rule was strictly obeyed, the number of distinct conformations would be reduced from 20^n to 5^n; the same point M_0 would correspond to a large number of primary sequences. Our calcuation would then lead to a much larger value (~ 30) for n_c.

We believe, however, that our earlier estimate for n_c is the correct one, for the following reason: the angles ϕ and ψ for each aminoacid do have some small deviations from the idealized values selected by Liquori. Thus, the many sequences, which, in a strict version of the code, would converge to the same point M_0, will in fact lead to a *cluster* of points M centered around M_0 (Fig. 2.4). Are these clusters well separated from each other, or do they overlap? The calculation of the previous section applies only when they overlap, since then the overall density of points M becomes uniform.

To answer this question, let us first estimate the size of one cluster. A spread ε in the angles for each aminoacid implies for the orientation of the final group A_{n+1} an uncertainty:

$$\delta\Omega_x \sim \delta\Omega_y \sim \delta\Omega_z \sim \varepsilon n^{1/2}.$$

As regards the position of A_{n+1}, since the overall loop size is of order $n^{1/2}b$, a change ε in the angles for *one* aminoacid will displace x, y, and z by $\sim \varepsilon n^{1/2}b$. The cumulative effects of n aminoacids will give a result $n^{1/2}$ larger, i.e.:

$$\delta x \sim \delta y \sim \delta z \sim n\varepsilon b.$$

Thus the volume of one cluster is approximately:

$$\Delta = \varepsilon^6 n^{9/2} b^3.$$

On the other hand, the number of clusters per unit volume in six-dimensional space, is given by the analog of equation (2):

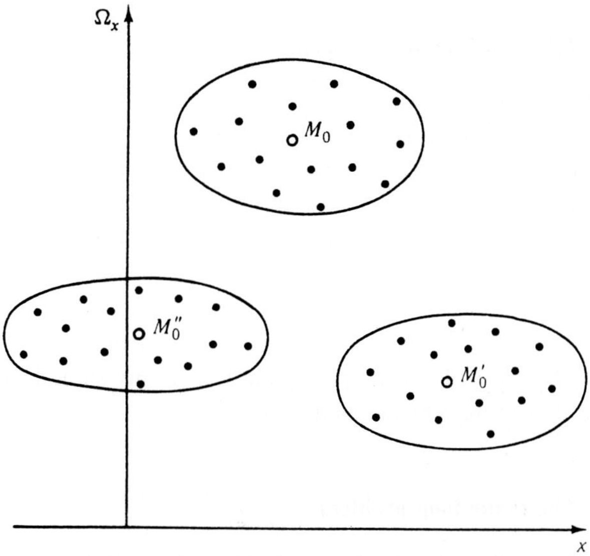

Fig. 2.4. Clustering effects in the (x,Ω_x) diagram. If the stereochemical code of Liquori was strictly obeyed, various primary sequences would give the same point M_0. Small deviations from the code spread these points in a cluster around M_0.

$$\tilde{\rho}_c = Cb^{-3}n^{-3/2}.$$

The overlap will be large if:

$$\tilde{\rho}_c \equiv C\varepsilon^{-6}n^3.$$

From the experimental distributions of points in the $(\phi\psi)$ plane for proteins of known structure (see, for instance, Ramachandran, 1968) we estimate $\varepsilon \cong 0.2$ rad (or 12°). Then $\tilde{\rho}_c\Delta$ becomes larger than 1 as soon as $n > 15$. Thus (if our guess for ε is correct), this 'clustering' effect will at most increase n_c by two units, and is not too important.

2.3.2 *Effects of the protein surface*

The receptor site has to be close to the surface, and the loops must reach the surface sometimes. As a model system, we may take a planar surface, and restrict the loop paths to one half space limited by this plane (the receptor site being treated as a point on the surface). The result of such a model is simply to modify slightly the coefficient C in equation (2), and the effect on n_c is very weak.

Of course, the surface will also influence the hydrophilic/hydrophobic balance of the loop sequence, and this should be included in a more refined calculation.

2.4 The many-loop problem

When the receptor site under consideration involves directly a number $p > 2$ of aminoacids (as it probably does in most cases), we have $p-1$ loops. Does the same value of n_c apply to all of them?

At first sight the answer would seem to be no, since the first loop restricts the amount of phase space available for the second one, etc. But we believe that in fact the answer is yes, for the following reasons.

(a) The Flory factor $\exp(-n)$ does contain the effect on one loop, of all the material present, and in particular of the other loops.

(b) It is of course true that, if p is comparatively large, many loops will have to converge towards the same small region defined by the active site: this implies a reduction in entropy since the loop will have to stretch radially as shown in Fig. 2.5. A reduction in entropy means an unfavorable factor. However, we shall now estimate this reduction and show that it is weak.

Assume for instance that the $p-1$ loops fill completely a spherical region of radius R around the active site. Apart from numerical coefficients, the volume of this region will be

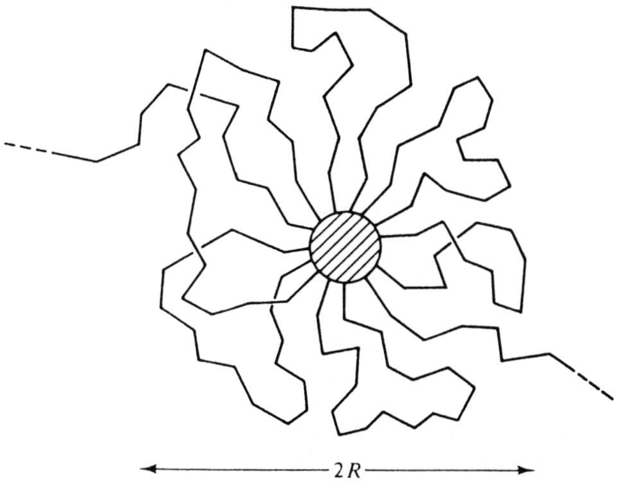

Fig. 2.5. When the number of loops is large, their paths near the active site are stretched, with a corresponding reduction in entropy.

$$R^3 \sim (n+1)(p-1)v \sim npv$$

where v is the average volume per aminoacid, and n the average chain length. Each loop is stretched over a distance of order R and suffers a reduction in entropy R^2/nb^2 (in dimensionless units). This corresponds to a reduction factor in the number of allowed conformations per loop, of order:

$$\exp(-R^2/nb^2) \sim \exp(-p^{2/3}n^{-1/3})$$

(assuming $v \sim b^3$). For the values of p and n of interest here, this factor is not much smaller than unity, and it may be omitted completely.

Thus, at least to a first approximation, our estimate of n_c should apply to the many-loop problem. Since there are $n-1$ loops, each of them involving at least n_c+1 aminoacids, the total number of aminoacids required to build up the receptor site may not be smaller than $(p-1)(n_c+1)$. Of course, this is only a lower limit: in particular, the requirements of stability, rigidity, and solubility will often impose longer chains.

References

Edsall, *et al.* (1966) *Biopolymers* **4**, (1966) 121.

Flory, P.J. (1961) *J. Polymer Sci.* **69**, 105, especially section 4.

Liquori, A.M. (1968) in A.D. Ketley (ed.) *The stereochemistry of macromolecules*, Dekker, New York, vol. 3, p. 287.

Monod, J. (1969) course, Collège de France.

Ramachandran, G.N. (1968) review in Rich and Davidson (eds.) *Structural chemistry and molecular biology*, Freeman, London, p. 77.

Stryer, L. (1968) *Ann. Biochem.* **37**, 25.

Afterthought: Minimum number of aminoacids required to build up a specific receptor with a folded polypeptide chain

This discussion is very naive. It ignores at least two basic requirements: a) the final, globular, protein, is more robust if it is compact (no voids inside, and a sharp boundary); b) the aminoacids at the outer surface should preferentially be hydrophilic (and the interior should be hydrophobic). However I feel optimistic about this calculation. If at one step, we do not have the complete freedom of choosing amongst twenty aminoacids, but only amongst the hydrophobic subgroup, we lose a factor of order 2. (This would roughly shift the magic number from 14 to 15.) On the other hand, many aminoacids have not one, but ~ 2 ground state conformations (of very similar energies), and this tends to bring us back to the original estimate.

This general type of approach — on the build up of functional proteins — has also been used, in a much more refined version, by K.A. Dill [*Biochemistry* **29**, 7133 (1990)].

THE JOURNAL OF CHEMICAL PHYSICS VOLUME 55, NUMBER 2 15 JULY 1971

Reptation of a Polymer Chain in the Presence of Fixed Obstacles

P. G. DE GENNES

Laboratoire de Physique des Solides, Faculté des Sciences, 91—Orsay, France

(Received 18 January 1971)

We discuss possible motions for one polymer molecule P (of mass M) performing wormlike displacements inside a strongly cross-linked polymeric gel G. The topological requirement that P cannot intersect any of the chains of G is taken into account by a rigorous procedure: The only motions allowed for the chain are associated with the displacement of certain "defects" along the chain. The main conclusions derived from this model are the following:

(a) There are two characteristic times for the chain motion: One of them (T_d) is the equilibration time for the defect concentration, and is proportional to M^2. The other time (T_r) is the time required for complete renewal of the chain conformation, and is proportional to M^3.

(b) The over-all mobility and diffusion coefficients of the chain P are proportional to M^{-2}.

(c) At times $t<T_r$ the mean square displacement of one monomer of P increases only like $\langle (r_t-r_0)^2\rangle =$ const $t^{1/4}$.

These results may also turn out to be useful for the (more difficult) problem of entanglement effects in unlinked molten polymers.

I. INTRODUCTION

The stochastic motions of a single polymeric chain dissolved in a solvent of low molecular weight are reasonably well understood in terms of a model set up by Rouse[1] and improved by Zimm.[2] On the other hand, the dynamical properties of concentrated (or molten) polymers are still poorly understood. An excellent review of the experimental situation has been given by Ferry.[3] The main features appear to be the following:

(1) The "creep compliance" $J(t)$ (i.e., the delayed response in strain induced by a step function increase in stress) displays a glasslike behavior at short times ($t\ll T'$), then an approximate plateau (rubberlike behavior $T'<t<T''$), and finally a liquidlike behavior ($t\gg T''$). The molecular mass dependence of the "terminal time" T'' has been measured on a few typical systems, and appears to be[3]

$$T''\sim M^{3.3}. \tag{I.1}$$

(2) In the liquidlike region $J(t)$ is linear in time:

$$J(t)\simeq J_e^0+(t/\eta).$$

The zero-frequency viscosity η depends strongly on M for large M, and follows the celebrated law[4]

$$\eta\sim M^{3.4}. \tag{I.2}$$

(3) The self-diffusion coefficient D for one polymeric chain moving among the others has been measured only in a few cases, and the dependence of D on M is not very well established. For polyethylene, McCall *et al.* have found[5]

$$D\sim M^{-5/2}. \tag{I.3}$$

Qualitative arguments leading to Eq. (I.2) for the viscosity have been given by Bueche[6] and Graessley[7]; but they lead to rather unsatisfactory exponents when applied to Eq. (I.1) ($T''\sim M^{4.5}$) or to Eq. (I.3) ($D\sim M^{-3.5}$). Another attractive approach to describe entanglement effects is based on the idea of a transient network[8]; the relaxational modes of a set of chains which are intercoupled by friction at some points have been worked out.[9] However, some of the statistical assumptions on which the model is based (e.g., the fact that entanglement takes place on the same monomer at all times) are still open to some doubt.

These observations have prompted us to try another approach: instead of a molten polymer, we focus our attention on a different system, where entanglements are also dominant, but where their effects can be understood in terms of a more precise statistical analysis. The system that we have in mind is made of a single, ideal, polymeric chain P, trapped inside a three-dimensional network G, such as a polymeric gel. A two-dimensional representation of this situation is shown on Fig. 1, where G reduces to a set of fixed obstacles $O_1O_2\cdots O_n\cdots$. The chain P is not allowed to cross any of the obstacles; however, it may still move in a wormlike fashion between them. The details of the model are given in Sec. II. In Sec. III we present an elementary calculation of the over-all chain mobility (or of the self-diffusion coefficient, which is related to the mobility by an Einstein relation). We find that both coefficients behave as M^{-2}. In Secs. IV and V we carry out a more detailed investigation of some microscopic properties: Brownian motion of one monomer, fluctuations of the end-to-end vector, etc. The two different times which emerge from these results are discussed in Sec. VI.

II. A MODEL FOR REPTATION

We consider a freely jointed chain of N "monomers." The positions of successive monomers are labeled $\mathbf{r}_1, \mathbf{r}_2, \cdots, \mathbf{r}_n, \cdots, \mathbf{r}_N$. The vector intervals

$$\mathbf{a}_n = \mathbf{r}_{n+1} - \mathbf{r}_n \simeq \partial\mathbf{r}/\partial n \tag{II.1}$$

are assumed statistically independent

$$\langle \mathbf{a}_n(t)\cdot\mathbf{a}_m(t)\rangle = \delta_{nm}a^2. \tag{II.2}$$

We assume that the length of the chain is very large when compared to the distance between neighboring links in the fixed network: then the possible chain motions are strongly restricted. We postulate that the

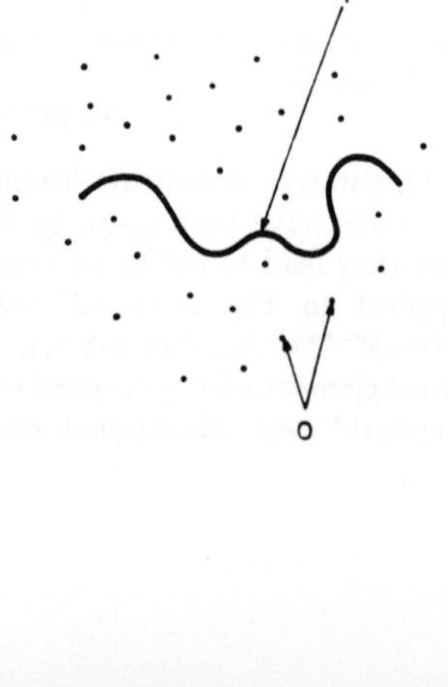

FIG. 1. The chain P is free to move between the fixed obstacles 0, but it cannot cross any of them.

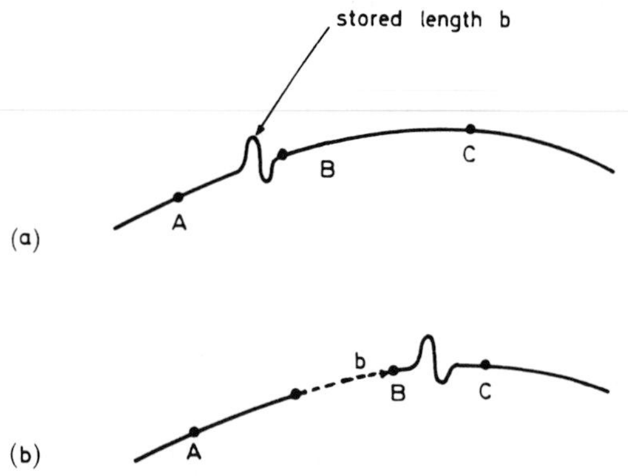

FIG. 2. A "defect" moves from A towards C along the chain. When it crosses monomer B, this monomer is displaced by an amount b.

only allowed motions correspond to the migration of certain "*defects*" along the chain. A qualitative picture for such a defect is shown on Fig. 2. The curvilinear interval between two monomers (n) and (m) would be equal to $(n-m)a$ in the absence of defects. But if we have ν defects in this interval, each of them storing an amount b of length, the curvilinear interval becomes

$$s_n - s_m = (n-m)a - \nu b. \tag{II.3}$$

For simplicity we take all defects with the same "stored length" b. (A distribution of b values can easily be included, but leads to no interesting change in the final results.) The most important consequence of Eq. (II.3) is that there is a conservation law on defect number: ν changes only when a defect crosses one of the end points (n or m) of the interval. This conservation law, plus an assumption of short-range forces (no backflow effects of the Zimm[2] type), leads to a one-dimensional diffusion equation for the gas of defects along the chain. To write down this equation, let us introduce the number ρ of defects per unit length of the extended chain. [$\rho = \nu/(n-m)a$ in the example of Eq. (II.3).] We shall constantly assume that ρ is small, i.e., that we have a dilute gas of defects. Let us also define a defect current J_n (=number of defects passing through point n per unit time). Then the conservation law is

$$(\partial\rho_n/\partial t) + (\partial/a\partial n)J_n = 0, \tag{II.4}$$

and J_n has the explicit form

$$J_n = \Delta\{-(\partial\rho/a\partial n) + (\rho/k_BT)\varphi_n\}. \tag{II.5}$$

Δ is the diffusion coefficient of the defects along the chain: It is a microscopic constant, characteristic of local jump processes, and is independent of the molecular mass M (for large M). The second term in Eq. (II.5) represents the drift of defects due to external forces $\mathbf{f}_n$ applied on the monomers. φ_n is the force per defect; to relate φ_n to $\mathbf{f}_n$ we write down the work performed when one defect moves by an amount $(a\delta n)$ along the chain: this means that δn monomers are displaced, each of them by an amount $\mathbf{b}$, where $\mathbf{b}$ is a vector of magnitude b, tangential to the chain. Thus

$$\delta n \mathbf{f}_n \cdot \mathbf{b} = a\delta n \varphi_n,$$

$$\varphi_n = a^{-1}\mathbf{b}\cdot\mathbf{f}_n. \qquad \text{(II.6)}$$

Equations (II.4) and (II.5) for the gas of defects must be supplemented by boundary conditions at both ends of the chain $n=0$ and $n=N$.[10] We assume that, at the ends, the density ρ relaxes very rapidly towards a constant equilibrium value $\bar{\rho}$,

$$\rho_0 = \rho_N = \bar{\rho}.$$

In the absence of external forces, the general solution of the diffusion equations (II.4) and (II.5) then takes the form

$$\rho_n - \bar{\rho} = \sum_p c_p \sin(p\pi n/N) \exp(-t/\tau_p), \qquad \text{(II.7)}$$

where p is a positive integer, labeling the various relaxation modes, c_p is a constant, and the relaxation times τ_p are given by

$$\tau_p = (1/\pi^2)[(Na)^2/\Delta p^2]. \qquad \text{(II.8)}$$

The longest among these relaxation times is

$$T_d \equiv \tau_{(p=1)} = \pi^{-2}(Na)^2/\Delta \qquad \text{(II.9)}$$

and it is proportional to the square of the molecular mass. We may say that T_d is the time for equilibration of the gas of defects.

When the defects move, the chain progresses, as shown on Fig. 4. The velocity of the nth monomer is related to the defect current J_n by

$$d\mathbf{r}_n/dt = \mathbf{b}J_n. \qquad \text{(II.10)}$$

Equation (II.10) must also be supplemented by boundary conditions. For instance, when a defect leaves the chain at the extremity (N), a new terminal arc (of length b) is set up and may take various orientations: in zoological terms, the head of the snake must decide which gate it will choose in the cross-linked network. We assume that this choice is at random, i.e., that the last vector $\mathbf{b}$ is completely uncorrelated to the preceding ones.

The reptation equations (II.4), (II.6), and (II.10) are significantly more complicated than the same equations for a free chain.[1] However, as we shall see in the following sections, it is possible to extract from them some definite predictions on the stochastic behavior.

III. ELEMENTARY CALCULATION OF THE OVER-ALL MOBILITY

Let us assume for the moment that the time T_r for complete renewal of the chain conformation is much larger than the equilibration time of the defects T_d. This inequality will be confirmed in Sec. V. When it holds, we can derive the chain mobility (defined as the ratio of the drift velocity to the applied force) by a very simple argument.

Consider the chain P in reptation under a constant (weak) force $\mathbf{F}$, applied along a certain direction (Z). The force per monomer is $\mathbf{f}_n = \mathbf{F}/N$ and the force on one defect is, according to Eq. (II.6),

$$\varphi_n = \mathbf{b}\cdot\mathbf{F}/Na. \qquad \text{(III.1)}$$

The vectors $\mathbf{b}$ are time dependent, but they change only on the scale T_r. On the other hand, the gas of defects reaches a steady state under external forces in a much shorter time T_d. Thus, we can treat the $\varphi_{n'}$ as constants, and write down the response of the gas of defects as a steady state conduction current $\bar{J}$:

$$J = \bar{\rho}\mu N^{-1} \int_0^N dn \varphi_n, \qquad \text{(III.2)}$$

where $\mu = \Delta/k_BT$ is the defect mobility. From Eq. (III.1) we get

$$J = (\bar{\rho}\mu/N^2a) \int \mathbf{F}\cdot\mathbf{b}dn$$
$$= [\bar{\rho}\mu b/(Na)^2] \int \mathbf{F}\cdot d\mathbf{l}, \qquad \text{(III.3)}$$

where $d\mathbf{l}$ is the line element of the chain. The integral

$$\int d\mathbf{l} = \mathbf{P} \qquad \text{(III.4)}$$

is the end-to-end vector, and we have for the average defect current

$$J = [\bar{\rho}\mu b F/(Na)^2] P_Z. \qquad \text{(III.5)}$$

The center of gravity $\mathbf{g}$ of the chain moves with a velocity

$$\dot{\mathbf{g}} = N^{-1} \int_0^N dn \dot{\mathbf{r}}_n.$$

We can transform $\dot{\mathbf{r}}_n$ by Eq. (II.10) and insert (III.5) as the current, obtaining

$$\dot{\mathbf{g}} = \frac{\bar{J}}{N} \int_0^N \mathbf{b} dn = \frac{Jb}{Na}\mathbf{P}. \qquad \text{(III.6)}$$

Averaging over the values of $\mathbf{P}$ we are left with one nonvanishing velocity component:

$$\dot{g}_Z = \mu_{\text{tot}} F,$$

where the over-all mobility μ_{tot} is explicitly given by

$$\mu_{\text{tot}} = \mu[\bar{\rho}b^2\langle P_Z^2\rangle/(Na)^3]$$
$$= \mu\bar{\rho}b^2/N^2a. \qquad \text{(III.7)}$$

Equation (III.7) may also be written in terms of the self-diffusion coefficient $D_{\text{tot}} = k_BT\mu_{\text{tot}}$. Both coefficients are seen to decrease like N^{-2} (or M^{-2}). This is to be compared with the case of a free Rouse chain, where $D_{\text{tot}} \sim M^{-1}$.

IV. DISPLACEMENTS OF ONE MONOMER

The quantity

$$\langle[\mathbf{r}_n(t)-\mathbf{r}_n(0)]^2\rangle=\Gamma(t) \qquad \text{(IV.1)}$$

provides a good description of the motions of one monomer. It turns out that $\Gamma(t)$ may be computed rather simply when t is *smaller than the equilibration time of the defects* T_d [provided that (n) is not too close from the ends (0) or (N)]. Then, end effects may be omitted and we may consider that the relevant portion of the chain (around n) worms its way inside a thin, permanent, undeformable tube (Fig. 3).

The position of the monomer (n) may then be defined (instead of using $\mathbf{r}_n$) by its curvilinear abscissa along the tube, $s_n(t)$. Let us discuss first the quantity

$$c(t)=[s_n(t)-s_n(0)]/b. \qquad \text{(IV.2)}$$

$c(t)$ is the number of defects which have passed from the "left" to the "right" of monomer n during time t, minus the number of those which went from right to left. The average $\langle c(t)\rangle$ is zero for symmetry reasons. On the other hand, $\langle c^2\rangle$ is that average number of defects which commuted either from right to left or from left to right, during the time t. Since in a time t diffusion takes place over a length $(\Delta t)^{1/2}$, this number will be of order $\bar{\rho}(\Delta t)^{1/2}$. The calculation given in Appendix A confirms this, and gives

$$\langle c^2(t)\rangle=(2/\pi^{1/2})\bar{\rho}(\Delta t)^{1/2}. \qquad \text{(IV.3)}$$

It is also possible to prove, by a standard Holtsmark–Markoff analysis[11] that the distribution $p_t(c)$ of c at a given time t is Gaussian, provided that the "number of migrating defects" $\bar{\rho}(\Delta t)^{1/2}$ is much larger than one:

$$p_t(c)=[1/(2\pi\langle c^2\rangle)^{1/2}]\exp[-c^2/2\langle c^2(t)\rangle]$$
$$[\bar{\rho}(\Delta t)^{1/2}\gg 1]. \qquad \text{(IV.4)}$$

We can now proceed and describe the spread in three-dimensional space of $\mathbf{r}_n(t)-\mathbf{r}_n(0)$. Two points separated by a curvilinear distance $s=bc$ have a mean square distance in space given by

$$\langle r_c^2\rangle_{\text{AV}}=|s|a=|c|ba \qquad \text{(IV.5)}$$

because the tube has the conformation of a random coil. Thus we may write

$$\langle[\mathbf{r}_n(t)-\mathbf{r}_n(0)]^2\rangle=ba\int dc\,p_t(c)\,|c|$$
$$=ba(2/\pi)^{1/2}[\langle c^2(t)\rangle]^{1/2}$$
$$\langle[\mathbf{r}_n(t)-\mathbf{r}_n(0)]^2\rangle=(2/\pi^{3/4})ba\rho^{1/2}(\Delta t)^{1/4}$$
$$(T_d\gg t\gg 1/\Delta\bar{\rho}^2). \qquad \text{(IV.6)}$$

Equation (IV.6) is one of our central results: It describes a very special type of Brownian motion. For comparison we shall write down the laws for systems which are more familiar:

(a) diffusion of one small molecule

$$\langle r^2\rangle\sim t.$$

(b) motion of one monomer in a free Rouse chain[12]

$$\langle r^2\rangle\sim t^{1/2}$$

($t\ll N^2W^{-1}$, where W is a fundamental jump frequency).

Equation (IV.6) shows that for a chain in reptation $\langle r^2\rangle\sim t^{1/4}$: The motion is extremely slow. For this reason Eq. (IV.6) cannot be tested by inelastic neutron scattering (at least in the present state of the art). On the other hand, this very slow creep might lead to observable effects in nuclear relaxation. This is discussed for one limiting case in Appendix C.

We end up this section with a word of caution: although the distribution $p_t(c)$ is Gaussian [when $\bar{\rho}(\Delta t)^{1/2}\gg 1$] the distribution of $\mathbf{r}=\mathbf{r}_n(t)-\mathbf{r}_n(0)$ is not Gaussian: It decays roughly like $\exp(-r^{4/3})$ at large r.

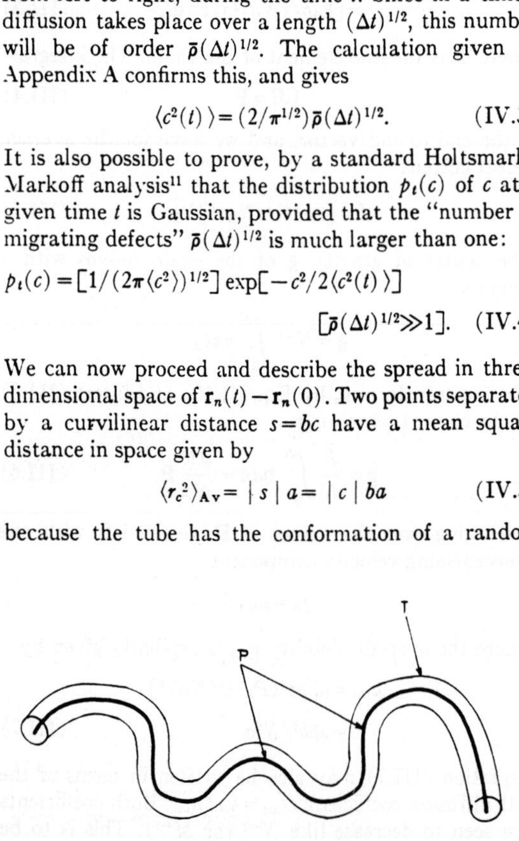

FIG. 3. An infinitely long chain P cannot move sideways: It is trapped in a thin tube T.

V. THE RENEWAL TIME T_r

Our aim in this section is to discuss the correlation for the end-to-end vector $\mathbf{P}=\sum_n \mathbf{a}_n$:

$$\langle\mathbf{P}(0)\cdot\mathbf{P}(t)\rangle=\phi(t). \qquad \text{(V.1)}$$

After deriving $\phi(t)$ we shall define a renewal time T_r as the time above which $\phi(t)$ becomes negligibly small.

It will turn out that T_r is much larger than T_d: This observation simplifies the analysis considerably. For times of order T_r we may forget completely the details of the reptation process inside the chain, and neglect in particular the fluctuations of its length $L=Na(1-\bar{\rho}b)\simeq Na$.

A. Relation between Vector Correlations and Migration along a Tube

In terms of the vectors $\mathbf{a}_n$, the function $\phi(t)$ [Eq. (V.1)] becomes

$$\phi(t)=\sum_{n,m}\langle\mathbf{a}_n(0)\cdot\mathbf{a}_m(t)\rangle$$
$$=a^2\sum_{n,m}G_{nm}(t). \qquad \text{(V.2)}$$

Our task is to derive the "vector correlation functions" $G_{nm}(t)$ in the limit $t\gg T_d$.

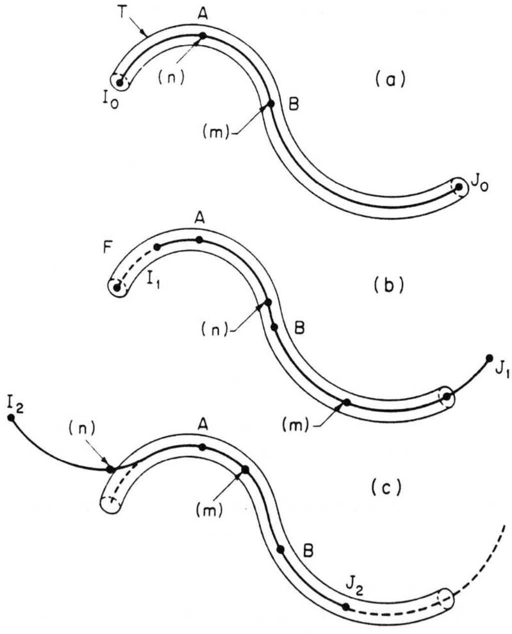

FIG. 4. Successive steps for the long time ($t\sim T_r$) motions of a chain inside a gel. (a) Initial position: The chain is restricted to a certain tube T. (b) First stage: The chain has moved to the right by reptation. (c) Second stage: The chain has moved to the left; the extremity I has chosen another path (I_1I_2). Note that, for this example, a certain fraction of the chain (I_1J_2) is still trapped in the initial tube at stage (c).

Let us consider first the chain P at the initial time $t=0$. It extends (on a length $\sim Na$) on a curvilinear path (I_0J_0) which we call the "initial tube" [Fig. 4(a)]. Let us call A the position of monomer (n) and B the position of (m), both taken at $t=0$. The initial tube has the conformation of an ideal coil: thus the orientations of $\mathbf{a}_n$ and $\mathbf{a}_m$ (both measured at the same time $t=0$) are uncorrelated for $n\neq m$. This may be written as

$$G_{nm}(0)=\delta_{nm}. \quad \text{(V.3)}$$

Now let us consider a later time t: The chain has moved back and forth and some portions of it are now out of the initial tube [Figs. 4(b), 4(c)]. We shall have nonvanishing correlations $G_{nm}(t)$ only if some part of the chain is still in the tube. For monomers (r) belonging to this part, we may still define a curvilinear abscissa $s_r(t)$, measured along the initial tube. Since, for the time scale ($\sim T_r$) of interest, changes in the chain length are negligible, we may write that all the monomers r have been displaced along the tube by the same amount $\sigma(t)$:

$$s_r(t)=s_r(0)+\sigma(t). \quad \text{(V.4)}$$

We get a nonvanishing contribution to $G_{nm}(t)$ if, and only if, the following conditions are satisfied:

(a) The monomer (m) must have moved to point A (because different line elements along the tube are uncorrelated).

(b) The vector $\mathbf{a}_m(t)$ must be in the direction imposed by the tube: That is, (m) must belong to the family of monomers (r) defined above. This demands that, at all times t' in the interval ($0t$), the extremities (I and J) of the chain have not passed through point A. (Figure 4 shows how the correlation is lost at a point F if I has passed F.)

Thus we must have

$$s_m(t)=s_n(0), \quad \text{(V.5a)}$$

$$s_n(0)-s_I(0)>\sigma(t')>s_n(0)-s_J(0). \quad \text{(V.5b)}$$

Finally, we may interpret $G_{nm}(t)$ as the probability for the random function $\sigma(t')$ to go from 0 (at time 0) to $s_n(0)-s_m(0)=(n-m)a$ (at time t), without ever exceeding the limits (V.5b).[12a]

B. Migration Along One Infinite Tube

The stochastic properties of $\sigma(t')$ depend only on the density fluctuations of the gas of defects, and are independent of the three-dimensional structure of the chain: To study $\sigma(t')$, it is enough to solve a restricted problem, where the chain is constantly confined to one unbranched tube of infinite length (Fig. 4).

The restricted problem is one dimensional, and is thus easily solved. The results, for the curvilinear abscissa s_n of one monomer, are the following:

(a) At times $t\ll T_d$ (and for n not too close to the end points) Eq. (IV.3) holds and we have

$$\langle[s_n(t)-s_n(0)]^2\rangle=(2/\pi^{1/2})\bar{\rho}b^2(\Delta t)^{1/2}.$$

(b) At times $t\sim T_d$ or $t>T_d$ a more detailed analysis, taking into account end effects, is required: It is described in Appendix B. The results become simple in the limit $t\gg T_d$:

$$\langle[s_n(t)-s_n(0)]^2\rangle=2D_ct \qquad (t\gg T_d). \quad \text{(V.6)}$$

Equation (V.6) describes the over-all diffusion of the chain along its fixed curvilinear path. The curvilinear diffusion coefficient D_c is given explicitly by

$$D_c=\bar{\rho}b^2\Delta/Na. \quad \text{(V.7)}$$

D_c (or the corresponding curvilinear mobility $\mu_c=D_c/k_BT$) is inversely proportional to the molecular mass: This is natural since the friction exerted by the chain (in uniform curvilinear motion) is proportional to its length.

The result (V.6) is independent of the index n: on the scale of times $t(\gg T_d)$ and of curvilinear distances ($\sim Na$) which is of interest here, the inner constrictions of the chain can be neglected, as was announced below Eq. (V.1): Eq. (V.6) is an equation for the over-all displacement σ:

$$\langle\sigma^2(t)\rangle=2D_ct. \quad \text{(V.6')}$$

It can also be shown that, when $t \gg T_d$, $\sigma(t)$ becomes a stochastic function with independent increments. This observation is crucial, because it is only for functions of their class that restrictions such as (V.5b) are tractable.

C. The Second Spectrum of Relaxation Times

Let us now restrict our attention to times $t \gg T_d$, and introduce a function $W(t, \sigma)$ giving the statistical weight for finding $\sigma(0)=0$ and $\sigma(t)=\sigma$. The function W is ruled by a standard diffusion equation

$$\partial W/\partial t = D_c(\partial^2 W/\partial \sigma^2) \qquad (t \gg T_d). \qquad \text{(V.8)}$$

We want to count in W only the random functions $\sigma(t')$ which remain inside the interval (of length Na) defined by Eq. (V.5b). This is obtained if we impose the boundary conditions

$$W[t', s_n(0)-s_I] = W[t', s_n(0)-s_J] = 0. \qquad \text{(V.9)}$$

Then W differs from G_{nm} only by a proportionality factor[13]:

$$G_{nm}(t) = aW[t, (n-m)a]. \qquad \text{(V.10)}$$

To find the explicit form of G_{nm} [satisfying the initial condition (V.3)] we then expand the solution of Eqs. (V.8) and (V.9) in a series of sine waves vanishing at both ends of the interval, and obtain

$$G_{nm}(t) = (2N)^{-1} \sum_p \sin(\pi p m/N) \sin(\pi p n/N) \times \exp(-t/\theta_p) \qquad (t \gg T_d), \qquad \text{(V.11)}$$

where p is again a positive integer, and the θ_p's represent a new set of relaxation times

$$1/\theta_p = \pi^2 p^2 (D_c/N^2 a^2) = p^2(1/T_r),$$

$$T_r = (Na)^2/\pi^2 D_c = (Na)^3/\pi^2 \bar{p} b^2 \Delta. \qquad \text{(V.12)}$$

Two remarks concerning T_r should be made at this point:

(a) T_r is proportional to N^3 or M^3, while T_d, as given by Eq. (II.9), is proportional only to N^2. Thus, for large N, $T_r \gg T_d$ as announced.

(b) T_d is independent of the density of defects ($\bar{p}$) and of the stored length per defect b. On the other hand, T_r is strongly dependent on both.

Finally, we insert Eq. (V.11) into (V.2) and derive explicitly the correlation function $\phi(t)$ for the end-to-end vector

$$\varphi(t) = \langle \mathbf{P}(0) \cdot \mathbf{P}(t) \rangle = (a^2/2N) \sum_p \sum_{nm} \exp(-t/\theta_p) \times \sin(p\pi n/N) \sin(p\pi m/N). \qquad \text{(V.13)}$$

Replacing the sums over n and m by integrals, we get $\phi(t)$ as a rapidly converging series

$$\phi(t)/Na^2 = (8/\pi^2) \sum_{(p\ \text{odd})} (1/p^2) \exp(-p^2 t/T_r). \qquad \text{(V.14)}$$

For $t < T_r$ $\phi(t) \to Na^2$. For $t \gg T_r$ $\phi(t)$ decays exponentially with time constant T_r: This shows that T_r is indeed the renewal time for chain conformations.

VI. DISCUSSION

A. Reptation in a Gel

We found two sets of relaxation times (τ_p and θ_p) for a chain in reptation. Each set has a structure reminiscent of a Rouse chain; however, the physical processes associated with both sets are very different:

(a) The first set (τ_p) corresponds to relaxation of the density of defects along the chain. The basic time for this set is $T_d \sim M^2$.

(b) After a time $t \sim T_d$ the chain is still essentially confined in the "tube" which held it at time $t=0$. Each monomer has then been displaced only by a small amount along the tube [$\delta s \sim b(\bar{p}aN)^{1/2}$].

(c) To disengage the chain completely from its initial tube we require a much longer time $T_r \sim M^3$. T_r may be interpreted as a dielectric relaxation time, if the monomers have a finite permanent dipole along the chain axis a_n.

Of course, at present, there are essentially no experimental data on the dynamics of free chains trapped inside a gel; but the self-diffusion coefficient D should be comparatively easy to measure. Our prediction is $D \sim M^{-2}$. The extra mechanical dissipation in the gel due to the presence of the chains P would also be of interest, but it would be extremely hard to measure it at the very low concentrations which are required to have independent chains. In any case the theoretical analysis of the viscosity η appears difficult, and the molecular mass dependence of η remains to be found.

What are the weaknesses of the model?

(1) At first sight our introduction of "defects" appears rather arbitrary. However, all our results are insensitive to the details of the reptation process: what is important is to take into account the sequestration in "tubes" and the fact that the total chain length is conserved.

(2) Back-flow effects have not been included. For a chain inside a conventional solvent, they are in fact very important[2]: However, in a gel, we expect that back flow around a moving object decays exponentially within a finite distance λ, as first shown by Debye[14]: thus our results are probably not too seriously affected by backflows, provided that the mean radius of the chain is much larger than λ.

(3) Excluded volume effects are not taken into account: they might modify slightly the exponents of the various laws. Also no special consideration has been

given to the knots which the chain may perform on itself. This is probably not too serious since these knots can also be unwound by our reptation process.

B. Conjectures on Polymer Melts

Obviously the time T_r is closely related to the "terminal time" derived from mechanical studies on concentrated or molten polymers. The M^3 dependence of Eq. (V.12) is in fact not very far from the experimental[3] ($M^{3.3}$) exponent.

It is also tempting to interpret T_d as the characteristic time for the glass–rubber transition. However, all these interpretations are merely guesses at the present stage: The transposition of our results (for one chain in a fixed gel) to a system of many mobile chains is extremely delicate and will require separate theoretical studies. If it turned out that the exponents are the same for both problems, we would come out with laws which are very different from the predictions by Bueche[6] and Graessley,[7] for both D and T_r.

ACKNOWLEDGMENT

The present author has greatly benefited from discussions with G. Agren, who is currently attempting to test some of these ideas by computer simulation.

APPENDIX A: MEAN SQUARE CURVILINEAR DISPLACEMENT OF ONE MONOMER

We consider a gas of defects with positions $m_1 \cdots m_i \cdots$ along an infinite chain P. The curvilinear abscissa of monomer (n) may be written as

$$s_n = \text{const} + b \sum_i \theta(m_i - n),$$

where

$$\theta(x) = \tfrac{1}{2}\,\text{sign}(x).$$

We then have

$$\langle c^2(t)\rangle = (1/b^2)\langle[s_n(t) - s_n(0)]^2\rangle$$

$$= \sum_{ij} \langle(\theta_{m_i(t)} - \theta_{m_i(0)})(\theta_{m_j(t)} - \theta_{m_j(0)})\rangle.$$

Different defects being uncorrelated, only the terms $i=j$ remain. The defects which contribute are those for which $\theta_{m(t)}$ is of opposite sign to $\theta_{m(0)}$. This leads to

$$\langle c^2(t)\rangle = a\bar{\rho} \sum_{m>n;\,m'<n} [p(m0 \mid m't) + p(m'0 \mid mt)],$$

where $p(m0 \mid m't)$ is the probability for one defect initially at m to move to m' in time t. To compute $\langle c^2\rangle$ it is convenient to differentiate it with respect to time and to make use of the diffusion equation

$$\partial p/\partial t = (\Delta/a^2)(\partial^2 p/\partial m^2)$$

$$= (\Delta/a^2)(\partial^2 p/\partial m'^2)$$

$$= -(\Delta/a^2)(\partial^2 p/\partial m \partial m').$$

Then the integrations on m and m' are easily done, and give

$$(\partial/\partial t)\langle c^2\rangle = 2\bar{\rho}(\Delta/a)\,p(n0 \mid nt).$$

Using the standard diffusion result

$$p(m0 \mid m't) = (2\pi u)^{-1/2} \exp[-(m-m')^2/2u^2],$$

$$(u^2 = 2\Delta t/a^2)$$

and integrating, one obtains Eq. (IV.3).

APPENDIX B: CURVILINEAR MOTIONS OF ONE MONOMER FOR A FINITE CHAIN IN AN INFINITE TUBE

Although the physical content is very different, the calculation of $\langle[s_n(t) - s_n(0)]^2\rangle$ is formally identical to the appendix of Ref. 15; thus we shall give here only the main lines of the argument. The equation for curvilinear motions, under applied forces f_n (we take f_n colinear to the tube at each point) is derived from Eqs. (II.5) and (II.6), and reads

$$\dot{s}_n \equiv \frac{\partial s_n}{\partial t} \equiv bJ_n = \Delta\left(\frac{\bar{\rho} b^2}{a k_B T} f_n + \frac{1}{a^2}\frac{\partial^2 s_n}{\partial n^2}\right).$$

The responses ($\dot{s}_n$) to a time dependent set of forces f_n may then be obtained by an expansion in Fourier series. The result has the causal form

$$\dot{s}_n(t) = \sum_m \int_{-\infty}^{t} dt' \beta_{nm}(t-t') f_m(t'),$$

where the response functions β_{nm} are given explicitly by

$$\beta_{nm}(t) = \frac{\bar{\rho} b^2 \Delta}{a} \{N^{-1}\delta_+(t) + (2/N) \sum_p \cos(\pi p n/N)$$
$$\times \cos(\pi p m/N)[\delta_+(t) - \tau_p^{-1} \exp(-t/\tau_p)]\},$$

where $\delta_+(t)$ is a delta function with its peak at $t=+0$; p is a positive integer. From the β's one derives correlation functions by the Kubo theorem

$$\langle \dot{s}_n(0) \dot{s}_m(t)\rangle = k_B T \beta_{nm}(t) \qquad (t>0)$$

and by two time integrations one obtains, for the diagonal term ($n=m$),

$$\langle[s_n(0) - s_n(t)]^2\rangle = \frac{2\bar{\rho} b^2 \Delta}{a}$$
$$\times \{N^{-1}t + (2/N) \sum_p \cos^2(\pi p n/N) \tau_p [1 - \exp(-t/\tau_p)]\}.$$

The second term in this equation is dominant at times $t<T_d$, and is then equivalent to Eq. (IV.3). The first term is the main contribution at large times ($t>T_d$) and is the only term kept in Eq. (V.6).

APPENDIX C: NUCLEAR RELAXATION BY REPTATION

To be specific, let us assume that the chain P carries some protons: their spin-lattice relaxation includes contributions from the modulation of their dipolar interactions by the motions of the chain.[16] Among these motions we shall consider only the slow reptations, omitting all the more rapid (and comparatively trivial) components such as methyl group rotation, etc. The reptation might indeed become dominant if the nuclear frequency ω_n is sufficiently low.

The spin–spin couplings which are modulated may involve either two protons of P [case (a)] or one proton of P and one fixed spin belonging to the obstacles[17] [case (b)]. In both cases we have to follow a moving object linked to a given monomer (n): a pair of protons (i and j) for case (a), or a single proton for case (b).

We shall now restrict our attention to a medium with *densely distributed obstacles* (i.e., a chain P moving inside a rubber), for the following reason: Consider a monomer (n) which, at time $t=0$, has a certain position and orientation. After a rather large time t, it has moved by reptation. The dipolar interaction of interest $\mathcal{H}_D(t)$ is then changed strongly, and becomes uncorrelated with $\mathcal{H}_D(0)$, except if by chance (n) has returned to its initial position (or within a small interval δc from it): The corresponding probability is $p_t(c=0)\delta c$. If the surrounding medium is rigid and dense, when monomer (n) returns to its initial position, it also recuperates its initial orientation: thus, the correlation function for the dipolar interaction $\mathcal{H}_D$ may be taken of the form:

$$\langle \mathcal{H}_D(0)\mathcal{H}_D(t)\rangle = \text{const}\, p_t(c=0).$$

Using Eq. (IV.4) (for large t) this gives

$$\langle \mathcal{H}_D(0)\mathcal{H}_D(t)\rangle = \text{const}\, t^{-1/4}.$$

The spin–lattice relaxation rate $1/T_1$ is proportional to the Fourier transform of correlation functions of this type, taken at the nuclear frequency ω_n or at the double frequency $2\omega_n$. Since

$$\int_0^\infty dt\, t^{-1/4} \cos\omega t = \text{const}\, \omega^{-3/4},$$

this leads to

$$1/T_1 = \text{const}\, \omega_n^{-3/4} \qquad (\text{for } \omega_n \to 0).$$

This exponent may also be derived, in case (b), by a different argument: One writes the relaxation rate $1/T_1$ in terms of a probability distribution

$$P(\mathbf{r}_i 0;\, \mathbf{r}_i' t;\, \mathbf{r}_j)$$

for finding spin (i) at $\mathbf{r}_i$ at time 0 and at $\mathbf{r}_i'$ at time t, in the presence of the (fixed) spin (j) with which it interacts. One then makes the approximation

$$P(\mathbf{r}_i 0;\, \mathbf{r}_i' t;\, \mathbf{r}_j) \to w(\mathbf{r}_i 0;\, r_i' t),$$

where $w(r_i 0;\, r_i' t)$ is the self-correlation function for which a second moment is given by Eq. (IV.6). What is needed is the function w for space intervals

$$\mathbf{r} = \mathbf{r}_i' - \mathbf{r}_i$$

which are comparable to an interatomic distance. An explicit calculation of w for such intervals (and large t) gives[18]

$$w(\mathbf{r}_i 0 \mid \mathbf{r}_i' t) = (3/2\pi rab)\, p_t(c=0).$$

Again this leads to a $|t|^{-1/4}$ dependence for the correlation function at large times.

To conclude this Appendix, we should emphasize that the $\omega_n^{-3/4}$ law for the relaxation rate refers only to one highly idealized limit: it may be unobservable in practice for various reasons:

(1) Effects related to the detailed monomer structure, to the chain rigidity, etc., have not been considered.

(2) In time intervals $t \sim \omega_n^{-1}$ the reptation process must involve many monomer units $[\bar{p}(\Delta t)^{1/2} \gg 1]$ for the asymptotic laws to be meaningful. This may be hard to obtain, even with high temperatures and low frequencies.

It is also not clear whether these results remain meaningful for the related, but different, problem of polymer melts. Here a recent experiment[19] indicates $1/T_1 \sim \omega_n^{-0.5}$ in the low-frequency limit.

[1] P. E. Rouse, J. Chem. Phys. **21**, 1272 (1953).
[2] B. H. Zimm, J. Chem. Phys. **24**, 269 (1956).
[3] J. D. Ferry, *Viscoelastic Properties of Polymers* (Wiley, New York, 1970), 2nd ed.
[4] A recent review on this field is G. C. Berry and T. G. Fox, Advan. Poly. Sci. **5**, 261 (1968).
[5] D. W. McCall, D. C. Douglass, E. W. Anderson, J. Chem. Phys. **30**, 771 (1959).
[6] F. Bueche, J. Chem. Phys. **20**, 1959 (1952). F. Bueche, *The Physical Properties of Polymers* (Interscience, New York 1962).
[7] W. W. Graessley, J. Chem. Phys. **43**, 2696 (1965) and **47**, 1942 (1967).
[8] A. S. Lodge, Rheol. Acta **7**, 379 (1968).
[9] A. J. Chompff, W. Prins, J. Chem. Phys. **48**, 235 (1968).
[10] Note that, in the limit of large N which is of interest here, we make no difference between $n=0$ or $n=1$. For the statement of boundary conditions leading to the modes of Eq. (II.7) it is more convenient to put the origin at $n=0$.
[11] For a review on the Holstein Markoff method, see for instance S. Chandrasckhar, Rev. Mod. Phys. **15**, 2 (1943).
[12] P. G. de Gennes, Physics **3**, 37 (1967).
[12a] We assumed here that, if monomer (m) returns to its initial position in the tube, it also recuperates the exact initial orientation. In fact, the arguments of Secs. V.B and V.C require only that (m) recuperates a finite fraction of its initial orientation. This fraction will appear as a time-independent scaling factor in Eq. (V.10).
[13] G_{nm} is a statistical weight per monomer while W is a weight per unit length.
[14] P. Debye, J. Chem. Phys. **14**, 636 (1946).
[15] E. Dubois Violette, P. G. de Gennes, Physics **3**, 181 (1967).
[16] See, for instance, A. Abragam, *Principles of Nuclear Magnetism* (Oxford U. P., London, 1961), Chap. 8.
[17] The fixed species might be a nuclear spin, or a paramagnetic impurity of spin S (acting through terms such as $S_Z I_+$ in the dipolar interaction).
[18] Again it must be emphasized that w is strongly non-Gaussian: imposing a Gaussian form, one would come out with a very different correlation function ($\sim |t|^{-3/8}$).
[19] R. Lenk, J. P. Cohen Addad, Solid State Commun. **8**, 1864 (1970).

Afterthought: Reptation of a polymer chain in the presence of fixed obstacles

Looking back at this paper, I am amazed at the slow course which we all followed after. The first active reaction was through the group of J. Ferry, who succeeded in trapping some chains in a compatible gel, and measuring their contribution to the complex elastic modulus. But in fact there was a much simple way to operate: inserting the test chains in a melt of (chemically identical) longer chains: the latter are so slow that they behave as a gel on the time scale of interest. It took us about four years to realise this simple point. Finally, the most delicate step is to go to a homodisperse melt, and to see if one chain is still trapped in an "effective gel" by its neighbours: the first detailed justification for this idea (with long enough chains $N \gtrsim 10\,N_e$) came from Doi and Edwards. There are still, however, some irresolved disputes between them and J. des Cloizeaux on "double reptation".

Coil-stretch transition of dilute flexible polymers under ultrahigh velocity gradients

P. G. De Gennes

Collège de France—11, place Marcelin-Berthelot, 75231 Paris Cedex 05—France
(Received 31 January 1974)

Because the hydrodynamic interactions are reduced by stretching, a solute polymer coil should unwind *abruptly* when a certain critical value of the velocity gradient is reached. Depending on the details of the velocity field ("longitudinal", or "transverse", or more complex gradients) this coil ↔ stretch transition may be continuous ("second order") or discontinuous ("first order"). In the latter case hysteresis should often be observed; a qualitative discussion of the associated relaxation times is given. Simple shear flow is an exceptional case, with no sharp transition. Some expected effects of the $C \leftrightarrow S$ transition on the mechanical behavior, on optical properties, and on chemical degradation in flow are briefly analyzed.

I. INTRODUCTION

An extremely large number of investigations on dilute, flexible polymers under various types of shear flow can be found in the literature. The effects involved are rather complex, but, they are well classified in various review papers (see for instance Ref. 1 and 2). We shall be concerned here mainly with possible distortions of the molecular shape, for which the most salient facts appear to be the following.

(1) For weak shear rates S, the relative distortions (as measured for instance by flow birefringence[3]) are essentially proportional to $\tau_0 S$, where τ_0 is the largest relaxation time of the unperturbed molecule,[4]

$$\tau_0 \cong 0.2\ \eta_S R_0^3/T \tag{I.1}$$

(R_0 is the unperturbed coil radius, η_S the solvent viscosity, T the temperature; we use units where Boltzmann's constant is unity).

(2) For larger S, at least three new effects come into play:

(a) For certain flows (in particular for "longitudinal" gradients as defined in Ref. 1) the linear theory of distortions predicts a divergence when $\tau_0 S > k$, where k is a numerical constant of order unity. The divergence announces a strongly stretched state, which may be of importance in connection with drag reduction by polymer additives, as suggested by Lumley.[5] Of course, the linear theory is not sufficient to describe properly the final state. Our aim here is to improve this description incorporating two effects (b) and (c) which turn out to be important. Effect (b) is rather trivial, but effect (c) is not.

(b) The elongation tends to saturate: The restoring force F which tries to maintain a compact chain becomes infinite if the chain is completely extended. In the "dumbbell model"[1] where the chain is represented by one single spring of extension $\mathbf{r}$, the restoring force is written as

$$F = -(3\,T\,\mathbf{r}/R_0^2)\,E(t), \tag{I.2}$$

where t is the extension ratio,

$$t = r/L = r/Za \tag{I.3}$$

(L is the extended length and Z is the number of monomers in one chain). The factor $E(t)$ is unity at small t, but diverges for $t \to 1$. Formulas for $E(t)$ in a simple chain have been quoted.[1] For realistic chains (with transgauche isomerism, etc.) the structure of $E(t)$ would become more complex, but the trend remains the same.

(c) *The hydrodynamic interactions between monomers decrease* when the molecule is stretched. With a compact coil, only the outer segments are subjected directly to the flow field.[6] But with an extended coil all segments are exposed. To understand this, consider a stretched molecule occupying a cylindrical domain of radius R ($\sim R_0$) and of length $\tilde{L} = Lt$ ($\sim L$) (Fig. 1). The number density of monomers/cm^3 inside the cylinder is

$$\rho = Z/\pi R^2 \tilde{L}. \tag{I.4}$$

The penetration depth δ of the flow field inside the cylinder can be calculated by the Debye–Bueche formula[6]:

$$\delta^2 = \eta_S/\rho f_1 = \pi R^2 L \eta_S / Z f_1, \tag{I.5}$$

where f_1 is the hydrodynamic friction coefficient for one monomer. We want to compare δ and R,

$$(\delta/R)^2 = \pi \eta_S \tilde{L}/Z f_1. \tag{I.6}$$

The quantity $f_1/\pi\eta_S$ has the dimension of length; for most practical situations, it is comparable to the monomer diameter a: the right hand side of Eq. (II.6) is thus

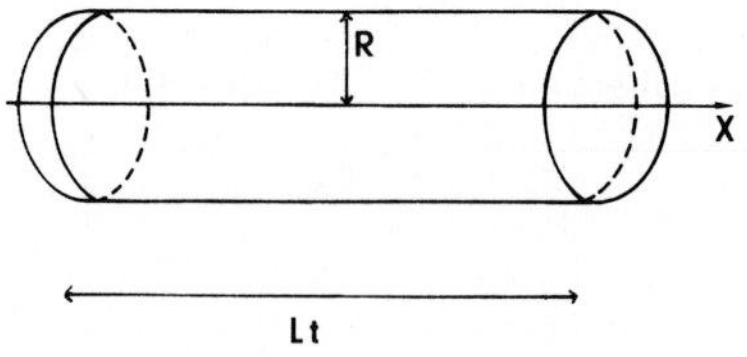

FIG. 1. Simple geometrical model for a flexible polymer stretched by a shear flow. The direction of stretching x will be defined as an eigenvector of the velocity gradient matrix in Sec. II. The radius R is comparable (slightly smaller) to the unperturbed coil radius R_0. The length is Lt, where t is smaller than unity. In the more detailed model of appendix A the radius R is allowed to vary with x.

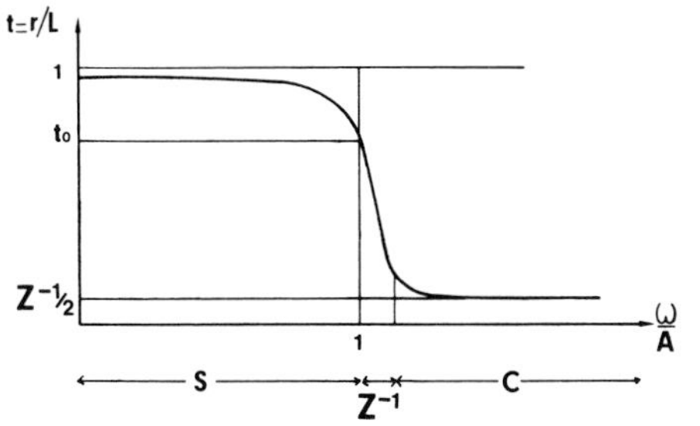

FIG. 2. Transition from a stretched state (S) to a coiled state (C) in two dimensional flow when the vorticity ω is increased. A is the symmetrized velocity gradient defined in the introduction, and is assumed large. The whole transition takes place in a relative interval $\Delta\omega/\omega \sim Z^{-1}$ (where Z is the number of monomers on one chain).

of order unity; the flow penetrates the entire cylinder. We conclude that the Rouse model[7] (which neglects all hydrodynamic interactions) becomes qualitatively valid at ultrahigh shear rates.

This effect has been known for a long time: it is for instance the source for the increase of viscosity at high gradients discussed in Ref. 8. However, the change in hydrodynamic interactions at high shear has some further consequences which do not seem to be recognized at present. Namely, if we increase S, the inner monomers become more exposed to the shear; the chain distortion increases, and the exposure is still further enhanced. Depending on the detailed structure of the flow, this feedback process may lead to two different behaviors, illustrated in Figs. 2 and 3.

The example shown in Fig. 2 is obtained with a two dimensional flow (v_x, v_z) described by the vorticity $\omega = \frac{1}{2}(\partial v_x/\partial z - \partial v_z/\partial x)$ and the pure deformation rate $A = \frac{1}{2} \times (\partial v_z/\partial x + \partial v_x/\partial z)$. A is assumed to be large and fixed, and ω is varied. When ω/A decreases and crosses the value 1, the end to end distance r of the chain increases monotonously but *steeply*: r switches from a value of order R_0 (coil) to a value of order L (extended length). We call this situation a *second order transition* between coil (C) and stretched (S) states.

The example shown in Fig. 3 is found in particular with longitudinal gradients (e.g., $v_x = Sx$, $v_z = -Sz$). Here the curve $r(S)$ folds back and in a certain range of S values, we have a *bistable equilibrium*. As will be shown in Sec. III, there is one critical value of $S = S^*$ at which the system (if it is operated very slowly) will switch from coil to stretch. We say that a first order transition takes place at $S = S^*$.

The importance of these abrupt transitions has not been fully appreciated in the past because the most usual case of pure "transverse" gradients (in the notation of Ref. 1) is special and shows only a weak transition. But in all other cases the transition is essential. We start here by a discussion of strong distortions (Sec. II) and then proceed with some simple examples of C–S transitions (Sec. III). At this stage, the arguments are essentially restricted to the simple "dumbbell" model. It might be feared that the sharp transitions become wiped out when the local distortions are considered: i.e., the ends of the chain tend to always remain coiled, while the central portion is most strongly stretched. Thus the $C \rightarrow S$ transition might proceed by progressive thinning of the end coils, without any sharp accident. We discuss this in an appendix (using a more complete "necklace" model) and find that the transition remains sharp. In Sec. IV we carry out a general discussion for an arbitrary velocity gradient $S_{kl} = \partial v_k/\partial x_l$ (except that the fluid is incompressible, $S_{kk} = 0$). In Sec. V we analyze some observable effects of the transition. Note that the present paper is restricted to *time independent* gradients. This is a serious restriction even for laminar flows[9] but a natural one for a first study.

All our discussion is qualitative in nature: we have not tried to calculate exactly all the numerical coefficients involved, and to predict precisely the shape of any experimental curve. Our aim is only to draw attention to certain possible new effects. If these effects do show up in some measurements, more refined numerical studies may be attempted. They would have in fact to include various minor corrections which are purposely omitted here (excluded volume, anisotropy of hydrodynamic interactions, etc.), and also the, more interesting, internal viscosity.

II. LARGE DISTORTIONS IN THE DUMBBELL MODEL

A. The Peterlin approximations

We shall follow closely the approximations and even the notation of Ref. 1. The distribution function for the end to end vector $\mathbf{r}$ will be called $\Phi(\mathbf{r})$. The associated current $\mathbf{J}$ contains three contributions: drift in the external velocity field $\mathbf{v} = \mathbf{S}\mathbf{r}$, motion under the restoring force $\mathbf{F}$ (Eq. 2), and Brownian motion with a diffusion coefficient D. (The effects of hydrodynamic interactions are incorporated into D.) Thus

$$\mathbf{J} = \Phi\,\mathbf{S}\mathbf{r} + D[(\Phi/T)\mathbf{F} - \nabla\Phi]. \tag{II.1}$$

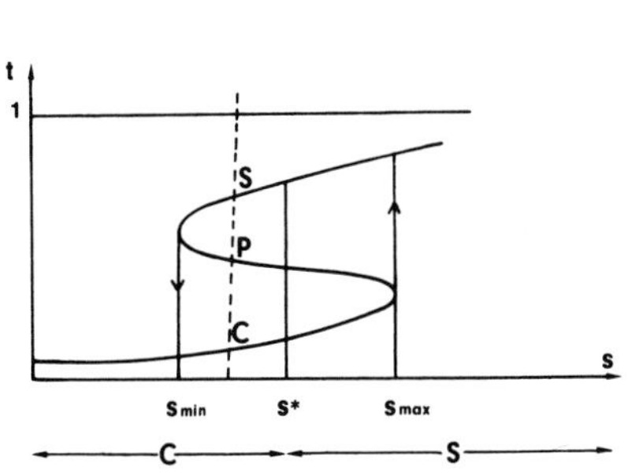

FIG. 3. $C \leftrightarrow S$ transition in a "longitudinal gradient." For a given shear rate S between the limits S_{min} (of the order of a Rouse frequency $1/\tau_R$) and S_{max} (of the order of a Zimm frequency $1/\tau_0$) there are three possible states of the molecule: a coil state C, an unstable state P, and a stretched state S. For reversible situations, the transition from S to C takes place at a particular value S^* of the shear rate. For most practical situations, hysteresis is expected, and should involve the cycle shown by arrows.

Conservation of probability imposes

$$\partial\Phi/\partial t + \operatorname{div} \mathbf{J} = 0. \qquad \text{(II.2)}$$

For the moment we shall be concerned only with stationary states where $\partial\Phi/\partial t = 0$. Also $S_{ii} = 0$. Then Eq. (II.2) may be rewritten as

$$\nabla^2\Phi + \frac{qE}{R_0^2}\Phi + \nabla\Phi \cdot \left[\frac{3E}{R_0^2}\mathbf{r} - \frac{1}{D}\mathbf{S}\cdot\mathbf{r}\right] = 0. \qquad \text{(II.3)}$$

As explained in the Introduction, both $\mathbf{r}$ and D depend on the elongation. But here, as usual, Eq. (II.3) will be solved in the Peterlin approximation where the factor E and D are treated as constants, and later adjusted self-consistently. Then the distribution $\Phi(\mathbf{r})$ remains a simple gaussian. Without constructing Φ explicitely, we can obtain all relevant averages through the following partial integrations:

$$\int x_i x_j \nabla^2\Phi\, d\mathbf{r} = 2\delta_{ij}\int \Phi d\mathbf{r} = 2\delta_{ij},$$
$$\int x_i x_j x_k (\partial\Phi/\partial x_k)\, d\mathbf{r} = -5\langle x_i x_j\rangle, \qquad \text{(II.4)}$$

where δ_{ij} is a Kronecker symbol, and the brackets $\langle\rangle$ represent an average over the distribution Φ. Multiplying (II.3) by $x_i x_j$ and using (II.4) we arrive at the following fundamental set of conditions:

$$\delta_{ij} - EC_{ij} + \tfrac{1}{2}\tau\sum_l (S_{il}C_{lj} + S_{jl}C_{li}) = 0, \qquad \text{(II.5)}$$

where

$$C_{ij} = 3\langle x_i x_j\rangle / R_0^2 \qquad \text{(II.6)}$$

and τ is a relaxation time of the molecule, dependent on the state of elongation, and related to the diffusion coefficient D by

$$\tau = R_0^2/3D(t). \qquad \text{(II.7)}$$

A plausible form for τ, as a function of the elongation parameter t, (justified in the Appendix), is

$$\tau(t) = \frac{\tau_R}{1 + u/t}, \qquad \text{(II.8)}$$

where τ_R is the Rouse relaxation time (proportional to Z^2).[7] The parameter u is essentially the ratio of the hydrodynamic radius to the end to end length a for one monomer unit, and is a constant of order unity. The relative contribution of the hydrodynamic interactions to the relaxation rate is measured by u/t. Two remarks may be useful in connection with Eq. (II.8):

(a) In the unperturbed coil ($t \sim Z^{-1/2}$) the relaxation time is

$$\tau = \tau_0 \cong (\tau_R/u) Z^{-1/2} = \text{const}\, Z^{3/2}$$

in agreement with the original calculation of Zimm.[4]

(b) For $t \sim 1$ we have $\tau \cong \tau_R$ as expected from our discussion of Sec. I.

Equation (II.5) must be supplemented by a self-consistency condition on the elongation parameter t, which we write, following Peterlin, as

$$\frac{r^2}{R_0^2} = \frac{1}{3}\sum_i C_{ii} = Zt^2. \qquad \text{(II.9)}$$

For a given set of velocity gradients S_{ij}, Eqs. (II.5, 8, 9) can then be solved to find the coefficients C_{ij}, t, and the effective relaxation time τ.

B. Limit of strong elongations

We shall now concentrate on the limit $t \gg Z^{-1/2}$. By Eq. (II.9) this implies that one (or more than one) eigenvalue of the symmetric matrix C_{ij} is much larger than unity. To investigate this situation, let us write down the set of Eqs. (II.5) in the frame (xyz) where $\mathbf{C}$ is diagonal. We get

$$1/C_{xx} = E(t) - \tau(t) S_{xx},$$
$$1/C_{yy} = E(t) - \tau(t) S_{yy}, \qquad \text{(II.10)}$$
$$1/C_{zz} = E(t) - \tau(t) S_{zz},$$

and

$$S_{xy}C_{yy} + S_{yx}C_{xx} = 0,$$
$$S_{yz}C_{zz} + S_{zy}C_{yy} = 0, \qquad \text{(II.11)}$$
$$S_{zy}C_{xx} + S_{xz}C_{zz} = 0.$$

Let us assume first that one, and only one, eigenvalue of $\mathbf{C}$ (e.g., C_{xx}) is large. (Physically this corresponds to a cigar shaped molecule elongated along x.)

Then we may neglect $1/C_{xx}$ in Eq. (II.10) and write

$$S_{xx} = E/\tau. \qquad \text{(II.12)}$$

Similarly, in the first and third equations (II.11), we see that, if C_{xx} is large,

$$S_{yx} = S_{yz} = 0. \qquad \text{(II.13)}$$

In other words, if the end to end vector $\mathbf{r}$ is along x, the velocity field

$$\mathbf{v} = \mathbf{S}\cdot\mathbf{r}$$

must also be along x; also because $S_{xx} = E/\tau$ is positive, $\mathbf{v}$ and $\mathbf{r}$ must be of the same sign. In mathematical terms, we may say that the (nonsymmetric) matrix S_{ij} must have one *real, positive,* eigenvalue. The corresponding eigenvector (x) then gives us the direction of elongation. (This property will be our starting point for the general discussion of Sec. IV).

Two remarks should be inserted at this point. (a) By combination of (II.10) it is possible to show that $1/C_{yy} + 1/C_{zz} = 3E(t)$. If (as usual) C_{yy} and C_{zz} are comparable, they must be of order $2/3E$: This tells us how the lateral dimension of the cigar decreases by stretching. (b) We could have reached Eq. (II.12) directly without using the complicated apparatus of the distribution function Φ, by writing a balance between two terms only; the hydrodynamic friction (proportional to $\tau\mathbf{S}\cdot\mathbf{r}$) and the elastic restoring force (proportional to $E\mathbf{r}$); in a strongly stretched state the diffusion terms can be omitted completely. The molecule is largely polarized and the fluctuations around the polarized state are of minor importance.[10]

Up to now we assumed that the $\mathbf{C}$ matrix had only one large eigenvalue. It may happen, however, that two of them are large (e.g., C_{xx} and C_{yy}), although not necessarily equal. This corresponds to a molecule shaped into an elliptical pancake. If this shape is achieved, we must have, by the analog of Eq. (II.12),

$$S_{xx} = S_{yy} = E/\tau \tag{II.14}$$

and also, through the second and third equation (II.11),

$$S_{zx} = S_{zy} = 0. \tag{II.15}$$

Comparing Eqs. (II.12, 13) with (II.14, 15) we see that the latter contain four conditions on the velocity field instead of three. Thus the "elliptical pancake" will be found only in exceptional cases, and we shall not discuss it any further.

Finally, it is easy to dispose of the case where the three constants C_{11} would be simultaneously large; this would imply

$$S_{xx} = S_{yy} = S_{zz} = E/C > 0 \tag{II.16}$$

but the incompressibility condition imposes $S_{xx} + S_{yy} + S_{zz} = 0$. Thus Eq. (II.16) is clearly ruled out. The only major type of distortion is the cigar shape.

III. EXAMPLES OF SHARP $C \leftrightarrow S$ TRANSITIONS

A. Couette flow as a marginal case

The most common example of flow corresponds to a "transverse gradient" or —in hydrodynamic language— to a Couette flow of the form

$$v_x = ez \tag{III.1}$$

(all other components of **v** vanishing). The flow (III.1) is a superposition of a pure rotation

$$\omega_y = \tfrac{1}{2}[(\partial/\partial z)v_x - (\partial/\partial x)v_z] = \tfrac{1}{2}e$$

and of a pure deformation

$$A_{xz} = \tfrac{1}{2}[(\partial/\partial z)v_x + (\partial/\partial x)v_z] = \tfrac{1}{2}e$$

of equal magnitude. We shall consider in the present section more general two dimensional flows of the form

$$v_x = ez, \quad v_z = \xi x, \tag{III.2}$$

where e is large ($e\tau_0 > 1$) while ξ is of arbitrary sign. The decomposition is now

$$\omega_y = \tfrac{1}{2}(e - \xi), \quad A_{xz} = \tfrac{1}{2}(e + \xi). \tag{III.3}$$

Lumley has noted that when $\omega_y < A_{xz}$ ($\xi > 0$) we may have strong distortions, while when $\omega_y > A_{xz}$ ($\xi < 0$) we always find a coil.[5] This shows that Couette flow is a marginal case. The property may easily be rederived in the terms of Sec. II: We have seen there that an elongated state is possible only when the velocity gradient matrix S_{ij} has a real, positive eigenvalue. In the present two dimensional problem **S** reduces to

$$\mathbf{S} = \begin{vmatrix} 0 & 0 & e \\ 0 & 0 & 0 \\ \xi & 0 & 0 \end{vmatrix}.$$

The eigenvalues are 0 and $\pm\sqrt{e\xi}$: There is a positive, real, eigenvalue only if $\xi > 0$.

Our aim here is to discuss the *transition* between the coil regime (for $\xi < 0$) and the stretched regime (for $\xi > 0$), and to show that this transition takes place in an interval $\Delta\xi$ which is very small. ($\Delta\xi/e \sim Z^{-1}$). The starting point is the set of Eqs. (II.5),

$$\begin{aligned} &1 - EC_{xx} + \tau e\, C_{zx} = 0, \\ &-EC_{xz} + \tfrac{1}{2}\tau(eC_{zz} + \xi C_{xx}) = 0, \\ &1 - EC_{zz} + \tau\xi C_{xz} = 0. \end{aligned} \tag{III.4}$$

Together with

$$C_{yy} = 1/E, \quad C_{yx} = C_{yz} = 0. \tag{III.5}$$

Solving the system (III.4) for the C's, we arrive at the following self-consistency equation:

$$\begin{aligned} Zt^2 &= \tfrac{1}{3}(C_{xx} + C_{yy} + C_{zz}) \\ &= E^{-1}\left[1 + \frac{1}{6}\frac{(e+\xi)^2\tau^2}{E^2 - e\xi\tau^2}\right]. \end{aligned} \tag{111.6}$$

In the region of interest, we may replace $(e+\xi)^2$ by e^2 and then solve for ξ, obtaining

$$\begin{aligned} \frac{\xi}{e} &= \left[\frac{E(t)}{\tau(t)e}\right]^2 - \frac{1}{6}\frac{1}{zt^2E(t) - 1} \\ &= f(t) - g(t). \end{aligned} \tag{III.7}$$

For $\xi = 0$ we recover a self-consistency condition discussed by Peterlin, giving $t = t_0(e)$. In the region $e\tau_0 > 1$ (where τ_0 is the Zimm relaxation time) the corresponding t_0 is much larger than $Z^{-1/2}$, i.e., we are in the stretched state. Note that

$$\begin{aligned} f(t_0) = g(t_0) &= \frac{1}{6}\frac{1}{Zt_0^2E(t_0) - 1} \\ &\cong \frac{1}{6Zt_0^2E(t_0)} \cong \frac{1}{Z}. \end{aligned} \tag{III.8}$$

Let us now discuss the variation of t with ξ at fixed e. The general aspect is shown in Fig. 3. Here, we want to insist particularly on the slope of the curve near $\xi = 0$, $t = t_0$. Thus we calculate the derivative

$$\begin{aligned} \frac{1}{e}\frac{d\xi}{dt}\bigg|_{t_0} &- f'(t_0) - g'(t_0) \\ &= f(t_0)\frac{\partial \log f}{\partial t} - g(t_0)\frac{\partial \log g}{\partial t}\bigg|_{t_0} \\ &= f(t_0)\left(\frac{\partial \log f}{\partial t} - \frac{\partial \log g}{\partial t}\right)\bigg|_{t_0} \end{aligned} \tag{III.9}$$

In the interval $Z^{-1/2} < t < 1$ the functions f and g are regular and smooth and their logarithmic derivatives are of order unity. Hence we may write

$$\frac{1}{e}\frac{d\xi}{dt}\bigg|_{t_0} \sim f(t_0) \sim \frac{1}{Z}. \tag{III.10}$$

Thus a change in ξ of order e/Z is enough to change t by a quantity of order unity: the transition from coil to stretch is extremely sharp. Note that here the slope $dt/d\xi$ is constantly positive. For this reason we call the transition *second order* (using a loose analogy with thermodynamics).

We conclude that a simple Couette flow (such as is

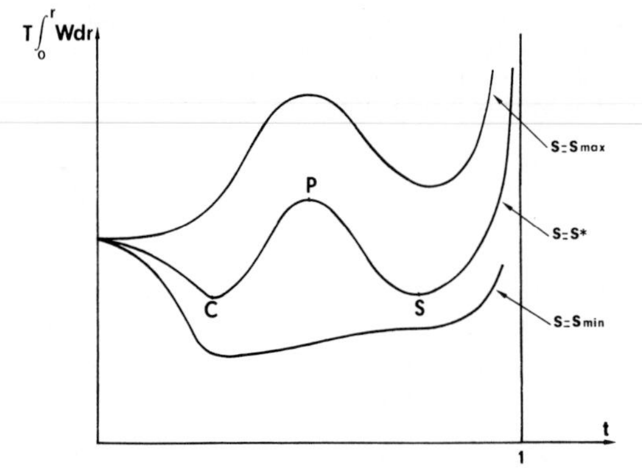

FIG. 4. Effective potential for the chain in a longitudinal gradient. In the region $S_{min} < S < S_{max}$ the potential has two minima. At $S = S^*$ these minima are of equal height.

found in laminar sublayers) with a solute polymer and a high shear rate ($e\tau_0 > 1$) is on the verge of instability from a structural point of view. This marginal character was not taken into account by Lumley in his discussion of drag reduction in turbulent flow: we shall come back briefly to this point in Sec. V.

B. Longitudinal gradients and first order $C \leftrightarrow S$ transitions

a. Bistable equilibria

Let us now consider briefly the behavior of a coil in a two dimensional longitudinal gradient, with a matrix **S** of the form

$$\mathbf{S} = \begin{vmatrix} S & 0 & 0 \\ 0 & -S & 0 \\ 0 & 0 & 0 \end{vmatrix} \qquad \text{(III.11)}$$

This case has been studied in detail in the literature, but usually without inclusion of variable hydrodynamic interactions.[1] Here we shall take them into account, using the dumbbell model and the form (II.8) for the elongation dependence of the relaxation times.

The matrix **S** has one positive eigenvalue (*S*). In the limit of strong elongations ($tZ^{1/2} > 1$) we may then use directly the simplified equation (II.12) and write

$$S = \frac{E}{\tau} = \frac{E(t)}{\tau_R}\left(1 + \frac{u}{t}\right) \quad (tZ^{1/2} > 1). \qquad \text{(III.12)}$$

This leads to the plot of *S* versus *t* which is shown in Fig. 3. For $t \lesssim \frac{1}{2}$ $E(t)$ is nearly equal to unity, and *S* is proportional to $1/t$. For $t \gtrsim \frac{1}{2}$ the factor $E(t)$ starts to increase [and diverges like $1/3(1-t)$]. Equation (III.12) must be supplemented by a more complete analysis [based on Eq. II.5], to deal with the region of small distortions ($t \sim Z^{-1/2}$): this gives

$$Zt^2 = 1 + \tfrac{1}{3}\tau_0 S \quad (Zt^2 \sim 1). \qquad \text{(III.13)}$$

On the whole the (S, t) relation has the aspect already shown in Fig. 3. The plot of *t* as a function of *S* is *multivalued* in a certain range of *S*, roughly given by

$$1/\tau_R < S < 1/\tau_0. \qquad \text{(III.14)}$$

In this range, for a given value of *S*, we fine three possible states of the molecule, described in Fig. 4 by the points *C*, *P*, and *S*. *C* represents a coil state, and is locally stable. *P* represents an unstable state, and is the analog of the maximum point in a static potential barrier. Finally *S* is associated with a stretched state.

b. The effective potential barrier

It is in fact possible to give a more precise meaning to a potential barrier in the present problem. Let us, for this purpose, return to the equation for the distribution function Φ discussed in Eqs. (II.1–3), and rewrite it as

$$\partial\Phi/\partial t = \operatorname{div} D[\nabla\Phi + \Phi W], \qquad \text{(III.16)}$$

where

$$\mathbf{W}(\mathbf{r}) = -\frac{1}{D(t)}\mathbf{v}(\mathbf{r}) + \frac{3E(t)\,\mathbf{r}}{R_0^2}$$
$$= -\frac{1}{D}\mathbf{S}\cdot\mathbf{r} + \frac{3E\mathbf{r}}{R_0^2}. \qquad \text{(III.17)}$$

Let us, for the moment, restrict our attention to longitudinal gradients. We may then write, for a stationary state ($\partial\Phi/\partial t = 0$) that $\operatorname{div}\mathbf{J} = 0$ is equivalent to $\mathbf{J} = 0$, or

$$\nabla\Phi/\Phi = -W.$$

The problem at hand (where, in the *P* and *C* states, there is strong one-dimensional stretching) is essentially one dimensional (along the stretching direction

$$\Phi(r) = \text{const}\exp\left(-\int^r W dr\right). \qquad \text{(III.18)}$$

Equation (III.18) shows that the function $T\int^r W dr$ may be interpreted as an effective potential barrier.[11] This barrier is shown on fig. 4. Of particular interest is the situation where the two minima (*C* and *S*) are of equal height, and thus of equal probability. This corresponds to

$$\Phi(r_S)/\Phi(r_C) = e^{\int_{r_C}^{r_S} W(r)dr} = 1$$

or

$$\int_{r_C}^{r_S} W dr = 0. \qquad \text{(III.19)}$$

This allows us to locate precisely the $C \rightarrow S$ transition, by the analog of a Maxwell construction on the function $W(r)$. The corresponding value of the velocity gradient will be called S^*.

c. Sharpness of the transition

To justify our description we must still show that, as soon as *S* departs slightly from S^*, one of the two probabilities $\Phi(r_S)$ or $(\Phi(r_C)$ becomes very small compared with the other. Then (and only then) we may indeed say that the transition at $S = S^*$ is sharp. The proof may be sketched as follows: to first order in $S - S^*$, we have

$$\ln\left|\frac{\Phi(r_C)}{\Phi(r_S)}\right| = (S - S^*)\int_C^S \frac{r\,dr}{D} \qquad \text{(III.20)}$$

$$\cong (S - S^*) r_S^2 / 2\bar{D} \, , \qquad \text{(III.21)}$$

where $\bar{D}$ is some average of D between the C and S states. We may write very roughly that $\bar{D}$ is related to the critical S^* by a relation of the form $R_0^2/\bar{D} \cong S^*$. (This will ensure at least that $R_0^2/\bar{D}$ is in between $1/\tau_R$ and $1/\tau_0$). Thus (omitting all coefficients of order unity) we arrive at

$$\ln \frac{\Phi(r_C)}{\Phi(r_S)} \cong \frac{S - S^*}{S} \frac{r_S^2}{R_0^2} = \frac{S - S^*}{S^*} Z t_S^2$$

$$\cong \frac{S - S^*}{S^*} Z \qquad \text{(III.22)}$$

(Since t_S, in the stretched state, is of order unity). We see that as soon as $(S - S_C)/S_C$ is larger than $1/Z$, one of the populations is completely negligible, as announced. We are really dealing with an abrupt transition.

Note that, to define strictly the population of one minimum one should in principle perform an integral $\int \Phi \, dr$ near this minimum, but the resulting corrections (related to the curvature of the potential energy) are unimportant from the point of view of Eq. (III.22).

d. Hysteresis effects

We have discussed the populations of the C and S state in a state of dynamical equilibrium. However, in a bistable system of this sort, we must obviously consider the possibility of *hysteresis*: If for instance, we start with $S < S^*$ and progressively increase S, the molecule may remain in the coil state, even when we reach $S > S^*$ simply because its relaxation time towards the stretched state is longer than the duration of the experiment. These relaxation times may be studied from the dynamical equation (III.16). The results can be qualitatively put in the simple form

$$1/T_{C\to S} \cong S^* \exp\left(-\int_C^P W dr\right),$$
$$1/T_{S\to C} \cong S^* \exp\left(-\int_P^S W dr\right). \qquad \text{(III.23)}$$

Here $T_{C\to S}$ is the relaxation time from the C state to the S state, etc. The formulas (III.23) are very similar to the usual relaxation rate equations for a system overcoming a static activation barrier. The prefactor (here indicated as S^*) has not been calculated accurately. The main effect comes from the exponentials. An order of magnitude estimate similar in spirit to Eq. (III.20) shows that for $S \sim S^*$

$$\int_C^P W dr \sim S^* r_P^2 / D$$
$$\sim (r_P/R_0)^2 \sim Z t_P^2 \gg 1 . \qquad \text{(III.24)}$$

Consider a specific example, with $Z = 10^4$ and $t_P \sim 0.1$ ($Z t_P^2 = 10^2$). Take for the Rouse relaxation time $\tau_R = 10^{-2}$ sec and for the Zimm relaxation time $\tau_0 = 10^{-4}$ sec. Assume that S^* (intermediate between τ_R^{-1} and τ_0^{-1}) is of order 10^{-1} sec^{-1}. There the hysteretical relaxation times, for $S \sim S_C$, would be of order $10^{-3} e^{100}$, i.e., of astronomical magnitude!

Of course, detailed predictions would require a more precise calculation of the integral (III.24), and this may reduce the exponent significantly. But the trend is clear: The system will be trapped in one of the minima of the effective potential, and will escape from it only when this minimum ceases to exist. Under oscillating fields of large amplitude, this should then lead to the set of hysteresis cycles—extending from S_{min} to S_{max} —shown in Fig. 3.

The above discussion on barriers and on hysteresis was restricted to the case of a purely "longitudinal" gradient ($\omega = 0$). We now argue that similar phenomena should occur even for $\omega \neq 0$, whenever the transition is of first order.

A molecule in the *coil* state rotates with angular velocity ω and is thus forced to expand and contract periodically. In the language of Eq. (II.1) this corresponds to $J \neq 0$ (although div $\mathbf{J} = 0$). In the stretched state, however, the situation is quite different. Let us still call $\mathbf{r}_S$ the elongation vector at which the drag and the elastic force are just balanced in the S state. Brownian motion spreads the representative points $\mathbf{r}$ in a finite region around $\mathbf{r}_S$. In this region, for $\omega \neq 0$, there is indeed a current $\mathbf{J}$. But this current describes a rotation *around* $\mathbf{r}_S$ (not around $\mathbf{r} = 0$): the molecule has small periodic motions around its average stretched conformation, but it does not shrink back periodically to the C state. Thus the barrier concept remains valid, and hysteresis should be observable for all first order transitions. The case where the $C \to S$ transition is weakly first order (namely, when $|\omega| \to A$ in the two dimensional example) should be particularly interesting, because the jumping frequencies may be more easily adjustable in this region.

IV. MORE GENERAL SHEAR FLOWS

A. Two dimensional incompressible flow

The most general flow of this type is (as mentioned in Sec. III) a superposition of a rotation ω_y and a pure deformation A_{xz}: It involves two independent parameters. For a given ω_y and A_{xz}, the molecule, in its dynamical equilibrium ($\partial\Phi/\partial t = 0$) may be either in the C state or in the S state. The borderline between the two is shown in Fig. 5. The straight portions M correspond to marginal Couette flow (second order $C \to S$ transition). The point L is associated with " a longitudinal" gradient: the curved arc around L describes a first order transition. This arc meets the second order line at a point T which might (loosely) be called a tricritical point. The behavior obtained when crossing T along one path (having constantly $\omega = A = G/2$ and increasing G) has been studied long ago by Peterlin[1]. From his calculations one finds that the elongation t shoots up with a *finite* slope as soon as G exceeds G_T,

$$G_T \, \partial t / \partial G \,|_{GT} \cong \text{unity}.$$

Thus the transition, as seen along this path, is less dramatic than in the examples of Sec. III. This is due to the special geometry of the chosen path (tangent to the transition curve).

The behavior along the first order arc around L has not been calculated in detail. Presumably the magnitude of the discontinuity in t at the transition decreases and vanishes when one moves from L towards T.

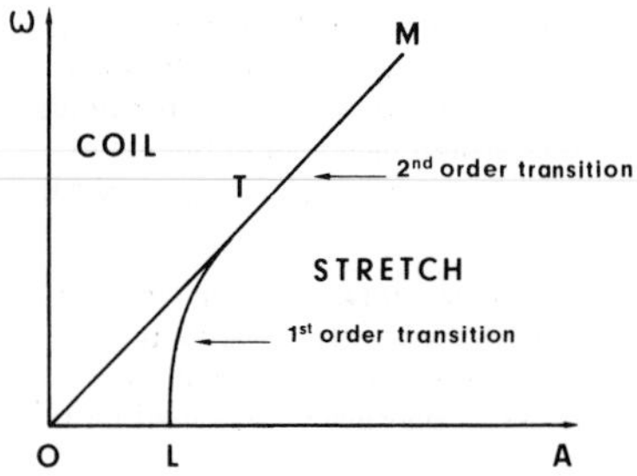

FIG. 5. Domains of stability of the stretched phase in incompressible two dimensional flows characterized by a pure deformation rate A and a vorticity ω [defined in Eqs. (III. 3)]. The usual studies on Couette flow are performed with ω constantly equal to A. In this case, one reaches the transition line *tangentially* at point T, and the transition is not very conspicuous. But any other path from the C region to the S region should show a sharp transition.

B. Three dimensional incompressible flow

Here the problem is geometrically more complex, and we shall reduce it as follows. We separate the velocity gradient tensor S_{ij} into its symmetrical part,

$$A_{ij} = \tfrac{1}{2}(S_{ij} + S_{ji}), \qquad \text{(IV.1)}$$

and the antisymmetrical part, associated to a rotation vector $\boldsymbol{\omega}$ (defined as in Sec. III). We use the frame $(x_1 x_2 x_3)$ which diagonalizes A_{ij}. The three eigenvalues A_{11}, A_{22}, A_{33} are of zero sum from the incompressibility condition. We assume that their absolute magnitude is *large* ($\gtrsim 1/\tau_0$); thus, in the absence of any rotation ($\boldsymbol{\omega} = 0$) we would surely be in the stretched state S. Now, keeping the A tensor fixed, we consider all possible states of rotation: Each state is associated with one point $(\omega_1 \omega_2 \omega_3)$ in our reference frame. We want to find what region, in this three dimensional space, is associated with the stretched phase.

The limit of the S domain can be obtained by returning to Eqs. (II.12, 13) and to the resulting conditions. In the stretched state the matrix **S** must have an eigenvalue S which is both *real* and *positive*. When we cross over to the coil state, this implies that either the eigenvalue becomes complex (an important possibility for a nonsymmetric matrix), or the eigenvalue goes through 0. These two possibilities will give rise to two distinct portions of the limiting surface, which we shall call, respectively, $\sum_d$ and $\sum_0$. (d stands for double root, since the occurrence of a complex pair of eigenvalues is always preceded by a double root).

Let us now write down the equation for the eigenvalues S,

$$\begin{vmatrix} A_1 - S & \omega_3 & \omega_2 \\ -\omega_3 & A_2 - S & \omega_1 \\ -\omega_2 & -\omega_1 & A_3 - S \end{vmatrix} = 0, \qquad \text{(IV.2)}$$

or explicitly

$$S^3 + PS + Q = 0, \qquad \text{(IV.3)}$$

where

$$P = \omega^2 - \frac{1}{2}\sum_i A_{ii}^2 , \qquad \text{(IV.4)}$$

$$Q = -A_{11}A_{22}A_{33} - \sum_i A_{ii}\,\omega_i^2 . \qquad \text{(IV.5)}$$

(a) Consider first the portion $\sum_d$, upon which Eq. (IV.3) has a *double root* S_d. This implies the condition

$$3S_d^2 + P = 0, \quad S_d^2 = -P/3 \qquad \text{(IV.6)}$$

and imposes that $P < 0$, or according to (IV.4)

$$\omega^2 < \frac{1}{2}\sum_i A_{ii}^2 .$$

Reinserting the value of S_d in Eq. (IV. 3) we obtain

$$S_d = 3Q/2P . \qquad \text{(IV.7)}$$

Since the eigenvalue of interest is positive, while $P < 0$, we must have $Q > 0$. Eliminating S_d between (IV. 6) and (IV. 7) we reach

$$27Q^2 + 4P^3 = 0. \qquad \text{(IV.8)}$$

Equation (IV.8), together with the restriction $Q > 0$, defines the surface $\sum_d$; it is a portion of a surface of degree 6 in the $(\omega_1, \omega_2, \omega_3)$ plane. The general aspect of $\sum_d$ is shown in Fig. 6 and 7 for two possible cases (Fig. 6: two of the eigenvalues of **A** are positive, Fig. 7: one only is positive).

(b) As explained earlier, there is another portion to the limiting surface, which we called $\sum_0$. This corresponds to a vanishing eigenvalue $S = 0$, or, from Eq. (IV.3) to the condition

$$-Q = \sum_i A_{ii}\,\omega_i^2 - A_{11}A_{22}A_{33} = 0. \qquad \text{(IV.9)}$$

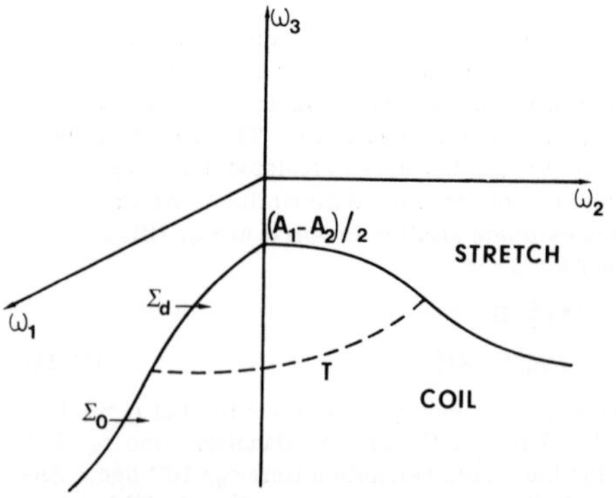

FIG. 6. Domain of stability of the stretched phase as a function of vorticity $\boldsymbol{\omega}$. The axes are the principal directions of the symmetrized shear rate tensor **A**. The figure is drawn for the case where $A_3 < 0 < A_1 < A_2$. The surface $\sum$ separates the stretch and coil region. Only one eighth of $\sum$ is shown for clarity (it should be completed by symmetry with respect to all mirror planes). On the portion $\sum_d$ the $C \leftrightarrow S$ transition is of second order. On the portion $\sum_0$ it is of first order.

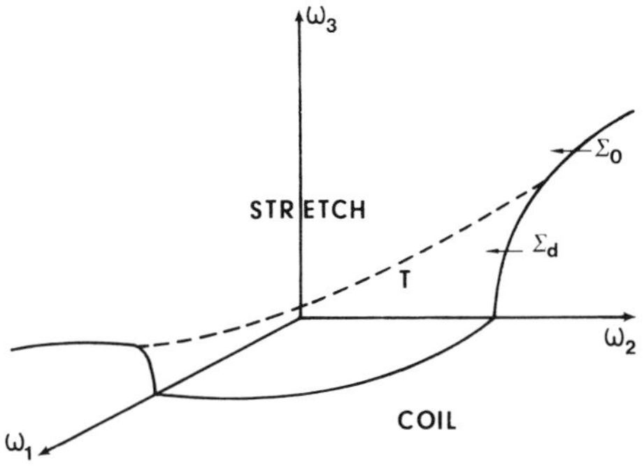

FIG. 7. Similar of Fig. 6, but with $A_2 < A_1 < 0 < A_3$.

Equation (IV.9) defines a hyperboloid. But $\sum_0$ does not cover the whole hyperboloid. To be sure that $S=0$ is the root of interest, we must investigate the two other roots; they correspond to

$$S^2 = -P. \tag{IV.10}$$

If $P>0$ they are pure imaginary, and $S=0$ is indeed the interesting root. But if $P<0$, there is one positive real root $(+\sqrt{-P})$ and this root will allow for a stretched state. Thus we conclude that $\sum_0$ is defined by Eq. (IV.9) plus the restriction $P>0$.

The surfaces $\sum_d$ and $\sum_0$ meet on the line T defined by the two equations $P=0$ and $Q=0$. Their over-all aspect is represented on Figs. 6 and 7. Their physical meaning appears to be the following:

The surface $\sum_d$ corresponds to "marginal" cases similar in nature to the Couette flow of Sec. (III.A) and thus to second order transitions.

The surface $\sum_0$ corresponds to a real eigenvalue which becomes small $(S\to 0)$ as for a "longitudinal" gradient. In actual fact the transition takes place not for $S=0$ but at some finite threshold $S=S^*$, as in Sec. (III.B). However, because we have assumed that the constants $|A_{ii}|$ are large, the distinction between $S=0$ and $S=S^*$ may be neglected. The transition is here of first order.

V. POSSIBLE CONSEQUENCES OF THE $C \leftrightarrow S$ TRANSITION

A. Orders of magnitude

We have defined the domain of existence of the stretched phase S. How can the $C\to S$ transition be observed? Of course, it takes place only at rather high values of the gradient $S=S^*$ where S^* is either equal to a Zimm relaxation frequency $1/\tau_0$, or to an intermediate between the Zimm and Rouse frequencies. Fundamental studies on the S state and on the transition will require a stable, simple, *laminar* flow. Consider for instance a Couette flow between two walls separated by a distance D. The associated Reynolds number is

$$\mathrm{Re} = \frac{\rho S D^2}{\eta_S} \sim \frac{5D^2}{R_0^3}\frac{\rho T}{\eta_S^2}, \tag{V.1}$$

where we have set $S\sim 1/\tau_0$ and used Eq. (I.1). (ρ is the density of the solvent.) To maintain a secure laminar flow we require $R \lessapprox 1500$.[12] This corresponds to

$$D \lessapprox 17 R_0^{3/2}\eta_S(\rho T)^{-1/2}. \tag{V.2}$$

Taking $R_0=300$ Å, $\eta_S=10^{-2}$ P, $\rho=1\text{g/cm}^3$ (water), and $T=300\,°\text{K}\cong 5.10^{-14}$ erg, this imposes $D<50\ \mu$. Thus to detect the $C\to S$ transition in water we need very thin samples —and for many experiments this may imply difficult signal to noise problems. However, the situation is improved if we go to more viscous solvents ($\eta_S \sim 1$ P).

B. Mechanical effects

To analyze the hydrodynamic consequences of the $C\to S$ transition we must first write down the contribution to the stress tensor σ_{ij} due to the solute molecules. In the dumbbell model this is simply

$$\sigma_{ij} = -n\langle r_j F_i\rangle, \tag{V.3}$$

where n is the number density of macromolecules and $\mathbf{F}$ is the restoring force defined in Eq. (I.2). The bracket denotes an average over the distribution function Φ. Note that, since $\mathbf{F}$ is radial ($\mathbf{F}\times\mathbf{r}=0$), the tensor σ_{ij} is symmetrical.

Prior to any discussion of the physical effects associated to the tensor σ_{ij} it may be useful to connect Eq. (V.3) to the energy balance in the solution. We shall write this down in the limit of strong elongations, which is particularly simple: As explained in the remarks preceding Eq. (II.14), Brownian motion is negligible in this limit; the diffusion currents can be omitted, and also the average symbol in (V.3). The balance of forces is

$$f(\mathring{\mathbf{r}} - \mathbf{v}(\mathbf{r})) = \mathbf{F}, \tag{V.4}$$

where $\mathring{\mathbf{r}} = d\mathbf{r}/dt$ and $f=T/D$ is the dumbbell friction coefficient. The work done (per sec) by the solvent on one macromolecule is

$$\begin{aligned}-\mathbf{v}\cdot\mathbf{F} &= f(\mathbf{v}-\mathring{\mathbf{r}})\,\mathbf{v},\\ &= f(\mathbf{v}-\mathring{\mathbf{r}})^2 - \mathbf{F}\cdot\mathring{\mathbf{r}}.\end{aligned} \tag{V.5}$$

We may relate $\mathbf{F}$ to the chain free energy $G(\mathbf{r})$.

$$\mathbf{F} = -\partial G/\partial\mathbf{r}. \tag{V.6}$$

Thus the work done by the solvent is

$$-\mathbf{v}\cdot\mathbf{F} = f(\mathbf{v}-\mathring{\mathbf{r}})^2 + dG/dt. \tag{V.7}$$

The first term on the right hand side represents frictional dissipation. The second term represents the work done by the solvent to change the internal state of the chain. It vanishes of course in steady state. The left hand side of Eq. (V.7), when counted per unit volume, becomes

$$-n\mathbf{v}\cdot\mathbf{F} = -n S_{ij} r_j F = \sigma_{ij}S_{ij}. \tag{V.8}$$

The last form is the classical expression for the work done by the fluid.

Let us now discuss the changes in the stress tensor σ_{ij} which occur when we have a $C\to S$ transition. By

Eq. (V.1), since $\mathbf{r}$ increases abruptly, we must have a corresponding change in certain components of $\boldsymbol{\sigma}$. In fact, if we define a local reference frame as in Sec. II, with the x axis along the direction of elongation, it is the component σ_{xx} (and this component only) which shows an important jump.

(a) In the case of a "longitudinal" gradient this implies a highly nonlinear relation between stress and shear rate of the type shown in Fig. 3. In most practical cases we expect that the relaxation between C and S states will be slow on the scale of the experiment, and thus that the stress/shear rate relation should show a hysteresis cycle. A word of caution should be inserted at this point: In magnetism a hysteresis cycle is defined by a curve $M(H)$, where the field H is the variable conjugate to the magnetization in a thermodynamic sense. On the other hand, the plot of Fig. 3 does not give the stress σ as a function of the conjugate strain, but as a function of the *time derivative* of the strain. Thus, for instance, the integral $\oint MdH$ around one magnetic cycle gives the heat dissipated. But the integral $\oint \sigma dS$ on the cycle of Fig. 3 does not have a similar meaning.

(b) With a "transverse" gradient (the usual Couette flow situation) the situation is somewhat more subtle. If, as is usually done, we define a viscosity through the ratio $\sigma_{xz}/A_{xz}=\eta$ (where X is the direction of flow, and Z the direction of the gradient, as in Sec. III.A). We do not expect any spectacular discontinuity at the $C \rightarrow S$ transition —and this is indeed what has been found up to now.[1,2] But if we study the component σ_{xx} of the stress, we do expect an anomaly.

To get a slightly more precise feeling about this effect, let us superimpose on a strong Couette flow $v_x^{(0)} = ez$ (with $e > 1/\tau_0$) a small fluctuation $\mathbf{v}^{(1)}$. As noticed by Lumley, the crucial parameter is the component $\partial v_z^{(1)}/\partial x$ of the gradient. If this is positive, the S phase is maintained. But if it is negative, the C phase is favored. We shall describe this situation by the following stress structure:

$$\sigma_{ij} = 2\eta\, A_{ij} + \sigma'_{ij}\,, \tag{V.9}$$

where η is the conventional viscosity of the solution, as defined above, and σ' describes the effect of the $C \rightarrow S$ transition: σ' has only one nonvanishing component. For small $\mathbf{v}^{(1)}$ we may write:

$$\sigma'_{xx} = \tilde{\eta}\ \partial v_z^{(1)}/\partial x + \text{const}\ . \tag{V.10}$$

Using Eq. (V.1) we see that $\tilde{\eta}$ is related to the slope of the plot $t(\xi)$ of Sec. III.A. Very qualitatively we may write

$$\begin{aligned}\tilde{\eta} &\sim \frac{nT}{R_0^2}\, \frac{R\,\partial R}{\partial \xi}\bigg|_{\xi=0} \\ &\sim nT\frac{R^2}{R_0}\,\frac{Z}{e} \\ &\sim nT\,\tau_0 Z^2\,. \end{aligned} \tag{V.11}$$

It is important to realize that $\tilde{\eta}$ is much larger than the usual increment $\delta\eta = \eta - \eta_S$ due to the polymer solute, since (again omitting all coefficients)

$$\delta\eta \sim nT\tau_0 \sim \tilde{\eta}/Z^2\,. \tag{V.12}$$

Equation (V.10), relating a diagonal component of $\boldsymbol{\sigma}$ to an off diagonal component of the velocity gradient, is very unusual in form. To get some impression of its practical effects, let us write down explicitly the equations of motion for the fluctuation $\mathbf{v}_1$ in the limit of low Reynolds numbers. For simplicity we consider only two dimensional motions, and have

$$\begin{aligned}\rho\ \partial v_{1x}/\partial t &= (\partial/\partial x)(\sigma'_{xx} - p) + \eta\nabla^2 v_{1x} \\ \rho\ \partial v_{1z}/\partial t &= -\ \partial P/\partial z + \eta\nabla^2 v_{1z}\,. \end{aligned} \tag{V.13}$$

Using the incompressibility condition div $\mathbf{v}_1 = 0$ to eliminate the pressure, and looking for variations of the form

$$\mathbf{v}_1(\mathbf{r}, t) = \mathbf{W}_1 \exp(i\mathbf{k}\cdot\mathbf{r} - \alpha\, t) \quad (\mathbf{k} = k_x, 0, k_z)\,, \tag{V.14}$$

we obtain an attenuation coefficient $\alpha(\mathbf{k})$ with the following structure

$$\alpha = \frac{\eta}{\rho}\,k^2 + \frac{\tilde{\eta}}{\rho}\,\frac{k_x^3 k_z}{k^2}\,. \tag{V.15}$$

Because of the large factor Z^2 present in Eq. (V.11), the $\tilde{\eta}$ correction can be quite significant. Depending on the relative sign of k_x and k_z, it may either increase or decrease the viscous damping! This remarkable asymmetry between wave vectors $\mathbf{k}$, which lie in the first or the second quadrant of the (x, z) plane, is of course due to the underlying Couette flow ($v_x^{(0)} = ez$).

It must be emphasized, however, that the above calculation, based on a linearized form [Eq. (V.10)] is valid only for infinitesimal perturbations $\mathbf{v}_1$. In most practical applications, a more complete nonlinear relation between σ_{XX} and ξ is required: it is closely related to the curve of Fig. 3.

(c) Let us end up this paragraph by some remarks concerning *turbulent flows*. The effects associated to polymer solutes in such flows are complicated, but two useful reviews linking the hydrodynamic and molecular points of view have been given by Lumley.[5] The strong nonlinearities of the $\sigma(\mathbf{S})$ relation, associated with the stretch-coil transition, are probably essential for all interpretations.

At any instant of time, certain regions of the fluid will be in a strongly deformed state. However, the location (and the shape) of these regions *cannot* be derived from the criteria of Sec. IV: This section dealt only with velocity gradients $S_{ij} = \partial v_i/\partial x_j$ which are time independent. When S_{ij} is modulated at frequencies comparable to $1/\tau$, the molecular response is delayed and becomes much more complex.

C. Optical detection in laminar flow

Flow birefringence has been studied with considerable detail at moderately high gradients, and mainly in planar Couette flow.[3] The quantity which is measured here is the anisotropic part of the dielectric tensor (at a certain optical frequency). For a rather broad class of dumbbell models, this is essentially proportional to

$$Q_{ij} = n\,(\langle r_i r_j\rangle - \tfrac{1}{3}\langle\mathbf{r}^2\rangle\,\delta_{ij})\,. \tag{V.16}$$

Comparing Q_{ij} and the stress tensor σ_{ij} as defined in Eq. (V.1) we see that they have very similar structures whenever the restoring force F is linear in the elongation ($E \cong 1$) [The experimental consequences of this relation are extensively discussed in Ref. 3]. To detect the $C \rightarrow S$ transition via Q_{ij} we should work with ultrahigh gradients (keeping in mind the restrictions on sample thickness discussed at the beginning of this section) and focus our attention on the diagonal components Q_{xx} along the direction of stretch.

(a) The simplest geometry corresponds to Couette flow between two flat plates. The main defect of this geometry is that, when increasing the shear rate, we move on the first bisector in the (ω, A) plot of Fig. 5 and we meet the transition line at point T *tangentially*. The transition is then less dramatic than usual. The Peterlin analysis[1] shows that the plot of Q_{xx} versus stress should have the aspect shown in Fig. 8. This particular circumstance probably explains why the concept of a $C \rightarrow S$ transition has not emerged yet from the experiments.

(b) Spectacular effects would be expected with *longitudinal gradients*. One possible geometry is the T shown in Fig. 9. The fluid (with an average viscosity of order 1 P) is sent through the central portion of the T and comes out on both sides. The dimensions are such that the main velocity gradients take place *in the plane* of the T. The optical beam is sent normal to this plane, and in the vicinity of the stagnation line. With this geometry, it should be possible to observe the stretched state.

Apart from the birefringence measurements which we have just discussed, it is of course conceivable to monitor the change in shape of the solute macromolecules from the angular distribution of the light scattering. But these experiments are rather delicate in all cases, and the size requirements on the sample would make them worse.

D. Chemical degradation in flow

It is known that solute macromolecules are significantly disrupted into smaller portions by strong turbulent flows.

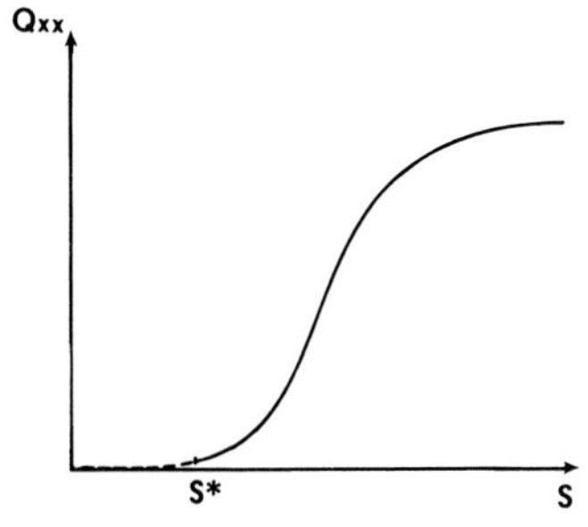

FIG. 8. Component Q_{xx} of the dielectric tensor along the flow lines (in simple shear flow) as a function of shear flow) as a function of shear rate: strong distortions should start at $S = S^* \, 1/\tau_0$ but they start quadratically in $(S - S^*)$: thus the transition is not very spectacular in Couette flow.

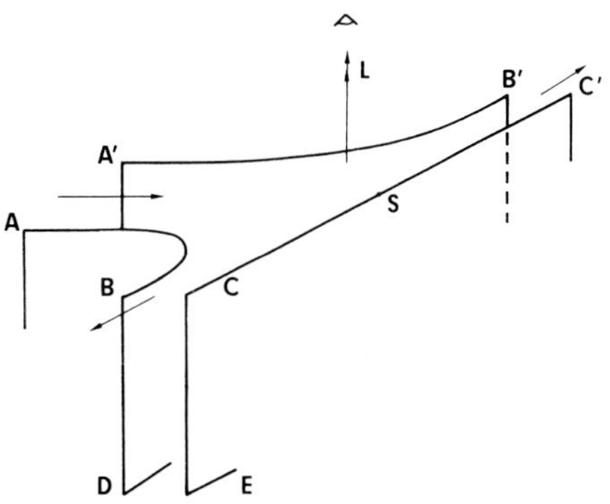

FIG. 9. Tentative geometry for the study of flow birefringence in high longitudinal gradients. A vertical light beam L probes the vicinity of the stagnation point S. The dimensions AA', BC, etc., must be small (of order 0.1 mm). The vertical dimensions (BD, etc.) must be somewhat larger to ensure that the velocity gradients are mainly in the horizontal plane. To maintain Reynolds numbers which are not too large, the viscosity of the solvent should be at least of order 1 P.

It may be that certain "aging effects" found in this situation, and often ascribed to intermolecular association,[12] are in fact due to chemical degradation. Unfortunately these experiments depend on rather refined aspects of the turbulence statistics, and no quantitative interpretation can be attempted at present.

Here we shall focus our attention on the simpler problem of degradation in *laminar* flow. Again, to reach the ultrahigh gradients of interest, while keeping the Reynolds number not too large, this implies experiments on thin channels, or with rather high solvent viscosities.

Let us first represent the binding energy of one particular link in the polymer backbone as a function of the distance x between the two atoms of interest (e.g., two neighboring carbon atoms in a polyethylene chain). Let us call this function $W_0(x)$(Fig. 10). We now add on this the contribution due to the tensile force **F** between segments [where F is defined by Eq. (I.2)] and obtain a new potential energy

$$U(X) = U_0(X) - FX . \qquad (V.17)$$

The correction FX is very small on the chemical scale. However, it has two effects (a) The minimum of U is shifted from U_{min} to U'_{min}. (b) A secondary maximum U_{max} appears at large X. The activation energy required to disrupt the link is

$$H(F) = U_{max} - U'_{min} . \qquad (V.18)$$

For small F, the shift $U'_{min} - U_{min}$ is of second order in F, and thus negligible. A more interesting contribution comes from U_{max}. Consider first the case where the energy $U_0(X)$ decreases exponentially at large X.

$$U_0 = - Ce^{-X/b} \quad (X \rightarrow \infty) . \qquad (V.19)$$

Then the position of the maximum X_{max} is given by

$$F = C(1/b)\, e^{-X_{max}/b}$$

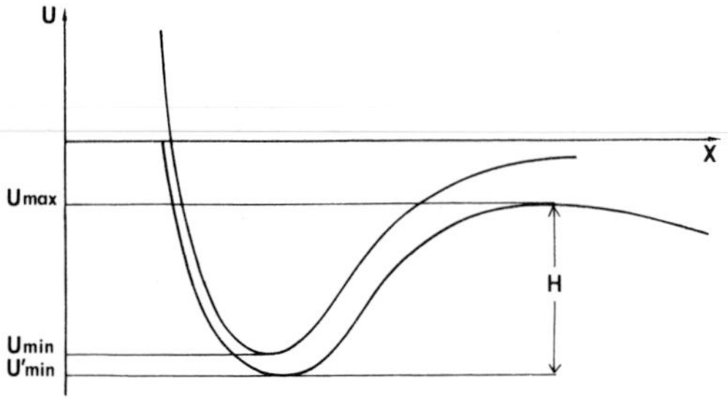

FIG. 10. Potential energy for one skeleton bond as a function of the interatomic separation X. The dotted curve is the unperturbed energy $U_0(X)$. The full curve includes the effect of an external tension F.

and the corresponding energy is

$$U_{\max} = -Fb\,. \qquad (\text{V}.20)$$

The main conclusion is that $U_{\max}$ is linear in F. Let us now investigate a slightly more realistic form of $U_0(X)$,

$$U_0(X) = -D/x^n\,, \qquad (\text{V}.21)$$

where $n = 6$ would be the natural candidate for neutral partners. A similar calculation then shows that

$$U_{\max} = -\text{const}\,(F)^{n/(n+1)}. \qquad (\text{V}.22)$$

In view of the large value of n, this is not very far from the linear law (V.20), and we shall retain (V.20) as our basic assumption. The activation energy is then of the form

$$H(F) = H_0 - Fb\,(+\text{order } F^2)\,, \qquad (\text{V}.23)$$

where b is related to the range of the attractive forces. We may now assume that the probability of rupture of any given bond, per unit time, is

$$W = \nu e^{-H/T}$$
$$= \nu e^{-H_0/T} e^{Fb/T}\,, \qquad (\text{V}.24)$$

where ν is a vibrational frequency.

Inserting Eq. (I.2) into (V.24) we obtain

$$W = \nu e^{-H_0/T} \exp(3bt/a)\, E(t)\,. \qquad (\text{V}.25)$$

The total probability for rupture of any bond along the chain is of order ZW.[13] The first dominant factor is of course the usual activation term $\exp(-H_0/T)$. (Note that the energy H_0 is not the energy required to disrupt a bond in the vacuum, but rather the energy of some bond-breaking reaction possibly involving the solvent.) Even so, the factor $\exp(-H_0/T)$ may be of order 10^{-20} or less at usual temperatures. This implies that degradation will be observable only when t is rather close to unity, so that $E(t) \sim 1/3(1-t)$ is then large. Thus we may write in practice

$$ZW \cong Z\nu \exp(-H_0/T) \exp[b/a(1-t)]. \qquad (\text{V}.26)$$

As tentative example, taking $Z = 10^4$, $\nu = 10^{12}$, $\exp(-H_0/T) = 10^{-20}$, $b \sim a/2$, and $t = 0.9$ we would find $(ZW)^{-1}$ of order of minutes. Of course all these estimates are very dubious. Other channels of degradation may be conceived; the asymptotic form of $E(t)$ at high t is model dependent, etc. However, it would be of great interest to have measurements on the rate of degradation in laminar flow, as a function of the shear rate S, and to compare these data to Eq. (V.26), together with the self-consistent values $t(S)$ which have been discussed in Sec. III, or can be found in Refs. (1 and 2). For instance in Couette flow, the analysis of Ref. 1, with variable hydrodynamic interactions, leads to the following form (omitting all numerical coefficients):

$$S\tau_0 \cong tE^{3/2}(t) \quad (tZ^{1/2} \gg 1). \qquad (\text{V}.27)$$

This can then be inserted in Eq. (V.25) to obtain a specific curve $W(S)$.

ACKNOWLEDGMENTS

The author has greatly benefited from a preprint, Ref. (5b), sent to him by Professor J. Lumley and from stimulating correspondences with Professor J. D. Ferry, Professor F. C. Frank, Professor A. Keller, and Professor W. Stockmayer.

APPENDIX: LOCAL PROPERTIES OF THE POLYMER IN A STRONGLY STRETCHED STATE.

We consider here a necklace model where successive beads have positions $\mathbf{r}_1 \ldots \mathbf{r}_n \ldots \mathbf{r}_Z$, and velocities $\mathring{\mathbf{r}}_1 \ldots \mathring{\mathbf{r}}_x \ldots \mathring{\mathbf{r}}_Z$. The equation of motion has the form

$$\mathring{\mathbf{r}}_n = \mathbf{v}(\mathbf{r}_n) + W\frac{\partial}{\partial n}(E\mathbf{a}_n) + \sum_m \frac{l}{r_{nm}}\, W\frac{\partial}{\partial m}(E\mathbf{a}_m). \qquad (\text{A}.1)$$

The first term is the external velocity field $\mathbf{v}$. The second term is the Rouse contribution, with

$$\mathbf{a}_n = \mathbf{r}_{n+1} - \mathbf{r}_n \cong \partial\mathbf{r}_n/\partial n\,. \qquad (\text{A}.2)$$

W is a microscopic frequency. Our notation here follows Ref. 14. The factor E is the nonlinear factor defined, as in Eq. (I.2), but for one unit (a_n). The third term in Eq. (A.1) describes the effect of hydrodynamic interactions, $r_{nm} = |r_n - r_m|$, and l is a length proportional to the hydrodynamic radius of one monomer. In principle we should add to (A.1) a random force. However, as explained in Sec. II, the resulting Brownian motion effects are negligible for our purposes in the case of strong elongations.

Let us assume that the stretched molecule has the shape of a cigar, with a cross section of radius R, and a number density of monomers ρ (note that both R and ρ may vary along the cigar axis). Replacing the sum $\sum_m$ by an integral we find

$$\sum_m (1/r_{nm}) \cong 2\pi\rho R^2[\log(L\bar{t}/R + \tfrac{1}{2}]\,l\,, \qquad (\text{A}.3)$$

where $L\bar{t}$ is the average length of the cigar. We shall replace the logarithmic factor by a constant and write

$$l\,[\ln(L\bar{t}/R) + \tfrac{1}{2}] = \tfrac{1}{2}au\,, \qquad (\text{A}.4)$$

where a is the extended length of one monomer and u is a number of order unity. We now restrict our attention to one dimensional problems (e.g., longitudinal gradi-

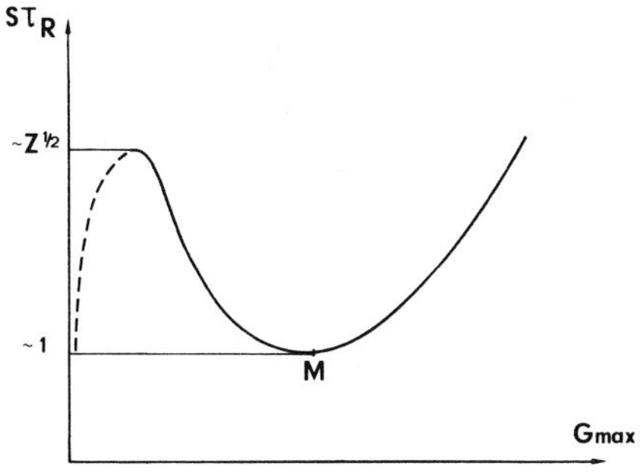

FIG. 11. Relation between the shear rates and the integration constant G_{max} for a necklace model under high longitudinal shear. For a given S in a certain range we have three solutions, i. e., a bistable equilibrium. This confirms the simpler analysis of Sec. III.

ents) where all vector indices may be omitted, and where we can write

$$a_n = t_n a \tag{A.5}$$

($\mathbf{a}_n$ being along the cigar axis). We must then have

$$\pi R^2 \rho a_n = 1. \tag{A.6}$$

Finally we are led to the following simplified form of the hydrodynamic interactions:

$$\sum_m \frac{l}{r_{nm}} W \frac{\partial}{\partial m} (E a_m) - \frac{u}{t_n} W \frac{\partial}{\partial n} (E a_n). \tag{A.7}$$

Differentiating (A.1) with respect to n, and using t_n instead of a_n as variable, we arrive at the final form

$$\mathring{t}_n = S t_n + W \frac{\partial}{\partial n} \left(f(t_n) \frac{\partial}{\partial n} [E(t_n) t_n] \right) \tag{A.8}$$

with

$$f(t) = 1 + u/t \quad (t \gg Z^{-1/2}). \tag{A.9}$$

For $t < Z^{-1/2}$ it is convenient (and correct) to take $f = 1 + uZ^{1/2}$. Equation (A.9) may be used to derive the approximate form (II.8) for the relaxation time of the dumbbell model. Here, however, we shall be concerned with the more detailed problem of finding the various t_n in a steady state, for a given shear rate S. Putting $\mathring{t}_n = 0$ in Eq. (A.8) we are left with a static problem which can be integrated explicitly. To do this we define the following auxiliary functions:

$$U(t) = \int_0^t f(t)\, d(Et), \tag{A.10}$$

$$G(t) = 2 \int_0^t t\, dU. \tag{A.11}$$

Then Eq. (A.8) becomes

$$(S/W) t + d^2 U/dn^2 = 0 \tag{A.12}$$

and has the first integral

$$[dU/dn]^2 = (S/W)(G_{max} - G), \tag{A.13}$$

where G_{max} is a constant, and is in fact the value of G which is realized at the middle of the chain. It is then possible to integrate (A.12), noting that $t = 0$ at both ends of the chain. This finally gives a relation between G_{max} and the number of monomers Z, in the implicit form

$$Z = 2\left(\frac{W}{S}\right)^{1/2} \int_0^{t_{max}} \frac{dU}{(G_{max} - G)^{1/2}}. \tag{A.14}$$

There are three main regimes.

(a) $t_{max} < Z^{-1/2}$. In this regime of small deformations, our starting point, Eq. (A.7), is not correct. Thus, we shall not discuss it any further.

(b) $Z^{-1/2} \ll t_{max} \ll 1$. Here, after suitable manipulation of Eq. (A.14), we get

$$S\tau_R = \text{const} \ln(Z^{1/2} G_{max})/G_{max}. \tag{A.15}$$

Thus S is a *decreasing* function of G_{max} in this regime.

(c) Finally for $t_{max} \lesssim 1$, Eq. (A.14) gives simply

$$S\tau_R = \text{const}\, G_{max}. \tag{A.16}$$

In this regime it can be shown that the maximum elongation t_{max} (found at the center of the chain) is given by a law of the form

$$1 - t_{max} = \text{const}/S\tau_R. \tag{A.17}$$

The over-all plot of S as a function of G_{max} has then the aspect shown in Fig. 11. The branch with negative slope corresponds to unstable solutions, and the branch with positive slope describes the stretched state: This confirms Fig. 3. For a given S one can then find G_{max} and construct $t(n)$ by a simple integration of (A.13). If one studies for instance the conformation at point M (i.e. when the shear rate is just large enough to allow for a stretched state) one finds a variation of $t(n)$ which is not very different from a parabola (t being maximum at $n = Z/2$ and vanishing at both ends). Thus, in the distorted chain, there is *not* a very well defined boundary between a central, stretched portion, and terminal portions which are coiled. This has also important dynamical consequences: Because $t(n)$ is not very different from a sine wave, the dynamics of any adjustment process will be controlled by the lowest modes, for which the dumbbell description is qualitatively correct.

[1]A. Peterlin, Pure Appl. Chem. 12, 273 (1966).
[2]A. Peterlin, Adv. Macromol. Chem. 1, 225 (1968).
[3]Janeshitz-Kriegl, Adv. Polymer Sci. 6, 170 (1969).
[4]B. H. Zimm, J. Chem. Phys. 24, 269 (1956).
[5]J. Lumley, (a) Ann. Rev. Fluid. Mech. 1, 367 (1969); (b) Macromol. Rev. (to be published).
[6]P. Debye and A. M. Bueche, J. Chem. Phys. 16, 573 (1948).
[7]P. E. Rouse, J. Chem. Phys. 21, 1272 (1953).
[8]A. Peterlin, J. Chem. Phys. 33, 1799 (1960). The initial de-

crease (for $\tau_0 S \gtrsim 1$) calculated in this paper is related to nonuniform chain expansion. It is dominated, however, for $\tau_0 S > 1$, by the global effect discussed above.

[9]The difficulty of realizing nontrivial laminar flows where each molecule sees a constant gradient has been pointed out by professor F. C. Frank (private communication).

[10]This ceases to be true, however, near a second order coil ⟷ stretch transition — for instance, if we consider a marginal case such as Couette flow.

[11]However, we must keep in mind the fact that a stationary state ($\partial\Phi/\partial t = 0$) is not a state of thermal equilibrium, since the velocity field $\mathbf{v} = \mathbf{S}\,\mathbf{r}$ always imposes some dissipation.

[12]See, for instance, H. D. Ellis, Nature 226, 352 (1970).

[13]This is a very rough estimate because the chain extension is not uniform.

[14]P. G. De Gennes, Physics 3, 37 (1967).

Afterthought: Coil-stretch transition of dilute flexible polymers under ultrahigh velocity gradients

Beautiful experiments, displaying the coil-stretch transition in longitudinal shear, have been performed with the 4 roller apparatus at Bristol: see for instance A. Keller and J. Odell, *Coll. Pol. Sci.* **263**, 181 (1985).

Simulations performed with a small number N of beads ($N \lesssim 50$) have *not* shown the transition. I tend to believe that this is related to the height of the barrier between the two states, which is increasing with some power of N: at low N the barrier is not much larger than kT, and the system commutes frequently from one state to the other: there is no sharp transition.

Solutions of Flexible Polymers. Neutron Experiments and Interpretation

M. Daoud,[1a] J. P. Cotton,[1a] B. Farnoux,[1a] G. Jannink,*[1a] G. Sarma,[1a] H. Benoit,[1b] R. Duplessix,[1b] C. Picot,[1b] and P. G. de Gennes[1c]

Laboratoire Léon Brillouin, Centre d'Etudes Nucléaires de Saclay, 91190 Gif-sur-Yvette, France; Centre de Recherches sur les Macromolécules, 67083 Strasbourg Cedex, France; and Collège de France, 11, Place M. Berthelot, 75005 Paris, France. *Received May 13, 1975*

ABSTRACT: We present small angle neutron scattering data on polystyrene (normal or deuterated) in a good solvent (carbon disulfide). All data are taken in the *semidilute* regime where the chains overlap strongly, but the solvent fraction is still large. We have measured the following. (a) The radius of gyration $R_G(c)$ for one deuterated chain in a solution of normal chains with concentration c. We find that $R_G^2(c)$ is proportional to the molecular weight and that R_G^2 decreases with concentration like c^{-x} where $x = 0.25 \pm 0.02$. (b) The screening length $\xi(c)$ (introduced, but not quite correctly calculated by Edwards) giving the range of the $\langle c(\mathbf{r})c(\mathbf{r}')\rangle$ correlations. We find $\xi \sim c^{-z}$, with $z = 0.72 \pm 0.06$. (c) The osmotic compressibility $\chi(c)$ (through the scattering intensity of identical chains in the small angle limit). From an earlier light-scattering experiment, we find $\chi \sim c^{-y}$ with $y = 1.25 \pm 0.04$. These results are to be compared with the predictions of the mean field (Flory–Huggins–Edwards) theory which are: R_G independent of c, $\chi \sim c^{-1}$, and $\xi \sim c^{-1/2}$ in the semidilute range. We show in the present paper that the measured exponents can all be interpreted in terms of a simple physical picture. The underlying basis is the analogy, recently found by Des Cloizeaux, between the semidilute system and a ferromagnet under an external field. However, in this paper, we emphasize mainly the polymer aspects. At short distances ($r < \xi$) the correlations are determined by excluded volume effects. At large distances ($\xi < r < R_G$) the chains are gaussian and the effective interaction between subunits of the same chain are weak.

I. Introduction

1. General Aims. For a long time, the main experiments on polymer solutions measured *macroscopic* parameters such as the osmotic pressure, or the heat of dilution. The resulting data for good solvents are rather well systematized by a mean field theory analysis due to Flory[2a] and Huggins.[2b] A precise description of the method and some of the data can be found in chapter 12 of Flory's book.[3] Recently, however, it has become possible to probe solutions more *locally* (i.e., at distances of order 20 to 500 Å) by small-angle neutron scattering.[4] Using the large difference between the scattering amplitude of protons and deuterons, many different measurements become feasible. For instance it is possible to study one *single-labeled chain* among other chains which are chemically identical, but not labeled.[5] In the present report we present two distinct series of neutron studies on polymer solutions: one with identical chains, and another one with a few labeled chains. These experiments prove that the mean field theory must be refined, and that a number of anomalous exponents occur; fortunately the theory of polymer solutions in good solvents has progressed remarkably in the last year mainly through the work of Des Cloizeaux.[6] In his original article, Des Cloizeaux was concerned mainly with thermodynamic properties. In the theoretical part of the present paper, we show that (a) his results can be derived directly from certain scaling assumptions and (b) his arguments can be extended to discuss local correlation properties. Our discussion is only qualitative, but it does account for the anomalous exponents, without the heavy theoretical background which is needed to read ref 6. (A simplified version of ref 6, requiring only a modest knowledge of magnetism and phase transitions, is given in the Appendix.) To reach a (hopefully) coherent presentation, we shall not separate the theory from the experiments, but rather insert the latter at the right point in the discussion.

2. Organization. This paper is organized as follows. Section II describes the experimental method. Section III contains a general presentation of the three concentration domains for polymer solutions, corresponding respectively to

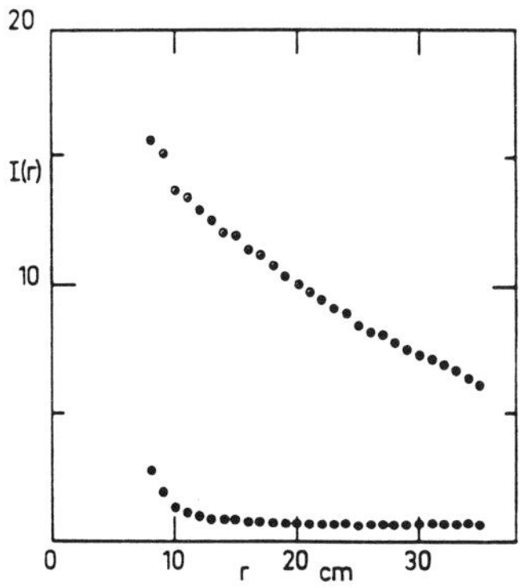

Figure 1. Scattered intensities $I(r)$ as a function of the distance r between a cell detector and the direct beam: r is a measure of the scattering angle θ, $r/L = \theta$, where L is the path length between sample and detector. Open circles: scattered intensity by a solution of 0.5×10^{-2} g cm^{-3} PSD and 95×10^{-2} g cm^{-3} PSH in CS_2. Closed circles: scattered intensity by a solution of 95×10^{-3} g cm^{-3} PSH in CS_2. The intensity variation at small r is due to the direct beam contamination.

separated chains, overlapping chains, and concentrated systems. Our interest is mainly on the second, "semidilute" regime. We discuss in section III the global thermodynamic properties in this regime. In section IV we consider the local structure of the correlation, and establish the distinction between a short-range behavior which is of the excluded volume type and a long-range behavior which corresponds essentially to gaussian chains; the critical distance ξ at which these two pictures merge is one crucial parameter of the theory and we report measurements of this length as a function of concentration. Finally, in section V, we discuss possible correlation experiments involving chains which carry a label at *one* definite position (for instance at the ends, or at the midpoint).

II. Experimental Technique

1. Contrast Factor in a Small-Angle Neutron-Scattering Experiment. The scattering of a radiation of wavelength λ by a medium with long-range correlations (i.e., greater than λ) gives rise to a central intensity peak about the forward scattering direction. The shape of this peak reflects the pair correlation function of the atoms in the sample. The intensity is determined by the interaction between radiation and atoms. In case of a neutron radiation, the interaction is of the neutron–nucleus type. It is characterized by a "coherent scattering length", a. The value of a varies in an impredictable manner[7] from nucleus to nucleus. For example, the hydrogen atom has a scattering length $a_H = -0.374 \times 10^{-12}$ cm, and the deuterium atom has $a_D = +0.670 \times 10^{-12}$ cm. This is the large difference responsible for the success of the labeling method[8,9] in the study of polymer configurations by small-angle neutron scattering. The labeling is an isotopic substitution, without chemical perturbation.

In a scattering experiment the intensity is recorded as a function of the scattering angle θ. The pair correlation function $S(q)$, for a given scattering vector

$$|\mathbf{q}| = 4\pi/\lambda \sin \theta/2 \tag{II.1}$$

is related to the coherent scattered intensity $I(q)$ by

$$I(q) = AK^2S(q) \tag{II.2}$$

where A is the apparatus constant and K the contrast factor or "apparent" scattering amplitude given by

$$K = (a_m - na_s) \tag{II.3}$$

a_m is the coherent scattering length for the monomer unit (i.e., the sum of the length associated with each atom), a_s is the coherent scattering length for the solvent molecule, and n is the ratio of the

Table I
Characteristics of Samples 1 and 2

Sample no.	Polymer	M_w	M_w/M_n	c^*, g cm^{-3}	λ_0, Å
1	PSD	114000	1.10	0.075	5.62
	PSH	114000	1.02		
2	PSD	500000	1.14	0.020	8.83
	PSH	530000	1.10		

partial molar volumes. Formula II.2 can be generalized to the problem of interest in the study of concentrated solutions, where a few labeled chains are dispersed in a solution of unlabeled chains and solvent. In this case

$$I(q) = A[K_D^2S_D(q) + K_H^2S_H(q) + 2K_DK_HS_{HD}(q)] \tag{II.4}$$

where the separate correlation functions for labeled and unlabeled polymers are respectively $S_D(q)$ and $S_H(q)$; S_{HD} is the interference term, K_D and K_H are the "apparent" scattering amplitudes defined by eq II.3 for the deuterated and protonated monomers, respectively. For polystyrene dissolved in carbon disulfide, the theoretical values[10] for these amplitudes are

$$K_D = 8.6 \times 10^{-12}\ \text{cm}$$

$$K_H = 0.3 \times 10^{-12}\ \text{cm}$$

It is clear, with these numbers, that the scattered intensity (eq II.4) is essentially proportional to the correlation function of the labeled chains; this assertion is true at about every practical concentration of deuterated and protonated chains, as will be seen from the concentration dependence of $S_D(q)$ and $S_H(q)$ in section IV.

Up to this point we have only discussed the "coherent" contribution to the scattered intensity. The presence of protons in the solution implies however a strong "incoherent" contribution to the scattered intensity, which is q independent. Because of the long-range monomer–monomer correlation, the weight of the coherent signal is of an order of magnitude greater than the incoherent signal. The two contributions can be compared in Figure 1. The upper curve corresponds to a concentration $c_D = 5 \times 10^{-3}$ g cm^{-3} of labeled chains and a concentration $c_H = 95 \times 10^{-3}$ g cm^{-3} of protonated chains, dispersed in carbon disulfide. The lower curve corresponds to the solution of protonated chains only ($c_H = 95 \times 10^{-3}$ g cm^{-3}). It is seen to contribute by an order of magnitude less.

2. Samples. Our experiments have been carried out with two kinds of polystyrene chains. (a) The first is deuterated polystyrene (PSD), obtained by anionic polymerization of perdeuterated styrene (kindly supplied by Mr. M. Herbert, CEA, Département des Radioéléments). The method of preparation and the characterization are detailed in ref 11. The molecular weights and polydispersity are given in Tables I and III (section IV). (b) The second kind of chain is hydrogeneous polystyrene, prepared by anionic polymerization (cf. Table I). The samples are mixtures of PSD and PSH, in proportions described in section IV, dispersed in carbon disulfide. This is a good solvent and has the advantage of being "transparent" to the neutron beam.

3. Apparatus. The wavelength of the incident neutron beam can be fixed in between $\lambda = 5$ Å and 10 Å. This range is very convenient for a study of polymer configuration and points out another advantage of neutron-scattering technique. It allows the measurement of the size of the polymer coil in the "Guinier" range, as in a light-scattering experiment,

$$qR_G < 1 \tag{II.5}$$

where R_G is the radius of gyration of the coil, and also the monomer–monomer correlation function in the so called "asymptotic" or "intermediate" range

$$1/R_G < q < 1/l \tag{II.6}$$

where l is the step length.

Measurements of R_G were performed with the small-angle neutron-scattering apparatus at the "Institut Laue Langevin" (Grenoble) (the "Institut Laue Langevin" has edited a Neutron Beam Facilities available for users, ILL, BP 156 Centre de tri 38042 Grenoble). The experimental setup is described in ref 11 and corresponds to the lines B and C of Table III of this reference. The incident beam has a broad wavelength distribution and the momentum transfer is defined here as inversely proportional to the mean quadratic wavelength λ_0

$$q = 4\pi/\lambda_0 \sin \theta/2 \qquad \text{(II.1')}$$

The values of λ_0 are given in Table I.

For the intermediate momentum range (eq II.6), the measurements were made on a two-axis spectrometer at the exit of a "cold" neutron guide of the EL3 reactor at Saclay, specially designed for small-angle scattering. A detailed description is also given in ref 11. The incident wavelength is 4.62 ± 0.04 Å (thus with a very narrow distribution as compared to the first apparatus) and the scattering vector range is

$$10^{-2}\ \text{Å}^{-1} \leq q \leq 10^{-1}\ \text{Å}^{-1}$$

III. Different Concentration Regimes and Global Properties

All our discussion will be restricted to the case of *good solvents*, where different monomers repel each other.

1. Dilute Limit. In this limit, the chains are well separated and behave essentially like a gas of hard spheres,[12] with a radius[43]

$$R_G \simeq N^{\nu} \qquad \text{(III.1)}$$

Here N is the polymerization index (the number of monomers per chain) and ν is an excluded volume exponent. We shall always use the Flory value[3] for ν; in the space dimension $d = 3$, $\nu = 3/5$. This value is confirmed by a large number of experiments[13] and by numerical studies.[14] Recent advances in the theory of phase transitions[15] have shown that it is instructive to consider not only the usual case $d = 3$, but also other possible dimensionalities ($d = 2$, for instance, corresponds to polymer films on a surface). The general Flory formula for $d \leq 4$ is

$$\nu = 3/(d + 2) \qquad \text{(III.2)}$$

Above four dimensions, ν sticks to the ideal chain value ($\nu = 1/2$).

The dilute regime corresponds to chain concentrations $\rho_p = \rho/N$ such that the coils are mostly separated from each other

$$\rho_p R_G{}^d = \rho_p N^{\nu d} \ll 1 \qquad \text{(III.3)}$$

In this limit the osmotic pressure π is given by a virial expansion[44]

$$\frac{\pi}{T} = \frac{1}{\bar{v}}\rho_p + \frac{N^2}{\bar{v}^2}A_2(0)\rho_p{}^2 + \ldots$$
$$= \frac{c}{M} + A_2(0)c^2 + \ldots \qquad \text{(III.4)}$$

where M is the molecular weight of the polymer chain. The quantity A_2 is a second virial coefficient. (The notation $A_2(0)$ recalls that it is defined here in the dilute limit.) As in a hard sphere system, $A_2(0)$ is proportional to the volume occupied by one chain

$$A_2(0) = \pi R_G{}^d/6M^2 \sim N^{\nu d - 2} \qquad \text{(III.5)}$$

Equations III.1, III.4, and III.5 have been amply verified[3] and we have nothing to add here. But eq III.4 and III.5 will be of use later for comparison with the behavior found at higher concentrations.

2. Semidilute Solutions. This regime sets in when $\rho_p R_G{}^d > 1$. The chains now interpenetrate each other and the thermodynamic behavior becomes completely different. In terms of the monomer concentration ρ, the semidilute regime is defined by the two inequalities

$$\rho^* \ll \rho \ll 1 \qquad \text{(III.6)}$$

where we have introduced a critical concentration

$$\rho^* = NR_G{}^{-d} = N^{1-\nu d} = c^*\bar{v} \qquad \text{(III.7)}$$

In our experiments ρ^* ranges between 10^{-3} and 8×10^{-2}. The changes occurring near $\rho = \rho^*$ have been studied by various methods and in particular by neutron scattering.[4] Here we shall be more concerned with the region above ρ^* defined by (III.6).

In this region the mean field picture could be presented as follows: for small ρ the free energy density is of the form

$$F = F_{\text{ideal}} + \frac{T}{2}\upsilon \int \rho^2(r)\frac{d^3r}{V} + 0(\rho^3) \qquad \text{(III.8)}$$

where F_{ideal} is the free energy for nonintersecting coils with their unstretched radius $R_G{}^0 = N^{1/2}$, and V is the volume of the sample. The parameter υ (with dimension volume) is a virial coefficient for the monomer–monomer interaction. One definition of υ follows the paper by Edwards.[16] In the Flory notation we would write

$$\upsilon = \frac{\bar{v}^2 m_0{}^2}{V_1}(1/2 - \chi_1) = \frac{\bar{v}^2 m_0{}^2}{V_1}\psi_1\left(1 - \frac{\Theta}{T}\right) \qquad \text{(III.9)}$$

where V_1 is the molar volume of the solvent and m_0 the molecular weight of the monomer. The constant χ_1 characterizes the solvent–solute interaction, ψ_1 is the entropy constant, and Θ is the Flory temperature.

We restrict our attention to good solvents where υ is positive. Returning now to eq III.8 the mean-field approximation amounts to replacing $\rho^2(r)$ by ρ^2. The osmotic pressure is related to F by the general formula

$$\pi = \rho_p{}^2\frac{\partial}{\partial \rho_p}\left(\frac{F}{\rho_p}\right) \qquad \text{(III.10)}$$

(since $1/\rho_p$ is the volume per chain and F/ρ_p is the free energy per chain). Using $\rho_p = \rho/N$, eq III.10 may also be written as:

$$\pi = \rho^2\frac{\partial}{\partial \rho}\left(\frac{F}{\rho}\right) \qquad \text{(III.10')}$$

Writing $F = F_{\text{ideal}} + 1/2 T\upsilon\rho^2$ in the mean-field approximation we get

$$\frac{\pi}{T} = \frac{c}{M} + \frac{1}{2}\frac{\upsilon}{m_0{}^2}c^2$$

The first term (derived from F_{ideal}) is in fact negligible for semidilute solutions. Thus the mean-field prediction is essentially

$$\frac{\pi}{T} = \frac{1}{2}\frac{\upsilon}{m_0{}^2}c^2 \qquad \text{(III.11)}$$

We notice that at a given concentration c in this range, the osmotic pressure is independent of the *chain* molecular weight. (In the dilute range, π/T is strongly dependent upon M (eq III.4).)

However, this formula omits a very important correlation effect; when one monomer is located at point **r**, all other monomers are repelled from the vicinity of **r**. Thus the average of $\rho^2(\mathbf{r})$ is expected to be smaller than ρ^2. (Also, as we shall see, the chains are still swollen in the semidilute region; this reacts on the entropy.) To include all these effects, let us assume that the osmotic pressure behaves like some powers of the concentration

$$\frac{\pi}{T} = \text{constant} \times c^m \qquad \text{(III.12)}$$

This particular scaling law can be, to some extent, demonstrated by a renormalization group technique[17] or related to the scaling properties of a magnetic system.[6] If we accept it, we can derive the value of the exponent m by asking that eq III.4 and III.12 give the same order of magnitude at the

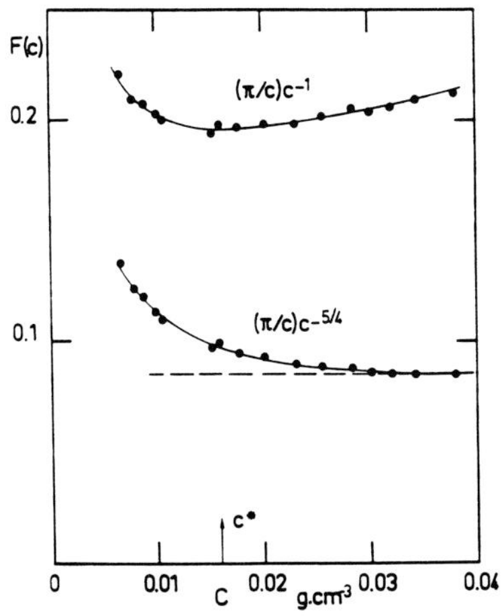

Figure 2. Representations of the osmotic pressure vs. concentrations. The data of ref 18 are divided by c^2 (open circles) and by $c^{9/4}$ (closed circles). The latter representation illustrates the proposition that in the semidilute range, the osmotic pressure behaves as $c^{9/4}$ rather than the mean field value c^2.

crossover between dilute and semidilute regimes ($c = c^*$)

$$\left.\frac{\pi}{T}\right|_{c^*} \simeq \frac{c^*}{M} \simeq \text{constant} \times (c^*)^m \qquad \text{(III.13)}$$

Inserting the value from eq III.7 for c^* gives

$$m = \nu d/(\nu d - 1) \qquad \text{(III.14)}$$

a value first derived in ref 6. For $d = 3$, and using $\nu = 3/5$, one finds $m = 9/4$, to be compared to the mean-field value $m = 2$. The difference is not very large but, as we shall see it, visible in careful experiments performed with sufficiently long chains (a large N is favorable, because it gives a small c^* and thus a wide range of semidilute systems).

(a) There are many data in this field[3] of osmotic pressure measurements. We shall discuss here the findings of Strazielle and coworkers,[18] who measured the pressures at the highest possible concentrations beyond c^* in order to detect deviations from the c^2 behavior. Figure 2 shows two representations of these data, respectively π/c^2 and $\pi/c^{9/4}$. The two curves illustrate the discussion about the exponent m. The figure is however not intended as a proof of eq III.12. The evidence which we consider as an experimental proof for the value of m is given in section IV.

(b) The heat of dilution, ΔH_1, and the specific heat of dilution $\overline{\Delta H}_1$, can be calculated as a function of c. A definition of ΔH_1 is[3]

$$\Delta H_1 = \zeta p_{12} \qquad \text{(III.15)}$$

where ζ is the change of energy for the formation of unlike contacts and p_{12} is the number of such pairs. This quantity is evaluated as follows. The probability that a site is occupied by a (polymer) segment is ρ. Formula III.12 tells us that the probability for another segment to be in an adjacent site is $\rho \times \rho^{5/4}$. In other words, the conditional probability for a segment to occupy a site, knowing that an adjacent site is occupied by another segment, is $\rho^{5/4}$. This quantity is smaller than ρ; there is a depletion effect around each site occupied by a polymer segment. The probability for unlike contacts is thus

$$\rho(1 - \rho^{5/4}) \sim p_{12}$$

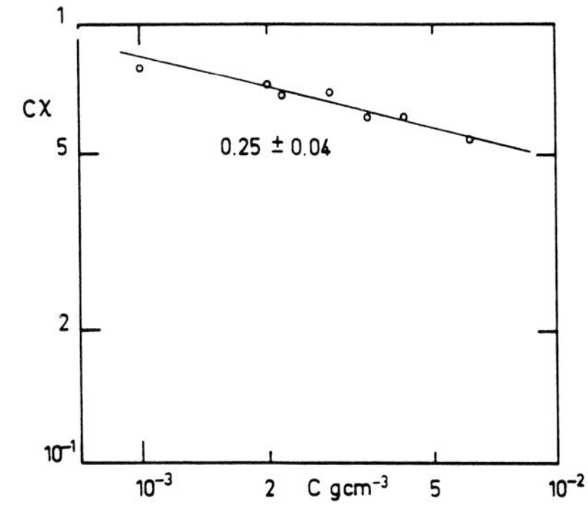

Figure 3. Log–log representation of the osmotic compressibility (multiplied by c) vs. concentration. The data are taken from ref 21. The extrapolation to $\theta = 0$ of the scattered intensities is made from the intermediate momentum range (eq II.6).

and the specific heat of dilution

$$\overline{\Delta H}_1 = \left.\frac{\partial H_1}{\partial \rho_s}\right|_{\rho} = \zeta \rho^{9/4} \qquad \text{(III.16)}$$

where ρ_s is the volume fraction of the solvent. Formula III.16 could explain deviations from the mean-field prediction: $\overline{\Delta H}_1/\rho^2$ = constant. The evaluation of the exponent m from existing ΔH_1 vs. ρ plots[19] yields however values in excess of 9/4. This is not meaningful, in so far as the molecular weights used in these experiments are very low. The semidilute range is thus here too narrow.

(c) Compressibility measurements through small angle scattering. These experiments were performed with solutions of identical (deuterated or protonated) chains. They measure the Fourier transform $S(q)$ of the monomer–monomer correlation function.

$$S(\mathbf{q}) = \int \langle \delta\rho(\mathbf{o})\delta\rho(\mathbf{r})\rangle e^{i\mathbf{q}\cdot\mathbf{r}} d^3\mathbf{r} \qquad \text{(III.17)}$$

where $\delta\rho(\mathbf{r}) = \rho(\mathbf{r}) - \rho$ represents the local fluctuation of concentration. In the limit of small q, a general thermodynamic theorem[20] imposes that $S(q)$ becomes proportional to the osmotic compressibility

$$\chi = \partial c/\partial \pi$$

$$\lim_{q \to 0} S(q) = Tc\chi \qquad \text{(III.18)}$$

The correlation function $S(q)$ is measured in the asymptotic, or intermediate, range defined by eq II.5, and the extrapolation at $q \to 0$ is made from this range. A typical plot of $S(q)$ at different concentrations is given in section IV.3. In order to compare the extrapolated values at different concentrations, some socalled "absolute" measurements must be performed in which the apparatus constant A in eq II.2 is determined with sufficient accuracy. This raises technical difficulties in neutron-scattering experiments. These difficulties do not appear in light-scattering experiments. We have therefore plotted values of ($c\chi$) as a function of c, obtained by the extrapolation described above, of light-scattered intensities[21] by very long chains ($M = 7 \times 10^6$), Figure 3. The details of this experiment are described in ref 21. We wish however to point out that the values $I(0)$ extrapolated from the "intermediate" momentum range (Figure 3) do not correspond to the values $I(0)$ reported in ref 21, which are extrapolated from the "Guinier range". The result is $\chi \simeq c^y$, where $y = 1.25 \pm 0.04$.

We conclude that the exponent $m = y + 1$ in eq III.12 is definitely larger than 2, and in fact close to 9/4 as expected

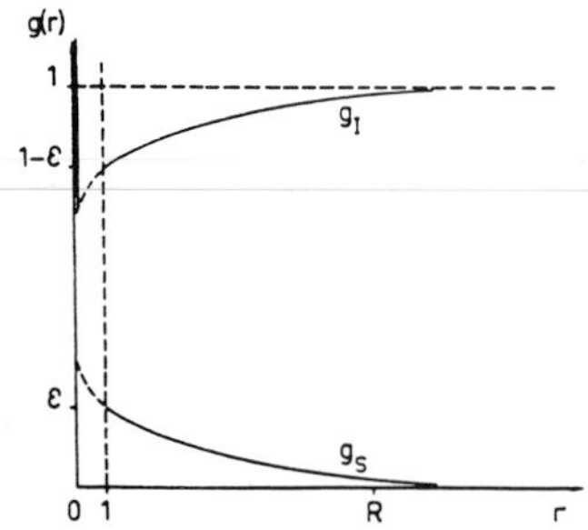

Figure 4. Schematic representation of the holes in the correlation functions $g_S(\mathbf{r})$ and $g_I(\mathbf{r})$. There are three distinct behaviors: (i) asymptotic ($r > R_R$), (ii) the hole ($r \sim R_R$), (iii) critical ($r < 1$).

from the scaling law. We may end up this paragraph by a qualitative remark, which will be of use later; the effective interaction energy $\frac{1}{2}vT\langle\rho^2(r)\rangle$ defined in eq III.8 is clearly smaller than it's mean field estimate $\frac{1}{2}vT\rho^2$, because of repulsive correlations between monomers. We may define a renormalized interaction constant by setting

$$v\langle\rho^2\rangle = \mathbf{v}\rho^2 \qquad \text{(III.19)}$$

Returning to eq III.10′ for the osmotic pressure and comparing between eq III.12, we see that the left-hand side is proportional to $\rho^m = \rho^{9/4}$. Thus **v** is proportional to

$$\mathbf{v} \simeq \rho^{m-2} \qquad (\rho^{1/4} \text{ for } d = 3) \qquad \text{(III.20)}$$

This renormalization was not included in the Edwards analysis of semidilute solutions.[22] We shall see in the next section that, when renormalization is taken into account, the Edwards calculation becomes compatible with our picture.

3. Concentrated Solutions. When ρ tends to one we find certain complications but also certain simplifications. On one hand, it is not possible any more to describe the interactions between monomers in terms of a simple coupling constant v; the system is less universal. However, there do remain universal properties on a large spatial scale. In an earlier series of experiments[11,23,24] we have shown that the radius of gyration for one deuterated polystyrene chain imbedded in normal chains was essentially the ideal chain radius $R_G \simeq N^{1/2}$. From a theoretical point of view, in the limit $\rho = 1$, we can neglect the fluctuations of ρ, the main effect of the monomer–monomer interactions in this limit is to preserve a constant density. The correlations present in that phase are nontrivial, however; the chains still repel each other and each chain is surrounded by a *correlation hole* (of radius $\sim R_G$) within which the concentration of other chains is slightly depleted. This hole can be discussed through a simple random phase approximation.[25] Here we shall present only qualitative features of the hole which are entirely model independent. Consider first the correlation function $\langle\rho(0)\rho(r)\rangle$ for all identical chains; since the density is constant ($\rho = 1$), this is equal to unity at all r. Let us split this correlation into two terms

$$1 = \langle\rho(\mathbf{0})\rho(\mathbf{r})\rangle = g_S(\mathbf{r}) + g_I(\mathbf{r}) \qquad \text{(III.21)}$$

where g_S represents the correlation inside our same chain and g_I the correlation between different chains. The shapes of g_S and g_I are shown in Figure 4. A detailed calculation of g_S and g_I could be obtained[45] by the technique of ref 25. The range of g_S is of the order R_G. The magnitude ϵ of g_S for a typical $r \sim R_G$ is such that the integral of g_S gives the total number of monomers on one chain

$$\int g_S(\mathbf{r})\mathrm{d}^3\mathbf{r} = N \qquad \text{(III.22)}$$

Thus $\epsilon R^d \sim N$, or $\epsilon \sim N^{1-d/2}$ ($\sim N^{-1/2}$ for $d = 3$). Note that $\epsilon \ll 1$. The correlation between *different* chains $g_I = 1 - g_S$ is thus slightly depressed (by an amount ϵ) in a region of radius R_G. This could be seen in principle by a study of mixtures between normal and deuterated chains; if we have a concentration ρ_D of deuterated monomers (the total concentration hydrogen + deuterium remaining equal to 1) we can measure a correlation function $\langle\rho_D(\mathbf{0})\rho_D(\mathbf{r})\rangle$. Since the mixture is ideal, we may write

$$\begin{aligned}\rho\langle\rho_D(\mathbf{0})\rho_D(\mathbf{r})\rangle &= \rho_D g_S(\mathbf{r}) + \rho_D^2 g_I(\mathbf{r}) \\ &= \rho_D[g_S + \rho_D(1 - g_S)] \\ &= \rho_D^2 + g_S(\mathbf{r})\rho_D(1 - \rho_D) \qquad \text{(III.23)}\end{aligned}$$

or in terms of the fluctuating part $\delta\rho_D$

$$\langle\delta\rho_D(\mathbf{0})\delta\rho_D(\mathbf{r})\rangle = \rho_D(1 - \rho_D)g_S(\mathbf{r}) \qquad \text{(III.24)}$$

For a given scattering vector q we would measure the Fourier transform $S_D(q)$ of this correlation. Let us restrict our attention for the moment to the *small* q limit. Making use of eq III.22 we have

$$S_D(q = 0) = \rho_D(1 - \rho_D)N \qquad \text{(III.25)}$$

The $\rho_D(1 - \rho_D)$ dependence is natural for an ideal mixture (as noted in ref 11). But another presentation of this result is interesting in terms of the number of deuterated chains per unit volume $\rho_{PD} = \rho_D/N$. We have

$$S_D(q = 0) = N^2[\rho_{PD} - N\rho_{PD}^2] \qquad \text{(III.26)}$$

We may express this in terms of a second virial coefficient $A_2(1)$ for the deuterated chains;[46] we have exactly

$$A_2(1) = (1/N)(\bar{v}/m_0) \qquad \text{(III.27)}$$

(cf. ref 11). This is to be compared with the virial coefficient $A_2(0)$ for the dilute limit (eq III.5). Note that

$$A_2(1)/A_2(0) \sim N^{1-\nu d} = N^{-4/5} \qquad \text{for } d = 3 \qquad \text{(III.28)}$$

Thus $A_2(1)$ is very small; indeed up to now, we have not been able to detect any deviation from a linear law for $S_D(q = 0)$ at small c_D. However, it is worthwhile to notice that eq III.26 for S_D applies to *all* concentrations ρ_D (for $\rho = 1$). Using the values of ρ_D of order $\frac{1}{2}$, it should be easy to check it. Equations III.25 and III.27 were derived in an earlier paper on liquid polystyrene.[11] We have rederived them for later uses in our discussion of $A_2(\rho)$ in the semidilute regime.

IV. Spatial Correlations in the Semidilute Regime

1. Correlations Inside One Chain. Let us again start from the concentration pair correlation function (which is measured in experiments on identical chains) and split it as in eq III.21 in two parts:

$$\langle\rho(\mathbf{0})\rho(\mathbf{r})\rangle = g_S(\mathbf{r}) + g_I(\mathbf{r}) \qquad \text{(IV.1)}$$

Again $g_S(\mathbf{r})$ will represent the intrachain contribution, which can be separated using deuterated labels. In the concentrated system ($\rho = 1$) the sum $g_S + g_I$ was constant. But in the semidilute systems g_S and g_I represent separate unknown functions.

Let us focus our attention first on $g_S(\mathbf{r})$. We may define it by the following recipe; at the origin (0) we put one particular monomer (the nth monomer) of a certain polymer chain P. Then we look at a neighboring point r and ask for the local concentration of monomers belonging to the same chain; this is $\rho^{-1}g_S(\mathbf{r})$. At small r we expect $g_S(\mathbf{r})$ to be dominated by excluded volume effects inside chain P. Using the scaling law[26,27] we are led to expect that

$$\frac{1}{\rho}g_S(\mathbf{r}) = \frac{p(r)}{r^d} \qquad \text{(IV.2)}$$

where $p(r)$ is the number of chain units which will give a coil of size r. Thus $p(r) \sim r^{1/\nu}$ and

$$\frac{1}{\rho} g_S(\mathbf{r}) = r^{-d+1/\nu} \simeq r^{-4/3} \quad \text{for } d = 3 \qquad \text{(IV.3)}$$

Equation IV.3, for an isolated chain, was written down long ago by Edwards.[16] This single chain behavior will apply whenever the local concentration $\rho^{-1}g_S(r)$ due to a chain P is larger than the average concentration ρ (due to other chains). The cross-over will thus occur at a certain characteristic distance ξ such that

$$\frac{1}{\rho} g_S(\xi) = \rho \qquad \text{(IV.4)}$$

Equation IV.3 then shows that

$$\xi \simeq \rho^{-\nu/(\nu d - 1)}$$

or

$$\xi = \rho^{-3/4} \quad \text{for } d = 3 \qquad \text{(IV.5)}$$

In a recent experiment,[5] the single chain behavior could be observed at concentrations substantially greater than ρ^*.

At the lower concentration limit ($\rho = \rho^*$) the length ξ is equal to the coil size $R_G(\rho^*) \sim N^\nu$. At higher concentrations ξ is smaller than $R_G(\rho)$. Ultimately, for ρ equal to unity ξ becomes of order one (i.e., comparable to the monomer size). In this limit all effects of the anomalous exponent ν are removed. This is why the simple random phase calculation of ref 25 is acceptable for molten polymers.

At distances larger than ξ, the chain P may be considered as a succession of statistical elements, with a number of monomers per element $p(\xi) \sim \xi^{1/\nu}$. The average square length of one element is ξ^2. Successive elements are screened out from each other by the other chains, and the overall behavior is thus expected to be gaussian. Thus the end-to-end distance $R(\rho)$ is given by

$$R^2(\rho) = \frac{N}{p(\xi)} \xi^2 \sim N\rho^{(1-2\nu)/(\nu d - 1)} \qquad \text{(IV.6)}$$

Equation IV.6 can also be derived from the Des Cloiseaux approach (see Appendix).

(a) Radius of Gyration. From relation IV.6, the c and M dependence of $R^2(c)$ is

$$R^2(c) \sim M_w c^{-1/4} \quad \text{for } d = 3 \qquad \text{(IV.7)}$$

where M_w is the molecular weight average.

In order to test the validity of this relation, we have measured the radii of gyration of the PSD chain dispersed in a PSH and CS_2 solution. Two sets of PSD, PSH, and CS_2 solutions have been used, corresponding to the two molecular weights given in Table I.

The data were obtained using the small angle apparatus at the I.L.L. with the set-up described in section II. The two values of the quadratic mean wavelength of the incident beam (given in Table I) are chosen to match, for each set of samples, the inequality $qR_G(c) < 1$ (II.1′).

The radius of gyration, $R_G(c)$, of a chain dispersed in a solution with chain concentration c is measured from the pair correlation of the labeled chains. Although the concentration c_D of labeled chains is small compared to c, there are interference terms which must be evaluated. This is achieved by extrapolation of the data at $c_D = 0$ while maintaining the total concentration c fixed. For each total concentration c, eight samples are studied; four solutions are made with the concentrations $c_D = 0.020, 0.015, 0.010$, and 0.005 g cm^{-3} of PSD (each containing a concentration c_H of PSH such that the sum $c_H + c_D$ remains equal to c). The four others are test samples which contain only the four

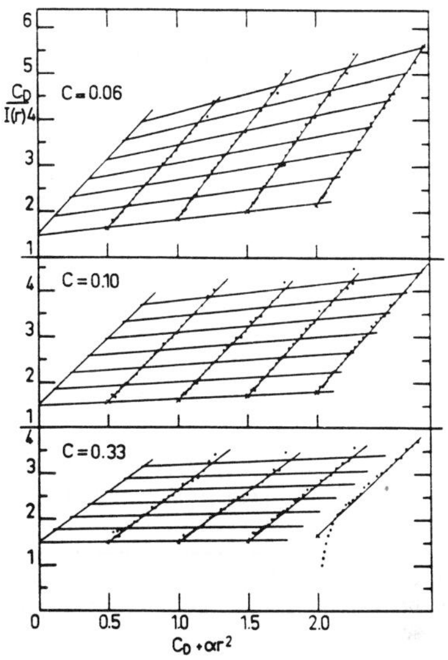

Figure 5. Zimm plots of the scattered intensities by deuterated chains of PSD, at three total concentrations (sample 1, Table I): (i) $c = 0.06$ g cm^{-3}, (ii) $c = 0.10$ g cm^{-3}, (iii) $c = 0.33$ g cm^{-3}. The length r is a measure of the scattering angle, $r/L = \theta$, where L is the path length between sample and detector. c_D is given in units of 10^{-2} g cm^3 and α is a scale factor.

concentrations $c - c_D$ of PSH. For each value of c_D the effective coherent intensity $I(q,c_D)$ or $I(r,c_D)$ is obtained by subtracting from the scattering intensity of the sample the one given by its test sample. This last operation is made in order to eliminate the proton incoherent background.[11]

The effective intensities $I(q,c_D)$ are extrapolated to $c_D = 0$, using the well-known relation of Zimm[28]

$$\frac{Kc_D}{I(q,c_D)} = \frac{1}{M_w}\left(1 + \frac{q^2 R_G^2(c)}{3}\right) + 2A_2(c)c_D \qquad \text{(IV.8)}$$

where K is a constant and $A_2(c)$ a coefficient which accounts for the pair interactions between labeled chains. This relation allows us to obtain the values of $R_G^2(c)$ and $A_2(c)$ using linear extrapolations to $c_D = 0$ and to $q^2 = 0$, respectively.

The method is applied to solutions of sample 1, for nine concentrations c from 0 to 1.06 g cm^{-3} (the bulk density) and to solutions of sample 2, for two concentrations ($c = 0.06$ and 0.2 g cm^{-3}).

Figure 5 shows three of these Zimm plots corresponding to the concentrations $c = 0.060$, 0.100, and 0.33 g cm^{-3} of solutions made with sample 1 ($M_w = 114000$). The experimental data are in good agreement with the Zimm relation if we neglect the curve obtained from the solution $c = 0.33$ g cm^{-3} ($c_D = 0.02$, $c_H = 0.31$) which has an anomalous variation. This last curve can be explained by a wrong determination of the c_H concentration for the test sample.

The slopes of the extrapolated lines at $c_D = 0$ and $\theta^2 = 0$ in Figure 5 indicate that $R_G^2(c)$ decreases more moderately than the coefficient $A_2(c)$, as concentration increases. This can be checked from the numerical values, given in Table II, of $R_G(c)$ and $A_2(c)$ determined with two sets of runs (respectively samples 1 and 2).

The third column of this table gives the thickness t of the containers used in the experiment. The optimal t for a given concentration was tested in an earlier experiment[29] in such a way that multiple scattering effects are avoided for a greatest possible scattered intensity. The case $c = 0$

Table II
Concentrations and Radii of Gyration

M_w	c, g cm^{-3}	t, mm	$R_G(c)$, Å(±5%)	$R_G^2c^{1/4}/M_w$	$A_2(c)$, 10^{-4} g^{-2} cm^3	$M_wA_2(c)$, g^{-2} cm^3
114000	0.00	10	137	0	6.21	70.8
	0.03	10	120	62	3.54	40.4
	0.06	10	117	60	1.02	11.6
	0.10	5	111	60	0.44	5.0
	0.15	5	104	59	0.29	3.3
	0.20	2	101	60	0.15	1.7
	0.33	2	95	59	0.19	1.0
	0.50	1	91	53	0	?
	1.06 (bulk)	1	82	59.5	0	?
500000	0.06	5	242	58	0.11	5.5
	0.20	2	226	68	0.03	1.5

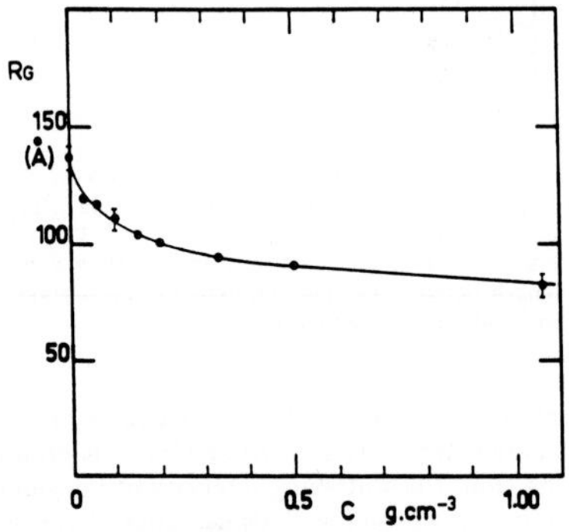

Figure 6. Radius of gyration R_G as a function of concentration, obtained from extrapolation of the Zimm plots, at nine different concentrations of PSH (sample 1).

(Table II) corresponds to the situation where chains are identical (PSD), i.e., there are no PSH chains. The values of R_G and A_2 obtained in these conditions are consistent with those obtained from light-scattering measurements.

Figure 6 shows the variation of the radius R_G as a function of c, determined with sample 1; $R_G(c)$ is continuously decreasing as the concentration increases.

In order to test the model of a gaussian chain, of which only the subunits (of length ξ) are submitted to the excluded volume effects, the log–log plot of $R_GM_w^{-1}$ as a function of c is given in Figure 7. The slope determined by this plot is 0.25 ± 0.02. It is in excellent agreement with the theoretical prediction of a $c^{-1/4}$ variation. The universality of this variation is confirmed by the data obtained from the samples of molecular weight 500000. (This representation yields however an artificial diminution of the experimental data dispersion, which appear if we look at the $R_G^2M_w^{-1}c^{1/4}$ values given in Table II.)

The theoretical predictions (eq IV.6) are given in terms of the quadratic end-to-end distance, R^2. The measured quantities are however the radii of gyration R_G^2

$$R_G^2 = \frac{1}{N}\sum_{i<j<N} \langle r_{ij}^2 \rangle \qquad \text{(IV.9)}$$

where r_{ij} is the distance between segments (i,j). The interpretation of Figure 7 is only meaningfull if the ratio R_G^2/R^2 does not vary as a function of concentration, for $c > c^*$. Since we have assumed that the chain is essentially "Gaussian" beyond c^*, we may expect that R_G^2/R^2 is effectively a constant. More precisely, the model described above tells us that

$$\langle r_{ij}^2 \rangle = |i-j|^{2\nu} \quad \text{for } |i-j| < n$$
$$\langle r_{ij}^2 \rangle = |i-j|\, n^{2\nu-1} \quad \text{for } |i-j| > n \qquad \text{(IV.10)}$$
$$n \simeq \rho^{-\nu/(\nu d-1)}$$

The calculation yields

$$\frac{R_G^2}{R^2} = \frac{1}{6} - x^3 \left| \frac{1}{6} - \frac{1}{(2\nu+1)(2\nu+2)} \right| \qquad \text{(IV.11)}$$

where $x = (\rho/\rho^*)^{1/(1-\nu d)}$.

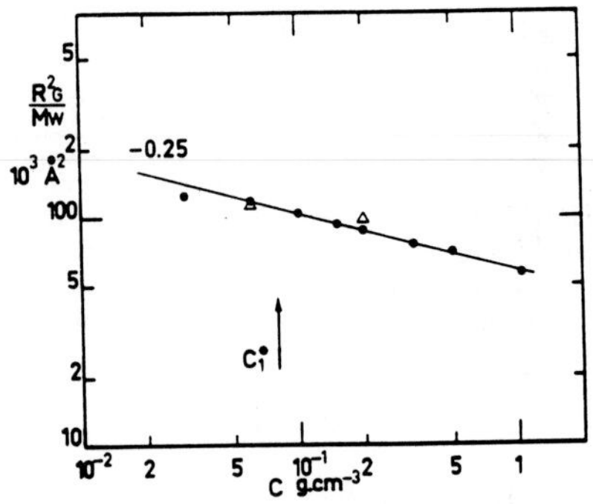

Figure 7. Log–log plot of $R_G^2M_w^{-1}$ as a function of concentration, for samples 1 (●) and 2 (△).

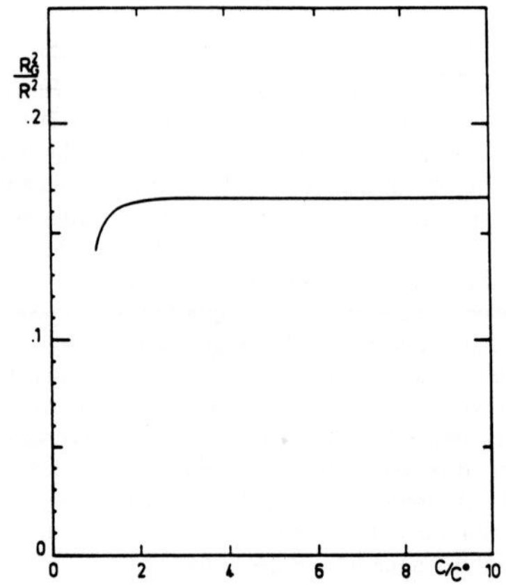

Figure 8. Calculated ratio R_G^2/R^2 as a function of c/c^*, in the interval $c/c^* \geq 1$ (eq IV.11).

The function R_G^2/R^2 is displayed in Figure 8 as a function of c/c^*. The figure shows clearly that this ratio is constant as soon as $c > c^*$. This result confirms a posteriori the experimental test described above. In the neighborhood of c^*, the departure from the $c^{-1/4}$ law should be more pronounced for R_G^2 than for R^2. In the limits of the experimental error, this departure does however not appear in Figure 7.

Relation IV.11 can only be used in the range $\rho^* \leq \rho \leq 1$. This raises naturally two questions, in connection with earlier remarks concerning the decrease of R_G as a function of concentration. The concentration c^* as defined in (III.7) is not a quantity that can be accurately determined either from the theoretical nor from the experimental point of view. Second, the mechanism which is responsible for the decrease of R_G^2 in the dilute range $c < c^*$ is not well understood.

(b) The Second Virial Coefficient. We may call the coefficient $A_2(c)$ in eq IV.8 the second virial coefficient of the deuterium fraction. The value of these coefficients is determined from the Zimm plots. The precision obtained on the evaluation of the $A_2(c)$ is not very satisfactory, and beyond $c = 0.4$ g cm^{-3} the experimental value is meaningless. In spite of this lack of precision, we have drawn the resulting $A_2(c)$ in a log–log plot as a function of concentration (Figure 9). Neglecting the data below c^*, it is seen that A_2 is a decreasing function of c, behaving roughly like $1/c$, and that A_2 is inversely proportional to the molecular weight (cf. eq III.27).

These variations of $A_2(c)$ may be explained as follow; since the mixture of deuterated and normal chains is ideal, we may write the relevant correlation function in the form

$$\langle \rho_D(\mathbf{0})\rho_D(\mathbf{r})\rangle = \frac{\rho_D}{\rho} g_S(\mathbf{r}) + \left(\frac{\rho_D}{\rho}\right)^2 g_I(\mathbf{r}) \qquad \text{(IV.12)}$$

where $g_S(\mathbf{r})$ is the self part defined in eq IV.1 and $g_I(\mathbf{r})$ the part corresponding to distinct chains. We are interested in

$$S_D(\mathbf{q}) = \int \langle \delta\rho_D(\mathbf{0})\delta\rho_D(\mathbf{r})\rangle e^{i\mathbf{q}\cdot\mathbf{r}} d^3\mathbf{r} \qquad \text{(IV.13)}$$

at the value $q = 0$

$$S_D(q = 0) = \int d^3r[\langle \rho_D(\mathbf{0})\rho_D(\mathbf{r})\rangle - \rho_D^2] = \frac{\rho_D}{\rho} I_S - \left(\frac{\rho_D}{\rho}\right)^2 I_I \qquad \text{(IV.14)}$$

where

$$I_S = \int g_S(\mathbf{r}) d^3\mathbf{r} = N\rho \qquad \text{(IV.15)}$$

$$I_I = \int [N^2\rho^2 - g_I(\mathbf{r})] d^3r \qquad \text{(IV.16)}$$

Returning to the case of identical chains, we can write the osmotic compressibility theorem (eq III.18) in the form

$$I_S - I_I = T\rho \frac{\partial \rho}{\partial \pi} \sim \rho^{m-2} = \rho^{-1/4} \qquad \text{(IV.17)}$$

where we have used eq III.12 for the osmotic pressure. Let us compare the right-hand side to I_S. Using eq III.7 and III.14 we have

$$\frac{\rho^{2-m}}{I_S} \simeq N^{-1}\rho^{(\nu d-2)/(\nu d-1)} = (\rho/\rho^*)^{2-m} \qquad \text{(IV.18)}$$

Thus, when $\rho \ll \rho^*$ we can, in a first approximation, neglect the right-hand side in eq IV.17 and put $I_S = I_I$ because the compressibility is weak in the semidilute regime. Returning to eq IV.14 we then have

$$S_D(q = 0) = N\rho_D(1 - \rho_D/\rho) \qquad \text{(IV.19)}$$

This we can write

$$\frac{\rho_D}{S_D(0)} \simeq \frac{1}{N} + \frac{\rho_D}{N\rho} \qquad \text{(IV.20)}$$

Comparing this with the Zimm relation (eq IV.8) we get

$$A_2(\rho) \sim 1/M\rho \qquad \text{(IV.21)}$$

Including the compressibility correction would change this to

$$A_2(\rho) \sim \frac{1}{M\rho}\left[1 - k\left(\frac{\rho^*}{\rho}\right)^{1/4}\right] \qquad \rho^* \ll \rho \ll 1 \; (d = 3) \qquad \text{(IV.22)}$$

where k is a numerical constant of order unity.

This form is in satisfying agreement with the data shown on Figure 9. However it must be emphasized that the leading term in this expression $(N\rho)^{-1}$ is essentially the consequence of a sum rule, and is not related to the anomalous exponents which occur in $\pi(c)$, in $R(c)$, and in $\xi(c)$. Thus the determinations of $A_2(c)$ remain of moderate interest.

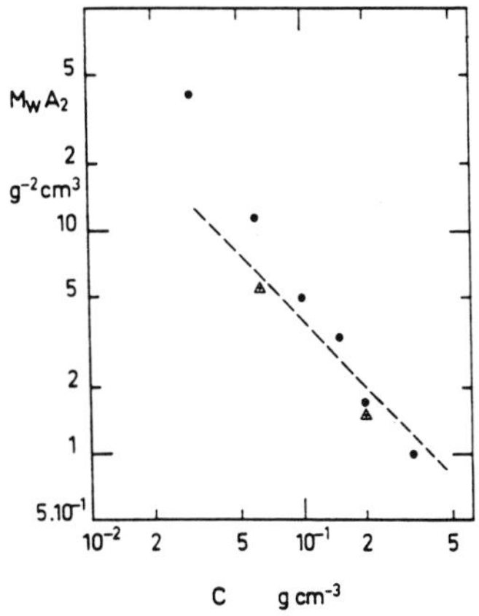

Figure 9. Log–log plot of the second virial coefficient of the deuterium fraction, $A_2(c)$, vs. concentration c of the protonated fraction. The straight line illustrates eq IV.21: sample 1 (●), sample 2 (△).

(c) Behavior of $g_S(\mathbf{r})$ for $\xi < r < R$. In this region we expect $\rho^{-1}g_S(r)$ to be of the ideal chain type, discussed long ago by Debye,[30] provided that the statistical unit is chosen according to the rules above. This is described most simply in terms of Fourier transforms. The most interesting range of q vectors corresponds to

$$1/R \ll q \ll 1/\xi \qquad \text{(IV.23)}$$

and in this range the Debye formula reduces simply to

$$\frac{1}{c} g_S(\mathbf{q}) = \frac{1}{c}\int g_S(\mathbf{r})e^{i\mathbf{q}\mathbf{r}}d_3r = cte\,\frac{12p(\xi)}{q^2\xi^2} \qquad \text{(IV.24)}$$

where the numerical constant depends on the coefficients chosen in eq IV.5 and IV.3. In what follow we shall normalize ξ so that this constant is unity. The spatial version of this formula is for $d = 3$

$$c^{-1}g_S(\mathbf{r}) = \frac{3p(\xi)}{\pi\xi^2 r} \qquad \text{for } \xi < r < R \qquad \text{(IV.25)}$$

One possible determination of ξ could be based on the comparison of scattering data for a single deuterated chain in a solution of protonated chains with eq IV.3 and IV.25 or with their Fourier transforms, the cross-over between the two terms occurring at $r = \xi$.

These types of data[5] indicate however that such a cross-over is not a quantity that can be measured with accuracy. In so far as it is observed, the characteristic distance r is small compared to the value of ξ obtained from measurements on *identical chains* in rather more favorable conditions. This will be described in the next paragraph.

2. Correlations between All Monomers. Let us now return to the full correlation function

$$\langle \rho(\mathbf{0})\rho(\mathbf{r})\rangle = g_S(\mathbf{r}) + g_I(\mathbf{r}) = g(\mathbf{r}) \qquad \text{(IV.26)}$$

where $g_S(\mathbf{r})$ is normalized following eq IV.15.

At small distance ($r < \xi$), $g(r)$ is dominated by $g_S(\mathbf{r})$ and described by eq IV.3. What happens at a larger distance? The first detailed approach to this problem is contained in an elegant paper by Edwards.[22] This is based on a random phase calculation, when the interaction v (eq III.9) is treated as small;[47] one essentially writes that the total correlation function $g(\mathbf{r})$ is the sum of a direct term $g_S(\mathbf{r})$, inside one chain, plus corrections involving two, three, . . . , n

Table III
Characteristics of the Samples in the Identical Chain Experiment

PSD Sample	M_w	M_w/M_n	c^*, g cm^{-3}	c, g cm^{-3}	t, mm	ξ, Å
A	5×10^5	1.14	0.02	0.04	10	29
B	5×10^5	1.14	0.02	0.075	10	18
C	5×10^5	1.14	0.02	0.15	5	10
D	1.1×10^6	1.17	0.01	0.025	10	41

chains.

$$g(\mathbf{r}) = g_S(\mathbf{r}) - \int g_S(\mathbf{r}_1) v g_S(\mathbf{r} - \mathbf{r}_1) d_3r_1 + \int g_S(\mathbf{r}_1) v g_S(\mathbf{r}_2 - \mathbf{r}_1) v g_S(\mathbf{r} - \mathbf{r}_2) d_3r_1 d_3r_2 - \ldots \quad \text{(IV.27)}$$

In terms of Fourier transform this simplifies to

$$S(\mathbf{q}) = \int d_3r g(\mathbf{r}) e^{i\mathbf{qr}} = g_S(\mathbf{q}) - v g_S{}^2(\mathbf{q}) + \ldots \quad \text{(IV.28)}$$

or

$$S(\mathbf{q}) = \frac{g_S(\mathbf{q})}{1 + v g_S(\mathbf{q})} \quad \text{(IV.29)}$$

Edwards used for $g_S(q)$ the Debye form, assuming that the chain was *ideal at all distances*, i.e.,

$$g_S(\mathbf{q}) \sim \frac{12Nc}{q^2 l^2} \quad \text{for } qR_G > 1 \quad \text{(IV.30)}$$

This gives

$$S(q) \sim \frac{12Ncl^{-2}}{q^2 + \xi_E{}^{-2}} \quad \text{(IV.31)}$$

where the "screening length" ξ_E is defined by

$$\xi_E = (12cvNl^{-2})^{-1/2} \quad \text{(IV.32)}$$

Two sets of experimental results are used in order to test the validity of these theoretical previsions. They both have been obtained in the intermediate q range from runs with the small angle neutron apparatus of Saclay. The first one, corresponding to semidilute solutions of protonated polystyrene dispersed in deuterated benzene, is largely described elsewhere.[10] The second is obtained from the two-axis spectrometer in the conditions described in section II. In this case the samples were solutions of identical deuterated polystyrene dissolved in carbon disulfide. The characteristics of these samples are given Table III.

The data of $c/S(q)$ for samples A, B, and C are shown in Figure 10. It is clear from these plots that the q dependence is correctly described by eq IV.31, i.e., $S(q)$ has the predicted Lorentzian form.

However, the data also show that the analysis of ref 22 must be amended.

(a) Figure 11 indicates a departure from the Lorentzian behavior beyond $q = 5.5 \times 10^{-2}$ Å^{-1}. The pair correlation becomes dominated by self-excluded volume effects, as predicted by eq IV.3. The cross-over is here clearly observed for the concentration $c = c_D = 2.5 \times 10^{-2}$ g cm^{-3} (sample D). The fact that it is not observed with samples A, B, and C (i.e., at higher concentrations) means that the momentum q^*, defining the cross-over boundary, is beyond the q range explored in this experiment.

(b) The intercept $S(q = 0)$ varies like $c^{-1/4}$ while in the Edwards calculation it retains it's mean field value ($S(0) = 1/v$ = constant). This is the compressibility anomaly which was described in section III. (The experimental evidence is taken here from the light-scattering experiment.)

(c) The experimental characteristic length ξ determined from the Lorentzian broadening of the scattered intensity is clearly not proportional to $c^{-1/2}$, but rather to c^{-z}, where

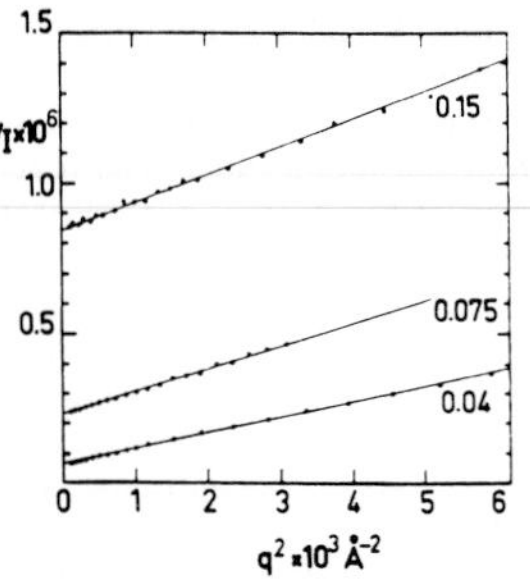

Figure 10. Inverse scattered intensities as a function of the squared momentum transfer. The intensities are expressed in arbitrary units for the concentrations A, B, C in Table III.

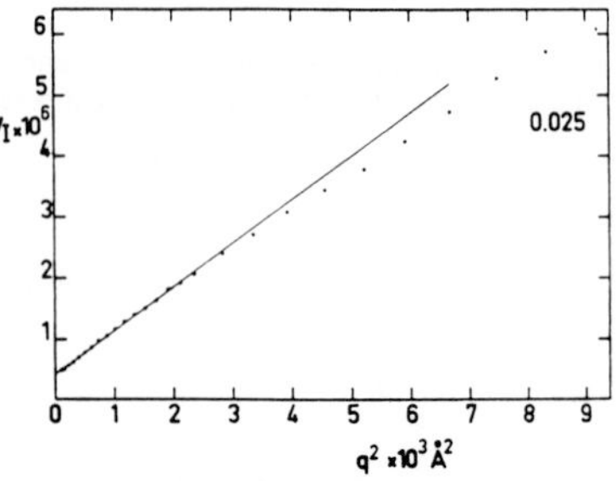

Figure 11. Inverse scattered intensity as a function of the squared momentum transfer. The concentration is here $c_D = 2.5 \times 10^{-2}$ g cm^{-3}, somewhat above c^*. The cross-over between the Lorentzian behavior and the isolated chain behavior is clearly seen at $q^2 = 5 \times 10^{-3}$ Å^{-2}. This value is about four times as high as the observed value ξ^{-2} (corresponding to concentration c_D).

$z = 0.72 \pm 0.06$ (Figure 12). The origin of this discrepancy is in fact simple. There are two flaws in the Edwards argument.

(i) The self-correlation function $g_s(\mathbf{r})$ is indeed of the Debye type for long distance, but the statistical element has a length ξ and a number of units $p(\xi)$, i.e., we must use eq IV.24 instead of eq IV.31 for $g_S(\mathbf{q})$.

(ii) The effective interaction, at small q vector, is reduced as explained at the end of section III; we must replace v by $\mathbf{v}$ (relation III.19). When these two corrections are included, we obtain:

$$S(\mathbf{q}) = \frac{12\rho Np(\xi)}{q^2\xi^2 + 12\mathbf{v}p(\xi)N\rho} = \frac{12\rho Np(\xi)\xi^{-2}}{q^2 + \kappa^2} \quad \text{(IV.33)}$$

Inserting relation III.20 for $\mathbf{v}$ and eq IV.5 for $p(\xi)$ we find that (apart from a numerical factor) $12\mathbf{v}p(\xi)c = 1$. Thus the corrected screening length is identical with ξ.

This coincidence is not accidental. From a renormalization group point of view[17] it expresses that there is only one characteristic length for a given c, in the limit of very long chains ($N \to \infty$).

It is perhaps of interest to recall that three independent experiments can be considered for the determination of the screening length ξ; namely, ξ^{-1} is defined (i) as the cross-over value q^{**}, in the single chain correlation function $S_D(q)$. (section IV1c); (ii) as the cross-over value q^*, in the segment correlation function, for a solution with identical chains; (iii) as the Lorentzian broadening κ of the scattered intensity in the range $q < q^*$ (eq IV.33).

The quantities q^{**}, q^*, and κ may differ by a numerical constant. However, they will have the same identical concentration dependence.

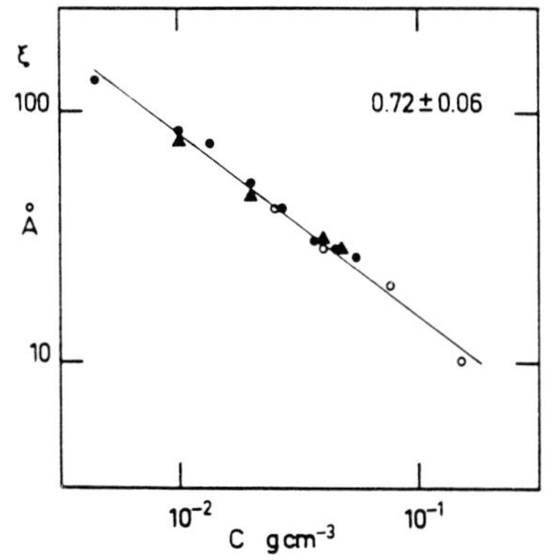

Figure 12. Log–log plot of ξ as a function of concentration. ξ is determined from the Lorentzian broadening ξ^{-2} of the scattered intensity curves (or abcissa intercept of the I^{-1} vs. q^2 representation). Closed symbols are taken from data of ref 10 for PSH in solution and deuterated benzene: closed circle $M_w = 2.1 \times 10^6$, closed triangle $M_w = 6.5 \times 10^5$. Open symbols are obtained from scattering by samples of Table III.

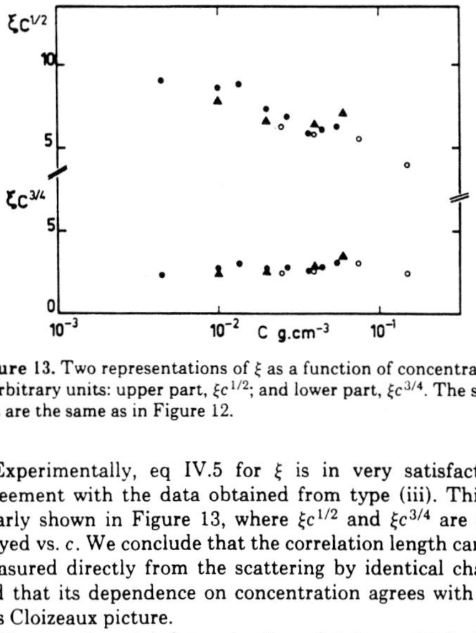

Figure 13. Two representations of ξ as a function of concentration in arbitrary units: upper part, $\xi c^{1/2}$; and lower part, $\xi c^{3/4}$. The symbols are the same as in Figure 12.

Experimentally, eq IV.5 for ξ is in very satisfactory agreement with the data obtained from type (iii). This is clearly shown in Figure 13, where $\xi c^{1/2}$ and $\xi c^{3/4}$ are displayed vs. c. We conclude that the correlation length can be measured directly from the scattering by identical chains and that its dependence on concentration agrees with the Des Cloizeaux picture.

The experimental determination of ξ from q^{**} (ref. 5) and q^* (Figure 11) does not yet lead to such a clear conclusion. We have however observed that the following inequality holds

$$q^{**} < \kappa < q^* \quad \text{(IV.34)}$$

V. Future Experiments with Chains Labeled at Specific Points

It is in some cases possible to label the chains (e.g., by selective deuteration) at the endpoints or at the midpoint. Experiments of this kind are currently being prepared in various groups. For heuristic reasons, it appears not inconsiderate to discuss here briefly the kind of correlation functions which will be measured in this way.

Let us choose, as an example, the case of chains labeled at *one* final point (for instance at one end). Let us call $f(r)$ the corresponding correlation function

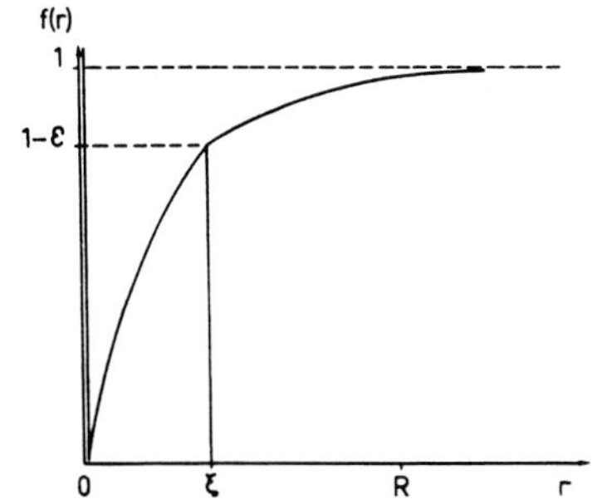

Figure 14. Pair correlation function for the labeled end of polymer chains in the semidilute regime.

$$f(\mathbf{r}) = \langle \rho_1(\mathbf{0})\rho_1(\mathbf{r})\rangle \quad \text{(V.1)}$$

where ρ_1 is the local concentration of the labeled monomer. The average ρ_1 is equal to $\rho_p = \rho/N$. We still have a compressibility theorem

$$\int [f(\mathbf{r}) - \rho_p^2] d^3r = T\rho_p \frac{\partial \rho_p}{\partial \pi} \quad \text{(V.2)}$$

(1) In the *dilute limit* the chains behave essentially like hard spheres. A plausible approximation for $f(\mathbf{r})$ would then be

$$f(\mathbf{r}) = \int d^3r_1 d^3r_2 p(\mathbf{r}_1) g_h(\mathbf{r}_1 - \mathbf{r}_2) p(\mathbf{r} - \mathbf{r}_2) \quad \text{(V.3)}$$

where $p(\mathbf{r}_1)$ would be the monomer distribution around the center of gravity of one coil, and $g_h(\mathbf{r}_1 - \mathbf{r}_2)$ would be the correlation function for a hard sphere gas, for which precise numerical data are available.[32] In particular, when the spheres become close packed ($\rho \to \rho^*$), the function g_h shows certain correlation peaks which might be reflected in $f(\mathbf{r})$. However, when ρ gets close to ρ^*, the coils begin to interpenetrate; the analogy with a gas of hard spheres breaks down, and all predictions become nebulous.

(2) Let us now go to the opposite limit $\rho = 1$. The correlation function $f(\mathbf{r})$ for this case is calculated explicitly in ref 22. It is very similar in shape to the function $g_I(r)$ of Figure 4 and shows a shallow correlation hole of size $R = N^{1/2}$. The Fourier transform $f(q)$ is a monotonously increasing function of the wave vector q; no peak is expected.

(3) Finally, let us consider the semidilute regime. Here we must discuss separately the region $r > \xi$ and $r < \xi$. For $r > \xi$, $f(r)$ is similar in shape to the high-concentration limit, with a properly scaled correlation hole extending to distances $r \sim R(c)$. For $r < \xi$, the correlation f is expected to drop abruptly (Figure 14). There is a strong excluded volume effect rejecting any other labeled monomer from the vicinity of the origin, where one monomer is present. This effect appears related to the terminal point distribution function $W_N(r)$ for an *isolated* chain starting from a fixed point 0. Des Cloizeaux[27] has shown that for small r

$$W_N(r) \simeq \left|\frac{r}{R_G}\right|^u \frac{1}{R_G{}^d} \quad \text{(V.4)}$$

where $R_G = N^\nu$ is the single chain radius, and u is another critical exponent. In terms of the γ exponent associated with the entropy of a single chain,[27] one expects

$$u = \frac{\gamma - 1}{\nu} \quad \left(\sim \frac{1}{3} \text{ for } d = 3\right) \quad \text{(V.5)}$$

The analog of $W_N(\mathbf{r})$ for our problem is $\rho_p^{-1}f(\mathbf{r})$. The analog of $1/R_G{}^d$ is the concentration of labeled points ρ_p. We expect that $\rho_p^{-1}f(\mathbf{r})$ will reach its unperturbed value ρ_p[48] for $r \sim \xi$.

Thus we are led to the following conjecture

$$\rho_P^{-1} f(\mathbf{r}) \simeq \left|\frac{r}{\xi}\right|^u \rho_P \qquad (r < \xi) \qquad \text{(V.6)}$$

Measurements in this domain are not easy. But, if they can be performed in the future, they may give direct information on the exponent γ.

Similar experiments can be considered with the chains labeled at their midpoint, or also with chains labeled at both ends. The correlation function $f_2(\mathbf{r})$ for the latter case is qualitatively reproduced on Figure 14.

VI. Concluding Remarks

1. Deviations from Mean Field. We hope to have shown that (a) semidilute solutions show significant deviations from the mean field theory and that (b) the deviations can be systematized in terms of one single scaling assumption (say for the osmotic pressure). We also conclude that the apparently very different points of view introduced by Des Cloizeaux and Edwards can be reconciled provided certain renormalizations (of the chain statistical unit, and of the interactions) are included in the Edwards picture.

Certain points remain obscure; the data on the heats of dilution do not fit with the scaling law, probably because they are not taken in a broad enough semidilute concentration range. Also the experimental behavior of the intrachain correlation function $g_S(\mathbf{r})$ (as measured on deuterated chains) does not show clearly the cross-over at $r = \xi$ which is expected. On the other hand, the cross-over observed for the segment pair correlation function, measured on identical chains, occurs at a distance r which is smaller than ξ. But the three independent neutron (plus the light scattering) data on the anomalous exponent do converge remarkably to the same value and are in good agreement with the Flory calculation of ν.

All our discussion has been very qualitative, and primarily oriented toward the determination of certain power laws. We have not investigated, experimentally or theoretically, the constant factors which enter in these power laws; comparing with the present experience in the field of phase transition,[34] we are let to believe that measurements of these prefactors will require long efforts.

2. Future Investigations with Poor Solvents. Another direction of interest is related to controlled changes of the *interaction constant* v. In all our discussion we have taken v to be positive and comparable to the monomer volume, as it is in a good solvent. When the solvent quality decreases, and we get close to a Θ point, the interaction constant and v become small and the next virial term in the free energy (eq III.8) must be included. One of us has argued recently that the region $T \to \Theta$ is related to a point in the magnetic analog.[33] This leads (for $T = \Theta$ exactly) to an osmotic pressure law in the semidilute regime (at $d = 3$), $\pi = c^3$, essentially identical with the mean field prediction for $T = \Theta$. For $T < \Theta$ the demixtion problem is still a challenge (see Appendix), but the complications introduced by polydispersity (even if the latter is weak) may forbid precise determinations of the associated critical exponents.

3. A Conjecture on the Rubber Elasticity of Entangled Chains. Returning to the good solvent situation, it may be appropriate to insert in these conclusions one remark and one conjecture. The remark concerns the osmotic pressure law $\pi \simeq Tc^{9/4}$ in the semidilute regime. It may be argued quite generally that (apart from a numerical coefficient) π/T measures the number of *contact points* between different chains (per unit volume). In a mean field picture this number is proportional to ρ^2. Here, as explained in section III, it is smaller than ρ^2.

This may possibly be related to an interesting physical parameter; if we measure the viscoelastic properties of a solution, at a finite but low-frequency ω, and if the chains are long enough (so that the relaxation rate is smaller than ω) we expect to find an elastic modulus E of the rubber type, due only to entanglements between different chains. A number of arguments suggests that E/T is proportional to the number of contact points; in mean field theory this would give $E = T\rho^2$. More complicated power laws have been proposed within mean field theory[35] but they are open to some criticism.[36] Our conjecture is that E is indeed proportional to the exact density of contact points, i.e., that

$$E = T\rho^{9/4} \qquad (\rho^* \ll \rho \ll 1)$$

The experimental situation is not too clear at present, but we hope to come back to this point in later studies.

4. Semidilute Behavior of Two-Dimensional Polymer Films. Monomolecular polymer films can be prepared on a liquid interface, or on a solid substrate. In principle, for $d = 2$ the deviations from mean field are much more glazing; for instance the osmotic pressure should behave like c^3 in good solvent conditions, rather than c^2. However, the range of semidilute concentrations is not very broad in two dimensions. Also, many experimental techniques (such as neutron scattering) become inapplicable for a single film. It would be necessary to work with a system of many superposed films, for instance, to incorporate solute polymer chains in a lamellar phase of lipid plus water,[37] each lamella behaving more or less like an independent two-dimensional sheet. Clearly these experiments belong to a rather remote future.

Acknowledgment. We wish to thank Des Cloizeaux, Boccara, Strazielle, Bidaux, and Mirkovitch for stimulating discussions over this subject; Decker, Herbert, and Rempp for their cooperation in the deuteration of the molecules; and Higgins, Nierlich, Boué, and Ober for preparing the data processing.

Appendix

The Magnetic Analog to Polymer Solution. An Introduction. In this appendix we present a new and perhaps simpler derivation[38] of the analogy found by de Gennes[26] and Des Cloiseaux[6] between the "polymer" problem and a "magnetic" problem.

1. A Few Basic Facts about Ferromagnets. The main features of ferromagnetic transitions are reviewed for instance in the book by Stanley.[39] Ferromagnetic order is described[49] by a magnetization vector $\mathbf{M}$ with a certain number n of independent componants: n may be equal to 3 (Heisenberg magnets), to 2 (planar), or to 1 (Ising magnets).

The average magnetization M of the magnet is a function of the temperature T and of the applied magnetic field H. In zero field and at high T, the magnetization M vanishes. However, below a certain critical temperature T_c there is a spontaneous magnetization $\mathbf{M}(T,H = 0)$ as shown in Figure 15. Coming toward the critical point from the high-temperature side ($T > T_c$, $H = 0$) the proximity of a transition is signalled by two main effects.

(a) The susceptibility $\chi = (\partial M/\partial H)|_{H=0}$ becomes large and diverges at T_c

$$\chi \sim \tau^{-\gamma} \qquad (\tau = (T - T_c)/T_c) \qquad \text{(A.1)}$$

γ is a certain critical exponent, depending only on d (the dimensionality) and on n (the number of components of $\mathbf{M}$). Approximate formulas for γ ($d = 3,n$) have now been worked out.[17]

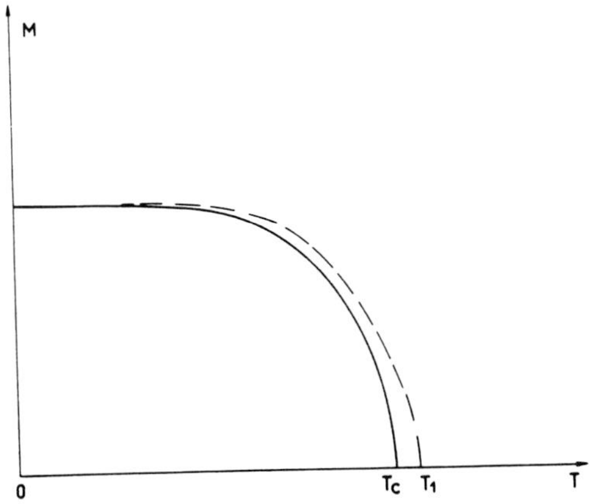

Figure 15. Average magnetization M as a function of temperature, in the absence of applied magnetic field (full line). In the vicinity of T_c, M behaves as $((T - T_c)/T_c)^{\beta}$. In the presence of an applied magnetic field, the relevant curve is the isometric line (broken line).

(b) The magnetization correlation function $\langle M(0)M(r)\rangle$ has the general shape shown on Figure 16. The spatial range of this function ξ is a function of the temperature T and diverges when T tends to T_c like

$$\xi \sim |\tau|^{-\nu} \tag{A.2}$$

where ν is another critical exponent.

The thermodynamical properties of the magnet can be described in terms of a free energy $F(H,\tau)$. It will be convenient for the following to separate in F the field independent part $F(0,\tau)$ and to write

$$\Delta F(H, \quad F(H,\tau) - F(0,\tau) \tag{A.3}$$

If we know ΔF we know the average magnetization through the simple rule

$$M = -\partial \Delta F/\partial H \tag{A.4}$$

In particular, at small H, and $\tau > 0$, ΔF can be expanded in the form

$$\Delta F = -(0.5)\chi H^2 \tag{A.5}$$

and then $M = \chi H$ as requested by the definition of the susceptibility. It is sometimes convenient to use M and τ (rather than H and τ) as independent variables in the thermodynamic analysis. One then introduces the function $\Gamma(M,\tau)$ such that

$$\partial\Gamma(M,\tau)/\delta M = H \tag{A.6}$$

It is related to ΔF by the formula

$$\Delta F(H,\tau) = \Gamma(M,\tau) - MH \tag{A.7}$$

Near T_c the dominant terms of $\Gamma(M,\tau)$ have the following structure (first proposed by Widom in connection with another critical point[40])

$$\Gamma(M,\tau) = |\tau|^{\nu d} g(x)$$
$$x = M/|\tau|^{\beta} \tag{A.8}$$

where β is the exponent defined in Figure 15 and $g(x)$ is a certain universal function (for fixed d and n). The great merit of this form is to reduce Γ to a function of one variable x. The structure of $g(x)$ is now known reasonably well[41] but we shall not need it in what follows.

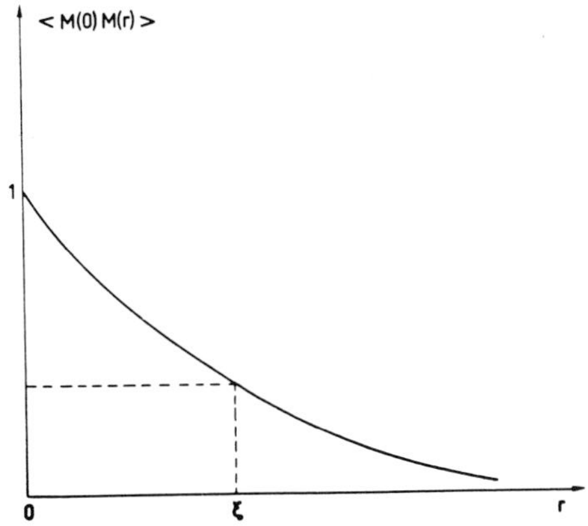

Figure 16. Magnetization correlation function.

2. Relation between the "Magnetic" Problem and the "Polymer" Problem. (a) Lemma. Let S be a "spin" with n components S_α ($\alpha = 1, 2, \ldots, n$), the length being chosen to be $n^{1/2}$, so that $\Sigma_\alpha S_\alpha^2 = n$. This implies that geometrical mean value

$$\langle S_\alpha^2 \rangle = 1 \quad \text{for any value of } \alpha \tag{A.9}$$

The constraint of the fixed length can be considered as a law of probability, whose characteristic function is defined by the geometrical mean value

$$f(k_\alpha) = \left\langle \exp\left(i \sum_\alpha k_\alpha S_\alpha\right) \right\rangle \tag{A.10}$$

It is clear that f is a spherical function

$$f(k_\alpha) = f(k) \qquad k = \left(\sum_\alpha k_\alpha^2\right)^{1/2}$$

This function obeys the differential equation

$$-\Delta^2 f = \left(\sum_\alpha S_\alpha^2\right) f = nf$$

which can be written for a spherical function as

$$-\left(\frac{d^2}{dk^2} + \frac{n-1}{k}\frac{d}{dk}\right) f = nf \tag{A.11}$$

The solution of this equation depends on two boundary conditions, which are given by the small k expansion of (A.10),

$$f(k_\alpha) = 1 - \tfrac{1}{2}\langle(\mathbf{k}\cdot\mathbf{S})^2\rangle + \ldots$$

that is, taking account of (A.9),

$$f(k) = 1 - \tfrac{1}{2}k^2 + \ldots$$

Now $f(k) = 1 - \frac{1}{2}k^2$ is a solution of (A.11) for $n = 0$.

Therefore, in the limit when n goes to zero,

$$f(k) = \langle \exp(i\mathbf{k}\cdot\mathbf{S})\rangle = 1 - \tfrac{1}{2}\sum_\alpha k_\alpha^2 \tag{A.12}$$

Note that in the general case the function $f(k_\alpha)$ defined by (A.10) is the generating function for mean values of the type $\langle (S_1)^{p_1}(S_2)^{p_2} \ldots (S_n)^{p_n}\rangle$, $p_1, \ldots, p_n$ being positive integers. It is clear that the latter mean value is proportional to the coefficient of $(k_1)^{p_1}(k_2)^{p_2} \ldots (k_n)^{p_n}$ in the expansion of $f(k_\alpha)$ in powers of k.

Hence, in the case of a zero-dimensional "spin", *all these mean values* vanish except the mean values $\langle S_\alpha^2\rangle = 1$ for any value of α,

(b) The Magnetic Model. We consider n component "spins" $\mathbf{S^R}$ located on the sites $\mathbf{R}$ of a lattice in a d-dimensional space, coupled by a nearest-neighbor interaction K. The length of the spins is taken as before to be $n^{1/2}$.

The Maxwell Boltzmann law of probability of the system reads

$$\frac{1}{Z}\exp\left(-\frac{\mathcal{H}}{T}\right) = \frac{1}{Z}\exp(T^{-1}\sum K_{\mathrm{RR'}}\mathbf{S^R}\cdot\mathbf{S^{R'}}) \quad (A.13)$$

where $K_{\mathrm{RR'}} = K$ for nearest neighbors and $K_{\mathrm{RR'}} = 0$ otherwise. Z is the partition function, which clearly will involve only angular integrations, because of the constraint of fixed length for the spins.

Z can be expanded as

$$Z = \left\langle \prod_{\mathrm{RNR'}\,\alpha}\left[1 + \frac{K}{T}S_\alpha{}^{\mathrm{R}}S_\alpha{}^{\mathrm{R'}} + \frac{1}{2}\left(\frac{K}{T}\right)^2 (S_\alpha{}^{\mathrm{R}}S_\alpha{}^{\mathrm{R'}})^2 + \ldots\right]\right\rangle_{\mathrm{A}}$$

where $\mathbf{RNR'}$ means that R and R′ are nearest neighbor sites, and $\langle\ \rangle_{\mathrm{A}}$ means the angular mean values. In the limit when $n \to 0$, thanks to the lemma (a), the expansion of $\exp(KS_\alpha{}^{\mathrm{R}}S_\alpha{}^{\mathrm{R'}}/T)$ does not need to be continued after the second term.

Terms of Z can be represented graphically; a typical term is a loop on the lattice

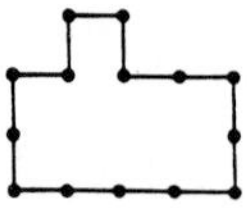

corresponding to a product of $KS_\alpha{}^{\mathrm{R}}S_\alpha{}^{\mathrm{R'}}/T$, where the index α *is fixed.* Moreover, since other mean values than $\langle S_\alpha{}^2\rangle$ are zero, one can pass only *one time* through a lattice point; so the limit $n = 0$ contains the *excluded volume*, the loop cannot intersect itself. Also, when there are many loops, one loop cannot intersect another one.

This contribution of the loop is easily seen to be $(K/T)^N$ where N is the number of segments. Now, the same loop could appear with any index α, so that the summation over α gives $n(K/T)^N$ which goes to zero when $n \to 0$.

So one sees that Z reduces to the value 1

$$Z = 1 \quad (A.14)$$

(c) The Problem of a Single Polymer Chain. The essential content of the lemma (a) is that for $n \to 0$ the only allowed graphs are *excluded volume* graphs that is, one cannot pass through a lattice point more than one time.

We now consider the correlation function for two spins $\mathbf{S}^{\mathrm{R}_1}$ and $\mathbf{S}^{\mathrm{R}_2}$ far apart.

Since $Z = 1$, we get

$$C(\mathbf{R}_1,\mathbf{R}_2) = \langle S_1{}^{\mathrm{R}_1}S_1{}^{\mathrm{R}_2}\rangle = \left\langle S_1{}^{\mathrm{R}_1}S_1{}^{\mathrm{R}_2}\prod_{\mathrm{RNR'}\,\alpha}\left[1 + \frac{K}{T}S_\alpha{}^{\mathrm{R}}S_\alpha{}^{\mathrm{R'}} + \ldots\right]\right\rangle_{\mathrm{A}}$$

The correlation is defined for a fixed component index (1) for the two spins.

The different contributions to $C(\mathbf{R}_1,\mathbf{R}_2)$ can be represented graphically. With the contribution of any loop being zero after summation over the index α, one can see that the only remaining graphs are all the excluded volume paths on the lattice joining the points $\mathbf{R}_1$ and $\mathbf{R}_2$. For each segment the term $KS_1{}^{\mathrm{R}}S_1{}^{\mathrm{R'}}/T$ is used, with the index $\alpha = 1$, so that for such a graph there is no summation on α. The graph

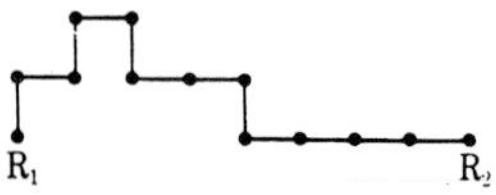

indeed directly represents a polymer configuration with the ends fixed at $\mathbf{R}_1$ and $\mathbf{R}_2$. All the configurations are counted, and only once.

This gives

$$C(\mathbf{R}_1,\mathbf{R}_2) = \langle S_1{}^{\mathrm{R}_1}S_1{}^{\mathrm{R}_2}\rangle = \sum_N \left(\frac{K}{T}\right)^N G(N,\mathbf{R}_1 - \mathbf{R}_2) \quad (A.15)$$

where N is the number of segments (or monomers) and $G(N,\mathbf{R}_1 - \mathbf{R}_2)$ is the number of configurations of a polymer of N monomers with its ends fixed at $\mathbf{R}_1$ and $\mathbf{R}_2$.

The result (eq A.15) has first been found by de Gennes.[26] For the magnetic problem, the correlation $C(\mathbf{R}_1 - \mathbf{R}_2)$ is known and behaves in the vicinity of T_c as

$$C(\mathbf{R}_1 - \mathbf{R}_2) \sim \frac{1}{|\mathbf{R}_1 - \mathbf{R}_2|^{d-2+\eta}} f\left(\frac{|\mathbf{R}_1 - \mathbf{R}_2|}{\xi}\right) \quad (A.16)$$

where η is an exponent of the magnetic problem, and f is a certain universal function (for fixed d and n). We recall that $\xi \sim |\tau|^{-\nu}$ (eq A.2). Comparing (A.15) and (A.16) one gets for large values of N, which appears as the conjugate variable of τ,

$$\left(\frac{K}{T_c}\right)^N G(N,\mathbf{R}_1 - \mathbf{R}_2) \sim N^{\gamma-1-\nu d}\, h\left(\frac{|\mathbf{R}_1 - \mathbf{R}_2|}{N^\nu}\right) \quad (A.17)$$

where h is a certain universal function. The exponent γ has been defined by (A.1). We have used in deriving (A.17) the "magnetic" scaling relation $\gamma = (2 - \eta)\nu$. Equation A.17 solves the one-polymer problem. The exponents appearing in (A.17) have been calculated for the magnetic problem as a function of d and n by renormalization group methods. Setting $n = 0$, the results agree rather well with the Flory estimate in three dimensions.[3]

(d) The Problem of Semidilute Polymers. We have shown above that the magnetic $n = 0$ problem solved the problem of a single chain. Des Cloizeaux[6] was the first to realize that the same problem with an additional applied magnetic field led to the understanding of the semidilute regime. Applying a magnetic field H along the (1) direction of the spins, the law of probability of the spin system becomes

$$\frac{\exp(-\mathcal{H}_1/T)}{Z} = \frac{1}{Z}\exp(T^{-1}[\sum K_{\mathrm{RR'}}\mathbf{S^R}\cdot\mathbf{S^{R'}} + H\sum S_1{}^{\mathrm{R}}]) \quad (A.18)$$

We have seen before that $Z(H = 0) = 1$. Thus we can write

$$\frac{Z(H)}{Z(0)} = \left\langle \prod_{\mathrm{RNR'}\,\alpha}\left(1 + \frac{K}{T}S_\alpha{}^{\mathrm{R}}S_\alpha{}^{\mathrm{R'}} + \ldots\right)\prod_{\mathrm{R}}\left(1 + \frac{H}{T}S_1{}^{\mathrm{R}} + \ldots\right)\right\rangle_{\mathrm{A}}$$

The expansion in powers of H contains obviously only even powers. The loops being prohibited, the only graphs which remain are "polymer" graphs. For instance the term in $(H/T)^2$ sums the contribution of all the *one* polymer configurations between two arbitrary ends $\mathbf{R}$ and $\mathbf{R'}$ with an arbitrary number N of monomers, the value of such a graph being $(K/T)^N(H/T)^2$. In general a term in $(H/T)^{2P}$ is obtained by drawing P excluded volume polymers of lengths N_1, N_2, . . . , N_P. Each possible configuration is counted once and only once. Let then N_P be the number of polymers, N_M the *total* number of monomers, then

$$\frac{Z(\mathrm{H})}{Z(0)} = \sum \left(\frac{K}{T}\right)^{N_M} \left(\frac{H}{T}\right)^{2N_P} U(N_M,N_P) \qquad \text{(A.19)}$$

where $U(N_M,N_P)$ is the number of configurations of N_P polymers with a *total* number of N_M monomers. So, provided that $n = 0$ in the magnetic problem, $Z(\mathrm{H})/Z(0)$ is the *grand partition function* of an assembly of polymers of variable length, defined with the convention that $2 \log H/T$ and $\log K/T$ are respectively the chemical potentials conjugate to the number of polymers N_P and the *total* number of monomers N_M. We can still write (A.19) in the form

$$\frac{Z(\mathrm{H})}{Z(0)} = \sum \exp[2N_P \log H - (N_M + 2N_P) \log T] U(\mathrm{N_M,N_P}) \qquad \text{(A.20)}$$

(we have set $K = 1$, which means that H and T are measured in units of K).

Let ρ_P and ρ be respectively the number of polymers and monomers per unit volume, and let us define ΔF (see eq A.3) by unit volume, then from (A.20) one gets

$$2\rho_P = -H \frac{\partial}{\partial H} \Delta F = HM = M \frac{\partial \Gamma}{\partial M} \qquad \text{(A.21)}$$

$$\rho + 2\rho_P = T \frac{\partial}{\partial T} \Delta F = T \frac{\partial \Gamma}{\partial T} \qquad \text{(A.22)}$$

The osmotic pressure π is given by a general theorem as

$$\pi = -\Delta F = -\Gamma + M \frac{\partial \Gamma}{\partial M} \qquad \text{(A.23)}$$

Finally the average number N of monomers per chain is defined by

$$N = \rho/\rho_P \qquad \text{(A.24)}$$

In the limit of large N, $\rho_P \ll \rho$ and can be dropped in (A.22). Finally, for reasons to come clear soon, the magnetic problem will be considered close to T_c, which allows us to replace $T(\partial/\partial T)$ by $\partial/\partial \tau$. With these remarks, eq A.21 to A.24 are now written as

$$\rho_P = \frac{1}{2} M \frac{\partial \Gamma}{\partial M} \qquad \text{(A.25)}$$

$$\rho = \frac{\partial \Gamma}{\partial \tau} \qquad \text{(A.26)}$$

$$\pi = -\Gamma + M \frac{\partial \Gamma}{\partial M} \qquad \text{(A.27)}$$

$$\rho/\rho_P = N \qquad \text{(A.28)}$$

The relations A.25 to A.28 are the fundamental equivalence relations between the magnetic problem and the polymer problem. Assume for the moment, as we shall show it later, that in the limit of large N one remains in the vicinity of T_c. Then we can use the Widom form (eq A.8) for $\Gamma(M,\tau)$, and we get

$$\left.\begin{aligned} \frac{\rho}{N} &= |\tau|^{\nu d} g_P(x) \\ \rho &= |\tau|^{\nu d - 1} g_c(x) \end{aligned}\right\} \longrightarrow \quad |\tau| N = \frac{g_c(x)}{g_P(x)}$$

$$\pi = |\tau|^{\nu d} g_\pi(x)$$

where $g_P(x)$, $g_c(x)$, $g_\pi(x)$ are universal functions related to $g(x)$ which we shall not write down explicitly. From the above relations it is easy to deduce the "scaled" equation of state, found first by Des Cloizeaux[6]

$$\pi N^{\nu d} = \phi(\rho N^{\nu d - 1}) \qquad \text{(A.29)}$$

valid for large N and high temperature of the polymer system (because we took into account only the excluded volume effect, that is the entropy). ϕ is a certain universal function. We can already see the existence of a "cross-over" concentration defined by $\rho^* N^{\nu d - 1} \sim 1$.

Our aim being to describe a situation where N is fixed, this imposes a certain relation between τ and M in the magnetic problem. The resulting line on a (M,τ) diagram will be called the *isometric* line; it is shown on Figure 15.

(i) The isometric line starts from a point T_1 on the T axis ($T_1 > T_c$). This region corresponds to $x = M/\tau\beta \ll 1$, and $g(x)$ has an analytical expansion in even powers of x. The use of (A.25), (A.26), and (A.28) allows us to show easily that $T_1 = T_c(1 + \gamma/N)$. This region corresponds to a close vicinity of T_c and also to $\rho \ll \rho^*$. It is associated with the dilute regime. Taking as a first approximation $g(x) \sim Ax^2$, one gets easily

$$\pi = \rho/N \qquad \text{(A.30)}$$

which means that $\phi(y) \sim y$ for $y \ll 1$. With the same approximation, (A.25) and (A.26) give that $N = \rho/\rho_P = \gamma/\tau$. The correlation length measures the length of the correlation between chain extremities; in this regime of separated chains, it coincides with the swollen coil radius

$$R \sim \xi \sim \tau^{-\nu} \sim N^{\nu} \qquad \text{(A.31)}$$

(ii) The isometric line then goes to an intermediate region with $T \sim T_c$ and M finite. This corresponds to concentrations ρ comparable to ρ^*.

(iii) Ultimately, the isometric line comes very close to the spontaneous magnetization curve. Here the reduced variable $x = M/|\tau|^{\beta}$ tends to a finite limit ($x \sim x_0$). This corresponds to the semidilute regime $\rho \gg \rho^*$, but as announced previously this situation happens still for small values of $|\tau|$ (τ is negative), of the order of $1/N$.

If $x = x_0$ one gets as a first approximation $\Gamma = |\tau|^{\nu d} f(x_0)$ and $\partial\Gamma/\partial M = 0$ on the spontaneous magnetization curve. Then from (A.26) and (A.27) one finds the fundamental result

$$\pi = A\rho^{\nu d/(\nu d - 1)} \qquad \text{(A.32)}$$

where the constant A is *independent of* N. Note that (A.26) gives $\rho \sim |\tau|^{\nu d - 1} f(x_0)$. The condition $\rho \gg \rho^*$ implies $|\tau|^{\nu d - 1} N^{\nu d - 1} \gg 1$, that is $|\tau| \gg 1/N$, which is not contradictory with $|\tau| \ll 1$ in the limit of large N. This completes the proof of our assumption of T being always close to T_c as the concentration varies. The result (eq A.32) implies that

$$\phi(y) \sim Ay^{\nu d/\nu d - 1)} \qquad \text{for } y \gg 1$$

On the whole the Des Cloiseaux analogy gives a very deep insight on the problem of polymer solutions. It suffers from one limitation, however; the problem which is solved is not a problem of monodisperse chains. There is a broad distribution of molecular weights. To be sure, we arrange that a certain number average N is fixed and independent of ρ. But the distribution itself does not retain exactly the same shape at all concentrations (and in particular in the vicinity of ρ^*). Thus the magnetic analogy is useful for qualitative purposes (to determine exponents) but probably less useful if one really wanted to compare the shape of a correlation function.[50]

3. Technical Remarks on the End-to-End Correlations. We shall briefly sketch here the derivation of eq IV.6 for the coil radius in semidilute solutions. The background in magnetism required for this step is slightly more elaborate; a list of references is given under ref 40.

As noticed by Des Cloiseaux[6] the correlations between

two ends of one same chain correspond to transverse correlations in the magnetic problem; if the applied field H is along the z axis, we look at $\langle M_x(0)M_x(r)\rangle$ or at it's Fourier transform $\langle |M_x(q)|^2\rangle$. This has the form[39]

$$\langle |M_x(q)|^2\rangle = \frac{A}{q^2 + \kappa_1^2} \qquad \text{(A.33)}$$

where A varies as ξ^η where η is another critical exponent, defined as in ref 39. The quantity κ_1^{-1} is a characteristic length,[51] and will give us the size of the coils. The quantity A/κ_1^2 is proportional to the transverse magnetic susceptibility which is simply[39]

$$\chi_\perp = M_0/H \qquad \text{(A.34)}$$

M_0 being the spontaneous magnetization. We can eliminate H through eq A.21 obtaining

$$\kappa_1^2 \simeq (A/M_0^2)\rho_P$$

Using the scaling law for ξ and M_0, and various relations between critical exponents, plus eq A.26 to eliminate τ in terms of ρ we get

$$\kappa_1^2 \simeq N^{-1}\rho^{(2\nu-1)/(\nu d-1)} \qquad \text{(A.35)}$$

an equation equivalent to (IV.7).

Similar arguments can be applied to the discussion of the long-range correlation hole (shown on Figure 14) in semidilute systems. What is considered here is the correlations between the ends of all chains which, as shown in ref 6, is related to longitudinal correlations or susceptibilities.

However, for a magnetic system with $n \neq 1$, these longitudinal susceptibilities are influenced by transverse fluctuations; an early discussion of this point can be found in ref 42. It is possible to show for instance that in three space dimensions the longitudinal correlation function contains a long-range term

$$\langle \delta M_z(0)\delta M_z(r)\rangle \simeq \frac{1}{M_0^2\xi^{2\eta}}\frac{1}{r^2}e^{-2\kappa r} \qquad \text{(A.36)}$$

This is the source of the correlation hole (extending to distances of order of R). More precise calculation of the hole cannot be attempted from this approach because of the polydispersity effect discussed above; the R.P.A. approach of ref 25 is probably preferable here.

References and Notes

(1) (a) Laboratoire Léon Brillouin, Centre d'Etudes Nucléaires de Saclay; (b) Centre de Recherches sur les Macromolécules; (c) Collége de France.
(2) (a) P. J. Flory, *J. Chem. Phys.*, **10**, 51 (1942); (b) M. L. Huggins, *J. Phys. Chem.*, **46**, 51 (1942).
(3) P. J. Flory, "Principles of Polymer Chemistry", Cornell University Press, Ithaca, N.Y., 1967.
(4) J. P. Cotton, B. Farnoux, and G. Jannink, *J. Chem. Phys.*, **57**, 290 (1972).
(5) B. Farnoux, M. Daoud, D. Decker, G. Jannink, and R. Ober, *J. Phys. (Paris) Lett.*, **36**, L35 (1975).
(6) J. Des Cloizeaux, *J. Phys. (Paris)*, **36**, 281 (1975).
(7) I. I. Gurevitch and L. V. Tarasov, "Low Energy Neutron Physics", North-Holland Publishing Co., Amsterdam, 1968.
(8) J. P. Cotton, B. Farnoux, G. Jannink, J. Mons, and C. Picot, *C. R. Hebd. Seances Acad. Sci., Ser. C*, **275**, 175 (1972).
(9) R. G. Kirste, W. A. Kruse, and J. Schelten, *Makromol. Chem.*, **162**, 299 (1972).
(10) J. P. Cotton, Thèse Université Paris VI, 1973; CEA, No. 1743.
(11) J. P. Cotton, D. Decker, H. Benoit, B. Farnoux, J. Higgins, G. Jannink, R. Ober, C. Picot, and J. Des Cloizeaux, *Macromolecules*, **7**, 863 (1974).
(12) W. H. Stockmayer, *Makromol. Chem.*, **35**, 54 (1960).
(13) C. Loucheux, G. Weil, and H. Benoit, *J. Chim. Phys.*, **43**, 540 (1958).
(14) C. Domb, J. Gillis, and G. Wilmers, *Proc. Phys. Soc., London*, **85**, 625 (1965).
(15) M. E. Fisher, *J. Chem. Phys.*, **44**, 616 (1966).
(16) S. F. Edwards, *Proc. Phys. Soc., London*, **85**, 613 (1965).
(17) S.-K. Ma, *Rev. Mod. Phys.*, **45**, 589 (1973).
(18) J. P. Cotton, B. Farnoux, G. Jannink, and C. Strazielle, *J. Polym. Sci., Symp.*, **42**, 981 (1973).
(19) G. Gee and W. J. Orr, *Trans. Faraday Soc.*, **42**, 507 (1946).
(20) P. A. Egelstaff, "An Introduction to the Liquid State", Academic Press, London, 1967.
(21) H. Benoit and C. Picot, *Pure Appl. Chem.*, **12**, 545 (1966).
(22) S. F. Edwards, *Proc. Phys. Soc., London*, **88**, 265 (1966).
(23) R. G. Kirste, W. A. Kruse, and K. Ibel, *Polymer*, **16**, 120 (1975).
(24) D. G. Ballard, J. Schelten, and G. D. Wignall, *Eur. Polym. J.*, **9**, 965 (1973).
(25) P. G. de Gennes, *J. Phys. (Paris)*, **31**, 235 (1970).
(26) P. G. de Gennes, *Phys. Lett. A*, **38**, 339 (1972).
(27) J. Des Cloizeaux, *J. Phys. (Paris)*, **31**, 715 (1970).
(28) B. H. Zimm, *J. Chem. Phys.*, **16**, 1093 (1948).
(29) J. P. Cotton, D. Decker, B. Farnoux, G. Jannink, and R. Ober, *Phys. Rev. Lett.*, **32**, 1170 (1974).
(30) P. Debye, *J. Phys. Colloid. Chem.*, **51**, 18 (1947).
(31) G. Jannink and P. G. de Gennes, *J. Chem. Phys.*, **48**, 2260 (1968).
(32) J. O. Hirschfelder, C. F. Curtis, and R. B. Bird, "Molecular Theory of Gases and Liquids", Wiley, New York, N.Y., 1954.
(33) P. G. de Gennes, *J. Phys. (Paris), Lett.*, **36**, L55 (1975).
(34) P. Heller, *Rep. Prog. Phys.*, **30**, 731 (1967).
(35) S. F. Edwards, *Proc. Phys. Soc., London*, **92**, 9 (1967).
(36) P. G. de Gennes, to be published.
(37) P. G. de Gennes, to be published.
(38) G. Sarma, Lecture Notes Saclay, 1974, unpublished.
(39) H. E. Stanley, "Introduction to Phase Transition and Critical Phenomena", Oxford University Press, London, 1971.
(40) B. Widom, *J. Chem. Phys.*, **43**, 3898 (1965).
(41) M. E. Fisher, *Rep. Prog. Phys.*, **30**, 615 (1967).
(42) P. G. de Gennes, "Magnetism", Vol. III, G. T. Rado and H. Suhl, Ed., Academic Press, New York, N.Y., 1963.
(43) To avoid cumbersome factors in the theoretical discussions, we shall consider our molecules as inscribed in a periodic lattice, as in the Flory-Huggins model, and we shall take the lattice parameter l as our unit of length. The concentration of polymer chains, ρ_p (number of chains per unit volume), is then a dimensionless number going from 0 (pure solvent) to $1/N$ (molten polymer). The concentration of monomers, ρ, is then equal to the polymer volume fraction. Whenever experimental results are discussed, the concentration c in g cm^{-3} is used. The relation between ρ and c is $\rho = c\bar{v}$, where $\bar{v}$ is the partial specific volume of the solute (monomer).
(44) We set the Boltzman constant equal to unity. T is the absolute temperature.
(45) $g_S(r)$ in the interval $1 < r < R$.
(46) The notation $A_2(1)$ recalls that this coefficient is measured at $\rho = 1$.
(47) Another, possibly more simple, presentation of the random phase method through a response function approach can be found in ref 31 and 25.
(48) The weak correlation hole effect is not important for this argument.
(49) We use here the standard notation in magnetism. Some symbols written in this appendix (M,f,g) do not correspond to the physical quantities discussed in the earlier sections. Others, such as ΔF, the free energy, γ, have the same meaning.
(50) (a) This difficulty does not occur for the single chain problem, where it is always possible to go back to a monodisperse situation by a suitable Laplace transform. The difficulty occurs with more than one chain: fixing the total number of monomers present does not fix exactly the length of one particular chain. (b) The polydispersity also blocks possible extensions of the problem to poor solvents; the problem is complicated by separation effects.
(51) We use the subscript 1 to avoid confusion with the length κ^{-1} in eq IV.33.

RIVISTA DEL NUOVO CIMENTO VOL. 7, N. 3 Luglio-Settembre 1977

Theoretical Methods of Polymer Statistics (*).

P. G. DE GENNES

Collège de France - 75231 *Paris Cedex* 05

(ricevuto il 20 Febbraio 1977)

1. – Long, flexible chains.

A large fraction of the materials which we use in everyday life are made of *long flexible chains*: a typical example would be polystyrene

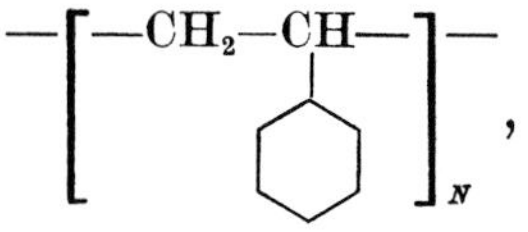

wheret he « polymerization index » N may reach values up to 10^5. The geometrical size a of the repeat unit (« monomer ») is of the order of 3 Å. For a general introduction to these materials, the reader is referred to the outstanding book of Flory [1]. If we do not look at the 3 Å scale, but rather at a broader scale (in the range (100÷300) Å, for instance), the polymer looks like a large coil

(*) Based on lectures delivered at the Symposium on « Interdisciplinary Aspects of Modern Physics » in Parma, Italy, May 3-7, 1976.

(fig. 1) and many features of these coils are *universal* (independent of the details of the monomer structure). It is these features which we want to emphasize here. The situation is somewhat similar to what we find in the physics of phase transitions. Near a critical point (τ_c) a critical system shows correlated regions which are of large size (typically, a few hundred angstroms

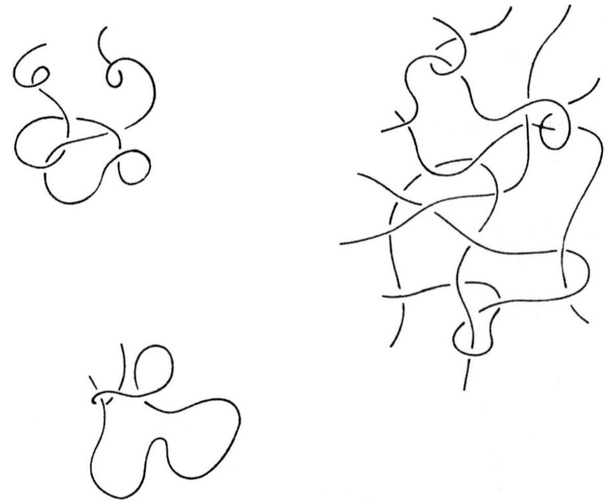

Fig. 1. a) b)

when the temperature shift from critical $\Delta\tau$ is of order 0.1 K). When these large regions occur, their properties become quite independent of the detailed properties of the atomic scale (*e.g.* of lattice structure for crystals): this gives then certain universal features.

More precisely, to have large correlated regimes near a critical point τ_c, we must have $\Delta\tau/\tau_c$ small. To have large coils in the polymer problems, we must have a large polymerization index N. This suggests a certain correspondence

$$\frac{1}{N} \leftrightarrow \frac{\Delta\tau}{\tau}\,. \tag{1.1}$$

As we shall see later, this correspondence does have a rigourous theoretical basis in many cases. Note that for $N \sim 10^5$ the polymer problem corresponds indeed to a system very close to τ_c—*i.e.* to a high degree of universality.

For a single chain floating in a solvent, the first attempt at modelization was based on an analogy with a simple random walk of N steps, each of equal length a, successive steps being uncorrelated. In this « ideal chain » model, the end-to-end vector of the chain R has a Gaussian distribution and a square mean:

$$R_0^2 = Na^2 \qquad \text{(ideal chain)}. \tag{1.2}$$

However, it has been soon realized that this model is rarely applicable to the single-chain problem. The usual situation is that the chain is dissolved in a « good solvent »: the monomer-solvent interactions are attractive and dominant. Then, if we bring two monomers in close contact, they both lose some of these monomer-solvent interactions and experience an effective *repulsion*. Of course, if we had put the two monomers exactly at the same place, they would also suffer from a large steric repulsion. Thus, in a good solvent, the sign of the interaction between monomers is always repulsive. This is a short-range repulsion, and for large-scale properties it can often be replaced by a delta-function form

$$U(12) = Tv(T)\,\delta(\boldsymbol{R}_{12}), \tag{1.3}$$

where T is the temperature (the Boltzmann constant is put equal to 1) and $v(T)$ has the dimensions of a *volume* in d-dimensional space. The real significance of $v(T)$ is that of a second virial coefficient between monomers. In practice, $v(T)$ is often called the *excluded-volume parameter*. Another presentation of the interaction is through the dimensionless coupling constant

$$u_0 = \frac{v(T)}{a^d}. \tag{1.4}$$

In a good solvent, u_0 is positive and finite. This implies essential departures from the ideal chain model. The repulsion forces the chain to swell to a radius $R \gg R_0$. A simple model for this situation is obtained with a chain inscribed in a periodic lattice (*e.g.* a cubic lattice of edge a): the case $v = a^d$ corresponds to a chain which cannot intersect itself, or equivalently to a *self-avoiding walk* (SAW) of N steps. Various numerical studies on SAW (as well as direct experiments on polymers, using light scattering, viscosities, etc.) indicate that for a SAW the radius R behaves like

$$R_{\text{SAW}} = aN^{\nu}, \tag{1.5}$$

where $\nu \simeq \frac{3}{5}$ for $d = 3$. (We shall come back to ν in sect. **2**.)

The next class of problems is obtained with polymer *solutions*, where a finite concentration c (monomer per unit volume) is reached. Polymer solutions (in good solvents) fall into three groups:

1) *Dilute solutions*, where different chains do not overlap (fig. 1*a*)). For c going to 0, one has a perfect gas of coils with coil concentration c/N. For slightly higher concentrations, the coils behave like a *dilute fluid of hard spheres* (for all dimensionalities $d < 4$). A certain feeling for this result can be obtained from the limit of small u_0: if we take the chains as nearly ideal, when two chains A and B overlap strongly the interaction between them is

$$\sum_{i\in A}\sum_{j\in B} Tv(T)\,\langle\delta(\boldsymbol{R}_{ij})\rangle \sim Tv(T)\frac{N^2}{R_0^d} \sim Tu_0 N^{2-d/2}\,, \tag{1.6}$$

where the sum $\sum_{i\in A}$ is on the monomers of the chain A, etc. We have estimated the monomer density in one coil as being $\sim N/R^d$, and we have used eqs. (1.2) and (1.4). Whenever d is smaller than 4, the repulsion (1.6) is much larger than T and the two coils cannot penetrate each other.

2) *Semi-dilute solutions* correspond to fig. 1*b*). Here the various chains do overlap, but the overall concentration c is still small with respect to the solvent concentration. The lower limit in c occurs when the chains are just in contact, and is thus comparable to the density in a single coil:

$$c^* = \frac{N}{R_{\text{SAW}}^d} \simeq N^{1-\nu d} a^{-d}\,. \tag{1.7}$$

For $d > 1$, $1-\nu d < 0$ and c is small: the semi-dilute regime, defined by

$$c^* \ll c \ll a^{-d}\,, \tag{1.8}$$

has a large span. For this regime the description of the interactions in terms of a single coupling constant $v(T)$ is valid, and thus we have a very universal behaviour.

3) *Concentrated solutions* ($ca^3 \sim 1$) are in a certain sense more complex. However, there are some important simplifications even in that limit. As first noticed by FLORY [1] long ago, the chains become ideal in this case! The argument can be summarized as follows: each monomer sees a repulsive potential $U(\boldsymbol{r}) = vTc(\boldsymbol{r})$ due to the other monomers. In a single coil $c(\boldsymbol{r})$ is a peaked function with a maximum near the coil centre. Thus there is a radial force $-\partial U/\partial r$ pointing outwards and swelling the chain. On the other hand, in the concentrated system, $c(\boldsymbol{r})$ is due to many different chains and is nearly constant in space (the fluctuations of c are very small). Thus, although the repulsive potential U is very large, the force vanishes and each coil is ideal. A more detailed description of this limiting case can be found in ref. [2].

This Flory theorem has been digested by the scientific community only rather slowly; a final experimental proof was given a few years ago by neutron scattering experiments: here a labelled (*e.g.* deuterated) chain is observed in a medium of normal (hydrogenated) chains and it is indeed found that it behaves like an ideal random flight [3].

In the following we shall be mainly concerned with semi-dilute solutions and we shall see how they interpolate between the dilute limit (swollen chains) and the concentrated limit (ideal chains). Section **2** gives a list of theoretical tools which have been used in this field, together with some specific applications. The discussion here is restricted to *static* properties in good solvents. More general cases and new problems are presented in sect. **3**.

2. – Calculational methods.

2'1. *Self-consistent fields.* – This method goes back very far in the past (KUHN, FLORY, HERMANS ...), but has been exploited fully by EDWARDS and co-workers and also by the Russian school. A general review is given in ref. [4] and we shall summarize the technique only very briefly. Whatever the detailed problem is (involving one chain, or many), the starting point is to assume a certain concentration profile $c(\boldsymbol{r})$ from which one extracts a repulsive potential

$$U(\boldsymbol{r}) = Tvc(\boldsymbol{r}) . \tag{2.1}$$

Then one computes the statistical weight $G_N(12)$ for an ideal polymer chain of N units, linking points (1) and (2), each unit being subjected to the potential $U(\boldsymbol{r})$. It turns out that this weight is very similar to the quantum-mechanical propagator: it obeys an equation of the Schrödinger form,

$$-\frac{\partial}{\partial N} G_N(12) = \mathscr{H} G_N(12) \qquad (N > 0) , \tag{2.2}$$

together with suitable boundary conditions. The operator $\mathscr{H}$ is explicitly

$$\mathscr{H} = -\frac{a^2}{6}\nabla^2 + \frac{1}{T}U(\boldsymbol{r}) . \tag{2.3}$$

For the assumed U one can thus construct the statistical weights G and finally derive a corrected expression for the local concentration $c_1(\boldsymbol{r})$. This can then be used to construct a new potential U_1, etc., and hopefully the procedure converges very much like in Hartree-Fock calculations on electron systems.

The greatest (and unusual) success of the self-consistent–field approach has been the calculation of the exponent ν in eq. (1.5) by FLORY [1]. The result is

$$\nu = 3/(d+2) \qquad (d \leqslant 4)\,. \tag{2.4}$$

For $d = 1$ this gives $\nu = 1$ as indeed expected for SAW. For $d = 2$, $\nu = \frac{3}{4}$ and, for $d = 3$, $\nu = \frac{3}{5}$, in very good agreement with most numerical and experimental data. Finally, for $d > 4$, $\nu = \frac{1}{2}$, *i.e.* the departure from ideality becomes weak.

On the other hand, the self-consistent–field method is *not* adequate for semi-dilute solutions: it leads to a description in which the chains are essentially uncorrelated and ideal in shape at all concentrations. Recent experiments using neutrons, osmometry and light scattering have shown that this is qualitatively incorrect [5]. We shall come back to this problem in subsect. **2**˙5.

2˙2. *Perturbation expansions.* – As usual in theoretical physics, it is tempting to assume first that the dimensionless coupling constant u_0 is small and to expand all observables in powers of u_0. This can be expressed in diagrammatic series, as shown on simple examples in fig. 2*a*. For instance the end-to-end radius is found to be of the form [6]

$$R^2 = R_0^2\left[1 + \text{const}\,\frac{W}{T} + \ldots\right], \tag{2.5}$$

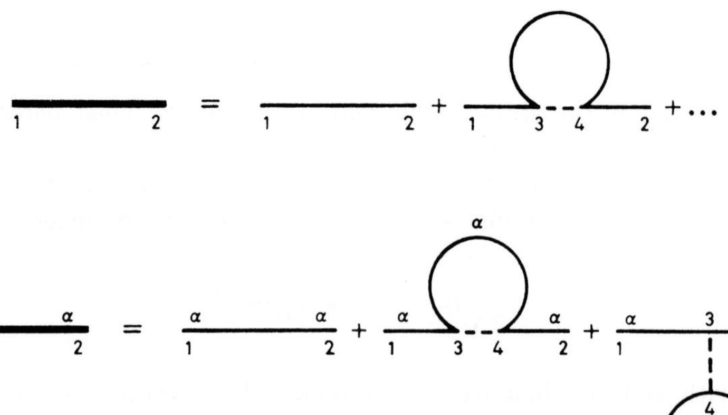

Fig. 2*a*.

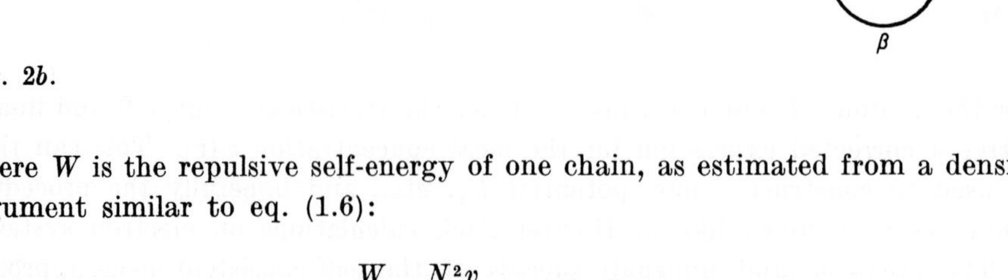

Fig. 2*b*.

where W is the repulsive self-energy of one chain, as estimated from a density argument similar to eq. (1.6):

$$\frac{W}{T} \sim \frac{N^2 v}{R_0^d} \sim u_0 N^{2-d/2}\,. \tag{2.6}$$

The expansion parameter (2.6) is large, even for small u_0, whenever $d < 4$, and thus the series is not very useful in practice. But it has a certain conceptual importance. Similar series can be obtained for the virial coefficient Ω between two coils (*) in the dilute regime, etc.

2·3. *Relation with the n-vector model.* – The diagrams of fig. 2*a* have a certain analogy with the diagrams for a Lagrangian field theory for an n-component order parameter $\boldsymbol{M}$, as occurs for instance in magnetic phase transitions ($\boldsymbol{M}$ being the magnetization). However, all the diagrams involving closed loops have no counterpart in the chain problem. Fortunately, each closed loop involves a summation on a certain dummy index for the components of $\boldsymbol{M}$, and gives a result multiplied by the number n of components. To dispose of the closed loops, it is enough to set $n = 0$. This was first noticed for the single-chain problem [7]. Later, a very elegant extension to the many-chain problem was derived by DES CLOISEAUX [8], and provided a considerable impetus for all studies on polymer solutions.

There has been a number of applications of the magnetic analogy. The first result was a comparison between the ε expansion results for the single-chain exponent ν, and the Flory formula (2.4). This showed that, although the Flory formula is excellent for all practical dimensionalities (1, 2, 3), it is not rigorous near $d = 4$. Recent reflexions on the success and limitations of the Flory approach can be found in ref. [9, 10].

But more fundamentally the main use of the $n = 0$ theorem has been to show how the *scaling properties* which are familiar in critical phenomena can be transposed to polymer statistics. We shall come back to this point in sect. **3**.

2·4. *Direct renormalization groups.* – The usual presentation of renormalization groups is based on a thinning of the number of degrees of freedom in the $\boldsymbol{k}$ space—or sometimes on « decimation methods » in an $\boldsymbol{r}$ space lattice. These techniques can, of course, be applied to polymers through the $n = 0$ theorem.

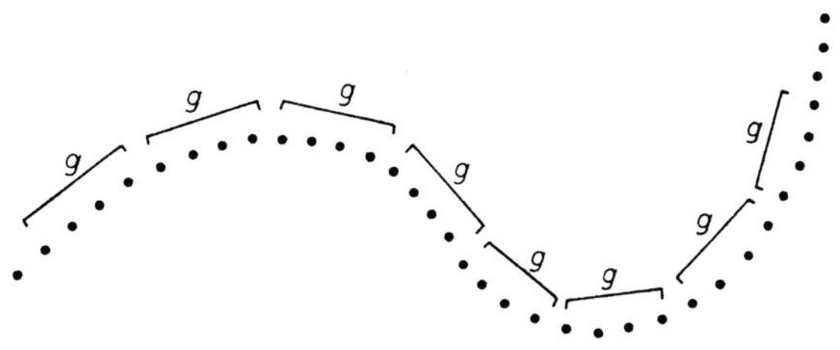

Fig. 3.

(*) Ω is not to be confused with the virial coefficient v between two *monomers*.

However, it is also instructive to operate in a different way, based on the one-dimensional structure of the chains. We shall briefly describe this approach, which can be called *decimation of the chemical sequence.*

The idea is shown in fig. 3. We start with N monomers of size a and reduced coupling constant u_0 (eq. (1.4)). We associate them into N/g consecutive subunits, each made of g monomers. By some method of calculation we then derive the average size a_1, and coupling constant u_1, for the subunits. We then iterate the process, going to N/g^m units of g^m monomers with size a_m and reduced coupling constant u_m.

The recurrence equations for a_m and u_m are

$$a_m = a_{m-1} g^{\frac{1}{2}}[1 + h(u_m)] , \tag{2.7}$$

$$u_m = g^{2-d/2} u_{m-1}[1 - k(u_m)] . \tag{2.8}$$

Consider first the size equation (2.7). If we were dealing with ideal coils, we would have of course $a_m = g^{\frac{1}{2}} a_{m-1}$. But here we must also include all interactions of pairs whose indices (i, j) on the chemical sequence differ by an amount smaller than g^m. When this is done, the subunits are swollen, and this is described by the function h. Perturbation expansions (2.5) and (2.6) indicate that, in fact, h depends only on the corresponding coupling constant u_m.

Now consider the recurrence relation (2.8) for the coupling constant, and start again from ideal chains. When the size of the units increases by a factor g, the number of interacting pairs between two units is g^2. The radius is multiplied by $g^{\frac{1}{2}}$ and thus v/a^d is multiplied by $g^{2-d/2}$. Now switch to interacting units: the number of effective pairs is reduced below g^2, because repulsive correlations decrease the chances of contact; also the radius a_m is larger than for ideal chains. All this points towards a negative correction $-k(u_m)$.

Now the crucial point is the following: when the process has been iterated a sufficient number of times, we are dealing with large subunits, and we know from the properties of dilute solutions (see eq. (1.6)) that two subunits *behave like impenetrable spheres.* This means that the excluded-volume parameter v_m becomes comparable to the subunit volume a_m^d, or equivalently that the reduced coupling constant

$$u_m = \frac{v_m}{a_m^d} \to u^* = \text{const} . \tag{2.9}$$

In renormalization group jargon, this tells us that the recurrence equation for u_n leads to a *fixed point.* But here the fixed point appears with a clear physical meaning.

Now let us focus our attention on large values of m, for which $u_m \sim u^*$. Then eq. (1.7) becomes a simple geometric series, since $h(u_m) \to h(u^*)$ is independent of m. Now we can write the form of the relation between overall size R and polymerization index N as

$$R = a_m f\left(\frac{N}{g^m}\right) \equiv a_{m+1} f\left(\frac{N}{g^{m+1}}\right). \tag{2.10}$$

This can be satisfied only if (*)

$$f(N) = \text{const}\, N^{\nu}, \tag{2.11}$$

$$\nu = \frac{\ln(a_{m+1}/a_m)}{\ln g} = \frac{1}{2} + \frac{\ln[1 + h(u^*)]}{\ln g}. \tag{2.12}$$

Equation (2.11) proves the scaling law. Equation (2.12) gives an explicit formula for ν provided that some numerical estimates of $h(u)$ and $k(u)$ have been constructed (the latter allows for an explicit determination of u^*).

This can be carried out by various procedures:

a) near $d = 4$, the fixed point u^* becomes small and the functions h and k can be derived from perturbation expansions. This must lead to the ε expansion which was already known through the $n = 0$ theorem [7].

b) for any $d < 4$, direct numerical calculations using a fixed small value of g could be carried out.

In practice, the exponent ν is so well described by the Flory form ((eq. (2.4)) that no great surprise is expected. However, more detailed information, such as prefactors in the scaling formula, could also be derived by this method.

One particular case has been found to have an interesting content. This occurs when the bare coupling constant u_0 is rather small (say $u_0 \sim 10^{-2}$ to 10^{-1}), or, in chemical terms, when the solvent becomes « poor ». Then, for the first few iterations, the constant u_m remains rather small, and it takes some time before u_m switches to the fixed-point value u^*. In physical terms, this means that there is a critical subunit size l such that at distances $r < l$ the chain behaves as ideal, while at distances $r > l$ it shows the excluded-volume effects. The value of l, or of the corresponding number of monomers P_l, can be estimated by a simple argument: *a*) since up to scales l the chain is nearly ideal, we must have $P_l \sim l^2/a^2$; *b*) since for larger scales the chain is strongly nonideal, the reduced coupling constant computed for size l must be of the order of unity. For $d = 3$ this means

$$\frac{vP_l^2}{l^3} \simeq 1 \tag{2.13}$$

(*) From a formal point of view, $\ln f(N)$ plotted *vs.* $\ln N$ might contain terms which are periodic with period $\ln g$, but repeating the argument for different choices of g eliminates this possibility.

and the result is thus

$$\frac{l}{a} \sim \frac{a^3}{v_0} (\gg 1) . \tag{2.14}$$

The existence of a cross-over between ideal-chain and swollen-chain behaviour at a wave vector $q_l \sim 1/l$ has been confirmed by recent neutron experiments of the Saclay group [11], and eq. (2.14) does apparently work.

Another interesting application of the decimation procedure has been found in connection with charged polymers, and will be described briefly in sect. **3**.

3. – Current applications.

3'1. *Semi-dilute solutions.* – The Des Cloiseaux analysis [8] has shown how to relate the properties of overlapping chains to a problem of magnetism under an external field. At present, however, detailed references to the magnetic problem do not give much insight, and the description of polymer solutions is mainly based on simple *scaling laws*. It turns out, in fact, that, knowing the single-chain exponent ν, one can account for most properties of the many-chain problem in good solvents ($u_0 \sim 1$). The spirit of the approach is sum-

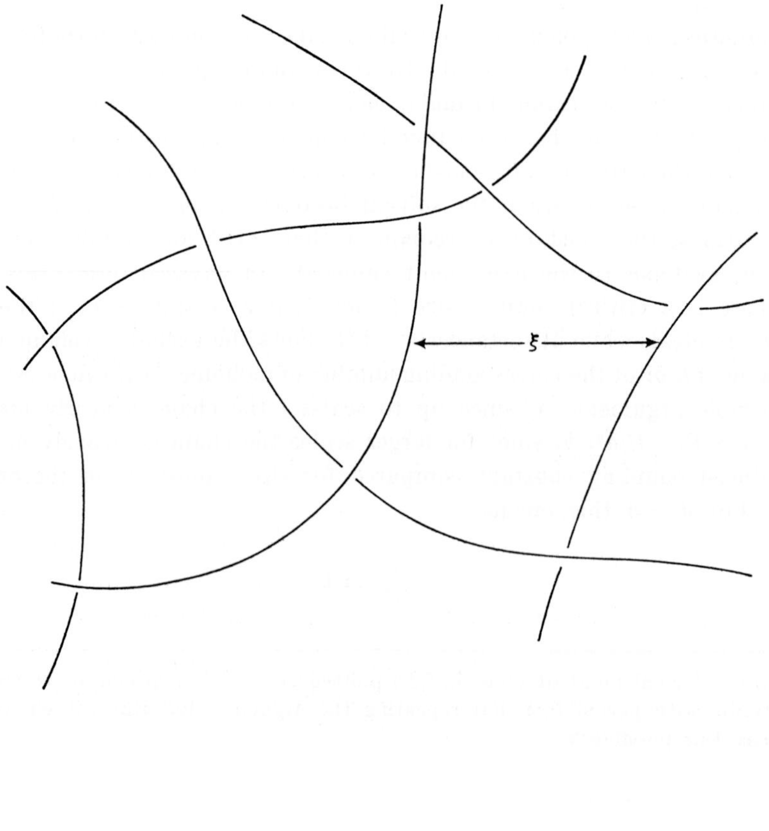

Fig. 4.

marized in fig. 4, where we see that an overlapping set of chains behaves like a « transient network » with a certain characteristic size ξ between contact points. The length ξ decreases clearly with increasing concentration, and has the following features:

a) At the lower concentration limit ($c \sim c^*$) ξ must become equal to the single-coil size R_{SAW}:

$$\xi \to R_{\text{SAW}} = aN^{\frac{3}{5}} \qquad (d = 3) .$$

b) At higher concentrations, the network structure depends on c, but not on N any longer. Thus ξ is independent of N.

This leads to the form

$$\xi = aN^{\frac{3}{5}}\left(\frac{c^*}{c}\right)^x , \tag{3.1}$$

where x is for the moment an unknown exponent. But noting that $c^* \sim N^{-\frac{4}{5}}$ (according to eq. (1.8) with $\nu = \frac{3}{5}$) and imposing condition *b*), we are led unambiguously to $x = \frac{3}{4}$. This is, in fact, in good agreement with neutron data [5]. On scales $r < \xi$, each chain behaves like an isolated (swollen) chain. On scales $r > \xi$ the chain may be discussed as a succession of noninteracting « blobs ». Each « blob » has a size ξ, contains a number g of monomers and obeys the swelling law $\xi/a \sim g^{\frac{3}{5}}$. Interactions between different blobs are screened out, and thus the chain is ideal at large scales. Now this picture has been checked by a great variety of experiments [5, 12].

3'2. *Extension to other dimensionalities.* – In two dimensions, correlation effects are more dramatic than in three. For this reason various efforts have been dedicated to restricted polymer systems:

a) Molecular films of long-chain polymers have recently been studied by OBER and VILLANOVE [13], by using refined techniques for low surface pressures $\Pi(c)$. The scaling prediction is here $\Pi \sim c^3$, while the mean field would. lead to $\Pi \sim c^2$. The experimental results show that $\Pi \sim c^y$, where $3 > y > 2.8$. The deviation $3 - y$ is probably due to secondary effects: when the pressure increases, the film thickness increases slightly.

b) There is special interest in the properties of polymer solutions confined inside small slits or pores, with applications in gel chromatography, oil recuperation, etc. A scaling theory has been constructed also for these cases, and it exhibits a number of unexpected regimes and cross-overs [14].

c) The problem of partial adsorption of a polymer on a weakly attractive surface is also of practical interest, and is the $n = 0$ analog of a ferromagnet with enhanced couplings near the surface [15].

3·3. *The case of poor solvents.* – In « poor » solvents the steric repulsions between monomers can be balanced by attractive interactions at slightly larger distances and, at a suitable temperature (called the Θ-point in the polymer literature), the excluded-volume parameter vanishes ($v = 0$ or $u_0 = 0$). This is a case in which the two-particle vertex of fig. 2 vanishes and in which the 3-particle vertex becomes essential: it is the $n = 0$ analog of a *tricritical* point in magnetism [16]. This analogy shows that for $d = 3$ the one-chain behaviour is nearly ideal. Logarithmic corrections (due to the marginal character of $d = 3$) have been worked out by STEPHEN [17]. For semi-dilute solutions, the cross-over between good-solvent and poor-solvent behaviour has been studied in theoretical [18] and experimental [11] papers of the Saclay group.

3·4. *Polyelectrolytes.* – Many polymers carry ionizable groups and in suitable solvents such as water they dissociate. A typical example is to have a backbone with a negative charge ($-e$) on each monomer and positive counterions floating in the solution. A few fundamental references on these systems are listed in [19]. Here we shall restrict our attention to the one-chain problem, assuming that the counter-ions are spread far away (negligible screening). More general situations are only partly understood [20].

Some feeling about the effects of charge can be obtained through the decimation method. Here we start with monomers of size a and charge $-e$, the interaction being of the form

$$U(r_{ij}) = \frac{e^2}{\varepsilon r_{ij}^{d-2}}, \tag{3.2}$$

where ε is the dielectric constant of the solvent, and we have written the Coulomb law for d dimensions (lumping all normalization factors into e^2). A suitable definition of the dimensionless coupling constant is

$$u_0 = \frac{e^2}{\varepsilon a^{d-2} T}. \tag{3.3}$$

Now we group the monomers into subunits of length g, g^2, ..., g^n, and construct recurrence equations for the size and for the couplings. The equation for sizes still has the structure of eq. (2.7)—though, of course, the detailed form of the correction h will differ. What is peculiar here is the equation for the charges: the charge of g units is *exactly* g times larger than the charge of one unit, and this implies that no other unknown function enters into the equations for u_m

$$u_m = u_{m-1} g^{[2-(d-2)/2]} [1 + h(u_m)]^{(2-d)}. \tag{3.4}$$

We can make this relation explicit in two simple limits, corresponding to weak coupling ($u_m \to 0$) and strong coupling ($u_m \to \infty$). In weak coupling the subunits are nearly ideal ($h \to 0$):

$$u_m/u_{m-1} = g^{3-d/2} \qquad (u_m \ll 1)\,. \tag{3.5}$$

In strong coupling the subunits are completely stretched ($a_m = g a_{m-1}$). This implies $1 + h = g^{\frac{1}{2}}$, and gives

$$u_m/u_{m-1} = g^{4-d} \qquad (u_m \ll 1)\,. \tag{3.6}$$

From the limiting properties (3.5), (3.6) one can reconstruct the general structure of u_m as a function of u_{m-1} for all values of u_m (fig. 5). Depending on the dimensionality d, we find three cases:

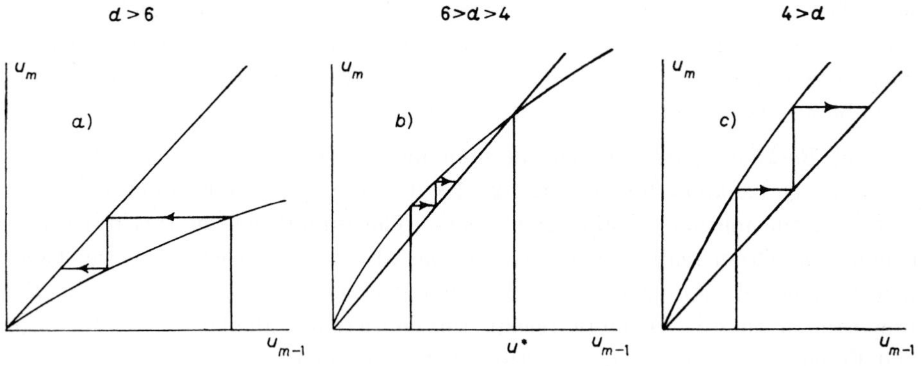

Fig. 5.

a) For $d \geqslant 6$ successive iterations lead to a fixed point with vanishing interactions $u^* = 0$: a long chain is then ideal.

b) For $4 < d < 6$ there is a finite fixed point $u_m \to u^*$. When u_m converges to the fixed point u^*, we must have

$$1 + h(u^*) = g^{6-d/2d-4}$$

and then

$$\frac{a_{m+1}}{a_m} = g^{2/(d-2)}\,, \qquad \nu = \frac{2}{d-2}\,. \tag{3.7}$$

Equation (3.7) has been derived first by PFEUTY and VELASCO [21] using the usual (k-space) version of renormalization groups.

c) For $d<4$ the iteration leads to $u_m \to \infty$ — *i.e.* to rigid chain behaviour. This occurs in particular for the physical case $d=3$. A single polyelectrolyte chain (in a pure solvent: no salts, no screening) should be fully stretched. Unfortunately, this limit of separate chains in a completely insulating solvent is very hard to reach experimentally: for instance the dissociation of water always implies a finite screening radius. Also the critical overlap concentration c^* becomes very small for extended chains: $c^* \simeq N/L^3$, where $L = Na$ is the full length. The single-chain problem can be studied only at $c < c^*$.

We end up by one special remark, concerning the single-chain problem with *weak charges* in 3 dimensions: if u_0 is very small (in practice, if there are only a few charges along the chemical sequence), the chain should appear coiled for studies at small r ($r < l_1$) and stretched for studies at large r ($r > l_1$).

3·5. *Dynamic scaling in polymer solutions.* – The dynamics of polymer chains is extremely complex. An introduction can be found in the book of Ferry [22] and in the review by GRAESSLEY [23]. Even restricting our attention to the simplest case (good solvent, semi-dilute system, friction between monomers and solvent dominant), we find two main difficulties; one is connected with *back flow*, the other with *entanglements*.

a) *Backflow effects.* When one monomer moves in the solvent, it creates a long-range backflow (decreasing like $1/r$ in the Stokes approximation) which reacts on the motion of other monomers. The importance of this dynamical coupling has been realized long ago. It has also become clear that the backflow is screened. An early work of Debye [24] led to a screening radius K^{-1} proportional to $c^{-\frac{1}{2}}$. A more recent self-consistent calculation of Edwards and Freed [25] (ignoring all repulsions between coils) gave a screening radius $K^{-1} \sim c^{-1}$. If this was still correct for interacting coils, we would have two separate characteristic lengths (ξ and K^{-1}) with different scaling properties, and the resulting behaviour would be very complex.

Fortunately this is not so: it turns out that « blobs » as defined in subsect. **3·1** represent the dynamical unit as well as the static unit. Thus K^{-1} scales like ξ and the scaling properties are simple [10]. These notions have been applied to the prediction of certain « gellike modes » in the long-wave-length limit: what is seen here (by light scattering) is concentration fluctuations of low wave vector k. They are expected to relax with a relaxation rate

$$\frac{1}{\tau_k} = D_c k^2 , \tag{3.8}$$

where the « collective diffusion coefficient » D_c scales like the self-diffusion coefficient of one blob [10]:

$$D_c \simeq \frac{T}{6\pi\eta_0 \xi} \sim c^{\frac{3}{4}} \tag{3.9}$$

(η_0 being the solvent viscosity). Recent photon beat experiments by ADAM and DELSANTI [26] do give a D_c increasing with concentration with an exponent of the order of 0.7.

For $k\xi > 1$, modes (3.6) cross over into *single-chain modes*, where now the width $\Delta\omega_k$ is independent of c. Thus we must have

$$\Delta\omega_k \simeq \frac{T}{\eta_0} k^3 . \tag{3.10}$$

Some data bearing on this limit have been obtained by light scattering (on coils of very large N) [27]. The experimental exponent z ($\Delta\omega_k \sim k^z$) is of order 2.9.

b) Apart from the backflow effects, there are great difficulties associated with the effects of entanglements (two chains cannot cross one another). Results (3.8)-(3.10) bear on regimes in which entanglements are simple or absent. But for certain mechanical properties (*e.g.* the static viscosity of the solution) entanglements are essential and their role is poorly understood.

* * *

The author has greatly benefited from discussions with M. DAOUD, J. DES CLOISEAUX, G. JANNINK, S. LIFSON, P. PINCUS, A. SILBERBERG. This article was written during a visit at the Weizmann Institute and at the Hebrew University, and the writer wishes to express his thanks to Prof. S. ALEXANDER for his hospitality on this occasion.

REFERENCES

[1] P. FLORY: *Principles of Polymer Chemistry* (Ithaca, N. Y., 1967).
[2] P. G. DE GENNES: *J. Physique*, **31**, 235 (1970).
[3] J. P. COTTON: *Macromolecules*, **7**, 863 (1974); R. KIRSTE: *Polymer*, **16**, 120 (1975).
[4] P. G. DE GENNES: *Rep. Prog. Phys.*, **32**, 187 (1969).
[5] M. DAOUD: *Macromolecules*, **8**, 804 (1975).
[6] See, for instance, H. YAMAKAWA: *Modern Theory of Polymer Solutions* (New York, N. Y., 1972).
[7] P. G. DE GENNES: *Phys. Lett.*, **38** A, 339 (1972).
[8] J. DES CLOISEAUX: *J. Physique*, **36**, 281 (1975). For these two references see also the appendix to ref. [5].
[9] J. DES CLOISEAUX: *Phys. Rev. A*, **10**, 1665 (1974).
[10] P. G. DE GENNES: *Macromolecules*, **9**, 587 (1967).
[11] J. P. COTTON: *Journ. Chem. Phys.*, **65**, 1101 (1976).
[12] P. G. DE GENNES: *Israel Journ. Chem.*, **14**, 154 (1975).
[13] R. OBER and R. VILLANOVE: to be published in *Journ. Colloid. Polymer Science.*
[14] M. DAOUD and P. G. DE GENNES: *J. Physique*, **38**, 85 (1977).
[15] P. G. DE GENNES: *J. Physique*, **37**, 1445 (1976).

[16] P. G. DE GENNES: *J. Physique Lett.*, **36**, 55 (1975).
[17] M. STEPHEN: *Phys. Lett.*, **53** A, 363 (1975).
[18] M. DAOUD and G. JANNINK: *J. Physique*, **37**, 973 (1976).
[19] See, for instance, *Polyelectrolytes*, edited by E. SELEGNY (Dordrecht, 1974).
[20] P. G. DE GENNES, P. PINCUS, R. VELASCO and F. BROCHARD: *J. Physique*, **37**, 1461 (1976).
[21] P. PFEUTY, R. VELASCO and P. G. DE GENNES: *J. Physique Lett.*, **38**, 5 (1977).
[22] J. FERRY: *Viscoelastic Properties of Polymers* (New York, N. Y., 1970).
[23] W. GRAESSLEY: *Adv. Pol. Sci.*, **16**, 1 (1974).
[24] P. DEBYE and A. BUECHE: *Journ. Chem. Phys.*, **16**, 573 (1948).
[25] S. EDWARDS and J. FREED: *Journ. Chem. Phys.*, **61**, 3626 (1974).
[26] M. ADAM, J. DELSANTI and G. JANNINK: *J. Physique Lett.*, **37**, 53 (1976).
[27] M. ADAM and J. DELSANTI: *J. Physique Lett.*, to be published.

C. R. Acad. Sc. Paris, t. 288 (9 avril 1979) Série B – 219

MÉCANIQUE DES FLUIDES. – *Écoulements viscométriques de polymères enchevêtrés.* Note (*) de **Pierre-Gilles de Gennes,** présentée par Paul Germain.

Lorsqu'il n'y a pas d'adsorption des chaînes sur les parois du tube, une solution concentrée de polymères flexibles doit montrer un effet de *glissement aux parois* : la loi de Poiseuille ne s'applique plus, lorsque le diamètre du tube D est inférieur à une longueur caractéristique (de l'ordre de 0,5 mm). Par contre, si le tube est rugueux, ou bien greffé avec des polymères, le glissement serait fortement réduit.

A concentrated polymer solution flowing in a clean capillary is predicted to show a significant slip *on the capillary wall, giving strong deviations to the Poiseuille law when the tube diameter* D *is smaller than a characteristic length (of order* 0.5 *mm). On the other hand, if the tube surface is rough, or grafted with polymer, the amount of slip should be much smaller.*

Les écoulements de polymères fondus ou concentrés jouent un rôle important pour divers phénomènes d'extrusion et de moulage. Ils sont compliqués par des effets viscoélastiques [1], et par certains effets non linéaires, qui peuvent engendrer des instabilités [2]. Dans la présente Note, nous considérons une situation limite bien plus simple : les écoulements dans un capillaire, en régime linéaire. Nous suggérons que ces écoulements sont anormalement sensibles à l'*état de surface* du capillaire.

Considérons pour commencer une paroi solide plane ($x=0$) et immobile, le liquide polymérique occupant le demi-espace $x>0$, et coulant avec une vitesse $v(x)$ dans la direction y. Incluons la possibilité d'un glissement, c'est-à-dire d'une vitesse à la paroi $v(0)$ différente de zéro. La contrainte $\sigma_{xy}=\sigma$ sera proportionnelle à cette vitesse :

(1) $$\sigma = kv(0).$$

La constante k a les dimensions d'une viscosité divisée par une longueur. La remarque cruciale est que (pour une paroi lisse, sans adsorption de polymères) la constante k n'est pas sensible aux effets d'enchevêtrement qui contrôlent la viscosité macroscopique η du liquide. On attend donc une constante k qui soit *indépendante de la longueur des chaînes,* et comparable à ce qu'elle est pour un système de monomères à même concentration [3].

Si nous nous plaçons un peu à l'intérieur du fluide (distance x faible par rapport au diamètre de l'écoulement) la contrainte $\sigma(x)$ est pratiquement égale à la contrainte au bord :

(2) $$\sigma(x) = \eta \frac{\partial v}{\partial x} = \sigma = kv(0).$$

La condition aux limites à utiliser pour décrire l'écoulement est donc de la forme mixte :

(3) $$v(0) = b \frac{\partial v}{\partial x},$$

où la « longueur d'extrapolation » b est définie par

(4) $$b = \eta / k.$$

En régime enchevêtré, la viscosité η croît très vite avec le nombre N de monomères par chaîne [1] $\eta = \eta_1 N^m$. Empiriquement l'exposant m est de l'ordre de 3,3, alors que le modèle de reptation suggère $m=3$ [4]. Dans tous les cas :

(5) $$b = \eta_1 k^{-1} N^m = b_1 N^m,$$

où b_1 est une longueur microscopique, et N^m un facteur très grand. Suivons par exemple le modèle de reptation [4] pour lequel

$$\eta = \eta_0 N^3 N_e^{-2}, \tag{6}$$

où N_e représente la distance entre enchevêtrements, et η_0 une viscosité microscopique. Alors $b = \eta_0 k^{-1} N^3 N_e^{-2}$. Pour $\eta_0 k^{-1} = 3$ Å, $N_e = 200$ et $N = 2000$ (ce qui correspondrait à un polystyrène fondu de masse 200 000) on attend $b = 60\,\mu$.

Les écoulements dans un capillaire de diamètre D comparable ou inférieur à $8\,b$ sont alors fortement modifiés. La relation entre débit (J) et gradient de pression (p') prend la forme

$$J = (-p')\frac{D^3 \pi}{16 k}\left[1 + \frac{D}{8 b}\right]. \tag{7}$$

Les mesures en capillaires fins sont souvent faites à des pressions p' un peu trop élevées et sont alors compliquées par des effets viscoélastiques non linéaires. Mais elles mériteraient d'être reprises et comparées à (7). Il faut insister toutefois sur les limitations du modèle physique :

(1) Si la paroi considérée comporte des ondulations perpendiculaires aux lignes de courant, le coefficient k est augmenté : par exemple si la surface se trouve non pas dans le plan $x=0$ mais à une distance $u(y)$ de ce plan

$$u(y) = u \sin(qy)$$

on trouve (par un calcul d'écoulement laminaire incompressible) une contribution à k de la forme :

$$k_{\text{ond}} = \eta\, q^3 u^2/2 \qquad (qu \ll 1). \tag{8}$$

Cette contribution est proportionnelle à η, donc importante, et peut l'emporter dans certains cas sur le coefficient k microscopique défini plus haut [5].

(2) Si la paroi est *greffée* avec des chaînes polymériques analogues (en longueur et en propriétés chimiques) à celles du liquide, il y a enchevêtrement entre chaînes mobiles et chaînes fixes, le coefficient k devient très élevé, et le glissement négligeable.

(3) Le même effet est attendu si les chaînes du liquide ont tendance à s'adsorber partiellement sur la paroi.

Au total il semble que les écoulements viscométriques de polymères enchevêtrés puissent fournir une méthode inattendue d'étude des états de surface.

(*) Remise le 9 avril 1979.

[1] J. D. FERRY, *Viscoelastic properties of Polymers*, 2ᵉ édition, Wiley, New York, 1970.

[2] C. J. PETRIE et M. M. DENN, *A.I. Chem. E. J.*, 22, 1976, p. 209.

[3] Cette estimation tient pour une solution concentrée. Pour une solution semi diluée, la constante k serait encore diminuée par l'existence d'une « couche de déplétion » : *voir* J. F. JOANNY, L. LEIBLER et P. G. DE GENNES, *J. Pol. Sc.* (à paraître).

[4] P. G. DE GENNES, *J. Chem. Phys.*, 55, 1971, p. 572; M. DOI et S. F. EDWARDS, *Faraday Trans.*, II, 74, 1978, p. 1789, 1802, 1818.

[5] Le cas de la chute d'une bille (de rayon R inférieur à b) dans le liquide est conceptuellement analogue au cas de l'écoulement près d'une surface ondulée. Malgré le glissement, on attend un coefficient de friction proportionnel à ηR, mais avec un coefficient inférieur au coefficient 6π de la loi de Stokes.

C.N.R.S., E.R.A. 542, Collège de France,
75231 Paris Cedex 05.

Afterthought: Ecoulements viscométriques de polymères enchevêtrés

Some early observations by Galt and Maxwell ("Modern Plastics", Dec 1964 issue) show slippage in a transparent extruder. In another geometry (plane/plane rheometer with a small gap), Burton, Folkes, Narm, and Keller [*J. Materials Sci.* **18**, 315 (1983)] found a strong slippage, (with extrapolation lengths in the range of 100 microns) for molten p.styrene on copper surfaces. But in these two cases we are dealing with rather large shear stresses, and it is not sure that we are actually in the linear regime discussed here.

In many practical cases one finds no slip [see J. Meissner, *Ann. Rev. Fluid Mech.* **17**, 45 (1985)]. The reason for this may be that even a few "reactive " points at the solid surface (each binding one chain) are enough to create a large surface friction: see F. Brochard, P.G. de Gennes and P. Pincus, *Comptes Rendus Acad. Sci. (Paris)* **314II**, 873 (1992).

On the other hand, a number of experiments indicate strong slippage at the interface between two incompatible polymer *melts*: for a discussion of this case see F. Brochard, P.G. de Gennes and S. Troian, *Comptes Rendus Acad. Sci. (Paris)* **310**, 1169 (1990).

Theory of Long-range Correlations in Polymer Melts

By P. G. de Gennes

Collège de France, 75231 Paris Cedex 05, France

Received 22nd October, 1979

We discuss first the small-angle scattering of neutrons by a homopolymer melt, when each chain is tagged at certain sites by deuteration. We then extend these considerations to block copolymers AB, assuming an AB interaction which is weakly repulsive. Starting from the homogeneous melt, we lower the temperature and determine first the locus of the spinodal instability towards microphase separation, and in particular the period of the incipient periodic structure. L. Leibler has recently constructed a Landau theory for the microphases, and a theoretical phase diagram (restricted to the vicinity of the melt) showing regions with lamellar, hexagonal or cubic structures. Finally we discuss the effect of a certain amount of disorder in the chemical sequence on the spinodal instability: in the most interesting case, where the disorder is the same in all chains, we find that disorder increases the spatial period and ultimately suppresses the trend toward microphases. These last considerations may be relevant for the design of certain nematic polymers.

CORRELATION IN HOMOPOLYMER MELTS

From the theoretical ideas of Flory[1] and from neutron experiments,[2] we know that the chains in a homopolymer melt are essentially ideal. It is important to realize, however, that there remain some interesting correlation effects in the melts, which can be seen in diffraction experiments provided that one uses tagged molecules.[3] This can be understood most directly for the case of homopolymer chains which are all tagged at *one end* only. Fig. 1 gives us a qualitative picture of the correlation $\gamma(\boldsymbol{r})$ between tagged sites: apart from normalization factors, we can think of $\gamma(\boldsymbol{r})$ as the concentration of tagged molecules at point r, when we have put one tagged molecule at the ori-

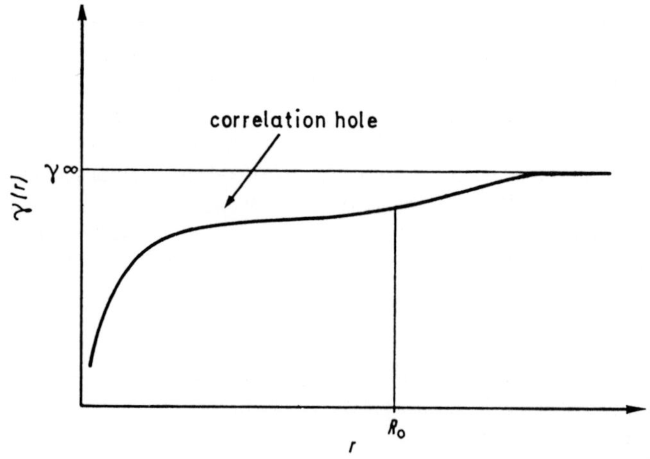

Fig. 1.—Qualitative plot of the correlation between " heads " of different molecules in a homopolymer melt. When one " head " is fixed at the origin, the space available for other coils around it is slightly decreased by the presence of the first coil: there is a correlation hole.

gin. At large r, $\gamma(\mathbf{r})$ becomes constant and equal to the average number of tags per cm^3. But when r is smaller than the average chain size $R_0 = N^{\frac{1}{2}}a$ (N = number of monomers per chain) a new effect appears: we know that one particular chain is present near the origin; thus the concentration of other chains must be slightly depleted. There is a *correlation hole*; the relative shift of $\gamma(\mathbf{r})$ in this region is rather small, since the density associated with the first chain is of order $N/R_0^3 \sim N^{-\frac{1}{2}}$. However, the correlation hole is visible in neutron experiments, because at small wavevectors ($qR_0 \sim 1$) there are no other physical phenomena contributing to the intensity: in particular the fluctuations of the overall density in the melt (controlled by a compressibility modulus) are very small when compared with the effects of the correlation hole.

A detailed theory of these correlation effects was constructed long ago,[3] and we shall summarize it only briefly here. The starting point is a set of identical chains, each with a sequence of monomers 1,2, . . . n, . . . N. Some of the monomers are deuterated, but they are all assumed to be identical as regards their mutual interactions: this appears to be a good approximation for most melts, although deviations may be found in polyethylene of very high molecular weight. We call $c_n(\mathbf{r})$ the concentration of monomers of rank n at point $\mathbf{r}$, and we are interested in the correlation functions

$$\langle c_n(0)c_m(\mathbf{r})\rangle - \langle c_n\rangle\langle c_m\rangle = kT\chi_{nm}(\mathbf{r}). \tag{1}$$

Note that the functions χ_{nm} introduced here have nothing to do with a Flory interaction parameter.[1] The notation expresses the fact, first shown by Yvon,[4] that $\chi_{nm}(\mathbf{r})$ may be visualized as a *non-local susceptibility*, giving the shift in $c_m(\mathbf{r})$ if a small perturbation, acting only on species (n), is applied at point 0. In the factor kT, T is the temperature and k is Boltzmann's constant.

The neutron experiments measure the Fourier transforms $\chi_{nm}(q)$ through a formula of the form:[5]

$$I(q) = \sum_{nm} \alpha_n \alpha_m \chi_{nm}(q)kT \tag{2}$$

where α_n is the coherent amplitude for species n, and q is the scattering wavevector. (In practice we control α_n by selective deuteration).

The calculation of the correlations χ_{nm} is based on the very same principles which were used by Debye and Hückel for polyelectrolytes. One puts one monomer (n) at the origin, and then looks at the self-consistent field built up in its neighbourhood. This idea was first applied to polymer *solutions* by Edwards.[6] One central assumption of the calculation is to assume that each chain is *nearly ideal* (just as Debye and Hückel assumed that in zero order, their ions behaved as an ideal gas). This assumption of ideality is not accurate for solutions, but is indeed correct for melts.

Another interesting feature of the melts is that we do not need to specify the interactions between monomers in any detail! At the long wavelengths of interest here, we may in fact describe them simply by imposing a constraint of constant density

$$\sum_n c_n(\mathbf{r}) = c = \text{constant}. \tag{3}$$

What is nice in this approach is that one then obtains a formula for the correlations which contains no adjustable parameter (except for the r.m.s. end-to-end distance $R_0 = N^{\frac{1}{2}}a$ of one chain), namely[3]

$$\chi_{nm}(q) = \chi^0_{nm}(q) - \frac{S_n(q)S_m(q)}{D(q)} \tag{4}$$

where $\chi^0_{nm}(q)$ is the intrachain correlation calculated for a gaussian coil

$$NkT\chi^0_{nm}(q) = \exp\{- q^2a^2 \, |n - m| \, /6\} \tag{5}$$

and the functions S_n and D are sums of χ^0 functions

$$\left.\begin{aligned} S_n(q) &= \sum_m \chi^0_{nm}(q) \\ D(q) &= \sum_n S_n(q) \end{aligned}\right\} . \tag{6}$$

In particular $D(q)$ is the ideal chain scattering function first computed by Debye.[7]

Although eqn (4) was derived from a specific Debye–Hückel type of calculation, it is slightly more general: the main assumption is that the self term χ^0_{nm} in eqn (4) is of the ideal chain type. For instance, with a chain where the first half ($1 \leqslant n \leqslant N/2$) is labelled and the second half is not, Benoit[8] has proven recently that the formula for $I(q)$ resulting from eqn (4) and (2) holds automatically if χ^0 is gaussian: the proof is based only on the incompressibility sum rules derived from eqn (3), namely:

$$\sum_m \chi^0_{nm}(q) \equiv 0 \tag{7}$$

plus the symmetry between the two parts of the chain.

What are the typical conclusions from eqn (4) and (2)? Let us start from the same simple case, where only one monomer ($n = 1$) is deuterated on each chain. The plot of $I(q)$ is then simple. At low q we have no intensity: the concentration of deuterated sites c_1 is simply $1/N$ times the overall concentration c, and the latter has no fluctuations in our model. At higher q values we do get a signal because, although c is constant, the chains can choose, in a given fluctuation, to put their end-to-end vectors (or " dipoles ") parallel: this creates a non-zero polarization $\boldsymbol{P}(\boldsymbol{r})$ and $c_1 = -\text{div}\, \boldsymbol{P}(\boldsymbol{r})$ can be finite. Ultimately, when qR_0 becomes much larger than unity, interference terms between different chains drop out from $I(q)$, and we are left with the scattering by an individual point-like object, which is constant.

Let us investigate next the case where the chain is tagged on a short sequence near the origin $n = 1, 2, \ldots n_e$. Then the low-q part of $I(q)$ is still exactly the same (except for a normalization factor) whenever $qR_e < 1$ where $R_e = n_e^{\frac{1}{2}}a$ is the size of the tagged portion. Thus $I(q)$ increases with q in this region. However, when we go to large q values ($qR_0 > 1$) the scattering now arises from independent particles of finite size (R_e) and the form factor of such a particle drops for $qR_e > 1$: thus we finally have a decrease in $I(q)$ at high q.

We conclude that there is a peak at intermediate q values. Peaks of this type have indeed been observed in many neutron experiments.[5] Originally, some experimentalists were tempted to associate them with an " incipient order " in the solution. However, this is not correct: the intensity at the peak is *smaller* than the intensity which would be due to uncorrelated scatterers with the same chemical sequence. The peak reflects the existence of a hole in the spatial correlations, and not of a hump.

We needed to recall this point before embarking upon the more complex case where our molecules are block copolymers, and where real segregation effects may become important: in this latter case, we shall find intensity peaks which are larger than the scattering due to independent objects, and which do represent incipient order.

To conclude with the homopolymer case: the neutron data on various types of partly labelled chains in melts do show that the Debye–Hückel formula (4) is a reasonable approximation in 3 dimensions. It should be emphasized, however, that in 2 dimensions (for chains confined to thin sheets in a lamellar system) the situation could be much worse: deviations from ideality might be more important. I do hope that some 2 dimensional situations will be studied in future experiments.

BLOCK COPOLYMERS

We assume now that all our chains are identical block copolymers made (for instance by anionic polymerization) with two types of monomers A and B. The chemical sequence will be described by indices $\sigma_1, \sigma_2 \ldots \sigma_n \ldots \sigma_N$ where:

$$\left.\begin{array}{ll}\sigma_n = +1 & (\mathrm{A}) \\ \sigma_n = -1 & (\mathrm{B})\end{array}\right\} . \tag{8}$$

There is a certain interaction between n and m which we shall write in the form:

$$V_{nm} = \tfrac{1}{2}kT\chi_{\mathrm{F}}(1 - \sigma_n\sigma_m). \tag{9}$$

Here χ_{F} is a Flory interaction parameter between A and B. Eqn (9) ensures that $V_{\mathrm{AA}} = V_{\mathrm{BB}} = 0$ while $V_{\mathrm{AB}} = kT\chi_{\mathrm{F}}$. In the simplest situations (to which we shall adhere here) V_{nm} is temperature independent and χ_{F} is proportional to T^{-1}. We assume χ_{F} to be positive and rather small (weak trend towards segregation).

The Debye–Hückel (or " random phase approximation ") result for the correlations in this case, has the form:[3]

$$\left[\chi^{-1}(q)\right]_{nm} = \left[\chi^{-1}_{\mathrm{homo}}\right]_{nm} + V_{nm}. \tag{10}$$

Here $[\chi^{-1}(q)]_{nm}$ is the inverse matrix to $\chi_{nm}(q)$ and χ_{homo} is the correlation matrix for homopolymers, given by eqn (4). It can be checked that eqn (10) is still compatible with the incompressibility sum rule.

Eqn (10) is less secure than eqn (4), because it implies assumption of ideal gaussian behaviour which might fail when the trend towards segregation is strong (χ_{F} large). However, certain qualitative arguments (restricted to the homogeneous melt, with A and B sequences of comparable length) do suggest that the assumption is still tolerable at temperatures just above microphase separation.

We shall first apply eqn (10) to the discussion of a relatively simple case, namely a periodic multiblock copolymer . . . A B A B . . ., each block having the same number d of monomers, and the same r.m.s. size $R_{\mathrm{A}} = R_{\mathrm{B}} = d^{\frac{1}{2}}a$. The complete chains are assumed to be very long, and this simplifies the analysis because end effects can be ignored: we may just as well replace the chain by a cyclic structure, and impose periodicity on the matrices χ_{nm}.

In this case it is rather simple to invert the matrix $\chi_{\mathrm{homo}}(q)$ and to write the inverse operator in differential form:

$$\chi^{-1}_{\mathrm{homo}}(q) = NkT(2x)^{-1}\left(-\frac{\partial^2}{\partial n^2} + x^2\right) \tag{11}$$

where $x = q^2a^2/6.$

Eqn (11) is easily checked on Fourier transforms with respect to the variable n, which here we treat as continuous. In general, eqn (11) would have to be supplemented by rather delicate boundary conditions for the operator on the right: but in our cyclic case this complication is removed. Also note that eqn (11) holds for all the vector space on which the matrix χ_{nm} acts, except for one vector (1 . . . 1 . . . 1): because of the incompressibility sum rule, $\chi^{-1}_{\mathrm{homo}}$ acting on this vector gives a divergent result.

We shall mainly focus our discussion here on the search for an instability premonitoring microphase separation: this will correspond to one particular eigenmode (ψ_n) of the matrix χ_{nm}, such that $\sum_m \chi_{nm}\psi_m = \infty$ or equivalently $\sum_m (\chi^{-1})_{nm}\psi_m = 0$. Using eqn (10) this gives:

$$-\frac{\partial^2\psi_n}{\partial n^2} + x^2\psi_n + \sum_m \frac{2x}{NkT} V_{nm}\psi_m = 0. \tag{12}$$

Inserting eqn (9) for the interaction matrix, and noting also that $\sum \psi_n = 0$ by the incompressibility sum rule, we can simplify eqn (12) greatly and obtain:

$$-\frac{\partial^2\psi_n}{\partial n^2} + x^2\psi_n - \frac{2x\chi_F}{N}\Gamma\sigma_n = 0 \tag{13a}$$

where the parameter

$$\Gamma = \sum_m \sigma_m\psi_m \tag{13b}$$

is independent of n. For a given sequence $\sigma_1 \ldots \sigma_n \ldots \sigma_N$ it is not hard to solve eqn (13*a*), find the structure of ψ_n for a given Γ, and finally impose the self-consistency condition (13*b*). For the periodic chemical sequence chosen in our example:

$$\left.\begin{aligned}\sigma_n = +1 \text{ for } & -d/2 < n < d/2 \\ -1 \quad & d/2 < n < 3d/2 \quad \textit{etc.}\end{aligned}\right\} \tag{14}$$

the adequate solution ψ_n is periodic and vanishes at the ends of each block. In the central block (A):

$$\left.\begin{aligned}\psi_n &= \cosh(xd/2) - \cosh(xn) \\ -d/2 &< n < d/2\end{aligned}\right\}. \tag{15a}$$

For the B blocks, ψ_n has a similar structure, but is reversed in sign:

$$\left.\begin{aligned}\psi_n &= -\cosh(xd/2) + \cosh x(n-d) \\ d/2 &< n < 3d/2\end{aligned}\right\}. \tag{15b}$$

From eqn (13*b*) one then finds

$$\Gamma = N[\cosh(xd/2) - 2(xd)^{-1}\sinh(xd/2)] \tag{16}$$

and comparing with eqn (13*a*) we arrive at the instability condition:

$$\chi_F = (x/2)[1 - \tanh(xd/2)/(xd/2)]^{-1}. \tag{17}$$

When plotted against x, the right-hand side of eqn (17) shows a maximum for $x = x^* \sim 3.21d^{-1}$. This means that the microphase period which tends to occur corresponds to a wavelength $2\pi/q^*$ which is just slightly smaller than the r.m.s. end-to-to end distance of one AB sequence. For χ_F smaller than the corresponding threshold, we expect a peak in the neutron scattering intensity at $q \cong q^*$.

We do not have many data on periodic multiblock systems to compare with these predictions. A different molecule which is easier to produce experimentally is the simple *diblock* (*AB*) *copolymer*. But here, on the theoretical side, the situation is slightly more complex. It has been worked out, however, in some detail by Leibler.[9] He gives plots of the incipient periodicity $2\pi/q^*$ for all values of the molar fraction f corresponding to $A_{Nf}B_{N(1-f)}$. Furthermore Leibler did not only compute the pair correlation functions $\langle c_n c_m\rangle$ but also some higher correlations $\langle c\, c\, c\rangle$ and $\langle c\, c\, c\, c\rangle$.

Starting from this, he can construct a Landau theory of the microphase separation of AB diblocks, at least for temperatures which are not too far below the liquidus. For $f \sim 0.5$ he predicts the onset of a simple lamellar phase. But for more dissymetric cases ($f \neq 0.5$), he finds a cascade of phases (isotropic, hexagonal, lamellar . . .).

These predictions are interesting: they supplement the earlier analysis of Helfand, which was concerned with a low-temperature regime where the spatial boundaries (between A rich and B rich portions) are rather sharp.

EFFECTS OF PARTIAL RANDOMNESS IN THE CHEMICAL SEQUENCE

The chemical sequence of a block copolymer is always expected to show some statistical deviations from its nominal structure. We can think of two main types of chemical disorder, which we call I and II.

Type I: here different chains in the same sample have different chemical sequences. This implies that the index σ_n defined in eqn (7) now has a certain average value $\bar{\sigma}_n$ which is intermediate between -1 and $+1$. In many practical cases the deviations $\sigma_n - \bar{\sigma}_n$ and $\sigma_m - \bar{\sigma}_m$ are expected to be uncorrelated. Then we need simply to replace the interaction V_{nm} of eqn (7) by an averaged interaction $\bar{V}_{nm} = kT\chi_{\mathrm{F}}(1-\bar{\sigma}_n\bar{\sigma}_m)$. The rest of the analysis is unchanged.

Type II: here all chains in the sample have the same sequence, but this sequence has statistical deviations from the model copolymer. This case is conceivably present in a periodic multiblock AB structure, prepared by anionic polymerization, if the successive time intervals for insertion of A or B monomers have a certain random distribution in length.

We shall discuss here the effects of type II disorder on one particularly simple case: namely we assume that we are in a nearly periodic multiblock structure, with indices σ_n which are random stationary variables, and a correlation function

$$\overline{\sigma_n\sigma_m} = \cos[\lambda(n-m)]\exp(-\alpha|n-m|). \tag{18}$$

Here $\lambda = \pi/d$ is related to the average periodicity ($2d$) of the chemical sequence, and α measures the randomness. We shall now see that eqn (18) suffices to give us the optimal wavevector q^*: returning to eqn (13) we find that:

$$\psi_n = \frac{\Gamma\chi_{\mathrm{F}}}{2N}\sum_m \sigma_m \exp(-x|n-m|) \tag{19}$$

and inserting this into the definition of Γ [eqn (13*b*)] we arrive at the self-consistency condition

$$\bar{\Gamma} = \overline{\frac{\Gamma\chi_{\mathrm{F}}}{2N}\sum_{nm} \sigma_n\sigma_m \exp(-x|n-m|)}. \tag{20}$$

On the r.h.s. of eqn (20) it appears plausible to decouple Γ from $\sigma_n\sigma_m$ (since Γ is a macroscopic quantity with fluctuations of relative magnitude $\sim N^{-\frac{1}{2}}$). This then leads to

$$1 = \frac{\chi_{\mathrm{F}}}{2}\sum_m \overline{\sigma_n\sigma_m}\exp - x|n-m| \tag{21}$$

and (always ignoring end effects) we obtain explicitly:

$$\chi_F^{-1} = \int_0^\infty dt \cos(\lambda t) \exp\{-(\alpha + x)t\}$$

$$\chi_F = \alpha + x + \frac{\lambda^2}{\alpha + x}. \quad (22)$$

(*a*) For $\alpha = 0$ (no disorder) this spinodal condition is qualitatively similar to eqn (17): the quantitative differences reflect the fact that σ_n is a succession of alternate steps for eqn (17) while it is a sinusoidal curve for eqn (22). The minimum in this case is obtained for $x = x^* = \lambda = 3.14\ d^{-1}$.
(*b*) For $\alpha > 0$ (increasing disorder) the whole curve (22) is shifted and the minimum is at $x^* = \lambda - \alpha$. When α becomes larger than λ the minimum remains at $q = 0$ (fig. 2).

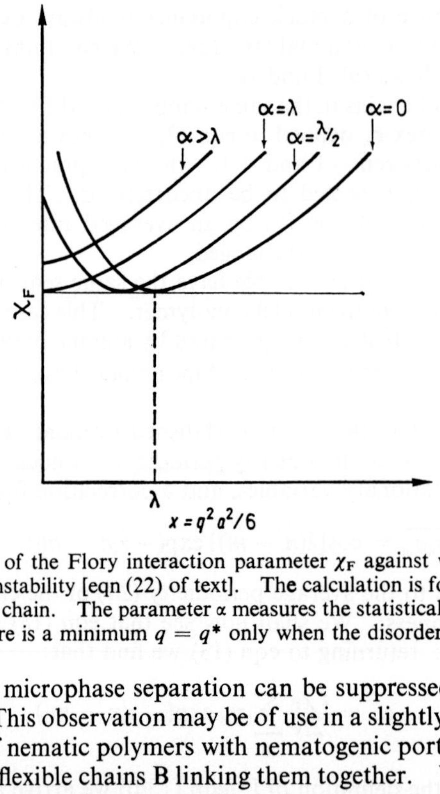

FIG. 2.—Qualitative plot of the Flory interaction parameter χ_F against wavevector squared at the onset of the microphase instability [eqn (22) of text]. The calculation is for a long sequence of equal blocks ABAB. . . on each chain. The parameter α measures the statistical departures from this ideal structure. Note that there is a minimum $q = q^*$ only when the disorder is not too strong ($\alpha < \lambda$).

We conclude that microphase separation can be suppressed by a certain amount of type II disorder. This observation may be of use in a slightly different field, namely for the preparation of nematic polymers with nematogenic portions A inserted in part of the backbone, and flexible chains B linking them together. When such systems are prepared, it often happens that the result at the temperatures of interest is not a nematic, but rather a smectic A (*i.e.*, a lamellar structure with alternating A rich and B rich portions). Eqn (22) suggests that, if a certain amount of randomness is introduced in the length of the B chains, we may well eliminate the spurious smectic phase. Of course in this case, the A sequence is not flexible, and the statistics may be more complex than in eqn (18), but the qualitative trends should be the same.

I am indebted to H. Benoit and L. Leibler for various discussions on the Debye–Hückel approach.

[1] P. Flory, *Principles of Polymer Chemistry* (Cornell Univ. Press, Ithaca, N.Y., 1953).
[2] D. G. Ballard, J. Schelten and G. D. Wignall, *Eur. Polymer J.*, 1973, **9**, 965; J. P. Cotton *et al.*, *Macromolecules*, 1974, **7**, 863; R. G. Kirste, V. A. Kruse and K. Ibel, *Polymer*, 1975, **16**, 120.

[3] P. G. De Gennes, *J. Physique*, 1970, **31**, 235.
[4] J. Yvon, *Rev. Scientifique* (Déc. 1939), p. 662.
[5] F. Boue *et al.* in *Neutron Inelastic Scattering 1977* (Int. Atomic Energy Agency, Vienna, 1978), vol. 1, p. 563.
[6] S. F. Edwards, *Proc. Phys. Soc.*, 1966, **88**, 265.
[7] P. Debye, *J. Phys. Colloid Chem.*, 1947, **51**, 18.
[8] H. Benoit, personal communication, Sept. 1979.
[9] L. Leibler, to be published.

Afterthought: Theory of long-range correlations in polymer melts

The first application of Debye Hückel methods to polymers is due to S.F. Edwards, *Proc. Phys. Soc. (London)* **88**, 265 (1966). This was dealing with semidilute solutions. The second paper, dealing with melts, was in *J. Physique* **31**, 235 (1970). But I decided to put in this compilation the (more recent) Faraday discussion because: (a) it starts with a good review of the early applications; (b) it opens the field of ordering in block copolymer melts: lately this has expanded considerably, in particular in the hands of L. Leibler. See the review by A. Halperin, M. Tirrell and T.P. Lodge, *Advances in Polymer Science* **100** (Springer, Berlin, 1991).

Reprinted from Macromolecules, 1984, *17*, 703

Tight Knots†

Pierre-Gilles de Gennes

Collège de France, 75231 Paris Cedex 05, France. *Received July 7, 1983*

ABSTRACT: We present a conjectural explanation for long-time memory effects in melts of crystallizable, linear polymers. During a crystallization process, chains that are knotted, and that are reeled in from *both ends*, might make very tight knots. These knots are expected to be very stable and to persist for long times after the sample is heated above the melting temperature. The density and the lifetime of tight knots should be sensitive functions of the original quenching temperature. Tight knots could also be generated, in the absence of any crystallization process, by strong longitudinal shear of a solution—especially with poor solvents.

I. Introduction

The natural time unit for entangled polymer melts of high molecular weight is the terminal time τ measured from viscoelastic data.[1,2] This time is a rapidly increasing function of the molecular weight M ($\tau \sim M^{3.3}$) and can become rather large (minutes) for high polymers ($M > 10^6$) even when they are well above their glass temperatures. A partial picture for τ has been provided by the reptation model.[3,4]

However, some strange things still happen at times $t \gg \tau$.

(a) The T_2 relaxation or protons in melts displays two distinct fractions associated respectively with fast or slow relaxation. The relative weight of the two fractions in a polyethylene melt appears to depend on the thermal history of the sample.[5] Memory effects are present after melting for many hours, in samples where the terminal time is smaller than 1 min.

(b) One can probe the organization of chains in a crystallizable melt by a rapid quenching below the crystallization temperature, followed by a measurement of the repeat period of the lamellae.[6,7] This again shows remarkably long relaxation times θ in the melt: θ is a rapidly increasing function of M, as is τ, but θ is considerably larger than τ; naive reptation cannot account for the long memory times.

In the present paper, we propose an explanation for the long times θ based on tight knots (section II). This is very tentative but may stimulate certain relevant experiments: we discuss them in section III.

II. Kinks and Knots

Figure 1 shows a tight knot made with a flexible aliphatic chain, preserving the tetrahedral carbon coordinations and forbidding overlap between neighboring C or H atoms but allowing for local conformations that can be neither trans nor gauche (arbitrary dihedral angles). With that set of constraints the minimal number N_m of carbon atoms required to make the knot is ~38. Imposing strict trans or gauche conformations, by applying the chain on a diamond lattice, we find a rather similar figure ($N_m \sim 36$).

Our first aim is to estimate the statistical weight for such a tight knot. We shall use a very qualitative approach based on the kink model, which was used as an image in early reptation work.[3] We picture a flexible thread with a certain average number ρ of kinks per monomer and a *stored length* b per kink. What characterizes the knot region is primarily the absence of kinks, or of stored length. Thus the probability P of finding a sequence of monomers $n, n + 1, ..., n + N_m$, in a tight knot may be estimated by the Poisson law

$$P = P_0 e^{-\rho N_m} \tag{1}$$

where P_0 is an unknown prefactor of order unity.

In a conventional melt of flexible chains, we expect $b \sim a$ (a being the monomer length) and $\rho \sim 1$. Then the probability P is exponentially small and tight knots are probably unobservable.

In a quenched system that is undergoing crystallization,[8] some chains may start to crystallize at two (or more) different points A and B. At each crystallization point the chain is reeled in with a certain tension ζ, which is simply related to the thermodynamics of undercooling

$$\zeta = \frac{\Delta F}{a} \simeq \frac{\Delta H}{a}\frac{T_m - T_Q}{T_m} \tag{2}$$

where ΔF is the free energy difference between crystal and melt and ΔH is the enthalpy of transition (per monomer). T_m is the melting point and T_Q is the quenching temperature. In steady state the tension ζ is applied all along the flexible portion AB. (In transient states some regions of the AB chain do not yet feel the tension, but we omit this complication in what follows.) We conclude that the density of kinks must be drastically reduced in the AB portion

$$\rho = \rho_0 \exp\left(-\frac{b\zeta}{kT_Q}\right) \tag{3}$$

$b\zeta$ being the work performed to destroy one kink. If we insert (3) in (1), we find a "doubly exponential" form, giving a very abrupt increase in P when $\zeta \sim kT_Q/b$.

When a tight knot is made, it becomes extremely difficult to undo even after melting and returning to $\zeta = 0$. The knot can disappear only through entry (in the knot region) of a certain amount of stored length; when this is provided, the usual reptation processes can start to work. We all know, from experience with macroscopic strings, that the insertion of some stored length in a tight knot is difficult to perform. Here, stored length will probably not enter in the form of a sharp kink, but rather by some very progressive distortion of many bonds. The activation energy E for this process is expected to be high and to be a rapidly increasing function of the original tension ζ. The calculation of $E(\zeta)$ will require a considerable effort of numerical analysis. But, at the present zero-order level, it may be plausible to think that the activation energy is linear in ζ

$$E(\zeta) = b'\zeta \tag{4}$$

where b' is another length comparable to a monomer size. The lifetime of the knots in the melt is expected to be of order

$$\theta = \tau \exp[E(\zeta)/kT] \gg \tau \tag{5}$$

† In spite of its obvious imperfections, we dedicate this work to Walter Stockmayer, whose advice, encouragement—and living example—have been a major help to our group during the past 15 years.

(a)

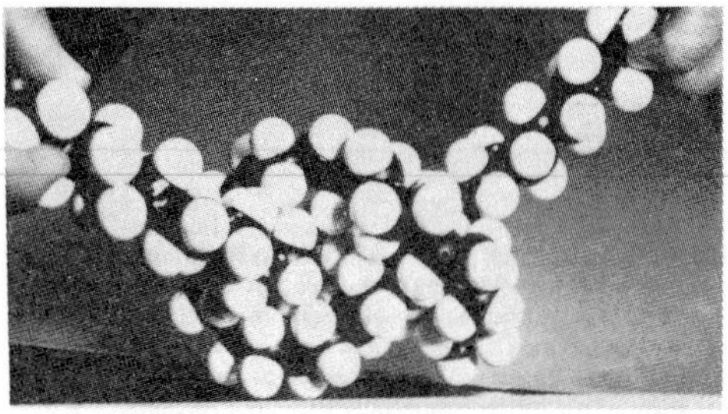

(b)

Figure 1. (a) Typical conformation for the simplest tight knot made with an aliphatic chain. The number of carbons involved in the "active" region is ~36–38. (b) Topology of the knot in (a).

For polyethylene quenched at room temperature,[14] eq 2 leads to $E(\zeta) \sim 0.5kTb'/a$. The precise value of b'/a is thus crucial, because it enters in eq 5 through an exponential. To explain the observed values of θ we need $b'/a > 10$. I do not have any precise argument for these high values of b'/a. I tend to believe that they reflect the *cooperativity* of a tight knot: to loosen it at one point requires extra freedom in another loop.

III. Concluding Remarks

(1) If they are high enough, the line tensions ζ present during a crystallization process may favor the formation of tight knots. These knots can be simple self-knots such as the one shown in Figure 1, but they can also be more complex objects involving two (or more) chains.

(2) Both the number of knots (eq 1 and 3) and the lifetime of the knots (eq 4 and 5) are expected to be strongly dependent on ζ and thus on sample history.

(3) Cumulative effects may occur: successive quench–melt cycles during time intervals smaller than θ could lead to rather high knot densities.

(4) The presence of one tight self-knot may slow down the opening of another knot on the same chain: there may be corrections to eq 5 describing knot/knot interactions.

(5) Apart from (4), a self-knot has few physical effects (it may reduce the crystallinity slightly). On the other hand, a tight knot of two chains transforms them into a star, with strongly enhanced relaxation times. A single star in a matrix of linear chains of comparable molecular weight is expected to display a mechanical relaxation

$$\tau_s \simeq \tau(N/N_e)^2 \qquad (6)$$

where N is the degree of polymerization and N/N_e is the number of entanglements per molecule.

(6) Tight knots may be generated, in the absence of any crystallization process, by *strong shear flows*. Consider, for instance, a single, flexible chain undergoing a coil–stretch transition in a longitudinal shear flow.[9,10] In the fully extended state the tensions in the middle of the chain (obtained by integrating the local friction forces along the whole chain length) are of order

$$\zeta \simeq \frac{kT}{a}\tau_R\dot{\gamma} \qquad (7)$$

where τ_R is the Rouse relaxation time (proportional to M^2) and $\dot{\gamma}$ is the shear rate. The tension ζ is often sufficiently strong to induce polymer degradation.[11] For our purposes, assume that the original coil carried a loose knot before stretching: then it may end up with a tight knot after stretching. Loose self-knots are probably rare in good solvents but more frequent in Θ solvents.[12,13] Thus this process may become important in poor solvents. Comparing (7), (3), and (1), we see that the minimal shear rate $\dot{\gamma}_m$ required to generate tight knots is of order

$$\dot{\gamma}_m \sim \frac{1}{\tau_R}\ln N_m \qquad (8)$$

(We usually expect $\dot{\gamma}_m < \dot{\gamma}_c$, where $\dot{\gamma}_c$ is the threshold for chemical degradation).

(7) We might also generate tight interchain knots in *semidilute* solutions (even with good solvents) under strong longitudinal shear flows: this may have interesting rheological consequences.

Acknowledgment. These ideas were developed during a very stimulating workshop at DSM. I thank Drs. Koningsfeld and Lemstra for their hospitality on this occasion. A Keller, A. Charlesby, M. Adam, F. Brochard, and M. Dvolaitsky have helped me in these matters but should not be held responsible for the very tentative ideas defined above!

References and Notes

(1) Ferry, J. D. "Viscoelastic Properties of Polymers", 2nd ed.; Wiley: New York, 1970.
(2) Graessley, W. *Adv. Polym. Sci.* **1974**, *16*.
(3) de Gennes, P.-G. *J. Chem. Phys.* **1971**, *55*, 572.
(4) Doi, M.; Edwards, S. *J. Chem. Soc., Faraday Trans. 2* **1978**, *74*, 1789, 1802, 1818. Graessley, W. *Adv. Polym. Sci.* **1982**, *47*, 1.
(5) Folland, R.; Charlesby, A. *Eur. Polym. J.* **1979**, *15*, 953. Charlesby, A. Oral presentation at the DSM Workshop (1983).
(6) Rault, J.; Robelin, E. *Polym. Bull.* **1980**, *2*, 273. Robelin, E. Thèse 3ème cycle, Univ. Paris 6 (1980).
(7) Cutter, D. J.; Hendra, P.; Sang, R. *Chem. Soc., Faraday Discuss.* **1979**, *68*, 320.
(8) See, for instance: (a) Wunderlich, B. "Macromolecular Physics"; Academic Press: New York, 1973. (b) Wunderlich, B. *J. Chem. Soc., Faraday Discuss.* **1980**, *68*. (c) Hoffman, J. D. In "Polyethylene 1933–1983"; Plastics and Rubber Institute: London, 1983; p D3-1.
(9) de Gennes, P.-G. *J. Chem. Phys.* **1974**, *60*, 5030.
(10) Mackley, M.; Keller, A. *Philos. Trans. R. Soc. London* **1975**, *278*, 29. Frank, F. C.; Mackley, M. *J. Polym. Sci.* **1976**, *14*, 1121. Crowley, D.; Frank, F. C.; Mackley, M.; Stephenson, R. *Ibid.* **1976**, *14*, 1111.
(11) A quantitative discussion of this point was recently given by A. Keller (Seminar at Cornell University (May 1983)).
(12) Brochard, F.; de Gennes, P.-G. *Macromolecules* **1977**, *10*, 1157.
(13) Vologodski, A.; Lakashin, A.; Frank-Kamenetskii, M. Ansaelevich, V. *Sov. Phys.—JETP (Engl. Transl.)* **1975**, *39*, 1059.
(14) I am very much indebted to a referee for this discussion.

Afterthought: Tight knots

Simulations on polymer knots have been carried out recently by M. Mansfield (to be published): they consider polyethylene, atactic polypropylene, and atactic poly (L butene). The knots were described by molecular dynamics; they were tightened by strong stretching, and then released. The basic conclusions is that tight knots are not stable — for these polymers. The question remains open for chains with more bulky side groups, like p. styrene.

Eur. Phys. J. E **1**, 93–97 (2000)

THE EUROPEAN PHYSICAL JOURNAL E
EDP Sciences

Viscosity at small scales in polymer melts

F. Brochard Wyart[1] and P.G. de Gennes[2,a]

[1] PSI, Institut Curie, 11 rue P. et M. Curie, 75231 Paris Cedex 05, France
[2] Collège de France, 11 place M. Berthelot, 75231 Paris Cedex 05, France

Received 19 April 1999

Abstract. Flows around small colloidal particles of diameter b, or in thin films, capillaries, etc., cannot always be described in terms of the macroscopic polymer viscosity. We discuss these features for entangled polymer melts, where two distinct regimes can be found: (a) the thin regime where b is smaller than the coil radius R_0, but larger than the diameter d_t of the Edwards tube; (b) the ultrathin regime, where $b < d_t$. We consider (i) non adsorbing particles, where slippage may occur between the melt and the solid surface; (ii) "hairy" particles, which carry some bound polymer chains. We obtain scaling predictions for mobilities of spheres, of needles, and of clusters of particles. We also discuss translational and rotational diffusion of needles.

PACS. 05.60.Cd Classical transport – 82.70.Dd Colloids – 83.70.Hq Heterogeneous liquids: suspensions, dispersions, emulsions, pastes, slurries, foams, block copolymers, etc.

1 A list of questions

Viscosity is a macroscopic concept. If we think of a polymer melt, we may describe weak, viscous flows in terms of one viscosity coefficient η_B, provided that the spatial scale b of the velocity variations is large compared to the size of the polymer coils R_0. Or equivalently, we can think in terms of a wave vector $q \sim 1/b$. The macroscopic regime corresponds to $qR_0 < 1$.

To extend this description at smaller scales, one natural tool is a scale dependent viscosity $\eta(q)$. This is defined by applying a spatially sinusoidal shear rate:

$$\frac{\partial v_x}{\partial z} = \dot{\gamma}(z) = \dot{\gamma} \sin qz \qquad (1)$$

and measuring the corresponding shear stress σ:

$$\sigma_{xz} = \sigma \sin qz,$$

(note that we are not imposing any modulation in time: we deal only with steady state properties):

$$\sigma = \eta(q)\dot{\gamma}. \qquad (2)$$

We shall discuss $\eta(q)$ in Section 2 for entangled melts and in Section 3 for unentangled chains. However, we have progressively noticed, over a long period (~20 years) that many practical problems, involving small objects and melts, cannot be expressed only in terms of $\eta(q)$. Our central example in this note is a colloidal particle (of small diameter b), moving in the melt.

[a] e-mail: pierre-gilles.degennes@espci.fr

These particles are important in many applications of plastics and rubbers: reinforcement, ignifugation, antioxydants, fungicides, etc. For these purposes it is often useful to know what is the order of magnitude of a diffusion coefficient in the melt, or equivalently of the friction coefficient f/V, relating an applied force f to a particle velocity V.

Friction depends on both the size and the shape of the particles; this leads us to a discussion of characteristic lengths. Two lengths are important in an entangled polymer melt: the coil radius $R_0 = N^{1/2}a$ (where N is the polymerisation index and a is a monomer size) and also the diameter of the Edwards tube $d_t = N_e^{1/2}a$, where N_e is the number of monomers between entanglements. Typically $d_t = 50$ Å and $R_0 = 200$ Å. Thus we must consider two microscopic domains:

(a) the *thin* regime, where the relevant scale satisfies

$$d_t < b < R_0; \qquad (3)$$

(b) the *ultrathin* regime $b < d_t$.

A third characteristic length ℓ shows up when we discuss microscopic friction at the solid/melt interface: slippage can occur. We have discussed this theoretically over many years [1,2]. We focus our attention here (mainly) on clean, smooth, non adsorbing, solid surfaces, over which slippage is important.

Slippage may be characterised by the Navier extrapolation length ℓ introduced in reference [1]. For non adsorbing particles ℓ is large (microns or more). This drives us to define two more regimes:

(c) if $R_0 < b < \ell$, we speak of *large* particles;

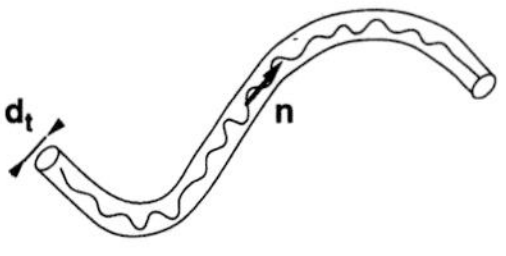

Fig. 1. A chain in a tube. The local alignment is defined by a unit vector **n**. Note that **n** vectors belonging to different tubes portions are uncorrelated.

(d) if $\ell < b$, we speak of *giant* particles: for the latter case, usual hydrodynamics (ignoring slippage) may be applied.

In Section 2, we concentrate on the thin regime: we argue that $\eta(q)$ is independent of q for entangled melts. This, however, does not rule out certain unexpected features of the mobility of needle-like particles.

Section 3 deals with ultrathin regimes, where certain features resemble Rouse dynamics (unentangled chains) while other features remain dominated by entanglements or by slippage.

These results are summarised, and to some extent, augmented in Section 4.

2 Thin particles ($d_t < b < R_0$)

2.1 Viscosity $\eta(q)$

Our formal starting point to study $\eta(q)$ is a Kubo formula expressing the viscosity in terms of stress-stress correlations:

$$\eta(q) = \frac{1}{kT}\int_0^\infty \mathrm{d}\mathbf{r}'\mathrm{d}t\langle\sigma_{xz}(\mathbf{r},u)\sigma_{xz}(\mathbf{r}'t)\rangle e^{i\mathbf{q}\cdot(\mathbf{r}'-\mathbf{r})}. \quad (4)$$

The stress σ is due to alignment inside one tube [3]. If $\mathbf{n}(\mathbf{r})$, a unit vector along the local tube axis (Fig. 1) and a^3 is the volume per monomer:

$$\sigma_{xz} = \frac{kT}{a^3}n_x n_z. \quad (5)$$

The crucial feature is that the tube, itself, is a random walk: the values of the alignment tensor $n_x n_z$ at different points are uncorrelated. The only contributions to (4) come from monomers which were originally sitting at one point ($\mathbf{r}$) in the tube, and which move along the tube, but return to point $\mathbf{r}$ at time t. Thus, the phase factor $\mathbf{q}\cdot(\mathbf{r}-\mathbf{r}')$ vanishes, and $\eta(q)$ is *independent of* q. This has a number of consequences, which will be discussed below.

2.2 Mobility of colloidal particles

2.2.1 Spheroids (Fig. 2a): from giant to thin

Consider first a sphere of radius b moving at velocity V in the melt. If we had no slippage, we would expect the standard Stokes result for the driving force f, with a q independent viscosity η_B:

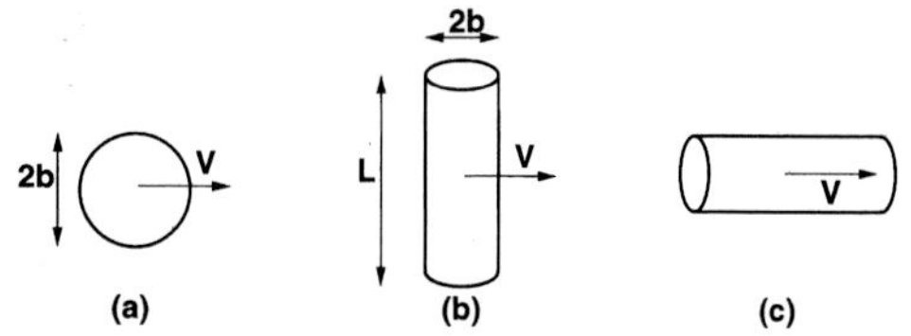

Fig. 2. Colloidal particles in a melt: (a) sphere, (b) needle moving in the "hard" direction, (c) needle moving in the "easy" direction.

$$f = 6\pi\eta_B bV. \quad (6)$$

However, with a non adsorbing sphere, we expect slippage on the surface [1] and f is reduced. This is similar to the problem of an air bubble in a fluid, when the bubble has no surfactant on its surface, and no surface shear stress. Then, instead of equation (6), we expect:

$$f = 4\pi\eta_B bV \qquad (d_t < b < \ell). \quad (7)$$

As explained in reference [1], this holds whenever the particle size b is much smaller than the Navier extrapolation length:

$$\ell \cong a\eta_B/\eta_1 \quad (8)$$

where η_1 is the viscosity of an equivalent liquid of monomers.

Thus, in our language, a giant particle follows equation (6), while a large particle follows equation (7).

2.2.2 Needles in the hard direction (Fig. 2b)

Here we have velocity gradients localised near the needle, of order V/b, extending over a volume $\sim Lb^2$. This gives a dissipation:

$$T\dot{S} = \eta_B(V/b)^2Lb^2 = fV \quad (9)$$

and thus:

$$f \cong \eta_B LV. \quad (10)$$

The exact numerical coefficient would depend on the presence, or absence, of slippage.

2.2.3 Needles in the easy direction (Fig. 2c)

It is useful here to decompose the friction into two parts:

- a frontal part, very similar to what we have for a sphere, giving a force f_1:

$$\frac{f_1}{V} \sim \eta_B b; \quad (11)$$

- a lateral part, due to a slippage velocity $\sim V$, acting on the surface monomers, each with a friction $\eta_1 a$. The number of these surface monomers is of order bL/a^2, and thus:

$$\frac{f_2}{V} = \eta_1 bL/a. \tag{12}$$

The ratio of the two forces is:

$$\frac{f_1}{f_2} = \frac{a\eta_{\mathrm{B}}}{L\eta_1} = \frac{\ell}{L} \tag{13}$$

and it is usually larger than unity. To keep in mind both contributions, ignoring again all coefficients, we can write, in the easy direction:

$$\frac{f}{V} = \eta_{\mathrm{B}} b\left(1 + \frac{L}{\ell}\right). \tag{14}$$

Notice that for $\ell > L$, there is a strong anisotropy in the friction. Comparing equations (10) and (11), we have:

$$\frac{f_{\mathrm{hard}}}{f_{\mathrm{easy}}} \sim \frac{L}{b}\,. \tag{15}$$

This is very different from what we have for giant needles (negligible slip), where standard hydrodynamics holds: then f_{hard} and f_{easy} are both proportional to $\eta_{\mathrm{B}} L$, and differ only by factor of 2.

Equation (15) implies that if an oblique force is applied to a needle, the resulting velocity is nearly parallel to the needle axis. Also, if we think in terms of diffusion coefficients, the only important diffusion is along the axis. (We come back to this point in Sect. 4.)

3 Ultrathin particles

3.1 Spheroids: the role of Rouse dynamics

An ultrathin sphere is expected to be very mobile: when the sphere moves, it does not have to disentangle any polymer chains. It must displace chain portions of size $\sim b$, by simple rearrangements within its own tube: here, we are dealing with Rouse dynamics [4].

The macroscopic viscosity of an assembly of Rouse chains, with a monomer concentration $c = a^{-3}$, is proportional to the polymerisation index N:

$$\eta_{\mathrm{Rouse}} \cong \eta_1 N. \tag{16}$$

If now we focus on smaller scales ($b < R_0$), what is involved is only a chain portion of size b, with a number of monomers:

$$n(b) \cong b^2/a^2. \tag{17}$$

Thus:

$$\eta_{\mathrm{Rouse}}\big|_b \cong \eta_1 n(b) \cong \eta_1 \frac{b^2}{a^2}\,. \tag{18}$$

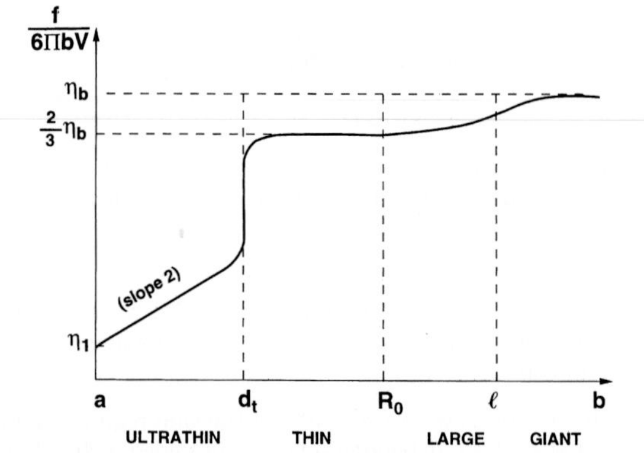

Fig. 3. Log log plot of reduced friction *versus* particle size for a colloidal sphere in a polymer melt.

The friction law for a sphere is then:

$$f/V \cong b\,\eta_{\mathrm{Rouse}}\big|_b = \eta_1 \frac{b^3}{a^2}\,. \tag{19}$$

This friction coefficient is very sensitive to the size b, but independent of the molecular weight of the chains.

In Figure 3, we have summarized our predictions for spheres in all regimes. Of course, for $b = a$, we return to a monomer friction $\eta_1 a$. Around $b = d_{\mathrm{t}}$, we have a dramatic jump of the friction. This is not a surprise: if we think of a permanent network, or equivalently of infinitely long chains (N large), our spheres get completely stuck when their size b is larger than the mesh size. Similar effects have been observed long ago on polymer solutions [5].

3.2 Ultrathin needles in the easy direction

Here again we deal with two distinct friction terms:

(a) frontal friction, due to the distortion of the flow near the ends of the needle, and identical in its scaling structure to what we had for spheres (Eq. (19));
(b) lateral friction, due to slippage with a relative velocity V, and described by equation (12).

Superposing these two terms, (and always ignoring all coefficients) we arrive at:

$$\frac{f}{V} = \frac{\eta_1 bL}{a}\left\{1 + \frac{b^2}{aL}\right\} \qquad (b < d_{\mathrm{t}}). \tag{20}$$

If L is not very large ($L < b^2/a$), frontal friction dominates. If L is large ($L > b^2/a$), lateral friction dominates.

3.3 Ultrathin needles in the hard direction

Assume that our needle moves laterally by some small distance u: it will first induce a Rouse response at scale b. But ultimately, some chains must reptate out of the

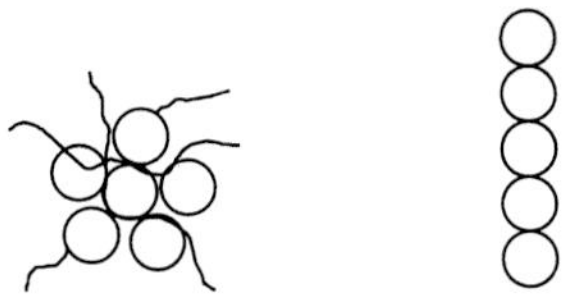

Fig. 4. Clusters of colloidal particles: a) a spheroidal cluster with some chains from the melt trapped in it, b) an elongated cluster

way; this was pointed out long ago [6]. The maximum displacement before relaxation is thus:

$$u = VT_{\mathrm{rep}} \tag{21}$$

where T_{rep} is the reptation time [4]. On this long time scale, the local Rouse response has been smeared out, and the relevant elastic modulus is simply the plateau modulus of the melt:

$$E \cong \frac{kT}{N_{\mathrm{e}}a^3} \cdot \tag{22}$$

This displacement u causes a deformation $\sim u/b$, and a stress $\sigma \sim Eu/b$. The resulting force is the product of the stress by the lateral area ($\sim Lb$). Thus:

$$f \cong LbEu/b = ET_{\mathrm{rep}}LV = \eta_{\mathrm{B}}LV \tag{23}$$

as already announced in reference [6].

Note again the extreme anisotropy of the friction factors (comparing Eqs. (20, 23)).

4 Extensions

4.1 Clusters (Fig. 4)

It may sometimes happen that we do not deal with well separated colloidal particles: instead they may associate in clusters, because of a lack of stability of the colloid.

(a) *Spheroidal clusters* should not always behave like a larger particle: if some melt chains can go through the clusters, they are stuck there for long times (T_{rep}), and cluster should behave like hairy particles (which are discussed below in Sect. 4.2).

(b) *Linear clusters* (needle shaped) may occur in special cases: for instance, if we deal with a magnetic colloid, embedded in a polymer melt, and prepared under a magnetic field, we generate needles along the field [7]. May we consider these needles as similar to the smooth needles discussed in Sections 2 and 3? The answer is yes for most friction problems, with an exception. If we deal with ultrathin needles ($b < d_{\mathrm{t}}$), moving along the easy direction, we find that the friction scales like the superposition of contributions from L/b Stokes spheres moving in a unentangled melt:

$$\frac{f}{V} = \frac{L}{b}\eta_{\mathrm{Rouse}}\Big|_b\, b = \eta_1 \frac{b^2 L}{a^2} \cdot \tag{24}$$

In the limit of large L, this is higher (by a factor b/a) than the friction for a smooth needle (Eq. (20)).

4.2 Hairy particles

Our discussion in Sections 2 and 3 was restricted to "naked" colloidal particles, which do not spontaneously bind any chain from the melt. What happens in the opposite limit? The answer is related to a number of experiments [8,9] and theoretical speculations [2,10] on the suppression of slippage, by grafted chains, at a solid/melt interface.

We can summarize the situation as follows. At low velocities V, the mobile chains in the melt are entangled with the "hair" and this dramatically reduces the slippage. Ultimately, a shell of thickness R_0 (the coil size) is comoving with the colloidal particle. Thus, even if the particle was a small sphere ($b < R_0$), the effective hydrodynamic radius ($b + R_0$), is relatively large: we end up with a friction coefficient $\eta_{\mathrm{B}}(b + R_0)$.

4.3 Memory of alignments for an ultrathin needle

From a friction coefficient $\zeta \equiv f/V$, we can extract a thermal diffusion coefficient $D = kT/\zeta$. Consider for instance an ultrathin needle, moving in the easy direction (Eq. (20)) and (for simplification) of long length $L > b^2/a$. We expect a relatively large diffusion coefficient:

$$D_{\parallel} = \frac{kTa}{\eta_1 bL} \tag{25}$$

independent of the molecular weight of the melt. The time required for complete renewal of the environment of our needle corresponds to a travel length L. We may call this time $\tilde{T}_{\mathrm{rep}}$, the *reptation time of the needle:*

$$\tilde{T}_{\mathrm{rep}} = \frac{L^2}{D_{\parallel}} = \frac{\eta_1 bL^3}{kTa} \cdot \tag{26}$$

During one move of length L, the orientation of the needle may tilt by a small angle θ_1. This was called the "reptation-rotation process" in reference [6]. At much longer times, $t \gg \tilde{T}_{\mathrm{rep}}$, the squared tilt angle is of order:

$$\theta^2 = \frac{t}{\tilde{T}_{\mathrm{rep}}}\theta_1^2. \tag{27}$$

We can define a memory time τ such that for $t \sim \tau$, $\theta \sim 1$. The scaling form of τ is:

$$\tau = \tilde{T}_{\mathrm{rep}}\theta_1^{-2}. \tag{28}$$

The scaling structure of $\theta_1$is non-trivial: in reference [6] we tentatively proposed $\theta_1 \sim d_{\mathrm{t}}/L$. This is the result expected for a periodic, cubic, network, with a rod nearly parallel to a (001) axis. If we believe this, and introduce a microscopic jump time for one monomer in the melt:

$$\tau_1 = \frac{\eta_1 a^3}{kT} \cdot \tag{29}$$

We arrive at:

$$\tau = \tau_1 \frac{L^5 b}{a^4 d_t^2} \cdot \tag{30}$$

Taking $\tau_1 = 10^{-11}$ s, $b = d_t = 50$ Å, $a = 3$ Å, the memory time is of order of seconds for $L = 100$ nm, and is of order of a day for $L = 1$ μm. This time may be important to predict the state of needles in a melt which have been oriented by extrusion or casting of the melt, and which may randomize their orientation before melt cooling.

4.4 Nanotubes and nanofilms

Is it conceivable to force a melt through a nanopore of radius $b < R_0$ – obtained for instance by particle track etching? At first sight, we might think that this probes viscosities at small scales. However, this guess is not very realistic. To avoid clogging, we want the pore surface to be non adsorbing: then the polymer flow will be a *plug flow*, with a pressure gradient ∇_p given by:

$$-\pi b^2 \nabla_p = V \eta_1 a \frac{2\pi b}{a^2} \cdot \tag{31}$$

Similar features are expected for polymer films in wetting processes: for a completely wettable surface, a droplet will spread by emission of a precursor film, which may be thin [11]: but with a smooth surface plug flows should dominate [12]. It may be possible to suppress the slippage, and impose a shear flow, by grafting the surface: the situation is then very complex.

4.5 General conclusion

The dynamics of colloidal particles embedded in a polymer melt raises a large number of questions. Samples are not easy to prepare, and the requirements of colloid stability are stringent. But the diffusion coefficients for ultrathin objects are not too small, and they could be measured by current techniques, such as Forced Rayleigh Scattering. The naive scaling laws proposed here may be of help to plan such experiments.

This reflection, initiated long ago (1976) was reactivated by a discussion meeting with J.F. Sassi, J.P. Marchand, L. Ladouce and Y. Bomal.

References

1. P.G. de Gennes, C.R. Acad. Sci. Paris **288**, 219-222 (1979).
2. F. Brochard, P.G. de Gennes, Langmuir **8**, 3033-3037 (1992).
3. M. Doi, S.F. Edwards, "*The theory of Polymer Dynamics*" (Oxford Science Pub., 1986).
4. For a general introduction on Rouse dynamics and on reptation, see P.G. de Gennes "*Introduction to Polymer Dynamics*" (Cambridge Univ. Press, 1990).
5. D. Langevin, F. Rondelez, Polymer **19**, 875 (1978).
6. P.G. de Gennes, J. Phys. France **42**, 473 (1981).
7. P. Pincus, P.G. de Gennes, Phys. Cond. Matter **11**, 189 (1970).
8. K.B. Migler, H. Hervet, L. Léger, Phys. Rev. Lett. **70**, 287 (1993).
9. V. Mhetar, L.A. Archen, Macromolecules **31**, 6639 (1998).
10. F. Brochard, C. Gay, P.G. de Gennes, Macromolecules **29**, 377-382 (1996).
11. P.G. de Gennes, Rev. Mod. Phys. **57**, 827 (1985).
12. F. Brochard, P.G. de Gennes, J. Phys. Lett. France **45**, 597 (1984).

C. R. Acad. Sci. Paris, t. 324, Série II *b*, p. 343-348, 1997
Solides, fluides : propriétés mécaniques et thermiques/*Solids, fluids : mechanical and thermal properties*

PHYSIQUE / *PHYSICS*

Un muscle artificiel semi-rapide

Pierre-Gilles de GENNES

Collège de France, 11 place Marcelin-Berthelot, 75231 Paris cedex 05, France.

Résumé. Dans sa version initiale (Katchalsky, 1949) un muscle artificiel est fait d'un gel gonflé par un agent chimique (pH, ions, *etc.*). Ceci souffre de deux défauts : la *lenteur* (conditionnée par la diffusion) et la *fragilité* (inhomogénéité des tensions en régime transitoire). Nous proposons ici un type de muscle différent, formé à partir d'une phase lamellaire de polymère tribloc *RNR* (*R* = élastomère, *N* = squelette nématogène). On réticule l'élastomère. Le spécimen composite résultant, exposé à un flash lumineux (dans une bande d'adsorption), peut transiter rapidement, et de façon homogène, d'une phase nématique à une phase isotrope. Il se contracte alors vite par propagation d'ondes de cisaillement depuis les extrémités. En revanche, le retour à l'état de basse température est lent, et accompagné de tensions inhomogènes ; mais la structure composite devrait résister assez bien à ces tensions.

A semi-fast artificial muscle

Abstract. *Katchalsky's (1949) idea for an artificial muscle was based on a gel swollen by a chemical effector (pH, ions,* etc.*). This scheme has two major defects: a)* long response times *(controlled by diffusion); b)* fractures *are often induced by the inhomogeneous mechanical tensions present at intermediate stages. We propose here another scheme, based on the lamellar phase of a triblock copolymer* RNR *(*R *= elastomer,* N *= nematogenic backbone). When the composite resulting from cross-linking the elastomer is exposed to a pulse of light (in a suitable adsorption band) it should go rapidly and homogeneously from the nematic to the isotropic state. It then quickly contracts (*via *acoustic waves from both ends). The reverse process (by cooling) is slow and inhomogeneous, but the composite structure should be rather robust.*

1. Principes

1.1. *L'idée des réseaux nématiques*

Pour améliorer les constantes de temps d'un éventuel muscle artificiel, Hébert, Kant et de Gennes (1996) ont récemment proposé de remplacer les gels gonflés de Katchalsky (1949) ou les gels à échange d'ions de Katchalsky et Zwick (1955) par un polymère nématique réticulé. Ici, la transition nématique–isotrope ($N \rightarrow I$) serait induite par un léger échauffement. L'idée de départ était que les

Note présentée par Pierre-Gilles de GENNES.

diffusivités thermiques sont bien supérieures aux diffusivités d'ions (ou de molécules de solvant). Toutefois, l'étude du front thermique a montré que, à cause de la chaleur de transition *N–I*) les temps de commutation doivent rester assez longs. Par ailleurs, il subsiste plusieurs difficultés.

– On doit garder un nématique *monodomaine*. Des réseaux monodomaines (PDMS-nématogène) ont été construits (*via* une méthode élégante, en deux étapes) par Kupfer et Finkelman (1991). Mais au cours des cycles, la préservation des monodomaines n'est pas évidente. On pourrait peut-être y parvenir par une méthode d'opposition (*fig. 1*), où l'élément actif est en série avec un caoutchouc banal.

– Au cours d'une transition ($I \to N$ ou $N \to I$) il se crée entre les régions *N* et *I* des inhomogénéités de contraintes mécaniques qui risquent d'engendrer des fractures, comme dans le gonflement isotrope d'un gel tel que l'ont observé Tanaka *et al.* (1987) et Matsuo et Tanaka (1992).

1.2. *Le muscle strié semi-rapide*

A partir de la *figure 1,* pour renforcer la structure, on est naturellement amené à envisager un composite « strié », représenté sur la *figure 2.* Ici, le polymère nématique n'est pas réticulé, mais apparaît comme la partie centrale d'un tribloc *RNR*. Par choix convenable de la longueur de la partie *R* (élastomère) on peut en principe s'arranger pour avoir ainsi une phase fluide de type lamellaire. Il doit être possible de réticuler ensuite les parties *R*, en gardant l'arrangement lamellaire.

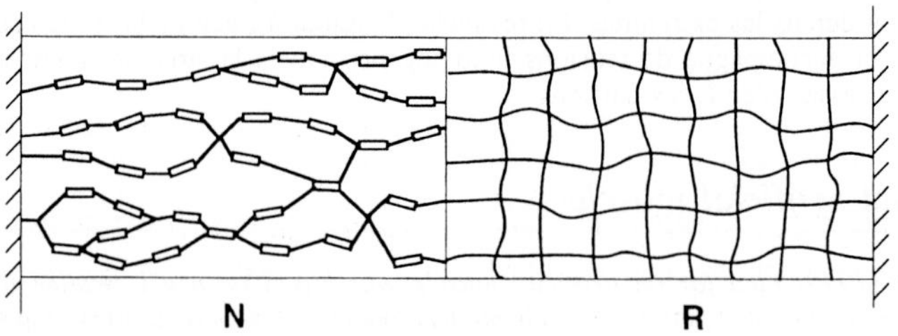

Fig. 1. – Un muscle artificiel basé sur un réseau nématogène (*N*) en série avec un caoutchouc (*R*). Le rôle de *R* est de maintenir (*N*) sous forme de monodomaine.

Fig. 1. – An artificial muscle based on a nematic network (N) *in series with a conventional rubber* (R). *The* (R) *piece helps to maintain* (N) *as a single domain.*

Nous examinons ici quelques propriétés de base d'un tel composite strié. Soulignons que le mot « strié » n'implique pas de similarité avec le muscle strié des vertébrés ou des insectes, qui est un objet très différent et bien plus sophistiqué !

2. Thermodynamique des composites

2.1. *Blocage des déformations*

L'épaisseur des lamelles est très inférieure à la largeur ℓ du spécimen. Dans ces conditions, les composantes transverses (e_{xx}, e_{yy}) du tenseur de la formation (*fig. 2*) sont les mêmes dans les régions *N* et *R*. A une bonne approximation, on peut supposer que les milieux *N* et *R* sont tous deux incompressibles. $e_{xx} + e_{yy} + e_{zz} = 0$. Alors on a nécessairement la même déformation axiale $e_{zz} = e$ dans *N* et dans *R*.

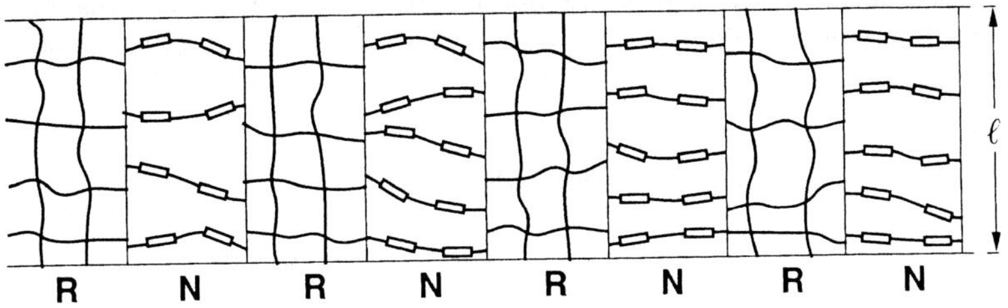

Fig. 2. – Le muscle « strié » fait à partir d'un copolymère *RNR* en phase lamellaire, avec réticulation des parties élastomères (*R*).

Fig. 2. – A 'striated muscle' based on a triblock copolymer RNR *in a lamellar phase – with suitable cross-linking of the elastomer part* R.

2.2. *Rappel de l'énergie libre pour un réseau nématogène* (N)

Nous avons jadis (de Gennes, 1975) proposé une forme simple d'énergie F pour un réseau nématogène.

L'idée est qu'il existe deux variables couplées : les déformations $e_{xx} = e$, et les orientations décrites par un tenseur symétrique de trace nulle $Q_{\alpha\beta}$. Ici, il suffit de considérer $Q_{xx} = Q$. L'énergie libre (par unité de volume) est alors :

$$F_N(e, Q) = F_0 + 1/2\, \mu_N e^2 + 1/2\, A(T)\, Q^2 + 1/6\, B(T)\, Q^3 + \ldots - \Lambda Q e \qquad (1)$$

Ici μ_N est un module élastique de cisaillement. Pour un nématogène réticulé, μ_N dépendrait du nombre de monomères entre points de réticulation. Pour notre composite, les chaînes N sont attachées seulement aux deux bouts, et ce nombre devient le degré de polymérisation X_N de la partie N :

$$\mu_N \sim \frac{kT}{X_N a^3} \qquad (2)$$

où a^3 est le volume par monomère. Dans l'énergie libre de l'équation (1) le coefficient de couplage Λ peut être positif et grand (pour des nematogènes sur le squelette) ou négatif et petit (pour des groupes nematogènes latéraux).

L'expression (1) est justifiée, au niveau du champ moyen, si Q et e ne sont pas trop grands. Elle a été utilisée par M. Warner *et al.*, (1994, 1995) qui lui ont trouvé certaines conséquences inattendues.

2.3. *Énergie libre du composite*

Il suffit ici d'ajouter un terme relatif à l'élastomère R (de volume Ω_R). Si Ω_N est le volume des régions N, on arrive à une énergie (pour le volume $\Omega = \Omega_R = \Omega_N$) :

$$F = F_N\, \Omega_N + 1/2\, \mu_R\, e^2\, \Omega_R \qquad (3)$$

Une conséquence amusante de la présence du composite est que les températures de transition T_{NI}dépendent des fractions en volume (de Ω_N/Ω). Ainsi la température spinodale de l'état isotrope $T_{IN}(S)$ correspond à :

$$A(T_{NI}(S)) = +\Lambda^2 \Omega_N /(\Omega_N \mu_N + \Omega_N \mu_R) \tag{4}$$

2.4. *Diagrammes*

On peut d'abord optimiser l'énergie libre de l'équation (3) par rapport à Q. L'énergie F résultante dépend encore de la déformation e, et de deux variables de contrôle : T (température) et σ (contrainte). L'optimum par rapport à e donne la déformation en fonction de T et σ.

Au-dessus du point d'ordre, et en contrainte nulle, le minimum de F correspond à $e = Q = 0$. En dessous du point d'ordre ($T = T_-$) y a une déformation spontanée (associée au point A de la *fig. 3*) e_A

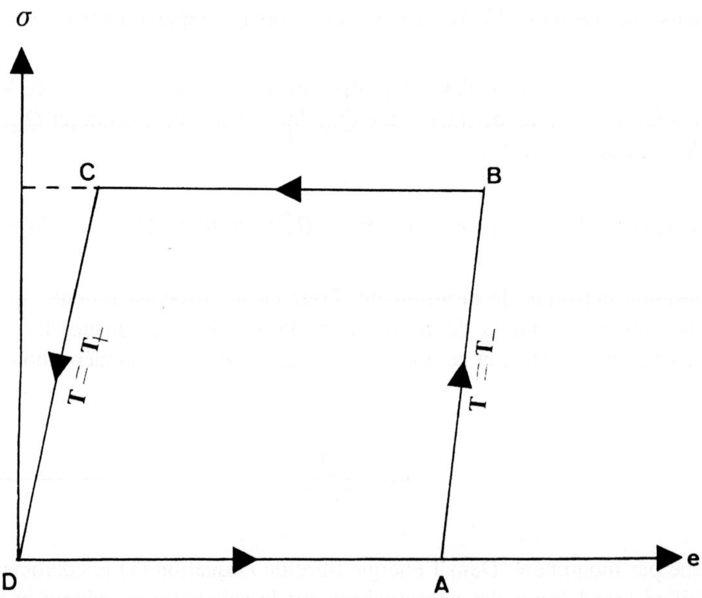

Fig. 3. – Cycle contrainte – déformation : l'étape rapide est l'étape d'irradiation – contraction $B \to C$.

Fig. 3. – Plot of stress versus deformation. The rapid stage corresponds to optical irradiation and contraction, along the path B → C.

$$e_A = \frac{\Lambda Q_A(T_-)\,\Omega_N}{\mu_N \Omega_N + \mu_R \Omega_R} \tag{5}$$

Lorsque T_- est peu inférieure à T_{NI}, on peut remplacer le paramètre d'ordre (en σ nul) $Q_A(T_-)$ dans l'équation (5), par sa valeur à la transition ($T = T_{NI}$), soit $Q(T_{NI})$ (qui est un nombre fixe de l'ordre de 0,44).

3. Cycle de travail

Il est représenté sur la *figure 3*.

– Partons du point A, à une température $T_- < T_{NI}$, et sous contrainte nulle.

– Élevons la contrainte sans changer la température : nous aboutissons au point B.

– Appliquons un pulse lumineux. Si le coefficient d'adsorption n'est pas trop fort, l'énergie injectée $\Omega \, \Delta U$ est répartie uniformément dans le spécimen. Celui-ci chauffe et atteint une température finale $T_+ > T_{NI}$. A cette température, l'ordre nématique est très faible (et induit seulement par la présence d'une contrainte). Donc le spécimen se contracte. (pour $\Lambda > 0$). Nous arrivons au point C.

Nous supposons ici que l'ordre nématique relaxe vite, et que la vitesse de contraction est limitée par la propagation acoustique de signaux dans le spécimen : comme dans un ruban de caoutchouc qui, à l'instant initial $t = 0_-$ est allongé (par une contrainte extérieure). Puis, à partir de $t = 0_+$, la contrainte est supprimée. Ce problème a été discuté par Hébert *et al.* (1996), en incluant les diverses sources d'amortissement acoustique. Mais, indépendamment de ces détails, on peut dire que le front de relaxation avance à partir des extrémités, avec la vitesse caractéristique du mode pair dans une plaque (Lord Rayleigh, 1889). Cette vitesse est double de celle (c) des ondes transversales dans le caoutchouc. (Le facteur 2 vient de ce que la déformation locale du ruban a deux composantes : axiale et transversale). Donc, le temps de relaxation mécanique d'un ruban de longueur L est :

$$\tau_e = \frac{L}{4\,c} \tag{6}$$

– Nous enlevons ensuite la contrainte, et passons du point C au point D.

– Finalement, nous refroidissons (lentement) le spécimen de la température T_+ à la température T_-, et nous retournons au point A : le cycle est complété.

4. Discussion

4.1. *Échelles de temps*

– *Relaxation de l'ordre nématique*

Pour des nématiques non polymères, le temps de relaxation de l'ordre, τ_N, est typiquement de 10^{-7} s. Pour des nématiques polymères à groupes nematogènes dans le squelette, il peut être plus long, peut-être de l'ordre de 10^{-5} s.

– *Relaxation élastique*

Si le modèle élastique μ du caoutchouc est de l'ordre de 1 atm. (0,1 MP) la vitesse $c = (\mu/\rho)^{1/2} \approx 40$ m/s. Donc notre muscle serait effectivement rapide dans sa phase de contraction.

Pour satisfaire aux conditions postulées dans la section **3** :

$$\tau_N < \tau_p < \tau_e$$

il faudrait (par exemple) des durées de pulse de l'ordre de 10^{-4} s.

– *Relaxation thermique*

Si l'épaisseur ℓ du muscle est de 1 mm, et la diffusivité thermique $D \sim 10^{-3}$ cm$^{2/s,}$ les temps d'équilibration thermique en phase isotrope seraient de l'ordre de :

$$\tau_T \sim \frac{1}{\pi^2} \frac{\ell^2}{D} \sim 1 \text{ s} \tag{7}$$

Le temps de refroidissement (avec transition $I \to N$) tel qu'il est discuté par Hébert *et al.* (1996) serait en gros aussi long (parce que la chaleur de transition $N \to I$ est faible).

4.2. *Dimensions spatiales*

– Pour avoir une structure robuste, il convient probablement d'avoir des épaisseurs comparables pour les lamelles N et R : $\Omega_N \sim \Omega_R$.

– Les polymères nématiques (préparés souvent par des réactions de condensation) ne sont pas très longs, par exemple 50 monomères de 20 Å chacun. On peut penser à des épaisseurs de lamelle dans la gamme de 100 nm.

– Si le polymère N a ses groupes nematogènes *dans le squelette,* le couplage orientation-déformation est fort, et les déformations e_A en phase N [(eq. 5)] devraient être de l'ordre de l'unité (pour $\Omega_N \sim \Omega_R$). Si au contraire, le polymère N est nématogène seulement par des groupes latéraux, e_A devrait être nettement plus faible.

4.3. *Remarque*

Un rapporteur anonyme nous a suggéré la possibilité d'un muscle *rapide* par l'usage de certaines photo isomérisations (par exemple *cis* $\to$ *trans*). On peut, dans des cas favorables, provoquer la transition *cis* $\to$ *trans* par irraditation à une longueur d'onde λ_1, et la transition inverse avec une autre longueur d'onde λ_2. Cette possibilité est très intéressante. Il y a une difficulté, d'ordre chimique : les transformations en cause se couplent facilement (en présence d'impuretés, d'oxygène, *etc.*) à des dégradations irréversibles. Ces réactions sont bloquées lorsque la molécule active est inclue dans un verre, mais plus dangereuses dans une matière souple (caoutchoutique). Le muscle rapide est concevable ; mais il n'est pas sûr qu'il soit très robuste.

Nous avons bénéficié de discussions avec J.-C. Daniel, M. Hébert, et R. Kant ; et des remarques innovantes du rapporteur mentionné ci-dessus.

Note remise le 13 décembre 1996, acceptée après révision le 3 février 1997.

Références bibliographiques

De Gennes P. G., 1975. *C. R. Acad. Sci (Paris),* B 281, p. 101-104.

Hebert M., Kant R. et de Gennes P. G., 1996. *Comptes Rendus du colloque de Saint Pétersbourg sur les polymères,* Plenum (à paraître).

Katchalsky A., 1949. *Experientia,* 5, p. 319-330.

Kupper J. and Finkelman H., 1991. *Makromol. Chem., Rapid Comm.,* 12, p. 717-726.

Matsuo F. S. et Tanaka T., 1992. *Nature,* 358, p. 482-485.

Rayleigh Lord, 1889. *Proc. London, Math. Soc.,* 20, p. 225-234.

Tanaka T., Sun S. T., Hirokawa Y., Katayama S., Kucera J., Hirose Y. et Amiya T., 1987. *Nature,* 325, p. 796-798.

Verwey G. C. et Warner M., 1995. *Macromolécules,* 28, p. 4303-4406.

Warner M., Bladon, P. et Terentiev E., 1994. *J. Phys. II,* 4, p. 91-100.

Afterthought: Un muscle artificiel semi-rapide

The title of this paper is ambitious! The hope is to have rapid contraction of the sample when it is heated. However, all the data on bulk "single liquid crystal elastomers" prepared by H. Finkelman *et al.*, suggest that the contraction is slow (~30'). But some hope remains for the nanostructure displayed in Fig. 2: they should have less defects in the nematic regions, and they might switch faster...

(revised version: Sept 1st, 2002)

Weak segregation in molten statistical copolymers

Pierre-Gilles de Gennes

Collège de France, 11 place M. Berthelot, 75231 Paris Cedex 05, France

Summary : we present a qualitative argument suggesting that a statistical (AB) copolymer should display a certain form of weak segregation, with A rich and B rich regions of size comparable to the coil radius. This should occur if the Flory parameter χ is larger than *unity* (much larger than for the phase separation of A and B of homopolymers) in agreement with detailed calculations by Frederikson, Milner and Leibler. We achieve a certain qualitative insight for the resulting microphase.

Keywords : statistical copolymers; segregation; polymer melts; microphase separation

Introduction

Two distinct homopolymers (A…A) and (B…B) are usually non miscible, when the product of the Flory parameter χ by the polymerisation index N is large ($\chi N > 2$).[1-3] This may be understood qualitatively through the work of transfer W of one A chain from a pure A medium to a pure B medium. This work is proportional to the chain length W = NU, where U is the enthalpy increase for the transfer of one A monomer ($U = \chi$ kT, where T is the temperature). Segregation will occur when W >> kT, or $\chi N >> 1$.

In this note, we consider the more complex problem of a melt based on a statistical linear polymer AB AA B… with N monomers, and a completely random chemical sequence. Here, the average enthalpy of mixing is trivial: what matters is the fluctuations of composition inside one chain. This question has been analysed recently by A. D. Litmanovich and coworkers[4] on a model with *two* chains in close contact: for this model, ref. [4] provides a detailed statistical analysis.

Our aim here is to extend the discussion to a real melt, in 3 dimensions, with many chains in interaction. This problem was solved long ago, at the level of the random phase approximation [5], by Frederikson, Milner and Leibler.[6] They set up a Landau theory, where the order parameter is the deviation from average of the A/B concentration. For our case, with no correlations along the chemical sequence, they find a spinodal threshold when $\chi > 2$.

Our main aim, in the present note, is to present a poor man's approach to this segregation problem. This is crude, but possibly helpful, because it gives a certain insight for the resulting mesophase.

Weak segregation

We assume that the sequence A...BB...A of each chain is generated independently. (This is very different from a melt of polypeptides obtained from one same protein, where all the chains have the same sequence). The fraction ϕ of monomers A on one chain has a gaussian distribution around $\phi = 1/2$, with a r.m.s. width

$$\Delta = N^{-1/2} \qquad (1)$$

We divide this distribution into two pieces:

(α) with $\phi > ½$

(β) with $\phi < - 1/2$

The (r.m.s.) excess fraction of A in the (α) group is Δ. In the (β) group it is $-\Delta$.

We call $\psi(r)$ the local volume fraction of the (α) chains (ranging from 0 to 1). The average Flory Huggins energy [1] f of the $\alpha\beta$ mixture (per monomer site) is given by:

$$f/kT = N^{-1} [\psi \ell n\, \psi + (1 - \psi)\, \ell n\, (1 - \psi)] - \chi\Delta^2 (\psi - ½)^2 \qquad (2)$$

This leads to a transition for:

$$\chi > 2N\Delta^{-2} \equiv 2 \qquad (3)$$

Spinodals

Let us now focus on the case where χ is slightly larger than 2. We start at time 0, with a completely disordered melt. Then we expect to observe a spinodal decomposition [5] with a certain caracteristic correlation length ξ_s. The formula for ξ_s may be obtained by adding to eq. (2) a term of order a^2 $(\nabla\psi)^2$ (where a is the size of one unit in the Flory picture). The coefficient of this gradient term is unaltered by our transformation from ϕ to ψ: in the language of ref. [5], the unperturbed correlation function for one chain is always the Debye function $g_D(r)$ [7].

The result (for χ slightly above 2) is:

$$\xi_s = a/3[2 - \chi]^{-1/2} \qquad (4)$$

This gives the size of the regions which should nucleate initially. Indeed, a recent simulation of Houdoyer and Muller [8] (with N ~ 20)shows A rich and B rich regions which may correspond to this.

At very long times, the α and β regions should grow into droplets, which swell by Ostwald ripening. But it is important to realise that the diffusion constants D should be slowed down by the weakness of the constant between A and B. We expect:

$$D \sim D_0 (\chi - 2) \Delta^2 \quad (5)$$

where D_0 is a diffusion constant for $\chi = 0$, which may be of the Rouse type (for small N) or of the reptation type (for large N).

Discussion

1) We see from eq. (3) that weak segregation should occur only for large values of the Flory parameter $\chi > 2$. This corresponds to very strong enthalpies of mixing. Thus all our discussion should be relevant only for some very special AB pairs, possibly using ionic components.

2) We should emphasize the weakness of the proposed segregation: the contrast between (α) and (β) regions is of order $\Delta << 1$, and the scattering intensities vary like $\Delta^2 \sim N^{-1}$.

3) We consider spatial variations in connection with eq. (4).But we can treat only wave vectors q which are small ($q R_0 < 1$, where $R_0 = N^{1/2}$ a is the size of one coil). At $q R_0 > 1$, the physically relevant value of Δ is $\Delta = g^{-1/2}$, where g is the number of monomers in a blob of size $q^{-1} = g = (qa)^{-2}$.

Acknowledgments: all this (conjectural) reflection was initiated by a discussion with Prof. A. Litmanovich. We also benefited from later exchanges with K. Binder, J. Houdayer and L. Leibler, and from an illuminating remark by F. Wyart.

[1] P. Flory, *Principles of Polymer Chemistry*, Cornell U. Press (Ithaca, NY), **1971**.

[2] R. L. Scott, *J. Chem. Phys*. **1949**, 17, 179.

[3] H.Tompa, *Trans. Faraday Soc*.**1949**, 45, 1142.

[4] A. D. Litmanovich, YA. V. Kudryavisev, YO. K.Kriksin, O. A. Kononenko, in *"Molecular order and mobility in polymer systems"*, Proceedings of the 4th International Symposium, St Petersburg **2002** (to be published).

[5] P. G. de Gennes, *Scaling Concepts in Polymer Physics*, Cornell U. Press (Ithaca , NY), 2nd printing, **1985**.

[6] G. Frederikson, S. Milner, L. Leibler, *Macromolecules*, **1992**, 25, 6341.

[7] P. Debye, J. Phys. Colloid. Chem., **1947,** 51, 18.

[8] J. Houdayer, M. Müller, *Europhys. Lett.* **2002**, 58, 660.

Part IV. Interfaces

C. R. Acad. Sc. Paris, t. 287 (9 octobre 1978) Série B – 207

PHYSIQUE DES COLLOÏDES. – *Phénomènes aux parois dans un mélange binaire critique.* Note (*) de **Michael E. Fisher** et **Pierre-Gilles de Gennes** présentée par M. René Lucas.

Au contact d'un mélange binaire critique, une plaque solide perturbe les concentrations sur une épaisseur de corrélation $\xi(T)$. Aux distances $z<\xi$, la perturbation décroît comme $z^{-\beta/\nu}$, ou β et ν sont deux exposants critiques ($\beta/\nu \sim 1/2$). Deux plaques identiques immergées dans le mélange, séparées par une épaisseur $D<\xi$, doivent s'attirer avec une énergie par centimètre carré proportionnelle à $1/D^2$. Cette attraction s'ajoute à celle de van der Waals, et pourrait jouer un rôle dans la stabilité de certains colloïdes.

In a near critical binary liquid mixture, the concentrations near a wall are perturbed over a distance of the order of the correlation length (ξ (T). At distances z smaller than ξ, the perturbation decreases, like $z^{-\beta/\nu}$, where β and ν are critical exponents ($\beta/\nu \sim 1/2$). Two identical plates in a mixture, separated by a thickness $D<\xi$, should attract each other, the interaction decreasing as in $1/D^2$. This attraction is superimposed on the van der Waals forces, and may play a role in the stability of certain colloids.

Au contact d'une paroi solide, un mélange binaire AB aura en général une concentration d'équilibre ([1]) Φ_s différente de la concentration en volume $\overline{\Phi}$. Si le mélange est proche d'un point critique de démixtion ($\overline{\Phi}=\Phi_c$, température $T=T_c$) cette modification n'est pas limitée au voisinage immédiat de la paroi, mais s'étend sur une portée [([2]), ([3])] :

$$(1) \quad \xi(T)=\xi_0 t^{-\nu}, \qquad t=\left|\frac{\Delta T}{T_c}\right|,$$

où $\xi(T)$ est la longueur de corrélation, $\Delta T = T - T_c$ l'écart au point critique. (Pour simplifier, nous discuterons seulement ici le cas $\overline{\Phi}=\Phi_c$ et le domaine monophasique.)

1. DEUX PAROIS PARALLÈLES IDENTIQUES. – Deux parois parallèles de surface A, séparées par une distance D ($\ll A^{1/2}$) sont immergées dans le fluide. On attend alors une partie singulière ΔF dans l'énergie libre, prévue par la théorie des lois d'échelle pour les systèmes finis ([4]) :

$$(2) \quad \frac{\Delta F}{kT} \approx A D\, t^{2-\alpha}\, W\left(\frac{D}{\xi};\frac{h_1}{t^{\Delta_1}}\right),$$

où $h_1=[(\mu_A-\mu_B)_S-(\mu_A-\mu_B)]/kT$ représente l'effet du mur comme une modification du potentiel chimique au contact des surfaces. L'équation (2) est construite pour h_1 petit, mais doit rester correcte même lorsque la valeur de Φ_S devient d'ordre unité. L'exposant $\alpha\,(\sim 0{,}1)$ décrit la divergence de la chaleur spécifique, alors que $\Delta_1\,(\sim 0{,}5)$ est un exposant de surface ([4]).

(*a*) pour $D \gg \xi$, l'énergie ΔF doit contenir un terme indépendant de D associé à une énergie interfaciale γ : ceci impose

$$\lim_{x\to\infty}\{W(x;y)\}=W_\infty + x^{-1}W(y),$$

$$(3) \quad \gamma-\gamma_r(T) \approx -kT\,\xi_0^{-2}\,t^{2-\alpha-\nu} \approx -kT\,\xi^{-2},$$

où $\gamma_r(T)$ est la partie régulière, et où nous acceptons la relation $2-\alpha=\nu d$ pour $d=3$

(*b*) on attend une *attraction* entre plaques. Dans la limite $t\to 0$ (où $D<\xi$), l'équation (2) donne une énergie de couplage par centimètre carré

$$(4) \quad U_{12} \cong -\left(\frac{kT}{\xi_0^2}\right)\left(\frac{\xi_0}{D}\right)^{(2-\alpha)/\nu-1} = -\frac{kT}{D^2}.$$

Dans le cas bidimensionnel, où la chaleur spécifique diverge logarithmiquement, ceci devient $U_{12} \approx k T (\ln D)/D$. Pour $T \neq T_c$ on retourne à la forme $U_{12} \approx -k T/\xi_0^2 \exp(-D/\xi)$. L'attraction (4) se superpose aux forces de van der Waals, et pourrait être mesurée mécaniquement : avec $D = 1\ \mu$ il faut réaliser $t < 10^{-4}$ pour avoir $\xi > D$.

2. Profil de concentration près d'un mur unique. – En dérivant l'équation (2) par rapport à h_1, et en prenant la limite $D \to \infty$, on trouve la loi pour la concentration Φ_S :

(5) $\Phi_S - \bar{\Phi} = m_S = t^{\beta_1} X_S(h_1 t^{-\Delta_1})$,

où $\beta_1 = 2 - \alpha - \nu - \Delta_1 \cong 0{,}8$. Pour un mur indifférent ($h_1 = 0$) on a alors $m_S \approx t^{\beta_1}$ en dessous du point critique. Nous généralisons l'équation ([5]) pour la valeur de $\Phi(z)$ à une distance z du mur

(6) $\Phi(z) - \bar{\Phi} = m(z) = t^{\beta_1} \left(\dfrac{z}{\xi_0}\right)^{\theta} X\left(\dfrac{z}{\xi}; h_1 t^{-\Delta_1}\right)$.

Pour $z = \xi_0$ et $T \to T_c$ l'équation (6) se réduit à l'équation (5) à condition de prendre $X(0; y) = X_c(y)$. Pour $z \to \infty$, $m(z) \to t^{\beta}$, où β est l'exposant de la courbe de coexistence. Ceci impose $\theta = (\beta_1 - \beta)/\nu$. La relation (6) impose à son tour :

(7) $m(z) = t^{\beta} Y\left(\dfrac{z}{\xi}; h_1 t^{-\Delta_1}\right)$.

(*a*) On pourrait éprouver l'équation (7) par ellipsométrie ([5]). Dans l'approximation la plus simple, l'indice local $n(z)$ varie comme $\Phi(z)$, et pour des longueurs d'onde $\lambda > \xi$ on mesure principalement

(8) $\displaystyle\int_0^{\infty} dz\,[n(z) - \bar{n}] \approx t^{\beta-\nu}$.

En principe, il faudrait inclure aussi les anomalies critiques de n ($\Delta n \sim m^{(1-\alpha)/\beta}$) mais comme $(1-\alpha)/\beta > 1$ ces effets sont minimes.

(*b*) Pour $t \to 0$ le profil tend vers une limite

(9) $m_c(z) = h_1^{\beta_1/\Delta_1} z^{\theta} X_c(h_1 z^{\Delta_1/\nu}) = z^{-\beta/\nu} Z(z h_1^{\nu/\Delta_1})$,

où $X_c(0)$ et $Z(\infty)$ sont finis et non nuls. Donc le profil critique décroît de façon universelle (pour $h_1 \neq 0$) et lente ($\beta/\nu \sim 1/2$). Aux courtes distances ($z \sim h_1^{-\beta/\Delta_1}$) les paramètres du mur (h_1) deviennent plus importants.

3. Profil entre deux plaques. – (*a*) On peut généraliser l'équation (7) à ce cas sous la forme

(10) $m(z) = t^{\beta} Y\left(\dfrac{z}{\xi}; \dfrac{D}{\xi}; h_1 t^{-\Delta_1}\right)$.

Ignorant encore les anomalies critiques de la constante diélectrique, ceci donne une anomalie de la *capacité* C entre plaques

(11) $\dfrac{C}{\bar{C}-1} \approx \left(\dfrac{\xi_0}{D}\right)^{\beta/\nu}$ $(D < \xi)$.

C. R. Acad. Sc. Paris, t. 287 (9 octobre 1978) Série B – 209

(*b*) Dans la région centrale le profil est la superposition de contribution des deux murs

$$(12)\quad m\left(\frac{D}{2}\right)\approx\left(\frac{\xi_0}{D}\right)^{\beta/\nu} V\left(\frac{D}{\xi}\right),$$

avec $V(0)=0(1)$ et $V(x\to\infty)\sim e^{-x}$.

(*c*) Par contre, près des murs, l'approximation de superposition est inacceptable. Le paramètre d'ordre est grand et les effets non linéaires sont dominants. Mais les propriétés *locales* d'un mur (spectroscopie infrarouge, ultraviolette, fluorescence, pK...) sont modifiées par la présence de l'autre mur.

Nous présentons ci-dessous une *conjecture* pour l'évaluation de ces effets à $T=T_c$, basée sur une énergie libre (par centimètre cube) de la forme

$$f(z)=\text{Cte}\, m^{\delta+1}\left[1+\left(\frac{\xi m}{m}\frac{dm}{dz}\right)^p\right],$$

où $\delta+1=(2-\alpha)/\beta$ et où $\xi_m=\xi_0 m^{-\nu/\beta}$ est une longueur de corrélation locale. La valeur de l'exposant p n'est pas essentielle pour la suite. Après minimisation de $\int_0^D f(z)\,dz$ on trouve une pente au contact du mur

$$\left.\frac{dm}{dz}\right|_D=\left.\frac{dm}{dz}\right|_{D=\infty}\left[1-\text{Cte}\left(\frac{\xi_0}{D}\right)^3\right].$$

Toutes les corrections aux propriétés locales devraient donc être d'ordre $(\xi_0/D)^3$.

4. Interactions entre grains colloïdaux. – L'attraction U_{12} implique un mécanisme nouveau de flocculation des colloïdes; par exemple, une suspension dans un mélange eau-phénol, stabilisée par la présence de charges sur les grains, pourrait devenir instable près du point critique eau-phénol.

Les considérations précédentes ne sont que qualitatives. Il faudrait ultérieurement préciser les fonctions d'échelle $Y(x;\ y;\ z)$, $Z(x)$, etc.; et aussi étendre la discussion aux concentrations différentes de la concentration critique $\overline{\Phi}$; étudier enfin les régimes biphasiques. Mais il paraît déjà possible d'envisager un programme expérimental sur la physique des interfaces solide/mélange binaire critique.

P.-G. de Gennes a bénéficié de discussions avec Mme L. Ter Minassian-Saraga. M. E. Fisher a bénéficié de l'aide de la National Science Fondation de la John Simon Guggenheim Memorial Foundation.

(*) Séance du 9 octobre 1978.

(1) Nous définissons les concentrations par les fractions molaires $\Phi=A/(A+B)$.

(2) H. E. Stanley, *Introduction to Phase Transitions*, Oxford University Press, 1970.

(3) M. E. Fisher, *Proc. Nobel Symp.*, 24, 1973, p. 16.

(4) M. E. Fisher, *J. Vac. Sc. Tech.*, 10, 1973, p. 665.

(5) F. L. McCrackin, *J. Res. Natl. Bureau of Standards*, 67 A, 1963, p. 363; D. Den Engelson, *J. Phys. Chem.*, 76, 1972, p. 3390.

Cornell University, Baker Laboratory,
Ithaca, New York 14853, *U.S.A.*;
Laboratoire de Physique de la Matière condensée,
11 *place Marcellin-Berthelot* 75231 *Paris Cedex* 05.

Afterthoughts: Phénomènes aux parois dans un mélange binaire critique

1) These scaling ideas have been tested by rigorous calculations in 2 dimensions: H. Au Yang and M. Fisher, *Phys. Rev.* **B21**, 3956 (1980).

2) I am now tempted to visualise the properties of the critical AB mixture, near a single wall, through a self-similar grid of concentration domains.

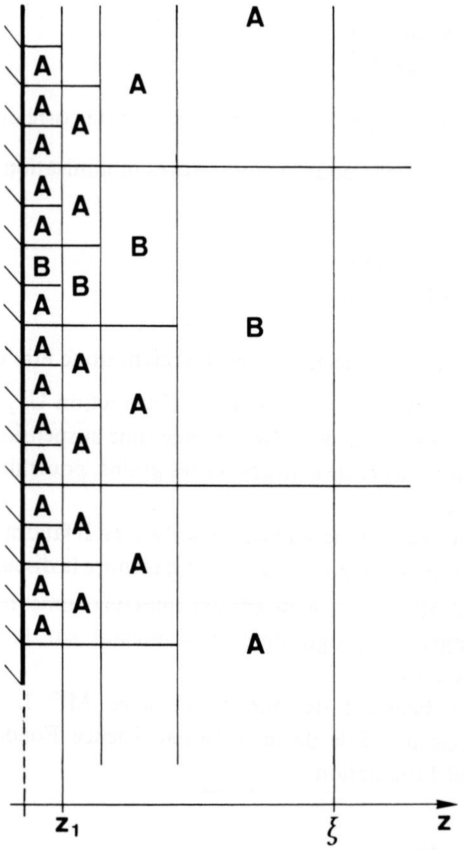

The grid holds from $z = z_1 = h_1^{-\beta/\Delta_1}$ up to $z = \xi$ (the correlation length). At the lower limit, the B fraction is small. At the upper limit, B/A = 1. The (tentative) rule which I use to construct the picture is:

a) Start from the first layer ($z = z_1$) with the correct B/A fraction [related to $m(z_1)$] to construct the second layer.

b) When we have an adjacent AA pair in the first layer, we put an A in the corresponding block of the second layer with probability of $\frac{1}{2}+\varepsilon$ or a B with a probability $\frac{1}{2}-\varepsilon$.

c) When we start with an AB pair, we put an A or a B with equal probability $\frac{1}{2}$.

d) When we start with a BB pair, we put an A with probability $\frac{1}{2}-\varepsilon$, and a B with probability $\frac{1}{2}+\varepsilon$.

e) We iterate: progressively the B fraction raises during the process, and the ratio B/A reaches unity when $z=\xi$.

For simplicity, the picture has $\varepsilon=\frac{1}{2}$. Scaling would predict $\varepsilon=\left(\frac{1}{2}\right)^{(\beta-\beta_1)/\nu}$. (I am indebted to J. Prost for a discussion of this point.)

C. R. Acad. Sc. Paris, t. 288 (11 juin 1979) Série B — 359

PHYSIQUE STATISTIQUE. — *Suspensions colloïdales dans une solution de polymères.* Note (*) de **Pierre Gilles de Gennes,** présentée par René Lucas.

Dans une solution semi-diluée de chaînes flexibles en bon solvant, on ajoute quelques sphères dures (rayon b). On considère ici le cas où il n'y a pas d'adsorption, et où b est plus petit que la longueur de corrélation ξ. On propose des lois d'échelle pour : (1) l'énergie de transfert W d'une bille depuis le solvant pur jusqu'à la solution de polymère; (2) le nombre n de monomères en contact direct avec une bille; (3) les interactions entre deux billes dues à la présence de polymères. On conclut que ces interactions ne suffisent pas à induire une précipitation des billes (contrairement à ce qui arrive pour $b \gg \xi$). Le cas des solutions diluées de polymère est aussi évoqué.

In a semi-dilute solution of flexible chains, with a good solvent, we add a few hard spheres of radius b (assuming no adsorption on the spheres). We choose b to be smaller than the correlation length ξ, and then propose scaling laws for: (1) the energy W of transfer (for one sphere) from the pure solvent to the polymer solution; (2) the number n of monomers in direct contact with one sphere; (3) the interaction between two spheres, due to the ambient polymer. We conclude that this interaction should not, by itself, induce a precipitation of the spheres. The case of dilute solutions is also discussed briefly.

(1) L'effet d'une paroi plane répulsive sur une solution de polymères a été discuté récemment du point de vue des lois d'échelle [1] : on attend une *couche de déplétion*, d'épaisseur comparable à la longueur de corrélation $\xi(c)$ du polymère [2] :

$$\xi(c) \cong a (ca^3)^{-3/4}, \tag{1}$$

où c est la concentration [3]; a, la taille d'un monomère. La contribution de la couche de déplétion à l'énergie interfaciale est d'ordre kT/ξ^2 et la concentration c_s au contact immédiat de la paroi est d'ordre ξ^{-3}.

(2) Ces considérations restent applicables aux parois faiblement courbées — et notamment au cas de sphères dures de rayon $b \gg \xi$. Dans ces conditions :

(*a*) une sphère en suspension a une énergie propre W (due aux polymères) de la forme

$$W \cong 4\pi b^2 (kT/\xi^2) \cong kT (b/\xi)^2 \qquad (b \gg \xi); \tag{2}$$

(*b*) deux sphères au contact ont des couches de déplétion qui se recouvrent sur un domaine de surface $\sim \xi b$ et il en résulte une interaction attractive [1] :

$$U_{12} \cong -(kT/\xi^2)\xi b \cong -kTb/\xi \qquad (b \gg \xi), \tag{3}$$

$|U_{12}|$ est grand devant kT. Donc, en l'absence de toutes autres forces, on prévoit qu'une suspension de billes *flocule*.

(3) Dans la présente Note nous examinons la limite opposée ($b < \xi$). Ici, la sphère est entourée d'une région de déplétion dont l'épaisseur n'est plus d'ordre ξ, mais plus petite (comparable à b) : les lois d'échelle sont très différentes.

(A) Le travail W requis pour extraire une sphère du solvant pur et l'amener dans la solution semi-diluée doit avoir la forme

$$W = kT f\left(\frac{b}{\xi}\right), \tag{4}$$

avec $f(1) \cong 1$ pour se raccorder à l'équation 2. Par ailleurs, aux échelles $< \xi$ qui nous concernent, il n'existe pas de longueur caractéristique autre que b. Ceci implique que W soit simplement proportionnel au nombre de monomères dans la région de déplétion ($\cong cb^3$) donc *linéaire en c*, d'où (par comparaison avec l'équation 1) :

$$W \cong kT\left(\frac{b}{\xi}\right)^{4/3} \quad (b \ll \xi). \tag{5}$$

L'équation (5) peut aussi être comprise par un calcul un peu plus détaillé, dans lequel la sphère n'est pas infiniment dure, mais plutôt est associée à un potentiel répulsif fini $U(\mathbf{r}) = U$ pour $r < b$. On peut calculer les contributions W_1 et W_2 à W qui sont respectivement du premier et du deuxième ordre en U :

$$W_1 = U c \frac{4\pi}{3} b^3, \tag{6}$$

$$W_2 = -\frac{1}{2}\frac{U^2}{kT}\iint_{r, r' < b} \langle \delta c(\mathbf{r})\, \delta c(\mathbf{r}') \rangle \, d\mathbf{r}\, d\mathbf{r}' \cong -\frac{U^2}{kT} cb^3 \left(\frac{b}{a}\right)^{5/3}. \tag{7}$$

Dans ces équations $\langle \delta c \, \delta c \rangle$ est une fonction de corrélation pour les fluctuations de c (en l'absence du potentiel U) pour laquelle nous avons utilisé la forme d'Edwards [4] :

$$\langle \delta c(\mathbf{r})\, \delta c(\mathbf{r}') \rangle \cong c a^{-5/3} |\mathbf{r} - \mathbf{r}'|^{-4/3}, \tag{8}$$

établie pour une chaîne unique, mais aussi valable ici [2] pour $|\mathbf{r} - \mathbf{r}'| \ll \xi$. On peut remplacer la série $W_1 + W_2 + \ldots$ par la forme approchée

$$W = \frac{U c (4\pi/3) b^3}{1 + \lambda (U/kT)(b/a)^{5/3}}, \tag{9}$$

où λ est une constante numérique. Si on retourne alors à la limite $U \to \infty$ (sphère dure) on aboutit à l'équation (5).

(B) La concentration immédiatement au contact de la sphère dure c_s est de la forme

$$c_s = \frac{1}{\xi^3} g\left(\frac{b}{\xi}\right), \tag{10}$$

où $g(1) \sim 1$ pour retrouver les propriétés de la section 1 quand $b > \xi$. Lorsque $b = a$, la sphère devient comparable à une molécule de solvant : pour une telle molécule la probabilité de contact avec un monomère est (pour une solution peu concentrée en polymère) simplement proportionnelle à c :

$$c_s(b = a) \cong c. \tag{11}$$

D'où la forme interpolée

$$c_s \cong \frac{1}{\xi^{4/3} b^{5/3}}. \tag{12}$$

On peut aussi traduire ce résultat au moyen du nombre n de monomères qui sont en contact direct avec la sphère

$$n \cong 4\pi b^2 a c_s \cong c b^{1/3} a^{8/3}. \tag{12'}$$

Le nombre n pourrait éventuellement être mesuré par des mesures de fluorescence, avec des groupes fluorescents fixés à la surface des sphères, et des pièges sur les chaînes.

C. R. Acad. Sc. Paris, t. 288 (11 juin 1979) Série B — 361

(C) Deux sphères voisines s'attirent, ici encore, quand leurs zones de déplétion se recouvrent, et l'interaction V entre elles est, au contact, de l'ordre de $-W$:

$$V(r_{12}) = -kT\left(\frac{b}{\xi}\right)^{4/3} h(r_{12}/b), \tag{13}$$

où r_{12} est la distance entre les centres, et h une fonction sans dimension (d'ordre unité) qui tombe à zéro pour $r_{12} \gg 2b$. Pour $b < \xi$ l'interaction V_{12} n'est que faiblement attractive, et les corrélations résultantes entre billes sont négligeables : il paraît donc possible de réaliser des suspensions du type considéré avec $b < \xi$, sans que la présence du polymère impose une floculation [5].

(4) Ces considérations peuvent être étendues au cas où la solution de polymère est très diluée (pelotes séparées). Dans ces conditions, une petite bille fixée au centre de gravité de la pelote à une énergie propre W_d obtenue à partir de l'équation (5) en remplaçant ξ par le rayon de Flory [6] $R_F \cong N^{3/5} a$ de la pelote isolée. Plus généralement, si la bille est à une distance r du centre de la pelote, elle subit un potentiel répulsif

$$V_{pb}(r) = kT\left(\frac{b}{R_F}\right)^{4/3} l(r/R_F), \tag{14}$$

où $l(x)$ est une autre fonction sans dimensions tombant à zéro pour $x > 1$. Ici encore, pour $b < R_F$, les interactions sont faibles, et on peut considérer que les billes pénètrent librement dans les pelotes. Cette remarque peut être intéressante en vue de certaines mesures de viscosité aux petites échelles spatiales, faites par sédimentation de billes dans une solution diluée [6].

Au total, les systèmes mixtes billes + pelotes statistiques ouvrent peut-être des voies expérimentales assez nouvelles, même dans le cas (étudié seul ici) où aucun phénomène d'adsorption n'intervient.

(*) Remise le 11 juin 1979.

[1] J. F. JOANNY, L. LEIBLER et P. G. DE GENNES, *J. Pol. Sc.* (à paraître).

[2] *Voir* par exemple M. DAOUD et coll., *Macromolecules*, 8, 1975, p. 804. Pour simplifier les lois d'échelle nous supposons que le solvant est *très bon* (dans la notation de Fory $\chi \ll 1/2$).

[3] Nous définissons c comme un nombre de monomères/cm^3.

[4] S. F. EDWARDS, *Proc. Phys. Soc.*, 93, 1965, p. 605.

[5] Bien entendu d'autres effets (comme les forces de Van der Waals entre sphères) peuvent déstabiliser la suspension.

[6] F. BROCHARD, Communication privée.

Laboratoire de Physique de la Matière condensée, Collège de France,
11, *place Marcelin-Berthelot*, 75231 *Paris Cedex* 05.

Conformations of Polymers Attached to an Interface

P. G. de Gennes

Collège de France, 75231 Paris Cedex 05, France. *Received April 10, 1980*

ABSTRACT: We discuss the conformations and the concentration profiles for long, flexible chains (N monomers per chain) grafted at one end on a solid surface (fraction of surface sites grafted σ). The chains are immersed either in a pure (good) solvent or in a solution of the same polymer (P monomers per mobile chain, volume fraction ϕ). It is assumed that the polymer does *not* adsorb on the wall surface. The zone occupied by the grafted chain may contain a large fraction of mobile P chains: we call this a mixed case (M), as opposed to the unmixed case (UM). Also the chains may be stretched (S) or unstretched (US). The combination of these two criteria gives four possible regimes. Using scaling laws, we locate the domains of existence of these four regimes in terms of the variables σ and ϕ. High σ values may be hard to reach by grafting but could be obtained with block copolymers at an interface between two immiscible solvents.

I. Introduction

Polymers grafted onto solid walls can be useful for many physicochemical applications:[1] wetting, adhesion,[2] chromatography,[3] colloid stabilization,[4] and biocompatibility[5] are typical examples. In the present paper, we discuss theoretically some conformation problems for grafted, flexible polymers immersed in *good* solvents. The situation which we have in mind is described on Figure 1: here, a set of linear chains (with N monomers per chain) is attached to a wall and immersed in a liquid which may be either a pure solvent or, more generally, a solution of the same polymer (with P monomers per chain). We assume that all chains are *uncharged*: this eliminates some important practical situations but is logical in view of the difficulties found in understanding polyelectrolyte conformations in solution. We also assume *no adsorption*: the chains are not attracted to the wall. The opposite case can be treated and has in fact been discussed in some limits[6] but is obscured by the increase in number of relevant parameters.

A global Flory–Huggins theory for the selective properties of a set of grafted chains with respect to solvent mixtures has recently been constructed.[7] Our aim here is somewhat different:

(a) We are mainly concerned with conformational properties and with spatial distributions.

(b) We wish to cover situations where the fraction σ of grafted sites on the solid surface is large (up to unity). Of course, these situations are not easy to achieve on a solid wall. But they may be reached with *monolayers of block copolymers* at an interface between two solvents.

(c) We allow for a liquid phase which contains mobile polymer chains (P chains), but we restrict our attention to P chains which are *chemically identical* with the N chains. This excludes a number of interesting chromatographic effects[7] but preserves some important physical questions: (i) the concentration (or volume fraction ϕ) of the mobile chains can be used to modify the properties of the grafted layer; (ii) the limit of a dense polymer melt ($\phi \to 1$) which is included here is of interest for certain applications, such as incompatible polymer mixtures doped by block copolymer additives.[8]

Clearly, the statistical problem raised by these rather complex interfaces is very delicate. However, it is possible to delineate the main *qualitative* features by comparatively simple arguments based on scaling laws.[9] We do this here in one limit, namely, when the system is athermal ($\chi = 0$ in the Flory–Huggins[10] notation). Also we restrict our attention to mobile chains which have lengths comparable to (or smaller than) that of the grafted chain ($P \lesssim N$).

There are three successive steps in our discussion. The case of grafted chains plus *pure solvent* is treated in section II. Here, most of the relevant scaling laws have already been constructed by Alexander:[6] we add only a few novel features, such as the concentration profile close to the wall. In section III, we discuss the opposite limit of a grafted wall in contact with a polymer melt. Of particular interest here is the progressive expulsion of the mobile chains from the grafted film when the graft density σ increases. Finally in section IV we deal with semidilute solutions: here we find that the regime of section II is prevalent when $\phi < \sigma^{2/3}$. In the opposite limit ($\phi > \sigma^{2/3}$) there are still two different possibilities, one with penetration of the grafted layer by the P chains and one without penetration.

The main conclusions are summarized in Figure 2, showing the different regimes which we expect for any given ϕ and σ. It should be immediately pointed out that the boundaries between successive regimes are not sharp: they correspond in reality to smooth crossovers.

II. Grafted Chains plus Good Solvent

A. Separate Coils. The limit of low σ is particularly simple (Figure 3): each chain occupies roughly a half-sphere with a radius comparable to the Flory radius for a coil in a good solvent[10]

$$R_F = N^{3/5}a \qquad \text{(II.1)}$$

where a is a monomer size (a is the mesh size in the lattice model of Figure 1). Here, the different coils do not overlap: thus we must have $\sigma a^{-2} R_F^2 < 1$ or

$$\sigma < N^{-6/5} \qquad \text{(II.2)}$$

Let us now discuss the *average* profile $\phi(z)$ for a random distribution of grafting points on the wall. We call z the

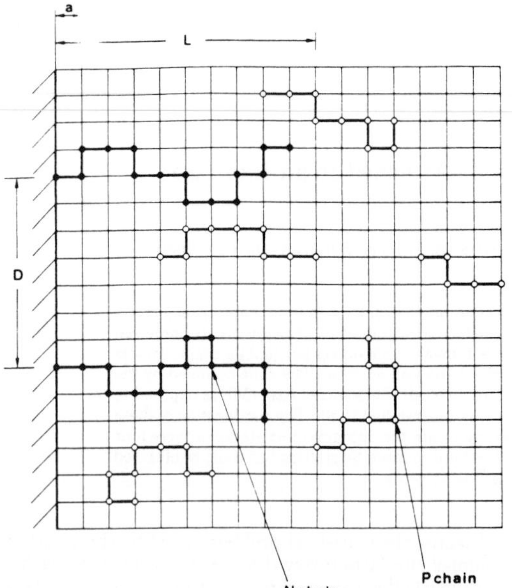

Figure 1. Lattice model for grafted chains (N monomers per chain) and mobile chains (P monomers per chain) in an athermal solvent. On this picture $N = 15$ and $P = 9$. In the text we are concerned with larger values ($N \gtrsim 10^3$ and $P \gtrsim 10^2$).

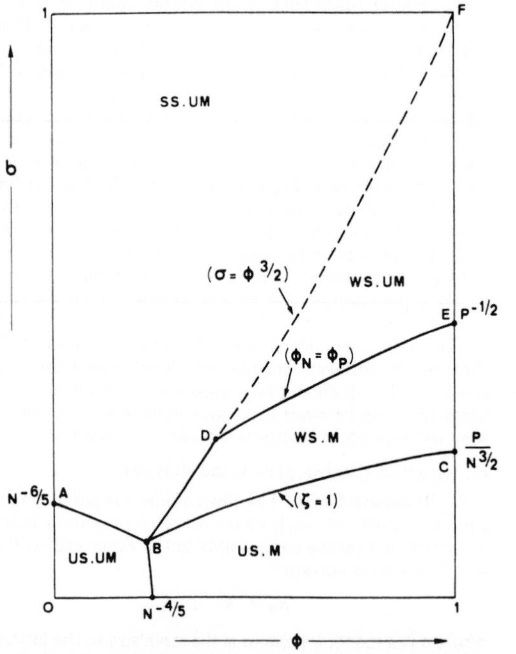

Figure 2. Various regimes for the grafted layer (density of graft points σ (in dimensionless units)) in the presence of a solution of P chains (volume fraction of P chains in the bulk ϕ). Symbols: S = stretched, SS = strongly stretched, WS = weakly stretched, US = unstretched, M = mixed (penetrated by the P chains), UM = unmixed. The figure is drawn for the case $N^{1/2} < P < N$ and is only qualitative.

distance to the wall and focus our attention on the interval $a \ll z \ll R_F$. At the upper limit ($z \sim R_F$) we expect a

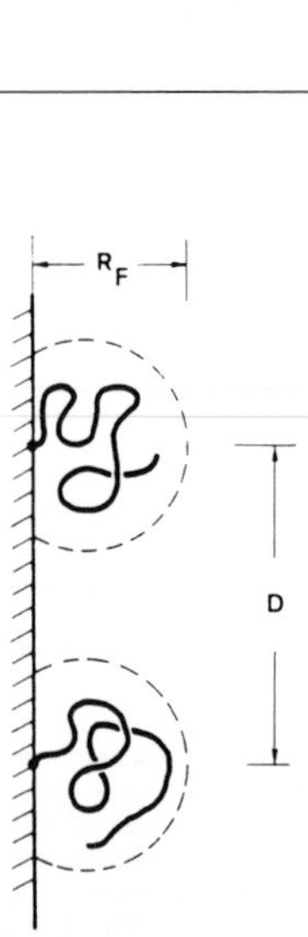

Figure 3. Low density of grafted points ($\sigma \to 0$) and pure solvent.

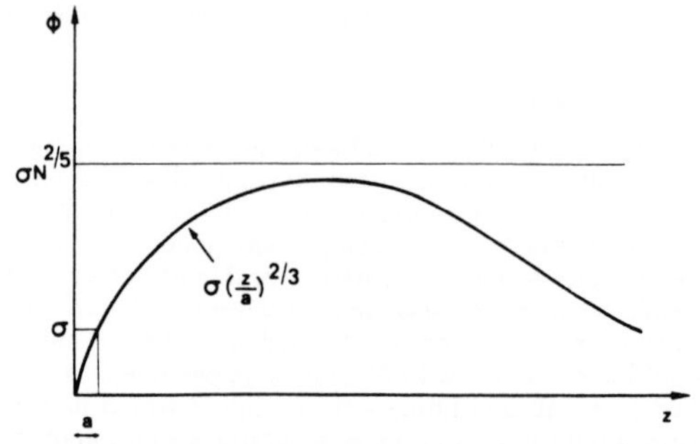

Figure 4. Average concentration profile for the situation of Figure 3.

concentration equal to the concentration inside a single coil (N/R_F^3) times the fraction of wall area occupied by coils $(\sigma/a^2)R_F^2$. Thus

$$\phi(z = R_F) \simeq N\sigma a/R_F = \sigma N^{2/5} \quad \text{(II.3)}$$

At the lower limit ($z \sim a$) we should return to the fraction of grafted points σ. Interpolating between these two ends by a power law

$$\phi(z) = \sigma(z/a)^m \quad \text{(II.4)}$$

(with an unknown exponent m) and imposing the condition (II.3), we arrive at

$$\sigma(R_F/a)^m = \sigma N^{2/5}$$

$$m = 2/3 \quad \text{(II.5)}$$

The resulting profile is very different from what we know for solutions of mobile chains.[11] Our conclusions for the profile are summarized in Figure 4. Note that for $z > R_F$ the concentration drops out very fast.

B. Overlapping Coils. This is the most interesting regime. It is obtained when $D < R_F$ or $\sigma > N^{-6/5}$. The general physical picture for this case (shown on Figure 5) is due to Alexander.[6] He has shown that the fundamental distance D of the problem is the average distance between grafted sites on the surface. In terms of the grafting fraction σ, we may define D through

$$D = a\sigma^{-1/2} \quad \text{(II.6)}$$

A grafted chain may be subdivided into "blobs" of linear size D, each of them containing a number g_D of monomers. Since at small scales ($r < D$) the correlations are dominated

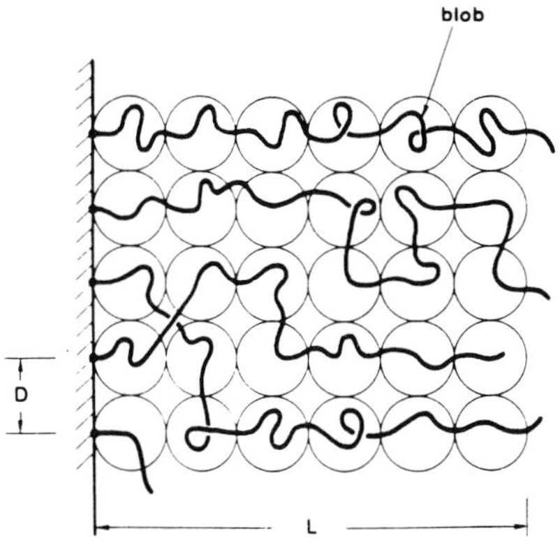

Figure 5. Strongly stretched situation for a grafted layer in a good solvent. The chains are mainly stretched along the normal to the wall.

by excluded-volume effects, the relation between g_D and D is of the form (similar to II.1)

$$a g_D^{3/5} = D \quad \text{(II.7)}$$

It is immediately seen that when $D \ll R_F$ (overlapping regime), we have $g_D \ll N$. In the region occupied by the grafted chains, the blobs act as hard spheres and fill space densely. Thus the polymer concentration is

$$a^{-3}\phi_N \simeq g_D/D^3 \quad \text{(II.8)}$$

Inserting (II.6) and (II.7) into (II.8) we get

$$\phi_N \simeq \sigma^{2/3} \quad \text{(II.9)}$$

The thickness L of the grafted layer is then immediately derived: the volume per grafted chain is LD^2 and this contains N monomers. Thus

$$\phi_N a^{-3} = N/(LD^2) = N/(La^2\sigma^{-1})$$

$$L \simeq N a \sigma^{1/3} \quad \text{(II.10)}$$

Note that L is proportional to the molecular weight. This corresponds to a regime of *stretched chains* (S).

It is also possible to write this thickness in the form

$$L \simeq (N/g_D)D \quad \text{(II.11)}$$

showing that the chain is mainly a linear string of blobs stretched along the normal to the wall. There is, however, a random-walk component parallel to the wall plane: as shown on Figure 5, each chain is expected to show a (root mean square) spread along these directions of magnitude

$$\Delta x \simeq (N/g_D)^{1/2} D \quad \text{(II.12)}$$

In this stretched regime the concentration profile is essentially flat, except for two adjustment regions at the end (Figure 6). The profile near the wall is still given by eq II.4 and II.5 but this equation applies only in an interval $a < z < D$. For $z > D$ we recover the concentration $\sigma^{2/3}$ of eq II.9.

To summarize, we find two extreme regimes, one unstretched and one stretched, depending of the graft density σ. In both regimes the concentration profile is predicted to show a depletion layer near the wall, although each chain must reach the wall.

III. Grafted Chains plus Polymer Melt

We now consider a less familiar case where the grafted chains (N chains) are in contact with a pure polymeric liquid (P chains). Thus at any distance z from the wall

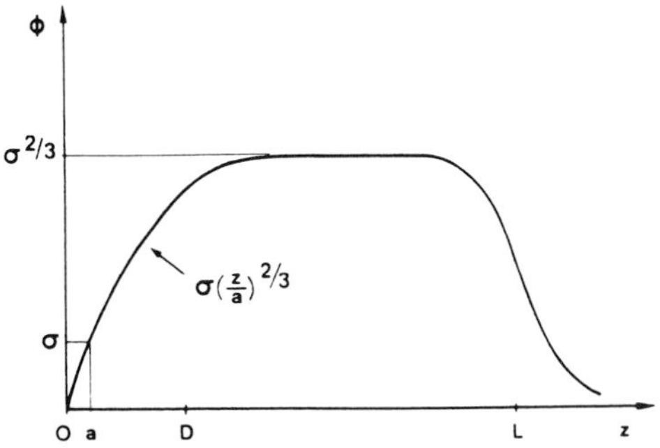

Figure 6. Concentration profile for a grafted layer immersed in a good solvent in the overlapping regime.

we have to define two volume fractions ϕ_N and ϕ_P, with $\phi_N + \phi_P = 1$.

A. Low-Density Limit. Here, the N chains do not overlap. One major difference with section II is that the N chains, being immersed in a sea of P chains, are *ideal* (provided that $P > N^{1/2}$).[9,10] For such a case, it is possible to construct the concentration profiles exactly. We shall not, however, insist on these technical points. The central fact is that each chain occupies a region of linear dimension $R_0 = N^{1/2}a$. The grafted chain concentration scales like

$$\phi_N(z) = \sigma z/a$$

$$a < z < R_0 \quad \text{(III.1)}$$

and reaches a maximum of order $\sigma N^{1/2}$ at R_0. This is correct since it corresponds to N monomers distributed in a region of volume D^2R_0.

The limit of validity of this regime is defined by $\sigma a^{-2} R_0^2 < 1$ or

$$\sigma < N^{-1} \quad \text{(III.2)}$$

Thus the density ϕ_N is always small ($<N^{-1/2}$). This is an important check of our starting assumption: mobile chains can screen out the interactions between the grafted chains only if ϕ_P dominates over ϕ_N.

We conclude that when eq III.2 holds, the grafted chains are mixed with the mobile chains and are unstretched (symbol US.M).

B. Onset of Stretching. If we increase the density σ of grafted points beyond the limit (III.2), the grafted coils begin to overlap, and this may lead to stretching. However, as we shall see, there is an intermediate domain where the chains overlap but where their interactions are still negligible: thus, up to a certain value σ_1 we still retain a grafted layer of thickness R_0. For $\sigma > \sigma_1$ the chains do stretch. Later, when we reach a second crossover region ($\sigma = \sigma_2$), the mobile chains are progressively expelled from the grafted layer.

The threshold σ_1 can be obtained from a consideration of the ζ parameter used in perturbation calculations of excluded-volume effects.[13] Essentially ζ measures the average repulsion energy inside one chain divided by the thermal unit kT. In our case the monomers have a "bare" excluded volume $v = a^3$ and a "screened" excluded volume $\tilde{v} = a^3P^{-1}$. (For a detailed discussion of screening by long chains, see ref 14 and 9.) The ζ parameter is then of order

$$\zeta = N\phi_N a^{-3}(a^3P^{-1}) \quad \text{(III.3)}$$

corresponding to N monomers, each of them feeling a repulsive potential proportional to the concentration ($\phi_N a^{-3}$) and to $\tilde{v}kT$. We estimate (III.3) by assuming that the chains are ideal and spread over a thickness R_0. This corresponds to $a^{-3}\phi_N = N/(D^2R_0)$ and gives

$$\zeta = \sigma N^{3/2}P^{-1} \quad \text{(III.4)}$$

We conclude that whenever σ is smaller than

$$\sigma_1 = PN^{-3/2} \quad \text{(III.5)}$$

the ζ parameter is small, the interactions are weak, and the ideal-chain assumption is correct. But when $\sigma > \sigma_1$ the chain must stretch.

C. Expulsion of the Mobile Chains. The calculations based on a ζ parameter are essentially limited to dilute N coils ($\phi_N \ll 1$). We shall now write down a slightly more general analysis, allowing for arbitrary values of ϕ_N and $\phi_P = 1 - \phi_N$. Our calculation here is extremely rough: it does not reach the scaling accuracy of the Alexander argument in section II but is based rather on a Flory type of self-consistency.

The starting point is a form of the free energy (per N chain) containing two physical contributions: (a) an entropy of mixing between P chains and N chains, which tends to swell the grafted chains, and (b) elastic terms for the grafted chains, which limit the swelling.

The usual expression for the entropy of mixing per lattice site in the Flory–Huggins scheme is $-\Delta S_{mix} = (1/N)\phi_N \ln \phi_N + (1/P)\phi_P \ln \phi_P$. However, for grafted N chains, we must drop out the first term, which is normally associated with translational freedom. Thus we shall write the corresponding free energy per grafted chain, F_{mix}, in the form

$$\frac{F_{mix}}{kT} = \frac{LD^2}{a^3}\frac{1}{P}\phi_P \ln \phi_P \quad \text{(III.6)}$$

where L is the thickness of the grafted layer (and is yet unknown) and LD^2 is the volume per grafted chain (LD^2/a^3 being the corresponding number of lattice sites).

To this we must add an elastic energy, which we take in the simplest form

$$\frac{F_{el}}{kT} = \frac{3}{2}\left(\frac{L^2}{R_0^2} + \frac{R_0^2}{L^2}\right) \quad \text{(III.7)}$$

The first term (dominant at large elongations) corresponds to ideal-chain behavior under stretch, while the second term gives the correct ideal-chain behavior under compression[15] (except for numerical coefficients). The whole expression is not perfect, but comparing with other known situations, we are led to hope that the inaccuracies in (III.6) and (III.7) cancel out to some extent.[9] Because of these difficulties, we shall omit all coefficients such as the factor $^3/_2$ in eq III.7.

We must now minimize the sum $F_{mix} + F_{el}$, keeping in mind the constraints

$$\phi_N = 1 - \phi_P = Na^3/(LD^2) \quad \text{(III.8)}$$

The result, expressed in terms of ϕ_N, has the form

$$kP\sigma^2[1 - (\phi_N^2/N\sigma^2)^2] = \phi_N \ln(1 - \phi_N) + \phi_N^2 \quad \text{(III.9)}$$

where k is a coefficient of order unity. The condition (III.9) is qualitatively displayed in Figure 7 as a plot of ϕ_N vs. $P\sigma^2$. There are three regions in the curve.

(a) When $\sigma < \sigma_1$ (σ_1 being defined in eq III.5), the relation reduces to $\phi_N^2 = N\sigma^2$ and this simply corresponds to $L \simeq R_0$, i.e., to unstretched chains.

(b) When $\sigma_1 < \sigma < \sigma_2$, where

$$\sigma_2 = P^{-1/2} \quad \text{(III.10)}$$

it is possible to expand the right-hand side of eq III.9 in powers of ϕ_N; then the leading term (of order ϕ_N^3) dominates. Physically this means that we have $L \gg R_0$ but still $\phi_N \ll 1$. We associate this with the symbol S.M.

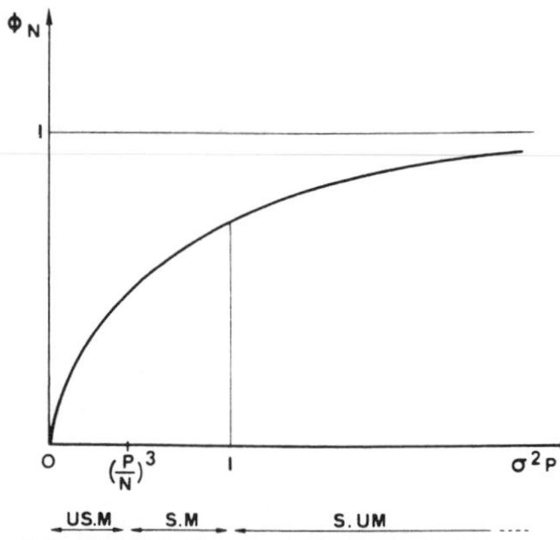

Figure 7. Relation between the grafted chain concentration (ϕ_N) and the density of grafted points (σ) for a grafted layer in contact with a polymer melt (mobile chains of P monomers per chain, with $P < N$) (qualitative plot).

(c) When $\sigma \gtrsim \sigma_2$, ϕ_N and ϕ_P become comparable: we find a progressive expulsion of the P chains, and we reach a regime UM.S. Ultimately for $\sigma \to 1$ the N chains are completely stretched and segregated from the melt.

It is of importance to note that

$$\sigma_1/\sigma_2 = (P/N)^{3/2} \leq 1 \quad \text{(III.11)}$$

Thus when all chains become equal in length, the intermediate regime b becomes evanescent. For $P > N$ we cannot use the simple form (III.6) of the mixing entropy: the P chains are too large, and a more detailed spatial calculation would be required.

IV. Semidilute Solutions

Let us start from the pure-solvent case of section II, assuming that the chains *do overlap* and build up a certain swollen layer. We now add mobile chains to this solvent, imposing a polymer volume fraction $\phi_P = \phi$ far from the wall. What happens to the grafted layer? Two regimes are clearly present, depending on the ratio ϕ/ϕ_{N0}, where $\phi_{N0} = \sigma^{2/3}$ was the concentration inside the grafted layer for the pure-solvent case (eq II.9).

A. Low Concentration of Mobile Chains. Strongly Stretched Regime. Here the layer is still essentially described by the analysis of section II. The layer thickness L is large and is given by eq II.10. The overall concentration profile does not differ much from Figure 6 (except for the presence of a small nonzero value at $z > L$).

We shall briefly discuss the *partition* of the mobile chains between the bulk solution and the layer. This can be calculated in detail by the propagator methods of Edwards,[16] adjusted for a case where blobs of size D are the basic unit. We give here a rough, but simplified calculation. Physically (for $P < N$) there are two regions of importance: one inner region, where standard partition arguments apply, and one outer region, where the chains penetrate partially (see Figure 8).

1. Outer Penetration Layer. Consider first the case of one mobile chain which has *partially* entered in the layer, ν monomers being inside and $P - \nu$ outside. Scaling theorems tell us that the work W_ν required to bring it from the low-ϕ solution outside to this situation is essentially kT per blob[9]

$$W_\nu = (\nu/g_D)kT \quad \text{(IV.1)}$$

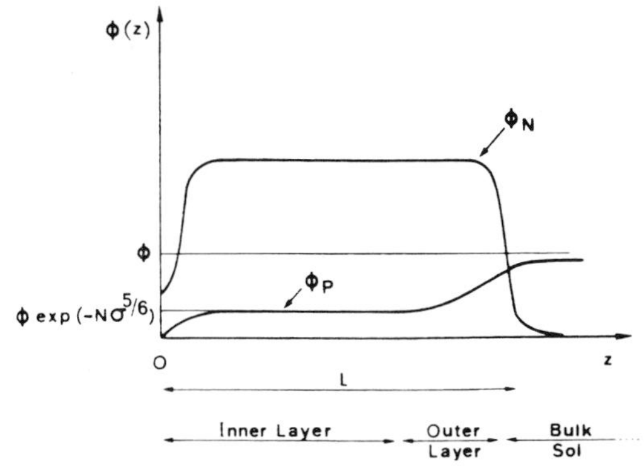

Figure 8. Concentration profiles in the strongly stretched regime.

(Note that there is no important contribution to W_ν originating on the other side by extraction from the bulk solution, since $\phi \ll \phi_{N0}$.) Omitting all prefactors, we write

$$\frac{\phi_P(z)}{\phi} \simeq \sum_{\nu=1}^{P} \exp\left(-\frac{W_\nu}{kT}\right) H\left[(L-z)^2 \Big/ \frac{\nu}{g_D} D^2\right] \qquad \text{(IV.2)}$$

The function $H(u)$ must be unity when u is small, i.e., when a random walk of ν blobs can reach point z. But $H(u)$ must be zero for the opposite case. It is natural (and in fact correct) to assume that $H(u) = \exp(-u)$. Then we can compute (IV.2) by saddle-point integration and find that the leading contributions come from

$$\nu / g_D = (L - z)/D \qquad \text{(IV.3)}$$

giving

$$\phi_P(z)/\phi \simeq \exp[-(L-z)/D] \qquad \text{(IV.4)}$$

The result (IV.4) is meaningful only when $\nu < P$ or, equivalently, in a certain outer layer

$$L - z < P/g_D = aP\sigma^{1/3} \qquad \text{(IV.5)}$$

Note that the thickness of the outer layer is *linear in P*: in this layer the mobile chains are stretched as much as the grafted chains.

2. Inner Region. When the inequality IV.5 is reversed, our eq IV.2 ceases to be valid: penetration occurs through mobile chains which are entirely immersed in the grafted layer; for these chains $\nu = P$ and there is no factor equivalent to $H(0)$. Thus we arrive at

$$\phi_P = \phi \exp(-N/g_D) \simeq \phi \exp(-N\sigma^{5/6})$$

This is the type of result which is discussed in ref 7.

B. Transition from Strong Stretching to Weak Stretching. Let us now assume that the bulk polymer concentration ϕ is larger than the threshold $\sigma^{2/3}$. We shall see that this leads to a *decrease* of the layer thickness L: we expect a (progressive) transition from strong stretching to weak stretching.

There are, in fact, two possible schemes for this process, and both of them can be physically obtained. We discuss them successively below.

1. Mobile Chain Dominance (Figure 9a). In this scheme, we reach a situation where the N chains are still a minority component, with $\phi_N \ll \phi_P$ in the grafted layer. Then ϕ_P is not very different from the bulk concentration ϕ. We say that, in this regime, the mobile chains are *dominant*. They impose, in particular, the blob size: this

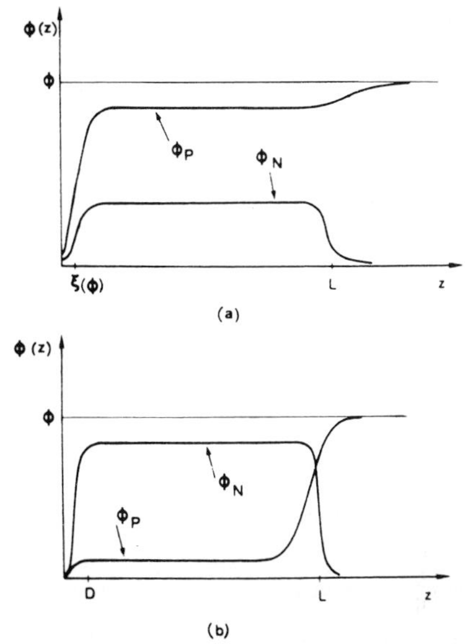

Figure 9. Two regimes of weak stretching: (a) mobile chain dominance $\phi_P \gg \phi_N$; (b) grafted chain dominance $\phi_P \ll \phi_N$.

becomes equal to the correlation length ξ associated with the overall semidilute solution

$$\xi = a\phi^{-3/4} \qquad \text{(IV.6)}$$

The corresponding number of monomers per blob is

$$g = \phi^{-5/4} \qquad \text{(IV.7)}$$

For all properties which involve scales larger than the blob size, we can transpose immediately our discussion of section III for melts, treating the solution as a *melt of nonoverlapping blobs*.[9] For instance, the ζ parameter of eq III.3 becomes

$$\zeta = \left(\frac{N}{g}\right)^2 \frac{\xi^3 g}{P} \frac{1}{D^2 R(\phi)} \qquad \text{(IV.8)}$$

where N/g is the number of blobs per grafted chain, ξ^3 is the bare excluded volume per blob, and g/P is the *screening factor*. (For a different discussion of these factors, see ref 12.) Finally $R(\phi)$ is the ideal size for a string of N/g blobs, each of size ξ

$$R^2(\phi) = (N/g)\xi^2 \qquad \text{(IV.9)}$$

It is essential to appreciate the difference introduced here by screening effects: for $\phi < \sigma^{2/3}$ we had an unscreened repulsion between blobs of size D, leading to very strong stretching. But for $\phi > \sigma^{2/3}$ (and if we are in a case of mobile chain dominance) the P chains penetrate in the grafted layer and screen out the interactions between the N chains: then the thickness of the grafted layer is significantly reduced.

Let us first write down the ζ parameter explicitly. Inserting (IV.6), (IV.7), and (IV.9) into (IV.8), we arrive at

$$\zeta = \sigma N^{3/2} P^{-1} \phi^{-7/8} \qquad \text{(IV.10)}$$

The condition $\zeta = 1$ corresponds to a crossover between unstretched and stretched chains. This defines a certain line BC in the (ϕ, σ) plane of Figure 2

$$\sigma = \phi^{7/8} P N^{-3/2} \qquad \text{(IV.11)}$$

Note that for $\phi = 1$ this agrees, as it should, with the corresponding result of section III.

Let us now assume that we are in the region $\zeta > 1$ (always maintaining our assumptions $\phi > \sigma^{2/3}$ and $\phi_P \gg \phi_N$) and compute the thickness L in this weakly stretched regime. The starting point is a free energy per chain, F, with the qualitative form

$$\frac{F}{kT} \simeq \frac{L^2}{R^2(\phi)} + \left(\frac{N}{g}\right)^2 \frac{\xi^3 g}{P} \frac{1}{LD^2} \quad \text{(IV.12)}$$

After optimization with respect to L this gives

$$L \simeq \xi \frac{N}{g}\left(\frac{\xi}{D}\right)^{2/3}\left(\frac{g}{P}\right)^{1/3} \quad \text{(IV.13)}$$

$$= \sigma^{1/3} N P^{-1/3} \phi^{-5/12} \quad \text{(IV.14)}$$

Note that this value does *not* cross over to eq II.10 when $\phi = \sigma^{2/3}$. The screening effect (described by $P^{-1/3}$ in eq IV.14) disappears when we return to the regime of section IV.1.

Let us now discuss the internal self-consistency of our approach. We assumed $\phi \simeq \phi_P \gg \phi_N$, but we can now compute the concentration due to grafted chains ϕ_N inside the grafted layer from eq IV.14. The result is

$$\phi_N = \sigma^{2/3}(P/g)^{1/3} = \sigma^{2/3} P^{1/3} \phi^{5/12} \quad \text{(IV.15)}$$

Our assumption of mobile chain dominance is correct only when $\phi_N \ll \phi$ or

$$\sigma < P^{-1/2} \phi^{7/8} \quad \text{(IV.16)}$$

This defines the line DE in Figure 2. Our next task is to find out what happens for σ values above this line.

2. Grafted Chain Dominance (Figure 9b). The second scheme, which can also be realized when $\phi > \sigma^{2/3}$, corresponds to $\phi_N \gg \phi_P$, i.e., to an unmixed (UM) situation.

This automatically implies

$$\phi_N = \phi \quad \text{(IV.17)}$$

The argument is the following: in the bulk solution each P chain sees a certain repulsive potential due to the other P chains, which we call $\mu_{rep}(\phi)$. Scaling imposes

$$\mu_{rep}(\phi) = kTPg^{-1} \simeq kTP\phi^{5/4} \quad \text{(IV.18)}$$

in the semidilute region, but we shall not need this precise form here. In the grafted layer, if the P chains are dilute, each of them sees a repulsive potential which is due only to the N chains and is the same function $\mu_{rep}(\phi_N)$. The equilibrium condition for the P chains then requires

$$\mu_{rep}(\phi) = \mu_{rep}(\phi_N) \quad \text{(IV.19)}$$

and this leads to (IV.17).

In this situation, we can easily find out the thickness L of the grafted layer: we must have $LD^2\phi_N = N$ and, using eq IV.17, this gives

$$L = \sigma N \phi^{-1} \quad \text{(IV.20)}$$

If we compare this form to eq IV.14, we see that the two predictions coincide when $\sigma = P^{-1/2}\phi^{7/8}$, in agreement with our definition (IV.16) of the crossover line DE (Figure 2). We conclude that above the line DE the second scheme is indeed realized: we have a weakly stretched, unmixed situation (WS.UM). One final remark concerns the crossover SS.UM $\leftrightarrow$ WS.UM on the line $\sigma = \phi^{3/2}$. If we compare eq IV.19 and II.10, we find that there is also a smooth crossover for $L(\sigma, \phi)$ on this boundary.

Thus the behavior of the grafted layer thickness—when we increase ϕ beyond the line $\phi = \sigma^{2/3}$—is very different in the two schemes. In the first scheme (line portion BD) we expect a rather abrupt collapse of the grafted layer. In the second scheme (line portion DF) the contraction is more gradual.

V. Conclusions

How can we test the simple ideas which have been presented above?

(1) For the pure-solvent limit, the thickness L of the grafted layer can be measured by hydrodynamic techniques[17] (investigating either the permeability of fine pores or the mobility of small spheres with a grafted surface). The main laws to be checked here are the linear dependence of L on molecular weight in the overlapping regime and the $\sigma^{1/3}$ dependence on graft density (eq II.10). It may be difficult to achieve σ values well beyond the overlap criterion ($\sigma > N^{-6/5}$): rather than bringing polymer chains to the surface and attaching them there, it may be preferable to synthesize them in situ, from an initiator site on the surface. Another approach employs a liquid interface, with amphiphilic block copolymers anchored at the interface, instead of a solid wall: here, each half of a diblock spreads in its preferred solvent and plays the role of a grafted chain. Our discussion may be useful in discussing the *detergent properties of block copolymers*.

(2) When some solute P chains are added, it is of interest to study their penetration in the grafted layer. One parameter that can be obtained by ellipsometry,[18] the local refraction index

$$n(z) = n_s + n_1[\phi_P(z) + \phi_N(z)] = n_s + n_1\phi(z) \quad \text{(V.1)}$$

measures only the global polymer concentration $\phi(z)$. A more sophisticated method would be based on neutron diffraction.[19] Using deuterated N chains, for instance, one can first (in principle) determine ϕ_N. Then deuterating the solvent, one obtains from a second experiment an independent combination of ϕ_N and ϕ_P. However, this requires large samples with a sizable surface-to-volume ratio and is clearly not an easy experiment.

(3) For the case of polymer melts, deuteration would again by very helpful. Another method would imply a slight generalization of our model: if the N chains and P chains are chemically different but compatible (as is known for a few favorable cases), our analysis should remain meaningful. Using two optical wavelengths, one could then, in principle, measure both ϕ_N and ϕ_P by ellipsometry.

Of course, in all these situations, we can also think of *local studies* by EPR, NMR, fluorescence, etc.[20] However, the interpretation of the resulting data is much more complex and is sometimes complicated by special effects due to the "label" molecule itself.

The essential conclusion is derived from Figure 2: in view of the many states predicted here it is absolutely necessary to start from a relatively *simple model system*; then there is hope for reaching a clear-cut classification and a good physicochemical picture for all states, provided that the parameters σ, N, P, and ϕ are varied systematically. On the other hand, all more delicate effects related to poor solvents, adsorption, etc. will probably not be understood fully before the end of this first step.

We particularly need model systems with *long chains* ($N \gtrsim 100$). Most practical systems have $N < 20$ (either in chromatography or for conventional detergents). However, a fundamental understanding of the amphiphilic properties of detergents is still lacking: the chains are too long to be analyzed by fully detailed molecular models, but

Macromolecules **1980**, *13*, 1075–1080

at the same time they are too short to display the general scaling laws discussed here. What is needed is a systematic effort based on block copolymers.

Acknowledgment. The author has greatly benefited from discussions with R. Audebert, R. Cantor, C. Taupin, and C. Quivoron.

References and Notes

(1) M. Rosoff, "Physical Methods in Macromolecular Chemistry", Vol. 1, B. Carroll, Ed., Marcel Dekker, New York, 1967, p 1.

(2) (a) R. Stromberg, "Treatise on Adhesion and Adhesives", Vol. 1, R. Patrick, Ed., Marcel Dekker, New York, 1967, p 1; (b) L. H. Lee, Ed., "Adhesion Science and Technology", Plenum Press, New York, 1975.

(3) M. Hennion, C. Picart, M. Caude, and R. Rosset, *Analysis*, **6**, 369 (1978).

(4) S. Ash, *Chem. Soc. Spec. Publ.: Colloid Sci.*, 1, 103 (1973); B. Vincent, *Adv. Colloid Interface Sci.*, **4**, 193 (1974).

(5) "The Chemistry of Biosurfaces", Vol. 1 and 2, M. Hair, Ed., Marcel Dekker, New York, 1971.

(6) S. Alexander, *J. Phys. (Paris)*, **38**, 983 (1977); see also P. G. de Gennes, "Solid State Physics", Seitz and Turnbull, Eds., Academic Press, New York, 1978, Suppl. 14, p 1.

(7) J. Lecourtier, R. Audebert, and C. Quivoron, *Macromolecules*, **12**, 141 (1979).

(8) R. E. Cohen and A. Ramos, *Macromolecules*, **12**, 131 (1979).

(9) For a general introduction, see P. G. de Gennes, "Scaling Concepts in Polymer Physics", Cornell University Press, Ithaca, N.Y., 1979.

(10) See P. Flory, "Principles of Polymer Chemistry", Cornell University Press, Ithaca, N.Y., 1971.

(11) J. F. Joanny, L. Leibler, and P. G. de Gennes, *J. Polym. Sci.*, **17**, 1073 (1979).

(12) P. G. de Gennes, *J. Polym. Sci., Symp.*, **61**, 313 (1977).

(13) H. Yamakawa, "Modern Theory of Polymer Solutions", Harper and Row, New York, 1972.

(14) S. F. Edwards, *Proc. Phys. Soc., London*, **88**, 265 (1966).

(15) E. Casassa, *J. Polym. Sci., Part B*, **5**, 773 (1967).

(16) See, for instance, P. G. de Gennes, *Rep. Prog. Phys.*, **32**, 187 (1969).

(17) See, for instance, M. Garvey, T. Tadros, and B. Vincent, *J. Colloid Interface Sci.*, **55**, 440 (1976).

(18) R. Stromberg, D. Tutas, and E. Passaglia, *J. Phys. Chem.*, **69**, 3955 (1965).

(19) E. Boue et al. in "Neutron Inelastic Scattering 1977", Vol. I, IAEA, Vienna, 1978, p 63.

(20) See, for instance, H. Hommel, L. Facchini, A. Legrand, and J. Lecourtier, *Eur. Polym. J.*, **14**, 803 (1978); H. Hommel, A. Legrand, J. Lecourtier, and J. Desbarres, *ibid.*, **15**, 993 (1979).

Afterthought: Conformations of polymers attached to an interface

This paper follows the pioneering reflection of S. Alexander (Ref. 6) who showed that mean field and scaling theories give the same result for the thickness of a wet brush in good solvents. The whole field has expanded considerably during the last few years: see the review by A. Halperin, M. Tirrell and T.P. Lodge, *Advances in Polymer Science* **100** (Springer, Berlin, 1991). But the present paper did introduce some basic notions: "mushrooms" or "brushes", "dry" or "wet" and the state diagram of Fig. 2.

For a long time the theoretical emphasis was mainly on brushes — but mushrooms may be more relevant in practice: a) they are easier to make; b) they can show rather spectacular mechanical effects, such as a coil-stretch transition in simple shear (F. Brochard, P. G. de Gennes, *Langmuir* — special issue in honor of Jean Perrin, 1992, to be published).

C. R. Acad. Sc. Paris, t. 290 (23 juin 1980) Série B — 509

PHYSIQUE STATISTIQUE. — *Sur une règle de somme pour des chaînes polymériques semi-diluées près d'une paroi.* Note (*) de **Pierre Gilles de Gennes**, Membre de l'Académie.

Dans une Note précédente [1] nous avions proposé une entropie de confinement S_c pour des chaînes entre deux parois répulsives, en milieu fondu ou semi-dilué. Nous analysons ici plus en détail le champ self consistent près d'une paroi. La « couche de déplétion » d'épaisseur ξ, présente près du mur [2] crée une force *attractive* qui s'oppose à l'effet répulsif de la paroi nue : la compensation est exacte et chaque chaîne doit montrer aux grandes échelles ($x > \xi$) un comportement de chaîne libre. L'entropie S_c est donc identiquement nulle.

In an earlier Note [1] *we proposed an entropy of confinement* S_c *for chains in a melt, or in a semi-dilute solution confined between two repulsive walls. We now investigate more closely the self-consistent field near one wall. The depletion layer* [2], *of thickness* ξ, *present near the wall, creates an* attractive *force which counteracts the repulsive effect of the bare wall. The cancellation is exact: each chain must show a on large scale* ($x > \xi$) *a macroscopic behavior which is not influenced by the wall. The entropy* S_c *thus vanishes identically.*

La question des conditions aux limites imposées par une paroi sur une chaîne polymérique *isolée* est assez bien comprise [3]. Pour une chaîne isolée idéale de N segments (avec une longueur statistique par segment a) le poids statistique $G_N(\mathbf{r}_0, \mathbf{r})$ correspondant à une chaîne allant de $\mathbf{r}_0$ à $\mathbf{r}$, est régi par une équation de type Schrodinger, écrite en premier par S. F. Edwards [4] :

$$-\frac{\partial G}{\partial N} = -\frac{a^2}{6}\nabla^2 G + \frac{U(\mathbf{r})}{kT} G, \tag{1}$$

où $U(\mathbf{r})$ est le potentiel agissant sur chacun des N segments. Dans nos problèmes de paroi, $U(r) \to U(x)$ dépend seulement de la distance x à la paroi, et devient infiniment répulsif pour $x \leq 0$. Par contre, pour $x > 0$, $U(x)$ peut avoir une partie attractive de portée l finie. On a alors trois possibilités :

(1) Si la partie attractive est faible, le potentiel $U(x)$ n'a pas d'état lié. Dans ces conditions, les fonctions d'onde associées $\varphi(x)$, c'est-à-dire les fonctions propres de l'équation

$$E\varphi = -\frac{a^2}{6}\frac{\partial^2 \varphi}{\partial x^2} + \frac{U(\mathbf{r})}{kT}\varphi \tag{2}$$

ont un comportement simple à basse « énergie » ($E \to 0$) :

$$\varphi(x \gg l) = (\text{Cte})(x + b). \tag{3}$$

La longueur d'extrapolation b est positive et finie dans le cas (1). Aux distances macroscopiques ($x \gg l$) la structure (3) revient à imposer au poids statistique G une condition de valeur nulle en $x = -b$. Pour deux parois séparées par une distance D grande devant b (mais plus petite que la taille naturelle $N^{1/2} a$ des chaînes) on a alors des modifications intéressantes de G, conduisant à une entropie de confinement [5] (par monomère).

$$S_c = -k\frac{\pi^2}{6}\left(\frac{a}{D}\right)^2. \tag{4}$$

(2) Si la partie attractive du potentiel U est suffisamment forte, il apparaît un *état lié* : physiquement, ce cas correspond à une adsorption de la macromolécule par la paroi [2]. Ici $b < 0$.

510 — Série B C. R. Acad. Sc. Paris, t. 290 (23 juin 1980)

(3) Le point de transition entre ces deux comportements correspond à $b=\infty$ (état lié d'énergie nulle). Dans un tel cas, la condition aux limites macroscopiques pour les poids statistiques G est une condition de *pente nulle*, et l'entropie S_c n'est plus présente.

Nous allons maintenant transposer cette discussion au cas d'une chaîne dans une solution *semi-diluée*, de concentration égale à *c* loin des parois. Notre discussion sera basée sur l'approximation du champ self-consistent [6]. Si $\psi(x)$ est le paramètre d'ordre local, relié à la concentration locale par $c(x)=\psi^2(x)$, il est régi par l'équation (2), mais ici $U(x)$ est la somme du potentiel extérieur U_{ext} et d'un potentiel self-consistent :

(5) $$U_{sc}(x)=k\,T\,v\,\psi^2 \qquad (v=\text{paramètre de volume exclu}).$$

Considérons par exemple le cas où U_{ext} se réduit à une paroi infiniment répulsive pour $x\leqq 0$. La solution du problème self-consistent pour ψ est alors connue

(6) $$\psi(x)=c^{1/2}\,th(x/\xi).$$

Elle décrit une déplétion [2] dans une couche d'épaisseur

(7) $$\xi=a/(3\,vc)^{1/2}.$$

Le potentiel U_{sc} est donc nul pour $x \lesssim \xi$ et tend vers une constante positive pour $x \gg \xi$: la portée *l* du potentiel est d'ordre ξ. Notons que la force associée à U_{sc} est une *attraction* vers le mur. Nous savons en fait quelle est la longueur d'extrapolation *b* pour ce potentiel. En effet, la solution (6) est la fonction d'onde d'énergie égale au minimum du spectre continu ($E_{min}=vc$). Or cette solution tend vers une constante pour $x \gg \xi$. Donc $b=\infty$ dans notre problème, et nous sommes *automatiquement* dans le cas (3).

La conclusion physique est que, dans une solution semi-diluée, près d'un mur, le champ self consistent dû aux modifications de concentration s'ajuste pour compenser exactement les effets du mur nu : la condition aux limites macroscopiques à appliquer aux poids statistiques est toujours une condition de pente nulle, et on n'attend *pas* d'interactions remarquables entre deux parois séparées par une distance $D \gg \xi$, contrairement à ce que nous avions prédit dans [1].

Cette conclusion paraît très générale; elle s'applique aussi au cas où les parois sont attractives : ici la densité est augmentée près des parois, et le champ self consistent développe une force répulsive qui compense l'attraction initiale. Nous sommes en présence d'une « règle de somme » un peu comparable, par sa généralité, à la règle classique de J. Friedel pour les impuretés dans les alliages métalliques [7].

(*) Séance du 2 juin 1980.

[1] P. G. de Gennes, *Comptes rendus*, 289, série B, 1979, p. 103.

[2] J. F. Joanny, L. Leibler et P. C. de Gennes, *J. Polymer Sc. (Physics)*, 17, 1979, p. 1073.

[3] P. G. de Gennes, *Rep. Prog. Phys.*, 32, 1969, p. 187.

[4] S. F. Edwards, *Proc. Phys. Soc. (London)*, 85, 1965, p. 613.

[5] E. F. Cassasa, *J. Polymer Sc.*, B 5, 1967, p. 773.

[6] *Voir* par exemple P. G. de Gennes, *Scaling Concepts in Polymer Physics*, Cornell Univ. Press, Ithaca, New York, chap. 9, 1979.

[7] *Voir* par exemple J. Friedel, *Nuovo Cimento Suppl.*, 7, 1958, p. 287.

Laboratoire de Physique de la Matière condensée,
Équipe associée au C.N.R.S., n° 542, *Collège de France*, 75231 *Paris Cedex* 05.

2294 J. Phys. Chem. 1982, 86, 2294–2304

FEATURE ARTICLE

Microemulsions and the Flexibility of Oil/Water Interfaces

P. G. De Gennes* and **C. Taupin**

Collège de France, 75231 Paris Cedex 05, France *(Received: November 30, 1981; In Final Form: February 1, 1982)*

The phase diagram of a system of oil + water + surfactant is usually dominated by a variety of regularly organized phases (lamellar, hexagonal, etc.) which are highly viscous. However, in some favorable cases, these organized phases are less stable than a certain "microemulsion" where no periodicity occurs. These microemulsions are much more fluid than the regularly organized phases. They often exist over a broad domain of concentrations. In some limiting cases a microemulsion is simply made of swollen micelles (of oil in water or water in oil). However, various experiments indicate that "bicontinuous" structures also occur. Our aim is to understand why a random structure of this type does not collapse into an ordered phase. The interface saturated by surfactant has a nearly vanishing surface tension; one essential parameter is then the elastic constant K describing the curvature elasticity of the interface. The "persistence length" ξ_K of the interface increases exponentially with K. This should have some important effects. (1) When K is above a certain critical value K_c the interfaces tend to stack or, more generally, to build up a periodic, stable, phase. (2) When K is below K_c the interface can become extremely wrinkled and the resulting gain in entropy is larger than the loss of energy due to the departure from a periodic array. This case $K < K_c$ would correspond to microemulsions. In this picture one major effect of cosurfactants (additives which favor the microemulsion phase) is to increase the flexibility of the layers.

I. Distinct Features of Microemulsions

Mixtures of oil and water are naturally unstable, but can be stabilized by addition of suitable "surfactants". These molecules contain a polar group (soluble in water) and an aliphatic tail (soluble in oil). They optimize their interactions by standing at the oil–water interface and decrease drastically the interfacial energy γ. (Similar reductions are observed at the air–water interface (see Figure 1).)

(1) In most cases this decrease of γ stops at some point; if we add a larger amount of surfactant (Figure 1c) the value of γ stays roughly constant and is definitely nonvanishing. The reason is that, beyond a certain limiting bulk concentration, the added surfactant does not go to the interface but prefers to stay in one of the bulk phases in the form of micelles (Figure 1c). With finite but small interfacial tensions γ, one can make *emulsions* of oil droplets in water (O/W) or of water droplets in oil (W/O). Here the droplets are rather large (10 μm) and they are metastable (they coalesce slowly).

(2) Some surfactants, however (or some surfactant *mixtures*: surfactant + "cosurfactant"), have a different behavior; by increasing their bulk concentration, it is possible to reach a state of *zero interfacial tension*, without being blocked by a previous micelle formation. A system of this sort will tend to increase the total area of interface between oil and water. This leads to highly divided systems; they fall into two classes.

(a) In most cases the oil and the water regions form a *periodic array* based on lamellae, or on rods, or on more complex objects.[1–4] These arrays or "macrocrystals" have been well characterized by Bragg peaks in low-angle X-ray crystallography, at least for relatively simple cases such as the soap–water systems.[2,3] Several orders of reflection have been observed in favorable cases. On a macroscopic scale, the periodic structures behave like weak solids, yielding to plastic flow, and giving very high apparent viscosities. This is enough to rule them out for many industrial applications.

(b) In a few favorable cases (e.g., for special surfactants) the interfacial sheets do not build up a regular array but are distributed at random. At low concentration of oil (or of water) this occurs in the form of swollen micelles (Figure 2a,c). But at intermediate concentrations the shapes are probably much more complex (Figure 2b).[5] In all these cases the system behaves like a *transparent* fluid of *low viscosity*; it is called a microemulsion.[6] Transparency is a consequence of the size of the oil (or water) regions (which is of order 100 Å, much smaller than an optical wavelength). The low viscosity expresses the fluid character of the overall structure; it is a favorable feature for most applications.

These observations suggest that two main conditions must be met for the successful achievement of a microemulsion: (i) the surfactants must prefer to remain at the oil–water interface rather than to form separate objects inside one of the bulk phase (as they do in Figure 1c), and (ii) the oil–water interface, saturated with surfactants, must not build up a periodic network; *the macrocrystal must melt*.

Unfortunately, it is very difficult to discuss quantitatively principles (i) and (ii). Our aim in the present paper is first to review some of the basic experimental facts leading to these principles (section II). Then we analyze

(1) P. Ekwall in "Advances in Liquid Crystals", Vol. 1, G. Brown, Ed., Academic Press, New York, 1971, p 1.
(2) V. Luzatti and F. Husson, *J. Cell. Biol.*, 12, 207 (1962).
(3) C. Madelmont and R. Perron, *Bull. Soc. Chim. Fr.*, 425 (1974).
(4) G. J. Tiddy, *Phys. Rep.*, 57, 1 (1980).
(5) The most recent and detailed images of the structure (obtained by freeze etching and electron microscopy) can be found in J. Biais, M. Mercier, P. Bothorel, B. Clin, B. Lalanne, and B. Lemanceau, *J. Microsc.*, 121, 169 (1981).
(6) L. M. Prince in "Surfactant Science Series", Vol. 6, Part I, K. J. Lissant, Ed., Marcel Dekker, New York, 1976, Chapter 3.

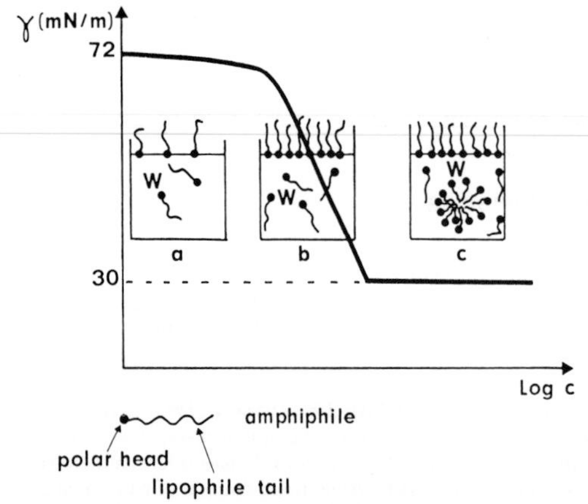

Figure 1. Interfacial behavior of amphiphiles at the air–aqueous solution interface: (a) the pressure in the dilute adsorbed film decreases progressively the surface tension γ; (b) the surface tension γ decreases abruptly and the film becomes compact; (c) when micelles appear the surfactant/water system is buffered; γ remains constant.

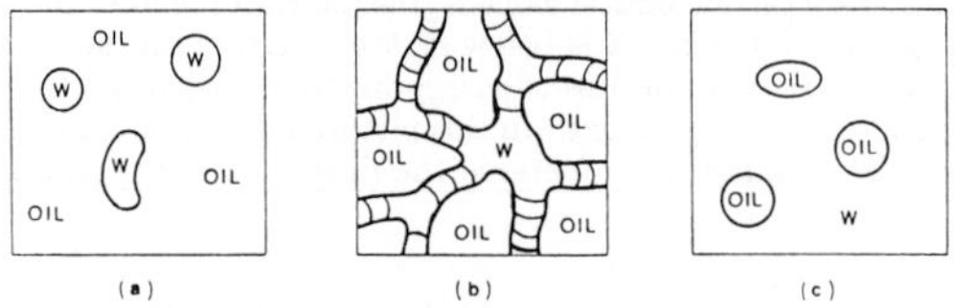

Figure 2. Structure of microemulsions as a function of the water-to-oil ratio (w/o): (a and c) swollen micelles (respectively w/o << 1 and w/o >> 1; (b) "bicontinuous" structures, first proposed by Scriven.[69] The example shown here can be described as a network of water tubes in an oil matrix. (In this figure the individual surfactant molecules are not shown; the surfactant film is represented as a continuous sheet.)

the main properties of a large saturated interface following the early ideas of Schulman. We extend his model, including in particular the entropy of a fluctuating interface (section III). In section IV we discuss the persistence length of the interface. In section V we arrive at the second principle and conclude that (ii) is satisfied if the interface is very *flexible*. Finally, in the Appendix we try to relate the flexibility to various types of measurements: magnetic birefringence, rheology, etc.

The ideas presented in this paper are conjectural and their predictive power is low; we are not able to guess what type of surfactant/cosurfactant mixture will give rise to high flexibilities. Our aim is simply to define certain important parameters, not to predict their actual value.

II. A Selection of Experimental Facts

1. The Ordered Phases. As mentioned in section I, the most common structure for concentrated systems involving water and/or oil plus surfactants is a periodic structure ("macrocrystal"). We discuss this now in some more detail.

a. Water/Surfactant Systems. Figure 3 gives the phase diagram of the system sodium laurate + water, which is a classic example.[3] When the laurate is dilute, we find micelles; but at higher concentrations (around room temperature) we find a hexagonal crystal of rods (middle phase) and a lamellar crystal of bilayers (neat phase).[7] In

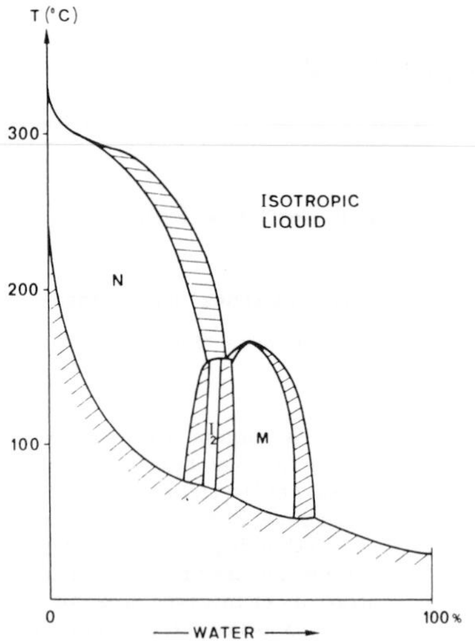

Figure 3. Phase diagram of the sodium laurate/water system (simplified plot, after Madelmont and Perron, ref 3). The "neat" (N) phase is lamellar, the "middle phase" (M) hexagonal, and the intermediate (I_2) phase is cubic. The shaded areas are two-phase regions.

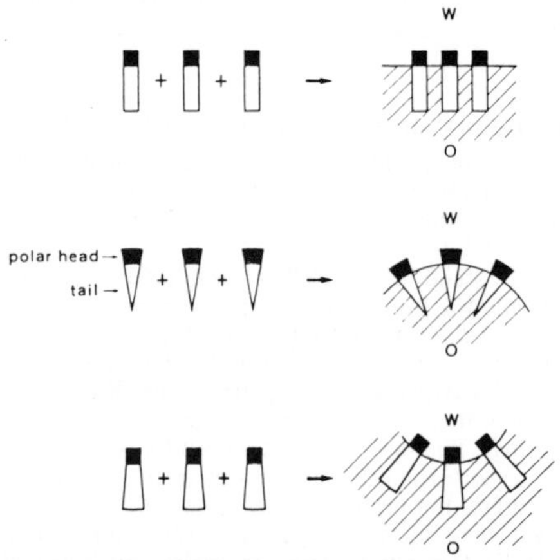

Figure 4. A naive steric model correlating the shape of the amphiphile to the spontaneous curvature of the interface.

many cases we also find (at low water content) inverted phases with a water sphere, or cylinder, surrounded by aliphatic regions. Above a certain temperature limit, the aliphatic chains in all these systems are liquidlike; we shall be concerned only with this regime, which is the relevant one for our purposes. Note that the macrocrystals are special: ordered and periodic at large scales (100 Å) but disordered and liquid like at the atomic level (3 Å). What are the stabilizing forces? One essential parameter is the area Σ of interface per surfactant molecule. In a zeroth order approximation, Σ is roughly constant for all surfactant concentrations. In an improved approximation it has been noted that Σ increases regularly with the water content.[2]

Another relevant parameter is the curvature $1/R$ of the interface counted as positive for a direct micelle and negative for an inverse micelle (here R is the radius of curvature). For ionic surfactants with a small polar head and an aliphatic tail, the curvature is largely controlled by

(7) The lamellar phase is smectic and is a liquid crystal rather than a crystal. But it is also highly viscous. We do not discriminate (here) between smectics and crystals.

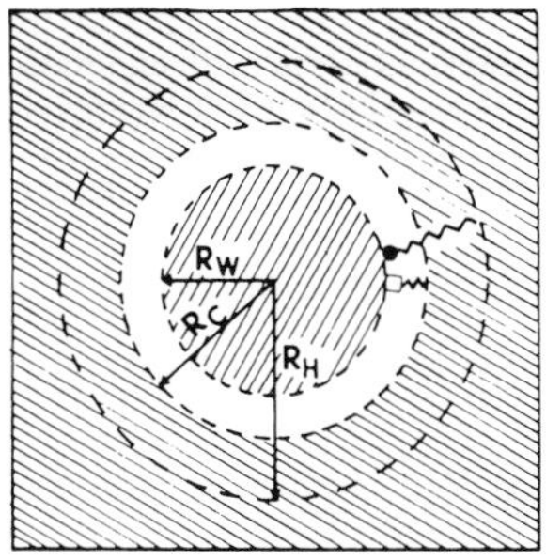

Figure 5. Structure of the interfacial film: R_w, radius of the water core; R_H, hydrodynamic radius (radius of a moving droplet as seen in viscosity and centrifugation experiments); R_c, corresponds to the discontinuity of chemical composition due to the limit of penetration of the oil in the film.

simple steric constraints. This is illustrated in Figure 4 and discussed by Mitchell and Ninham.[8]

Let us call V the volume of a surfactant molecule; the ratio $3V/\Sigma$ defines a natural radius of curvature R_0. This is the radius of a dense, spherical micelle made with p molecules; the total area is $4\pi R_0^2 = p\Sigma$ and the total volume is $(4\pi/3)R_0^3 = pV$. However, the surfactant chains have an extended length l_0 and the micellar structure is acceptable only if l_0 and R_0 are comparable.[9] Similarly for cylinders $l_0 \sim 2V/\Sigma$ and for lamellar structures $l_0 \sim V/\Sigma$. Thus by forming the parameter $l_0\Sigma/V$ one can, to some extent, guess the type of aggregate which will show up preferentially.[10]

On the whole the discussion of ref 8 helps us to understand why a particular macrocrystal form tends to occur (i.e., lamellae vs. rods). But the establishment of a well-defined periodicity between the lamellae, or the rods, depends on the existence of long-range forces between them: attractive (van der Waals) or repulsive (electrostatic, steric, plus the short-range Marčelja repulsion[11]). In any case we end up with a material which is not liquid and is highly viscous.

b. Water + Oil + Surfactant. We have but few X-ray or neutron data on systems with oil, water, and surfactant. But from the mechanical properties (also sometimes from observations on the optical birefringence), it usually appears that most of the concentrated regions of the phase diagram again corresponds to macrocrystals.[12]

2. Microemulsions. If we choose special surfactants[6] or preferably if we add to the surfactant a *cosurfactant* (usually an alcohol ranging from C_4 to C_6), we can obtain, in a broad domain of concentrated mixtures,[14] a transparent, nonviscous liquid which has been called a microemulsion by Schulman.[15] For some reason, macrocrystals are suppressed.

By various methods (electron microscopy,[16,18] X-ray,[16] or neutron scattering,[17] etc.) we know that the constituent objects are small (~100 Å)—much smaller than the droplets of an ordinary emulsion (10 μm). The main features are the following:

a. Area and Curvature. The interfacial area per surfactant Σ can be inferred from measurements of the droplet size, at least in cases where the droplets are well defined in shape (e.g., spherical with radius R) and reasonably monodisperse. Consider for instance a dilute system of water droplets in oil, with ν droplets/cm^3, giving a volume fraction of water ϕ_W and associated with n_s surfactants/cm^3. These parameters are linked by the relations

$$n_s = \nu\frac{4\pi R^2}{\Sigma} \quad \text{(II.1)}$$

$$\phi_w = \nu\frac{4\pi R^3}{3} \quad \text{(II.2)}$$

giving

$$\phi_w/n_s = \tfrac{1}{3}\Sigma R \quad \text{(II.3)}$$

Thus a measurement of R gives Σ directly.

At large R, Σ *tends toward a well-defined limit* Σ^*; this appears to be a rather general rule, valid for systems with a single surfactant as well as for the more frequent cases with surfactant + cosurfactant.[19] We shall see in section III how this relates to the pioneering ideas of Schulman.

Another interesting aspect of these data is that they tell us what is the curvature $1/R$ of the interface under given constraints (given ϕ_w and n_s). Here the situation is quite different from what we had in the water/surfactant systems. One given microemulsion may show a broad spectrum of curvatures depending on ϕ_w and n_s. Indeed a number of microemulsions transform without any apparent discontinuity from water/oil to oil/water.[20] Thus the average curvature is not a leading feature of their stability. (We shall give a theoretical discussion of curvature effects later in section III).

b. Structure of the Interface. Several detailed structural studies have been performed more recently on microemulsions.[17-23] We focus here on the studies of the interfacial film. The composition and thickness of the film can be deduced from hydrodynamic measurements and from neutron scattering. The local state of the surfactant in the film is known by various means, e.g., EPR on spin-labeled amphiphiles, NMR relaxation, fluorescence depolarization.

(i) *Neutron scattering* allows one to perform "variable contrast experiments"[24] which are not feasible with X rays.

(8) D. J. Mitchell and B. W. Ninham, *J. Chem. Soc., Faraday Trans. 2*, 77, 601 (1981).
(9) This statement is very approximate (the chains do not converge radially toward the center of the micelle) but it gives an upper limit which is not unrealistic for conventional surfactants.
(10) The weakest point in this discussion is the precise definition of l_0. For molecules which are not too long, it may be nearly correct to define l_0 by the length of the fully extended chain. But for block copolymers, where the chains are often far from full extension, l_0 would become a function of Σ.
(11) S. Marcelja and N. Radic, *Chem. Phys. Lett.*, 42, 129 (1976).
(12) Exceptional objects, such as the lyotropic nematics made of rod-like (or platelike) micelles,[13] are not included in our discussion.
(13) J. Charvolin, A. M. Levelut, and E. T. Samulski, *J. Phys. Lett.*, 40, L 587 (1979).
(14) It is important to observe that, in many systems, microemulsions dominate a large fraction of the phase diagram, extending *continuously* from the oil-rich to the water-rich side.
(15) T. P. Hoar and J. H. Schulman, *Nature (London)*, 152, 102 (1943).
(16) J. H. Schulman, W. Stockenius, and L. M. Prince, *J. Phys. Chem.*, 63, 1677 (1959).
(17) M. Dvolaitzky, M. Guyot, M. Lagües, J. P. Lepesant, R. Ober, C. Sauterey, and C. Taupin, *J. Chem. Phys.*, 69, 3279 (1978).
(18) J. Biais, M. Mercier, P. Bothorel, B. Clin, P. Lalanne, and B. Lemanceau, *J. Microsc.*, 121, 10 (1980).
(19) See, for instance, H. F. Eicke and J. Rehak, *Helv. Chim. Acta*, 59, 2883 (1976).
(20) B. Lindmann, N. Kamenka, I. Kathopoulis, B. Brun, and P. G. Nilson, *J. Phys. Chem.*, 84, 2485 (1980); F. Larche, J. Rouvière, P. Delord, B. Brun, and J. L. Dussossoy, *J. Phys. Lett.*, 411, 437 (1980).
(21) One difficulty may occur, however; the behaviors of an ionic surfactant in H_2O and in D_2O are not exactly identical (see, for instance, S. Chiou and D. Shah, *J. Colloid Interface Sci.*, 80, 49 (1981)).
(22) A. M. Cazabat and D. Langevin, *J. Chem. Phys.*, 74, 3148 (1981).
(23) R. Ober and C. Taupin, *J. Phys. Chem.*, 84, 2418 (1980).

The principle is explained in ref 17; it is based on the difference between the scattering amplitudes of hydrogen and deuterium. From plots of the scattered intensity vs. H/D fraction in the oil (and in the water) one can ultimately extract the *composition of the interfacial film.* The "film" is defined here as the region for which the scattering amplitude is unaffected by isotopic substitution in the oil (or in the water).[21]

(ii) *Hydrodynamic experiments* measure for instance the mobility of water droplets in an oil matrix; this can be done either by sedimentation[17] or by photon beat methods.[22] What is measured here is a certain "hydrodynamic thickness" of the film (see Figure 5).

(iii) *Local probes in the interface* can also be used; there are many possibilities here (fluorescent dyes, etc.). Some of the most detailed information comes from *spin labels,* modified surfactant molecules carrying a nitroxide radical on the aliphatic chain.[25] From the EPR spectrum of the label one can extract two parameters: an *order parameter* S measuring the alignment of the label along the normal to the interface, and a *correlation time* τ_c giving some estimate of the motional frequencies in the film.[26]

The main conclusions from all these sources appear to be following:

The cosurfactant adsorbs strongly on the interface. Typically one finds two cosurfactant molecules per surfactant.[27]

The order parameter S measured by spin labels in the middle of the chain, in the presence of the cosurfactant, is *very small* ($S < 0.1$).[28,29] This is to be contrasted with the ordered phases; for instance, in the lamellar phases of amphiphile + water, S is of order 0.3–0.4. This points toward a special *fluidity of the interface* in microemulsions. The same qualitative conclusion is also obtained from NMR studies of correlation times[30] and by various indirect procedures: e.g., the addition of a cosurfactant often increases the interpenetration between the surface films of two adjacent droplets, and increases the coalescence rates.

The interface is not seriously modified when the oil/water ratio is modified and in particular when structural inversion occurs (oil/water → water/oil).[31]

III. Existing Models

1. The Saturated Interface (*Schulman*).[32] *a. Principle.* Let us start with a single interface of arbitrary shape separating the oil from the water; the total area of interface is A. It contains a number n_s of surfactant molecules, each of them covering an area $\Sigma = A/n_s$. We wish to discuss first a simple view where (i) the surfactant is insoluble in bulk oil or water, (ii) interactions between different portions of the interface are negligible, and (iii) curvature energies are omitted. We are then led to a free energy of the form

$$f = f_{\text{bulk}} + \gamma_{\text{ow}}A + n_s G(\Sigma) \qquad \text{(III.1)}$$

where the second term corresponds to the bare (i.e., without surfactant) interface (with an interfacial tension γ_{ow}) and $G(\Sigma)$ is a surfactant free energy, depending on the area Σ and containing in particular the effect of surfactant/surfactant repulsions. The Langmuir surface pressure of the film is

$$\Pi(\Sigma) = -\partial G/\partial \Sigma \qquad \text{(III.2)}$$

and the actual interfacial tension is

$$\gamma = \gamma_{\text{ow}} - \Pi \qquad \text{(III.3)}$$

If we minimize (III.1), at fixed n_s, with respect to Σ we obtain the condition

$$0 = \mathrm{d}f/\mathrm{d}\Sigma = n_s(\gamma_{\text{ow}} - \Pi) = n_s\gamma \qquad \text{(III.4)}$$

Thus the system will adopt a well-defined area per surfactant which we call Σ^*; it is defined by the implicit equation

$$\Pi(\Sigma^*) = \gamma_{\text{ow}} \qquad \text{(III.5)}$$

The state $\Sigma = \Sigma^*$ will be called the *saturated state.* It can be reached only if other possible states of the surfactant (such as pure surfactant micelles in water, or in oil) are of higher free energy. (See again Figure 1c for a counter example).

In the saturated state we are dealing with a system of zero surface tension, as shown by eq III.4. The area A is entirely defined by the number of surfactants available

$$A^* = n_s\Sigma^* \qquad \text{(III.6)}$$

b. The Case of Block Copolymers. The ideas described above are essentially due to Schulman[32] and have experienced successive periods of fashion and of rejection. But, in our view, they represent a first, necessary step in the discussion. The second step is to observe that, in many systems, the saturated state cannot be reached because, at surface densities Σ^{-1} smaller than $(\Sigma^*)^{-1}$, the surfactant may prefer to remain inside one of the bulk phases, either in micelles or in more complicated forms. (Recall that in (a) above we started with the surfactant being insoluble in the bulk.)

A semiquantitative study of these points has been carried out for a special class of detergents:[33] diblock copolymers, with an aliphatic chain and a polar chain welded together (e.g., polyethylene–polyoxyethylene). This type of molecules is an extension (toward high molecular weights) of the more common "nonionic surfactants". From a theoretical point of view, these long-chain objects are attractive because (i) they can be treated by well-known procedures of polymer statistics and (ii) the resulting theoretical laws give definite predictions concerning the dependences on molecular weight. For instance, with a diblock (AB) of $N_A(N_B)$ monomers in the A(B) portions, one ends up with a formula for $\Sigma^*(N_A,N_B)$. In the simplest case of a symmetric diblock ($N_A = N_B = N$) the prediction for large N is of the form

$$\Sigma^* \sim N^{6/11} \qquad \text{(III.7)}$$

One interesting feature of the Cantor calculations[33] is that a (rough) comparison is made between the surfactant at an interface and the surfactant in a micelle (on either side, oil or water). The conclusion (for large N, and $N_A \sim N_B$) is that the interfacial situation is favored, even at $\Sigma = \Sigma^*$; the repulsions between neighboring surfactant molecules on the interface are weaker than the repulsions in the case of a micelle.

(24) H. B. Stuhrmann, *J. Appl. Crystallogr.*, **7**, 173 (1974).
(25) M. Dvolaitzky and C. Taupin, *Nouv. J. Chim.*, **1**, 355 (1977).
(26) W. L. Hubbell and H. M. Mc Connell, *J. Am. Chem. Soc.*, **93**, 314 (1971).
(27) However, if the microemulsion is near a limit of stability (highly salted systems) the ratio cosurfactant/surfactant is often much reduced.[23]
(28) M. Dvolaitzky, C. Sauterey, and C. Taupin, Communication presented at the Third International Conference on Surface and Colloid Science, Stockholm, 1979.
(29) One interesting exception is found with the "constrained" films of unstable microemulsions.[28]
(30) A. M. Bellocq, J. Biais, B. Clin, P. Lalanne, and B. Lemanceau, *J. Colloid Interface Sci.*, **70**, 524 (1979).
(31) M. Dvolaitzky, R. Ober, and C. Taupin, *C. R. Acad. Sci. Paris*, **296**, II, 27 (1981).
(32) J. H. Schulman and J. B. Montagne, *Ann. N.Y. Acad. Sci.*, **92**, 366 (1961).
(33) R. Cantor, *Macromolecules*, **14**, 1186 (1981).

Thus it appears feasible to comply with principle (i) of section I by using diblock copolymers, provided that one block has a strong preference for oil, and the other for water.[34]

2. *Limitations of the Schulman Argument.* The discussion of the saturated interface in paragraph II.1a gives a simple feeling for the balance of forces. But it must be refined in practice. We list some of the major factors below.

a. Entropy Effects. The interface is very soft and may be bent randomly at large scales (100 Å); the associated entropy is very small (since we are dealing typically with one degree of freedom per (100 Å)^2) but it is nevertheless relevant, since all other contributions to the free energy are also small. These entropy effects have been mentioned first by Ruckenstein and Chi.[35] Talmon and Prager[36] later showed that, in the absence of any other interaction effects, the entropy term may impose certain phase transitions; when the amount of surfactant available is decreased, they predict that the microemulsion separates into *two phases* (oil rich and water rich). We give a modified version of their argument in section III.3.

b. Curvature energies are weak and do not affect the local properties (e.g., the area per surfactant Σ) very much. However they influence the large-scale properties and the phase equilibria. We give a discussion of the two basic curvature parameters (and of their influence) in section III.4.

c. Electrostatic energies play a role in the stability of swollen micelles of ionic surfactants. Some specific long-range effects have been considered in a recent review.[37] For droplets larger than the Debye screening length, the electrostatic terms are not singular and may be lumped into the interfacial energy $G(\Sigma)$. They also contribute to the curvature parameters.

d. Interactions between different droplets (or other shapes) tend to favor ordered structures or even to promote droplet coalescence. This will be reviewed briefly in section III.5.

3. *Entropy Effects for Flexible Interfaces.* We discuss these effects in a model which is related to, but somewhat different from, the original proposal by Talmon and Prager.[36] We observe first that the interface must have a certain *persistence length* ξ_K: (i) it is essentially flat[38] at scales smaller than ξ_K (we shall justify this concept and estimate ξ_K later in section IV); (ii) consecutive "pieces of interface", with an area ξ_K^2, have independent orientations. A rough but convenient model is then obtained by dividing all space into consecutive cubes, each of linear size ξ_K. Each cube is either filled with oil or with water. The overall proportion of cubes filled with oil (water) is called ϕ_0 (ϕ_w). Two adjacent cubes will have no interface, and no energy, if they are of the same type. But if they are different, we must count a free energy contribution $\gamma\xi_K^2$, γ being the interfacial tension (III.3). For the moment, we do not assume $\gamma = 0$. Rather, we say that a given chemical potential μ_s of the surfactant imposes a certain area per surfactant Σ_s, through the thermodynamic condition (obtained from differentiating (III.1) with respect to n_s and using (III.2) and $\Sigma_s \equiv A/n_s$)

$$\mu_s = G(\Sigma_s) + \Pi(\Sigma_s)\Sigma_s \qquad \text{(III.8)}$$

(34) The case of polyoxyethylene is more complex. POE is soluble in water (at not too high temperatures) but has also a certain compatibility with oil.

(35) E. Ruckenstein and J. Chi, *J. Chem. Soc., Faraday Trans. 2*, 71, 1690 (1975).
(36) Y. Talmon and S. Prager, *J. Chem. Phys.*, 69, 2984 (1978).
(37) J. Overbeek, *Faraday Discuss., Chem. Soc.*, 65, 7 (1978).

(38) For the moment we assume no spontaneous curvature.

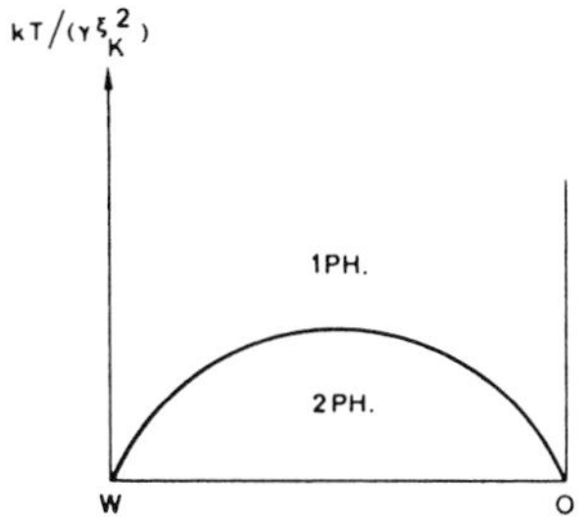

Figure 6. Phase diagram in the modified Talmon–Prager model. At high surfactant contents (low γ) the microemulsion is stable at all water fractions. At lower surfactant contents, a phase separation occurs. The present model does not include the spontaneous curvature of the interface associated with Brancroft's rule. Then (and only then) the plot is symmetrical.

The result is then a certain $\gamma(\Sigma_s)$ which depends ultimately on μ_s.[39]

We have now reduced the statistics of the interface to a "lattice gas model". Clearly, the description is very crude; the oil (or water) regions in a microemulsion do not look like an assembly of cubes. But the lattice gas model keeps some essential features of a random surface. Also, the resulting statistical behavior is well-known. When the coupling between adjacent cubes ($\gamma\xi_K^2$) is weaker than kT, or more precisely when

$$\gamma < \gamma_c = \alpha kT/\xi_K^2 \qquad \text{(III.9)}$$

(where $\alpha = 0.44$ for a simple cubic lattice), we expect a single phase with the oil and water mixed down to the scale ξ_K. But when inequality III.9 is reversed, we may have *phase separation.* The phase diagram is shown in Figure 6.

A number of significant properties emerge from the model and are probably of more general validity:

(a) The values of γ involved are weak. As shown by (III.9) the range of interest is $\gamma \sim kT/\xi_K^2$. The persistence length ξ_K is expected to be rather large, and thus γ should be small; we are not very far from the Schulman criterion.

(b) Phase separation occurs not because of specific interactions between the droplets, but purely because of a balance between interfacial entropy and interfacial energy.

(c) The region of phase separation corresponds to $\gamma > \gamma_c$, i.e., low surfactant content. Consider for instance a small water fraction ϕ_w in the form of droplets inside an oil matrix. If R is the radius of the droplets (assumed spherical and monodisperse) and ν the number of droplets/cm^3, we have (cf. eq II.1–II.3)

$$\nu(4\pi R^2)\Sigma_s^{-1} = n_s$$

$$\nu(4\pi/3)R^3 = \phi_w$$

giving

$$R = 3\phi_w/(n_s\Sigma_s)$$

Thus we would expect that a sufficiently low n_s (surfactants/cm^3) can lead to very large R. But this is never observed! In practice the droplet sizes of all microemulsions remain rather small. A low n_s leads to phase separation between the oil-rich microemulsion and a (nearly pure) water phase.

(d) In practice one often observes more complex phase diagrams. In particular, certain microemulsions can coexist

(39) Alternatively, we could start from a situation with a fixed number of surfactant molecules (assuming that they all lie at the interface) and introduce γ as a Lagrange multiplier.

simultaneously with an oil phase and a water phase. The above lattice gas model generates two-phase equilibria only. However, the model is highly degenerate. Small perturbations on the structure of the free energy (induced by curvature effects or other corrections) might lead to three-phase equilibria. A first attempt in this direction is described in ref 36. Some difficulties are pointed out in ref 40.

4. Curvature Effects. The Schulman description ignores all energies associated with the curvature of the interface. Indeed, for many problems involving fluid/fluid interfaces, curvature energies represent only a very minor correction, and the interfacial energy γ dominates the behavior. Here, however, we deal with interfaces where $\gamma \to 0$, and curvature effects become relevant.

a. Rigidity and Spontaneous Curvature. These two basic ingredients have been defined most clearly in a paper by Helfrich.[41] For a curvature $1/R$ we expect an energy contribution per unit area of the form

$$F = \gamma - \frac{K}{R_0 R} + \frac{K}{2R^2} \qquad \text{(III.10)}$$

here $1/R_0$ is the spontaneous curvature and can be of either sign (we count R_0 as positive when the trend is toward direct micelles). The parameter K has the dimensions of an energy and may be called the rigidity of the interface.[42] Equation III.10 holds only if R and R_0 are much larger than the interfacial film thickness L. How can we estimate K and R_0?

(i) For ionic surfactants the steric considerations of Ninham and Mitchell[8] may give an estimate of the spontaneous curvature. More detailed computer studies on aliphatic chains anchored at a curved interface have been carried out recently in Bordeaux.[43] All these studies assume that the interfacial shell occupied by the surfactant tails is free of oil. This may be correct for the relatively short chains of conventional surfactants.

(ii) The addition of a cosurfactant may act strongly on $1/R_0$ and also on K. For instance, in the (different but related) soap water systems, Charvolin and Mely showed that a certain mixture of C_{18} and C_{10} soaps could give a cubic phase which is not present with the pure C_{14} soap.[44] The cubic phase is believed to be an array of rod portions (with positive curvature) related by branching regions (with negative curvature). More recently, Hendricks[45] took data on another system and showed the curvature trends induced by an exchange between sulfate polar heads and alcohols.

From the theoretical side, we can think of at least two effects of the cosurfactant related to the curvature: a simple wedge effect and a concentration effect.

In the simple "wedge effect", changes in the average parameters Σ, l_0, and V for the mixed interface react on $1/R_0$. Ninham and Mitchell[8] propose that the main role of the cosurfactant is to change R_0 by this process. Our opinion is different; many alkanols can adsorb in the interfacial film and modify R_0, but only a few are efficient to induce microemulsions. Also, many microemulsions are continuously stable when we change the composition from oil rich ($R > 0$) to water rich ($R < 0$). Thus the role of the cosurfactant cannot be entirely reduced to its effect on R_0.

In the concentration effect, the cosurfactant may migrate (for instance) toward the regions of strong curvature of the interfacial film; this has been discussed by Helfrich.[41,46] It leads to a weaker K value and may facilitate structures which include regions of both positive and negative curvature.

(iii) For films of nonionic surfactants, Robbins has considered curvature effects in some detail.[47] We shall give here only a simplified presentation of his ideas, rephrased in the language of eq III.10. He proceeds in three steps. First, using steric considerations, he estimates the optimal radius R for a spherical droplet. Second, by an independent argument (analyzing the stresses on the two sides of the surfactant film) he estimates K/R_0. (The sign of K/R_0 is actually related to a classic rule of Bancroft[48] for ordinary emulsions; the trend is to have the best solvent of the amphiphile on the *outside* of the droplet.) In a third step, Robbins writes the free energy F_d of one droplet in the form

$$F_d = \gamma(4\pi R^2) - \frac{K}{R_0}(4\pi R) + 2\pi K + \lambda\frac{4\pi R^3}{3}$$

where the first three terms derive from eq III.10, while the last term is related to the chemical potential of the inner constituent (e.g., water, if we are dealing with inverse micelles). He then restricts his attention to a microemulsion which is in equilibrium with one solvent (e.g., w/o in equilibrium with water) and argues that $\lambda = 0$ at this point. He then derives the optimal radius R by minimization of F_d (with $\lambda = 0$) and reaches the condition

$$\gamma = \frac{K}{2R_0 R} \qquad \text{(III.11)}$$

Inserting his estimates for K/R_0 and for R, he can now predict γ; γ is the interfacial energy/cm^2 for a flat interface ($R = \infty$), i.e., γ represents the interfacial tension between the w/o microemulsion and pure water.

This procedure works remarkably well; both the estimates of R and γ are quite good for a number of nonionic surfactants. There are, however, certain ambiguities in the scheme. In particular, it is not clear why the steric arguments of step 1 should give R rather than R_0. Our own preference would be more phenomenological; namely, to take R and γ from direct measurements and to derive from these the spontaneous curvature constant R/R_0, using the Robbins' eq III.11.

All the discussion assumes droplets which are not far from spherical and rather monodisperse. These two conditions often seem to be met in practice. But we should keep in mind that, whenever $R > \xi_K$, the shapes must be strongly nonspherical.

(iv) For diblock copolymers (which are nonionic surfactants with relatively long chains, allowing for rather simple statistical calculations), Cantor has given explicit expressions for R_0 and K.[33]

Consider for instance the following (idealized) case: each chain (A) (B) is in an excellent solvent (oil for A, water for B); also for simplicity assume the same monomer size (a) for (A) and (B). Then the condition of weak curvature imposes that N_A/N_B be close to unity

(40) J. Jouffroy, P. Levinson, and P. C. De Gennes, to be published.
(41) W. Helfrich, *Z. Naturforsch. C*, **28**, 693 (1973).
(42) Alternatively, we shall sometimes define the *flexibility* of the film by the inverse (K^{-1}) of the rigidity.
(43) B. Lemaire and P. Bothorel, *Macromolecules*, **13**, 311 (1980).
(44) J. Charvolin and B. Mely, *Mol. Cryst. Liquid Cryst. Lett.*, **41**, 209 (1978).
(45) Y. Hendricks and J. Charvolin, *J. Phys.*, **42**, 1427 (1981).
(46) W. Helfrich, *Phys. Lett. A*, **43**, 409 (1973). A detailed calculation of the rigidity K for a mixture of two types of block copolymers has been performed by one of us (P. G. De Gennes, lectures at Collège de France, 1982, unpublished).
(47) M. L. Robbins, "Micellization, Solubilization and Microemulsions", Vol. 2, K. Mittal, Ed., Plenum, New York, 1977, p 273.
(48) W. D. Bancroft and C. W. Tucker, *J. Phys. Chem.*, **31**, 1680 (1927).

$$N_A = N(1 + \epsilon) \qquad N_B = N(1 - \epsilon)$$

and the parameters in eq III.8 are predicted to obey the following scaling laws:

$$R_0 = L/(6\epsilon)$$

$$L = 2Na(a^2/\Sigma^*)^{1/3} = (\text{constant})N^{9/11}a$$

$$K = (\text{constant})\gamma_0 L^2 = (\text{constant})kTN^{18/11} \qquad \text{(III.12)}$$

(In eq III.12 we have ignored prefactors containing powers of the dimensionless parameter $\gamma_{ow}a^2/kT$, which turns out to be of order unity in many practical cases.[49] The essential features are as follows:

The thickness of the surfactant layer is nearly proportional to the chain length (varying like $N^{9/11}$ at large N). This is remarkable because we are still far from close packing (Σ^* is much larger than A^2) as is clear from eq III.7.

The spontaneous curvature ($1/R_0$) is directly proportional to the dissymmetry (measured by ϵ).

The elastic constant K becomes enormously larger than kT when N is large. (This, as we shall see, may be unfavorable from the microemulsion point of view.)

b. Impact of Curvature Energies on the Phase Diagrams. (i) The rigidity parameter K controls the persistence length ξ_K. We shall discuss this in section IV, where we find that ξ_K increases exponentially with rigidity. This in turn may react on phase separation properties, since (as explained in III.3) one essential parameter is $\gamma/(kT\xi_K^2)$.

(ii) The spontaneous curvature also reacts on the phase diagram. In the original Talmon–Prager study,[36] a very special view was taken, according to which interfaces of finite curvature $\pm 1/R_0$ were equally favored with respect to the flat interface. This may then lead to three-phase equilibria where these three possibilities compete.

If we keep the (more realistic) view, according to which only one curvature sign (dictated by the Bancroft rule) is favored, we come to slightly different conclusions. The only case where predictions are available is the case of weak spontaneous curvature ($|\xi_K/R_0| < 1$); the main effect is a distortion of the phase diagrams.[40] The phase (o/w or w/o) which is privileged by Bancroft's rule increases its domain of existence.

For instance, at high salinity, many ionic microemulsions show a phase equilibrium with nearly pure water. In the low salt systems, the Coulomb repulsions between polar heads tend to give a positive curvature (i.e., as in direct micelles). This then leads, from our discussion, to a natural phase equilibrium between droplets (of oil in water) and pure oil. On the other hand, in high salt concentration, the Coulomb forces are suppressed, and the steric repulsions between aliphatic tails dominate, favoring negative curvature; then if we have droplets of water in oil, they can coexist with pure water.

In nature more complex phase equilibria may occur;[47] in particular, one may find coexistence between *three phases*: nearly pure oil and water, plus a microemulsion. But the "augmented Talmon-Prager model" of ref 40 does not seem to account for these three-phase equilibria; interaction effects have to be invoked here.

5. Long-Range Interactions. Up to now we omitted all effects related to possible interactions between different droplets or, more generally, between different portions of the interface. But it is clear that these interactions must

(49) Also we replaced the mean field exponents of Cantor by the (slightly different) scaling exponents: see S. Alexander, *J. Phys.*, 38, 983 (1977).

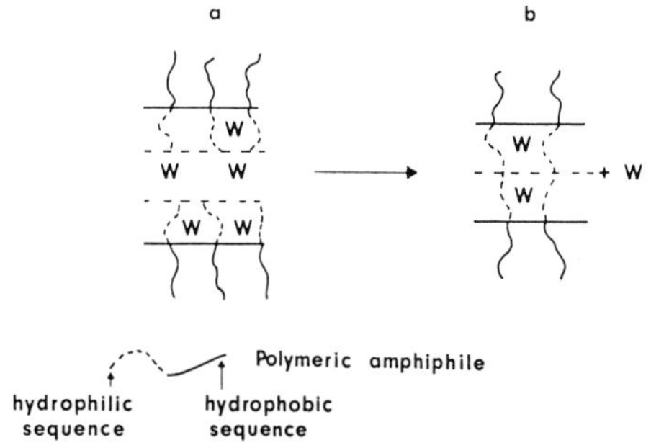

Figure 7. Phase separation induced by long-range van der Waals forces. (a) We start with a lamellar phase of a polymeric amphiphile, where the hydrophilic portion is swollen by water, and the hydrophobic portion by oil. (b) van der Waals attractions (e.g., between two neighboring hydrophobic regions) tend to squeeze out the water, but the process stops when the surfactants from these two layers come into contact (steric repulsion).

play a role in most cases (i.e., provided that we are not dealing with dilute droplets). This has been approached from two main directions: (a) through a study of ordered phases (lamellar, etc.); (b) through measurements and theories involving the interactions between droplets in disordered phases.

a. Stability of the Lamellar Phases. The simplest systems allowing for a study of these interactions are the ordered lamellar phases. For soap/water and lecithin/water systems, Parsegian assessed the role of the long-range van der Waals forces and of certain repulsion forces (electrostatic, steric). More recently, a (less-refined but more transparent) discussion was given by Huh[50] for the case of interest here, namely, when we have successive layers of oil and water separated by surfactant films. There are, however, some points where we disagree with Huh.

(i) He assigns a finite surface tension γ to each layer; we tend to believe that $\gamma = 0$ for any lamellar system of this type (the free energy is stationary with respect to Σ).

(ii) His analysis of thermal fluctuations in the lamellar phase is open to some doubt. Because of (i) he has stabilizing terms due to interfacial tension which, we think, are not realistic. We shall come back to this fluctuation problem in section V.

An improved discussion of thermodynamic stability (against phase separation) has been given in ref 33 for the lamellar phases of oil + water + block copolymer. When the long-range van der Waals attractions are significant, a swollen lamellar phase, containing sheets of pure oil and pure water, should be thermodynamically unstable with respect to a system of "wet bilayers" as shown on Figure 7.

In the final state each half of bilayer is still wetted with its preferential solvent (water on the part shown in the figure) but the water film has become thinner (as required by the long-range van der Waals forces). The thinning stops when the two copolymer sheets are put into contact; then steric repulsions between the chains become rapidly dominant. (For charged systems, electrostatic repulsions play a similar role.)

One interesting consequence of this discussion is the prediction of *three-phase equilibria*: oil + water + wet bilayers. The composition corresponding to the wet bilayer is easily obtained from standard statistical arguments on

(50) Chun Hu, *J. Colloid Interface Sci.*, 71, 408 (1979).

block copolymers. The changes in this composition which are predicted by the theory when the length of the chains (or the quality of the solvents) is modified, resemble closely what is found experimentally on three phase equilibria involving microemulsion + oil + water. A future aim for theoretical work should thus be to enlarge the Cantor discussion from lamellar phases on to more disordered systems, such as a "bicontinuous" microemulsion where both the oil and water regions are connected over macroscopic distances.

Our approach in section V will not be to try and analyze these complex disordered phases but rather to search for *fluctuation instabilities* of the corresponding lamellar phases.

b. Interactions between Droplets. When two spherical droplets (of radius R) are separated only by a small gap h, the van der Waals attraction V between them is of order

$$V(h) = -AR/(6h) \qquad (h < R) \qquad \text{(III.13)}$$

where A is the Hamaker constant (and is often comparable to kT). Light-scattering[51] and neutron[23] experiments on microemulsions can give the osmotic compressibility. Assuming a specific model with monodisperse, undeformable droplets, one then estimates the interactions V. In a number of cases V is indeed strong, especially for large droplets as predicted by (III.13). The distance of closest approach h may also be rather small (because the soft interfacial films from the two spheres interpenetrate each other slightly). These effects have been analzyed recently on one typical system (still using an empirical value for h);[52] with surfactant and cosurfactant, the agreement between osmotic data and theoretical estimates for A and h is rather good.

Very often the van der Waals attractions between droplets may lead to a *liquid–gas transition* where each droplet plays a role similar to that of one argon atom in fluid argon.[53] This could provide an explanation for certain phase equilibria between *two microemulsions* which have been observed in systems of well-adjusted salinity[54] or temperature.[55] However, it is not at all sure that this picture is always right: (i) the actual concentration of the minority phase at criticality is often much smaller than predicted by the liquid–gas analogy; (ii) in many cases the denser system of droplets probably becomes bicontinuous, and the description should be refined accordingly—returning to the end of paragraph III.5a.

IV. The Role of Flexibility

1. Comparison with Red Blood Cells. In eq III.10 two fundamental constants are associated with a saturated interface: the spontaneous curvature $1/R_0$ and the curvature elastic constant K. These two parameters have been introduced first, in connection with surfactant films, by Helfrich,[41] with applications to vesicles and red blood cells.

In fact, in one of the existent models for red blood cells,[56] the surface area of the cell is assumed to adjust and to minimize the overall free energy; thus, in this model, the surface has $\gamma = 0$. One major success of the model has been to explain the *scintillation* of red blood cells and their complex space–time correlations. The general idea is that, with $\gamma = 0$, we have giant fluctuations of the cell shape, which become observable (under phase contrast) with an optical microscope.

Our approach to microemulsions in section IV and V will follow similar lines. We start in section IV with a discussion of the fluctuations of a single interface. In section V we modify this discussion to include (as far as we can) the interactions between different pieces of the divided structure.

2. Statistics of a Random Interface. Let us assume now that our interface has a negligible spontaneous curvature ($1/R_0 \to 0$) and is close to a certain reference plane (xy). The distances between the plane and the interface will be called $\zeta(xy)$. The curvature is then

$$\frac{1}{R} = \frac{\partial^2\zeta}{\partial x^2} + \frac{\partial^2\zeta}{\partial y^2} \equiv \Delta_\perp \zeta \qquad \text{(IV.1)}$$

The free energy (III.10) becomes

$$f_K = \int \tfrac{1}{2}K(\Delta_\perp\zeta)^2 = \sum_q \tfrac{1}{2}Kq^4|\zeta_q|^2 \qquad \text{(IV.2)}$$

where we have gone to two-dimensional Fourier transforms

$$\zeta_q = \int \mathrm{d}x\, \mathrm{d}y\, \zeta(xy) \exp[i(q_x x + q_y y)] \qquad \text{(IV.3)}$$

We shall be mainly interested in the local orientation of the surface, defined by a unit vector $\mathbf{n}$ normal to it

$$n_x = -\partial\zeta/\partial x \qquad n_y = -\partial\zeta/\partial y \qquad n_z \simeq 1 \qquad \text{(IV.4)}$$

For small fluctuations $\delta\mathbf{n} = (n_x,n_y)$ we can write

$$|\delta\mathbf{n}_q|^2 = q^2|z_q|^2 \qquad \text{(IV.5)}$$

Applying the equipartition theorem to all modes in eq IV.3 we obtain the thermal average of these fluctuations

$$\langle|\delta n_q|^2\rangle = kT/(Kq^2) \qquad \text{(IV.6)}$$

where T is the temperature and k is Boltzmann's constant. We can now look at the angular correlations between two points (0 and $\mathbf{r}$) on the surface

$$\theta^2(r) = \langle|\delta\mathbf{n}(0) - \delta\mathbf{n}(r)|^2\rangle = \sum_q 2[1 - \cos(\mathbf{q}\cdot\mathbf{r})]\langle|\delta n_q|^2\rangle \qquad \text{(IV.7)}$$

$$= \frac{kT}{\pi K}\int_0^{1/a} [1 - J_0(qr)]\,\frac{\mathrm{d}q}{q} \qquad \text{(IV.8)}$$

where $1/a$ is a high q cutoff—a microscopic length related to the detergent size. $J_0(x)$ is a Bessel function; the factor $1 - J_0(x)$ is essentially equal to 1 for $x >> 1$ and to 0 for $x << 1$. If we omit uninteresting constants, the result is thus

$$\langle\theta^2\rangle = \frac{kT}{\pi K}\ln\left(\frac{r}{a}\right) \qquad \text{(IV.9)}$$

For small θ we may also present it in the form

$$\langle\cos\theta\rangle \simeq \left\langle 1 - \frac{\theta^2}{2}\right\rangle \simeq \exp\left(-\frac{\langle\theta^2\rangle}{2}\right) = \left(\frac{a}{r}\right)^{kT/(2\pi K)} \qquad \text{(IV.10)}$$

The law (IV.10) shows an exponent $kT/(2\pi K)$ which is continuously variable with T, a frequent feature of two-dimensional fluctuations.[57] We have derived it here only for θ small or not too large r. The question of its validity

(51) A. Calje, W. Atgerof, and A. Vrij in "Micellization, Solubilization and Microemulsions", K. Mittal, Ed., Plenum, New York, 1977, p 779.
(52) B. Lemaire, D. Roux, and P. Bothorel, *J. Phys. Chem.*, to be published.
(53) C. Miller, R. Hwan, W. Benton, and Tomlinson Fort, Jr., *J. Colloid Interface Sci.*, **61**, 554 (1977). See also E. Ruckenstein, *Chem. Phys. Lett.*, **57**, 517 (1978).
(54) K. Shinoda and H. Kunieda, *J. Colloid Interface Sci.*, **75**, 601 (1980).
(55) K. Shinoda and S. Friberg, *Adv. Colloid Interface Sci.*, **4**, 281 (1975).
(56) F. Brochard and J. F. Lennon, *J. Phys*, **36**, 1035 (1975).
(57) Phase transition in surface films, J. G. Dash and J. Ruvalds, Ed., "NATO Advances Study Series (Physics)", Plenum Press, New York, 1980.

for larger r values is not solved at present. For our purposes, however, we know enough with eq IV.10. In particular we can define a *persistence length* ξ_K by the following procedure: at distances r smaller than ξ_K the angle θ is small on the average, while at distance $r > \xi_K$ it is large. Choosing for instance $\langle\cos\theta\rangle = 1/e$ as the cross over value, we obtain from eq IV.10

$$\xi_K = a\exp[2\pi K/(kT)] \qquad \text{(IV.11)}$$

Thus the persistence length ξ_K is extremely sensitive to the value of the rigidity constant K. If, following Helfrich[41] we assume that a simple monolayer (without any cosurfactant) has a rigidity comparable to that of a thermotropic liquid crystal, we arrive at values $K \sim 10^{-13}$ erg, corresponding to $2\pi K/(kT) \sim 12$. In such a case ξ_K is exponentially large, $\xi_K \sim 10^3 a$, and the interface is *stiff*. On the other hand, if, by addition of a suitable cosurfactant, we can decrease K by a factor of 5 ($K \sim 2 \times 10^{-14}$ erg) then $\xi_K \sim 10a \sim 100$ Å and the interface, observed at scales r larger than 100 Å, is strongly wrinkled. Clearly this distinction must play an important role for the selection of disordered (rather than ordered) structures.

V. The Multisurface Problem

1. The Average Distance between Sheets. We now consider a situation where the interface is present in all the sample volume Ω. Numerically, this can be characterized by an *amount of surface per unit volume* which we call $1/d$ (since it has the dimensions of an inverse length). If the number of surfactant molecules per cm^3 is n_s, and if they are all located at an interface, we have

$$1/d = n_s\Sigma \qquad \text{(V.1)}$$

Physically d represents a certain average distance between consecutive sheets of the interface; for instance, if we had a lamellar structure, $2d$ would be the repeat period (each period containing *two* interfaces).

If our sheets were completely ideal, i.e., if there were no interaction between them, each could show a persistence length ξ_K; it would be rigid at short scales ($r < \xi_K$) and wrinkled at large scales ($r > \xi_K$). However, all sheets interact. As mentioned in section III, we have long-range van der Waals attractions and also repulsive forces: electrostatic forces in the water phase, and steric forces appearing whenever the aliphatic tails from two neighboring sheets begin to overlap.[33]

All these forces tend to restore order in the sheet system; in the following paragraphs, we try to give a qualitative analysis of these very complex effects.

2. Steric Effects. Let us start with noninteracting sheets; it is probably correct to visualize any sheet as a system of adjacent platelets, each with a certain typical size ξ_K and area ξ_K^2. Let us think of them as independent units. Each platelet has a number ξ_K^2/Σ^* of detergent molecules, and thus the number of platelets per unit volume is

$$c_p = n_s\frac{\Sigma^*}{\xi_K^2} = \frac{1}{d\xi_K^2} \qquad \text{(V.2)}$$

If different platelets cannot intersect each other, they may tend to stack and form a nematic phase of flat objects (Figure 7);[58] this type of liquid has indeed been observed recently with suitable organic molecules and is currently called a *discotic* phase.[58] Clearly, the discotic phase can exist only if the distance d between platelets is not too large. We can make this statement slightly more quantitative by a transposition of the Onsager argument concerning the nematic alignment of rodlike molecules.[59] In the present case we may say that each platelet is associated with an interaction volume of order ξ_K^3 (Figure 8). Whenever two platelets have a finite overlap of their interaction volumes, they are strongly correlated in their orientation. Thus the criterion for nematic order is of the form

(58) J. Billard in "Chemical Physic Series", Vol. II, Springer, Berlin, 1980, pp 383, 395.

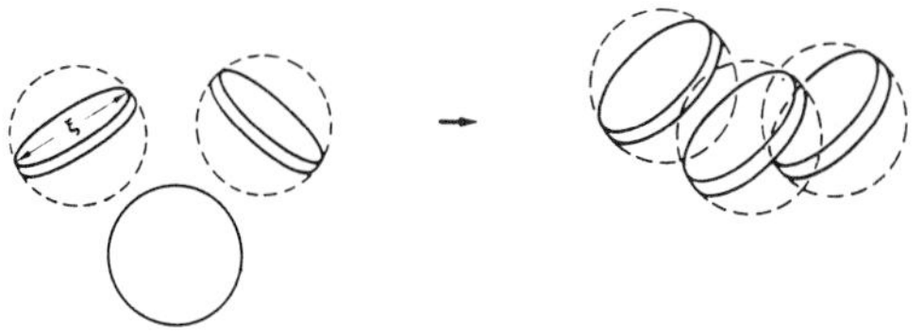

Figure 8. The stacking of disks: (a) at low concentration the disks are disordered; (b) as soon as their "envelope spheres" (dashed) overlap significantly they tend to stack. A "discotic phase" appears.

$$c_p\xi_K^3 >> 1 \qquad \text{(V.3)}$$

Returning now to (IV.2) we see that there are two limits: (a) if $\xi_K > d$ the sheets tend to be parallel to each other; (b) if $\xi_K < d$ the sheets are wrinkled and can build up an isotropic, disordered liquid phase; the discussion of section III.3 becomes relevant.

Thus for a given surfactant concentration n_s, there is in this case a critical value of the rigidity constant $K = K_c$ corresponding roughly to

$$a\exp[2\pi K_c/(kT)] = d = 1/(n_s\Sigma^*) \qquad \text{(V.4)}$$

where we have used eqs IV.11 and V.1.

3. Long-Range Interaction Effects. Our approach here will start from an ordered lamellar phase, adding fluctuations to the average order, and looking for their amplitude. This line of thought has already been followed by Huh,[48] but there are important differences—already mentioned in section III. Huh's interfacial energy γ is different from zero (in fact, for the fluctuation calculation, he uses the bare oil/water surface tension). In the present paper we start from a saturated interface with $\gamma = 0$ and thus the fluctuations are automatically much more important.

We use the analogue of the Einstein model for lattice vibrations of solids.[60] Namely, we look at the motions of one interface in the presence of neighboring layers which are fixed at their equilibrium position. Again we call the displacement of our layer $z(xy)$ and look at the associated free energy. This is an augmented version of eq IV.2 where we include a potential energy term $\frac{1}{2}U''z^2$ (per unit area). Here U'' is the curvature of the potential due to the neighboring layers and is positive (stable equilibrium). Thus we write

$$F = \tfrac{1}{2}K(\nabla_\perp^2 z)^2 + \tfrac{1}{2}U''z^2 \qquad \text{(V.5)}$$

$$= \tfrac{1}{2}\sum_q (Kq^4 + U'')z_q^2$$

$$= (K/2)\sum_q (q^4 + \xi_u^{-4})z_q^2 \qquad \text{(V.6)}$$

where we have introduced a second characteristic length

$$\xi_u = (K/U'')^{1/4} \qquad \text{(V.7)}$$

(59) L. Onsager, *Ann. N. Y. Acad. Sci.*, **51**, 627 (1949).

(60) Quoted in C. Kittel, "Introduction to Solid State Physics", 3rd ed., Wiley, New York, 1967, p 168.

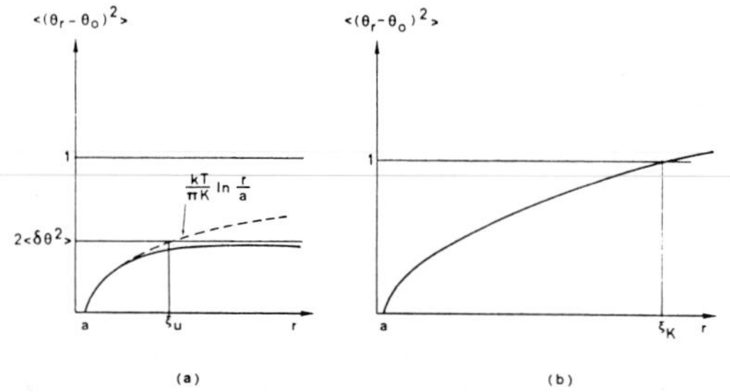

Figure 9. Variation of the orientational correlation between two portions of one same interfacial film as a function of the distance between them in the case of interacting flexible layers: (a) the potential U due to neighboring layers limits the angular fluctuations; (b) the potential U is negligible.

We can then repeat the calculation of orientational correlations; the results are shown in Figure 9. Equation IV.9 still holds for $r < \xi_u$. But for $r > \xi_u$ the average $\langle \theta^2 \rangle$ reaches a finite limit

$$\langle \theta^2(r \to \infty) \rangle = 2\langle \delta \mathbf{n}^2(0) \rangle \simeq \frac{kT}{\pi K} \ln\left(\frac{\xi_u}{a}\right) \quad \text{(V.8)}$$

Of course eq V.8 is meaningful only if the angular deflections are small ($\theta^2 < 1$). This condition leads to the following two cases:

(a) if $\langle \theta^2 \rangle$, as given by eq V.8, is small, the whole scheme is consistent and ξ_u is the relevant characteristic length. We can then compute the average squared displacement of one point on the sheet.

$$\langle z^2 \rangle = \sum_q \langle z_q^2 \rangle = \sum_q \frac{kT}{K(q^4 + \xi_u^{-4})}$$

$$= \frac{kT}{8K}\xi_u^2 = \frac{kT}{8(KU'')^{1/2}} \quad \text{(V.9)}$$

The result (V.9) can be understood in simple terms; a correlated unit of size ξ_u, area ξ_u^2, when displaced by an amount z, has a potential energy $(1/2)U''z^2\xi_u^2$. Setting this equal to $^1/_2kT$ we get eq V.9 (except for the coefficient). We can now decide if the ordered phase is stable by comparing $[\langle z^2 \rangle]^{1/2}$ with the layer thickness (of order d). If

$$K < (kT)^2/(64U''d^4) \quad \text{(V.10)}$$

we conclude that a disordered state is preferable.

(b) If the quantity $\langle \theta^2 \rangle$ as given by eq V.8 is larger than unity, the plot of Figure 9 is cutt off at $\xi = \xi_K$ rather than at $\xi = \xi_u$. Here ξ_K is the relevant characteristic length and the system is clearly not ordered.

VI. Concluding Remarks

1. Summary of Main Ideas. We propose that an oil + water + surfactant system can generate microemulsions in a large part of the phase diagram when two principles are satisfied

(i) The surfactant is able to saturate the oil interface rather than building up pure micelles inside one of the two phases.

(ii) The interface is highly fluid (low K) and the long-range interactions are weak enough so that any "macrocrystal" structure must melt. Already in the pioneering work of Schulman[61] the cosurfactant was pictured as "increasing the disorder of the interfacial film". This was fully confirmed by the measurements on spin labels described in section II. What we tried to show here is that (a) the fluidity is absolutely needed to maintain a microemulsion phase in the central region of the phase diagram and (b) the fluidity is quantitatively measured by a certain rigidity constant K or by the resulting persistence length.

We also noted that the spontaneous curvature $1/R_0$ plays a significant role. This brings us back to principle (i). If R_0 is too small, the surfactant always prefers to make slightly swollen micelles, rather than to build a large interfacial film; the water solubility is low.

It is relatively simple to control the spontaneous curvature; it depends on steric effects and on solvent qualities (hydrophilic–lipophilic balance). But it is much harder to control the interfacial rigidity K. In the Appendix we list a few experimental methods which may be of interest here.

2. Low Interfacial Tensions. Tertiary oil recovery depends crucially on the use of surfactants which lead to very low interfacial tensions between an oil-rich and a water-rich phase. Indeed, with some practical surfactants, low tensions ($<10^{-3}$ cgs) have been detected;[62] to reach these low values usually requires careful adjustment of one control parameter (e.g., of the relative length of oil chains vs. surfactant chains). It is only for a narrow set of values of the control parameter that low tensions are obtained. Is there a profound relation between these special situations and microemulsion properties? The full answer is not known at present, but we may present some relevant remarks, related to our discussion in section III.

We know *two* limiting cases where small interfacial tensions can be expected:[63,64] (a) phase equilibria between a microemulsion (m) and a (nearly) pure phase, e.g., (water in oil)/(water); and (b) phase equilibria near a critical point.

In case (a) the interface is *narrow*; it is essentially composed of a single surfactant film. The interfacial tension γ_{mw} is dominated by this film and remains finite even when the water fraction ϕ_w in the microemulsion tends toward zero.[64] In the modified Talmon–Prager model of section III, we would expect $\gamma \sim kT/\xi_K^2$ for this case.

In case (b), on the other hand, low surface tensions are obtained when two coexisting phases become nearly identical.[65] Here the interfacial region becomes thick, as in any critical phenomenon. Detailed experiments on one ionic system—using light scattering, photon beats, and measurements of the interfacial tension $\gamma_{mm'}$—indicate that the interfacial thickness diverges exactly like the bulk correlation length ξ when the critical point is reached.[65] The scaling laws for ξ and $\gamma_{mm'}$ resemble those which are known for liquid–gas transition. But this does not tell us much about the objects which are involved (droplets, bicontinuous structures, or worse) because, in case (b), the interface (of thickness ξ) is much larger than the size of an individual "object". Note that the Talmon–Prager model does lead to a critical point with $\xi \to \infty$ and $\gamma_{mm'} \to 0$. In this model, near criticality, the microemulsion is bicontinuous.

The interfacial tension $\gamma_{mm'}$ becomes particularly small in case (b) when the system is close to critical; thus we do

(61) J. Schulman and M. Mc Roberts, *Trans. Faraday Soc.*, **42B**, 165 (1946).

(62) J. Cayias, R. Schechter, and W. Wade, *J. Colloid Interface Sci.*, **59**, 31 (1976).

(63) A. M. Bellocq, D. Bourbon, B. Lemanceau, and G. Fourche, *J. Colloid Interface Sci.*, in press.

(64) A. Pouchelon, D. Chatenay, J. Meunier, and D. Langevin, *J. Colloid Interface Sci.*, **82**, 418 (1981).

(65) A. M. Cazabat, D. Langevin, J. Meunier, and A. Pouchelon, *J. Phys. Lett.*, **43**, L89 (1982). A. M. Cazabat, D. Langevin, J. Meunier, and A. Pouchelon, *Adv. Colloid Interface Sci.*, in press. G. Fourche, A. M. Bellocq, and S. Brunetti, *J. Colloid Interface Sci.*, in press.

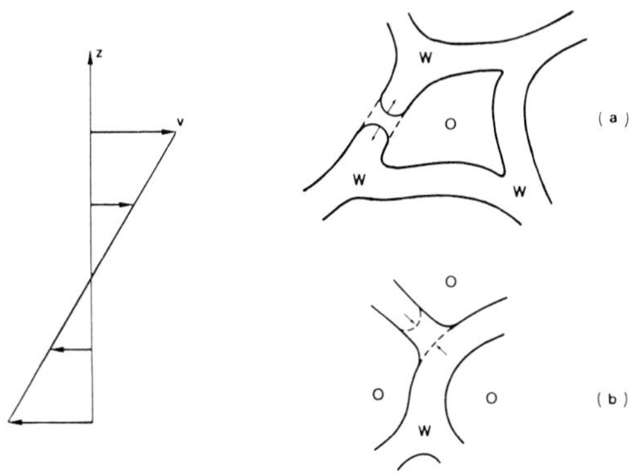

Figure 10. Effects of a shear flow on bicontinuous structures: rupture or coalescence.

have at least two processes generating small interfacial tensions. In practice low surface tensions are often found in a portion of the phase diagram where *three phases* coexist (oil rich/microemulsion/water rich). In this region one benefits from a superposition of effects (a) and (b).

Acknowledgment. This review was written for a session of the French cooperative action on microemulsions (Greco Microemulsions, director P. Bothorel). It has benefited from exchanges of preprints and ideas with the groups in Bordeaux, Paris, Montpellier, and Strasbourg, especially with M. Dvolaitzky and D. Langevin. We also thank J. Israelashvili, B. Ninham, and W. Kanzig for related discussions.

Appendix. Other Physical Effects Dependent on the Flexibility

1. Magnetic Birefringence of Isolated Droplets. If the detergent has a certain amount of diamagnetic anisotropy, it will tend to align in a magnetic field H. This will lead to a certain magnetic birefringence of the droplets, which is currently under study in Grenoble.[66] A discussion of this effect must include two opposite situations: (a) when the droplet radius R is smaller than ξ_K, we are dealing with weak deformations of a spherical droplet; (b) when R is larger than ξ_K the droplets have large fluctuations in their shape, and the field can align the spontaneously deformed objects.

a. The weak distortion limit is relatively easy to compute.[46] The magnetic birefringence depends both on the rigidity K and on the spontaneous curvature $1/R_0$.

Large values of the birefringence would require small values of K; but, if we assume that K is small, the calculation breaks down; the droplet shape fluctuates strongly.

b. The Strong Fluctuation Limit. If $R >> \xi_K$ the droplet is not spherical in the absence of the field. In many cases we could assume that it has a quadrupolar deformation of order unity. The so-called Cotton–Mouton constant (magnetic birefringence/H^2) is then modified; the effect of a weak field $\mathbf{H}$ is primarily to align the distorted droplets, putting their larger dimension along $\mathbf{H}$. The field now competes not with elastic energies but with the thermal agitation, which tends to disorient the droplets. The net result is to maintain a formula for the magnetic birefringence which is not very different from case (a) above but where we make the replacement $K \rightarrow$ (constant)kT. Thus, in this limit, the existence of a significant magnetic birefringence may indicate that we are dealing with flexible interfaces but the value of the Cotton–Mouton constant should become independent of K.

On the whole, the magnetic birefringence is interesting, but complex; it may not be very sensitive to K in the region of utmost interest. Also it does not discriminate directly between fluctuating, fluid droplets, and another extreme case, elongated rigid micelles, which also give a large effect.

2. Dynamical Processes. a. Flow Properties. (i) Isolated droplets of water in oil (or the reverse) lead naturally to a low shear viscosity η. But it is much more surprising to find that more concentrated microemulsion, which (from their transport properties) are most probably *bicontinuous*, also display low η values! The shear must impose dramatic changes to the interface, some of which are tentatively illustrated in Figure 10. We are probably dealing with microscopic coalescence and rupture processes which are very sensitive to the local fluidity of the film, but the relation between them is obscure.

(ii) Apart from the viscosity, measurements of the *flow birefringence* would be interesting (and possibly related to the magnetic birefringence discussed above).

(iii) An interesting experiment has been attempted by Shah and co-workers;[67] this amounts to forcing a microemulsion through very small, calibrated, pores (nucléopores of diameters as low as 150 Å). The results are not spectacular; the overall permeability does not show any anomaly when the size of the microemulsion droplets is increased and becomes larger than the pore diameter. There may be various explanations of this fact, depending on the wetting properties of the pore surface, but one possible line of thought is the following: If the droplets are very flexible, as postulated by us, they can be forced through pores which are smaller than their own size. We believe that these experiments should be pursued with different surface treatments for the pores.

b. Electrical Properties. The *average* conductance of microemulsions has been studied as a function of water content and shows abrupt changes near a certain threshold. This onset of conductance is not purely a geometrical, static effect. One approach to it, showing the importance of time constants, has been constructed under the name of "percolation brassée" (stirred percolation).[68]

Studies on the superimposed *noise* might be instructive and have been started recently; indeed, if bicontinuous microemulsions exist, they will probably show, even in the absence of any flow, coalescence and rupture processes somewhat similar to those of Figure 10, but due to Brownian motion rather than to shear. On major aim for future work should be to display these dynamical effects. One characteric frequency involved would be the relaxation rate of a deformed droplet of radius R

$$1/\tau \sim K/(6\pi\eta R^3)$$

(66) C. Meyer, G. Maret, and P. Poggi, private communication.

(67) D. O. Shah and S. Chou, Department of Energy Report BETC 0008-5, 1979.

(68) M. Lagües, *J. Phys. Lett.*, **40**, L331 (1979).

(69) "Micellization, Solubilization, and Microemulsion", K. Mittal, Ed., Plenum, New York, 1977, p 277.

C. R. Acad. Sc. Paris, t. 301, Série II, n° 10, 1985 675

PHYSIQUE DES SURFACES ET DES INTERFACES. — *Transitions de monocouches à molécules polaires.* Note de **David Andelman, Françoise Brochard, Pierre-Gilles de Gennes**, Membre de l'Académie, et **Jean-François Joanny.**

Lors d'une transition (par ex. liquide L/gaz G) dans une monocouche de Langmuir, la phase la plus concentrée (L) a un moment dipolaire moyen (vertical) assez élevé, et révélé par les mesures de potentiel de contact. Nous montrons que les interactions entre dipoles (faibles, mais à longue portée) doivent bouleverser l'interprétation du diagramme de phase : il apparaît une phase « supercristal » avec un arrangement périodique des régions L et G, et des mailles très variables selon la température et la force des dipôles.

SURFACE AND INTERPHASE PHYSICS. — Monolayer transitions with polar molecules.

At a transition (e.g. liquid L/Gas G) in a Langmuir monolayer, the more concentrated phase (L) has a non zero average dipole moment (along the vertical direction) which is commonly detected by measurements of the contact potential. We show here that the weak, long range (r^{-3}) interactions between dipoles change drastically the interpretation of the phase diagram: in a significant part of the traditional coexistence region, we expect a supercristal phase, where the L and G regions build a spatially periodic arrangment, with periods $\sim 10^3$ Å.

I. Dipôles et énergies. — Les molécules amphiphiles qui forment une monocouche à la surface de l'eau [1] portent en général un moment dipolaire $\boldsymbol{\mu}$ non nul. La moyenne de ces moments est dirigée verticalement (Oz) et donne une densité de polarisation à deux dimensions :

$$m = \Sigma^{-1} \langle \mu_z \rangle \qquad (\Sigma = \text{surface par tête}). \tag{1}$$

Le moment dépend fortement de la concentration φ :

$$\varphi = \Sigma_0/\Sigma \qquad (\Sigma_0 = \text{surface compacte}). \tag{2}$$

On détermine classiquement $m(\varphi)$ par l'intermédiaire du potentiel de contact V (2). Par exemple, si les moments $\boldsymbol{\mu}$ sont immergés dans la partie aqueuse, et si celle-ci a partout la constante diélectrique macroscopique ε, on a :

$$V = 4\pi m \varepsilon^{-1}. \tag{3}$$

Gardons ce modèle et considérons maintenant une situation ou la concentration φ varie de point en point dans le plan de la couche :

$$\varphi = \varphi_0 + \sum_q \delta\varphi_q e^{i\mathbf{q}\cdot\mathbf{r}}, \qquad \mathbf{r} = (x, y). \tag{4}$$

Alors un calcul électrostatique montre que, à q petit, l'énergie dipolaire D est modifiée par :

$$\delta D = -\frac{8\pi}{\varepsilon(\varepsilon+1)} [m'(\varphi_0)]^2 \sum_q |q| . |\delta\varphi_q|^2 = -\frac{\varepsilon}{16\pi(\varepsilon+1)} \left[\frac{m'}{m}\right]^2 V^2(\varphi_0) \sum_q |q| . |\delta\varphi_q|^2, \tag{5}$$

$$= -k\,TK(\varphi) \sum_q |q|\, \delta\varphi_q^2, \tag{6}$$

δD est négatif, et linéaire en q, ce qui caractérise une interaction répulsive en r^{-3}. Le coefficient K se relie assez directement au potentiel de contact V (φ). En général K est petit, mais la forme singulière de (6) doit néanmoins avoir des effets que nous analyserons ci-dessous.

II. Statistique des transitions. — Pour fixer les idées, nous décrivons une transition liquide-gaz, à l'approximation des solutions régulières, avec une énergie libre de mélange par « site » (de surface Σ_0) F(φ) :

(7) $$\frac{F(\varphi)}{kT} = \frac{1}{N}\varphi \ln \varphi + (1-\varphi)\ln(1-\varphi) + \chi\varphi(1-\varphi).$$

La facture N^{-1} dans le premier terme est réminiscent de la théorie de Flory Huggins [3] pour les systèmes polymère solvent : $N(>1)$ serait le nombre de sites Σ_0 occupés par un amphiphile couché sur l'interface. L'usage de cette forme est ici imparfait, puisque les molécules se redressent en phase liquide, mais le facteur N permet de rendre compte, à l'ordre 0, de la forte dissymétrie des courbes de coexistence ([4], [5]). Le paramètre $\chi>0$ décrit l'enthalpie de mélange; il incorpore les interactions attractives (dominantes) entre molécules, et aussi l'effet moyen des répulsions dipoles-dipoles.

Plaçons-nous au voisinage du point critique (φ_c, χ_c) :

$$\varphi = \varphi_c + \Psi(\mathbf{r}),$$
$$\chi = \chi_c + \frac{1}{2}\Delta.$$

L'énergie libre (7) se développe sous la forme :

(8) $$\frac{F-F_c}{kT} = -\Delta/2\,\Psi^2(r) + \frac{1}{4}B\,\Psi^4(\mathbf{r}).$$

(Il n'y a pas de termes en Ψ^3 lorsque χ est indépendant de φ.) Il faut ajouter à (8) les termes de variation spatiale F_g :

(9) $$\frac{F_q}{kT} = \Sigma_0 \sum_q \left[-K(\varphi)\,q + \frac{1}{2}L(\varphi)\,q^2\right] |\Psi_q|^2.$$

Ici L(φ) provient des interactions attractives entre proches voisins. La fonctionnelle (9) fait apparaître un vecteur d'onde optimal :

(10) $$q^* = K(\varphi)/L(\varphi).$$

Une équation analogue à (10) a été obtenue en premier par Garel et Doniach [6] à propos d'un problème similaire : film ferromagnétique avec moments normaux au plan du film, et incorporation des interactions dipolaires.

En prenant V = 0,1 V, les équations (5), (6), (11) donnent $(q^*)^{-1} \sim 1\,000$ A. La période spatiale est donc assez grande. Ceci est lié à la faiblesse des interactions dipolaires, mesurées par le paramètre sans dimension :

(11) $$\eta = q^*(\varphi_c)\,\Sigma_0^{1/2} = KL^{-1}\,\Sigma_0^{1/2}.$$

(*a*) *Près du point critique*, la monocouche tend à réaliser une phase spatialement modulée, de maille $2\pi/q^*$, que nous appellerons un *supercristal*. L'ordre peut-être smectique (un seul vecteur q^*) triangulaire (trois vecteurs q^*), etc. La stabilité comparée des solutions smectique et triangulaire est discutée dans la référence [6]. Près de l'axe du domaine de coexistence ($\Psi=0$) c'est la phase smectique qui est stable. Près de la courbe de coexistence, c'est une phase triangulaire formée d'un réseau régulier d'îlots (à peu près circulaires).

(*b*) *A plus basse température*, le calcul à une composante q^* n'est plus valable : on peut raisonner en termes de *domaines* L ou G, séparés par des interfaces minces (d'épaisseur ξ). Pour le cas d'une structure « zébrée » (smectique : bandes L et G alternées) on trouve par sommation de Poisson sur l'équation (7), un gain d'énergie

C. R. Acad. Sc. Paris, t. 301, Série II, n° 10, 1985 677

dipolaire par centimètre carré :

$$D = \frac{4}{\pi M} k T K (\varphi_L - \varphi_G)^2 \log_c (M/\xi) \tag{12}$$

(où M est la période des zébrures, et où on a pris $\bar{\varphi} = \varphi_c$ exactement).

On sacrifie par contre une énergie interfaciale :

$$F_{int} = \frac{2\gamma}{M}, \tag{13}$$

où γ est l'énergie de 1 cm d'interface L/G. En optimisant (12+13) on arrive à :

$$M \cong \xi \exp \left| \frac{\pi\gamma}{2 k T K (\varphi_L - \varphi_G)^2} \right|, \tag{14}$$

M devient exponentiellement grand dès que :

$$\gamma \Sigma_0 \lesssim k T \eta (\varphi_L - \varphi_G)^2. \tag{15}$$

Dans l'approximation de champ moyen des équations (8) et (9) ceci correspond à $\eta \gtrsim \Delta^{1/2}$.

III. Conclusions. — 1. Toute transition de monocouches formées avec des molécules polaires doit faire intervenir une phase *supercristal*, formées d'îlots qui se repoussent par interactions dipolaires.

2. L'importance de l'effet dépend de la valeur de η [équation (11)] : η est sans doute petit dans les cas usuels ($\eta < 10^{-2}$) mais il augmente si : (*a*) les dipoles sont forts (ex : lécithines) et peu immergés; (*b*) le liquide support a une constante diélectrique ε moins élevée que celle de l'eau (en gros $\eta \sim \varepsilon^{-2}$). Par contre, pour les surfactants non ioniques, η doit être très faible.

3. L'échelle de taille des îlots est $\Sigma_0^{1/2} \eta^{-1}$ près du point critique, mais augmente exponentiellement à basse température. Il faut rapprocher ceci de certaines observations sur les figures de croissance de films de phosphatidyl-choline à la transition liquide-solide [7] = on y voit des structures périodiques à l'échelle de quelques dizaines de microns. Notons toutefois qu'il ne s'agit pas d'un état d'équilibre (les tailles croissent dans le temps) et que les anisotropies spécifiques de la phase solide peuvent peut-être jouer aussi un rôle [8].

4. Les détails du diagramme de phase sont sûrement plus complexes que notre discussion ne le suggère. Par exemple, Garel et Doniach discutent la *fusion* éventuelle du supercristal [6]. Ceci n'est probablement pas observable à l'équilibre thermique à petit (ou les îlots sont gros et leur interactions fortes, $\sim \eta^{-1}$). Mais des bruits non thermiques (mécaniques...) peuvent détruire l'arrangement périodique.

5. Mais, indépendamment de ces détails, il est clair que les effets dipolaires doivent être inclus dans la discussion de la plupart des transitions de monocouches. Il est même peut-être possible que certaines transitions « expansé/condensé » puissent s'interpréter par ce mécanisme : en effet, dans les régions où existe un supercristal, la dérivée de la pression de Langmuir par rapport à la surface $\partial\Pi/\partial\Sigma$ n'est pas nulle.

Nous avons bénéficié de discussions sur ces questions avec D. Canell, M. W. Kim, W. Helfrich et H. McConnell.

Remise le 17 juin 1985.

RÉFÉRENCES BIBLIOGRAPHIQUES

[1] *Voir* par exemple : G. GAINES, *Insoluble monolayers at liquid gas interface*, Wiley, N.Y., 1966.

[2] M. PLAISANCE et L. SARAGA, *Comptes rendus*, 270, série C, 1970, p. 1269-1272.

[3] P. FLORY, *Principles of polymer chemistry*, Cornell University Press, 1971.

[4] G. A. HAWKINS et G. BENEDEK, *Phys. Rev. Lett.*, 32, 1974, p. 524.

[5] M. W. KIM et D. CANELL, *Phys. Rev. Lett.*, 35, 1975, p. 885.

[6] T. GAREL et S. DONIACH, *Phys. Rev.*, *B* 26, 1982, p. 325.

[7] R. M. WEIS et H. MCCONNELL, *Nature*, 310, 1984, p. 5972; H. MCCONNELL, L. TAMM et R. M. WEIS, *Proc. Nat. Acad. Sc. U.S.A.*, 81, 1984, p. 3249.

[8] M. MARDER, H. L. FRISCH, J. S. LANGER et H. MCCONNELL, *Proc. Nat. Acad. Sc.*, 81, 1984, p. 6559.

[9] L. LANDAU et I. M. LIFSHITZ, *Statistical physics*, Pergamon, 1958.

Collège de France, Laboratoire de Physique de la Matière condensée,
75231 *Paris Cedex* 05.

Advances in Colloid and Interface Science, 27 (1987) 189–209 189

POLYMERS AT AN INTERFACE; A SIMPLIFIED VIEW

P.G. de GENNES

Collége de France, 75231 Paris, Cedex 05, FRANCE

CONTENTS

I. ABSTRACT

An elementary description of polymers adsorbed, or terminally attached, at an interface is given. Discussion is restricted to linear, flexible, neutral chains in good solvents. Terminally attached systems, in the "brush" regime, have weak fluctuations and are thus amenable to a simple mean field description. On the other hand, the adsorbed, diffuse layers have large fluctuations and cannot be described in mean field terms. Adsorbed layers have a certain **self-similar structure** which allows for a compact interpretation of most fundamental data (ellipsometry, hydrodynamic thickness). Self-similarity has been proven most explicitly by recent neutron experiments of Auvray and Cotton which were made on long chains with variable contrast techniques. Recent advances on the **kinetics** of exchange between adsorbed and free polymer are also briefly described.

II. INTRODUCTION

Many colloidal systems are made of grains, which tend to flocculate because of certain long range van der Waals attractions. If we add certain soluble polymers which adsorb on the grain surface, the adsorbed layers on two adjacent grains usually repel each other -- the colloid is stabilized.

A recent reference giving an excellent pragmatic overview on this type of "steric stabilization" is a book by D. Napper (ref. 1). The internal structure of the adsorbed layers is discussed in a detailed experimental text (ref. 2). In the present review, we try to extract a few fundamental facts from the vast body of data which is presented in Refs. 1 and 2. We focus our attention on certain **modeling systems**: linear, flexible, neutral chains in good solvents.

The model chain for studies in organic non-polar solvents is polystyrene (PS). Polydilethylsiloxane (PDMS) is also of interest and has one advantage: it remains fluid at room temperature, even in concentrated solutions, thus avoiding certain glassy features which occur in adsorption layers of PS (see Part IV). If we want to work in water, the main examples of neutral flexible changes available are polyoxyethylene (POE) and polyacrylamide (PAA).

A. Adsorption versus depletion

The polymer of interest will be adsorbed (or attached terminally) to a surface which is often a polystyrene latex. In some cases, the surface can be a liquid (emulsion droplets) or even a gas (the free surface).

It is essential to distinguish between the following regimes:

1. Adsorption layer. If the surface prefers the polymer to the solvent, we have adsorption (Fig. 1). In most practical cases, the free energy gained when a monomer is brought from the bulk solution to the wall is relatively large (i.e., it is comparable to the thermal energy kT); this is called strong adsorption.

2. Depletion layer. If the surface prefers the solvent to the polymer, the chains avoid the vicinity of the surface; a polymer solution then shows a depletion layer near the surface (Fig. 2) (ref. 4,5).

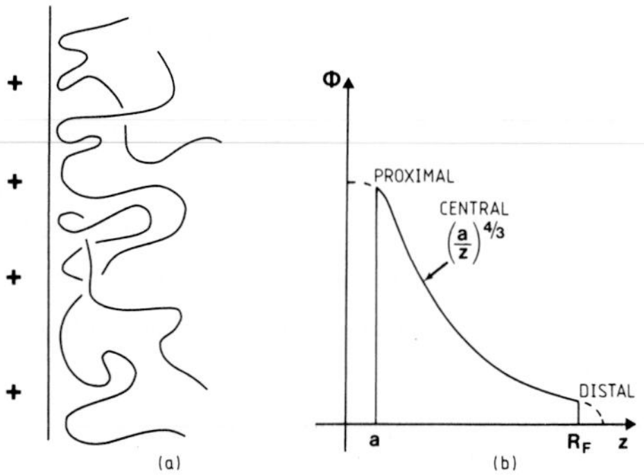

Fig. 1. Multichain adsorption from a good solvent: a) qualitative aspect of the diffuse layer; b) the concentration profile, $\phi(z)$. Note the three regions: 1) proximal (very sensitive to the details of the interactions); 2) central (self-similar); and 3) distal (controlled by a few large loops and tails).

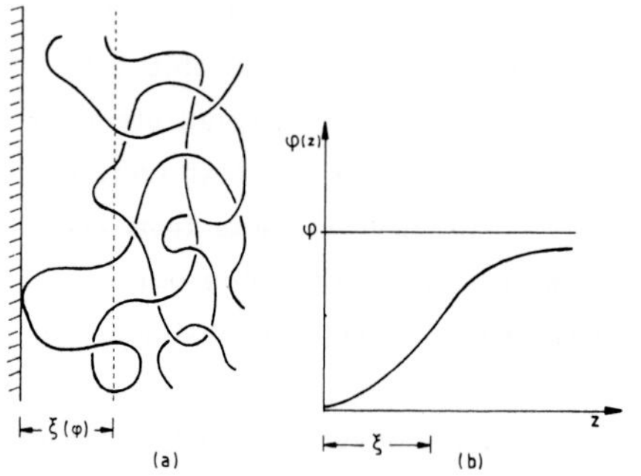

Fig. 2. A depletion layer near a repulsive wall. The thickness of the layer is the correlation length, $\xi(\phi)$.

In practice, the most important regime corresponds to "semi-dilute" solutions (where the coils overlap strongly but the polymer volume fraction, ϕ, is still small); the depletion layer then has a characteristic thickness, $\xi(\phi)$ which is a decreasing function of ϕ (to be discussed in more detail later). Direct measurements of the depletion layer thickness have been performed recently (ref. 6).

The difference between adsorption and depletion is illustrated in Fig. 3, where we show the surface tension γ of polymer solutions as a function of the concentration, ϕ. In the adsorption case, there is an immediate drop of $\gamma(\phi)$ at $\phi = 0$ due to the build- up of the adsorbed layer, which occurs even in very dilute solutions. In the depletion case, $\gamma(\phi)$ increases very slowly with ϕ; at low ϕ, the polymer never touches the surface.

Let us quote one example: many polymers like PS adsorb from organic solvents on a glass surface. However, if the glass is silanated (i.e., covered with short aliphatic tails, terminally attached by an Si-O bond to the glass), we often obtain a depletion layer. In the present review, we shall be mainly concerned with polymers **bound** to a surface.

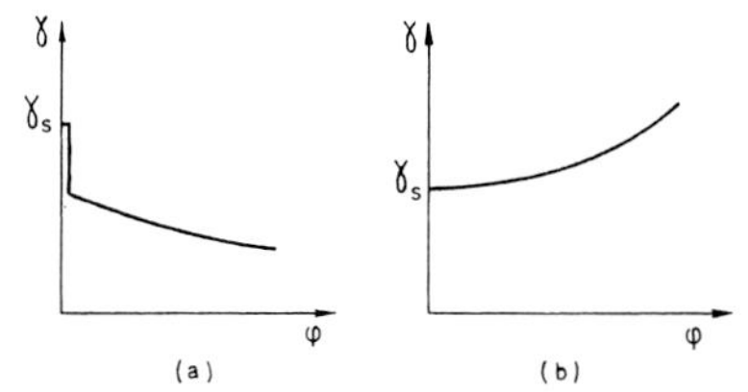

Fig. 3. Surface tension, γ, of a polymer solution (volume fraction, ϕ); a) adsorptive case -- as soon as some polymer is added, the tension drops abruptly; b) depletion case -- the interface remains nearly pure at low ϕ and γ is relatively unaffected. A rule of thumb to decide if a particular polymer/solvent pair falls into case a or b is to compare $\gamma(0)$ (pure solvent) and $\gamma(1)$ (pure polymer). If $\gamma(0) > \gamma(1)$, we expect to find case a.

B. Three modes of fixation

1. Adsorption. Adsorption is one of the simplest methods available for colloid protection (ref. 1,2).

2. Grafting. In situations where the polymer does not adsorb spontaneously, one can sometimes attach chemically the end of the chain to the surface (Fig. 4). In good solvent conditions, the chains repel each other and the grafting reaction terminates when the chains are disposed like adjacent "mushrooms" (Fig. 4a). In some favorable cases, one can push the density of grafted points to higher values; the result is the "brush" of Fig. 4b, with a relatively thick region of constant concentration.

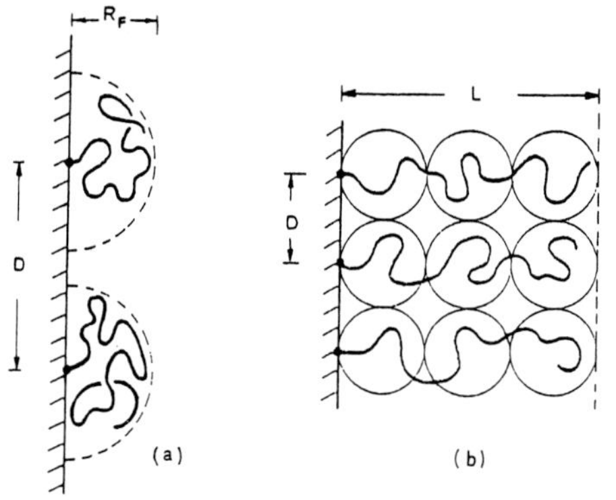

Fig. 4. Two types of grafted surfaces: a) low grafting density -- the distance between heads D is larger than the coil size R_F (this is the "mushroom" regime); b) high grafting density $D < R_F$) (the "brush").

3. Use of block copolymers (ref. 1). If a chain is made of one insoluble part A (the "anchor") and one soluble part B (the "buoy"), it will often

happen that A precipitates near the surface while B builds up an external brush (Fig. 5). For instance, in water, A may be an aliphatic chain while B is made with PE or polyacrylamides. This method is extremely useful but the formation process of the adsorbed layer is delicate and poorly understood.

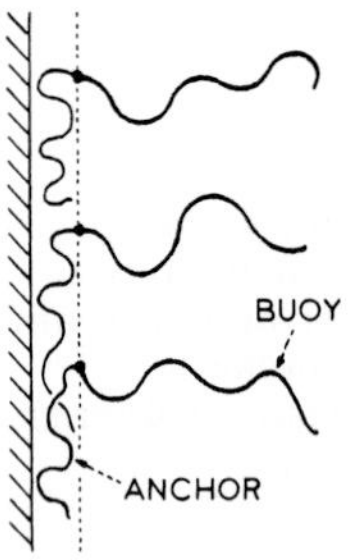

Fig. 5. Terminally attached chains via block copolymers AB. The "anchor" A precipitates against the wall while the "buoy" B protrudes toward the solution.

C. Experimental and theoretical tools

Originally, the main observable in a polymer layer was the total coverage, Γ (number of monomers/cm^2 of surface). Currently, we learn more about the layer thickness by two techniques: 1) hydrodynamics; and 2) ellipsometry.

1. Hydrodynamic measurements. For instance, a spherical grain of radius R, moving with a velocity V, suffers a friction force, $6\pi\eta(R+e_H)V$ where e_H is a certain hydrodynamic thickness of the adsorbed layer.

2. Ellipsometry. A flat bare surface usually reflects light of all polarizations. However, at one special incidence (the Brewster angle), with light polarized in the plane of the incidence, the reflectance coefficient vanishes. This cancellation is suppressed as soon as the surface is covered by an adsorbed layer (ref. 7). The residual reflectance at the Brewster angle is mainly dependent on the total coverage, Γ (i.e., the number of monomer units/cm^2 in the layer). There is a correction term which does depend on the detailed concentration profile c(z) in the layer. A complete study (ref. 8) shows that what is measured is the first moment, $\bar{z}$, of the profile. The ellipsometric thickless, e_ℓ, is thus:

$$e_\ell = 2\bar{z} = 2\Gamma^{-1}\int_0^\infty c(z)\, z dz ,$$

where z is the distance to the wall; the factor 2 can be understood from the trivial case of a uniform layer $c(z) = c_0(z<\ell)$ where $\bar{z} = \ell/2$.

The hydrodynamical measurements are simple and can be performed on colloidal particles. The ellipsometric method requires a neat flat surface and is thus more restricted. We shall see later that they supplement each other rather nicely.

More sophisticated techniques using evanescent waves (ref. 6) or neutron scattering (ref. 9,10) are just beginning to be used.

3. Forces between surfaces. Thanks to the work of J. Israelachvili (ref. 10), it is now possible to measure the forces between two mica plates separated by very small controlled distances h (10 Å < h < 10^4 Å). This technique is extremely well adapted to the case where each plate is covered with a polymer layer; the pioneer in this field was J. Klein (ref. 11-15). We shall discuss some of his results in Part III.

4. Theoretical tools. For many years the structure of polymer layers was derived by some form of self-consisted field theory where each chain feels an average potential with a short range part due to the wall and a long range repulsive part proportional to the concentration profile, c(z). This existed in two versions: 1) one, analytic and compact, but restricted to the main features due to Jones and Richmond (ref. 16); and 2 the other, more numerical but very detailed, due to Scheutens and Fleer (ref. 17). It became progressively recognized, however, that the mean field approach is not reliable for **adsorbed** layers, which have large fluctuations. A completely different scheme, incorporating the correlations in a rigorous fashion, has been constructed from general scaling laws (ref. 18). In particular, this approach showed that adsorbed layers have a remarkable property called self-similarity which is discussed in Part II-B. The scaling analysis is not precise (it does not predict the numerical coefficients in any formula) but it does provide an improved insight; most of our discussion will be based on it.

5. Scaling analysis. The starting point of the scaling analysis is the current understanding of bulk polymer solutions with overlapping coils (the so called "semi-dilute" solutions [ref. 19]). An idealized representation of such a solution is shown in Fig. 6. The most characteristic feature is the mesh size, ξ, which is a decreasing function of ϕ:

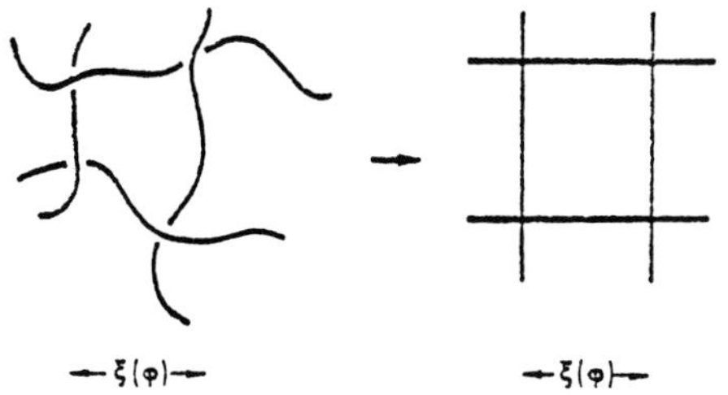

Fig. 6. A polymer solution (volume fraction ϕ) idealized as a "grid" with the same mesh size, $\xi(\phi)$.

$$\xi = a\phi^{-3/4} , \qquad (1)$$

where a is a monomer size (~3A). ξ is typically in the range 0 - 100 A. Another important parameter is the number g of monomers in one subunit of size ξ. The scaling structure of g is:

$$g = \phi^{-5/4}. \qquad (2)$$

Knowing the relations 1 and 2, one can predict most properties of a bulk solution (ref. 19). For instance, the osmotic pressure $\Pi(\phi)$ is simply:

$$\Pi = kT/\xi^3 \quad . \qquad (3$$

We shall see in the following se ctions that the structure of many adsorbed or grafted layers in good solvents can also be described very compactly in terms of ξ and g.

III. STRUCTURE OF ONE LAYER

A. Adsorbed chains

1. The "self-similar grid". Both the hydrodynamic and ellipsometric measurements show that the thickness of an adsorbed layer is large (hundreds or thousands of Angströms). This is indeed what is expected from the theory of an adsorbed layer at equilibrium. The basic clue is provided by Fig. 7, showing what we call the "self-similar grid" structure of the layer. At any distance, z, from the wall, the local mesh size is equal to z. This statement gives us the structure of the profile (ref. 18,19):

$$\xi(\phi(z)) = z \quad . \qquad (4)$$

Using Eq. 1, we find:

$$\phi(z) = (a/z)^{4/3} \; . \qquad (5)$$

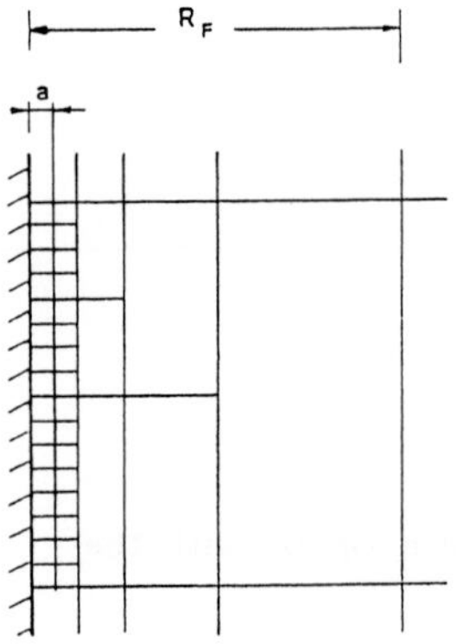

Fig. 7. An adsorbed polymer layer represented as a "self-similar grid". At any distance, z, from the wall, the local mesh size is equal to z.

Thus, the profile decreases very slowly. Of course, the self-similar structure must break down at very small z and also at very large z. It is thus important to state what the two limits ((z_{min} and z_{max}) are: 1) for the (usual) case of strong adsorption, the lower limit is a monomer size:

$$z_{min} = a \; ; \qquad (6)$$

for the (usual) case of an adsorbed layer facing nearly pure solvent, the upper limit is the size of one coil in bulk solution, which we call R_F (the Flory radius):

$$z_{max} = R_F = aN^{3/5} \; , \qquad (7)$$

where N is the number of monomers per chain. Note that the relation between R_F and N is identical to the relation between g (Eq. 2) and ξ; $\xi = ag^{3/5}$. Because N is often large (10^3 - 10^4), the interval (z_{min}, z_{max}) is large and the self-similar features of Fig. 7 are indeed meaningful.

2. Relation to experimental data. The total coverage, Γ, predicted from Eq. 5 is:

$$\Gamma = \int_a^\infty \frac{\Phi(z)dz}{a^3} \simeq \frac{1}{a^2} \; , \qquad (8)$$

(where a^3 is a monomer volume). We see that Γ corresponds roughly to one full monolayer of dense polymer. However, this should not mislead us; there are long loops and tails in the structure, extending up to the higher cut-off R_F. A detailed hydrodynamic analysis shows that one single loop (or tail) is enough to perturb the flow of solvent near the wall very significantly (ref. 18,20,21). The conclusion is that the hydrodynamic thickness, e_H, should be comparable to the coil size, R_H:

$$e_H \sim aN^{3/5}. \qquad (9)$$

Eq. 9 is supported by certain data on polstyrene latices covered by POE (in water) (ref. 22) but some other experiments (on the same system!) give a higher exponent ($e_H \sim N^{0.8}$) (ref. 23).

We have seen that the elllipsometric thickness, e_e, should be given:

$$e_e \sim \Gamma^{-1} \int_a^{R_F} \frac{\Phi(z)}{a^3} zdz \; .$$

This integral is also very sensitive to the cut-off R_F:

$$e_e \sim a^{1/3} R_F^{2/3} \sim aN^{2/5}. \qquad (10)$$

Eq. 10 is in excellent agreement with data by Kawaguchi et al. (ref. 24) which were taken before any theoretical prediction. Note the remarkable difference between e_H (Eq. 9) and e_ℓ (Eq. 10). With these very diffuse profiles, **each experimental method may measure a different thickness.**

The most detailed proof of self-similarity in adsorbed layers has been obtained by neutron scattering. Early experiments by Cosgrove and coworkers (ref. 2) had proven the existence of a diffuse layer. However, they were performed with rather small chains and in scattering conditions which did not allow for a precise check of Eq. 5.

More recently, Auvray and Cotton studied polydimethylsiloxane adsorbed on silica in a good solvent (ref. 25). They used the powerful method of "variable contrast", and ultimately measured the Fourier transform:

$$\phi_s(q) = \int_0^\infty \sin(qz)\ \phi(z)\ dz$$

They found that $\phi(q)$ obeys a power law $\phi_s(q) \sim q^\alpha$ in a broad interval of q's (from 10^{-2} A^{-1} to 0.1 A^{-1}. This proves that $\phi(z)$ is a power law (i.e., that we do have a self-similar structure). Furthermore, the exponent α is found to be equal to 0.35 ± 0.1 which is very close to the scaling prediction derived from Eq. 5 (α = 1/3).

B. Terminally attached chains.

We restrict our attention to chains which are not adsorbed on the surface. (In the opposite case, with adsorption, we return to Fig. 7). If the grafting level is low, we have separate "mushrooms", each with a size ~R_F. If the grafting density is high, we have a "brush". The scaling structure of the brush in good solvents had been deciphered first by S. Alexander (ref. 26). He points out that the crucial parameter is the average distance, D, between attachment points. The brush builds up a region of uniform concentration, ϕ (Fig. 8) and **the mesh size in the brush is equal to D:**

$$\xi(\Phi) = D\ . \qquad (11)$$

The chains in the brush are stretched out. As shown in Fig. 4b, they can be described as a linear sequence of subunits, each of size D and number of monomers g_D (Eq. 2):

$$g_D = \phi^{-5/4} = (D/a)^{5/3}\ .$$

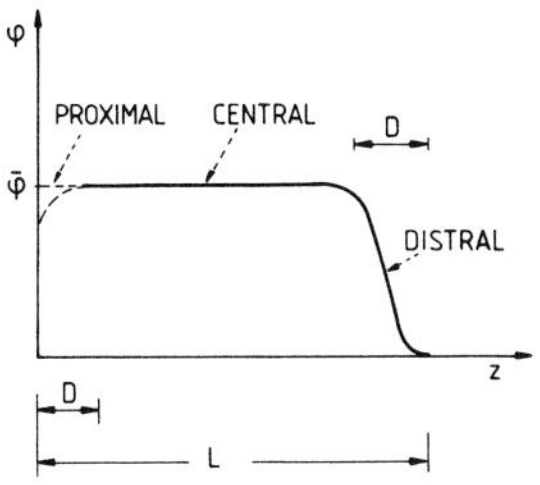

Fig. 8. Theoretical concentration profile inside a brush. In the central region, the profile is flat; in the prosimal region, the profile depends on the details of the monomer-wall interaction.

The overall thickness of the brush is then:

$$L = \frac{N}{g_D} D = Na \left(\frac{a}{D}\right)^{23} . \tag{12}$$

Unfortunately, at present we do not have very clearcut data on these "brushes". The brush regime requires a high density of attached chains ($D \ll R_F$) and this is not achieved easily in grafting. It may be reached, however, with block copolymers.

C. <u>More general situations</u>. All of our presentation of adsorption layers (or of brushes) assumed a good solvent and no additives. If we decrease the quality of the solvent, the structures tend to become more compact. For adsorbed layers, this regime has been computed in some detail by Klein and Pincus. If we add some free polymer (chemically identical to the grafted chains to a "brush", the structure also tends to shink and many different regimes can occur (ref. 28). One basic source of shrinkage is a "screening effect"; the mobile polymers tend to screen out the repulsive interfactions between monomers from the brush.

IV. INTERACTIONS BETWEEN TWO PARALLEL PLATES

A. <u>Plates coated with adsorbed polymer</u>

The basic experiment is idealized in Fig. 9. In Stage 1, two mica plates are incubated with polymer solutions at large distance, h. In a second stage, h is reduced to some small prescribed value (monitored by Fabry Perot interferometry) such that the two diffuse layers overlap ($h \lesssim 2R_F$). Then the forces are measured.

There have been some long disputes even concerning the **<u>sign</u>** of the force. The most detailed experiments (by J. Klein and coworkers [ref. 11-15]) have been performed with POE (of molecular weights in the range 10^5 - 3.10^5 daltons) in water at room temperature (i.e., in a rather good solvent). They show that in slow (hopefully reversible) conditions the force between the

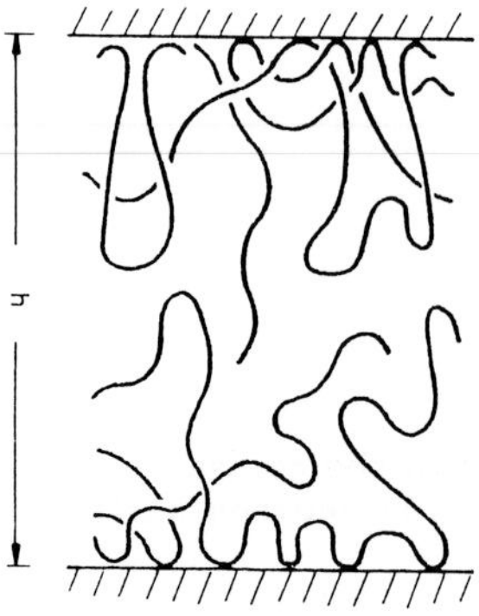

Fig. 9. Two adsorbed layers brought into contact, but before any bridging.

plates is **repulsive** at all measurable distances. What are the theoretical predictions? It turns out that the conditions of the experiment have to be stated in detail (ref. 18): 1) if the equilibrium was complete, allowing for some polymer to flow out of the gap when the two plates are squeezed together, one expects attraction (adjustable layers should lead to adhesion); and 2) in practice, the polymer moves only very slowly along the mica and a more plausible condition is to impose constant coverage (fixed Γ) on both plates. Indeed, certain optical measurements in the gap (which can be done using another set of Fabry Pérot fringes) suggest that this condition is satisfied for the Klein experiments.

At fixed Γ, one expects the force to be repulsive and to be given by the osmotic pressure at the mid-point (Fig. 10). Since the mesh size at this point is h/2, we expect from Eq. 3 that the force per unit area is:

$$F \sim (\text{constant})\ kT/(h/2)^3 \sim kT/h^3\ , \tag{13}$$

or that the interaction energy U (per cm^2) scales like:

$$U \sim kT/h^2\ (a \ll h \ll R_F). \tag{14}$$

This is compatible with the Klein data, but the exponent remains to be established in more detail.

An interesting complication is brought in by bridging (Fig. 11); after some time of contact, the chains from one plate may establish bridges with the other plate. The final equilibrium number, n_b, of bridges per cm^2 can be directly understood from the self-similar picture of Fig. 10. The spacing between possible bridges is of the order h/2; thus the scaling structure of n_b must be:

$$n_b \sim 1/h^2\ (h < R_F)\ . \tag{15}$$

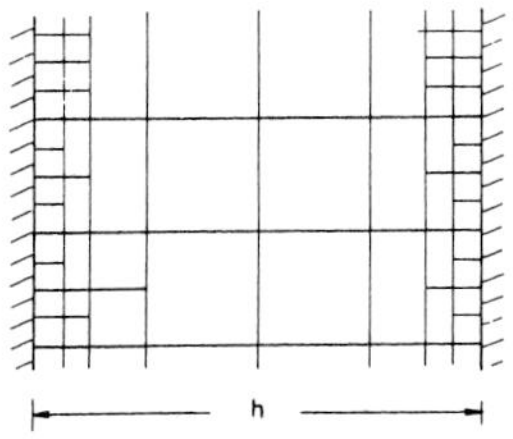

Fig. 10. The self-similar grid for two adsorbed layers in contact. The maximum mesh is of the oreder h/2 where h is the gap.

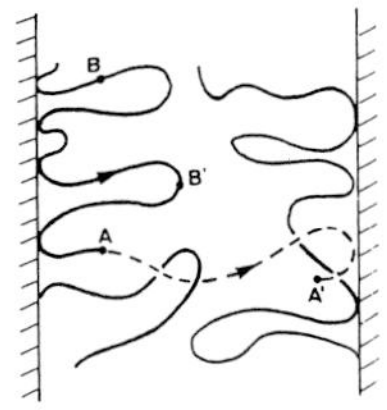

Fig. 11. The bridging process; one chain end moves from A to A' while the other end moves from B to B'.

Eq. 14 can be translated into a prediction for the number of bridges between two spherical grains of radius R_{grain} separated by a gap, $h<R_F<<R_{grain}$. The total number is then:

$$n_{total} \sim hR_{grain}\, n_b \sim \frac{R_{grain}}{h} \qquad (16)$$

How can we detect the bridges? If we pull the grains apart very slowly (remaining constanty in thermal equilibrium), the bridge should retract without any great effect on the forces. However, most practical velocities of separation are fast compared to the bridging kinetics and then the bridges should act as elastic springs, keeping the grains togehter. Scaling arguments predict a bridging energy $\sim kT/h^2$ which is very similar to Eq. 14 but with a reversed sign. Thus, adhesive properties may tell us something about the bridges. We shall come back to the kinetics aspects in Part IV.

A (possibly related) striking experimental effect is obtained under conditions of "starvation" (ref. 14). Here the plates are incubated at large h (always in good solvent) but only for a short time; the coverage, Γ, is lower than its equilibrium value. Then when the two plates are brought together, a strong attractive force shows up at intermediate distances. In these conditions, we expect a fast bridging process.

B. Plates with terminally attached chains

Experiments using the Israelachvili machine with block copolymers (anchor + buoy) have been performed (ref. 29). However, the exact state of these chains (their surface density, etc.) are still to be described. At present, we can only give the theoretical predictions, corresponding to the "mushroom and "brush" regimes of Fig. 4a,b.

1. Interactions between brushes. These interactions have been computed by many methods; in particular, self-consistent theory (ref. 30,31) and, more recently, scaling arguments (ref. 18). The two brushes come into contact when $h = h_c = 2L$ where L (Eq. 12) is linear in N. At $h < h_c$, the two brushes are squeezed against each other but they do **not** interdigitate (Fig. 12).

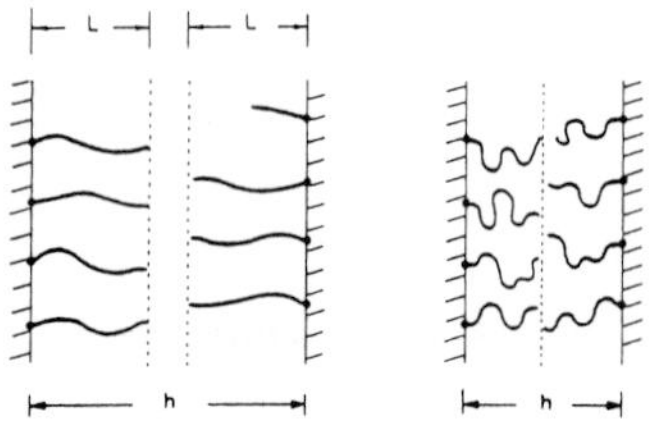

Fig. 12. Two brushes: a) before contact $h < h_c$; b) under compression $h < h_c$; the two brushes do not intermix very much.

The polymer concentration inside each brush increases $\phi = \Phi h_c/h$. This gives two contributions to the force: 1) the osmotic pressure inside each brush increases; and 2) the elastic restoring forces (tending to thin out the brush) decrease. In the single brush problem, the osmotic and elastic terms balance each other exactly at the equilibrium thickness (L). In the compressed brush (thickness h/2), they do not and the result is a force/cm^2:

$$F \cong \frac{kT}{D^3} \left[\left(\frac{h_c}{h}\right)^{9/4} - \left(\frac{h}{h_c}\right)^{3/4} \right] \quad (h < h_c) \quad . \tag{17}$$

Here, D is always the average distance between attachment points. The first term in the bracket is the osmotic term; the second one is the elastic term. At strong compressions, the osmotic term should dominate completely.

2. Interactions between dilute "mushrooms". Here the interaction shows up first at a critical gap thickness $h = h_c \sim R_F$. Thus, h_c is now a weaker function of $N(h_c \sim N^{3/5})$. At $h \ll h_c$, the chain is squeezed in a thin gap and scaling arguments predict a confinement energy per chain (ref. 19):

$$U_1 \cong kT \left(\frac{R_F}{h}\right)^{5/3} = NkT\left(\frac{a}{h}\right)^{5/3} \quad . \tag{18}$$

The energy per cm^2 of plates in U_1/D^2 and the corresponding force is:

$$F \simeq \frac{kT}{D^2 h_c} \left(\frac{h_c}{h}\right)^{8/3} \quad (h < h_c) \ . \tag{19}$$

The recent data by Tirrell and coworkers (ref. 29) on copolymers (A = polyvinylpyridine; B = polystyrene) give overall forces, F, which increase more weakly than these predictions (Eq. 19 or Eq. 17) at small h. However, we need to check whether D is independent of N in these experiments.

V. DYNAMICAL PROBLEMS

A. Basic results on exchange between adsorbed layers and solutions.

Can a molecule leave an adsorbed layer and go into solution, or the reverse? Two groups of observations are important here.

1. Stability under washing. If we prepare an adsorbed layer of long chains (large N) by slow incubation and then wash it with pure solvent, we usually find that the layer does not redissolve at all. The classical interpretation of this fact was the following: A finite fraction, f, of the monomers is in direct contact with the wall; f is measured by various tachniques and, in particular, by nuclear resonance, the spectrum of the f fraction being solid-like and broad, while the spectrum of the complement 1-f is liquid-like and narrow (ref. 2). Typically, f $\sim$ 0.1 to 0.8. The binding energy of the chain to the wall is of the order fNU_b where U_b (the adsorption energy) is then reduced by the repulsive monomer/monomer interactions in the good solvent. This repulsive energy is itself a finite fraction, ψ, of the above estimate, $\psi f N u_b$ with $\psi \sim$ 1/2 to 1/3, depending on the detailed model. The net result is a binding energy $f(1 - \psi)NU_b$. The Boltzmann factor for removing a chain would then be of the order:

$$\tau \sim \exp(-0.1\ NU_b/kT) \sim \exp\ (-N/10). \tag{20}$$

With N values above 10^3, this Bolzmann factor is extremely small and this could explain the stability under washing. We shall show, however, that this argument is utterly wrong, as suggested by the experiments quoted below.

2. Allowed exchanges at finite bulk concentrations. It has been known for some time that an adsorbed layer, when faced with bulk solution, can exchange some molecules with the solution. The first hint came from situations where the adsorbed chains are relatively short, while the solution chains are longer (ref. 32); then the long chains displace the short one.

A more detailed proof was obtained in 1985 by isotopic exchange measurements performed in Strasbourg (ref. 33.). Here one starts with radioactively labeled chains covering a set of grains and achieving a coverager, Γ^*. Then the system is washed with a solution of chemically identical unlabeled chains (concentration, ϕ_b). Over periods $\lesssim$ 1 day, there is a net loss of radioactivity, ruled by a kinetic equation of the form:

$$\frac{d\Gamma^*}{dt} = - k\Gamma^*\phi_b \quad . \tag{21}$$

As pointed out in Ref. 33, this resembles a second order process:

(bound chain)* + free chain → (free chain)* + bound chain .

However, we know that, for these model systems, there is no reason for a (1,1) association; thus, Eq. 21 raises an interesting question. The current answer is described in Part IV-B. Note that Eq. 21 is compatible with the aforementioned stability of adsorbed layers with respect to pure solvent; when $\phi_b \to 0$, the rate (21) vanishes.

B. <u>Tentative interpretation (ref. 34)</u>.

1. <u>Absence of very high energy barriers</u>. To extract a chain from the adsorbed layer, we need not overcome the huge barrier described by Eq. 20. Physically, this may again be understood, thanks to the self-similar representation of Fig. 7. When one particular coil wants to move (from the surface outwards) in the self-similar grid, it meets a dense repulsive grid only in the early steps; later, the grid is quite open. The final result (ref. 34) is that the barrier factor is not given by Eq. 20 but is a much weaker function of N:

$$\tau = N^{-x} \quad (x \sim 0.3) \quad . \tag{22}$$

With $N \sim 10^3$, $\tau \sim 0.1$; exchanges can occur rather freely.

2. <u>The saturation condition</u>. An important feature of the adsorbed layer is the strong repulsive interaction between adjacent chains. A single chain is very firmly adsorbed; adding a second chain, we find a little less binding energy. Ultimately, adding 1,2...p chains, we can fill the adsorbed layer up to an equilibrium coverage Γ. However, if we insist on adding more chains, their free energy becomes higher than the bulk solution level and they "spill out" fast. Thus, there is a strong restoring force, tending to maintain Γ constant. When this feature is injected into the adsorption and desorption rates, one finds that chains can escape from an equilibrated adsorbed layer **<u>only if</u>** some chains from the solution immediately come and maintain Γ nearly

constant. Thus, in the Strasbourg experiments, it is natural that the rate for escape ($- d\Gamma^*/dt$) be proportional to the outer concentration, ϕ_b, as observed in Eq. 21. This condition of nearly constant Γ is what we call the **satruation condition**. It is somewhat similar to the condition of electrical neutrality which is met in an ion exchanger; the exchange can liberate its own mobile ions (e.g., Na^+) only if some other positive ions are available in the ambient bath.

To summarize, we understand that the energy barriers are weak and also that the saturation condition enforces rate equations which resemble second order kinetics (Eq. 21). There remain many delicate points, however. To predict the factor k in Eq. 21, we need a detailed description of the crawling motions of chains in the barrier region. This crawling is complicated by another effect which is present with many polymer systems; it is described below.

C. The glassy layer

An important feature of adsorbed layers has been discovered recently by Cohen-Stuart and coworkers (ref. 39). They studied adsorbed layers of conventional polymers (POE, PVA, etc.) at the **free surface** of a solvent. Performing delicate rheological measurements on this thin layer, they found that it shows a non-zero yield strength. The adsorbed layer is not fluid but behaves like a (weak) solid crust.

The most plausible interpretation of this effect (ref. 36) is based on the observation that these polymers, in thin bulk phase at room temperature, are in a **glassy state**. In solution, they have a lower glass transition temperature, $T_g(\phi)$, but we know from Eq. 5 that the first few monomer layers near the surface are quite dense. In this region, we are probably dealing with a glass. This slows down all creeping motions of the chains.

Future experimental work on the dynamics should probably focus first on selected polymer systems which remain liquid even at room temperature in their bulk phase (elastomers or polydimethylsiloxane). However, even with these systems, there is a danger that the adsorbing surface (when it is solid) still favors a certain "freezing" of the motions in the first layer.

D. Kinetics of bridging: a few conjectures

Let us start from Fig. 11 and consider two overlapping layers which are brought at a distance $h < 2R_F$ at time 0. Qualitatively, one may think of bridging in terms of exchange of polymer between one layer (e.g., the left layer) and an equivalent solution (replacing the layer on the right-hand side). The concentration of the equivalent solution would then be the concentration at midpoint.

This scheme allows a description of bridging in terms of rate equations similar to Eq. 21. Again, there is no high barrier due to polymer repulsions and the whole line of thought can be transposed (ref. 34). However, the same difficulty remains; the magnitude of the rate constants will often be controlled by the glassy features of the inner sublayers and there is one (possibly important) complication if we want to compare the rate constants at different values of the gap, h: when h is decreasd, the concentration in the inner sublayers raises slightly. This in turn increases the glassy features.

E. Breathing modes of a polymer layer

Let us return to a single plate carrying a polymer layer and restrict our attention to **small deformations** of the layer. What are the collective modes of these systems? They fall into two classes (Fig. 13): transverse (shear modes) and longitudinal breathing modes.

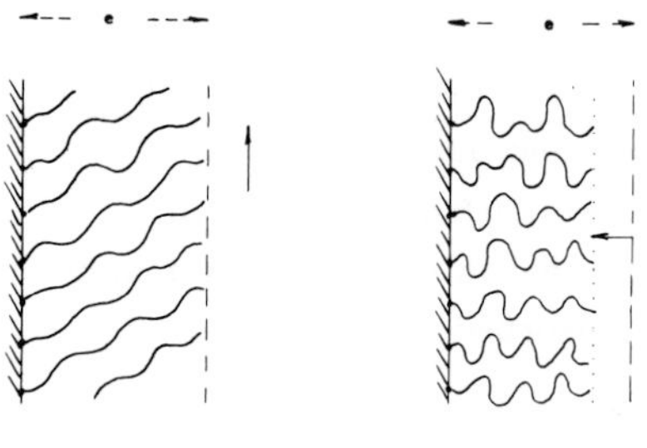

Fig. 13. Collective modes of a brush: a) the lowest shear mode; b) the lowest breathing mode.

1. Terminally attached chains. For terminally attached chains, in the "brush" regime, the low frequency collective modes are very similar to the modes of a **gel** with suitable boundary conditions. The transverse modes are similar to sound waves but are damped through contact with the bulk solvent. The longitudinal modes are diffusive with a diffusion coefficient $D_{coop} \sim kT/6\pi\eta_o D$ (η_o being the solvent viscosity).

2. Adsorbed layers. For adsorbed layers, the breathing dynamics are somewhat more complex. Writing down a balance between elastic and viscous forces, one can construct the modes in an inhomogeneous continium. The result is a (surprisingly) simple result; the n'th mode has a relaxation rate:

$$\frac{1}{\tau_n} = (\text{const.})(\frac{n^2}{T_z} \quad (n = 1,2,\ldots) \quad , \tag{23}$$

where T_z is the Zimm relaxation time of one free coil:

$$T_z = (\text{const.}) \frac{R_F^3 \eta_o}{kT} \quad . \tag{24}$$

Eq. 23 is similar in form to the simple result for a free ideal chain; however, it has in fact a much more delicate content. The chains are not assumed to be ideal and backflow is included.

How can we probe the breathing modes? Inelastic neutron scattering does not appear adequate. The available spatial scales q^{-1} for inelastic measurements are too small (<30 Å); the signals would be dominated by an inner region which is not very universal. Optical methods may be better. In some favorable cases, one could possibly measure the fluctuations of the center of gravity $\bar{z}(t)$ of the adsorbed layer. Scaling laws for the corresponding correlation function have been constructed.

VI. CONCLUDING REMARKS

A. Complementarity of the two theroetical approaches.

Our discussion was based only on scaling ideas (rather than mean field ideas) for two reasons: 1) the intrinsic weakness of self-consistent field ideas for strongly fluctuating systems such as adsorbed layers; and 2) a tutorial aspect; general trends can be memorized more efficiently from the scaling laws than from numerical calculations.

On the other hand, we do not wish to underestimate the mean field approach: 1) for grafted layers and also for systems with attached polymer + free polymer, the self-consistent calculations are indispensable; and 2) evern for adsorbed layers, the mean field calculations (ref. 17) give us the best available information on short chains (where self-similarity is not significant), on the effects of long loops and tails, on marginal solvent system (intermediate between good and θ), etc.

Clearly, the two techniques (mean field and scaling) supplement each other. A challenge for the future is to construct a suitable generlization of self-consistent fields which would lead to the right exponents in the self-similar regime (Eq. 5). An attempt at this is described in Ref. 18 but the energy function which is used here is too crude and too arbitrary.

B. Equilibrium or non-equilibrium?

All the available theoretical pictures are based on some assumption of thermodynamic equilibrium; in some cases, this can be a constrained equilibrium (e.g., the problem of two interacting plates at constant coverage Γ) but the structure of one adsorbed layer (in good solvents) is always assumed to be independent of the sample history.

This may deviate from reality in certain systems, especially when the inner adsorbed layer tends to be glassy. Let us think a little more about the incubation process; a plate is exposed to a dilute solution. When a first chain is adsorbed, all theoretical calculations tell us that it makes a flat **pancake**, not a fluffy layer. If we let the incubation proceed, then competition between different chains begins to show up. But since the early stages corresponded to a flat carpet, if a dense layer tends to be glassy, it will be very difficult for a newcomer to reach the adequate number of binding sites on the surface. We may be rather far from thermodynamic equilibrium. Unfortunately, the early stages of incubation are difficult to probe experimentally, but a lot remains to be done in this field.

VII. LIST OF SYMBOLS

a	size of monomer
D	distance between grafted sites
D_{coop}	cooperative diffusion coefficient
e_e	ellipsometric thickness of adsorbed layer
e_H	hydrodynamic thickness of adsorbed layer
F	force per unit area between plates
g	number of monomers in one mesh volume
g_D	number of monomers in a volume D^3
h	gap between two interacting plates
k	Boltzmann constant
L	thickness of brush
N	number of monomers/chain
n	mode index
n_b	nb of bridges/cm^2
n_t	total number of bridges between two grains
q	scattering wave sector
R_F	size of one coil in solution (good solvents)
T	absolute temperature
τ	Barrier factor
T_g	glass transition temperature
T_z	Zimm relaxation time
U	interaction energy between plates
z	distance from adsorbing wall
Γ	number of monomers adsorbed/cm^2
Γ^*	nb of radioactive monomers/cm^2
γ	surface tension of solution
γ_S	surface tension of pure solvent
η_0	viscosity of solvent

ξ correlation length of a polymer solution

π osmotic pressure

τ_n relaxation time of n'th mode

Φ volume fraction of polymer

$\Phi_S(q)$ Fourier transform of adsorbed profile

VIII . ACKNOWLEDGEMENTS

This presentation started from a long series of discussions with A. Silberberg and P. Pincus. It also benefited greatly from discussions with M. Cohen-Stuart, R. Varoqui and L. Auvray.

IX. REFERENCES

1 D. Napper, Polymeric Stabilization of Colloidal Dispersions, Academic Press, 1983.
2 M. Cohen-Stuart, T. Cosgrove, and B. Vincent, Advances Colloid Interface Science, 24(1986)143.
3 T. Hayter, G. Jannick, F. Brochard and P.G. de Gennes, J. Physique (Lett.), 41L(1980)451.
4 S. Asakura and F. Oosawa, J. Chem. Phys., 22(1954)1255.
5 J.F. Joanny, L. Leibler and P.G. de Gennes, J. Polymer Sci., 17(1979)1073.
6 C. Allain, D. Ausserre and F. Rondelez, Phys. Rev. Lett., 49(1982)1694.
7 R. Azzam and N. Bashara, Ellipsometry and Polarized Light, North Holland, 1977.
8 J.C. Charmet and P.G. de Gennes, J. Opt. Soc. Am., 73(1983)1777.
9 K. Barnett, T. Cosgrove, T. Crowley, J. Tadros and B. Vincent, in The Effects of Polymers on Dispersion Stability, J. Tadros, ed., Academic Press, 1982, p. 183.
10 L. Auvray, Comptes Rend. Acad. Sci. (Paris), 302(1986)859.
11 J. Israelachvili and G. Adamans, Faraday Trans. I, 74(1978)975.
12 J. Klein and P. Luckham, Nature, 300(1982)429.
13 J. Klein and P. Luckham, Nature, 308(1984)836.
14 J. Klein, Adv. Colloid Interface Sci., 16(1982)101.
15 J. Klein and P. Luckham, Faraday Trans. I, 80(1984)865
16 J.S. Jones and P. Richmond, Faraday Trans. II, 73(1977)1062.
17 J.Scheutens and G. Fleer, in The Effects of Polymers on Dispersion Properties, J. Tadros, ed., Academic Press, 1982, p. 45.
18 P.G. de Gennes, Macromolecules, 14(1987)1637; 15(1982)492.
19 P.G. de Gennes, Scaling Concepts in Polymer Physics, Cornell University Press, second printing, 1985.
20 P.G. de Gennes, Compt. Rend. Acad. Sci. (Paris), 297-II(1983)883.
21 M. Cohen-Stuart, F. Waajen, T. Cosgrove, B. Vincent and T. Crowley, Macromolecules, 17(1984)1825.
22 T. Kato, K. Nakamura, K. Kawaguchi and A. Takahashi, Polymer J., 13(1981) 1037.
23 T. Cosgrove, T. Crowley, M. Cohen-Stuart and B. Vincent, in Polymer Adsorption and Dispersion Stability, ACS Symposium, 240(1984)147.
24 M. Kawaguchi and A. Jarahashi, Macromolecules, 16(1983)1465.
25 L. Auvray and J.P. Cotton, Macromolecules (to be published).
26 S. Alexander, J. Physique (Paris) 38(1977)983.
27 J. Klein and P. Pincus, Macromolecules, 15(1982)1129.
28 P.G. de Gennes, Macromolecules, 13(1980)1069.
29 G. Hadzioannou, S. Patel, S. Granick and M. Tirrell, J. Am. Chem. Soc., 108(1986)2869.
30 A. Dolan and S.F. Edwards, Proc. Roy. Soc. London, A337(1974)509.
31 A. Dolan and S.F. Edwards. Proc. Roy. Soc. London, A343(1975)427.

32 C. Thies, J. Phys. Chem., 70(1976)3783.
M. Cohen-Stuart, J. Scheutens and G. Fleer, J. Polymer Sci. (Phys.), 18(1980)559.
K. Furusawa, K. Yamashita and K. Konno, J. Colloid Interface Sci., 86(1982) 35.
33 E. Pfefferkorn, A. Carroy and R. Varoqui, J. Polymer Sci. (Phys.), 23(1985) 1997.
34 P.G. de Gennes, C.R. Acad. Sci. (Paris), 301-II(1985)1399.
35 M. Cohen-Stuart, Colloids and Surfaces, 17(1986)91.
36 K. Kremer, J. Physique, 47(1986)1269.

Afterthought: Polymers at an interface; a simplified view

The scaling description of brushes which is described here is right, but the flat profile drawn on Fig. 8 is not correct. Milner, Witten and Cates, in a series of elegant papers, have shown that the profile is parabolic, and that the chain ends have a broad distribution of distances to the wall. This is very different from what we have in a surfactant layer, where the chains are strongly extended and the chain ends are all very close to the outer surface. A common treatment has been constructed by Milner *et al.*: they show that in the surfactant case, the self-consistent potential is still (nearly) parabolic: but the relation between potential and concentration is highly nonlinear, and this does restitute the surfactant features.

Shear-Dependent Slippage at a Polymer/Solid Interface

F. Brochard[†] and P. G. de Gennes*[‡]

PSI, Institut Curie, 11 rue Pierre et Marie Curie, 75231 Paris Cedex 05, France, and Collège de France, 75231 Paris Cedex 05, France

Received July 1, 1992

We discuss shear flows of a polymer melt near a solid surface onto which a few chains (chemically identical to the melt) have been grafted. At low shear rates $\sigma < \sigma^*$ we expect a strong friction, analyzed in ref 1.[1] Above a certain critical shear σ^* the grafted chains should undergo a coil stretch transition. In the stretched state, they are not entangled with the melt, and a significant slippage is expected when $\sigma > \sigma^*$. This transition may be important in the processing of polymers, where a few chains from the melt can be bound on an extruder wall and play the role of the grafted chains.

I. Principles

When a polymer melt flows along a solid surface (Figure 1), under a shear stress σ, there may exist a nonzero flow velocity, V, at the surface. The ratio $k = \sigma/V$ is the friction coefficient. Equivalently one may describe the flow pattern in terms of a slippage length

$$b = V/\frac{dV}{dy}\Big|_0 = \eta/k \tag{1}$$

where η is the melt viscosity.

(1) In ideal conditions, with a smooth solid surface, and no chains attached to it, one expects that the friction k is comparable to what it is in a fluid of monomers: $k = k_m$. On the other hand the viscosity η of an entangled melt is huge, and b can become very large ($\sim$100 μm).[2] There are some experiments which do suggest strong slippage: (a) observations in a transparent extruder;[3] (b) rheological studies on molten polystyrene in a plate–plate geometry with small gaps;[4] (c) studies on thin films in wetting or dewetting processes;[5] (d) measurements on multilayer extrusions[6] (where we probe a number of melt/melt interfaces).

(2) However, in most usual situations, no slippage is observed.[7] This may be due to various effects: (a) a thin layer of polymer near the wall may be *glassy*; (b) some chains from the melt can bind strongly to special sites on the solid surface. Thus the solid behaves really like a fluffy carpet, and this suppresses slippage. We recently investigated a model example, where a small number of chains (ν chains per unit area) is grafted on the solid.[1,8] We found that in the low velocity regime ($V \to 0$) slippage is strongly suppressed; the slippage length b being reduced to

$$b_0 \simeq (\nu R_0)^{-1} \tag{2}$$

where $R_0 = Z^{1/2}a$ is the coil size of the grafted chains (Z is the number of monomers per grafted chain and a is a monomer size). Equation 2 gives b values which are small

Figure 1. An idealized view of shear flow near a surface, assuming that the viscosity of the liquid (η) is the same at all scales. The shear stress σ is the same at all distances y; the shear rate $S = dv/dy$ is also constant. There is a finite velocity at the surface $V(y=0) \equiv V = Sb$.

($\sim$10 nm); thus, in all cases where (through a glassy layer or through local attachment) some chains are bound to the wall, we expect no slippage in slow flows, in agreement with ref 7.

(3) In the present paper, we investigate stronger flows, and their effect on a weakly grafted layer ("weakly" meaning the "mushroom" regime, where different grafted chains do not overlap, $\nu R_0^2 < 1$). The opposite case of a strong "brush" ($\nu R_0^2 \gg 1$) under strong flows was discussed in detail by Alexander and Rabin.[9] However, one of us (P.G.) pointed out a year ago that the mushroom regime might be more interesting, become the grafted chains will undergo a "coil stretch transition" under flow.[10] At this time, a computer simulation (or an ideal grafted chain plus solvent) exhibited the transition.[11]

We realized progressively that the central problem is different: Instead of a simple solvent, we must have a polymer melt (chemically identical to the grafted chain), then the discussion becomes relevant to many practical systems, such as extruders, where the solid surface may often attach a few chains. Also, on the theoretical side, we shall see that the coil/stretch transition induced by polymer flow is more spectacular, because the grafted chains *disentangle* at the transition point; the friction in the disentangled regime is a Rouse friction,[12] much weaker than the entangled friction.

In section II, we consider a simplified system, with no wall and a chain of Z units hooked at one end onto a fixed point (Figure 2). The chain is immersed in a polymer melt (N units per chain) flowing at a velocity V, and it exhibits a transition at a certain critical velocity V^*. We constantly assume $N > Z$.

[†] PSI, Institut Curie.

[‡] Collège de France.

(1) Brochard, F.; de Gennes, P. G.; Pincus, P. *C.R. Seances Acad. Sci.* **1992**, *314*, 873–878.

(2) de Gennes, P. G. *C.R. Acad. Sci.* **1979**, *288B*, 219.

(3) Galt, J.; Maxwell, B. *Modern Plastics*; McGraw-Hill: New York, 1964.

(4) Burton, R.; Folkes, M.; Karm, N.; Keller, A. *J. Mater. Sci.* **1983**, *18*, 315.

(5) Redon, C. Private communication.

(6) Miroshnikov, A. *Vysokomol. Soedin., Ser. A* **1987**, *29*, 579–82.

(7) See the following reviews: (a) Meissner, J. *Polym. Test.* **1983**, *3*, 291. (b) Meissner, J. *Annu. Rev. Fluid Mech.* **1985**, *17*, 45.

(8) de Gennes, P. G. *Simple Views on Condensed Matter*; World Scientific: London, 1992.

(9) Rabin, I.; Alexander, S. *Europhys. Lett.* **1990**, *13*, 49–54.

(10) de Gennes, P. G. *J. Chem. Phys.* **1974**, *60*, 5030.

(11) Mavrantzas, V. G.; Beris, A. N. *J. Rheol.* **1992**, *36*, 175–213.

(12) For a general presentation of this concept, see de Gennes, P. G. *Introduction to polymer dynamics*; Lezioni Lincee; Academia Nazionale dei Lincei: Rome, 1990; pp 1–57.

0743-7463/92/2408-3033$03.00/0

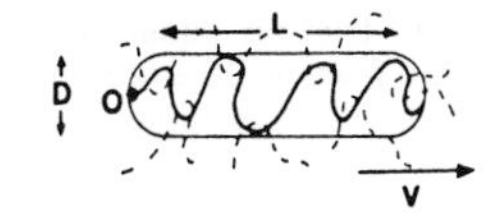

Figure 2. A tethered chain—attached by one end at a fixed point O—immersed in a moving polymer melt, of velocity V. When V is not too low ($V \sim V^*$) the chain elongates significantly.

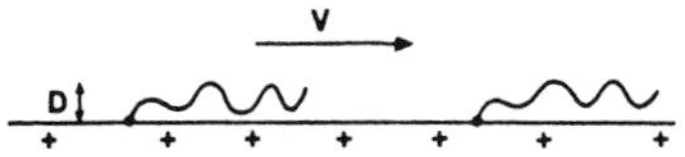

Figure 3. A grafted chain under shear flow in a melt. Where the shear stress σ exceeds a low threshold σ^* (eq 22), the chain is strongly stretched.

In section III, we investigate the chain plus wall (Figure 3). Here, what is imposed is not the velocity at the surface V, but the shear rate S or, equivalently, the shear stress $\sigma = \eta S$. In this situation it will turn out that, beyond threshold, the slippage length b is velocity dependent, and thus the velocity at the surface $V = bS$ is not proportional to S; thus the discussion is slightly more complicated.

Possible consequences, limitations, and extensions of these ideas are listed in section IV.

All our discussion is restricted to the level of scaling laws; in most formulas, we ignore numerical coefficients and replace the equality signs by the symbol $\simeq$. To find exact coefficients in the entangled regime would represent a very heavy program.

II. One Tethered Chain in a Flowing Melt

1. Friction in the Entangled Regime. Our chain is pictured in Figure 2 as an elongated object of length L and diameter D. The elastic force F or the hook is derived from standard discussions on ideal chains[12,13]

$$F = \frac{3kT}{R_0^2} L \simeq \frac{kT}{D} \tag{3}$$

(a) At very low velocities ($L < R_0$), our chain is a spheroid of size R_0, and we know from ref 1 that the friction law, relating F and V, is (surprisingly) similar to a Stokes law

$$F \simeq \eta R_0 V \tag{4}$$

(b) How is eq 4 modified when we have a strong elongation ($L > R_0$)? Let us extend the argument of ref 1 and first count the number m of matrix chains which are entangled with the (Z) coil. The volume of the elongated object is $\Omega \sim D^2 L$. Each of the m chains has a number g of monomers in the volume Ω, and crosses it over a length $\sim D$. Thus $g \sim D^2/a^2$. The volume Ω is mainly filled with matrix chains, and thus

$$\Omega \sim a^3 m g \tag{5}$$

where a^3 is a monomer volume. Thus

$$m = L/a \qquad (L \gg R_0) \tag{6}$$

We now estimate the dissipation $T\dot{S}$ due to the m chains reptating following the ideas of refs 1 and 14. Each of them has a curvilinear velocity $v \sim (N/N_e)V$ (where N_e is the entanglement distance) and

$$T\dot{S} = m \zeta_1 N v^2 \tag{7}$$

where ζ_1 is a friction coefficient for a monomer. By use of eq 6 this can be cast in the form

(13) Pincus, P. *Macromolecules* **1976**, *9*, 386–391.
(14) de Gennes, P. G. *MRS Bull.* **1991**, *20–21*.

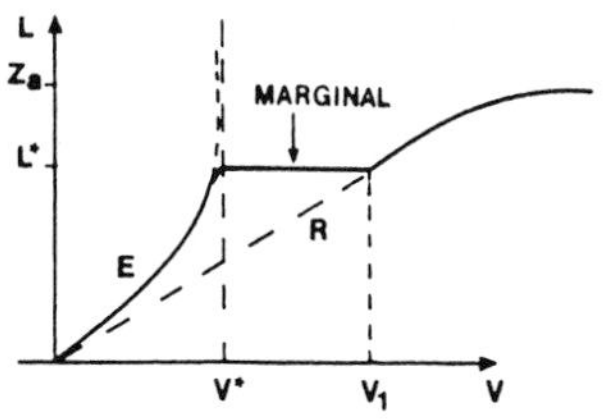

Figure 4. Elongation L versus flow velocity V for a tethered chain in a melt: (E) = entangled regime; (R) = Rouse regime (disentangled).

$$T\dot{S} \simeq \eta V^2 L \tag{8}$$

where

$$a^{-1} \zeta_1 \frac{N^3}{N_e^2} = \eta$$

is the reptation viscosity of the melt. Equating $T\dot{S}$ to the product of force by velocity FV we arrive at the basic formula

$$F \simeq \eta L V \qquad (L \gg R_0) \tag{9}$$

(c) To interpolate between the low elongation case (eq 4) and the high elongation case (eq 9), we shall use the following formula

$$F = \eta V (R_0^2 + L^2)^{1/2} \tag{10}$$

Other interpolatioins could be used—they are all equivalent at our scaling level.

2. Critical Velocity V^*. Equating the elastic force (eq 3) and the viscous force (10), we obtain a relation between elongation L and velocity V of the form

$$V/V^* = \frac{L}{(R_0^2 + L^2)^{1/2}} \tag{11}$$

where we have defined

$$V^* \simeq \frac{kT}{\eta R_0^2} = \frac{kT}{\eta Z a^2} \tag{12}$$

Typically for $\eta = 10^4$ P, $Z = 1000$, $a = 3$ Å, $V^* = 0.05$ μm/sec. The plot of $L(V)$ is shown on Figure 4. In our approximation L diverges for $V = V^*$. Of course, this divergence is formally suppressed by the finite extensibility of the Z chains. But we shall see that the physical cut off is different.

3. Marginal State. (a) The preceding discussion assumed that the Z chains are entangled with the surrounding melt. This requires that the number, g, of monomers spanning the cross sectional diameter D, be larger than the entanglement length N_e. The limiting point corresponds to

$$D = D^* \simeq N_e^{1/2} a \tag{13}$$

$$L = L^* \simeq R_0^2 / D^*$$

Whenever $N_e \ll Z$, this elongation L^* is much larger than R_0, and from the plot of Figure 4, we see that the elongation L^* is reached at a velocity practically equal to V^*.

(b) What happens if we impose a velocity V somewhat larger than V^*? At first sight we might think that the Z chain disentangles completely. However, if we go to this regime, we expect a friction force much weaker than eq 9. From the Rouse model we would write the friction as a sum of independent contributions from each monomer

$$F = Z\zeta_1 V \quad \text{(Rouse)} \tag{14}$$

Equating (14) to the elastic force (3), we would obtain a much shorter elongation L_R (where R stands for Rouse)

$$L_R \simeq \frac{Z\zeta_1 R_0^2 V}{kT} = L^* \frac{V}{V_1} \tag{15}$$

where we have introduced another characteristic velocity

$$V_1 = \frac{kTL^*}{Z\zeta_1 R_0^2} = \frac{kT}{Z\zeta_1 D^*} = V^* \frac{\eta}{\eta_1 N_e^{1/2}} \tag{16}$$

The quantity $\eta_1 \propto a^{-1}\zeta_1$ is something like the viscosity of a liquid of monomers. Typically $N_e^{1/2} \sim 10$ and $\eta/\eta_1 \sim 10^4$. Thus $V_1 \gg V^*$; there is a large interval of velocities $V^* < V < V_1$ where we can say the following: If the Z chain disentangles, it returns to a length L_R much smaller than L^*, or to a diameter $D > D^*$. Then it must entangle again!

Thus we are led to expect that, at fixed velocity in this interval, the Z chain stays in a *marginal state*, with $D = D^*$ and $L = L^*$. Ultimately, at $V > V_1$, we should go to a disentangled state, with $L = L_R > L^*$. The overall plot of elongation L versus velocity then looks as shown on Figure 4, with a long plateau.

It is nearly equivalent to plot the force F as a function of the conjugate variable V, since the force is proportional to L (eq 3) in the domain of interest ($L \sim L^* \ll Za$). Note finally that if the *force F* is monitored, we expect an abrupt jump of velocity (from $V = V^*$ to $V = V_1$) when F reaches a threshold value

$$F^* \simeq \frac{kTL^*}{R_0^2} = \frac{kT}{N_e^{1/2} a} \tag{17}$$

III. Grafted Chains in Strong Shear Flows

1. General Program. We now return to the situation depicted in Figure 2, with ν grafted chains per unit area, and a low coverage ($\nu R_0^2 < 1$). The imposed shear stress is

$$\sigma = \eta S = k_m V + \nu F \tag{18}$$

Here $S = dv/dz$ is the shear rate. The term $k_m V$ describes the weak friction due to monomer wall interactions

$$k_m \sim \zeta_1 a^{-2} \sim \eta_1 a^{-1}$$

where η_1 is something like the viscosity of a fluid of monomers. The second term νF is proportional to the elastic force F present on the stretched Z chains, given by eq 3. The main point here is that the Z chains stretch near the wall very much as they do in the infinite matrix of section II. The formulas for the diameter D and the length L remain valid (within coefficients). In particular we retain the friction eqs 9 (in the entangled regime) and 14 (in the Rouse regime). The meaning of V is now the slippage velocity at the surface. It will turn out that the slip length b is always much larger than the chain diameter D; this implies that the velocity profile is essentially flat at the scales probed by the grafted chain

$$\frac{V(z=D) - V}{V} = \frac{SD}{V} = \frac{D}{b} \ll 1$$

Thus, for our purposes, the grafted chains see a uniform flow field.

Our program is then the following: (a) we fix the shear stress σ (or equivalently the macroscopic shear rate $S = \sigma\eta^{-1}$); (b) using the friction equations (10 or 14), we eliminate F and find from eq 18 a relation between σ and V, from which we extract the slippage length

$$b(\sigma) = V/S = \eta V/\sigma$$

2. Simple Limits. It may be useful to first recall certain limiting cases: (i) *At very small σ* the friction equation is eq 4; inserting this into (18), we find a friction coefficient

$$k_0 = k_m + \nu\eta R_0 \qquad (\sigma \to 0) \tag{19}$$

(Note that $k_0 \simeq \nu\eta R_0$ since the k_m term is usually negligible here). The corresponding slippage length is

$$b_0 = \frac{\eta}{k_0} = \frac{\eta}{k_m + \nu\eta R_0} \simeq \frac{1}{\nu R_0} \tag{20}$$

in agreement with eq 2.

(ii) *In the marginal state* ($L = L^*$) the force F has a fixed value F^* (eq 17) and

$$\sigma = k_m V + \nu F^* \tag{21}$$

Here again the k_m term is usually negligible and σ is essentially constant

$$\sigma = \sigma^* \equiv \nu F^* = \frac{\nu kT}{N_e^{1/2} a} \tag{22}$$

In this regime the slippage length $b(V)$ is increasing linearly with V

$$b(V) = \eta V/\sigma^* \tag{23}$$

(iii) *In the disentangled state* $F(V)$—given by eq 14—is linear in V and the slippage length is now independent of V

$$b = b_\infty = \frac{\eta}{k_m + \nu\zeta_1 Z} \tag{24}$$

Here k_m is not negligible, and in fact is dominant

$$k_m + \nu\zeta_1 z \sim \frac{\eta_1}{a}[1 + \nu Za^2] = \frac{\eta_1}{a}[1 + \nu R_0^2] \simeq \frac{\eta_1}{a} = k_m \tag{25}$$

since we constantly assume a mushroom regime ($\nu R_0^2 \ll 1$). Thus $b_\infty \simeq \eta/k_m$ is nearly equal to its value for an ideal wall. Note that the ratio between the two extreme slippage lengths is very large

$$\frac{b_\infty}{b_0} \sim \frac{\eta}{\eta_1} Z^{-1/2} \gg 1 \tag{26}$$

3. Slippage Length versus Velocity. If we call $L(V)$ the elongation deduced from section II, we may write eq 18 in the form

$$k_m V + \frac{\nu\, kT\, L(V)}{R_0^2} = \sigma = \eta S = \eta V/b(V) \tag{27}$$

and this gives

$$b(V) = \frac{\eta}{k_m + \nu \dfrac{kT}{R_0^2}\dfrac{L(V)}{V}} \tag{28}$$

The results are shown in Figure 5. At $V = 0$, we start with a relatively small value $b = b_0$. When V increases, eq 10 shows us that $F(V)/V$ increases; thus $b(V)$ decreases and reaches a minimum at $V = V^*$ where $L = L^*$. This value is

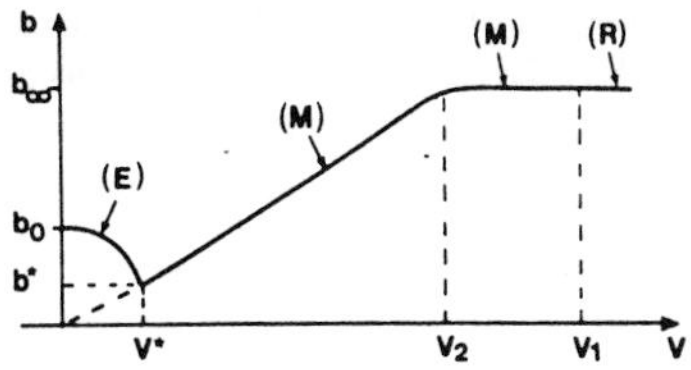

Figure 5. Slippage length versus surface velocity for a weakly grafted surface exposed to a polymer melt (chemically identical to the grafted chains): (E) = entangled regime; (M) = marginal regime; (R) = Rouse regime (disentangled).

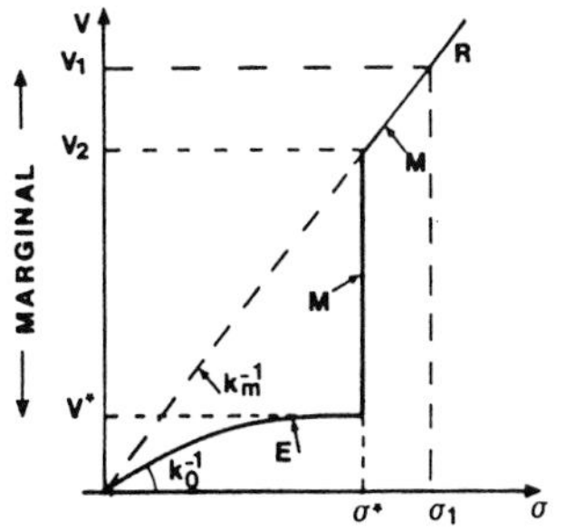

Figure 6. Surface velocity versus surface shear for a grafted surface exposed to a melt. Note the jump in velocities at $\sigma = \sigma^*$.

$$b = b^* = \frac{\eta V^*}{F^*} = \frac{N_e^{1/2}}{\nu Z a} \qquad (29)$$

We can check that $b^* < b_0$ since $Z \gg N_e$. Beyond $V = V^*$, we have a marginal regime with $L = L^*$ and b linear in V (eq 23). At this poiint an interesting deviation from section II shows up; the upper velocity limit of the linear law $b(V)$ is not $V = V_1$ as defined in eq 16, but is a smaller velocity V_2. When $V = V_2$, the slippage length b crosses over continuously from the marginal form (eq 16) to the ideal surface value $b = b_\infty$; at this moment the dissipation due to the grafted chains becomes negligible. Using eq 28 with $b(V_2) = b_\infty$, we arrive at

$$V_2 = \frac{\nu kT}{\eta_1 N_e^{1/2}} = V_1 \nu R_0^2 < V_1 \qquad (30)$$

We should emphasize that the marginal state is still realized in the interval $V_2 < V < V_1$, but in this region, it has no visible effects on the slippage length.

4. Velocity V versus Stress σ. This is again derived from eq 27 and shown in Figure 6. The stress $\sigma(V)$ increases linearly at low V ($V \ll V^*$), and nearly diverges at $V = V^*$. A cut off is provided by the marginal state $L = L^*$, giving $\sigma = \sigma^*$ (eq 22). In the interval $V^* < V < V_2$, σ is nearly constant ($\sigma = \sigma^*$). At higher velocities $V > V_2$, σ is linear in V ($\sigma \sim k_m V$).

IV. Concluding Remarks

1. Critical Stress σ^* and Critical Shear Rate S^*. The major conclusion from section III is that slippage is restored above a low stress threshold σ^* (eq 22). It is important to realize that the associated shear rate $S^* = \sigma^* \eta^{-1}$ is *not* the inverse of a reptation time for the Z chain. In fact it is related to the reptation time $T_R(N)$ of the *matrix* chains[12]

$$T_R(N) = \frac{N^3}{N_e} \frac{\eta_1 a^3}{kT} \qquad (31)$$

The relation between $T_R(N)$ and S^* is (from eq 22)

$$S^* T_R(N) \sim \nu a^2 N_e^{1/2} \qquad (32)$$

2. Role of Polydispersity. All our discussion assumed a highly peaked distribution of chain lengths (Z) for the grafted chains. This may hold if the surface is grafted with great care. On the other hand, we mentioned the possibility that certain solid surfaces can bind a few matrix chains at one (or more) unit along the chain. This would then generate a very broad distribution of Z. We call this $\nu(Z)$; $\nu(Z)$ is a number distribution, i.e. the number of chains (or loops) with Z units per unit area. This is normalized by

$$\sum_Z \nu(Z) = \nu \qquad (33)$$

The jump in $V(\sigma)$ shown in Figure 6 will clearly be attenuated by polydispersity. We shall now discuss this briefly, assuming that we are in the transition regime, i.e. all chains are entangled or marginal. At a given V, we can then write down the stress by splitting the grafted population into two parts:
(a) coils for which $V < V^*$, or equivalently $Z < Z(V)$, where

$$Z(V) \equiv V_0/V \qquad V_0 \simeq kT/(\eta a^2) \qquad (34)$$

For these coils we use the friction equation (10).
(b) coils with $V > V^*$ or $Z > Z(V)$, where we have a marginal state, and a force $f^* = kT/D^*$ independent of Z. This leads to

$$\sigma = \eta V \sum_{Z=N_e}^{Z(V)} \nu(Z) \frac{a Z^{1/2}}{[1 - (Z/Z(V))^2]^{1/2}} + \sum_{Z(V)}^{\infty} \nu(Z) \frac{kT}{D^*} \qquad (35)$$

Let us choose for instance a *flat distribution* extending from $z = 1$ to $z = P$ (with $P \gg N_e$)

$$\nu(Z) = \begin{cases} P^{-1}\nu & (1 < Z < P) \\ 0 & (Z > P) \end{cases} \qquad (36)$$

Inserting (36) in (35) we find two regions:
(a) for $V < V_0/P$, no molecule is marginal and σ increases originally linearly in V

$$\sigma \sim \eta P^{1/2} a \nu V$$

(b) for $V > V_0/P$, some grafted chains are marginal and

$$\frac{\sigma}{\nu \kappa T} \simeq \frac{1}{P^{1/2} a} \left(\frac{V_0}{PV}\right)^{1/2} + \frac{1}{D^*}\left(1 - \frac{V_0}{PV}\right) \qquad (37)$$

The second term on the right-hand side of eq 37 is due to the marginal chains, and is rapidly dominant. Thus the $\sigma(V)$ plot has lost the exact plateau of Figure 6, but σ does reach σ^* as soon as $V \gg V_0/P$, i.e. when V is larger than the characteristic V^* of the longer chains; in practice we still expect a large plateau.

3. Coil Stretch versus Debonding. If we observe a sharp transition between nonslippage to slippage in increasing shear stresses, σ, may we conclude that this is the coil stretch transition discussed in the present paper? We must be quite sure that the Z chains are not torn out from the surface.

(a) If we are dealing with *grafted chains*, it is easy to check whether the plot of $b(\sigma)$ is reversible or not.

(b) If we are dealing with a *nongrafted surface*, onto which a few chains from the matrix do bind (with a distribution of loop sizes extending up to some maximum length P), then we must compare forces. In the marginal state, the force experienced by the bound chains is $f^* = kT/(N_e^{1/2}a)$. We must compare this to the minimum force for tearing out one chain/surface bond, which we call f_B. Our discussion will hold only if $f_B > f^*$.

Acknowledgment. This work was initiated by discussions with H. Hervet and L. Léger. It was performed at the NATO ASI workshop on Interfacial Interactions in Polymeric Composites (June 1992). We wish to thank Professor G. Akovali for his hospitality on this occasion and X. Olympos for various exchanges.

Reprinted from Macromolecules, 1996, 29. 377

Slippage of Polymer Melts on Grafted Surfaces

F. Brochard-Wyart,† C. Gay,‡ and P.-G. de Gennes*,‡

PSI, Institut Pierre et Marie Curie, 11 rue P. et M. Curie, 75005 Paris, France, and Collège de France, 11 place Marcelin Berthelot, 75231 Paris Cedex 05, France

Received May 31, 1995; Revised Manuscript Received October 11, 1995®

ABSTRACT: We study the slippage of a highly viscous polymer melt (P monomers per chain) on a solid substrate grafted by a few smaller chains in the mushroom regime (N monomers per chain, grafting density ν). The friction is provided by the sliding motion of the P chains of the "skin" (thickness $R_p = P^{1/2}a$) which are entangled with the tethered chains. At low grafting densities, only a fraction of the P chains in the skin are coupled to the N chains, and the friction on the mushrooms is additive. Above a threshold ν_c, all P chains of the skin are trapped, and the low-velocity friction becomes independent of the grafting density. Above a certain threshold slippage velocity $V^*(\nu)$, the N chains are strongly stretched and reach a "marginal state", corresponding to a constant shear stress. We expect that for $\nu > \nu_c$, $V^*(\nu)$ increases linearly. Depending on N, P, ν, and V, we predict a cascade of regimes, where the N chains may be ideal, stretched, or "marginal", while the trapped chains may be ideal or stretched and progressively disentangle from the N chains.

I. Introduction

Entangled polymers do not flow like usual liquids. One of us[1] predicted that polymer melts slip on a "smooth, passive" surface. More generally, the slippage is characterized by the extrapolation length b, defined by the distance to the wall at which the velocity extrapolates to zero. The ratio between the shear stress σ at the S/L interface and the surface velocity V defines a friction coefficient k:

$$\sigma = kV \tag{1}$$

The polymer melt has a huge viscosity η_p. By equating the two forms of the shear stress,

$$\sigma = \eta_p \left.\frac{dv}{dz}\right|_0 = kV \tag{2}$$

one can relate the extrapolation length b to k:

$$b = \eta_p/k \tag{3}$$

For *ideal conditions*, with an ungrafted surface, $k = k_0 = \eta_0/a$, where η_0 is a monomer viscosity and a a molecular size. The viscosity η_p is huge, $\eta_p \cong \eta_0 P^3/N_e^2$, where N_e is the number of monomers per entanglement. Equation 3 leads to

$$b_\infty = a\frac{P^3}{N_e^{\,2}} \tag{4}$$

(typically, $b = 10\ \mu$m for $P \approx 10^3$ and $N_e \approx 10^2$).

Some time ago, we studied *semiideal conditions*, where a few long N molecular chains ($N < P$) have been grafted on the solid surface, in the mushroom regime ($\nu < 1/R_N^2$, where $R_N = N^{1/2}a$ is the coil size of the grafted chains). We found[2] that these few chains lead to a huge friction in the low-velocity regime ($V \to 0$):

$$\sigma = [k_0 + \nu\eta_p R_N]V \tag{5}$$

At most practical values of ν, the second term dominates and b reduces to

$$b_0 \cong (\nu R_N)^{-1} \tag{6}$$

Equation 6 gives b values which are extremely small ($b_0 \sim 100$ Å). Slippage is suppressed at small velocities.

Experiments[3] have shown that slippage is progressively *restored* above a critical velocity V^*. We have interpreted these results[4] by a coil−stretch transition of the N chains, which disentangle from the melt.

In refs 2 and 4, we showed the dissipation is due to the sliding motion of the P chains, which are entangled with one grafted chain. However, above a threshold ν_c, one P chain is entangled with more than one mushroom, and the sliding motion of a melt chain releases many entanglements. Our aim here is to include this cooperative effect, which will decrease the effective dissipation per chain. In section II, we go back to low grafting densities where the effect of the grafted chains is additive. In section III, we look at higher grafting densities, at which the cooperative effect described above plays a role.

The main difficulty, encountered in section II, is related to the number X of mobile chains which are trapped by a single mushroom. It turns out that X depends on a very detailed description of entanglements, as pointed out in ref 5. Opposite limits correspond to (a) binary entanglements (where only *two* chains build up a constraint) and (b) collective entanglements (where a number $\sim N_e^{1/2}$ of chains are required to construct one constraint, following an interesting idea of Kavassalis and Noolandi.[6]

The "binary" assumption is described in detail in ref 7. It is somewhat simpler to explain: for this reason, we use it in our discussion of section II. Fortunately, the distinction between (a) and (b) drops out when we come to the case of interest in this paper, namely, higher grafting densities, ν, where *all* the mobile chains which touch the surface are trapped, independently of a detailed model of entanglements.

We also simplify section II by assuming that the grafted chains are not exceedingly long ($N < N_e^2$). The opposite limit ($N > N_e^2$) is described in Appendix A. Appendix B shows the explicit link with experimental parameters.

† PSI, Institut Pierre et Marie Curie.
‡ Collège de France.
® Abstract published in *Advance ACS Abstracts,* December 1, 1995.

0024-9297/96/2229-0377$12.00/0

II. One Single Chain and the Low Grafting Density Limit

This case will be important at small grafting densities $\nu < \nu_c$. It can also be observed directly by manipulating one single chain immersed in a melt, using optical tweezers.[8] Let us first return to the calculation of the friction F_v experienced by one single tethered chain in the limit $V \to 0$.[2] When the mobile chains (P) move at velocity V with respect to the N grafted chain, the P chains which are entangled have to slide.[5] To move one entanglement a distance D^*, each P chain must slide over its tube length $L_t = Pa^2/D^*$. The tube velocity of the P chain is then $V_t = (P/N_e)V$. The friction force per sliding chain is

$$f_v = \eta_0 aP\left(\frac{P}{N_e}\right)^2 V = \eta_p aV \tag{7}$$

where η_p is the viscosity of the P melt.

The total friction force acting on the grafted chain is then given by

$$F_v = Xf_v = Xa\eta_p V \tag{8}$$

where X is the number of trapped chains: i.e., P chains entangled with the mushroom.

Different forms have been proposed for X.[2,4,5,7] Here, we shall simply summarize the results of the "binary entanglement model" described in ref 7, assuming $P \geq N$. There are two upper bounds for X:

(a) X cannot be larger than the total number of P chains which intersect the (N) "mushroom": the corresponding volume is $N^{3/2}a^3$ and each intersecting P chains puts $\sim N$ monomers in this region. Thus the total number of P chains concerned is $N^{3/2}/N = N^{1/2}$.

(b) X cannot be larger than the number of constraints experienced by the N chain, namely N/N_e.

Ultimately, for binary entanglements, it turns out that X is the smaller of these two bounds.[7] Thus

$$\left.\begin{aligned} X &= N/N_e \quad (N < N_e^2) & (9a)\\ X &= N^{1/2} \quad (N > N_e^2) & (9b)\end{aligned}\right\}$$

In most practical cases, $N_e \gtrsim 100$ and $N < N_e^2$. Then, the friction force in eq 8 is proportional to N: we call this the *Rouse regime*. There may be some (rare) cases where $N > N_e^2$. Then, the friction force in eq 8 should be proportional to the radius of the mushroom $N^{1/2}a = R_n$ and to the melt viscosity: it is similar (at the level of scaling laws) to the Stokes friction on a hard sphere of radius R_N. We call it the *Stokes regime*.

In what follows, we focus our attention on the Rouse regime. The opposite limit $N > N_e$ is discussed in Appendix A.

The friction form at low velocity is thus, from eqs 5, 8, and 9a,

$$\begin{cases} F_v = \dfrac{N}{N_e}a\eta_p V & (10a)\\ \sigma = \nu F_v = \nu \dfrac{N}{N_e}a\eta_p V & (10b)\end{cases}$$

This corresponds to a slippage length

$$b = \frac{N_e}{\nu aN} \tag{11}$$

The linear regime described by eq 10 will hold when $F_v < kT/R_N$, or equivalently at velocities $V < V_1$, where

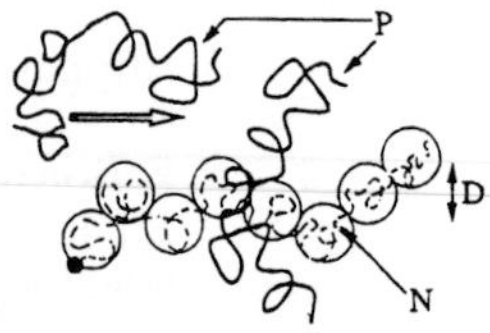

Figure 1. A tethered chain (N monomers) is elongated into a sequence of blobs of size D. Some of the melt chains (moving with velocity V) are entangled with it. In the marginal state ($D = D^* = aN_e^{1/2}$), each blob contains N_e monomers.

$$V_1 = \frac{kTN_e}{a^2\eta_p N^{3/2}} \tag{12}$$

V_1 can be found by two methods: (i) by equating the friction force F_v and the elastic force required to get an elongation $\sim R_N$:

$$\frac{N}{N_e}a\eta_p V_1 \equiv kT\frac{R_N}{Na^2}$$

(ii) by considering the coil relaxation time in the melt flow. If δR is the end-to-end vector of the grafted chain, we can write down the force balance:

$$kT\frac{\delta R}{Na^2} = \frac{N}{N_e}a\eta_p(V - \delta\dot{R})$$

As a matter of fact, if the coil is elongating ($\delta\dot{R} > 0$), monomers experience a flow at a reduced speed. We calculate therefrom a typical relaxation time:

$$\frac{1}{\tau_{rel}} = \frac{kT}{\eta_p a^3}\frac{N_e}{N^2}$$

The threshold velocity V_1 for large elongations ($\delta R > R_N$) is then given by $V_1 = R_N/\tau_{rel}$.

At higher velocities, the grafted chains are elongated into a cigar shape (as assumed in ref 4) or more precisely into a trumpet shape (as shown in ref 5), because the cumulative friction on the chain increases from the free end to the attachment point. Both descriptions are essentially equivalent in their final results. Here, for simplicity, we use the simpler picture, where the N chain is deformed into a cigar of diameter D, length $L = Na/D$, under the viscous friction force due to the sliding motion of the P chains entangled with it.

Elongation always favors a Rouse regime. (This is shown in detail for binary entanglements.) Thus, even for $V > V_1$, we keep $X = N/N_e$ for the distorted chain, and eqs 10 and 11 remain valid. The diameter of the cigar is derived from the Pincus law:[9]

$$D = kT/F_v = R_N\frac{V_1}{V} \tag{13}$$

When V increases, D decreases, and ultimately we reach the *marginal regime* where $D = D^* = aN_e^{1/2}$ (the tube diameter) (Figure 1). At this point, the N chain is on the verge of disentangling from the melt. But, if it did so, the drag force would decrease, D would increase, and entanglements would reappear. Thus we stick to $D = D^*$, and the shear stress σ reaches a plateau value:

$$\sigma^* = \nu kT/D^* \quad (14)$$

The corresponding velocity is

$$V_0^* = \frac{kTN_e^{1/2}}{Na^2\eta_p} \quad (15)$$

When $V > V_0^*$, σ sticks to the value σ^*, and the slippage length is linear in V:

$$b = \frac{\eta V}{\sigma^*} \qquad (V > V_0^*) \quad (16)$$

What happens at velocities $V \gg V_0^*$? First, the number of trapped chains $X(V)$ decreases. The friction per chain in the marginal regime is constant and equal to kT/D^*. The friction per sliding chain increases linearly with velocity V. Thus

$$X(V) = X(0)V_0^*/V$$

Ultimately, at a certain (high) velocity V^{**}, the melt chains decouple from the (N) chains. This process may be understood as follows:

The tube velocity of one (P) chain disentangling from the (N) chain is of order

$$V_t = \frac{L_t}{D^*}V \quad (17)$$

where $L_t = PD^*/N_e$ is the tube length. Equation 17 ensures that when the N chain has moved by a distance D^* with respect to the fluid, the P chain has released one constraint (and thus moved by $\sim L_t$ inside its tube). This removal of constraints is balanced by the establishment of new entanglements. The trapped P chains will be deeply perturbed if their relaxation time $T_{rep}(P)$ becomes longer than the tube evacuation time:

$$T_{rep}(P) > L_t/V_t = D^*/V \quad (18)$$

The nature of the perturbed state has recently been discussed by Ajdari for a related problem:[10] it may in fact lead to stick slip instabilities. From the present point of view, we conclude that there is a limiting velocity:

$$V^{**} = D^*/T_{rep} \quad (19)$$

Using standard relations for entangled media, we can write a relation between T_{rep} and the P melt viscosity η_p, namely

$$\eta_p = \frac{kT}{N_e a^3}T_{rep} \quad (20)$$

where $kT/N_e a^3$ is the plateau modulus of the melt. Combining eqs 19 and 20, we arrive at

$$V^{**} = \frac{kT}{D^* a\eta_p} \quad (21)$$

Notice that the pulling force $a\eta_p V^{**}$, acting on a P chain entangled with the N chain, is just equal to the marginal value kT/D^*: the P chains become marginal at $V = V^{**}$. The number of trapped chains per grafted chain also happens to be equal to unity at this stage: $X(V^{**}) = 1$.

III. Higher Grafting Densities

(1) From Separate Mushrooms to Cooperative Behavior. The central question here is to find how many P chains are entangled with the mushrooms: we call them *trapped chains*. Their number (per unit area) is designated by ν_p.

(a) At very low grafting densities ($\nu \to 0$), we must simply add up the contributions of independent mushrooms, and this gives (in the binary entanglement model)

$$\nu_p = X\nu = \frac{N}{N_e}\nu \qquad (N < N_e^2) \quad (22)$$

(b) At some critical value of the grafting density ($\nu = \nu_c$), the number of trapped chains saturates. All the trapped chains are confined in a certain "skin" of thickness $R_p = aP^{1/2}$ near the solid surface. The maximal number of trapped chains is thus

$$\nu_{pm} = \frac{1}{Pa^3}R_p = a^{-2}P^{-1/2} \quad (23)$$

and (by comparison with eq 22) this defines

$$\nu_c = \frac{N_e}{Na^2P^{1/2}} \quad (24)$$

In order to understand how ν_p saturates for $\nu > \nu_c$, it is also instructive to calculate ν_p as the total number of available entanglement points with the grafted chains $\nu N/N_e$, divided by the number of C of constraints that each entangled melt chain makes with grafted chains; C is evaluated by appropriate counting. This viewpoint will be discussed separately. Note that when $P > N_e^2$, the quantity $\nu_c Na^2 = \nu_c R_N^2$ is smaller than unity: the mechanical effects of the grafted chains are not additive, *although the mushrooms do not overlap*!

(c) It is of some interest to compare this discussion with another feature, related to the statistics of the grafted chains: at a certain grafting density ν_b, we cross over from mushrooms to extended chains. In this "brush" regime, the free energy per N chain has the scaling form[11]

$$\frac{F_N}{kT} = \frac{L_b^2}{Na^2} + \frac{N\phi}{P_{eff}} \quad (25)$$

The first term describes the entropic elasticity of deformed chains: L_b is the thickness of the brush. The second term describes repulsive interactions between grafted monomers. The volume fraction inside the brush is

$$\phi = \nu\frac{Na^3}{L_b} \quad (26)$$

The repulsive interactions inside the brush are screened by the P matrix:[11] hence the factor P_{eff} in eq 25. If the P chains were small ($P < N$), we would simply have $P_{eff} = P$. In our case ($P > N$), it is legitimate to take $P_{eff} = N$ (the effective length of the P chain portions which penetrate). To discuss ν_b, we are interested in the onset of the brush regime—i.e., when the second term of eq 25 (estimated via eq 26 with an unperturbed thickness

$L_b \sim N^{1/2}a$) becomes of order unity, extension starts. This leads to

$$a^2 \nu_b = N^{-1/2} \tag{27}$$

Comparing ν_b (eq 27) and ν_c (eq 24), we see that

$$\frac{\nu_b}{\nu_c} = \frac{(NP)^{1/2}}{N_e} \tag{28}$$

This ratio is always larger than unity (since $N > N_e$ and $P > N_e$). Thus, upon increasing ν, the dynamical coupling occurs before the onset of a brush.

(2) Dynamical Behavior of the *N* Chains in the Coupled Regime ($\nu > \nu_c$).

(a) Onset of Distortions on the *N* Chains. At low V, each trapped P chain contributes a shear force $\eta_p a V$. The number of trapped chains per unit area is $P^{-1/2}a^{-2}$, and the resulting stress is

$$\sigma = P^{-1/2}a^{-2}\eta_p a V \tag{29}$$

corresponding to a length $b \sim P^{1/2}a$ (the thickness of the trapped layer).

The force F_v on one N chain is such that $\sigma = \nu F_v$. When this force becomes equal to kT/R_N, the N chain begins to distort. Using eq 29, this gives a crossover velocity

$$V_1 = \frac{\nu\, kTP^{1/2}}{\eta_p N^{1/2}} \tag{30}$$

Note that eqs 12 and 30 coincide at the crossover point $\nu = \nu_c$.

(b) Marginal Regime of the *N* Chains. The velocity $V^*(\nu)$ at which the N chains become marginal is derived directly from the force balance, writing the stress in two forms:

$$\sigma = \nu \frac{kT}{D^*} = P^{-1/2}a^{-2}\eta_p a V^* \tag{31}$$

Thus V^* is linear in ν:

$$V^*(\nu) = V_0{}^* \frac{\nu}{\nu_c} \tag{32}$$

where $V_0{}^*$ is the value for a single mushroom (eq 15).

(c) Decoupling of the *P* Chains. For $V > V^*$, the number of coupled chains decreases and as explained in section II, the P chains are seriously affected at velocities $V > V^{**}$, where V^{**} is given by eq 21 and is independent of ν. We have no deep knowledge of what happens at $V > V^{**}$, but we suspect that the P chains are torn out from the grafted chains and that a completely novel regime follows: this will require a separate study. The various regimes (a, b, c) are summarized in Figure 2.

It must also be emphasized that, above a certain velocity $V_{1p} = kT/a^2\eta_p P^{1/2}$ (which is reached before V^{**}), the trapped chains are also deformed: their density ν_p has to be counted somewhat differently. However, we find that this does not lead to any deep alteration of the scaling laws. Thus the regimes found above do extend up to V^{**}.

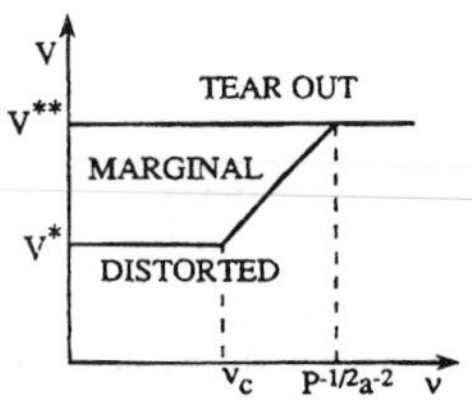

Figure 2. Various regimes for a weakly grafted solid (grafting density ν) under shear flow (slippage velocity V). Here, the surface is grafted with moderately long chains ($N_e < N < N_e^2$).

IV. Concluding Remarks

(1) The main surprise is the correlated behavior between mushrooms which occurs at grafting densities $\nu > \nu_c$. In this regime, (a) all chains of the skin are coupled and (b) the grafted polymer becomes denser and stronger upon increasing ν, while the number of trapped melt chains saturates at $\nu > \nu_c$. Thus the interface chains become less deformed (at fixed V), or equivalently the characteristic velocities V_1 (for deformation) and V^* (for marginal behavior of the grafted chains) increase. Ultimately, we should go back to the case of strong brushes in strong flows. A related problem has been discussed by Alexander and Rabin[12] and by Barrat.[13] However, their analysis applied for a liquid of small molecules ($P = 1$) facing a wet brush while, here, we are talking of a melt ($P > N$) facing a dry brush.

(2) The marginal velocity $V^*(\nu)$ in eq 33 in the cooperative regime is independent of the model chosen for entanglements: "binary" or "collective" (or other).

(3) We assumed long mobile chains ($P > N$). But the problems at small P ($P < N$) are also interesting; they will be discussed separately.

(4) At very high N ($N > N_e^2$), a single mushroom (at low ν) has a Stokes behavior rather than a Rouse behavior, and the diagram of Figure 2 becomes more complex. This is discussed in Appendix A. However, once again, the main formulas—in particular, $V^*(\nu)$ in eq 32—are quite independent of the precise choice of X for a single mushroom.

Acknowledgment. We have greately benefited from discussions with A. Ajdari, H. Hervet, J. F. Joanny, L. Léger, E. Raphaël, and especially J. L. Viovy, who patiently explained to us, over many years, the difference between binary entanglements and collective entanglements.

Appendix A: Grafted Surface with Ultralong Chains

When $N > N_e^2$, two main differences show up:

(1) Friction on One Single Chain ($\nu \to 0$): Zimm to Rouse Behavior. The number $X(V{=}O)$ of coupled chains is now $N^{1/2}$. The friction forces the grafted chain as the P melt moves at velocity V ($V \to 0$) is a Stokes friction:

$$F_v = N^{1/2}\eta_p a V = \eta_p R_{0N} V \tag{A1}$$

The chain starts to be deformed at a velocity V_1:

$$\eta_p R_{0N} V_1 = \frac{kT}{R_{0N}}$$

i.e.

$$V_1 = \frac{kT}{a^2 \eta_p N} \tag{A2}$$

As in the case of smaller chains, V_1 can be found by considering the relaxation time coming from the equation $kT\delta R/Na^2 = N^{1/2} a \eta_p (V - \delta \hat{R})$.

For $V > V_1$, the chain undergoes an abrupt coil–stretch transition[5,14] and the deformation L increases exponentially ($L \sim \exp(V/V_1)$). At a velocity V_c close to V_1, the friction becomes of the Rouse type: the chain can be pictured as a string of blobs, size D, containing each $g_D = D^2/a^2$ monomers. If $g_D > N_e^2$, the friction on each blob is a Stokes friction. If $g_D < N_e^2$, the friction becomes a Rouse friction ($X(V > V_c) = N/N_e$), and we go back to the previous discussion: thus the threshold velocity $V_0{}^*$ to enter the marginal regime is still given by eq 15.

(2) "Locking" Transition of the Skin at Increasing Velocity. At zero velocity, the grafted chains are unperturbed and the number of coupled P chain is

$$\nu_p = \nu N^{1/2} \tag{A3}$$

All the skin is coupled when $\nu_p = 1/P^{1/2}a^2$, i.e., when the grafting density reaches

$$\nu a^2 = \nu_{c2} a^2 \equiv \frac{1}{(NP)^{1/2}} \tag{A4}$$

Between V_1 and V_c, each chain is deformed and traps an increasing number of free P chains. Above V_c, X saturates at a constant value N/N_e, and the threshold density to couple to the whole skin is again

$$\nu_c a^2 = \frac{N_e}{NP^{1/2}} \tag{A5}$$

The threshold velocity $\tilde{V}_1$ to deform the chains in the cooperative regime is given by

$$\frac{1}{P^{1/2} a^2} \eta_p a \tilde{V}_1 = \nu \frac{kT}{R_{0N}}$$

i.e.

$$\tilde{V}_1 = V_1 \frac{\nu}{\nu_{c2}} \tag{A6}$$

The marginal velocity, for $\nu > \nu_c$, is still given by eq 32. The different regimes are shown in Figure 3.

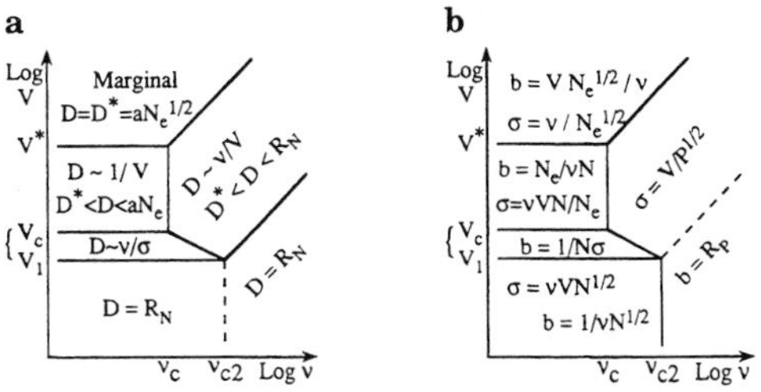

Figure 3. Slippage velocity V versus grafting density ν diagram for a long chain grafted surface ($N > N_e^2$): grafted chain deformation (a) and melt flow parameters (b). The $V_1 < V < V_c$ region describes an abrupt transition in terms of slippage velocity; however, all quantities vary continuously with the shear stress σ. Dimensional factors (a, kT, η_p) have been ignored.

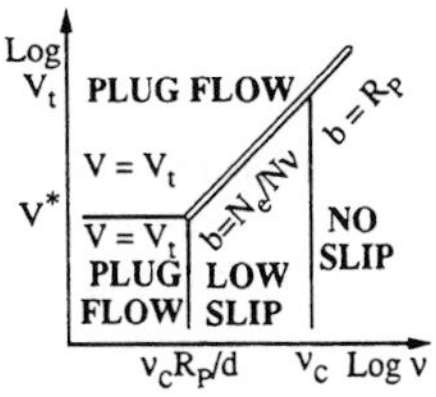

Figure 4. In some experiments, the polymer sample top velocity V_t is monitored. For moderately long grafted chains ($N < N_e^2$), the V_t versus ν diagram is shown. Slippage velocity V and extrapolation length b display strong variations across the double line; on the other hand, the shear stress σ is continuous. Dimensional factors (a, kT, η_p) have been ignored.

Appendix B: Experimental Conditions

The situation discussed in the present paper may be tested by various experimental methods. One of them, ref 3, involves a moving plate (at a distance d from the surface of interest). At the moving plate, there is no slippage, and the velocity has an imposed value V_t. The slippage velocity V, shear rate $\dot{\gamma}_p$, and extrapolation length b can be measured by optical means. The shear stress is $\sigma = \eta_p \dot{\gamma}_p$. We have

$$V = V_t \frac{b}{b + d} \tag{B1}$$

and

$$\dot{\gamma}_p = \frac{V_t - V}{d} = \frac{V_t}{b + d} \tag{B2}$$

We therefore expect regime transitions when b becomes comparable to d:

$$b > d \begin{cases} V = V_t & \text{(B3)} \\ \sigma = \eta_p \dfrac{V_t}{b} & \text{(B4)} \end{cases}$$

$$b < d \begin{cases} V = V_t \dfrac{b}{d} & \text{(B5)} \\ \sigma = \eta_p \dfrac{V_t}{d} & \text{(B6)} \end{cases}$$

We can now predict the different regimes in terms of the top velocity V_t and the grafting density ν. They are presented for short grafted chains ($N < N_e^2$) in Figure 4. We review now the behavior of observable quantities under these circumstances.

A plug flow ($V \simeq V_t$: the whole sample moves at roughly the same speed with respect to the grafted surface) should be observed in two regions: 1. For $V_t < V^*$ and $\nu < \nu_c R_p/d$, the observable parameters vary as follows:

$$\begin{cases} V = V_t \\ b = \dfrac{N_e}{N} \dfrac{1}{\nu a} \\ \sigma = \dfrac{N}{N_e} \nu V_t a \eta_p \end{cases} \tag{B7}$$

2. For $V_t > V^*$ and $V_t > (\nu d/N_e^{1/2}) kT/a\eta_p$ (marginal state $D = D^*$), we have

$$\begin{cases} V = V_t \\ b = N_e^{1/2}\dfrac{V_t}{\nu}\dfrac{a\eta_p}{kT} \\ \sigma = \dfrac{\nu}{N_e^{1/2}}\dfrac{kT}{a} \end{cases} \quad \text{(B8)}$$

Slippage is lowered in a transition region: 3. For $\nu_c R_p/d < \nu < \nu_c$ and $V_t < (\nu d/N_e^{1/2})kT/a\eta_p$, we have

$$\begin{cases} V = \dfrac{N_e}{N}\dfrac{V_t}{a\nu d} \\ b = \dfrac{N_e}{N}\dfrac{1}{\nu a} \\ \sigma = \dfrac{V_t}{d}\nu_p \end{cases} \quad \text{(B9)}$$

Eventually, slippage is visible only at the molecular scale ($b = R_p$); i.e., it is macroscopically suppressed: 4. For $\nu > \nu_c$ and $V < (\nu d/N_e^{1/2})kT/a\eta_p$, we get

$$\begin{cases} V = V_t\dfrac{R_p}{d} \\ b = R_p \\ \sigma = \dfrac{V_t}{d}\eta_p \end{cases} \quad \text{(B10)}$$

References and Notes

(1) de Gennes, P.-G. *C. R. Acad. Sci. (Paris)* **1979**, *228B*, 219.
(2) Brochard, F.; de Gennes, P.-G.; Pincus, P. *C. R. Acad. Sci. (Paris)* **1992**, *314*, 873.
(3) Migler, K.; Hervet, H.; Léger, L. *Phys. Rev. Lett.* **1993**, *70*, 287.
(4) Brochard, F.; de Gennes, P.-G. *Langmuir* **1992**, *8*, 3033.
(5) Ajdari, A.; Brochard, F.; de Gennes, P.-G.; Leibler, L.; Viovy, J. L.; Rubinstein, M. *Physica A* **1994**, *204*, 17.
(6) Kavassalis, T.; Noolandi, J. *Phys. Rev. Lett.* **1987**, *59*, 2674; *Macromolecules* **1989**, *22*, 2709–2720.
(7) Brochard-Wyart, F.; Gay, C.; de Gennes, P.-G. *J. Phys. II Fr.* **1995**, *5*, 491–495.
(8) Perkins, T. T.; Smith, D. E.; Chu, S. *Science* **1994**, *264*, 819.
(9) Pincus, P. *Macromolecules* **1976**, *9*, 386.
(10) Ajdari, A. *C. R. Acad. Sci. (Paris)* **1993**, *317II*, 1159–1163.
(11) See, for instance: de Gennes, P.-G. *Scaling Concepts in Polymer Physics*; Cornell University Press: Ithaca, NY, 1979.
(12) Rabin, Y.; Alexander, S. *Europhys. Lett.* **1990**, *13* (1), 49.
(13) Barrat, J. L. *Macromolecules* **1992**, *25*, 832-834.
(14) Brochard-Wyart, F. *Europhys. Lett.* **1993**, *23*, 105.

MA950753J

Afterthought: Slippage of polymer melts on grafted surfaces

At the moment (1997), the most detailed data on this problem are due to E. Durliat [E. Durliat *et al.*, *Europhys. Lett.* **38**, 383 (1997); L. Léger, H. Hervet, G. Massey and E. Durliat, *J. Phys.* **37**, 7719 (1997)], with grafted wafer supports facing a PDMS melt. He finds that the plateau stress σ^* varies with the graft density ν, and isolates 3 regimes —

a) At very low ν, the physics is dominated by a few chains from the melt which stick to some untreated points on the surface. This regime was studied before by G. Massey: see L. Léger *et al.* in "Rheology for polymer melt processing" (Vol. 5 of Rheology Series) Elsevier, 1996.

b) At intermediate grafting densities, σ^* increases linearly with ν as predicted in the present paper.

c) At higher ν, $\sigma^*(\nu)$ decreases: Durliat interprets this as a transition from a "wet brush" to a "dry brush": this transition is discussed in "Conformations of polymers attached to an interface," p. 235 of the present book.

C. R. Acad. Sci. Paris, t. 323, Série II *b*, p. 473-479, 1996
Physico-chimie/*Physical chemistry*

Injection threshold for a star polymer inside a nanopore

Françoise BROCHARD-WYART and Pierre-Gilles de GENNES

F. B.-W. : Physico-Chimie des Surfaces et Interfaces, Institut Curie, 11, rue Pierre-et-Marie-Curie, 75231 Paris CEDEX 05, France ;
P.-G. de G. : Physique de la Matière Condensée, Collège de France, 11, place Marcelin-Berthelot, 75231 Paris CEDEX 05, France.

Abstract. A linear, flexible polymer (in dilute solution with a good solvent) enters a pore (of diameter D smaller than its gyration radius R) only when the suction flux J is larger than a threshold value $J_{cl} \cong kT/\eta$ (T: temperature; η: solvent viscosity). We discuss here the case of an f arm star polymer ($f >> 1$). The results suggest that permeation through nanopores may provide an interesting characterization of mixtures containing linear and branched polymers with the same overall molecular weight.

Seuil d'injection pour un polymère en étoile dans un nanopore

Résumé. *Un polymère linéaire flexible (en solution diluée dans un bon solvant) ne pénètre dans un pore (de diamètre* D *inférieure à son rayon de gyration* R*) que si le flux d'entraînement* J *est supérieur à une valeur seuil* $J_{cl} \cong kT/\eta$ *(*T *: température ;* η *: viscosité du solvant). Nous étendons ici la discussion au cas d'une étoile à* f *branches* $(f >> 1)$. *Les résultats suggèrent que la perméation dans des nanopores peut être une méthode utile pour caractériser des mélanges de polymères linéaires et branchés de même masse moléculaire globale.*

Version française abrégée

Un polymère flexible enfermé dans un tube étroit (non adsorbant) voit son entropie réduite, et son énergie libre augmentée d'une quantité E_{conf}.

a) La perméabilité statique est réduite par un facteur $\exp(-E_{\text{conf}}/kT)$.

b) Si l'on ajoute un courant d'aspiration J le polymère ne passe que pour J supérieur à un certain courant critique J_c.

Note présentée par Pierre-Gilles de GENNES.

F. Brochard-Wyart and P.-G. de Gennes

Ces propriétés ont été étudiées jadis pour des chaînes linéaires en bon solvent (Guillot *et al.*, 1985 ; de Gennes, 1988). Nous les discutons ici pour des étoiles (au niveau des lois d'échelle). Les mécanismes régissant a) et b) sont différents.

a) Pour la perméabilité statique la configuration dominante est « symétrique » (*fig. a*) : une moitié des bras de l'étoile est en avant, l'autre en arrière.

b) Le courant critique J_c est, lui, donné par des configurations *dissymétriques* (*fig. b*).

Ces deux modes de perméation sont analysés ici : la mesure la plus intéressante paraît être celle de J_c, en mesurant J_c pour deux valeurs du diamètre de tube D, on devrait déterminer à la fois la longueur (N) et le nombre (f) des bras.

I. General aims

Today, with the use of suitable metallocene catalysts, the plastics industries produce polyolefins, where the overall molecular weight and the branching level can both be varied. The characterization of these complex mixtures has been amply discussed in a recent DSM-Cambridge symposium. The two main pieces of information come from: *a*) measurements of the complex shear modulus $\mu(\omega)$ as a function of frequency; *b*) rheological studies in shear flow and in longitudinal flow. However, in view of the many unknowns involved, there is a great need for other characterization methods, which should be sensitive to branching.

The aim of this Note is to propose one such method, based on permeation studies, performed with dilute solutions of the polymer mixture, and using very thin pores.

By particle track etching, it is possible to construct nanopores (commercially known as Nuclepores) with a diameter D ranging from nanometres to micrometres. It is also possible (by a suitable choice of solvent) to arrange that polymer solute do *not* absorb on the pore surface (Guillot and Rondelez, 1981). This situation has been studied in some detail, both experimentally (Guillot *et al.*, 1985) and theoretically (de Gennes, 1988) for *linear*, flexible coils in good solvents. With dilute polymers, the main results are as follows: whenever the pore diameter D is much smaller than the coil size R, the chains do not penetrate in static conditions. But if we impose a certain pressure drop between the two ends of the pore, or equivalently a certain solvent flux J (volume/sec.) we can force permeation, provided that J is larger than a certain threshold:

$$J_{c1} = \text{const.}\, kT/\eta \qquad (1)$$

where T is the temperature, and η the solvent viscosity. The threshold is independent of D (for $D << R$) and independent of the polymerization index N of the coil. For this reason, permeation studies on solutions do not give any information on the molecular weight distribution in the solution.

What do we expect for branched species? We are far from knowing how to cope with this problem in general. However, the situation becomes simpler if we deal with star polymers: f arms, each of them containing N monomers. In section II, we discuss the confinement of a star in a tube. In section III, we analyse the effect of a flow current J on the confined structure assuming a "symmetrical mode" of entry (shown on figure *a*). In section IV we analyse certain non-symmetric modes, for instance, with one arm forward (*fig. b*), and find that they are often more favourable. Our discussion is restricted to the level of scaling laws, but it does predict a strong difference between stars and linear polymers.

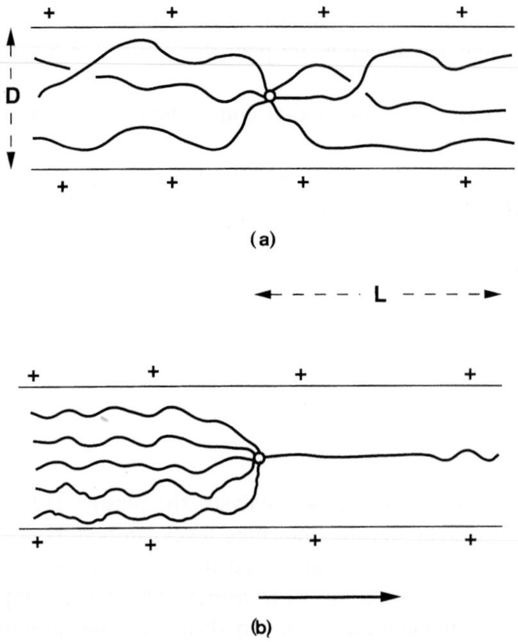

A star trapped in a nanopore. The case shown corresponds to $f=6$. (*a*) Symmetric mode. (*b*) One arm forward.

Une étoile confinée dans un tube. Exemple avec f=6. (a) Mode symétrique. (b) Un bras en avant.

II. Confined stars

Let us, for simplicity, restrict our discussion to stars in a very good (athermal) solvent, where the excluded-volume parameter $v \sim a^3$ (a: monomer size). Let us assume also that the stars have relatively long arms $N >> f^{1/2}$ (the opposite limit would yield a compact nodule).

In a bulk solution, the conformation of the star has been discussed by Daoud *et al.* (1982). The overall radius follows the scaling law:

(2) $$R = aN^{3/5} f^{1/5}$$

which can be obtained from a simple extension of the Flory argument (Flory, 1971). In Daoud and Cotton (1982), this result is associated with a more refined picture of the self-similar structure.

Let us now proceed to confined stars:

1) The organization of a star trapped in a narrow *slit* has been studied in detail by Halperin and Alexander (1987). They find: *a*) an interior region which is still 3-dimensional; *b*) an intermediate region, where the Daoud-Cotton blobs are still 3d, but the overall organization is 2d; *c*) an external region where the Daoud-Cotton blobs are 2d.

2) We expect the confinement in a *tube* to be much simpler (see *fig.*). There should again exist an inner region of size D. However, it has a negligible weight: most of the star arms are simply stretched. Let us assume for the moment that the distribution of arms between "backward" and "forward" corresponds to equal number: $\widetilde{f} = f/2$ arms in each direction. We call this the symmetrical mode.

For a tube diameter D smaller than R, the length L of the tube occupied by the star should then follow a simple scaling law of the form:

(3) $$L(D) = D\left(\frac{R}{D}\right)^{u}$$

where u is (for the moment) unknown. Equation (3) ensures that for $D = R$, we return to $L = R$. Another condition to be imposed is that when D reaches its minimum allowed value:

(4) $$D_{\min} = a^2\tilde{f} \qquad (\tilde{f} >> 1)$$

the star reduces to a compact bundle of stretched chains:

(5) $$L(D = D_{\min}) = Na$$

Inserting equation (5) into equation (2) leads to $u = 5/3$. And the powers of $\tilde{f}$ showing up on both sides of equation (3) are then automatically equal.

The polymer volume fraction inside the confined star is:

(6) $$\phi = \frac{\tilde{f}Na^3}{2LD^2} \cong \left(\tilde{f}\frac{a^2}{D^2}\right)^{2/3}$$

and the corresponding correlation length is $\xi = D\tilde{f}^{-1/2}$. From this one can work out the free energy of the confined star is:

(7) $$E = kT\frac{LD^2}{\xi^3} \cong kT\,N\left(\frac{a}{D}\right)^{5/3}\tilde{f}^{11/6}$$

$E(\tilde{f})$ is a *convex* function: this ensures that the optimal partitioning between "forward" and "backward" arms corresponds to equal numbers ($\tilde{f} = f/2$).

The confinement force F_{conf} (the force required to push the star in the tube) is the product of the internal osmotic pressure (Halperin, Alexander, 1987) $kT\xi^{-3}$ by the tube cross-section:

(8) $$F_{\mathrm{conf}} \cong \frac{kT\,D^2}{\xi^3} = \frac{kT}{D}\tilde{f}^{3/2}$$

We shall now see how this force can be balanced by hydrodynamic flow.

III. Suction in the symmetrical mode

We assume that half of the arms lie in the forward direction, as shown on figure *a*. To describe suction we shall follow the approach of de Gennes (1988). Let us assume that the star has penetrated over a length l ($< L$) inside the tube. The confinement energy is $F_c\,l$. However, we also have a hydrodynamic force F_h : following the classical picture of semi-dilute solutions in good solvents, the force per blob of size ξ is of order:

(9) $$\eta\xi V = \eta\xi J D^{-2}$$

where V is the flow velocity, and $J \cong VD^2$ the corresponding current.

The number of blobs inside the region of length l is lD^2/ξ^3, and the overall hydrodynamic force is:

(10) $$F_h(l) = \frac{lD^2}{\xi^3}\eta\xi J D^{-2} = \frac{lJ\eta\tilde{f}}{D^2}$$

The overall energy required to squeeze in a length l is then:

$$E = F_{\text{conf}}l - \int_0^p dx F_h(x) = F_{\text{conf}}l - \frac{1}{2}lF_h(l) \tag{11}$$

This has a maximum for:

$$l = l^* = \frac{F_{\text{conf}} D^2}{J\eta\tilde{f}} \tag{12}$$

corresponding to an energy barrier:

$$E^* = \frac{1}{2}F_{\text{conf}}l^* = \frac{(kT\tilde{f})^2}{J\eta} \tag{13}$$

We expect permeation when $E^* \sim kT$. This gives a critical current:

$$J_c = \frac{kT}{\eta}\tilde{f}^2 \tag{14}$$

IV. Suction in an asymmetric mode

1) *One arm forward*

The symmetric mode of section III gives the lowest energy barrier, but need not give the lowest critical current! We show this now by going first to the opposite extreme: a non-symmetric mode, with *one* arm forward, and $(f-1)$ arm backward (*fig. b*). At currents $J > J_c$, the forward arm will be able to enter. A detailed discussion of stretching for a tethered chain under uniform flow has been given by one of us (Brochard-Wyart, 1994). There are a number of regimes. For our purposes, the "stem-flower" regime is expected to be dominant (at currents $J \tilde{>} J_{c1}$). This essentially means that the forward arm is completely stretched, with a length $L \sim Na$. The scaling structure of the drag force is then simply:

$$F_h = \eta V L \cong \eta N a J / D^2 \tag{15}$$

At the threshold current $(J = J_c)$ this must balance the confinement force of equation (8), taken with $\tilde{f} = f - 1 \sim f$. The result is:

$$J_c = \frac{kT}{\eta} f^{3/2} D/(Na) \qquad \left(\frac{D}{a} f^{-1/2} < N < \frac{D}{a} f^{3/2}\right) \tag{16}$$

We must have $N < f^{3/2}D/a$ to ensure that $J_c > J_{c1}$ so that the forward arm has indeed been able to enter. On the other hand equation (16) must still be compared with the critical current for the symmetrical mode [equation (14)]. When $N > D/af^{-1/2}$ the asymmetric mode wins: this explains the inequalities quoted after equation (16).

2) *Optimal number of forward arms*

Let us now consider more general asymmetric modes, with a certain number ϕ of forward arms, and $f - \phi$ backward arms. It will turn out that $\phi << f$ in most practical cases, but that $\phi > 1$ at the onset of permeation.

If we want a number ϕ of arms to enter in the forward direction, the current J must be above the threshold value deduced from equation (14) when we replace $\widetilde{f}$ by ϕ.

$$J/J_{c1} > \phi^2 \tag{17}$$

Once this criterion is satisfied, we can repeat the argument of equations (15), (16) but the drag force is multiplied by ϕ, and equation (16) is replaced by:

$$J/J_{c1} = f^{3/2}\frac{D}{Na}\phi^{-1} \tag{18}$$

Comparing the conditions (17) and (18) we arrive at a threshold:

$$\phi = f^{1/2}\left(\frac{D}{Na}\right)^{1/3} \tag{19}$$

$$J_c/J_{c1} = f\left(\frac{D}{Na}\right)^{2/3} \tag{20}$$

Equations (19), (20) require $N < D/af^{3/2}$: then we have $J_c > J_{c1}$ and $\phi > 1$. If $N < D/af^{3/2}$ we expect $J_c = J_{c1}$ and the experiment brings no useful information.

V. Concluding remarks

1) As shown by equation (20), the critical current J_c should be sensitive to the amount of branching (*i.e.*: to f) whenever $f > (Na/D)^{2/3}$. Thus permeation measurements may usefully supplement rheological measurements for the characterization of branched polymers.

2) One unexpected feature is that, in the regime described by equation (20) the critical current decreases when the length N of the arms increases: the forward arms provide more hydrodynamic drag.

3) From a more applied side, we face at least three problems:

a) the possibility of *clogging* (by a few molecules of high f);

b) the limits in mechanical strength of the polymer. The drag force on one forward arm is of order:

$$\eta LV = \frac{Na\,kT}{D^2}\frac{J_c}{J_{cl}} = \frac{kT}{a}\left(\frac{a}{D}\right)^{5/3}N^{1/3}f \tag{21}$$

This should be compared with the force required for bond rupture, which scales roughly like U/a (where U is the energy of a covalent bond). Thus we must have:

$$f < \left(\frac{D}{a}\right)^{5/3}N^{-1/3}\frac{U}{kT} \tag{22}$$

c) the mechanical strength of the permeation plate. From Poiseuille's law (but ignoring all coefficients) we expect a pressure difference between the two sides of the form:

$$p \cong \frac{kTL_m}{D^4} \frac{J_c}{J_{cl}}$$

where L_m is the total length of the nanotube. Clearly we need thin membranes (L_m = a few microns) backed by a robust structure (e.g. a millipore filter).

4) Of course, the stars are a very primitive–but simple–model system. Another branched system of special interest is a *comb* polymer: the case of combs is currently analysed by Gay. For statistically branched polymers, the confinement problem is much more complex (Vilgis *et al.*, 1994). The analysis of threshold currents for this case is under way.

Acknowledgments. This work was initiated by discussions during a DSM/Cambridge meeting at the Isaac Newton (1996). We benefited from discussions with T. Mc Leish, S. Milner, G. Benedek, E. Raphaël and C. Gay.

Note remise le 25 mai 1996, acceptée après révision le 6 août 1996.

References

Brochard-Wyart F., 1995. *Europhysics Letters*, 30, pp. 387-392.
Daoud M. and Cotton J.-P., 1982. *J. Physique (Paris)*, 43, pp. 531-538.
Flory P., 1971. *Principles of Polymer Chemistry*, Cornell University Press.
Gennes P.-G. de, 1985. *Scaling Concepts in Polymer Physics*, Cornell University Press, 2nd Printing.
Gennes P.-G. de, 1988. *Transport, Disorder and Mixing*, **Guyon E.** Ed., Kluwer (Dordrecht, Netherlands), pp. 203-213.
Guillot G. and Rondelez F., 1981. *J. Applied Physics*, 52, 12, pp. 5155.
Guillot G., Léger L. and Rondelez F., 1985. *Macromolecules*, 18, pp. 2531-2537.
Guillot G., 1987. *Macromolecules*, 20, pp. 2600-2606.
Halperin A. and Alexander S., 1987. *Macromolecules*, 20, pp. 1146-1152.
Vilgis T. A., Haronska P. and Benhamou M., 1994. *J. Physique*, 4, pp. 2187-2196.

Macromolecules **1996,** *29,* 8379–8382 **8379**

Injection Threshold for a Statistically Branched Polymer inside a Nanopore

C. Gay,[†] P. G. de Gennes,*[,†] E. Raphaël,[†] and F. Brochard-Wyart[‡]

Labo. Matière Condensée, Collège de France, 11 pl. M. Berthelot, 75231 Paris Cedex 05, France, and PSI, Institut Curie, 11 rue P. & M. Curie, 75005 Paris, France

Received July 1, 1996; Revised Manuscript Received September 17, 1996[⊗]

ABSTRACT: A nonadsorbing, flexible polymer (in dilute solution with a good solvent) enters a pore (of diameter D smaller than its natural size, R) only when it is sucked in by a solvent flux, J, higher than a threshold value, J_c. For linear polymers $J_c \sim kT/\eta$ (where T is the temperature and η the solvent viscosity). We discuss here the case of statistically branched polymers, with an average number, b, of monomers between branch points. We find that there are two regimes, "weak confinement" and "strong confinement", depending on the tube diameter. By measuring J_c in both regimes, we should determine *both* the molecular weight and the number b.

I. Introduction

Thanks to suitable metallocene catalysts,[1] it is now possible to produce polyolefins with adjustable average molecular weights, $M = XM_0$ (X, polymerization index; M_0, monomer molecular weight), and with adjustable average levels of branching. But standard rheological measurements are not quite sufficient to characterize the resulting complex mixtures. This led us recently to propose another, complementary method of characterization, based on permeation studies using nanopores.[2] The discussion in ref 2 was restricted to a very simple family of branched objects, namely star molecules. We found that stars can be sucked in a narrow pore when the solvent flux J inside the pore exceeds a certain threshold, J_c(star):

$$J_c(\text{star}) \cong \frac{kT}{\eta} f \left(\frac{D}{Na}\right)^{2/3} \qquad (1)$$

$$(\text{whenever } f > (Na/D)^{2/3})$$

where f is the number of arms in the star ($f \gg 1$), and $N = X/f$ is the arm length. k is the Boltzmann constant, a is the monomer size, T is the temperature, and η is the solvent viscosity. (Equation 1 should hold when the molecules do *not* adsorb on the pore walls.) The critical current predicted by eq 1 is very sensitive to the amount of branching (i.e. to f).

In the present paper, we present a theoretical discussion (at the level of scaling laws) for the more usual case: a *statistically branched* polymer. In section II, we discuss the statistics of confined chains. The main idea here is based on what we call the "Ariadne length" of a cluster. The principle was discovered first by Vilgis et al.,[3] using a slightly different language, in relation with what is called the spectral dimension, d_s of clusters (see Appendix A for a discussion of d_s). Our approach in section II is based on a simple Flory argument, and predicts in fact the value $d_s = {}^4/_3$, which is currently recognized to be an excellent approximation.[4] In section III, we set up the hydrodynamics and compute the critical current. Section IV extends the discussion to cases of "weak branching", where the number of monomers between adjacent branch point has a large average value b: this is the most useful case in practice. Section V analyses all the results.

For readers inclined to somewhat more mathematical discussions, the case of branched objects with more general spectral dimensions d_s is studied in Appendix B.

II. Statistics of Branched Clusters

(1) Overall Size in Dilute Solution, *R*. An interesting approach, based on a Flory type of calculation, was set up by Lubensky and Isaacson;[5] a lucid description (incorporating many physical phenomena) was set up by Daoud and Joanny.[6]

An *ideal* branched structure (with no steric interactions) would have a size $R_0 \sim X^{1/4}a$.[7,8] If we now incorporate excluded volume (with a volume per monomer of a^3), we can write a coil energy $F(R)$ depending on the size R as follows:

$$F(R) \cong kT\left[\frac{R^2}{R_0^{\,2}} + \frac{X^2a^3}{R^3}\right] \qquad (2)$$

Here, the first term is an elastic energy, and the second term is the effect of intermonomer repulsions (Xa^3/R^3 is the internal volume fraction). Optimizing (2) with respect to R, we arrive at:

$$R \cong X^{1/2}a \qquad (3)$$

Some verifications of eq 3 have been obtained (on dilute solutions of branched polymers) by M. Adam et al.[9] In the following, we shall be concerned with pores of diameter $D \ll R$.

(2) Flory Argument in a Tube. Let us now modify eq 2 for a confined polymer, extended over a length, L (Figure 1). The allowed volume is now LD^2, and we have

$$\frac{F}{kT} = \frac{L^2}{X^{1/2}a^2} + \frac{X^2a^3}{LD^2} \qquad (4)$$

Optimizing with respect to L, we find

$$\frac{L}{D} = X^{5/6}\left(\frac{a}{D}\right)^{5/3} \equiv \left(\frac{R}{D}\right)^{5/3} \qquad (5)$$

[†] Collège de France.
[‡] Institut Curie.
[⊗] Abstract published in *Advance ACS Abstracts,* November 1, 1996.

S0024-9297(96)00941-2 CCC: $12.00

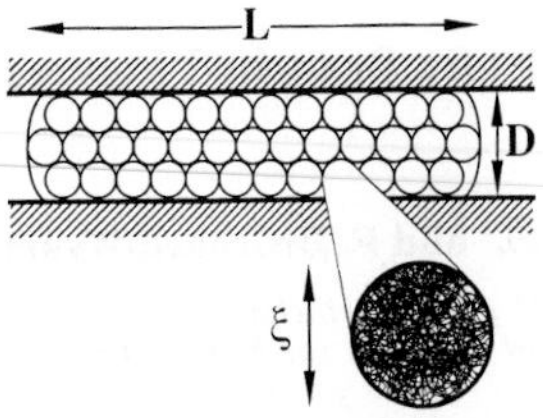

Figure 1. A branched polymer forced in a pore of diameter D smaller than its natural size, R. The interior of the polymer is a semidilute solution of correlation length ξ.

and the internal concentration is

$$\phi = \frac{Xa^3}{LD^2} = \left(\frac{D_{\min}}{D}\right)^{4/3} \tag{6}$$

The diameter

$$D_{\min} = aX^{1/8} \tag{7}$$

corresponds to maximum squeezing ($\phi = 1$).[10] Of particular interest is the corresponding value of L:

$$L_{\max} = aX^{5/6}\left(\frac{a}{D_{\min}}\right)^{5/3} \tag{8a}$$

Ariadne helped Theseus through the Minoan labyrinth, by giving him a reel of thread, which kept a track of his march. l_A represents the length of the shortest path, from the starting point, to the monster which is to be killed—in our context, the thread distance between two arbitrary points on the cluster (see Appendix A). In the squeezed polymer, assuming that there are *no loops*, the corresponding path becomes completely stretched, and thus:

$$l_A = L_{\max} = aX^{3/4} \tag{8b}$$

As usual for semidilute solutions in good solvents, we can think of the squeezed polymer as a compact stacking of blobs, each with a diameter (ξ) and a number of monomers (g). The relation between g and ξ inside one blob is derived from the size of a single cluster (eq 3):

$$\xi = ag^{1/2} \tag{9}$$

Writing that $\phi = ga^3/\xi^3$ (compact arrangement) and comparing with eq 6, we ultimately find the correlation length:

$$\xi = a\left(\frac{D}{a}\right)^{4/3} X^{-1/6} \tag{10}$$

For weak confinement ($D = R$) we recover $\xi = R$, and for very strong confinement ($D = D_{\min}$) we have $\xi \sim a$.

III. Suction into the Tube

We now force our polymer through the pore and assume that a certain length (y) of the squeezed structure has entered, as shown in Figure 2. The free energy required for this may be written as

$$F = F_c y - \int_0^y dy'\, F_h(y') \tag{11}$$

where F_c is a force resulting from confinement, while F_h is a hydrodynamic force. We can write simply

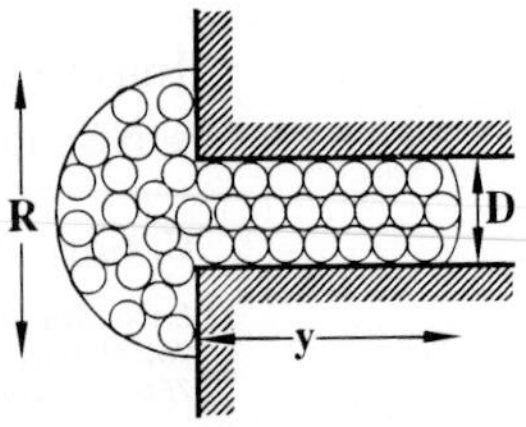

Figure 2. The entry process for the branched polymer: only a certain length, y, has penetrated in the tube.

$$F_c = \Pi D^2 \tag{12}$$

where Π is the osmotic pressure of the squeezed object, given by the usual scaling law,

$$\Pi = kT/\xi^3 \tag{13}$$

and D^2 is the cross sectional area of the tube. The hydrodynamic force is equivalent to a Stokes force per blob,

$$F_h \cong \eta\, \xi\, V \frac{D^2 y}{\xi^3} = \frac{\eta J y}{\xi^2} \tag{14}$$

where $V = J/D^2$ is the local solvent velocity and D^2y/ξ^3 is the number of blobs inside the pore.

Returning now to the energy, $F(y)$ (eq 11), we see that it is a maximum for

$$y = y^* = \frac{F_c \xi^2}{J\eta} \tag{15}$$

corresponding to an energy barrier,

$$E^* = \frac{1}{2} F_c y^* = \frac{(kT)^2}{J\eta}\left(\frac{D}{\xi}\right)^4 \tag{16}$$

Aspiration occurs when $E^* \sim kT$; this gives a critical current,

$$J_c \cong \frac{kT}{\eta}\left(\frac{D}{\xi}\right)^4 \tag{17}$$

Equations 16 and 17 are a natural extension of the result for the "symmetric mode" of stars (ref 2). For the stars in this mode, we had $f/2$ branches in parallel, each occupying a cross section $\xi^2 = 2D^2/f$, and the barrier energy was proportional to $(f/2)^2/J$, as in eq 16.

We can now insert the results of section II on the correlation length (eq 10) , and we find:

$$J_c = \frac{kT}{\eta}\left(\frac{R}{D}\right)^{4/3} \tag{18}$$

$$= \frac{kT}{\eta} X^{2/3} (a/D)^{4/3} \tag{19}$$

IV. Extension to Weaker Branching

Our discussion assumed a very high density of branching. In a more chemical language, if we make our polymer via condensation reactions, using a mixture of two and three functional units, we were assuming that the initial concentrations of these species were comparable.

In practice, we often operate with a much smaller fraction of three functional unit species, and the average number of difunctional monomers between two adjacent branch points (b) is much larger than unity. Then, our formulas have to be adjusted as follows.

(1) An unconfined polymer in dilute solution will have a size

$$R = aX^{1/2}b^{1/10} \tag{20}$$

and an Ariadne length

$$l_A = aX^{3/4}b^{1/4} \tag{21}$$

One may check these exponents by noticing that if $b = X$ (i.e.: when we return to linear polymers), we have the standard values $R = aX^{3/5}$ (the Flory radius) and $l_A = aX$).

(2) The length L of the confined polymer is still given by the second form of eq 5

$$L = D\left(\frac{R}{D}\right)^{5/3} = a\left(\frac{a}{D}\right)^{2/3} X^{5/6} b^{1/6} \tag{22}$$

where R is now taken from eq 20. At maximum stretching, we must have $L = l_A$, and this corresponds to a size

$$D_{\min} = X^{1/8} b^{-1/8} a \tag{23}$$

(3) There is one complication, however: there are two regimes, depending on the size of the correlation length (ξ) when compared to the size (ξ_b) of one linear piece of b monomers. In good solvents, ξ_b is given by the Flory law:

$$\xi_b = ab^{3/5} \tag{24}$$

(a) When the tube diameter D is larger than a certain crossover value, D^*, we have $\xi > \xi_b$.We call this *"weak confinement"*. In this regime, the results of section II can then easily be transposed, using blobs of b monomers (and size ξ_b) instead of monomers. (This leads, in particular, to a derivation of eq 22). The correlation length is

$$\xi = a\left(\frac{D}{a}\right)^{4/3} X^{-1/6} b^{-1/30} \tag{25}$$

(b) The crossover occurs when $\xi = \xi_b$. Using eq 25, we find that the corresponding tube diameter is

$$D^* = aX^{1/8} b^{19/40} \tag{26}$$

(Conversely, if one type of nanopore (fixed diameter D) is being used to separate a polydisperse mixture, it is useful to rewrite eq 26 in terms of a critical molecular weight: $X^* = (D/a)^8 b^{-19/5}$).

Note that (from eqs 23 and 26)

$$\frac{D^*}{D_{\min}} = b^{3/5} > 1 \tag{27}$$

(c) In the interval $D^* > D > D_{\min}$, we have a new regime of *strong confinement*. We retain the same law for L (eq 22), but the blob structure is different. There are many blobs between two adjacent branch points, and the correlation length ξ versus volume fraction ϕ follows the classical law for semidilute solutions of *linear* polymers $\xi = a\phi^{-3/4}$. The concentration ϕ can be derived from

$$\phi = \frac{Xa^3}{LD^3} \tag{28}$$

Using eq 22 for L, we then arrive at a very simple result:

$$\xi = a\frac{D}{D_{\min}} \tag{29}$$

(4) Knowing these static properties, we can now return to the critical current, J_c (eq 17), and we find

$$\frac{\eta}{kT}J_c = \begin{cases} X^{2/3}(a/D)^{4/3}b^{2/5} & (D > D^*) \quad (30a) \\ (X/b)^{1/2} & (D < D^*) \quad (30b) \end{cases}$$

In the second regime (strong confinement), J_c is independent of the tube diameter.

V. Concluding Remarks

(1) The critical current $J_c(X)$ of branched polymers ($J_c \sim X^{2/3}$) is a signature of their Ariadne length (or equivalently of their spectral dimension, as defined in the appendices). For instance, if we had not a statistically branched object but a spheroidal clump taken from a three-dimensional gel ($X \sim R^3$), we would expect $J_c \sim X^{4/3}$.

(2) If we return to the general formulas (30a,b) for statistically branched clusters with weak branching levels ($b \gg 1$), we see that one measurement of J_c in each regime ($D > D^*$ and $D < D^*$) should allow us to determine *both* the molecular weight (X) and the chemical distance between branch points (b). Thus the permeation may be a rather powerful characterization method; it may also have separation potentialities.

Acknowledgments: This program was initiated by discussions at the DSM meeting on Polymer Rheology (Isaac Newton Institute, Cambridge, May 1996). We are especially thankful for the remarks of T. McLeish and S. Milner during this symposium.

Appendix A: Fractal and Spectral Dimensions of Branched Objects

(1) The *fractal* dimension of our clusters describes the relation between the size, R and the polymerization index, X:

$$X \cong \left(\frac{R}{a}\right)^{d_f}$$

For our branched systems in good solvents, $d_f = 2$ (eq 3)

(2) The *spectral* dimension of a cluster, d_s, depends on its chemical formula (describing linear sequences and branch points), but is independent of the exact conformation of the polymer (e.g. its linear segments may be rigid or flexible, d_s will be the same). The concept was introduced by Alexander and Orbach,[4] and exploited by Rammal and Toulouse.[11] Here, we shall present it qualitatively, using an acoustic model as a tool.

Let us assume that our bonds can propagate sound, with a velocity c measured along the chemical sequence: the transit time between adjacent, bonded, monomers is a/c. We choose an "emitter" site (one particular monomer) and we send a signal at time 0. After a certain time, t, all the monomers which have a curvilinear distance to the emitter smaller than ct, have received the signal. Their number is called $m(t)$. Clearly, $m(t)$ increases with time. For "self similar" situations (i.e. when $m < X$), it is correct to postulate a power law for $m(t)$:

$$m(t) = \left(\frac{ct}{a}\right)^{d_s} \tag{A1}$$

The exponent d_s is characteristic of the chemical structure: a linear polymer has $d_s = 1$ (be it flexible or rigid). A sheet (like a graphite layer) has $d_s = 2$, even if it is crumpled, etc.

After a certain time, t^*, our whole cluster (X monomers) has received the acoustic signal. t^* is defined by

$$X = \left(\frac{ct^*}{a}\right)^{d_s} \tag{A2}$$

The length

$$l_A = ct^* \tag{A3}$$

is what we call the Ariadne length. From eq A2, we see that

$$l_A = aX^{1/d_s} = aX^{3/4} \tag{A4}$$

Comparing this with eq 8b for our statistically branched clusters, we get $d_s = 4/3$.

(3) We now derive the cluster conformation (L and ξ, eqs 5 and 10) using the Ariadne length.

(a) Vilgis et al.[3] assumed that the maximum stretched length of a confined cluster would scale like l_A (eq A4). Using the scaling form derived from the Flory argument for the length of the confined object (eq 5), they find the corresponding minimum tube diameter, D_{min} (eq 7).

(b) Conversely, the maximum stretching ($L_{max} \equiv l_A$) is due to confinement; it is thus reasonable to assume $\phi \sim 1$ in this situation (which defines some tube diameter $D \equiv D_{min}$, eq 7). Assuming the existence of a unique scaling law from the unperturbed regime $L \sim R \sim D$ to the maximum stretching ($L = l_A$, $D = D_{min}$), we recover eq 5.

(c) The same scaling argument as in b can be used to provide a description of the confined fractal in terms of blobs. Let us assume that confining the fractal into a tube of diameter D leaves its structure unperturbed at small length scales, i.e. that blobs of a certain size $\xi(D)$ retain their unaltered structure. Renormalizing these blobs as monomers, we can look at the confined object as a cluster in a tube of minimum diameter: $D \equiv D_{min}(\xi(D))$. More precisely, we have $L \equiv \xi(X/x)^{1/d_s}$ (similarly to eq A4), and $LD^2 \equiv (X/x)\xi^3$ (i.e., $\phi_\xi \sim 1$), where $x = (\xi/a)^2$ is the mass of one blob (eq 3). The resulting blob size is given by eq 10.

Appendix B: Injection Threshold for Statistically Branched Polymers of Arbitrary Spectral Dimensions

(1) A useful picture for the Ariadne length, l_A, is the following. The fractal object, made of monomers, can be parametrized[3] by a function $\vec{R}(x)$, where $\vec{R}$ is the monomer position in real space ($d = 3$) and x is a vector in the discrete, d_s − dimensional, parameter space $\{1, 2, ..., N\}^{d_s}$. The parameter space describes the object connectivity, independently of how it is embedded in real space (loose or dense, crumpled or stretched). It is clear from this picture that N represents the maximum distance between two points in the object, i.e., $l_A \cong Na$. (This is obvious with the particular cases $d_s = 1$, linear polymer, and $d_s = 2$, crumpled sheet). Obviously, the fractal molecular weight is also $X = N^{d_s}$.

(2) A generalized Flory approach yields the object radius of gyration, without use of the ideal size of a branched polymer $R_0 = aX^{1/4}$ (eq 2). Indeed, the Edwards Hamiltonian[12] can be generalized[3] as

$$\frac{F}{kT} = \int d^{d_s}x(\nabla_x\vec{R}(x)^2) + \int d^{d_s}x \int d^{d_s}x'(a^3\,\delta^3(\vec{R}(x) - \vec{R}(x'))) \tag{B1}$$

The corresponding mean-field Flory free energy,

$$\frac{F}{kT} = \frac{R^2N^{d_s}}{a^2N^2} + \frac{a^3N^{2d_s}}{R^3} = \frac{R^2}{a^2X^{(2-d_s)/d_s}} + \frac{X^2a^3}{R^3} \tag{B2}$$

gives the correct radius of gyration in good solvent, $R = aX^{(2+d_s)/5d_s}$ (when $d_s = 4/3$, we recover eq 3).

(3) The conformation of the fractal confined in a tube can be worked out in the same way as in the main text. The Flory approach gives directly the exponent $5/3$ for the dependence of L (eq 5). The assumption that the Ariadne length is the maximum stretched length[3] is made clear through the parametrization (paragraph B1): since a monomer cannot be stretched beyond extension a, the fractal cannot be stretched beyond $L_{max} \cong Na = l_A$. From this we deduce, as in paragraph A3, another derivation of the scaling law for L (eq 5) and the blob size $\xi = D(D/R)^w$ (compare with eq 10), where $w = (d_s - 1)/(2d_s/d_f + 1 - d_s)$ (or $w = 5(d_s - 1)/3(3 - d_s)$), using the fractal dimension derived from the Flory argument, $d_f = 5d_s/(2 + d_s)$. Note that $w = 0$ for linear polymers ($d_s = 1$, $d_f = 5/3$, $\xi = D$), and $w = 1/3$ for statistically branched polymers ($d_s = 4/3$, $d_f = 2$; see eq 10).

(4) The critical solvent current for the injection of such a fractal into a nanopore is derived in the same way as for a statistically branched polymer (section III):

$$J_c = \frac{kT}{\eta}\left(\frac{D}{\xi}\right)^4 = \frac{kT}{\eta}\left(\frac{a}{D}\right)^{4w} X^{4w/d_f} \tag{B3}$$

References and Notes

(1) See the review: Guyot, A. In *Matériaux Polymères (Arago 16)*; Masson éd. (Paris) 1995; Chapter *II*, pp 94−101.
(2) Brochard-Wyart, F.; Gennes, P. G. (de) *C. R. Acad. Sci., Paris* **1996**, *323II*, 473−479.
(3) Vilgis, T. A.; Haronska, P.; Benhamou, M. *J. Phys. II Fr.* **1994**, *4*, 2187−2196.
(4) Alexander, S.; Orbach, R. *J. Physique Lett.* **1982**, *43*, L625−L631.
(5) Isaacson, J.; Lubensky, T. C. *J. Phys. Lett.* **1980**, *41*, L469−L471.
(6) Daoud, M.; Joanny, J. F. *J. Phys.* **1981**, *42* , 359−1371.
(7) Zimm, B.; Stockmayer, W. H. *J. Chem. Phys.* **1949**, *17*, 1301.
(8) Gennes, P. G. (de) *Biopolymers* **1968**, *6*, 715.
(9) Adam, M.; Delsanti, M.; Munch, J. P.; Durand, D. *J. Physique* **1987**, *48*, 1809−1818.
(10) The minimum diameter (eq 7) was introduced a few years ago by Vilgis (*J. Phys. II*, Fr. **1992**, *2*, 2097−2101) as a basis for a separation process through mere diffusion of the branched polymer inside the tube (no solvent flow): only those polymers, the minimum diameter of which is smaller than the tube diameter, can pass through (on exponentially long time scales).
(11) Rammal, R.; Toulouse, G. *J. Phys. Lett.* **1983,** *44*, L13−L22.
(12) Edwards, S. F. *Proc. Phys. Soc. (London)* **1965**, *85*, 613; **1966**, *88*, 265.

MA960941P

Faraday Discuss., 1996, **104**, 1–8

Introductory Lecture

Mechanics of soft interfaces

Pierre-Gilles de Gennes

Laboratoire de Physique de la Matière Condensée, Collège de France, 11, place Marcelin-Berthelot, 75231 Paris Cedex 05, France

This (partial) review discusses: (1) the bursting of liquid films; (2) reactive wetting and (3) the slippage of polymer melts against walls. In all three cases, I try to explain a few basic features only, using very simple terms. Some uncertainties, and some challenges, are presented.

This meeting considers 'complex fluids'; I like this subject, but I do not like the words. With students, I find that they are scared by the word ' complex'. I prefer 'soft matter'. In practice, as we know, this means polymers, liquid crystals, surfactants, and so on. The word 'soft' is not just a pleasant adjective: it implies materials which have 'large response functions'.[1]

Soft matter is basic for many industrial purposes. The mechanics of interfaces is a good example: I want to show its practical interest along the way.

Of course, my selection of topics is rather arbitrary, based as it is on personal interests; I shall concentrate on three areas.

Bursting of soap films

We have all played with soap bubbles. The death of a bubble has been studied by many authors.[2] One of the great masters here was Karol Mysels, together with a remarkable theorist, S. Frankel.

Fig. 1 shows the growth hole in a thick water film after the hole has been started by a spark.

To predict the velocity V, the simplest approach is by Culik (1962).[3] It is based on the model of Fig. 2, where we assume that the water from the central part of the film was collected into a small rim. Ahead of the rim, we postulate that there is no motion (and thus no deformation of the film). This can be checked by interferometric measurements of the film thickness, ahead of the rim.

Culik then writes Newton's law in the form:

$$\frac{\mathrm{d}P}{\mathrm{d}t} = \frac{\mathrm{d}}{\mathrm{d}t}(MV) = F = 2\gamma \tag{1}$$

where P is the momentum (per unit length of rim), M the rim mass (again per unit length) and F the corresponding pulling force, related to the surface tensions, γ, on both sides. In the present case, the velocity V turns out to be time independent, and $\mathrm{d}P/\mathrm{d}t$ comes from the change of mass:

$$\frac{\mathrm{d}M}{\mathrm{d}t} = \rho e V \tag{2}$$

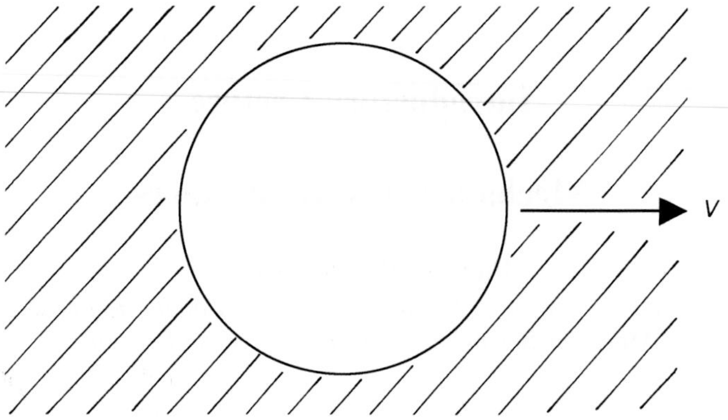

Fig. 1

where ρ is the density and e the initial film thickness.

Combining eqn. (1) and (2) we arrive at:

$$V = V_{\text{Culik}} = \left(\frac{2\gamma}{e\rho}\right)^{1/2} \tag{3}$$

All this looks very innocent, but it is not. The crucial point is that the work done by capillarity is:

$$\int FV \, dt = MV^2$$

and not the standard kinetic energy $1/2MV^2$.

We can get a better feeling for this strange observation if we return to a classic high school exercise, based on a conveyer belt (Fig. 3). Here, some sand is poured on the belt, and is therafter transported at a prescribed horizontal velocity V. The force required is by eqn. (1) $F = V \, dM/dt$. Let us look at the collision of one grain with the (rough) belt surface, idealised in Fig. 4.

In the reference frame of the belt, the grain hits the solid at a horizontal velocity $-V$, and then stops: the collison is strongly inelastic. Thus, for the conveyor belt, we should write:

$$\text{work} = MV^2 = \tfrac{1}{2}MV^2|_{\text{kinetic}} + \tfrac{1}{2}MV^2|_{\text{dissipation}}$$

Using similar arguments, Mysels was led to assume a strong dissipation in the rim, possibly by viscous flow.

Another (earlier) model was set up by Dupré in 1867,[4] and echoed by Lord Rayleigh

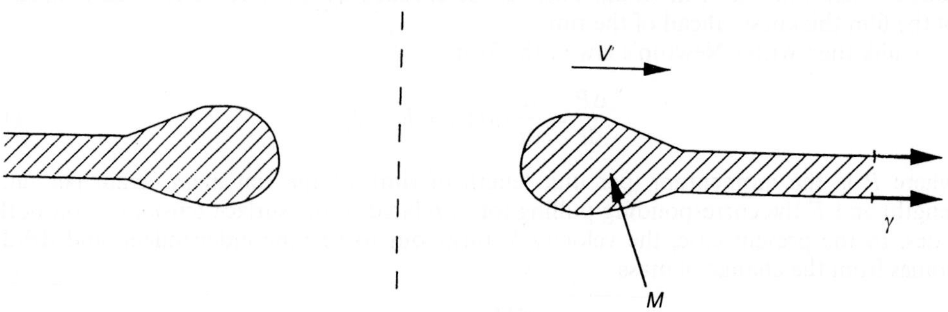

Fig. 2

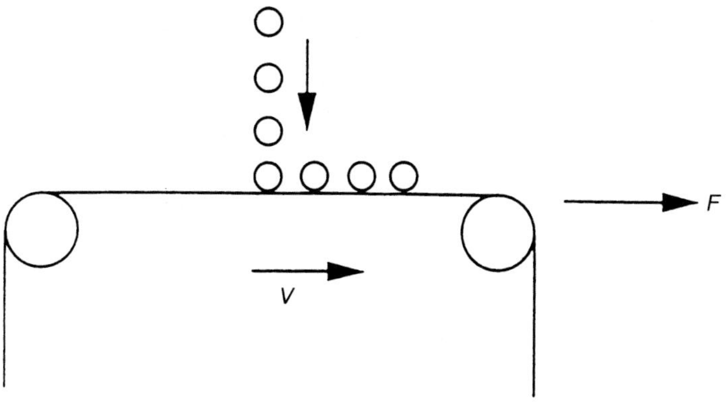

Fig. 3

in 1891.[5] It assumes no loss (work = $1/2MV^2$). But, in the conveyor belt analogy, this would mean particles which rebound, and travel faster than the rim.

My (proposed) way out of the paradox was different. If we sit in the reference frame of the rim, we have a jet of liquid, coming with velocity $(-V)$ from the right, and hitting a fixed wall. We have seen something of this sort, observing the jet from a fawcett hitting the bottom of a basin. The water escapes from the collision point by moving out (fast) along the wall.

For our soap films, I thus proposed that: (a) the rim breaks into droplets (possibly by the Rayleigh instability); (b) these droplets escape by moving at a velocity $\pm V$ normal to the film plane (in the rim reference frame). Half of them go up, half go down. The momentum balance of Culik remains correct.

Returning to the laboratory reference frame, the droplet should escape at 45° from the film plane, with velocities:

$$V_{\parallel} = V_{\perp} = V \tag{4}$$

The energy balance is:

$$\text{work} = 1/2MV_{\parallel}^2 + 1/2MV_{\perp}^2 = MV^2 \tag{5}$$

Up to now, as far as I know, nobody has seen droplets escaping during film rupture! Prof. Frens and his group argue that the extra dissipation occurs in the schock front.

The crucial parameter is the surface pressure head of surfactant Π_1. There is a natural limit to Π_1: when $\Pi_1 = \gamma_{wo}$ (the surface tension of pure water) the surface tension vanishes. Assuming this value for Π_1, the Hugoniot conditions (for mass and

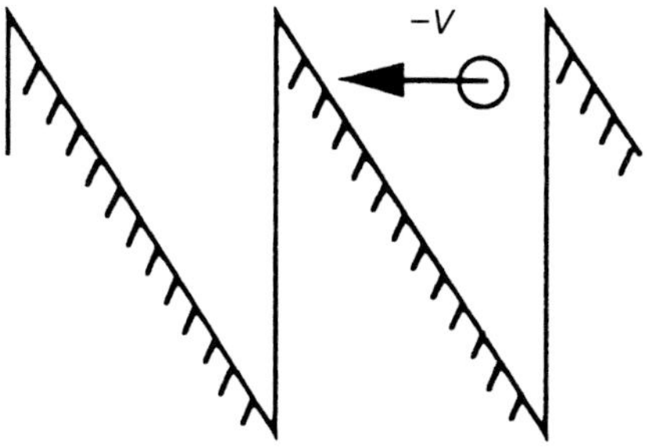

Fig. 4

momentum transfer) give a rim velocity:

$$V^2 = V_{\text{Culik}}^2\left(1 - \frac{e}{e_1}\right)$$

where e_1 is the (increased) water thickness of the 'aureole'. The ratio e_1/e is equal to Σ_0/Σ_1, where $\Sigma_0(\Sigma_1)$ is the area per polar head in the rest state (in the rim). Energy conservation then gives:

$$(\Sigma_0 + \Sigma_1)[\Pi(\Sigma_1) - \Pi(\Sigma_0)] - 2\int_{\Sigma 1}^{\Sigma 0} \Pi \, d\Sigma = e\Sigma_0 C_p \, \delta T$$

where C_p is the specific heat of water and δT is the temperature rise in the water. $\delta T \sim 1/e$ is small, and the curve $\Pi(\Sigma)$ may thus not be very different from the isotherm. Then writing $\Pi(\Sigma_1) = \gamma_\omega$ fixes Σ_1 and e_1.

All this can apply only to thick films. For thin films (<1000 Å) the Reynolds numbers become small, and viscosity effects come in. Also, if we go towards black films, the surfactant becomes the majority component: this is analysed in the paper by Frens and co-workers in this Discussion.

Dewetting: Bursting of supported films

Here, we force a liquid film (*e.g.* water) on a surface which does not like to be wetted (*e.g.* polyethylene). Dry patches nucleate and grow, with a certain velocity V. In certain well controlled cases, using silanized surfaces, one then finds a simple law:[7]

$$V = (\text{const})\,\frac{\gamma}{\eta}\,\theta_e^3; \quad \theta_e \ll 1 \tag{6}$$

γ is again the surface tension, η the viscosity, θ_e is the equilibrium contact angle, Eqn. (6) holds when there is again a rim, and when the losses are dominated by hydrodynamic friction in this rim. (There may be other cases where molecular losses very near the contact line are important, especially at high θ_e.)

What is the origin of the factor θ_e^3 in eqn. (6)? I try to explain in Fig. 5, where I focus on the border, *B*, between rim and film at rest. (The other end, *A*, gives comparable contributions.) At point *B*, there is a driving force (in the Young sense):

$$F = \gamma(1 - \cos\theta) \approx \gamma\,\frac{\theta^2}{2} \tag{7}$$

We have to balance this against a dissipation in the wedge which can be written in the form:[7]

$$F = zV = \text{const.}\,\frac{\eta}{\theta}\,V \tag{8}$$

The presence of θ in the denominator reflects what I call the 'plumber's theorem': dissipation is strong in thin pipes. Combining eqn. (8) and (7), and guessing that θ is (within a numerical factor) comparable to θ_e, we get eqn. (6).

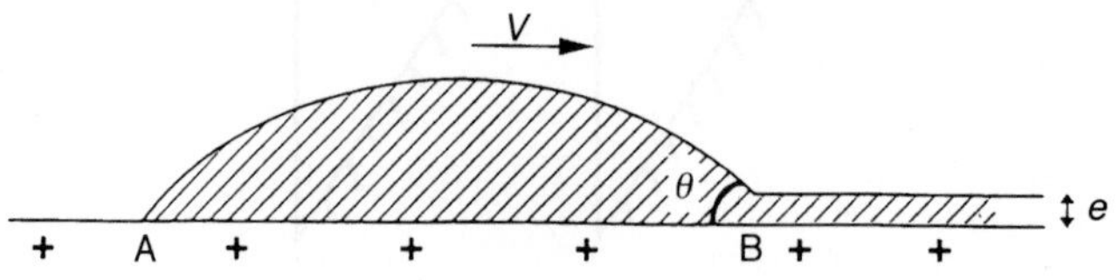

Fig. 5

There are many variants to this dewetting process, which I shall not describe. Some of them are of industrial importance: drying inside a washing machine; offset four colour printing; protecting plants *via* insecticides; hydroplaning of cars, *etc*. Here, I will just mention another type of bursting film, studied recently by G. Debregeas *et al.*[8], where the physics is completely different. Here, we deal with a freely suspended film of a linear polymer melt (*e.g.* poly(dimethylsiloxanes) of molecular weight *ca.* 10^6).

These films are rather stable (even for thickness, e, of order 10 μm or less) without any surfactant. But if we initiate a hole, the hole grows. The scenario is widely different from Fig. 2: there is no rim, but there is a radial velocity field extending up to a large distance. The radius of the hole does not increase linearly with time, but exponentially! A simple viscoelastic model for this is proposed in ref. 8.

Reactive wetting

Ondarçuhu was planning to graft chlorosilanes on a glass surface. He put a drop of solution, containing the chlorosilane, on the glass; the drop started to run at a well defined velocity! Sometimes it would change its direction (probably *via* dust particles on the glass) but it never crossed its own path. Ondarçuhu and Domingues dos Santos then studied the process in detail.[9]†

The basic scheme is simple.[10] The grafting reaction makes the surface less wettable: the drop escapes from the reacted region, which is hostile. You can summarise the process as follows. If the reaction has proceeded for a time t, the solid/liquid interfacial tension is modified by a term:

$$F = \gamma_1[1 - \exp(-t/\tau)] \tag{9}$$

where τ is a reaction time. [The exponential form of eqn. (9) is a consequence of a first-order reaction and τ^{-1} is proportional to the silane concentration c.] The modification described by eqn. (8) acts as a driving force in the Young sense to push the droplet. It induces a velocity V and we can often assume:

$$F = zV \tag{10}$$

where z is a friction coefficient, which will depend on the geometry. (Do we have a free drop on a surface, or a drop trapped in a capillary. *etc*?) If the drop has a certain size L, the reaction time is:

$$t = L/V \tag{11}$$

Let me choose the specific case where $t < \tau$. Then, eqn. (9) simplifies and becomes:

$$F = \gamma_1 t/\tau \tag{12}$$

Combining eqn. (1–12) we obtain:

$$V^2 = \frac{\gamma_1}{z}\frac{L}{\tau} \tag{13}$$

For the case of a droplet on a solid surface, z is known to be independent of L.[10] Then, eqn. (13) predicts $V \sim L^{1/2}$ and $V \sim c^{1/2}$, if we vary the silane concentration c. Both predictions are reasonably well observed.[9]

Let me mention one possible extension of these ideas. Suppose that we do not have a drop, but a lump of swollen gel, containing the same sort of reactive solution. My expectation is that it will also start to move.[11] If the gel is in the form of a little pill, it should

† During the Discussion meeting, I learnt about some earlier work on running droplets by C. Bain, G. Burnett-Hall and R. Montgomerie, *Nature (London)*, 1994, **372**, 414! Unfortunately, funding for this subject has not been forthcoming.

glide at very low velocities (μm s^{-1}). However, if the gel is in the form of a sphere, it should roll, and if the sphere is not too small (say 0.1 mm) the velocities should be quite sizeable. I do hope that somebody will try this.

Slippage of polymer melts

In mechanics, we are taught that 'velocity fields are continuous': *e.g.* if we shear a fluid near a solid surface, which is immobile, the fluid velocity at the wall vanishes. This is essentially correct for simple fluids, but not necessarily true for polymer melts (Fig. 6).

If the solid wall is ideal (no polymer chain bound to it) we would write a relation between shear stress σ and slip velocity V at the wall in the form:

$$\sigma = kV \tag{14}$$

where k describes a microscopic friction between the monomers and the wall: k is then independent of chain length. Inside the fluid, we have the standard viscous relation.

$$\sigma = \eta \frac{dV}{dz} \tag{15}$$

where η is again the viscosity. This leads to an 'extrapolation length':

$$b = \frac{\eta}{k} \tag{16}$$

In these melts, η is large (because of chain entanglements) while k remains small. Thus, b can be quite large (hundreds of μm). On an ideal surface, slippage should be significant.

Most usual walls (in rheometers, viscometers, *etc.*) are not ideal, and show no slippage. This is probably due to some chains being bound firmly to the wall, and blocking the surface flow.

Systematic experiments were carried out during the past four years by Hervet, Léger and co-workers.[12,13] They prepared silanated surfaces where only a few sites (of bare silica) can bind the polymer chains (silicones in their case).

They found that, in a system like this, slippage is velocity dependent. At very low V ($V < V^*$ with $V^k \approx 1$ μm s^{-1}.) b is small (no slip). Beyond V^*, there is a large velocity domain, where b increases continuously. In this regime, the surface stress is constant $\sigma = \sigma^*$. How can we explain this?

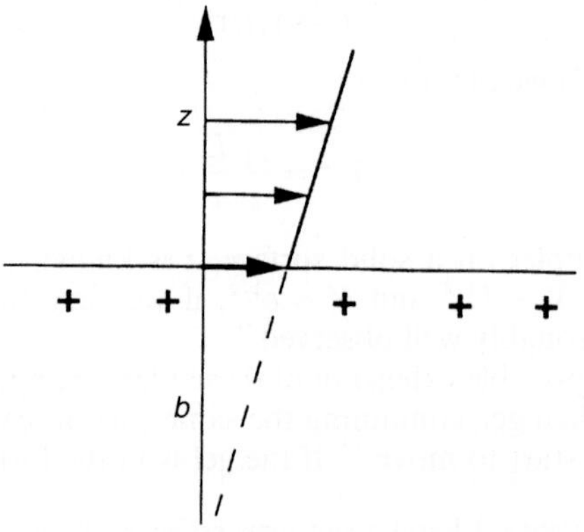

Fig. 6

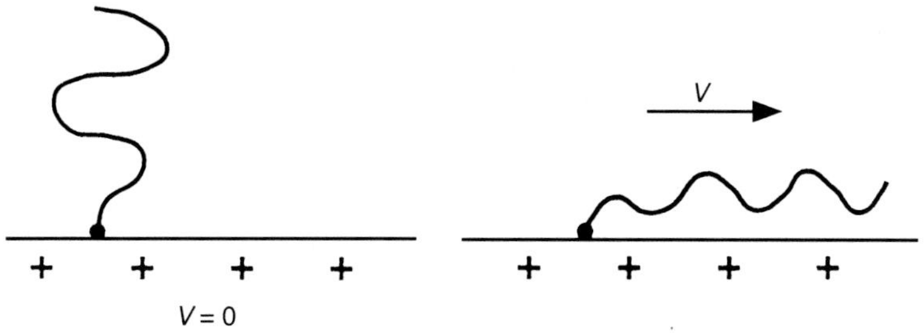

Fig. 7

We might be tempted to think that, at $V = V^*$, the bound chains are torn out from the surface. But this could give an abrupt jump in b around $V = V^*$, which is not observed. Thus we are led to another model (Fig. 7) which I call 'wind in the willows'.

When the melt velocity is large enough, the willows bend so much that their vertical is comparable to the diameter of the 'Edwards tube' defining entanglements. Then we are in a situation where a chain is confined in a linear pipe: it is on the verge of being disentangled. We call this the marginal regime. In the present case, the marginal regime has a certain intrinsic stability, if the willow stretches more, it decouples and suffers less from the flow. This encourages the willow to return towards the upright position.[14]

Let me now mention a related, but different situation. Here, we talk about molten polyethylene (PE) pushed through a thin steel tube. These experiments have been performed over many years, both in industry and in academia. Recently, Wang and co-workers in Cleveland[15] showed that in these systems, there is again a transition from a no slip state to a slip state, with a very large b (*ca.* 1 mm). This is clearly related to what I described above, but there are some important differences. In these polyethylene experiments, we see a trend towards stick slip instabilities; the most natural picture for this involves a shear stress σ which, in a certain range of V, becomes a decreasing function of V. With the 'wind in the willows', we do not have this: $\sigma(V)$ is flat. One possible model for the polyethylene/steel system is based on the 'arches' of Fig. 8.

The arches must be due to PE chains which are bound to many points at the surface. In this case, the mobile PE chains glide in and out the arch. It was pointed out by Ajdari that this would naturally lead to instabilities.[16]

Concluding remarks

This talk has ignored some of the most important chapters of 'soft interfaces': especially the area of nanotribology which will be tackled in some of our later discussions (see the paper by Klein and co-workers in this Discussion). Here some important empirical features emerge, but they are not easy to systematise by simple arguments. Another area

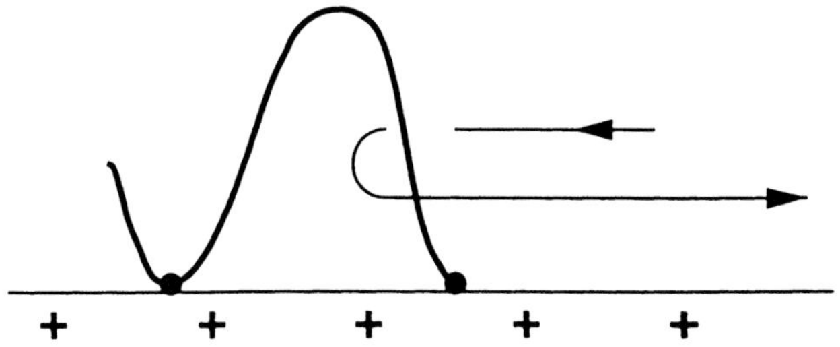

Fig. 8

overlooked here is adhesion. My impression of this is that we do understand some features of adhesion on really soft materials (weakly cured rubbers, tackifiers). But we are far from a clear picture for glassy adhesives (such as epoxy resins) because their distortions near the fracture tip are hindered by cross links. We may need a later Faraday Discussion on these matters.

I am deeply indebted to K. Mysels, who introduced us to many of these problems, to Prof. Frens, who instructed me on the shock wave dissipation in soap film rupture, and to F. Brochard-Wyart and her team, who produced many of the results and interpretations described here.

References

1 *Fragile Objects*, Springer, New York, 1996.
2 (*a*) W. Ranz, *J. Appl. Phys.*, **30**, 1950, (*b*) W. Mc Entee and K. Mysels, *J. Phys. Chem.*, 1969, **73**, 3018.
3 F. E. Culik, *J. Appl. Phys.*, 1960, **31**, 1128.
4 A. Dupré, *Ann. Chim. Phys.*, 1867, **4**, 194.
5 Lord Rayleigh, *Nature* (*London*), 1891, **44**, 249.
6 S. Frenkel and K. Mysels, *J. Phys. Chem.*, 1969, **73**, 3028.
7 F. Brochard and P. G. de Gennes, *Adv. Colloid Sci.*, 1992, **39**, 1.
8 G. Debregeas, P. Martin and F. Brochard-Wyart, *Phys. Rev. Lett.*, 1995, **75**, 3886.
9 F. Domingues dos Santos and T. Ondarçuhu, *Phys. Rev. Lett.*, 1995, **75**, 2972.
10 F. Brochard and P. G. de Gennes, *C. R. Acad. Sci.* (*Paris*), 1995, **321 II**, 285.
11 P. G. de Gennes, *C. R. Acad. Sci.* (*Paris*), 1996, submitted.
11 P. G. de Gennes, *C. R. Acad. Sci.* (*Paris*), 1996, submitted.
12 K. Migler, H. Hervet and L. Léger, *Phys. Rev. Lett.*, 1993, **70**, 287.
13 (*a*) G. Massey, Thèse U. Paris 6, 1995; (*b*) L. Léger, H. Hervet, Y. Marciano, M. Deruelle and G. Massey, *Isr. J. Chem.*, 1995, **35**, 65.
14 (*a*) F. Brochard and P. G. de Gennes, *Langmuir*, 1992, **8**, 3033. (*b*) F. Brochard, C. Gay and P. G. de Gennes, *Macromolecules*, 1996, **29**, 377.
15 P. A. Drda and S. Q. Wang, *Phys. Rev. Lett.*, 1995, **75**, 2698.
16 A. Ajdari, *C. R. Acad. Sci.* (*Paris*) *II*, 1993, 317, 1159.

Paper 6/06869C; *Received* 1st *October*, 1996

Langmuir **1996,** *12,* 4497–4500 4497

Soft Adhesives†

P. G. de Gennes

Collège de France, 11 place Marcelin Berthelot, 75231 Paris Cedex 05, France

Received October 17, 1995. In Final Form: February 1, 1996®

We attempt a simple theoretical picture for the strong adhesion properties of weakly cross linked rubbers. These ideas are then extended to zero crosslinking -ie the problem of tack for a linear polymer.

I. General Aims

Most fracture processes inside a glue are dominated by dissipation near the fracture tip; this occurs for instance with polymers which craze, where most of the separative energy (per unit area) G is due to a precursor craze.[1,2] However, there is another possibility, which is important for very soft materials, in practice, for weakly cured rubbers; inside the network, we have many chains which are free and many chains which are tied at one end only. In cases like this, the low frequency modulus μ_0 (related to the network) is small. But the high frequency modulus μ_∞ (which contains the effects of the entangled free chains and of the dangling ends) is high. Typically, we can achieve:

$$\lambda = \mu_\infty/\mu_0 \sim 100 \qquad (1)$$

It was shown by Gent and Petrich[3] that poorly cross-linked systems of this type have a very anomalous curve $G(V)$ (adhesive energy/velocity) as shown on Figure 1. The anomaly disappears upon further "curing" (when the network is more cross-linked and μ_0 raises to become comparable to μ_∞).

Here, we are concerned only with what Gent and Petrich call the *first transition*: from a soft rubber (modulus μ_0 due to a few cross links) to a hard rubber (modulus μ_∞ dominated by entanglements). Both states are assumed to be fluid. The second transition (from hard rubber to a glass) will not be discussed here. We also restrict our attention to bulk fracture and do not consider any detailed interfacial features, such as transitions between cohesive and adhesive modes.

In section II, we construct a simple picture of viscoelastic fracture in a weakly cross-linked rubber.[4] The major idealization which we impose is to assume that mechanical relaxation inside the network is described by a *single* relaxation time τ. Of course, this is very crude; the dangling ends have a wide distribution of length, etc. But the one time approximation allows an understanding of what happens in space, far behind the fracture tip.

In section III, we proceed to the more general case where there are many relaxation times. Various relations between the fracture energy $G(V)$ and the complex modulus $\mu(\omega)$ have been proposed, often assuming that ω and V are related via a fixed length l ($\omega = V/l$). We arrive here at a formula which is quite different, where $G(V)$ is an *integral* over many frequencies (eq 21). This equation

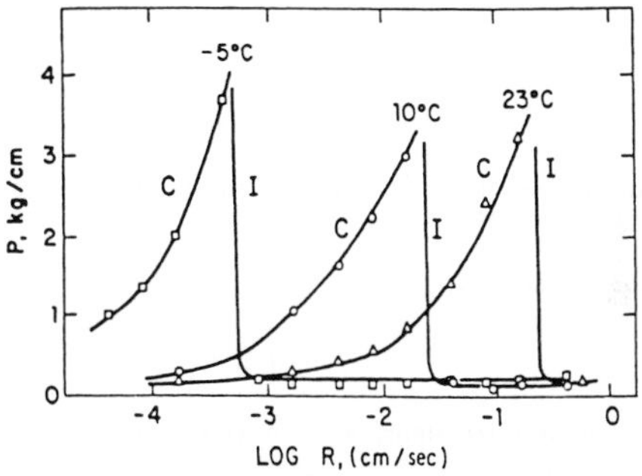

Figure 1. Peel force P versus peel rate R for an un-cross-linked butadiene–styrene rubber adhering to a PET polyester film after ref 3. The symbols C and I denote cohesive failure and interfacial failure, respectively.

will be derived here but will not be discussed in detail; the subject is so vast that a separate publication will be required.

Section IV discusses the fracture energy $G(V)$ for un-cross-linked melts, where $\mu_0 = 0$. This, as we shall see, can lead to very large fracture energies and is clearly an important component of *tack*. Of course, our discussion is still very far from real materials, such as pressure sensitive adhesives; in these systems, the *second* transition of Gent and Petrich (to a glassy state) plays a major role. But, once again, and independently of the detailed process involved in the transition, we believe that we can give a simple insight.

In all this text, our approach is mainly based on *scaling* laws; we gain in simplicity, but we ignore all numerical coefficients.

II. Bulk Fracture: The "Viscoelastic Trumpet"

Let us start with some remarks on the viscoelastic features.

The simplest formula for the complex modulus $\mu(\omega)$ as a function of frequency (ω) is the following:

$$\mu(\omega) = \mu_0 + (\mu_\infty - \mu_0)\frac{i\omega\tau}{1 + i\omega\tau} \qquad (2)$$

Usually, with one relaxation time τ, we think that we have to cope with two regimes $\omega\tau > 1$ and $\omega\tau < 1$. Here, in fact, when $\lambda = \mu_\infty/\mu_0 \gg 1$, we have *three* regimes:

(i) At very low $\omega\mu = \mu_0$, we expect a *soft solid*.

(ii) When $1 > \omega\tau > \lambda^{-1}$, we can approximate

$$\mu(\omega) \sim (\mu_\infty - \mu_0)i\omega\tau = i\omega\eta \qquad (3)$$

The modulus is purely imaginary; we are dealing with a

† Presented at the Workshop on Physical and Chemical Mechanisms in Tribology, held at Bar Harbor, ME, August 27 to September 1, 1995.

® Abstract published in *Advance ACS Abstracts,* Sept. 15, 1996.

(1) Brown, H. *Macromolecules* **1991**, *24*, 2752.
(2) de Gennes, P. G. *Europhysics Lett.* **1991**, *15*, 191.
(3) Gent, A.; Petrich, R.; *Proc. R. Soc. London* **1969**, *A310*, 433.
(4) de Gennes, P. G. C. R. *Hebd. Seances* Acad. Sci. **1988**, *307*, 1949.

S0743-7463(95)00886-9 CCC: $12.00

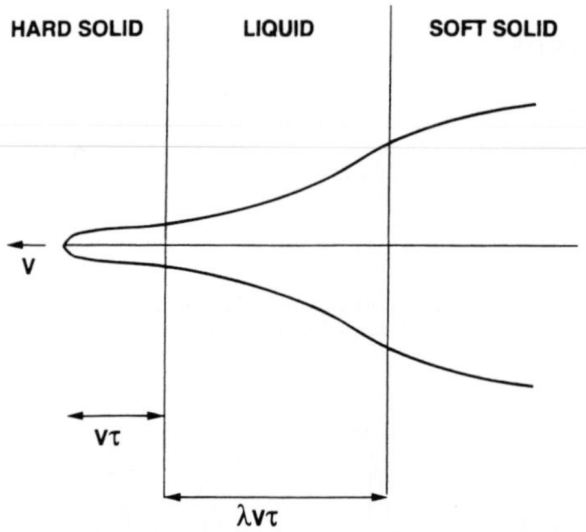

Figure 2. Viscoelastic "trumpet", fracture profile when a single relaxation time is present.

liquid of viscosity

$$\eta = (\mu_\infty - \mu_0)\,\tau \sim \mu_\infty \tau \tag{4}$$

(iii) At high frequencies ($\omega\tau > 1$) we recover a *strong solid* ($\mu = \mu_\infty$).

What are the consequences of this on a moving fracture? When the fracture velocity V is not too small, so that $V\tau$ is larger than the size of the fracture tip, we can think in terms of a continuum, and we find three regions in our rubber (Figure 2).

Near the fracture tip, we are discussing small spatial scales, or short times, and we have a strong solid.

At intermediate distances, $V\tau < r < \lambda V\tau$, we have a liquid, giving a large dissipation.

At higher distances $r > \lambda V\tau$, we have a soft solid.

You might think that the effects in the liquid zone are not very important, because the stresses here are relatively small—we are far from the fracture tip. On the other hand, because r is large, we have a huge volume of fluid-like behavior and the overall dissipation is increased. It will turn out that this volume factor dominates.

Although we are talking about a complex viscoelastic medium, the scaling law for the stress as a function of distance is still simple, and equivalent to what we have in a simple elastic medium[5]

$$\sigma(r) \sim K_0/r^{1/2} \tag{5}$$

where the factor K_0 is associated with the adhesion energy G_0 due to local processes near the tip; again ignoring all coefficients, we have the standard relation from fracture mechanics

$$G_0 \cong K_0^{\,2}/\mu_\infty \tag{6}$$

Three remarks concerning the stress distribution:

(i) Equation 5 is simply a scaling law, ignoring all specific details for different components of the stress; for instance, on the fracture tip, the normal stress component σ_{zz} must vanish identically—but the other components still follow the scaling form (5).

(ii) Why does this simple form remain valid in a viscoelastic medium? The equations of motion, in terms of the density ρ and the local velocity v reduce to

$$\rho\,\frac{Dv}{Dt} = -\nabla\sigma \cong 0 \tag{7}$$

where we set $Dv/Dt = 0$, because we are dealing with velocities much smaller than the sound velocity; inertial terms are negligible. Thus the equations are simply $\nabla\sigma = 0$ and are the same for any viscoelastic medium. The stress components must also satisfy certain compatibility conditions (because they derive from a displacement field). But these geometric conditions, again, are the same for any viscoelastic medium.

(iii) In an elastic medium of modulus μ, with a stress field described by eq 5, the displacement field u is simply related to the stress by the scaling law

$$\sigma = \mu \nabla u \tag{8}$$

and this imposes

$$u = (K_0/\mu) r^{1/2} \tag{9}$$

If we now take the scaling structure of the product (σu) along the fracture (where u now measures the opening of the fracture), we find that it is independent of distance

$$\sigma u = K_0^{\,2}/\mu = G_0 \tag{10}$$

where G_0 is the corresponding adhesion energy. For our more complex problem, we shall again find the separation energy G from the product (σu), calculated at larger r (in the soft medium); measuring at large distances we incorporate all dissipation effects.

Let us now return to Figure 2 and find out the overall shape of the fracture profile $u(x)$. In the strong solid region, we have a classical parabolic shape

$$u = K_0/\mu_\infty x^{1/2} \tag{11}$$

and $\sigma u = G_0$. These equations hold provided that x is larger than the (microscopic) size l of the zone, where nonlinear processes are dominant. (We expect $l \lesssim 100$ Å for our materials). On the other hand, the strong solid region is restricted to $x < V\tau$. Thus, our discussion holds, provided that $l \ll V\tau$.

In the liquid zone, σu is not constant. The scaling law relating u to σ is based on a viscous stress

$$\sigma = \eta \frac{d}{dx}\left(\frac{du}{dt}\right) = \eta\, V \frac{d^2u}{dx^2} \tag{12}$$

and with $\sigma \sim x^{-1/2}$ (eq 5) this gives $u \sim x^{3/2}$. Thus the product σu increases linearly with x

$$\sigma u = x/V\tau G_0 \tag{13}$$

When we reach the soft solid region ($x = \lambda V\tau$), we find

$$\sigma u = \lambda G_0 \tag{14}$$

and this gives us the overall adhesion energy G

$$G/G_0 = \lambda = \mu_\infty/\mu_0 \tag{15}$$

A more rigorous derivation of eq 15 has been given by Hui.[5] The result deserves a number of comments. Note first that the viscoelastic corrections give a *multiplicative* effect; they do not add up a constant term to G_0. Second, we see that the ratio G/G_0 can be very large if the material

(5) See, for instance: Lake, G.; Thomas, A. In *Engineering with rubber*; Gent, A., Ed.; Hanser: Munich, 1992.

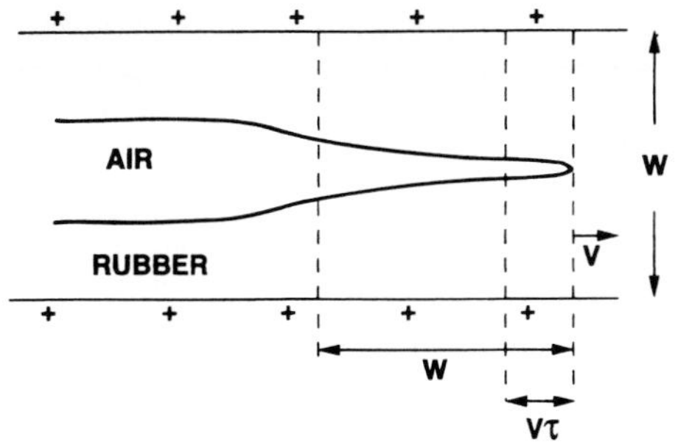

Figure 3. Truncation of the viscoelastic trumpet in a thin slice of adhesive material.

is very poorly cross-linked ($\mu_0 \to 0$). This also explains why the enhancement in G disappears upon further curing ($\mu_0 \to \mu_\infty$).

All this discussion held for a thick glue in the cohesive mode (bulk fracture). But very often the glue is in the form of a thin slab (thickness W) as shown in Figure 3. The opening process is then cut off at distances $x \sim W$. Thus when $\lambda V\tau > W$, the dissipation zone is restricted in size and we have

$$G = G_0 W/V\tau \tag{16}$$

In this regime, $G(V)$ is a decreasing function; the pulling force drops if the velocity increases. This often generates mechanical instabilities in the fracture process. We believe that this instability is the source of the peak in the experimental $G(V)$ curve observed by Gent and Petrich.

III. Other Types of Relaxation

(1) Sol/Gel Transitions. What happens when we have not one single relaxation time, τ, but a broad distribution of relaxation times, giving a complex form to the dissipative modulus $\mu(\omega)$? A natural attempt, proposed long ago, assumes that $G(V)$ is proportional to the modulus $\mu''(\omega)$ measured at a frequency $\omega = V/l$, where l is some characteristic length of the rubber network. However, this is not correct in general, as pointed out in particular by Barber et al. on simple examples.[6] There is one case where it holds, namely, when the network is *just at its gelation point*, at the transition between isolated clusters and one infinite cluster plus finite molecules. At this point, the elastic modulus $\mu(\omega)$ is given by a power law

$$\mu'(\omega) = \mu''(\omega) \cong \omega^{\alpha} \tag{17}$$

where α is a certain critical exponent. In this case, a simple extension of the above argument (with a distribution of relaxation times which is fixed by α) does show that $G \sim V^{\alpha}$. This could appear as a rather exceptional situation; however, since this region near the connectivity threshold is the region of long τ and large G, many practical materials may be purposely chosen to lie in this region.

(2) General Relation between $G(V)$ and $\mu(\omega)$. Our starting point here is the dissipation $T\dot{S}$ (per unit length of the fracture line).

$$T\dot{S} \equiv V\,G(V) = \int \mathrm{d}x\,\mathrm{d}y\,\sigma\,\dot{\gamma} \tag{18}$$

where σ is the local stress and $\dot{\gamma}$ the local share rate. As point out before, σ retains its unperturbed scaling structure (eq 5). The deformation γ can be related to σ through the complex modulus $\gamma = \sigma/\mu(\omega)$, where ω is the local sweeping frequency

$$\omega = V/r \tag{19}$$

Thus, $\dot{\gamma} = i\omega\sigma/\mu(\omega)$. We see that our scaling estimates all depend only on r; thus we may replace the integral

$$\int \mathrm{d}x\,\mathrm{d}y \to \int 2\pi r\,\mathrm{d}r \to \mathrm{const} \int r\,\mathrm{d}r$$

Collecting all these results, we arrive at

$$V\,G(V) \propto \int r\,\mathrm{d}r \frac{\sigma^2}{\mu(\omega)} i\omega \cong \int r\,\mathrm{d}r \frac{\sigma^2}{\mu(\omega)} i\omega = K_0^2 V \int \frac{\mathrm{d}\omega}{\omega} \mathrm{Im}\left(\frac{1}{\mu(\omega)}\right) \tag{20}$$

where Im(f) denotes the imaginary part of f. Remembering that the bare fracture energy is $G_0 = K_0^2/\mu_\infty$, we end up with

$$\frac{G(V)}{G_0} \cong \mu_\infty \int_0^{\omega_{\max}} \frac{\mu''(\omega)}{[\mu'(\omega)^2 + \mu''(\omega)^2]^{1/2}} \frac{\mathrm{d}\omega}{\omega} \tag{21}$$

For bulk fracture, the cut off frequency $\omega_{\max}$ is

$$\omega_{\max} = V/l \tag{22}$$

where l is the size of the nonlinear region near the fracture tip. (Typically $l \sim 10$ to 100 Å). Two remarks concerning these formulas are as follows:

(a) When there is a single relaxation time τ, the domain of important frequencies in eq 21 is

$$\frac{1}{\lambda\tau} < \omega < \frac{1}{\tau}$$

and in this region

$$\mathrm{Im}(1/\mu) \sim 1/\omega\tau\mu_\infty$$

One can check that eq 15 is recovered provided that $V\tau \gg l$.

(b) Equation 21 is closely similar to certain formulas derived much more rigorously by Bowen and Knauss.[7] One of the forms which they propose is

$$\frac{G(V)}{G_0} = \mu_\infty D(t) \tag{23}$$

where $D(t)$ is a compliance

$$D(t) \cong \int \mathrm{d}\omega\, \mathrm{e}^{i\omega t} \frac{1}{i\omega\,\mu(\omega)} \tag{24}$$

and t is a cut off time $t = l/V$. The role of $\mathrm{e}^{i\omega t}$ in eq 24 is precisely to cut off the integral at $\omega t \sim 1$; thus eqs 21 and 24 are very similar. Neither of them is rigorous, because the exact nature of the cut off is fuzzy.

IV. Extension to "Tack"

Certain polymer melts (un-cross-linked) are sticky. For instance, if we establish contact between a fluid elastomer and a metal, we find that to separate them we need energies G which are high—much higher than the van der Waals energy.

(6) Barber, M.; Donley, J.; Langer, J. S. *Phys. Rev.* **1989**, *A40*, 366.

(7) Bowen, J.; Knauss, W. *J. Adhes.* **1992**, *39*, 43.

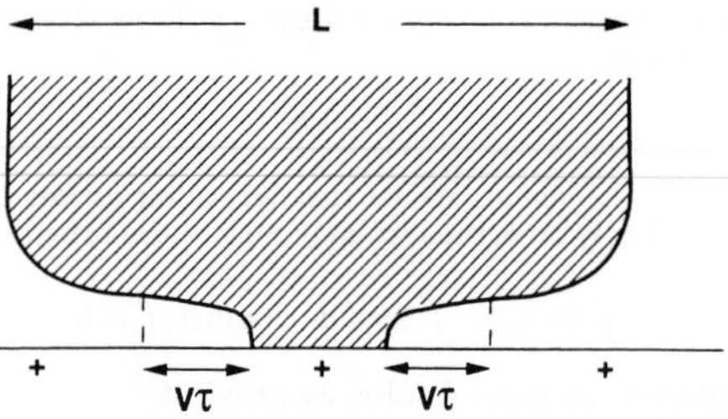

Figure 4. First model for early separation of a tacky material on a flat surface.

It is tempting to relate this remarkable effect to the viscoelastic properties of the elastomer.[8] The proposed mode of separation is shown in Figure 4. A fracture propagates along the solid, with velocity $V \sim L\theta^{-1}$, where L is the contact size and θ the separation time. We again have a liquid zone, ranging from $x = x_1 = V\tau$ up to L. The adhesion energy is

$$G = \sigma u|_{x=L} = G_0 \frac{L}{x_1} = G_0 \frac{\theta}{\tau} \tag{25}$$

Thus if $\theta \gg \tau$, we expect a large enhancement. A number of comments should be made here:

(i) Another presentation of the result (eq 25) is

$$G = G_0 \frac{\mu_\infty}{|\mu(\omega)|} \tag{26}$$

where $\mu(\omega)$ is the analog of eq 2 for a viscoelastic liquid ($\mu_0 = 0$)

$$\mu(\omega) = \frac{i\eta\omega}{1 + i\omega\tau} \approx i\eta\omega \qquad (\omega\tau \ll 1) \tag{27}$$

and $\omega = \theta^{-1}$ is the separation frequency. In the form (26) this tack property is a natural generalization of eq 15.

(ii) It is important to observe that the enhancement factor θ/τ is large if the elastomer has a rather low molecular weight (low τ). Of course, there is a limitation to this: if we choose chains which are so short that they do not entangle, μ_∞ drops down, and we lose the enhancement factor μ_∞/μ_0 in eq 15.

(iii) Our material is liquid; at very large times a liquid collapses under its own weight. The collapse time t_c may be estimated as follows. The shear rates are of order $1/t_c$ and the shear stresses are η/t_c. They must balance the gravitational pressure, which for a sample of size L is of order $\rho g L$ (ρ = density, g = gravitational acceleration). This gives a maximum value for the enhancement factor

$$\frac{G}{G_0}\Big|_{max} = \frac{t_c}{\tau} \approx \frac{\mu_\infty}{\rho g L} \approx 10^3 \tag{28}$$

It may be, however, that the accelerations involved in separation process are $g_{eff} \gg g$ and that g_{eff} should be used in this estimate. Note on eq 28 that if μ_∞ becomes small (with chains smaller than the entanglement length) the enhancement disappears.

(iv) A nice feature of eq 25 is that it is independent of the size L of the sample; thus the result may hold even if the contact zone breaks up into a bundle of elastomer fibrils (Figure 5).

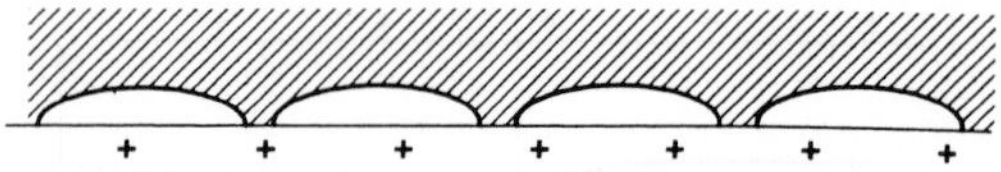

Figure 5. Separation of a tacky material, a more realistic picture.

On the whole, we have an attractive (but tentative) picture of tack based on viscoelastic effects; this should be compared to careful experiments at variable θ.

The main practical use of tack is in *rapid adhesion*: when we must bind objects very fast (e.g., assembling books on a production line). The materials used are commonly known as "pressure sensitive adhesives". They are based on (nearly) un-cross-linked chains (plus some additives). The adhesive joint is prepared by applying the glue under a certain weak pressure. What is the role of this pressure? It improves the contact on a rough solid surface. There are some "hollow" regions where contact is not initially realized and must be encouraged by the pressure p. Early descriptions of this filling process were based on a purely viscous model, where the adhesive flows toward the hollow regions like a liquid. Recently, Creton and Leibler[9] have pointed out that the viscoelastic features are essential and that the length of time required for good contact can be better understood on that basis.

Acknowledgment. This work has been initiated, and constantly stimulated, by discussions with A. Gent. I also benefited greatly from exchanges with C. Creton, L. Léger, and L. Leibler.

LA950886Y

(8) de Gennes, P. G. *C. R. Hebd. Seances Acad. Sci.* **1991**, *312II*, 1415.

(9) Creton, C.; Leibler, L. *J. Polym. Sci.; Polym. Phys. Ed.*, in press.

ELSEVIER

Physica A 278 (2000) 32–51

PHYSICA A

www.elsevier.com/locate/physa

Transient pores in stretched vesicles: role of leak-out

F. Brochard-Wyart[a,*], P.G. de Gennes[b], O. Sandre[a]

[a]*Physico-Chimie Curie UMR 168 CNRS/Institut Curie, Section de Recherche 11, rue Pierre et Marie Curie, 75248 Paris Cedex 05, France*

[b]*Physique de la Matière Condensée, Collège de France 11, place Marcelin Berthelot, 75231 Paris Cedex 05, France*

Received 2 November 1999

Abstract

We have visualized macroscopic transient pores in mechanically stretched giant vesicles. They can be observed only if the vesicles are prepared in a viscous solution to slow down the leak-out of the internal liquid. We study here theoretically the full dynamics of growth (driven by surface tension) and closure (driven by line tension) of these large pores. We write two coupled equations of the time evolution of the radii $r(t)$ of the hole and $R(t)$ of the vesicle, which both act on the release of the membrane tension. We find four periods in the life of a transient pore: (I) exponential growth of the young pore; (II) stop of the growth at a maximum radius r_m; (III) slow closure limited by the leak-out; (IV) fast closure below a critical radius, when leak-out becomes negligible. Ultimately the membrane is completely resealed. © 2000 Elsevier Science B.V. All rights reserved.

1. Introduction

Forcing the passage of molecules, or genes, through a cellular membrane is a central problem of drug delivery. Model experiments based on vesicles represent a first step in this direction. A variety of physical techniques have been proposed to increase the permeability of lipid bilayers:

* Corresponding author. Fax: +331-4051-0636.
E-mail address: brochard@curie.fr (F. Brochard-Wyart)

PII: S0378-4371(99)00559-2

Notation

d	membrane thickness
E	surface stretching modulus
K_b	Helfrich bending constant
Q	leak-out flux
r	pore radius
r_i	pore radius at nucleation
r_c	pore radius at zero tension
r_L	characteristic radius of leak-out
r_m	radius at maximum (II)
r_{23}	pore radius at cross-over between (II) and (III)
r_{34}	pore radius at cross-over between (III) and (IV)
R	vesicle radius
R_i	initial vesicle radius
R_0	vesicle radius at zero tension
V_L	leak-out velocity
V_3	slow closure velocity limited by leak-out (III)
V_4	fast closure velocity at end (IV)
η_2	lipid viscosity
η_s	surface viscosity
η_0	viscosity of solution
σ	surface tension
σ_0	surface tension before pore opening
τ	rise time of pore growth (I)
$\mathfrak{T}$	line tension

(a) raising the osmotic pressure inside a vesicle induces a surface tension σ, and the permeability (of dyes, or of spin labels) is significantly increased [1],

(b) if a bilayer becomes stuck on a porous surface (e.g. with a cationic bilayer and a negative surface) it tends to invade the pores; this again creates a tension σ and an extra permeability [2],

(c) instead of a porous medium it is possible to use a patched solid surface, some regions being attractive, and some being neutral or repulsive: a vesicle stuck on this surface becomes more permeable, or even breaks [3], (d) electroporation is a classical technique [4,5]: with voltage drops of the order of 1 V across the bilayer, some large pores can be induced.

In a recent series of experiments [6] some of us have studied giant vesicles where the tension σ could be established by two different ways:

(i) adhesion on an attractive surface, which generates flattened vesicles,

(ii) irradiation of a vesicle (doped with suitable fluorescent dyes). For some (mysterious) reason, the equilibrium area per polar head under illumination is slightly reduced, and a tension σ appears (within minutes). An effect of the same kind was reported for unlabeled vesicles submitted to laser tweezers [7]. This optically induced tension is very convenient, because it allows to work in a simple (spherical) geometry.

Another important trick introduced in Ref. [6] amounts to replace water by a more viscous mixture of water and glycerol (typical viscosity $\eta_0 = 30$ cP). This allows to observe the formation, growth, and ultimate closure of *large pores* (radius $r \approx$ a few microns). The role of glycerol is explained in Ref. [6]: a large pore allows for a certain leak-out of the inner compartment which has been known for some time [5,8]. Leak-out reduces the tension σ, and leads to pore closure. But with glycerol this leak-out becomes very slow, and we can watch large pores persisting over long times (s). Some of the salient features of growth and closure in the slow leak-out limit have been modelized in Ref. [6].

Our aim in the present paper is to make the theoretical discussion more general, in two directions:

(a) we discuss fast leak-out as well as slow leak-out,

(b) it is assumed in Ref. [6] that the growth and the late stages of closure were dominated by the internal viscosity η_2 of the lipid layer. This was correct for the experiments at hand, but in some other cases, all steps may be governed by the solvent viscosity η_0. This limit is discussed here in the appendix.

In Section 2 we construct the dynamical equations describing the pore size $r(t)$ and the overall radius $R(t)$ of the vesicle – allowing for fast or slow leak-out. We display some numerical result on $r(t)$ for a few typical cases. In Section 3 we concentrate on the slow leak-out regime: here, it is possible to derive explicit formulae describing the four major time intervals involved: growth (I), bloom (II), leak-out (III) and ultimate closure (IV). Some of the physical questions which appear are discussed in Section 4.

2. Transient pores

2.1. Counting the areas

The geometry is shown in Fig. 1. We constantly assume that the pore area (πr^2) is much smaller than the overall area of the envelope sphere ($4\pi R^2$). We call R_i the radius of the original vesicle under initial tension (σ_0) and R_0 the radius in zero tension (σ=0).

The lipid area A_L differs from $4\pi R_0^2$ because tension stretches our system

$$A_L = 4\pi R_0^2 \left(1 + \frac{\sigma}{E}\right) , \tag{1}$$

where E is a two-dimensional modulus. In the regime of interest here (weak tensions) E is not related to the van der Waals interactions in the lipid. It is in fact controlled by the unfolding of "wrinkles" (fluctuations) of the surface [9].

$$E = \frac{48\pi K_b}{R_0^2} \frac{K_b}{kT} , \tag{2}$$

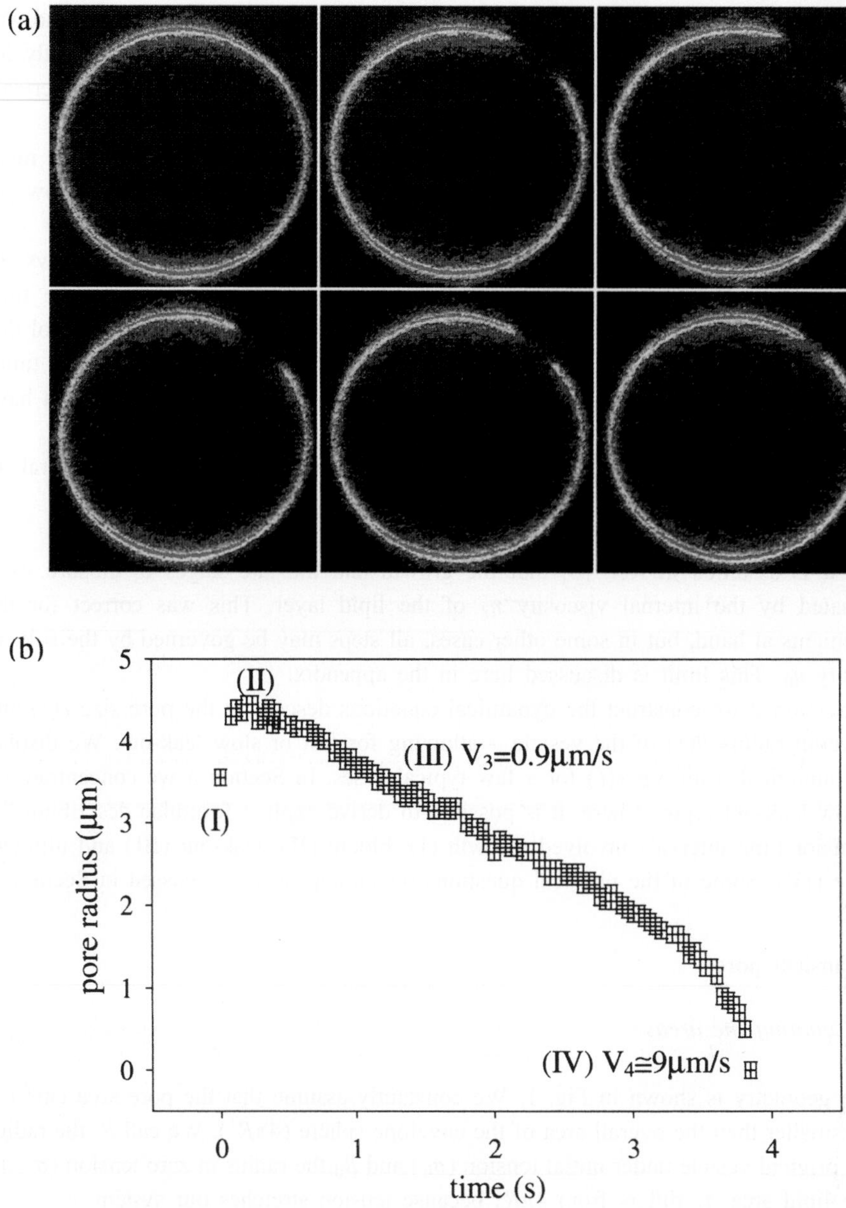

Fig. 1. Typical experiment (a) time sequences of a transient pore in a giant vesicle stretched by intense illumination; the membrane lipid is DOPC labeled as described in Ref. [6]; the solvent is a 71 : 29 w/w mixture of glycerol and water, with a measured viscosity $\eta_0 = 32 \pm 0.4$ cP; the solutes are sucrose 0.1 M inside, glucose 0.1 M outside; (b) corresponding measurement of pore radius versus time, showing four stages in the pore dynamics.

where K_b is the Helfrich bending constant [10] and kT is the thermal energy. The overall distribution of areas is ruled by

$$4\pi R^2 = A_L + \pi r^2 = 4\pi R_0^2\left(1+\frac{\sigma}{E}\right) + \pi r^2 \,. \tag{3}$$

It is sometimes convenient to define a critical pore radius r_c which corresponds to the complete relaxation of the tension ($\sigma = 0$) in situations of zero leakage ($R = R_i$). Thus,

$$r_c^2 = 4R_0^2\frac{\sigma_0}{E} \,. \tag{4}$$

2.2. Rate equations

(a) Changes of the pore radius $r(t)$ are driven by two forces:
(i) the tension σ (favoring expansion),
(ii) the line energy $\mathfrak{J}$ of the pore (favoring closure).
When the internal viscosity of the bilayer is dominant [11], this leads to

$$2\eta_2 d\frac{\dot{r}}{r} = \sigma - \frac{\mathfrak{J}}{r} \,, \tag{5}$$

where d is the lipid thickness. In most macroscopic physical systems, $\mathfrak{J}/r$ is a negligible correction. But here it is important, because the tensions σ are unusually small in vesicles.

(b) Leak-out provides a way of changing the overall solvent volume, as explained in Refs. [5,6,8]. The leak-out flux is

$$Q = \text{const}\; V_L r^2 \,,$$

where V_L is the leak-out velocity. The shear stresses involved in this outward flow are of order $\eta_0 V_L/r$ where η_0 is the solvent viscosity. They are balanced by the Laplace pressure $p = 2\sigma/R$. The result is

$$Q = \frac{2\sigma}{3\eta_0 R} r^3 = -4\pi R^2 \dot{R} \,. \tag{6}$$

On the whole, we now have three equations (3), (5) and (6) for three unknowns (r, R, σ).

2.3. Non-dimensional form

In the following, we consider small variations of the vesicle radius $R = R_0(1+\delta)$ and $R_i = R_0(1+\delta_0)$ at $t = 0$, with $\delta, \delta_0 \ll 1$. Then the critical pore radius introduced in Eq. (4) is given by $r_c^2 = 8\delta_0 R_0^2$. The two radii can be expressed in reduced units: $\tilde{r} = r/r_c$ and $\Delta = (\delta_0 - \delta)/\delta_0$. The latter represents the drop of the vesicle radius due to leak-out: Δ varies from 0 to 1 as R decreases from R_i to R_0. The reduced time is $\tilde{t} = t/\tau$ with

$$\frac{1}{\tau} = \frac{\sigma_0}{2\eta_2 d} \,. \tag{7}$$

Introducing the reduced surface tension $\tilde{\sigma} = \sigma/\sigma_0$, Eq. (3) rewrites

$$\tilde{\sigma} = 1 - \tilde{r}^2 - \Delta . \tag{8}$$

The differential equations (5) and (6) are normalized as

$$\frac{d\tilde{r}}{d\tilde{t}} = \tilde{r}\tilde{\sigma} - \tilde{\mathfrak{I}} , \tag{5'}$$

$$\frac{d\Delta}{d\tilde{t}} = \tilde{\sigma}\frac{\tilde{r}^3}{\tilde{r}_L} . \tag{6'}$$

System (5′) and (6′) contains only two adjustable parameters

$$\tilde{\mathfrak{I}} = \frac{\mathfrak{I}}{\sigma_0 r_c} \tag{9}$$

and

$$\tilde{r}_L = \frac{3\pi\eta_0}{8\eta_2 d}\frac{R_0^2}{r_c} . \tag{10}$$

The boundary conditions are $\tilde{r}(\tilde{t}=0) = \tilde{r}_i$ and $\Delta(\tilde{t}=0) = 0$.

The radius $r_L = \tilde{r}_L r_c$ is the upper pore size determined by leak-out first introduced in Ref. [6]. The two extreme cases are (i) instantaneous release for $r_L = 0$: pore never opens (ii) gelified content for $\tilde{r}_L \to \infty$: there is no leak-out at all, and $\Delta(\tilde{t}) = 0$ all the time. Then Eq. (5′) reduces to $d\tilde{r}/d\tilde{t} = \tilde{r}(1-\tilde{r}^2)$ and can be solved analytically, as in [6]

$$\ln\frac{\tilde{r}}{\tilde{r}_i} - \frac{1}{2}\ln\frac{1-\tilde{r}^2}{1-\tilde{r}_i^2} = \tilde{t} . \tag{11}$$

2.4. Numerical simulations

In the following, we take as a conjectural value $\tilde{r}_i = 0.15$ for the initial radius, which comes from the fits of Eq. (11) with experimental data on vesicles that burst irreversibly [6]. We vary $\tilde{r}_L$ in a broad range (10^{-2}–10^4) as a model of the thickening of the ambient liquid. Simulations are performed using the simple Newton algorithm with a variable time step $\Delta\tilde{t}$ and an iteration number N ranging from 4000 to 160 000 for the largest $\tilde{r}_L$. We show here typical results:

(i) *Fast leak-out* ($\tilde{r}_L < 10$): For the lowest values $\tilde{r}_L = 10^{-2}$–10^{-1}, we see "aborted" pores with a dynamics in three stages only: growth (I), maximum (II) and fast closure (IV). The slow closure regime (III) appears clearly for $\tilde{r}_L \geqslant 1$, as seen in Fig. 2(a). It becomes broader when $\tilde{r}_L$ increases further. An intriguing result is that the final drop of the vesicle radius Δ_∞ (or equivalently the surface tension at end $\tilde{\sigma}_\infty$) is not a monotonous function of $\tilde{r}_L$. Instead, Δ_∞ has a maximum ($\tilde{\sigma}_\infty$ has a minimum) near $\tilde{r}_L = 10^{-1}$: see Fig. 2(b) and (c). A look to the curves of the outward flow in Fig. 2(d) helps to understand this. The peak value of Q decreases monotonously with $\tilde{r}_L$, but the peak shape is not symmetrical: the left

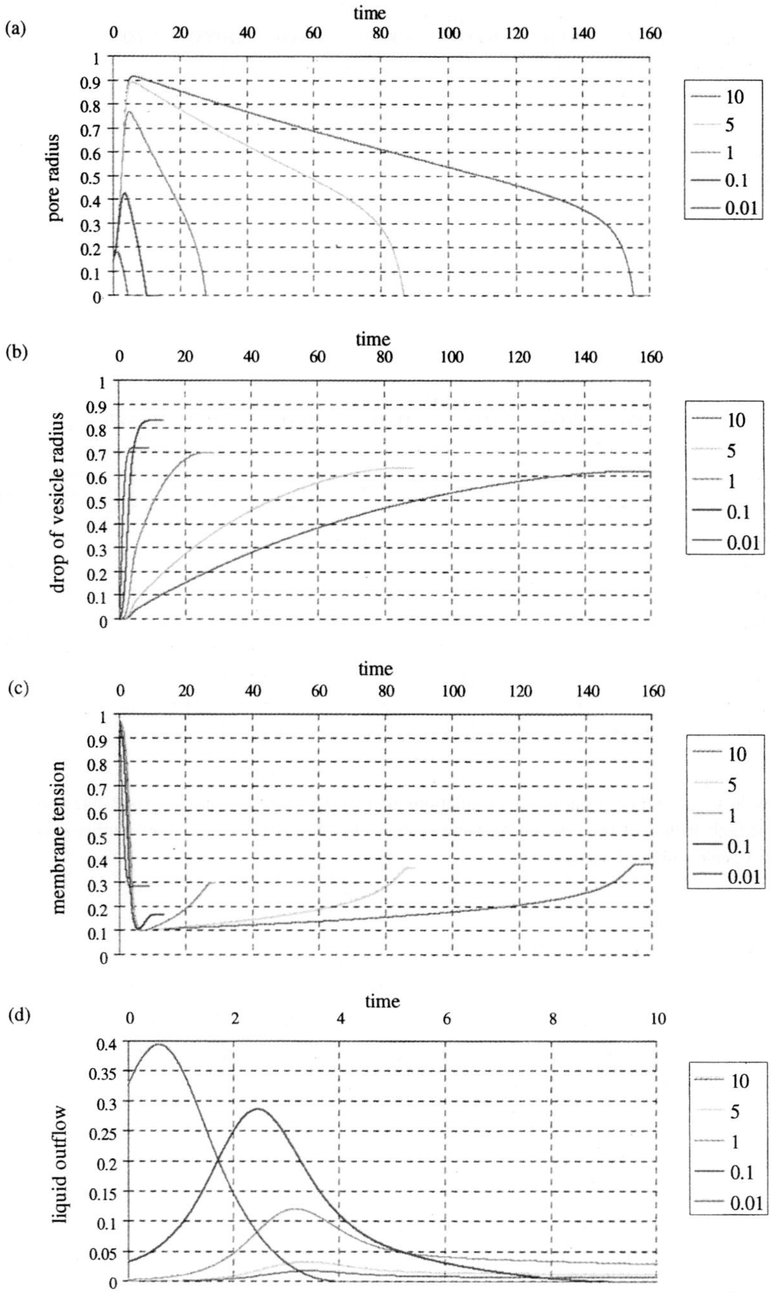

Fig. 2. Fast leak-out regime – numerical simulations of Eqs. (5′) and (6′) for constant line tension $\tilde{\mathfrak{I}} = 0.1$ and different low values of the leak-out parameter $\tilde{r}_L = 0.01–0.1–1–5–10$: (a) pore radius $\tilde{r}$ versus time $\tilde{t}$; (b) drop of vesicle radius Δ versus time $\tilde{t}$; (c) surface tension $\tilde{\sigma}$ versus time $\tilde{t}$; (d) liquid outflow Q in the early times.

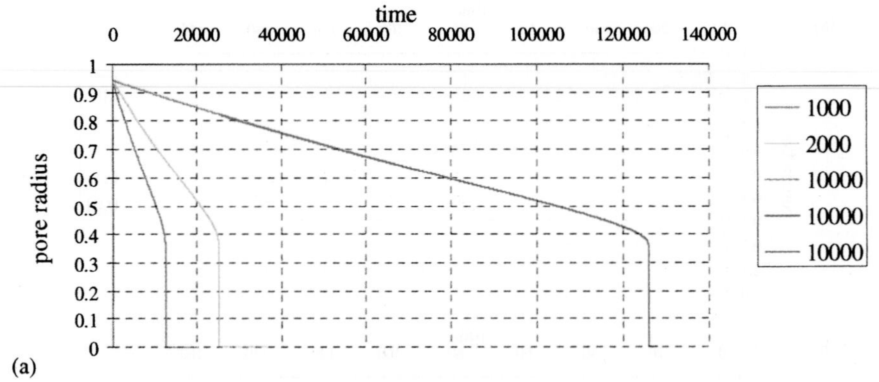

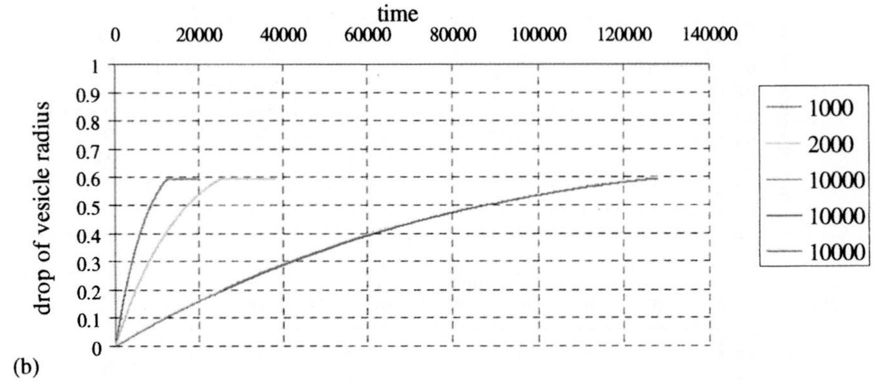

Fig. 3. Slow leak-out regime – numerical simulations of Eqs. (5′) and (6′) for constant line tension $\tilde{\Im} = 0.1$ and different high values of the leak-out parameter $\tilde{r}_L = 10^3$–2×10^3–10^4: (a) pore radius $\tilde{r}$ versus time $\tilde{t}$; (b) drop of vesicle radius Δ versus time $\tilde{t}$.

tail is lacking at $\tilde{r}_L = 10^{-2}$, while there is a wide spread right tail for $\tilde{r}_L \geqslant 1$. The release of the vesicle content and the relaxation of surface tension being proportional to the integral of Q, they are both maximum at $\tilde{r}_L = 10^{-1}$, which corresponds to a nearly symmetrical peak with maximum area. This means that there is a solvent viscosity that maximizes the delivery of the vesicle content: this effect could be useful for drug delivery by artificial vesicles and endocytosis processes involved in cells.

(ii) *Slow leak-out* ($\tilde{r}_L \gg 1$): In that case, the slow closure (III) takes much more time than the three other stages of the long-lived pores. The closure velocity V_3 (defined as the slope of the curves in Fig. 3(a)) decreases when $\tilde{r}_L$ increases, for a given value of $\tilde{\Im}$. But the radius $\tilde{r}_{34}$ at the crossover between slow (III) and fast (IV) closures does not vary ($\tilde{r}_{34} \approx 0.4$ on that example). The final value of Δ (and hence of the vesicle radius) is also constant, as seen in Fig. 3(b): in the slow leak-out case, raising the solvent viscosity increases the pore lifetime, but it does not change the efficiency of the content release. However, one should keep in mind that Eq. (5) stands for a viscous dissipation dominated by the flows of lipid. For a

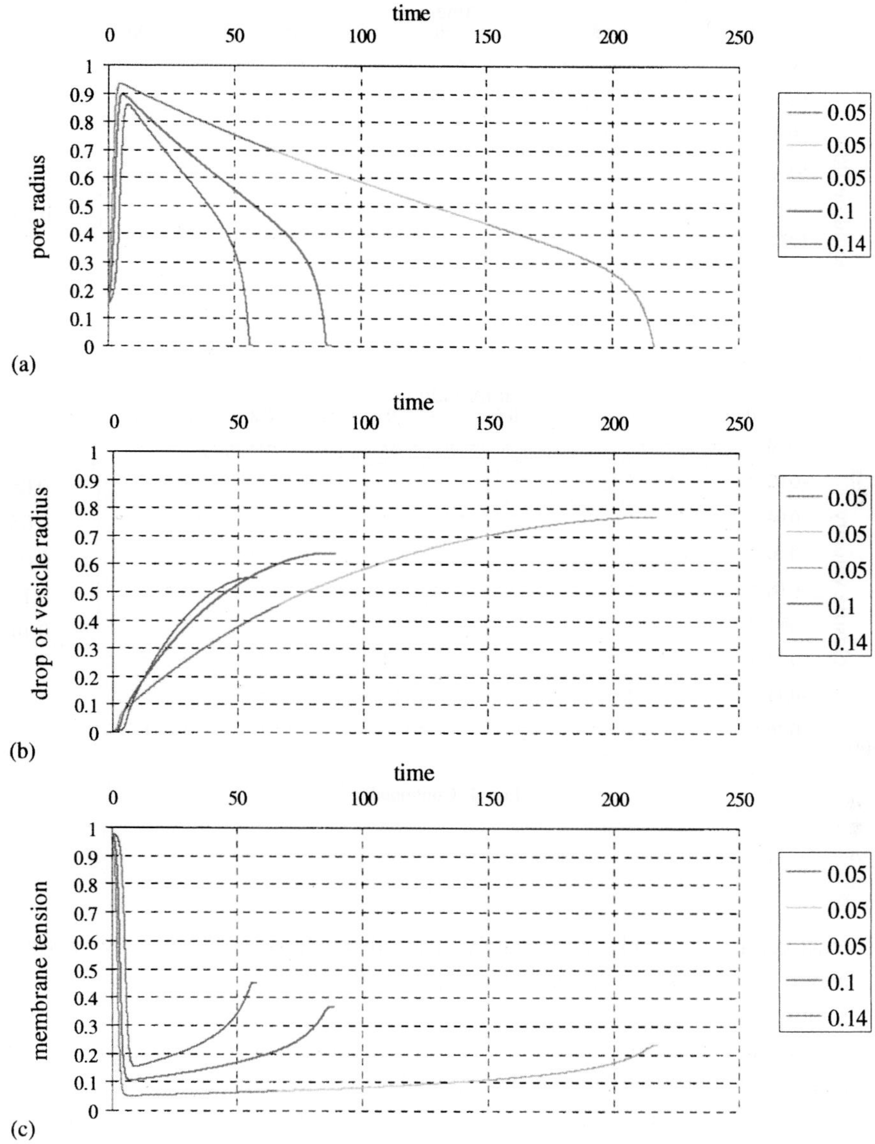

Fig. 4. Influence of line tension – numerical simulations of Eqs. (5′) and (6′) for constant leak-out parameter $\tilde{r}_L = 5$ and three different line tensions $\tilde{\mathfrak{I}} = 0.05–0.1–0.14$: (a) pore radius $\tilde{r}$ versus time $\tilde{t}$; (b) drop of vesicle radius Δ versus time $\tilde{t}$; (c) surface tension $\tilde{\sigma}$ versus time $\tilde{t}$; (d) liquid outflow Q in the early times; (e) closure velocity $\dot{\tilde{r}} < 0$ versus pore radius $\tilde{r}$.

very viscous content ($\tilde{r}_L \to \infty$), a more appropriate model is developed in the appendix.

(iii) *Influence of the line tension*: The first role of the line energy $\tilde{\mathfrak{I}}$ is to limit the pore nucleation. If $\tilde{\mathfrak{I}} > \tilde{r}_i$ (or $\mathfrak{I} > \sigma_0 r_i$), the force acting on the pore edge is always

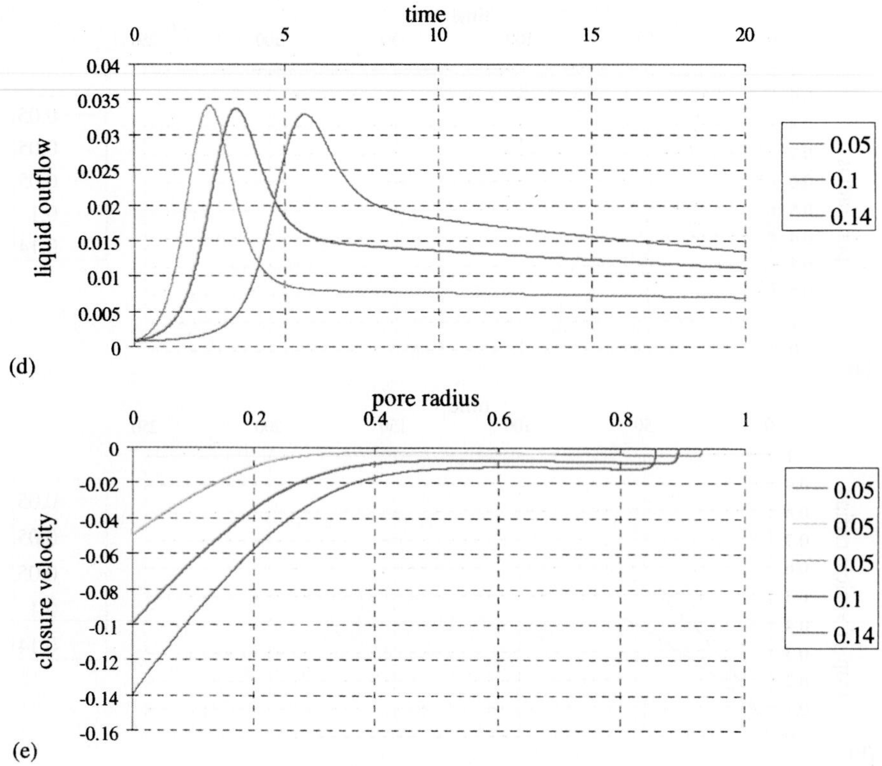

Fig. 4. Continued.

negative, and it shrinks immediately. The other effects are studied by varying $\tilde{\Im}$ among three values: 0.05, 0.10 and 0.14 (< 0.15).

We chose $\tilde{r}_L = 5$, an intermediate value between fast and slow leak-out, which fits rather well our experimental results (that will be presented in a forthcoming paper). It is clear on the curves of Fig. 4 that, at higher line tension: (a) the pore reaches a smaller maximum radius $\tilde{r}_m$ and closes faster; (b) the decrease of the vesicle radius (and so the released volume) is smaller; (c) there is less relaxation of the surface tension at end; (d) the peak of the outward flow occurs at a later time. This last observation means that, for $\tilde{r}_L < 10$, the line energy $\tilde{\Im}$ plays a role even in the growth regime (I): it slows down the pore opening which cannot be described simply by Eq. (11).

The two closure regimes (III) and (IV) depends also directly on $\tilde{\Im}$. The plot of the closure velocity (i.e., the negative values of $\dot{\tilde{r}}$) versus pore radius (e) shows two parts: – a plateau between $\tilde{r}_m$ and $\tilde{r}_{34}$ corresponding to regime (III) at almost constant velocity denoted V_3 – an acceleration of the closure, with a final velocity $\dot{\tilde{r}}_\infty = \tilde{\Im}$. This fast regime (IV) will be discussed in the next section.

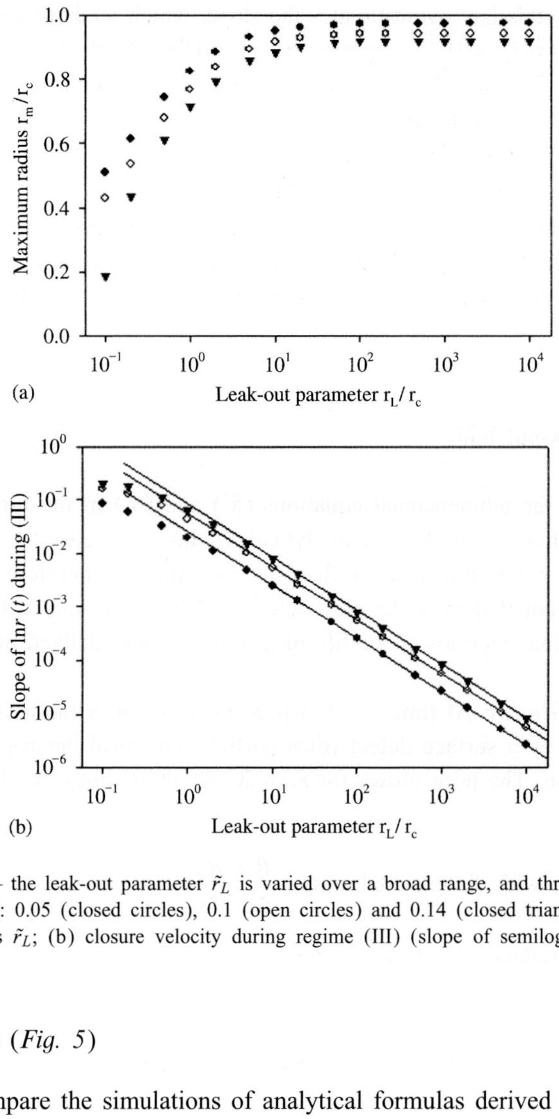

Fig. 5. Scaling laws – the leak-out parameter $\tilde{r}_L$ is varied over a broad range, and three values of the line tension $\tilde{\mathfrak{I}}$ are studied: 0.05 (closed circles), 0.1 (open circles) and 0.14 (closed triangles): (a) maximum pore radius $\tilde{r}_m$ versus $\tilde{r}_L$; (b) closure velocity during regime (III) (slope of semilogarithmic plot $\ln \tilde{r}(\tilde{t})$ versus $\tilde{t}$) versus $\tilde{r}_L$.

2.5. *Scaling laws* (*Fig. 5*)

In order to compare the simulations of analytical formulas derived in Section 3, we analyze here the variation of some characteristic features of the curves with the two control parameters $\tilde{r}_L$ and $\tilde{\mathfrak{I}}$:

(a) *The maximum pore radius $\tilde{r}_m$ versus $\tilde{r}_L$*: $\tilde{r}_m$ starts to increase with $\tilde{r}_L$, then it saturates at a level that depends on $\tilde{\mathfrak{I}}$ only.

(b) *The slow closure velocity versus $\tilde{r}_L$*: Although the pore radius $\tilde{r}(\tilde{t})$ appears linear in time during regime (III), a better linear regression is obtained when we plot $\ln \tilde{r}(\tilde{t})$ versus $\tilde{t}$. The slope is then plotted as a function of $\tilde{r}_L$ for the three values of $\tilde{\mathfrak{I}}$. We find a scaling law that is valid from $\tilde{r}_L = 10\text{–}10^4$ (over three decades):

$$\left.\frac{\mathrm{d}(\ln \tilde{r})}{\mathrm{d}\tilde{t}}\right]_{(\mathrm{III})} = (0.53 \pm 0.3)\frac{\tilde{\mathfrak{I}}}{\tilde{r}_L} \,. \tag{12}$$

For smaller $\tilde{r}_L$, this law overestimates the slope, which tends to a value that depends on $\tilde{\mathfrak{I}}$ only. Below $\tilde{r}_L = 10^{-1}$, regime (III) is undistinguished from the final closure at velocity $\dot{\tilde{r}}_\infty = \tilde{\mathfrak{I}}$ (the difference between $\tilde{\mathfrak{I}}$ and the saturation levels in Fig. 5(b) comes from the semi-logarithmic plot).

(c) *The surface tension at the end* $\tilde{\sigma}_\infty$ *versus* $\tilde{\mathfrak{I}}$: For $\tilde{r}_L > 10$, $\tilde{\sigma}_\infty$ is independent on $\tilde{r}_L$ and fits the scaling law $\tilde{\sigma}_\infty = 1.89\tilde{\mathfrak{I}}^{2/3}$.

In the following section, we derive analytical developments of the solutions $\tilde{r}(\tilde{t})$ for the "slow leak-out limit" $\tilde{r}_L \gg 1$. Interestingly, we have shown here numerically that, in practical, the condition $\tilde{r}_L > 10$ is sufficient to ensure the validity of the scaling laws.

3. The slow leak-out limit

We can solve the adimensional equations (5′) and (6′) in the limit of slow release $\tilde{r}_L \gg 1$. This condition simplifies the analytical resolution because the characteristic time of leak-out $\tau_L \cong \tilde{r}_L\tau$ is much larger than the rise time τ. Therefore the initial growth of the pore is decoupled from the slow release of the encapsulated liquid. We analyze in this limit the four regimes of the life of a transient pore defined on the experimental curve (Fig. 1).

Regime (I): *Growth*: At time $t = 0$, a pore nucleates at a radius r_i. This nucleation may be induced by a surface defect (dust particle, chemical heterogeneity, etc.) or by thermal activation. The pore grows for $\tilde{r}_i > \tilde{\mathfrak{I}}$. At short times, we have the following approximations:

$$\begin{matrix} \Delta = 0 \\ \tilde{\sigma} = 1 \end{matrix} \quad \text{or with dimensional units} \quad \begin{matrix} R = R_i, \\ \sigma = \sigma_0. \end{matrix} \tag{13}$$

Eq. (5′) then reduces to $\mathrm{d}\tilde{r}/\mathrm{d}\tilde{t} = \tilde{r}$, i.e.,

$$\tilde{r} = \tilde{r}_i \mathrm{e}^{\tilde{t}}, \quad r \cong r_i \mathrm{e}^{t/\tau}. \tag{14}$$

In the limit $\tilde{r}_L \gg 1$, the solution over the whole regime I verifies $\Delta = 0$, and the pore radius is given by Eq. (11) as in the case $\tilde{r}_L \to \infty$.

Regime (II): *Maximum radius*: The hole grows up to $\tilde{r}_m$, and starts to decrease. At its maximum size, $\dot{\tilde{r}}_m = 0$ leads to a first relation

$$\tilde{r}_m \tilde{\sigma}_m = \tilde{\mathfrak{I}}, \quad r_m \sigma_m = \mathfrak{I}, \tag{15}$$

where $\tilde{\sigma}_m = 1 - \tilde{r}_m^2 - \Delta_m$. We can estimate Δ_m by using a scaling solution of Eq. (6′):

$$\Delta_m \cong \tilde{r}_m^3 \frac{\tilde{\sigma}_m}{\tilde{r}_L} = \tilde{r}_m^2 \frac{\tilde{\mathfrak{I}}}{\tilde{r}_L}, \quad R_i^2 - R_m^2 \cong 4r_m^2 \frac{\mathfrak{I}}{\sigma_0 r_L}. \tag{16}$$

In the limit $\tilde{r}_L \gg 1$, $\tilde{r}_m$ becomes close to one, and $\Delta_m \ll 1$. Eq. (15) leads to

$$\tilde{\mathfrak{I}} \cong \tilde{r}_m(1 - \tilde{r}_m^2), \tag{17}$$

i.e.,

$$\tilde{r}_m \cong 1 - \frac{\tilde{\mathfrak{I}}}{2}, \quad r_m \cong r_c - \frac{\mathfrak{I}}{2\sigma_0}. \tag{18}$$

The parameter $\mathfrak{I}/\sigma_0$ has a simple meaning: it is the nucleation radius of a pore (in conditions of ideal nucleation) as derived from a pore energy $\pi r^2\sigma_0 - 2\pi r\mathfrak{I}$. Indeed, this radius is small (< 1 μm).

The numerical simulations show that the maximum pore radius fits well Eq. (17) as soon as $\tilde{r}_L > 20$ with less than 1% uncertainty Fig. 5(a). The linearized form (18) becomes a good approximation for small values of $\tilde{\mathfrak{I}}$.

The curvature $\ddot{\tilde{r}}_m = -\tilde{r}_m\dot{\Delta}_m = -\tilde{r}_m^4\tilde{\sigma}_m/\tilde{r}_L = -(\tilde{r}_m^3/\tilde{r}_L)\tilde{\mathfrak{I}}$ tends to zero like $\tilde{r}_L^{-1}$ when $\tilde{r}_L \to \infty$.

Regime (*III*): *Quasistatic leak-out* ($r_{23} > r > r_{34}$): As $r(t)$ starts to decrease, $\sigma(r,R)$ grows: a release of the inner liquid is then necessary in order to maintain an almost zero surface tension. More precisely, the pore closes if the driving force is negative, i.e.,

$$\tilde{r}\tilde{\sigma} < \tilde{\mathfrak{I}}, \quad r\sigma < \mathfrak{I}\,. \tag{19}$$

We shall see that inequality (19) is very nearly an equality: in regime (III) the driving force of Eq. (5′) is nearly zero $\sigma \cong \mathfrak{I}/r$ or

$$\frac{\sigma - \mathfrak{I}/r}{\sigma} \ll 1. \tag{20}$$

We may say regime (III) is *quasistatic*. A detailed argument for this follows: We calculate the derivative of Eq. (5′):

$$\ddot{\tilde{r}} = \dot{\tilde{r}}(1 - \tilde{r}^2 - \Delta) - \tilde{r}(2\tilde{r}\dot{\tilde{r}} + \dot{\Delta}). \tag{21}$$

In this regime, the closure velocity is very slow, and we can assume $\tilde{r}\tilde{\sigma} \cong \tilde{\mathfrak{I}}$ to evaluate $\dot{\Delta} \cong (\tilde{r}^2/\tilde{r}_L)\tilde{\mathfrak{I}}$. Eq. (21) becomes

$$\ddot{\tilde{r}} \cong \dot{\tilde{r}}\left(\frac{\tilde{\mathfrak{I}}}{\tilde{r}} - 2\tilde{r}^2\right) - \frac{\tilde{r}^3}{\tilde{r}_L}\tilde{\mathfrak{I}} \cong 0. \tag{22}$$

We set $\ddot{\tilde{r}} = 0$ because the scale time variation for leak-out is $\tilde{r}_L$ times larger ($\ddot{\tilde{r}}/\dot{\tilde{r}} \cong 1/\tilde{r}_L \ll 1$): the closure is at almost constant velocity. If $\tilde{\mathfrak{I}} \ll 2\tilde{r}^3$ (or $r > r_{34}$, which defines the limit of regime (III), where the pore is not too small so that leak-out has a significant effect), the first term between parentheses in Eq. (22) is negligible. This leads to

$$2\dot{\tilde{r}} + \frac{\tilde{r}}{\tilde{r}_L}\tilde{\mathfrak{I}} \cong 0. \tag{23}$$

Introducing r_{23} and t_{23} as the cross-over values between regimes (II) and (III), the solution is

$$\ln\frac{\tilde{r}}{\tilde{r}_{23}} = -\frac{\tilde{\mathfrak{I}}}{2\tilde{r}_L}(\tilde{t} - \tilde{t}_{23}), \quad \ln\frac{r}{r_{23}} = -\frac{2\mathfrak{I}}{3\pi\eta_0 R_0^2}(t - t_{23}). \tag{24}$$

This formula is very close to the scaling law deduced from Fig. 5(b). A linearized form of Eq. (24) is $\tilde{r} \cong \tilde{r}_{23}[1 - \tilde{\mathfrak{I}}/2\tilde{r}_L(\tilde{t} - \tilde{t}_{23})]$. Taking into account that $r_{23} \cong r_m$, we find an approximate expression for the closure velocity during regime (III)

$$V_3 \cong \frac{2\mathfrak{I} r_m}{3\pi\eta_0 R_0^2}. \tag{25}$$

Regime (IV): Fast closure ($r < r_{34}$): When the size of the hole becomes too small, leak-out is nearly suppressed. The radius r_{34} at cross-over between regimes (III) and (IV) has already been introduced, and verifies

$$2\tilde{r}_{34}^3 \cong \tilde{\mathfrak{I}}, \quad 2r_{34}^3 \cong \frac{\mathfrak{I} r_c^2}{\sigma_0}. \tag{26}$$

The vesicle has almost reached a constant radius R_∞ and Eqs. (5′) and (6′) become

$$\dot{\tilde{r}} = \tilde{r}(1 - \tilde{r}^2 - \Delta_\infty) - \tilde{\mathfrak{I}}, \tag{27}$$

$$\dot{\Delta} \cong 0. \tag{28}$$

Assuming that $\tilde{r}\tilde{\sigma} \cong \tilde{\mathfrak{I}}$ is still valid at the cross-over enables to estimate the surface tension at the end

$$\tilde{\sigma}_\infty = 1 - \Delta_\infty \cong \frac{\tilde{\mathfrak{I}}}{\tilde{r}_{34}} + \tilde{r}_{34}^2 \cong 3\tilde{r}_{34}^2 \cong 1.89\tilde{\mathfrak{I}}^{2/3}. \tag{29}$$

This relation (also verified on the numerical simulations) shows that the vesicle recovers a non-negligible surface tension at the end of the pore closure. For example, above 40% of the initial tension is maintained for $\tilde{\mathfrak{I}} = 0.10$. This could explain why series of successive pores are observed in the experiments.

The fast closure of the pore is governed by Eq. (27), which rewrites

$$\dot{\tilde{r}} = \tilde{r}(3\tilde{r}_{34}^2 - \tilde{r}^2) - 2\tilde{r}_{34}^3. \tag{30}$$

We do not propose an analytical solution that would be valid for the whole duration of regime (IV). Instead, we just look at the final stage, when the equation is linear

$$\dot{\tilde{r}} \cong 3\tilde{r}_{34}^2(\tilde{r} - \tfrac{2}{3}\tilde{r}_{34}). \tag{31}$$

This is a good approximation after a time t_4 that is defined arbitrarily such as

$$\tilde{r}(\tilde{t}_4) = \frac{\tilde{r}_{34}}{3}. \tag{32}$$

Then the integration leads to

$$\ln\left(1 - \frac{3\tilde{r}}{2\tilde{r}_{34}}\right) = 3\tilde{r}_{34}^2(\tilde{t} - \tilde{t}_4) - \ln 2 \tag{33}$$

which can be linearized as

$$\tilde{r} \cong \frac{2\ln 2}{3}\tilde{r}_{34} - \tilde{\mathfrak{I}}(\tilde{t} - \tilde{t}_4). \tag{34}$$

Therefore it appears that the final closure of the pore is at constant velocity

$$V_4 \cong \frac{\mathfrak{I}}{2\eta_2 d}. \tag{35}$$

This linear regime is valid over a range that is not negligible, since $\tilde{r}(\tilde{t}_4) \cong 0.26\tilde{\mathfrak{I}}^{1/3}$ (for example $r(t_4) \cong 0.12 r_c$ for $\tilde{\mathfrak{I}} = 0.10$).

4. Conclusions

(1) *A thickening of the solution increases the size of transient pores*: The cross-over between "fast" and "slow" leak-out is ruled by $r_L \cong \eta_0 R_0^2/\eta_2 d$. If $\tilde{r}_L = r_L/r_c < 1$, the blooming of pores (regime II) occurs for a radius $r_m \cong r_L$: pores abort before reaching the equilibrium size r_c, because enough surface tension is relaxed by the fast release of the vesicle content. If $\tilde{r}_L > 1$, a full blooming of pores is allowed: slowing down leak-out leads to mature pores, now at almost their equilibrium radius. After blooming, pores close slowly (regime III), before dying abruptly (regime IV). In the limit of slow leak-out ($\tilde{r}_L \gg 1$), the dynamical equation for $r(t)$ can be derived by a simple scaling argument. Line tension closes the pore with a net power that equilibrates the entropy production per unit time

$$- \mathfrak{I}\dot{r} = T\dot{S} = \eta_0 \left(\frac{V_L}{r}\right)^2 r^3 . \tag{36}$$

The viscous losses are due to flows of solution through the pore. The leak-out velocity has already been estimated as $V_L = r\sigma/\eta_0 R$. The closure is quasi-stationary and we can use approximation (20) to get $V_L \cong \mathfrak{I}/\eta_0 R$. Thus Eq. (36) rewrites

$$- \mathfrak{I}\dot{r} = \frac{\mathfrak{I}^2}{\eta_0 R^2} r . \tag{37}$$

This simple expression contains the salient features of regime (III): the logarithm of the pore radius decreases linearly in time, with a slope in agreement with Eq. (24). Of course the exact prefactor cannot be obtained by this scaling argument.

(2) Looking back at Fig. 1, our theoretical picture allows us to derive with a good accuracy two important parameters of the membrane: the surface viscosity $\eta_s = \eta_2 d$ and the line energy of a pore. Indeed the ratio between the velocities in the two linear regimes (III) and (IV) contains η_s as the only unknown:

$$\frac{V_4}{V_3} = \frac{3\pi R_0^2}{4 r_m} \frac{\eta_0}{\eta_s} . \tag{38}$$

The values $R_0 = 14.5$ μm, $r_m = 4.5$ μm, $\eta_0 = 3.2 \times 10^{-2}$ N s m^{-2} (=32cP), $V_3 = 0.9$ μm/s and $V_4 \approx 9$ μm/s are measured. Thus we calculate $\eta_s \approx 3.5 \times 10^{-7}$ N s m^{-1} ($=3.5 \times 10^{-4}$ surfacePoise). The thickness of a phospholipid bilayer being $d \approx 3.5$ nm, we predict the bulk viscosity of 1,2-dioleoyl-sn-glycero-3-phosphocholine (DOPC): $\eta_2 \approx 100$ N s m^{-2} ($=10^5$ cP). We calculate also (from V_3 only) $\mathfrak{I} \approx 0.6 \times 10^{-11}$ N. The usual measurements of line energies $\mathfrak{I}$ are extremely delicate [4,5,8,12], and $\mathfrak{I}$ is expected to be sensitive to certain lipid impurities. We have studied one extreme case, based on the well-known effect of cholesterol to stiffen the bilayers. Our preliminary results show that the line energy is three times larger for a membrane containing DOPC and 30 mol% cholesterol.

(3) The presence of glycerol may modify the membrane characteristics and the two-dimensional mobility of lipids. Vesicles of DOPC containing a hydrophobic fluorophore were studied by time-resolved fluorescence spectroscopy [13]. The rotational motion of the molecular probe was slowed down at high glycerol content (in a 91:9 w/w mixture of glycerol and water, the rotation time becomes three times larger than in water). This result tells us that a slight viscous thickening of the lipid is observed, but no dramatic effects are reported. This is crucial because the maximum size of transient pores in the fast leak-out limit is $r_L \cong (\eta_0/\eta_s)R_0^2$. If we assume no dependence of η_s on the glycerol content, the size of transient pores in pure water are expected to be 30 times smaller: 100–200 nm. These pores open and close before reaching a visible size in optical microscopy. But their presence explains certain anomalous permeabilities observed for stretched adhesive vesicles [14]. They have also been observed in the membrane of human erythrocytes [15,16]. The authors used a very clever setup to freeze the cells with a controlled time delay after electroporation, and observe them later by electron microscopy. Despite statistical scatter among the cells, they reached a good insight of the pore dynamics. They divided it into three stages: (i) within 3 ms, pores expand rapidly up to 20–40 nm in diameter (ii) between 20 and 220 ms they saturate at a maximum diameter of 100–160 nm (iii) after 1–5 s almost all the pores have shrank. This scenario looks similar to ours except that the scales of times and pore sizes are about 30 times smaller.

(4) We assumed that, at all times, σ is uniform all over the vesicle. This means that no rim surrounds the hole. This is the signature of plug flows extending over large distances, which have been observed in the bursting of polymer "bare" films (i.e., with no surfactants, [11]). An experimental evidence of this assumption is that no bright circle around the pore is visible in fluorescence microscopy. This simplification may be lost when the membrane viscosity is no more dominant. Our experiments on liquid/liquid dewetting have shown that indeed a rim appears when dissipation in the bulk liquid overcomes dissipation in the film [17].

Note added in proof

While this paper was in review, we became aware that Eq. (15) in H. Isambert, Phys. Rev. Lett. 80 (1998) 3404–3407 is an equivalent form of our Eq. (11) for the pore radius in case of negligible leak-out.

Appendix. Transient pores in low-viscosity membranes

We now study the opposite limit, when the viscous dissipation in the membrane is small compared to the dissipation of the backflows induced by the pore opening in the

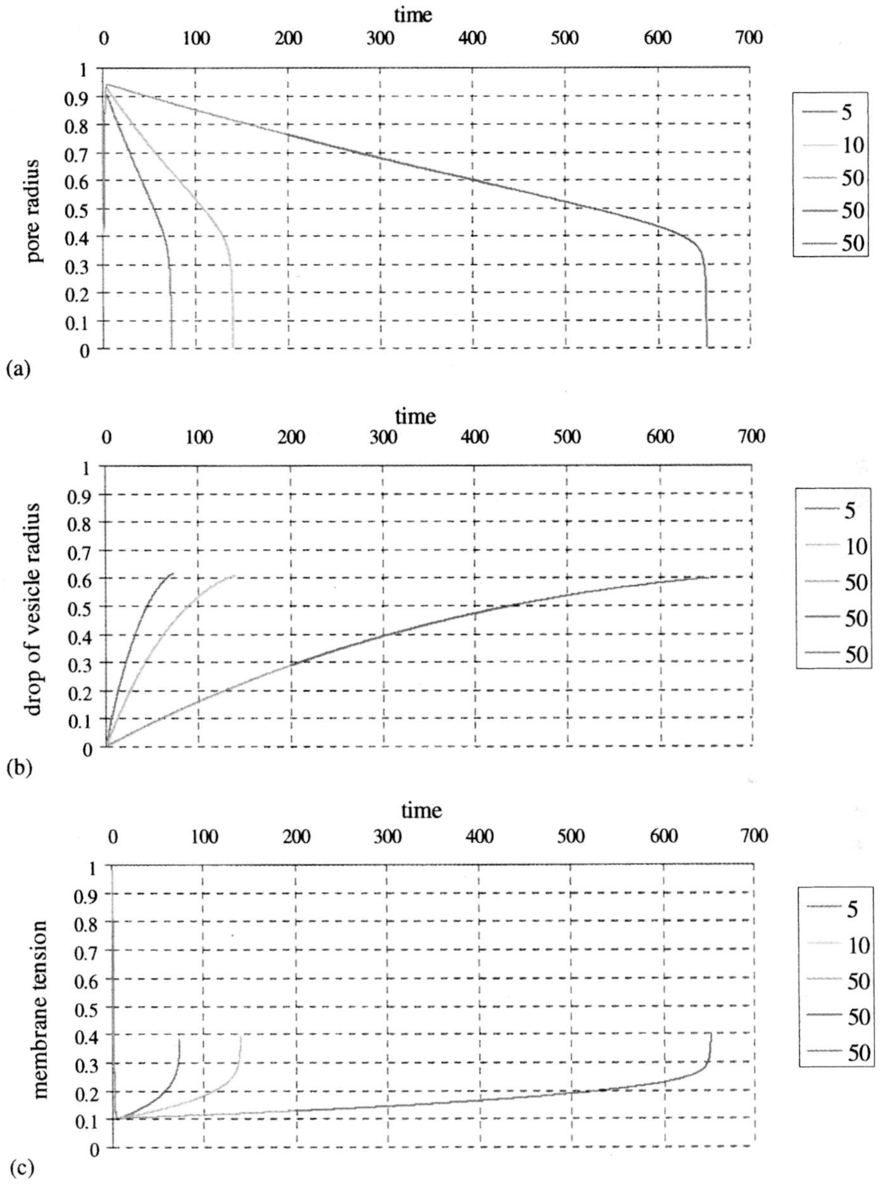

Fig. 6. Dissipation dominated by backflows – numerical simulations of Eqs. (A.5) for constant line tension $\tilde{\mathfrak{I}} = 0.1$ and different values of the leak-out parameter $r_L' = 5$–10–50: (a) pore radius $\tilde{r}$ versus time $\tilde{t}$; (b) drop of vesicle radius Δ versus time $\tilde{t}$; (c) surface tension $\tilde{\sigma}$ versus time $\tilde{t}$; (d) liquid outflow Q in the early times; (e) enlarged view of the growth (I); (f) enlarged view of the final stage (IV).

surrounding liquid. The extension of Eq. (5), incorporating a mixed dissipation both in the membrane and in the solvent may be written as

$$\eta_0 r\dot{r} + 2\eta_2 d\dot{r} = r\sigma - \mathfrak{I}. \tag{A.1}$$

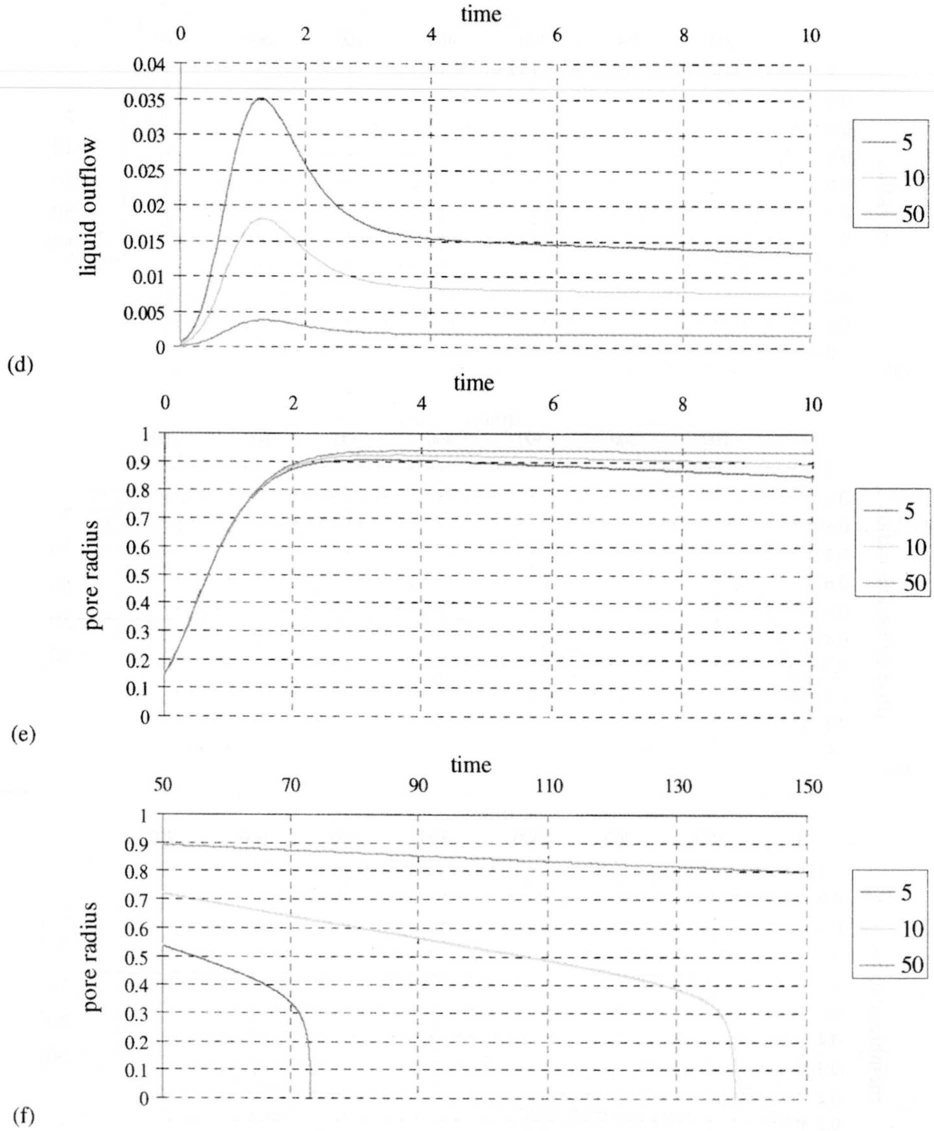

Fig. 6. Continued.

The first term is the contribution of the viscous dissipation in the solvent. It can be justified by the following scaling agrument. The entropy production per unit time is $T\dot{S} \cong \eta_0(\dot{r}/r)^2 r^3$, where r^3 is the volume of the flows occurring through the pore. The transfer of surface and line energies into viscous losses leads to Eq. (A.1). Note that for the backflows in the solvent, we have only a scaling form and we do not know the exact numerical coefficient. This is not true for the second term, because dissipation due to radial plug flows in the membrane was calculated exactly.

We study here the case where solvent dissipation is dominant. This implies $\eta_0 r \gg \eta_2 d$. Eq. (A.1) then becomes

$$\eta_0 r\dot{r} = r\sigma - \mathfrak{T}, \tag{A.2}$$

where

$$\sigma = \sigma_0\left(1 - \frac{r^2}{r_c^2} - \frac{R_i^2 - R^2}{R_i^2 - R_0^2}\right).$$

The other rate equation (describing leak-out) is not modified

$$-4\pi R^2\dot{R} = \frac{2\sigma}{3\eta_0 R}r^3. \tag{A.3}$$

Eqs. (A.2) and (A.3) with initial conditions $r(t=0) = r_i > \mathfrak{T}/\sigma_0$ and $R(t=0) = R_i$ describes the full life of transient pores in that limit. At short times the pores grow at constant velocity $\dot{r}_g = \sigma_0/\eta_0$, where σ_0 is the stress before hole opening. We can define a characteristic time of growth

$$\tau' = \frac{r_c}{\dot{r}_g} = \frac{r_c\eta_0}{\sigma_0}. \tag{A.4}$$

Eqs. (A.2) and (A.3) can be written in adimensional units $\tilde{r} = r/r_c$, $\tilde{t}' = t/\tau'$, $\Delta = (\delta_0 - \delta)/\delta_0$ and $\tilde{\sigma} = \sigma/\sigma_0$ as

$$\tilde{r}\frac{d\tilde{r}}{d\tilde{t}'} = \tilde{r}\tilde{\sigma} - \tilde{\mathfrak{T}},$$

$$\tilde{\sigma} = 1 - \tilde{r}^2 - \Delta,$$

$$\frac{d\Delta}{d\tilde{t}'} = \tilde{\sigma}\frac{\tilde{r}^3}{r_L'}, \tag{A.5}$$

where

$$\tilde{\mathfrak{T}} = \frac{\mathfrak{T}}{\sigma_0 r_c} \quad \text{and} \quad r_L' = \frac{R_0^2}{r_c^2}.$$

Because the same viscosity is involved in the pore expansion and in solvent release, η_0 does not show up in the controlled parameter. In addition the pore can reseal only if $r_c < R_0$ (otherwise the vesicle bursts irreversibly). Therefore in practice $r_L' > 1$, which means that when dissipation in backflows dominates, the pores are always in the slow leak-out regime. Pores will open at constant velocity $\dot{r}_g$ (I) up to a large size r_m (II). Then they close slowly during step (III) which is identical to the previous case, with a closure velocity $V_3 \cong 2\mathfrak{T} r_m/3\pi\eta_0 R_0^2$. The cross-over radius r_{34} between slow and fast closure is unchanged. But the fast regime (IV) is entirely different. Eq. (A.1) shows that $r^2 \cong \mathfrak{T}/\eta_0(t_\infty - t)$, i.e., the slope as $r \to 0$ becomes infinite. The line tension can still be deduced, from the curvature of $r(t)$ near $r = 0$. Numerical simulations of Eq. (A.5) are shown in Fig. 6 for several values of r_L' (=5, 10, 50).

References

[1] M. Dvolaïtsky, P.G. de Gennes, M.-A. Guedeau, L. Jullien, C.R. Acad. Sci. Paris 316 (II) (1993) 1687–1690.

[2] M.-A. Guedeau-Boudeville, L. Jullien, J.-M. di Miglio, communicated by P.G. de Gennes, Proc. Natl. Acad. Sci. USA 92 (1995) 1–3.

[3] A.-L. Bernard, M.-A. Guedeau-Boudeville, O. Sandre, S. Palacin, J.-M. di Miglio, L. Jullien, Langmuir, accepted for publication.

[4] W. Harbich, W. Helfrich, Z. Naturforsch. A 34 (1979) 1063–1065.

[5] D.V. Zhelev, D. Needham, Biochim. Biophys. Acta 1147 (1993) 89–104.

[6] O. Sandre, L. Moreaux, F. Brochard-Wyart, communicated by P.G. de Gennes, Proc. Natl. Acad. Sci. USA 96 (1999) 10591–10596. Commentary by P. Pincus, p. 10 550.

[7] R. Bar-Ziv, T. Frisch, E. Moses, Phys. Rev. Lett. 75 (1995) 3481–3484.

[8] J. Moroz, P. Nelson, Biophys. J. 72 (1997) 2211–2216.

[9] F. Brochard, P.G. de Gennes, P. Pfeuty, J. Phys. (France) 37 (1976) 1099–1103.

[10] W. Helfrich, Z. Naturforsch. C 28 (1973) 693–703.

[11] G. Debrégeas, P. Martin, F. Brochard-Wyart, Phys. Rev. Lett. 75 (1995) 3886–3889.

[12] L.V. Chernomordik, M.M. Kozlov, G.B. Melikyan, I.G. Abidor, V.S. Markin, Y.A. Chizmadzhev, Biochim. Biophys. Acta 812 (1985) 643–655.

[13] L.B.-Å. Johansson, B. Kalman, G. Wikander, A. Fransson, K. Fontell, B. Bergenstål, G. Lindblom, Biochim. Biophys. Acta 1149 (1993) 285–291.

[14] A.-L. Bernard, M.-A. Guedeau-Boudeville, J.-M. di Miglio, L. Jullien, Langmuir, submitted for publication.

[15] D.C. Chang, in: Chang, Chassy, Saunders, Sowers (Eds.), Guide to Electroporation and Electrofusion, Acad. Press, San Diego, 1992, pp. 9–27.

[16] D.C. Chang, Biophys. J. 58 (1990) 1–12.

[17] F. Brochard-Wyart, G. Debregeas, R. Fondecave, P. Martin, Macromolecules 30 (1997) 1211–1213.

2416 *Langmuir* **2001,** *17,* 2416–2419

"Young" Soap Films

P. G. de Gennes*

Collège de France, 11 place M. Berthelot, 75231 Paris Cedex 05, France

Received November 3, 2000. In Final Form: December 27, 2000

If we pull out rapidly a metallic frame out of a surfactant solution, we arrive at a "young" soap film with relatively simple features, as noticed first by Lucassen. The weight of the film is equilibrated by a vertical gradient of surface tension. At each level, the local solution concentration $c(z)$ equilibrates with the local monolayers, of surface concentration $\Gamma(z)$. A detailed analysis of the young films was started by us in 1987. We present here an approach which is more illuminating: (a) the concentration profiles decay exponentially at large heights, with a characteristic length $\lambda \sim$ meters; (b) the surface is protected up to a thickness h_m larger than λ; (c) we also review the dynamic requirements. The surfactant must reach the surface in a time shorter than the free fall time of a pure water film. This discussion explains (to some extent) the compromise which is achieved in practice by good foaming agents.

I. Introduction

Can a given surfactant produce a strong foam? there are some qualitative rules. We know, for instance, that a strongly insoluble surfactant cannot cover the surface of a rapidly growing bubble, while a soluble surfactant may succeed—by diffusion from the bulk solution. We would like to make this more precise—to characterize a surfactant solution by a few control parameters, which tell us what are its activities.

One of the methods for generating soap films is provided by a mechanical egg beater. Or, more scientifically, by pulling rapidly a metallic frame out of a surfactant solution (Figure 1a), as described in the book by Mysels et al.[1]

We might first think of pulling the film at constant vertical speed. This is a version of a classical Landau Levich problem.[2] But the quality of the surfactant plays only a minor role in this process:[3] the surfactant is pulled up at (essentially) the same speed as the solution, and no fresh surface is generated.

A more relevant situation was invented by Lucassen.[4] Here, the film is pulled abruptly, and its height h is comparable to (or larger than) the horizontal span of the reservoir. Thus, we must create a fresh surface by pulling the frame, and migration of surfactant from the water phase to the surface is dominant.

This situation corresponds to what Lucassen[4] called a "young film". We discussed some aspects of the young films in a note.[5] But we missed some important points: (a) We chose as our central variable the local surface tension. It turns out that it is more illuminating to study first the concentration $c(z)$ at all levels in the film. (b) We did not discuss the dynamical features which tell us if the young film can indeed be made—or not. (c) Some serious misprints occurred in ref 5.

Here, we return to the static structure of the young films: in section II we deal with concentrations c below

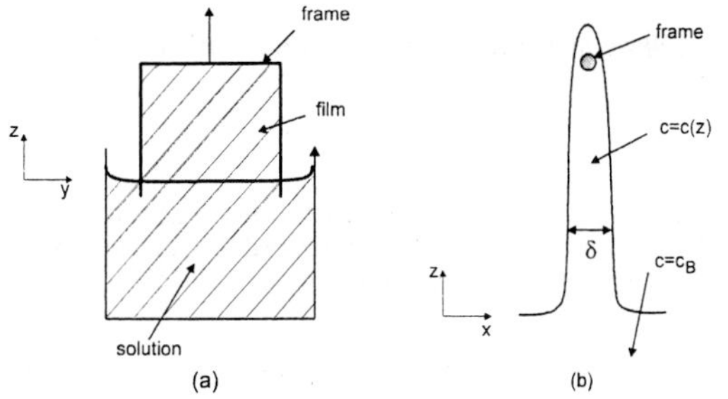

Figure 1. "Young" film pulled out from a surfactant solution: (a) general setup; (b) detailed cross section of the film.

the critical micelle contentration (c^*). We find two characteristic heights: both are relevant for a discussion of film stability. In section III we extend this to $c > c^*$. In section IV we consider the major time constants involved: the young film must achieve its protective surface before falling.

II. Profile of a Young Film ($c < c^*$)

A. Basic Equations. The film is drawn vertically (along z) and has a certain width $\delta(z)$ (Figure 1b). We assume that, in each interval dz, there is a rapid equilibration between solution (concentration $c(z)$) and surface (with monolayer concentration $\Gamma(z)$). The total number of surfactants was originally $c_B\delta$ (where c_B is the bulk concentration in the reservoir) and remains the same. Thus, we must have

$$c_B\delta = \delta c + 2\Gamma \tag{1}$$

We assume that equilibrium has been achieved, and this may be written in the differential form

$$\Gamma^{-1}\,d\Pi = d\mu(c) \tag{2}$$

where $\Pi = \gamma_0 - \gamma$ is the Langmuir pressure, describing the shift of the surface tension from γ_0 (for pure water) to γ (at concentration c) and $\mu(c)$ is the chemical potential of the surfactant in solution.

These equations must be supplemented by a requirement of mechanical equilibrium: the weight of the film must be balanced by Marangoni forces

* To whom correspondence should be addressed. E-mail address: pierre-gilles.degennes@espci.fr.

(1) Mysels, K.; Shinoda, K.; Frankel, S. *Soap films*; Pergamon Press: London, 1959.
(2) Levich, V. *Physicochemical Hydrodynamics*; Prentice Hall: Upper Saddle River, NJ, 1962.
(3) Quéré, D.; de Ryck, A.; Ou Ramdane, O. *Europhys. Lett.* **1997**, *37*, 305. Ou Ramdane, O.; Quéré, D. *Langmuir* **1997**, *13*, 2911.
(4) Lucassen, J. *Anionic Surfactants (Surfactant series)*; Marcel Dekker: New York, 1981; Chapter 6.
(5) de Gennes, P. G.; *C. R. Acad. Sci. (Paris)* **1987**, *305*, 9.

10.1021/la001538l CCC: $20.00
Published on Web 03/24/2001

$$\frac{-2\,d\Pi}{dz} = \rho g \delta \qquad (3)$$

(ρ = water density, g = gravitational acceleration).

B. Concentration Profile. Combining eqs 1–3, we arrive at

$$\frac{-2\,d\Pi}{dz} \equiv -2\frac{\alpha\Pi}{dc}\frac{dc}{dz} = \frac{\rho g 2\Gamma}{c_B - c} \qquad (4)$$

In this section, we focus our attention on the dilute case $c < c^*$. Then

$$\mu = kT \ln c + \text{constant} \qquad (5)$$

and eq 4 becomes

$$-\frac{dz}{\lambda} = \frac{dc}{c_B}\left(\frac{c_B - c}{c}\right) \qquad (6)$$

where we have introduced a characteristic length λ such that

$$\rho g \lambda = c_B kT \qquad (7)$$

λ measures the osmotic pressure of the surfactant in terms of hydrostatic heights. Note that our definition of λ differs from that of ref 5. Equation 6 integrates to

$$\frac{z}{\lambda} = \frac{c - c_B}{c_B} + \ln\left(\frac{c_B}{c}\right) \qquad (8)$$

This concentration profile is shown in Figure 2a. Two interesting limits are

$$\frac{c_B - c}{c_B} = \left(\frac{2z}{\lambda}\right)^{1/2} \quad (z < \lambda) \qquad (9)$$

$$\frac{c}{c_B} = e^{-z/\lambda} \quad (z > \lambda) \qquad (10)$$

C. Thickness Profile. We can translate $c(z)$ into a thickness profile $\delta(z)$ using eq 1. In most of the region $c < c^*$, the monolayer concentration Γ is nearly constant (adsorption has taken place abruptly at very low c's). Then we have

$$\delta = \frac{2\Gamma_B}{c_B - c} \qquad (11)$$

where Γ_B is $\Gamma(c{=}c_B)$. The thickness $\delta(z)$ described by eq 11 is a decreasing function of c. When $z > \lambda$, we have $c \ll c_B$, and δ reaches a simple limit $\delta = 2l$, where

$$l = \Gamma_B/c_B \qquad (12)$$

The length l plays an important role in foaming problems: l is the minimal thickness of solution required to transfer surfactant from bulk to surface, achieving the required surface concentration Γ_B.

D. Vertical Variations of the Surface Tension γ. We know $c(z)$, and we know (from classical plots for usual surfactants) the surface tension $\gamma(c)$: thus, we may construct $\gamma(z) = \gamma_0 - \Pi(z)$. The general aspect of this curve is shown in Figure 2b.

For $z > \lambda$ ($c \ll c_B$), we may write

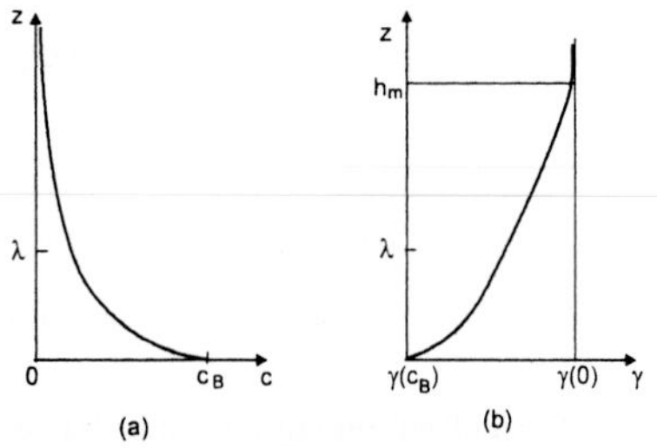

Figure 2. Structure of a young film when the bulk concentration (c_B) is smaller than the cmc (c^*): (a) concentration profile; (b) variation of surface tension with height.

$$\frac{d\gamma}{dz} = -\frac{d\Pi}{dz} = \frac{\rho g \Gamma}{c_B - c} \cong \frac{\rho g \Gamma_B}{c_B} = \text{constant} \qquad (13)$$

Thus, $\gamma(z)$ increases linearly with z, up to a certain height (h_m), where we return to a bare surface ($\Gamma \sim 0$, $\gamma = \gamma_0$). This corresponds to

$$h_m \frac{d\gamma}{dz} \sim \gamma_0 - \gamma_B \equiv \Pi_B$$

or

$$h_m = \lambda \frac{\Pi_B}{kT\Gamma_B} \qquad (14)$$

Note that h_m is larger than λ, because Π_B is much larger than an ideal gas pressure. The height h_m is an absolute limit of stability for our young film. (But, of course, other instabilities may occur in the regime of high z, low δ.)

III. Extension to Higher Concentrations $c_B > c^*$

A. Structure of the Chemical Potential at $c \gg c^*$. At concentrations $c > c^*$, most of the surfactant is in a micellar form (concentration c_M). Only a small fraction (c_1) is present in a monomer form. Let us call N the number of surfactants per micelle. Then a rough but convenient description (with N fixed) of the micelle/monomer equilibrium may be written in the form

$$c_M = \frac{c_1^N}{c^{*N-1}} \qquad (15)$$

or for the total concentration

$$c = c_1 + c_M = c_1\left\{1 + \left(\frac{c_1}{c^*}\right)^{N-1}\right\} \qquad (16)$$

The chemical potential is still

$$\mu = kT \ln c_1 + \text{constant} \qquad (17)$$

Let us consider only the limit $c \gg c^*$. Then

$$\frac{c}{c^*} \sim \frac{c_M}{c^*} = \left(\frac{c_1}{c^*}\right)^N \qquad (18)$$

and

$$\mu = kT\left\{\ln c^* + \frac{1}{N}\ln\frac{c}{c^*}\right\} \qquad (19)$$

$$\frac{d\mu}{dc} = \frac{kT}{Nc} \qquad (20)$$

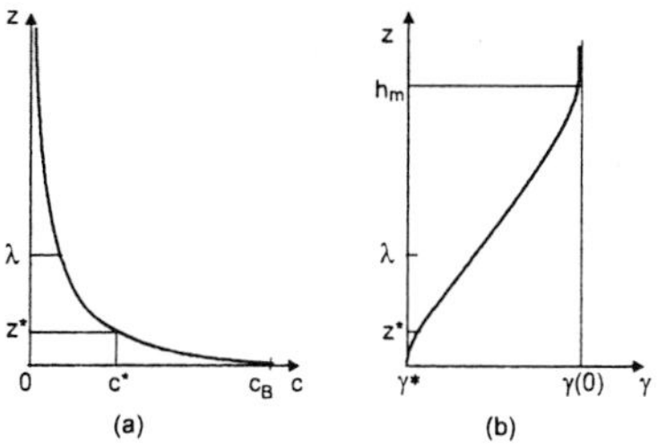

Figure 3. Structure of a young film when $c_B \gg c^*$: (a) concentrations; (b) surface tensions.

B. Concentration Profile. Returning to eq 4, we see from eq 20 that for $c \gg c^*$ we should replace ρg by $N\rho g$. Or, equivalently, the characteristic length in the high concentration regime is

$$\tilde{\lambda} = \frac{\lambda}{N} \tag{21}$$

This is much smaller than λ, since N is large ($\sim$60).

The resulting plot of $c(z)$ is shown in Figure 3a. We start at $z = 0$ from a concentration $c_B \gg c^*$. Then, upon increasing z, the concentration decreases, and when $c \ll c_B$, we have simply

$$c(z) = c_B \exp\left(-\frac{Nz}{\lambda}\right) \quad (z < z^*) \tag{22}$$

This holds up to a height z^* where $c = c^*$:

$$z^* = \frac{\lambda}{N} \ln \frac{c_B}{c^*} \tag{23}$$

At $z > z^*$, we return to the low concentration regime (eq 6), and ultimately, for $z > \lambda$, we have

$$c(z) \sim c^* \exp\left(-\frac{z}{\lambda}\right) \tag{24}$$

C. Surface Tensions. In all the region $c > c^*$ or $z < z^*$, the surface tension keeps its minimum value γ^*. Above $z < z^*$ we return to the constant slope regime of eq 13. The maximum height allowable h_m is still given by eq 14, as shown in Figure 3b.

Note the inequalities

$$h_m > \lambda > z^* \tag{25}$$

IV. Discussion

A. Summary of Static Predictions. (a) In a young film, we expect a concentration $c(z)$ which decays exponentially at large z, with a characteristic length λ (eq 7). λ is typically a few meters. This statement holds for bulk concentrations c_B which may be lower or higher than c^*.

(b) The film thickness at high z is simply $\delta = 2l$, where $l = \Gamma_B/c_B - (c < c^*)$ or $l = \Gamma_B/c^*$ for $c > c^*$.

(c) The film is bare and completely unstable at heights $z > h_m$, where h_m is given by eq 14 and is larger than λ.

(d) On the whole, going to bulk concentrations, $c_B > c^*$ does not produce great alterations: most static film properties should be close to what they are for $c_B = c^*$.

B. Characteristic Times. (1) A water film with no surfactant will collapse under its own weight in a time t_c governed by gravitational flow with the acceleration g:

$$t_c \sim \sqrt{\frac{h}{g}} \tag{26}$$

where h is the height of the film. For $h = 10$ cm, $t_c \sim 0.1$ s.

(2) What is the time required for the surfactant to reach the surface? We consider here first the (relatively simple) case where $c_B < c^*$: all the transport is via monomers. We also assume that there is no barrier at the surface opposing the surfactant adsorption. Then, we must transfer the surfactant by diffusion (coefficient D) over a distance l defined in eq 12. The diffusion time is roughly

$$t_d = \frac{l^2}{D} = \frac{\Gamma_B^2}{c_B^2 D} \tag{27}$$

The condition of formation of a young film at $c < c^*$ is essentially that t_d be shorter than t_c. To get small t'_d values, we must use concentrations c_B which are as high as possible (but remain below c^*).

(3) The extension of these ideas to $c_B > c^*$ is quite delicate. Micelles provide a big reservoir of surfactant, but the delivery is slow.

(a) Micelles can emit monomers which are then absorbed by the fresh surface: this process is fast (10^{-6} s) but inefficient, because each micelle cannot give more than two or three monomers (beyond that, the micellar free energy rises fast).

(b) Micelles can split in two, and the corresponding characteristic time τ_{split} is relatively long (10^{-3} s). After splitting, the parts end up as monomers and can feed the surface. This process has been documented by Shah.[6]

(c) The micelles may unfold by direct contact with the surface. The energy barriers for processes b and c are comparable—maybe the barrier for process c is slightly lower.

If, for simplicity, we focus on process b, the conclusion is that if we have $\tau_{split} < t_c$, the young film should be realizable. The opposite limit ($\tau_{split} > t_c$) would occur only for very long surfactant chains.

V. Conclusions

(a) To achieve young films of height h, we need $h_m > h$, where h_m is given by eq 14. As shown by this equation, high h_m corresponds to high surface pressures (or $\gamma \ll \gamma_0$) in the present solution. To achieve this, our surfactants must have an aliphatic tail which is not too short.

(b) The surfactant must migrate fast to the water/air interface. For $c \sim c^*$, this implies that $l^* = \Gamma^*/c^*$ must be relatively small. The critical micelle concentration c^* should not be too small. The surfactant tails should not be too long. The diffusion time of the surfactant over the length l^* must be shorter than the time of free fall of the film (with no surfactant forces).

Thus, a good foaming agent results from a compromise: this does correspond to the empirical rule stating that the HLB should be in the range 7–9.

A final remark: two questions arise in connection with the force balance (eq 3): (1) How is the vertical force transmitted from the water to the monolayers? (2) What about the horizontal balance of forces (question raised by a referee)? Answer: the pressure p in the water is independent of height and equal to the atmospheric pressure. This ensures the horizontal equilibrium. There is however a bulk gravity force in the water, inducing a (very slow) Poiseuille flow downward between two fixed

(6) Oh, S. G.; Shah, D. O. *Langmuir* **1991**, 7, 1316.

walls. The velocity field $\vec{v}$ is ruled by

$$\eta\nabla^2\vec{v} - \nabla p + \vec{\rho}g = 0$$

with $\nabla p = 0$.

If we compute the resulting shear stress from the water on the two lateral walls, we find that the weight force is transmitted exactly to the walls.

Acknowledgment. We have benefited from useful advice by D. Langevin and M. Schott.

LA001538L

Langmuir **2002**, *18*, 3413–3414 3413

On Fluid/Wall Slippage

P. G. de Gennes

Collège de France, 11 place Marcelin Berthelot, 75231 Paris Cedex 05, France

Received November 2, 2001. In Final Form: January 22, 2002

Certain (nonpolymeric) fluids show an anomalously low friction when flowing against well-chosen solid walls. We discuss here one possible explanation, postulating that a gaseous film of small thickness *h* is present between fluid and wall. When *h* is smaller than the mean free path *l* of the gas (Knudsen regime), the Navier length *b* is expected to be independent of *h* and very large (micrometers).

1. Introduction

The standard boundary condition for fluid flow along a wall is a no slip condition. If the wall is at rest, the tangential fluid velocity at the wall vanishes. The validity of this postulate was already checked in pioneering experiments by Coulomb.[1]

However, the spatial resolution available to Coulomb was limited. In our days, we characterize the amount of slip by an extrapolation length *b* (the Navier length): the definition of *b* is explained in Figure 1. The length *b* can be related to the surface friction coefficient *k* defined as follows: The shear stress σ induces at the wall a surface velocity v_s:

$$\sigma = kv_s \qquad (1)$$

Equating this to the viscous shear stress in the fluid (of velocity η), we get

$$\sigma = kv_s = \eta\left|\frac{\mathrm{d}v(z)}{\mathrm{d}z}\right| \qquad (2)$$

where $v(z)$ is the fluid velocity, increasing linearly with the distance from the wall *z*. When compared with Figure 1, we see that eq 2 gives

$$b = \eta/k \qquad (3)$$

For most practical situations, with simple fluids (made of small molecules with a diameter *a*), we expect a very small Navier length $b \sim a$: this has been verified in classical experiments, using a force machine under slow shear.[2] However, we know a certain number of anomalous cases with large *b*:

(1) A polymer melt, facing a carefully prepared surface, with grafted chains which are chemically identical to the melt: here, when σ becomes larger than a certain critical value σ^*, the Navier length jumps to high values (~50 μm).[3] This can plausibly be explained in terms of polymer dynamics:[4] at $\sigma > \sigma^*$, the grafted chains are sufficiently elongated by the flow to disentangle from the moving chains.

(2) With water flowing in thin, hydrophobic capillaries, there is some early qualitative evidence for slippage.[5,6]

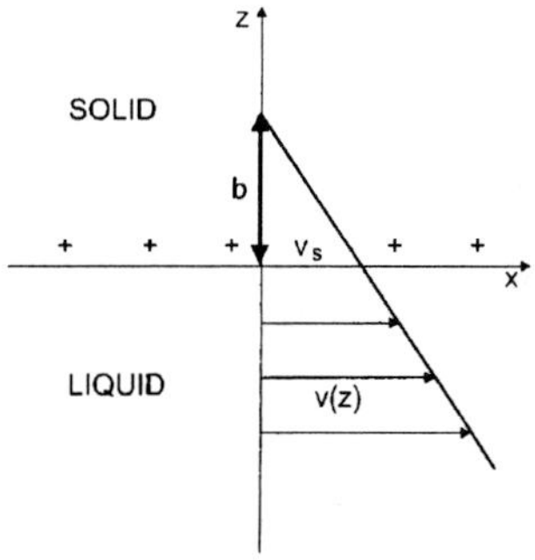

Figure 1. Definition of the Navier length *b* for simple shear flow near a solid wall.

(3) The role of the hydrophobic surfaces has been analyzed theoretically by J. L. Barrat and co-workers using analysis plus simulations.[7,8] They show that the first layer of "water" molecules is depleted in the presence of a hydrophobic wall: this can lead in their case to $b \sim 15$ molecular diameters in a typical case, for a contact angle $\theta = 150°$, corresponding to strong hydrophobicity.

(4) Recent experiments by Hervet, Léger, and co-workers are based on a local photobleaching technique (using evanescent waves) which probes the velocity field $v(z)$ within 50 nm of the surface.[9] They studied hexadecane flowing on sapphire: with bare sapphire, no slip was detected. On a fully grafted (methylated) sapphire, the slip length was found to be large (175 nm) and independent of σ in the observed range. This is a surprise: (a) we do not expect any hydrophobic effect here; (b) even if it was present, the *b* values are much larger than the Barrat–Bocquet estimates.

(5) Similar (as yet unpublished) results have been obtained by Zhu and Granick (using a force machine) and by Tabeling (using etched microcapillaries).

The results of refs 4 and 5 are unexpected and stimulating. This led us to think about unusual processes which could take place near a wall. In this Letter, we discuss one (remote) possibility: the formation of a gaseous film at the solid/liquid interface.

The source of the film is unclear: when the contact angle is large ($\theta \to 180°$), a type of flat bubble can form at the surface with a relative low energy. But this energy is still high when compared to the thermal energy *kT*.

(1) Coulomb, C. A. Mémoires relatifs à la physique; Société Française de Physique, 1784.
(2) See the book: *Dynamics in confined systems*; Drake, J., Klafter, J., Kopelman, R., Eds.; Materials Research Society: Pittsburgh, PA, 1996.
(3) Léger, L.; Hervet, H.; Pit, R. In *Interfacial properties on the submicron scale*; Fromer, J., Overney, R., Eds.; ACS Symposium series 781; American Chemical Society: Washington, DC, 2000.
(4) Brochard-Wyart, F.; de Gennes, P. G. *Langmuir* **1992**, *8*, 3033.
(5) Churaev, N.; Sobolev, V.; Somov, A. *J. Colloid Sci.* **1984**, *97*, 574.
(6) Blake, T. *Colloids Surf.* **1990**, *47*, 135.
(7) Barrat, J. L.; Bocquet, L. *Phys. Rev. Lett.* **1999**, *82*, 4671.
(8) Barrat, J. L.; Bocquet, L. *Faraday Discuss.* **1997**, *112*, 119.
(9) Pit, R.; Hervet, H.; Léger, L. *Phys. Rev. Lett.* **2000**, *85*, 980.

10.1021/la0116342 CCC: $22.00
Published on Web 03/30/2002

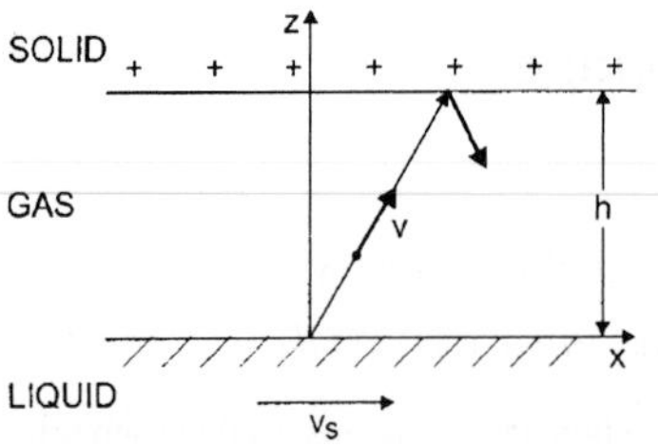

Figure 2. A gas molecule leaves the liquid with the velocity $\vec{V}$, and transmits (on the average) a momentum mV to the solid surface.

We first considered flat vapor bubbles generated at the solid fluid interface by thermal fluctuations. But the typical thickness of these bubbles turns out to be very small, of order $(kT/\gamma)^{1/2}$ (where T is the temperature and γ the surface tension), even when θ is close to 180°. Thus, this brings us back to the Barrat picture, with deplection in one first layer of the liquid.

Another possible source is an external gas dissolved in the liquid, up to metastable concentrations. This would then nucleate bubbles preferentially near the wall if $\theta > 90°$.

In section 2, we simply assume that there exists a uniform gas film of thickness h, larger than the molecule size a, but smaller than the mean free path l in the gas. We calculate the Navier length b for this case and find large values. The possible meaning of this result is discussed in section 3.

2. A Gas Film in the Knudsen Regime

The situation is shown in Figure 2. In the Knudsen regime, gas molecules travel freely in the film and hit the boundaries. We assume that there is no specular reflection on either boundary. Then, a molecule leaving the liquid will have a Gaussian velocity distribution for the tangential component v_x, with the peak of this Gaussian at velocity v_s (the translational velocity at the liquid surface). On the average, it will transmit a momentum mv_s to the solid. The number of such hits per second is ρ/mv_z, where ρ is the gas density, m the molecular mass, and v_z the normal component of velocity.

This leads to a shear stress

$$\sigma = mv_s \frac{\rho}{m} \bar{v}_z = \rho v_s v_z \qquad (4)$$

with

$$\bar{v}_z = \int_0^\infty \frac{1}{(2\pi)^{1/2} v_{th}} v_z e^{-v_z^2/2v_{th}^2} \, dv_z = v_{th}/(2\pi)^{1/2} \qquad (5)$$

where $v_{th}^2 = kT/m$.

Equation 4 gives us a precise value of the friction coefficient k in eq 2, and from it, we arrive at a Navier length

$$b = -h + \frac{\eta}{\rho \bar{v}_x} \cong \frac{\eta}{\rho v_z} \qquad (h < l) \qquad (6)$$

(The $-h$ term is due to the distance between liquid surface and solid surface; it is completely negligible in practice.)

If we choose typical values, $\rho = 1$ g/L, $v_{th} = 300$ m/s, and $\eta = 10^{-2}$ P s, we find $b = 7$ μm. Thus, a gas film can indeed give a very large slip length. Our calculation assumed complete thermalization at each particle/boundary collision. If we had some nonzero reflectance (especially on the solid surface), this would increase b even more.

3. Conclusions

When gas films with thickness h in the interval $a \ll h \ll l$ are present in a flow experiment, they may indeed generate a strong slippage, independent of the film thickness. But the process which could generate such films remains obscure. If, for some reason, the liquid entering the channel was supersaturated with a certain gas, a pressure drop in the channel could indeed promote the release of gas bubbles. If the solid surface is not very wettable ($\theta > 90°$), the bubbles should preferentially nucleate at the liquid/solid interface (as they do in a glass of champaign). Then, we would have to postulate that the shear flow transforms them into a flat film, provided that the shear stress is strong enough to flatten them. Indeed, in the Zhu–Granick experiments, there is a threshold stress, but it is much smaller than what would result from a compromise between shear stress and surface tension for small bubbles.

It is worth emphasizing that the amount of gas required to lubricate the solid liquid contact is very small: if D is the macroscopic width of the shear flow, the weight fraction of gas required is

$$\psi = \frac{\rho}{\rho_L} \frac{h}{D} \qquad (7)$$

where ρ_L is the density of the liquid. Typically $\psi \sim 10^{-5}$.

On the whole, we cannot present a complete explanation of the anomalous Navier lengths based on gas films. But the films may possibly show up in some instances; e.g., if the liquid is purposely oversaturated with gas, then eq 6 may become useful.

Acknowledgment. I benefited from stimulating discussions with S. Granick, L. Léger, H. Hervet, and F. Wyart-Brochard.

LA0116342

Afterthought: On fluid/wall slippage

The anomalous slip of certain (simple) liquids against some controlled surfaces is a challenge. At this moment (2002) I tend to believe that this slip is due to deviations from the ideal picture. For instance, with water facing a hydrophobic surface, a minute amount of surfactant reaching the water/gas interfaces can lead to complete wetting by an air film. Also, a certain amount of ronghness may favor the presence of gas bubbles, as pointed out by E. Charlaix.

Part V. Wetting and Adhesion

Wetting: statics and dynamics

P. G. de Gennes

Collège de France, Physique de la Matière Condensée, 11 Place Marcelin-Berthelot, 75231 Paris Cedex 05, France

The wetting of solids by liquids is connected to physical chemistry (wettability), to statistical physics (pinning of the contact line, wetting transitions, etc.), to long-range forces (van der Waals, double layers), and to fluid dynamics. The present review represents an attempt towards a unified picture with special emphasis on certain features of "dry spreading": (a) the final state of a spreading droplet need not be a monomolecular film; (b) the spreading drop is surrounded by a precursor film, where most of the available free energy is spent; and (c) polymer melts may slip on the solid and belong to a separate dynamical class, conceptually related to the spreading of superfluids.

CONTENTS

I. INTRODUCTION

Many practical processes require the spreading of a liquid on a solid. The liquid may be a paint, a lubricant, an ink, or a dye. The solid may either show a simple surface or be finely divided (suspensions, porous media, fibers). Water, for instance, may be sucked into a porous soil, because it tends to wet the solid components of the soil. Tertiary oil recuperation also involves the penetration of water into the channels of a porous rock, which were originally filled mainly by oil. The "flotation" of ores is based on selective wetting properties for the ore particles.

In spite of their importance, these processes are still poorly understood.

(1) All interfacial effects are very sensitive to contaminants and to physical modifications of the surface (e.g., steps, dislocations, if we are dealing with a crystalline solid); this may explain why certain basic experiments (e.g., spreading a single small droplet on a flat solid surface) have been fully carried out only recently.

(2) The solid/liquid interfaces are much harder to probe than their solid/vacuum counterpart; essentially all experiments making use of electron beams become inapplicable when a fluid is present. A few sensitive techniques may still be applied specifically to the interface (fluorescence, EPR, etc.), but they are often restricted to very specific examples. Similar limitations occur with the electrochemical data.

(3) On the theoretical side, 180 years after the pioneering work of Young and Laplace, a number of basic capillary problems are just beginning to be solved.

(a) The physicochemical parameters controlling the *thermodynamic wettability* of solid surfaces were clarified through the long, careful efforts of Zisman (1964) and others (Fowkes, 1964; Padday, 1978), but the deviations from thermodynamic equilibrium are just beginning to be understood. Here I shall insist particularly on two such deviations—the hysteresis of contact angles, due to the pinning of the contact line on localized defects, and the regimes of "dry spreading," where the final state of a spreading droplet is not necessarily a monomolecular layer, but may be a film of greater thickness. These relatively novel aspects are explained in Sec. II.

(b) The transition from "*complete wetting*" to "*partial wetting*" (defined in Fig. 1 below), first predicted in 1977 (Cahn, 1977; Ebner and Saam, 1977), has become an active field of research and dispute (Sec. III).

(c) The *dynamics* of spreading is delicate: a pioneering paper (Huh and Scriven, 1971) suggested a singularity in the dissipation, which provoked many discussions. Recently, a useful distinction has appeared between simple fluids, in which the liquid spreads by a rolling motion (Dussan and Davis, 1974), and polymer melts, which often tend to slip on a solid surface (Brochard and de Gennes, 1984). These two regimes (and the corresponding removal of singularities) are presented in Sec. IV.

Our discussion does not include the *local* structure of the interfaces—the arrangement of the atoms, or molecules, at the 3-Å scale near a boundary. For fluid/fluid interfaces, this aspect is, in fact, well reviewed in a recent book (Rowlinson and Widom, 1982). For solid/fluid interfaces our knowledge is still rather limited. In the present paper, the emphasis will be on behavior at somewhat larger scales (say 30 to 300 Å), where long-range forces (van der Waals, electrostatic, etc.) become essential, control many practical features, and give rather universal properties.

II. CONTACT ANGLES

A. Thermodynamic equilibrium

1. Angles and energies: the Young condition

When a small liquid droplet is put in contact with a flat solid surface, two distinct equilibrium regimes may be found: partial wetting [Figs. 1(a) and 1(b)] with a finite contact angle θ_e, or complete wetting ($\theta_e=0$) [Fig. 1(c)].[1] In cases of partial wetting, the wetted portion of the surface is delimited by a certain *contact line* $\mathscr{L}$ (which, for our droplets, is a circle).

The situation near the contact line is represented more precisely in Fig. 2. Here we are dealing with a macroscopic wedge, and the line $\mathscr{L}$ is normal to the plane of the figure. Three phases are in contact at the line: the solid S, the liquid L, and the corresponding *equilibrium vapor* V. Each interface has a certain free energy per unit area γ_{SL}, γ_{SV}, and γ_{LV} (the latter, for simplicity, will simply be called γ).

These parameters describe adequately the energy content of the interfaces in the far field (far from $\mathscr{L}$). In the vicinity of $\mathscr{L}$, the structure is much more complex and depends on a detailed knowledge of the system (examples of the complications that may occur are shown in Fig. 3). There is a "core region" around the nominal position of $\mathscr{L}$, where the complications occur. It is possible, however, to relate θ_e to the far-field energies γ_{ij} *without any knowledge of the core.* This was one of the (many) discoveries of the British scientist Thomas Young (1773–1829).

The basic idea is to write that, in equilibrium, the energy must be stationary with respect to any shift (dx) of the line position. In such a shift, (a) the bulk energies are unaffected (since the pressure is the same in the liquid and in the vapor); (b) the core energy is unaffected—the core is simply translated; and (c) the areas of the far-field interfaces (for a unit length of line) are increased, respectively, by dx (for S/V), $-dx$ (for S/L), and $-\cos\theta_e dx$ (for L/V). Hence the condition

$$\gamma_{SV}-\gamma_{SL}-\gamma\cos\theta_e=0\ . \tag{2.1}$$

[1]The subscript (e) in θ_e refers to an *equilibrium* property.

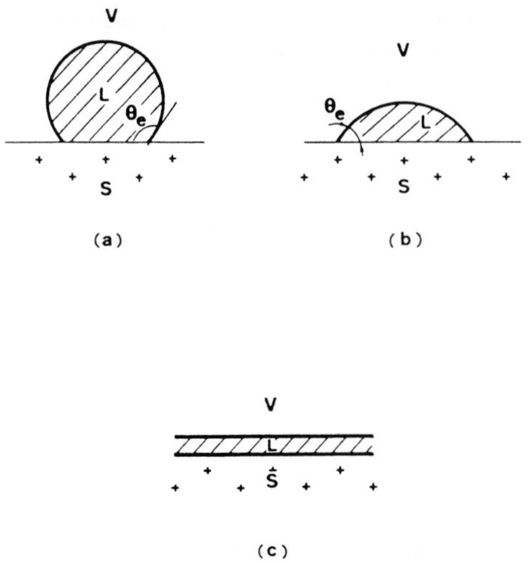

FIG. 1. A small droplet in equilibrium over a horizontal surface: (a) and (b) correspond to partial wetting, the trend towards wetting being stronger in (b) than in (a). (c) corresponds to complete wetting ($\theta_e = 0$).

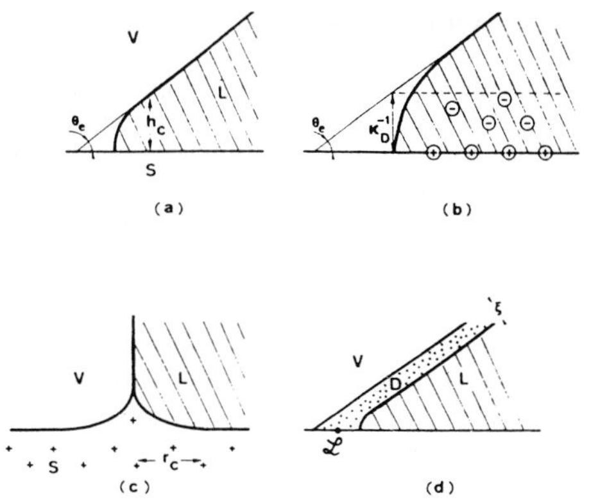

FIG. 3. Various types of core structures for the triple line. (a) Effect of attractive van der Waals forces. For $\theta_e \ll 1$, the profile is hyperbolic and the height h_c of the perturbed region is of order a/θ_e (where a is an atomic length). (b) A charged solid surface wetted by salty water (screening length κ_D^{-1}). (c) Effect of the finite deformability of the solid. The width r_c of the deformed region is $r_c \sim \gamma/E$ (γ, surface tension of the liquid; E, Young's modulus of the solid). For simplicity the special case $\theta_e = \pi/2$ has been drawn. (d) In the vicinity of a liquid/vapor critical point, the L/V interface becomes diffuse (thickness ξ) and the triple line has a core of radius $r_c \sim \xi$.

Equation (2.1) shows that the angle θ_e is entirely defined in terms of thermodynamic parameters: measurements on θ_e give us certain information on the interfacial energies. Usually, we know $\gamma_{LV} \equiv \gamma$ by separate measurements. Thus we are left with two unknowns (γ_{SL}, γ_{SV}) and only one datum (θ_e). But it is only the *difference* $\gamma_{SV} - \gamma_{SL}$ which is relevant for experiments involving the liquid.

2. Spatial scales for the definition of a contact angle

Equation (2.1) was derived for a wedge (planar interfaces in the far field). For many practical applications

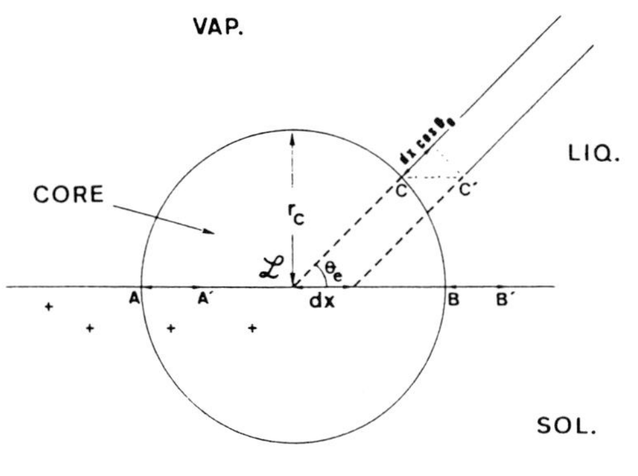

FIG. 2. Translation of a liquid wedge (triple line $\mathcal{L}$) by an amount dx. The energy is unchanged in this process, and this leads to the Young equation (2.1).

(such as the droplets of Fig. 1), some weak curvatures are superimposed.

(a) The liquid/vapor interface may have a total curvature $C = R_1^{-1} + R_2^{-1}$, and this is associated with a certain pressure difference between liquid and vapor:[2]

$$p_L - p_V = \gamma C \ . \tag{2.2}$$

The angle θ_e is still well defined in this case, provided that the radii of curvature (R_1, R_2) *are much larger than the size (r_C) of the core region.*

(b) The line $\mathcal{L}$ itself may be curved, and, in this case, a displacement of the line modifies the core energies. Again, this leads to measurable effects only when the radius of curvature of the line is not too large, when compared to the core size r_C [see, for instance, Platikhanov *et al.* (1980)]. In many practical examples $r_C < 100$ Å. Thus, for most macroscopic experiments, where the droplets or the capillaries have sizes $R \sim 1$ mm, all curvature corrections are negligible. A measurement of θ_e at a distance r from the line, where

$$r_C \ll r \ll R \ ,$$

should give a well-defined θ_e, independent of r.

[2]Equation (2.2) is also due to Young (1804) and was rediscovered independently one year later by Laplace.

3. Practical determinations of θ_e

The angle θ_e can be obtained (a) from a direct photograph, (b) through the reflection (or deflection) of rays by the liquid prism of Fig. 2, (c) by interferential techniques (Callaghan *et al.*, 1983), especially for small θ_e, (d) from the rise of a liquid column in a fine capillary (Fig. 4); for a general discussion of the various capillary effects, see the classical book by Bouasse (1924) and the recent tutorial article of Guyon *et al.* (1982).

In actual experiments the main difficulty is to avoid a certain *pinning of the triple line* $\mathscr{L}$ on defects of the solid surface. This pinning leads to a hysteresis of the contact angles, which can very seriously obscure the determination of θ_e. Clearly, to avoid pinning one requires solid surfaces that are smooth and chemically homogeneous, but the question is, what level of smoothness do we require to reduce the uncertainty in θ_c below a prescribed limit $\Delta\theta$? A partial answer to this question is given in Sec. II.C.

4. Special features of complete wetting

Equation (2.1) gives $\cos\theta_e$ as a function of interfacial energies. The special case

$$\gamma=\gamma_{SV}-\gamma_{SL}$$

leads to $\cos\theta_e=1$ or $\theta_e=0$ (complete wetting). At first sight, this situation appears rather exceptional. In fact, it is not, because $\gamma_{SL}+\gamma$ can never be larger than γ_{SV} (in thermodynamic equilibrium). If it were larger, this would imply that the free energy of a solid/vapor interface (γ_{SV}) could be lowered by intercalating a liquid film of macroscopic thickness (energy $\gamma_{SL}+\gamma$). The equilibrium solid/vapor interface then automatically comprises such a film, and the true γ_{SV} is identical to $\gamma_{SL}+\gamma$, i.e., we have complete wetting in this regime.

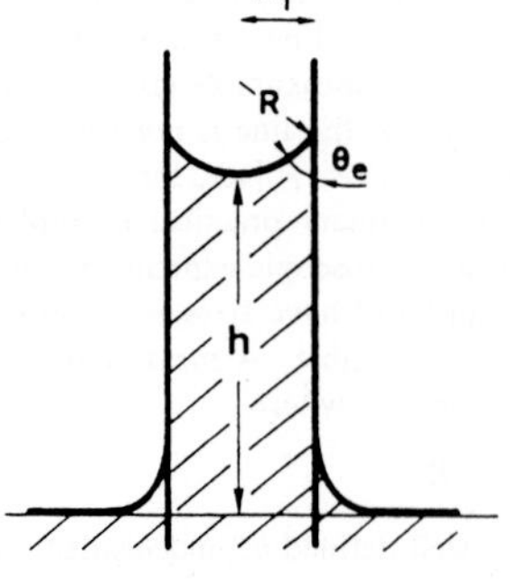

FIG. 4. Capillary rise: in a thin capillary the meniscus is a spherical cap of radius $R=R_1/\cos\theta_e$. The Young-Laplace capillary pressure $2\gamma/R$ is balanced by the hydrostatic component ρgh (ρ, density difference between liquid and vapor, g, gravitational constant). Thus a measurement of h gives θ_e.

On the other hand, if we deal with nonequilibrium situations, we may have a solid/vapor interfacial energy γ_{SO} that is larger than $\gamma_{SL}+\gamma$. The difference

$$S=\gamma_{SO}-\gamma_{SL}-\gamma \tag{2.3}$$

is called the spreading coefficient. Physically, γ_{SO} is associated with a "dry" solid surface, while γ_{SV} is associated with a "moist" surface. For some systems the difference is enormous. With water on metallic oxides, $\gamma_{SO}-\gamma_{SV}\sim300$ ergs/cm^2; with organic liquids on oxides, $\gamma_{SO}-\gamma_{SV}\sim60$ ergs/cm^2. On the other hand, with organic liquids on molecular solids, the difference (as seen via contact angles) is perceptible only when the liquid is very light and volatile. For instance, with normal alcanes on a Teflon surface, the angles θ_0 (on dry Teflon) and θ_e (in equilibrium with the vapor) are found to differ only when the carbon number n of the alcane is ≤ 5 (Zisman, 1964).

The importance of the spreading coefficient (2.3) for practical purposes was first recognized by Cooper and Nuttal (1915) in connection with the spraying of insecticides on leaves. Large positive S favors the spreading of a liquid.

There remains, however, a fundamental ambiguity in cases where the experimentalist observes complete spreading on macroscopic scales: he cannot tell whether $S=0$ or $S>0$. For the "moist" case, we know that $\theta_e=0$ (or that the corresponding $S_{\text{moist}}\equiv0$) because the system "locks in" at this value, as explained at the beginning of this section. But for the "dry" case there is no "lock-in" process, and we expect that complete spreading will usually be associated with a positive S.

How large is S? We shall see in Sec. II.D that the answer may sometimes be obtained by probing the thickness of the ultimate wetting film achieved in spreading: the smaller the S, the larger the equilibrium thickness, in qualitative agreement with the trend noticed by Cooper and Nuttal.

B. Wettability

Our aim in this section is to understand qualitatively how the contact angle θ_e depends on the chemical constitution of both the solid S and the liquid L. The basic reference here is still the review by Zisman (1964).

1. High-energy and low-energy surfaces

Let us discuss the solid first. From studies on the bulk cohesive energy we know that there are two main types of solids, (a) hard solids (covalent, ionic, or metallic), and (b) weak molecular crystals (bound by van der Waals forces, or in some special cases, by hydrogen bonds). A similar classification is found from the solid/vacuum surface energies (Fox and Zisman, 1950). Hard solids have "high-energy surfaces" ($\gamma_{SO}\sim500$ to 5000 ergs/cm^2), while molecular solids (and also molecular liquids) have "low-energy surfaces" ($\gamma_{SO}\sim50$ ergs/cm^2).

2. Standard behavior of high-energy surfaces

Most molecular liquids achieve complete wetting ($S \geq 0$) with high-energy surfaces. Let us try to understand this in simple terms, assuming that hard bonds control γ_{SO}, while van der Waals interactions control the liquid/solid energies (no chemical binding between liquid and solid). This amounts to writing for the solid/liquid energy

$$\gamma_{SL} = \gamma_{SO} + \gamma - V_{SL} \quad (V_{SL} > 0) . \tag{2.4}$$

Here the term $-V_{SL}$ describes the attractive van der Waals (VW) interactions between solid and liquid near the surface. Equation (2.4) can be understood if we progressively bring into contact the regions S and L; when they are well separated (by an empty slab), the energy is $\gamma_{SO} + \gamma$; when we establish contact, we recover the energy $-V_{SL}$.

Very similarly, by bringing into contact two *liquid* portions, we start with an energy 2γ, and end up with zero interfacial energy,

$$0 = 2\gamma - V_{LL} \quad (V_{LL} > 0) , \tag{2.5}$$

where $-V_{LL}$ represents the LL attractions. Using Eqs. (2.4) and (2.5), we find that the spreading parameter S, defined in Eq. (2.3), is equal to

$$S = -2\gamma + V_{SL} = V_{SL} - V_{LL} ,$$

and the condition of complete wetting ($S > 0$) corresponds to

$$V_{SL} > V_{LL} . \tag{2.6}$$

It is also possible to translate (2.6) in terms of the dielectric polarizabilities α_S (α_L) for the solid (liquid). To a first approximation the VW couplings between two species (i) and (j) are simply proportional to the product of the corresponding polarizabilities

$$V_{ij} = k\alpha_i\alpha_j , \tag{2.7}$$

where k is (roughly) independent of (i) and (j). Then the condition (2.6) reduces to

$$\alpha_S > \alpha_L . \tag{2.8}$$

Thus high-energy surfaces are wetted by molecular liquids, not because γ_{SO} is high, but rather because the underlying solid usually has a polarizability α_S much higher than the polarizability of the liquid. Of course, these considerations are very rough (the frequency dependence of the α's should be taken into account), but they still provide us with some guidance.

3. Low-energy surfaces and critical surface tensions

Low-energy surfaces can give rise to partial or to complete wetting, depending on the liquid chosen. In a complex situation like this, it is natural to choose a series of homologous liquids (for instance, the n-alcanes) and to study how they wet a given solid.

In some cases we find complete wetting for the whole series. This occurs, for instance, for liquid alcanes against solid polyethylene. But in other cases we find a finite contact angle θ_e, varying within the homologous series. A useful way of representing these results is to plot $\cos\theta_e$ versus the surface tension γ of the liquid. (An example is shown in Fig. 5.) Although, in many cases, we never reach $\cos\theta_e = 1$, i.e., we never reach complete wetting, we can extrapolate the plot down to a value $\gamma = \gamma_C$, which would correspond to $\cos\theta_e = 1$. The details of the extrapolation produced differ from author to author (in the pioneering work of Zisman a linear extrapolation was used), but this is not essential.

In general, we would expect γ_C to depend on the solid S, but to depend also on the liquid series L. However, when dealing with simple molecular liquids (where van der Waals forces are dominant), Zisman observed that γ_C is essentially independent of the nature of the liquid, and is a characteristic of the solid alone. Typical values are listed below:

	ergs/cm^2
Nylon	46
P. Vinyl chloride	39
P. Ethylene	31
PVF_2	28
PVF_4	18

If we want to find a molecular liquid that wets completely a given low-energy surface, we must choose a liquid of surface tension $\gamma < \gamma_C$. Thus γ_C may be called a "critical surface tension" and is clearly the essential parameter for many practical applications.

Can we relate γ_C to some simple physical parameters of the solid? This has been attempted by various authors

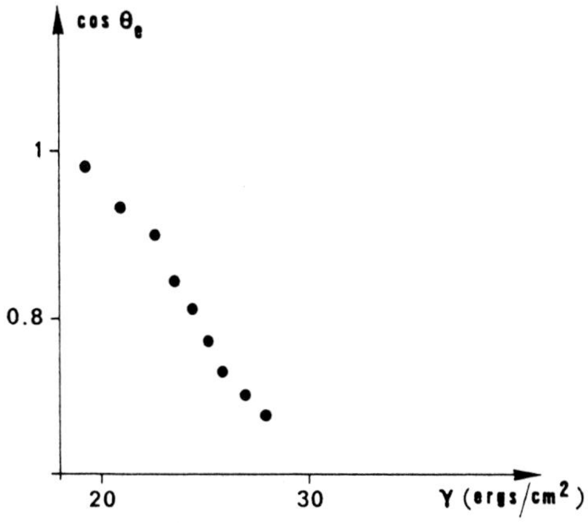

FIG. 5. A typical Zisman plot (cosine of equilibrium angle θ_e vs surface tension) for a polytetrafluoroethylene (Teflon) surface in contact with liquid n-alkanes (after Fox and Zisman, 1950). The critical surface tension γ_c for this system is ~ 18 ergs/cm^2.

(Girifalco and Good, 1957; Fowkes, 1962; Good, 1964). In the present review, we shall present only a naive argument, following the simple "van der Waals model" of Eqs. (2.4)–(2.7) and always assuming that the polarizability γ_V of the vapor is negligible. This amounts to writing

$$\cos\theta_e = \frac{\gamma_{SV}-\gamma_{SL}}{\gamma} \cong \frac{\gamma_{SO}-\gamma_{SL}}{\gamma} \cong \frac{V_{SL}-\gamma}{\gamma} = \frac{2\alpha_S}{\alpha_L}-1 . \quad (2.9)$$

When we compare different chemicals within the same homologous series, we vary α_L. For instance, with alcanes, the polarizability of the terminal (CH_3) groups is higher than the polarizability of the $-CH_2-$ groups: shorter alcanes have larger α_L values. The value of α_L at which $\cos\theta_e$ extrapolates to 1 is

$$\alpha_{LC}=\alpha_S . \quad (2.10)$$

If we prefer to work in terms of surface tensions $\gamma=\frac{1}{2}V_{LL}=\frac{1}{2}k\alpha_L^2$, we may write Eq. (2.9) in the form

$$\cos\theta_e = 2\left[\frac{\gamma_C}{\gamma}\right]^{1/2}-1 , \quad (2.11)$$

$$\gamma_C=\tfrac{1}{2}k\alpha_S^2 . \quad (2.12)$$

Equation (2.12) does show that γ_C depends only on the properties of the solid and is an increasing function of its polarizability. Numerically, Eq. (2.12) is not good, and various lines of improvement have been pursued.

(a) In practice many forces contribute to the solid/liquid interactions—dipolar, hydrogen bonds, etc. Thus one adds more terms in the decomposition (2.4), each force giving its contribution to V_{SL} and to V_{LL} (see, for instance, Good, 1964). In such a case γ_C may depend slightly on the nature of the liquids chosen to define it.

(b) Even when VW forces only are present, the simple expression (2.7) of the interaction in terms of some average polarizabilities is too primitive. More precise theories incorporate the frequency dependence of the polarizabilities, following the lines of the Lifshitz calculation of VW forces (Owens *et al.*, 1978). We shall not insist on these points, but mention that there exists (at least) a third group of corrections.

(c) The density distribution and the pair correlations in the liquid are modified near the solid surface, and these modifications may be quite different from what they are at the free surface of the liquid. For instance, recent work by Israelashvili (1982) and others shows that the forces between two closely spaced (20 Å) solid surfaces, through a liquid, are often oscillatory in sign, suggesting a one-particle density function in the liquid, which oscillates as a function of the distance to the solid. Computer work on these structures has begun (see, for instance, Snook and Van Megen, 1979,1980), but some more time will be required before we extract from it some really general rules and trends.

Let us now return to the practical aspects, and comment upon the values of γ_C that have been listed above.

(a) The system of high γ_C (nylons, PVC) are those most wettable by organic liquids. They carry rather strong permanent dipoles.

(b) Among systems that are controlled by VW interactions, we note that CF_2 groups are less wettable (↔ less polarizable) than CH_2 groups. In practice, many protective coatings (antistain, waterproofing, etc.) are based on fluorinated systems.

(c) It is possible to study specifically the wetting properties of terminal groups CF_3- or CH_3- by depositing a surfactant monolayer on a polar solid surface (Fig. 6). For CH_3 groups $\gamma_L=24$ ergs/cm^2, and for CF_3, γ_C is amazingly small ($\sim$6 ergs/cm^2). For more details on all these fascinating questions, we again refer the reader to the beautiful review by Zisman (1964).

C. Contact angle hysteresis

1. Experiments

The determination of the thermodynamic contact angle requires very clean experimental conditions. In many practical situations, one finds that the triple line $\mathcal{L}$ is pinned, and immobile, not only for $\theta=\theta_e$ but whenever θ lies within a finite interval around θ_e,

$$\theta_r < \theta < \theta_a . \quad (2.13)$$

The angle θ_a (advancing angle) is measured when the solid/liquid contact area increases, while θ_r (receding an-

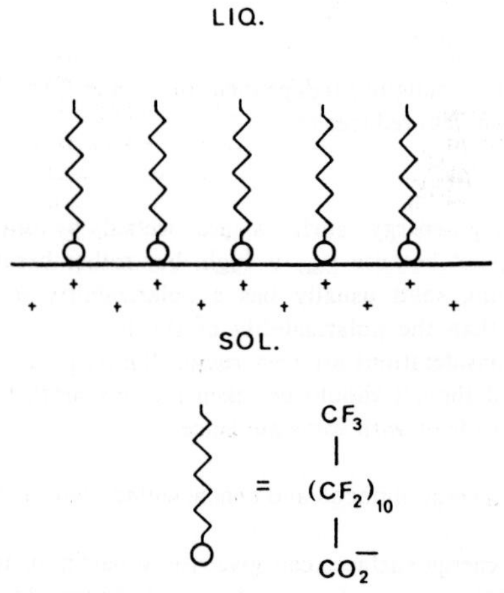

FIG. 6. Idealized structure of a surfactant monolayer attached to a polar solid. The particular example chosen provides one of the least wettable surfaces ever found.

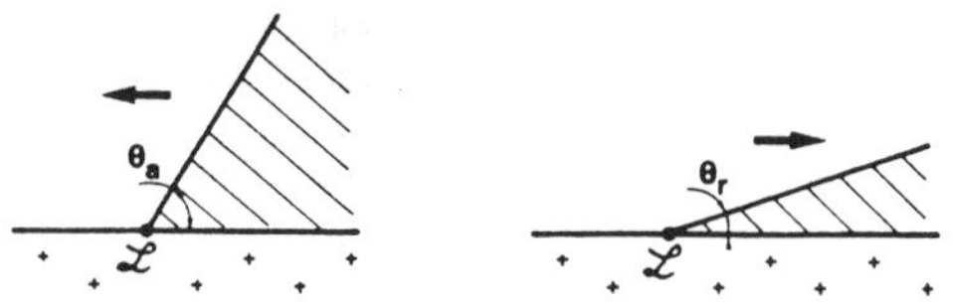

FIG. 7. Definition of the advancing (θ_a) and receding (θ_r) angles for a liquid on a nonideal solid surface.

gle) is measured when the contact area shrinks (Fig. 7). The interval $\theta_a - \theta_r$ may be 10° or more for surfaces that have not been specially prepared.

What is the source of this hysteresis? Three major causes have been invoked.

(a) *Surface roughness.* Early observations by Trillat and Fritz (1937) showed that the triple line $\mathcal{L}$ was easily trapped when parallel to a system of grooves. Among the more recent experiments, those by Dettre and Johnson (1964) deserve special mention because they were performed with a series of solid surfaces of increasing roughness. A typical set of data is shown in Fig. 8. It exhibits a remarkable, nonmonotonous variation of θ_r with the degree of roughness, to which we shall return later. Further systematic studies were carried out by Mason (1978).

(b) *Chemical contaminations*, or inhomogeneities, in the solid surface may also play an important role. Some of the experiments of Dettre and Johnson (1964) were made with glass beads immersed in paraffin wax, and the differences in wettability between glass and paraffin may have contributed to the hysteresis. But systematic studies of purely chemical effects at a smooth surface are still lacking.

(c) *Solutes* in the liquid (surfactants, polymers, etc.) may deposit a film on the solid surface, and the presence

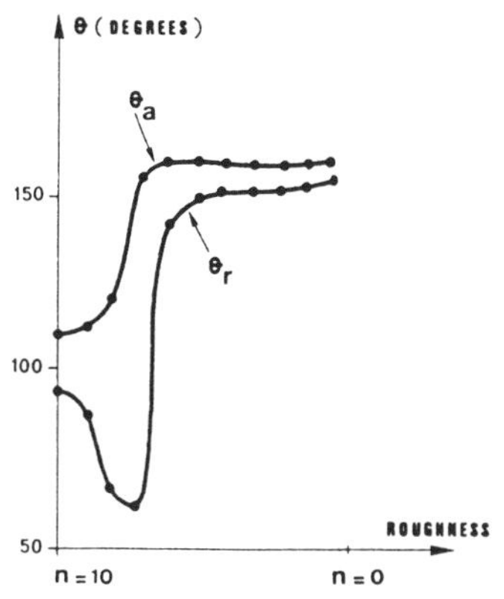

FIG. 8. Advancing and receding angles for water on fluorocarbon wax: a rough surface is obtained by spraying the wax. It is then made smoother by heating in an oven. The numbers n on the horizontal scale (0,1.0,10) refer to the number of successive heat treatments. Notice the abrupt jump of θ_r between $n=6$ and $n=7$ (after Dettre and Johnson, 1964).

or absence of the film can, in some cases, lead to hysteretic effects. In many cases the film, once formed, is stuck on the solid surface. See, for instance, Chappuis (1984).

2. Models with parallel grooves

Early discussions on the effects of surface roughness were restricted to *periodic* surfaces—for instance, with a parallel set of grooves (Johnson and Dettre, 1964; Mason, 1978; Cox, 1983). These systems have some reality—a classical example is a phonograph record (Oliver *et al.*, 1977).

When the triple line $\mathcal{L}$ is parallel to the grooves, it may have a number of pinned positions (described in Fig. 9), and it is possible to compute numerically the magnitude of the energy barriers between two such positions. Some aspects of these calculations are very artificial (the energy barriers are proportional to the total length of line

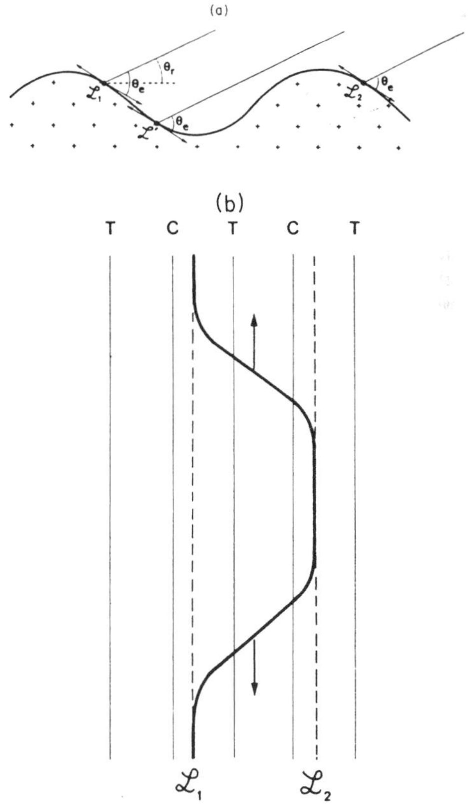

FIG. 9. (a) Equilibrium positions of a contact line $\mathcal{L}$ (normal to the sheet) on a system of grooves. $\mathcal{L}_1$, $\mathcal{L}_2$, are locally stable, while $\mathcal{L}'$ is unstable. θ_e is the thermodynamic contact angle. θ_r is the macroscopic angle. (b) The creep process for a contact line $\mathcal{L}$ moving from position $\mathcal{L}_1$ to position $\mathcal{L}_2$. C stands for "crest" and T for "trough."

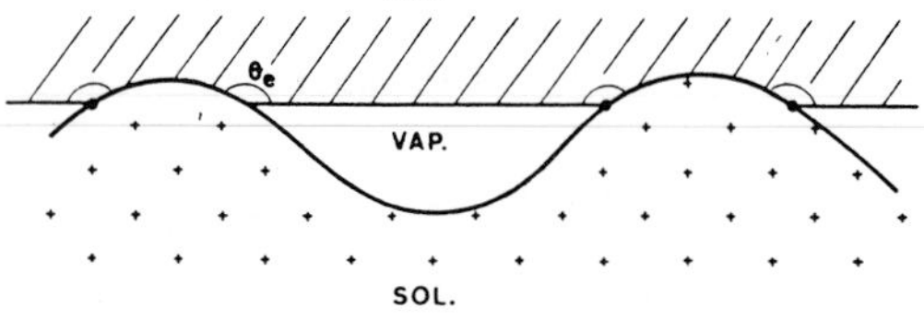

FIG. 10. A "composite" structure, with vapor bubbles trapped between liquid and solid, when the solid has deep, parallel grooves.

involved), but some aspects are instructive. For instance, when the grooves are rather deep, it may happen that vapor bubbles remain locked in the troughs and are covered by the liquid (Fig. 10). The resulting "composite structures" are then predicted to display much smaller barriers. The minimum of θ_r are a function of roughness, observed in various systems by Dettre and Johnson (1964), has been interpreted along these lines: When we increase the roughness, we first find a normal increase of the barrier heights and a corresponding decrease of θ_r; but when the troughs become deep enough, we obtain a composite structure, with weaker barriers, and θ_r increases.

Note, finally, that the groove systems show an extremely strong anisotropy. When the line $\mathcal{L}$ is parallel to the grooves, it is pinned. When $\mathcal{L}$ lies (on the average) at a certain angle ψ from the grooves, it has the local structure shown in Fig. 11, and $\mathcal{L}$ can be displaced continuously

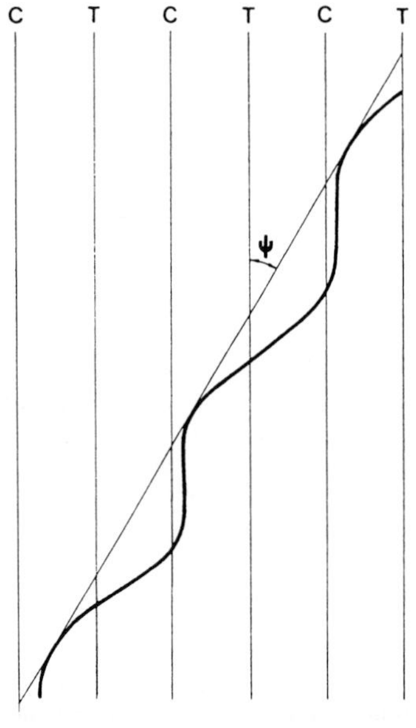

FIG. 11. A triple line $\mathcal{L}$ at an oblique angle (ψ) from a system of grooves (the crests and troughs are marked C and T, respectively). The overall pattern can be translated along the groove direction without any energy change (no pinning).

without any pinning (Cox, 1983). Experiments have been carried out for the special case $\psi=90°$ and indeed show no pinning (Mason, 1978).

In fact, the cascade of "jogs" displayed in Fig. 11 also gives us a hint about the physical processes that take place when $\mathcal{L}$ is parallel to the grooves ($\psi=0$). At $\psi=0$ the line wants to jump from one crest ($\mathcal{L}_1$) to the next ($\mathcal{L}_2$), as explained in Fig. 9(a). But the optimal method of doing this is not an overall jump (which would correspond to a huge barrier energy). What should really happen (in an infinite sample, with no edge effects) is shown in Fig. 9(b): nucleation of two "jogs" of opposite sign, followed by a glide of each jog along the grooves, until the ($\mathcal{L}_2$) strip spans the whole crest. Thus the physical barrier energy is independent of sample size, and is related to the nucleation and depairing of two adjacent jogs.

This statement holds when the grooves are infinitely long (or close on themselves, as they would in a capillary with grooves normal to the axis). If the grooves have a finite length (e.g., on a grooved plate), then a single jog may easily nucleate at the end of the groove and sweep through it. This process was discovered in recent numerical studies by Garoff and Schwartz (private communication).

3. Random surfaces

a. Weak fluctuations

A natural extension of the groove models is the case of surfaces that have a double periodicity, e.g., two orthogonal sets of grooves (Cox, 1983). However, it is clear that the major physical problem in this case corresponds to a random surface (random shape, or random chemical composition). This situation is, of course, more difficult. A first step, to make it simpler, is to focus on cases of *weak fluctuations*. To explain what this means quantitatively, let us start with a flat surface, but allow for chemical contamination. This will be described in terms of the local interfacial energies $\gamma_{SV}(x,y),\gamma_{SL}(x,y)$ at the point (x,y) on the surface. What matters is the difference $\gamma_{SV}-\gamma_{SL}$, or, more accurately, the fluctuating part

$$h(x,y)=\gamma_{SV}-\gamma_{SL}-\langle\gamma_{SV}-\gamma_{SL}\rangle \,, \tag{2.14}$$

where the angular brackets denote a space average. The local contact angle $\theta(x,y)$ at point (x,y) is given by the Young condition (2.1),

$$\gamma\cos\theta=h+\langle\gamma_{SV}-\gamma_{SL}\rangle \,, \tag{2.15}$$

while the unperturbed angle θ_0 is ruled by

$$\gamma\cos\theta_0=\langle\gamma_{SV}-\gamma_{SL}\rangle \,. \tag{2.16}$$

For small h we may thus write

$$\theta-\theta_0=-\frac{h(x,y)}{\gamma\sin\theta_0} \,, \tag{2.17}$$

and the condition for weak fluctuations is $|\theta-\theta_0|\ll\theta_0$ or, equivalently, $|\theta-\theta_0|\ll\sin\theta_0$, imposing

$$|h(x,y)| \ll \gamma \sin^2\theta_0 \ . \tag{2.18}$$

We shall assume that Eq. (2.18) holds in most of our discussion. This automatically eliminates certain interesting features (e.g., the "composite structures" mentioned earlier), but many important nonlinear effects are still present in this limit. [Remark: A further (convenient) simplification to be made here is to require that θ_0 itself be small, $\theta_0 \ll 1$. All calculations become simpler in this limit, without losing much physical content.]

Having defined weak fluctuations for chemical contamination, let us now turn to the case of surface roughness. Here the height of the interface at point (x,y) differs from the ideal value by a correction $u(x,y)$. We assume that the slopes $\varepsilon_x = \partial u/\partial x$, $\varepsilon_y = \partial u/\partial y$ are small. A systematic analysis of the effects to order ε^2 has been carried out by Cox (1983). Here we shall restrict our attention to the lowest significant order (ε). Let us define our axes (x,y) in the plane of average interface, so that the average contact line $\mathcal{L}$ is parallel to x. The "liquid side" is chosen to be the half-plane $y<0$ (Fig. 12). Then the major effect at any point is the rotation of the local surface, along an axis parallel to x, by an angle ε_y. The liquid/vapor interface makes an angle θ_0 with the tilted surface, but makes an angle

$$\theta = \theta_0 + \varepsilon_y = \theta_0 + \partial u/\partial y \tag{2.19}$$

with the average boundary plane (Fig. 12). Comparing Eqs. (2.19) and (2.17), we see that *the surface roughness problem and the chemical contamination problem coincide* (to first order in ε), provided that we set

$$-h(x,y) \leftrightarrow \gamma\theta_0 \partial u/\partial y \tag{2.20}$$

(for $\theta_0 \ll 1$).

The detailed statistical features of the random function $h(xy)$ depend on the particular system under consideration. In what follows, we shall restrict our attention to cases where $h(xy)$ is a random noise function with an amplitude h and a *finite* correlation range ξ. This is probably adequate for many types of chemical contamination and for certain forms of surface roughness (e.g., induced by abrasion). But the assumption may break down for certain special systems. For instance, as pointed out by Huse (private communication), if the solid is a glass, and if it has retained the thermal fluctuations of the surface it had as a melt, the surface $u(xy)$ is "rough" in the particular sense of statistical mechanics, and exhibits some anomalous long-range correlations.

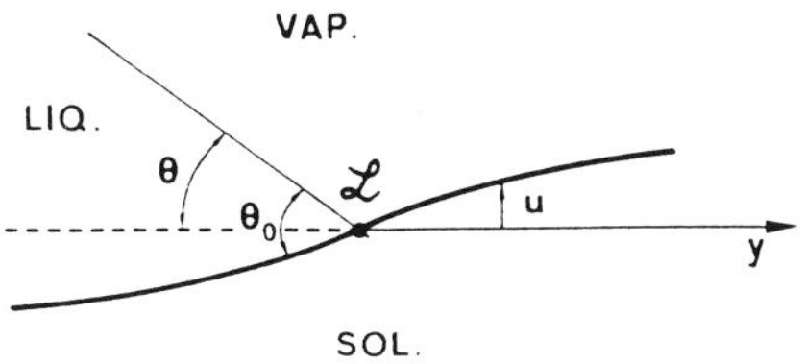

FIG. 12. Effect of a local tilt of the surface on the apparent contact angle θ.

b. The wandering triple line

Having defined the irregularities of the surface through a certain random function $h(x,y)$, we wish to understand how these irregularities react on the shape of the triple line $\mathcal{L}$. This has been attempted independently by Pomeau and Vannimenus (1984) and by Joanny and de Gennes (1984). The two approaches supplement each other, the first being more rigorous and the second giving certain physical insights.

A first step is to study a deformed line $\mathcal{L}$ [specified by a position $y=\eta(x)$ on the average surface] and to construct the elastic energy of the line. At first sight one might think of a line tension $\mathcal{T}$, giving an energy

$$f_{\rm el} = \int \tfrac{1}{2}\mathcal{T}\left[\frac{d\eta}{dx}\right]^2 dx = \tfrac{1}{2}\mathcal{T}\sum_q q^2 |\eta_q|^2 \ , \tag{2.21}$$

where we have introduced the Fourier transform η_q of $\eta(x)$. The form (2.21) however, is wrong, and should be replaced by

$$f_{\rm el} = \tfrac{1}{2}\gamma\theta_0^2 \sum_q |q|\,|\eta_q|^2 \ . \tag{2.22}$$

Physically, the usual $|q|$ dependence for a mode of wavelength $2\pi/q$ expresses the fact that the line distortion perturbs the liquid/vapor interface on a thickness q^{-1}. Integrating a capillary energy (proportional to q^2) over this thickness, we get Eq. (2.22).

Let us now add the inhomogeneities described by $h(x,y)$. They contribute an energy

$$f_i = \int dx \int_{\eta(x)}^{\infty} dy\, h(x,y) \ . \tag{2.23}$$

We may equivalently say that $h = -\delta f_i/\delta\eta(x)$ is the local force f acting on the line $\mathcal{L}$,

$$f(x) = h[x,\eta(x)] \ . \tag{2.24}$$

We must now balance the elastic force [linear in the displacements $\eta(x)$] against the random force (2.24). But the random force is itself a (nonlinear) function of η. This point, emphasized by Pomeau and Vannimenus, makes the discussion quite delicate. Here, we shall use an illuminating presentation, due to Huse (1984), which parallels a classic idea of Imry and Ma (1975), improved later by Grinstein and Ma (1983) for the discussion of random field effects in ferromagnets. We consider a piece of line of macroscopic length l, which is pinned at both ends,

$$\eta(x=0) = \eta(x=l) = 0 \ . \tag{2.25}$$

Let us look for the ground-state energy of the line, assuming that it is characterized by a fluctuation amplitude

$$\eta(x) \sim W \quad (0 < x < l) \ . \tag{2.26}$$

The corresponding elastic energy is derived from Eq. (2.22) with $q \sim l^{-1}$, and is

$$f_{\rm el} \sim l \tfrac{1}{2}\gamma\theta_0^2 l^{-1} W^2 \cong \tfrac{1}{2}\gamma\theta_0^2 W^2 \,. \qquad (2.27)$$

The energy f_i associated with the random force h can be estimated simply in two limits.

(i) If $W<\xi$ the line meets l/ξ uncorrelated inhomogeneities, each with random forces $\pm h$. The overall force is of order $\sqrt{l/\xi}h$, and the energy is

$$f_i \cong -Wh\sqrt{l/\xi} \,. \qquad (2.28)$$

When this is added to Eq. (2.27), we find an optimum displacement

$$W \sim \frac{h}{\gamma\theta_0^2}\sqrt{l/\xi} \,. \qquad (2.29)$$

This law has been quoted by Pomeau and Vannimenus (1984) and Joanny and de Gennes (1984a), but it is restricted to $W<\xi$, or, equivalently,

$$h<\gamma\theta_0^2\sqrt{\xi/l} \,. \qquad (2.30)$$

(ii) If $W>\xi$, the line, when moving from its unperturbed position ($\eta=0$), has swept a ribbon of area Wl, containing Wl/ξ^2 uncorrelated inhomogeneities. The resulting energy is

$$f_i \sim -h\xi^2(Wl/\xi^2)^{1/2} \sim -h\xi\sqrt{Wl} \;, \qquad (2.31)$$

and the optimum W, obtained by minimization of $f_i+f_{\rm el}$, is

$$W \sim \left[\frac{h}{\gamma\theta_0^2}\right]^{2/3} l^{1/3}\xi^{2/3} \,. \qquad (2.32)$$

For most practical purposes, this second case is adequate, and the Huse formula (2.32) should hold. Taking $l=1$ mm,[3] $h=\gamma\theta_0^2$, and $\xi=1$ μm, we get $W=10$ μm.

c. Line pinning

In the preceding section we considered only one (optimal) shape for the contact line $\mathscr{L}$. However, to describe hysteresis, we must compare different shapes. To understand the competition between two shapes, let us first consider a single "defect," following the arguments of Joanny and de Gennes (1984a). The word "defect" means a perturbation $h(x,y)$ localized near a particular point (x_d,y_d) and with finite linear dimensions $\Delta x \sim \Delta y = d$. Typical forms are shown in Fig. 13. The contact line $\mathscr{L}$ may have more than one equilibrium position near such a defect. In certain regimes it can become "anchored" to the defect as shown in Fig. 14. Far from the defect, the line coincides with $y=y_L$. Just on the defect ($x=x_d$), the line is shifted and reaches a certain value of $y=y_m$.

An essential parameter is the total force f_1 exerted by the defect on the line,

[3]This is an upper limit. Beyond that size gravitational energies come into play.

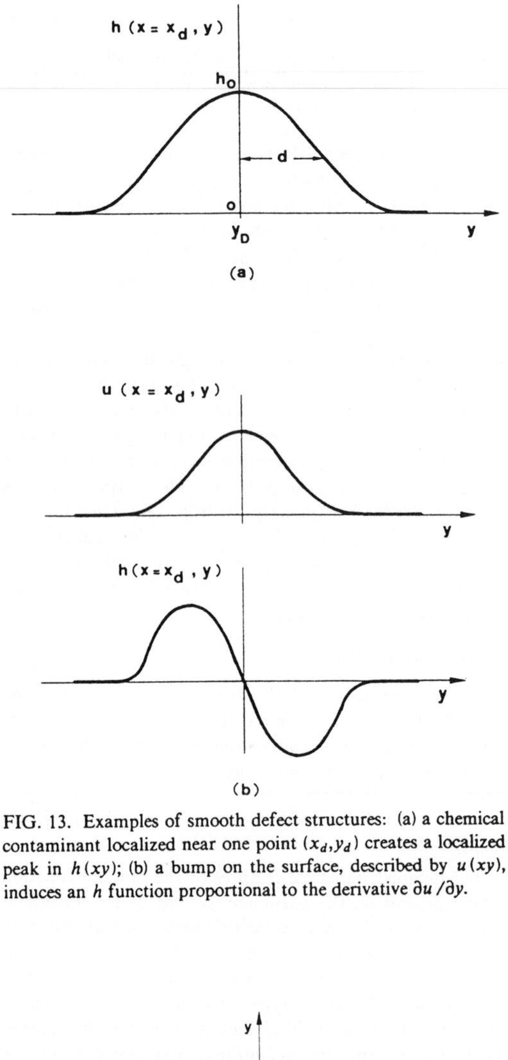

FIG. 13. Examples of smooth defect structures: (a) a chemical contaminant localized near one point (x_d,y_d) creates a localized peak in $h(xy)$; (b) a bump on the surface, described by $u(xy)$, induces an h function proportional to the derivative $\partial u/\partial y$.

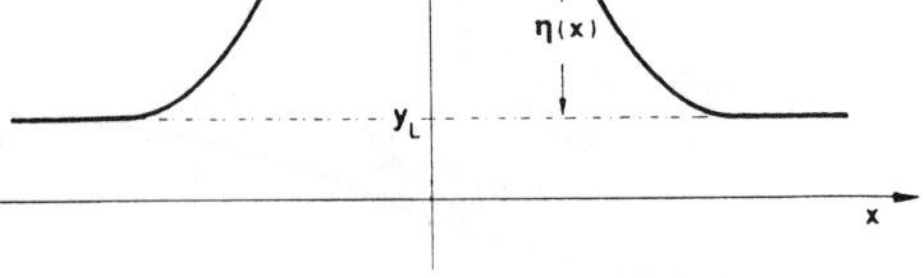

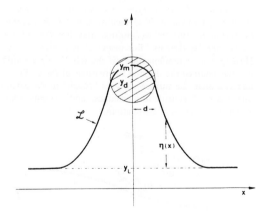

FIG. 14. A contact line $\mathscr{L}$ anchored on a defect. The defect is restricted to a small region (of diameter d), but the line is perturbed much farther out.

$$f_1=\int_{-\infty}^{\infty} dx\, h[x,y_L+\eta(x)] \; . \tag{2.33}$$

The integral (2.33) is dominated by the central region $[x \sim x_d,\ \eta(x) \sim y_m - y_L]$ and will be approximated by the simpler form

$$f_1(y_m)=\int_{-\infty}^{\infty} dx\, h(x_1 y_m) \; . \tag{2.34}$$

For a given defect structure, $f_1(y_m)$ is then a known function of $y_m - y_d$ (Fig. 15). A simple example, to which we shall sometimes refer, is a Gaussian defect

$$h(x,y)=h_0 \exp\left[-\frac{(x-x_d)^2+(y-y_d)^2}{2d^2}\right] . \tag{2.35}$$

In this case the force $f_1(y_m)$ is also Gaussian:

$$f_1(y_m)=(2\pi)^{1/2} h_0 d \exp\left[-\frac{(y_m-y_d)^2}{2d^2}\right] . \tag{2.36}$$

Let us now consider the line tip $(x = x_d, y = y_m)$. The line here is in equilibrium under two forces, the force f_1 defined in Eq. (2.34), and an elastic restoring force, which tends to bring y_m back to the unperturbed line position y_L. This elastic force can be derived from the elastic energy (2.22). It has the simple Hooke form

$$f_{el}=k(y_L-y_m) \; , \tag{2.37}$$

where k may be called the *spring constant of the line* and is given by

$$k=\frac{\pi\gamma\theta_0^2}{\ln(l/d)} \; . \tag{2.38}$$

Here l is a long-distance cutoff (for the single-defect problem, l would be the total length of line available), and d is always the defect size. The nice feature of Eq. (2.38) is that k is nearly independent of all defect properties. The balance of forces then gives the fundamental equation

$$k(y_m-y_L)=f_1(y_m) \; , \tag{2.39}$$

which is solved graphically in Fig. 15. When the strength of the defect [measured by h_0 in Eq. (2.35)] is small, there is only one root y_m for any specified y_L; we have no hysteresis. On the other hand, when the strength h_0 is

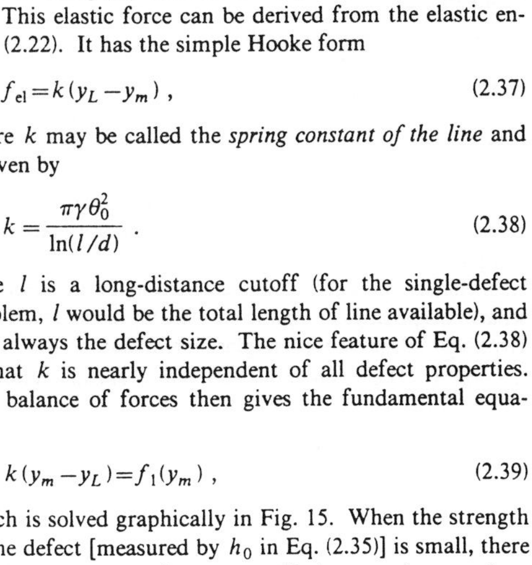

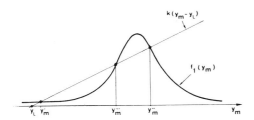

FIG. 15. Equilibrium positions for the anchoring point $(y = y_m)$ of the line on the defect. The position of the line far from the defect is imposed $(y = y_L)$. For a given y_L there may be three equilibrium positions; two of these (y_m', y_m'') are locally stable.

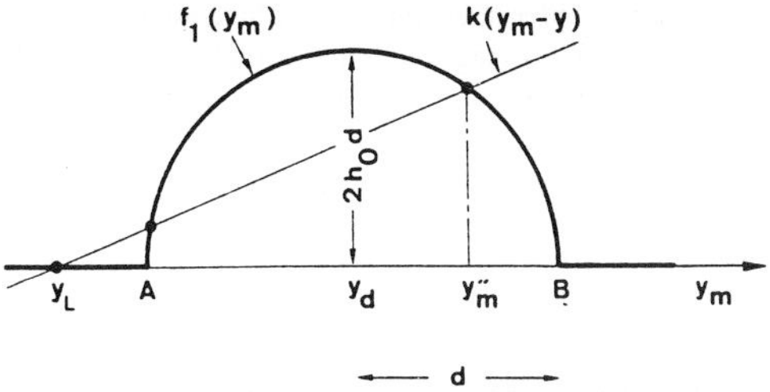

FIG. 16. An example of a "mesa" defect. The function $h(x_g)$ is zero, except in a circle of radius d around the point $(x_d y_d)$, where it is constant $(h = h_0)$. Then the force $f_1(y_m)$ has the aspect shown. Even for very weak h_0 there exist two competing equilibrium positions (at y_m'' and y_L), when y_L is just to the left of point A: hysteresis is always present.

beyond a certain threshold, we can find three roots (for a certain interval of y_L values); then we expect hysteresis.

This brings us to a very important conclusion for "regular" defects, i.e., for cases when $f_1(y_m)$ is a smooth function [with a finite derivative $f_1'(y_m)$]. We see that weak perturbations create strictly no hysteresis: to have a good determination of the thermodynamic angle θ_e, we do not need an ideal surface; we need only a surface with irregularities below a certain threshold.

The case of "mesa" defects (where the function h has step discontinuities) is completely different (Fig. 16). With mesa defects we can have hysteresis even for very small h functions. (Mesa structures can be obtained, for instance, with fatty acids on glass; see Brockway and Jones, 1964.) The mesa case was the only one considered by Pomeau and Vannimenus (1984). For this reason, some of their conclusions on the magnitude of the macroscopic hysteresis are somewhat specialized.

Up to now we have discussed only a single defect. It is not too hard, however, to extend the arguments to a *dilute* system of defects, and to produce detailed formulas for the hysteresis parameters θ_a and θ_r (Joanny and de Gennes, 1984a). The only nontrivial point in this extension is a renormalization of the spring constant k [Eq. (2.38)]: here the cutoff length l becomes the average distance between defects, as seen by the line $\mathscr{L}$. If n is the number of defects/cm^2,

$$l \rightarrow (nd)^{-1} \; .$$

There is good hope of comparing these predictions with experiments performed on *controlled defects,* with sizes in the 10-μm range.[4] They can be prepared by film deposition, using the many techniques currently in use in microelectronics. One can arrange to have either diffuse edges ("regular" defects) or sharp edges ("mesa" defects), and one can purposely locate these defects at random on a

[4]At much smaller sizes (below 1000 Å) the barriers between the equilibrium positions may be overcome by thermal agitation.

given surface. Measurements of contact angles in such systems should provide much more information on the basic laws of hysteresis.

d. Effects of inhomogeneities in situations of complete wetting

The above discussion was concerned with cases where $S<0$ (partial wetting), allowing for finite contact angles. What happens in the opposite case, where S is positive (complete wetting) but varies from point to point? This has been discussed recently (de Gennes, 1984e). It turns out that the final state of a droplet in dry spreading can be quite complex. The regions of high S are wet, while the regions of low S are not. The final thickness $\bar{e}$ of a *macroscopic* droplet adjusts itself so that the dry "islands" are just at their percolation threshold (i.e., at the onset of a continent). These considerations also suggest that many hysteretic effects could take place in dry spreading, but (to the author's knowledge) we do not have any experimental observations in this regime.

D. Wetting films and contact lines

1. Role of long-range forces

All our previous discussion dealt with macroscopic scales (larger than 1 μm). We now want to investigate smaller scales (say from 30 Å to 1 μm), where a continuum picture is still applicable, but where certain long-range forces become relevant, mainly van der Waals (VW) forces for organic liquids, or double-layer forces for water. (Classical reviews on these forces were given long ago by Dzyaloshinskii *et al.*, 1961; Overbeek and Van Silfhout, 1967; and Lyklema, 1967). Let us call $P(\zeta)$ the long-range tail of the energy/cm^2 of a flat liquid film of thickness ζ, lying on the solid. It is related to the celebrated "disjoining pressure" $\Pi(\zeta)$ introduced by Deryagin (1940) and reviewed in Deryagin (1955; see also Deryagin and Churaev, 1976), via $\Pi=-dP/d\zeta$,

$$P(\zeta)=\int_{\zeta}^{\infty} d\zeta'\Pi(\zeta') \,. \tag{2.40}$$

A lucid discussion of our present knowledge of the function $\Pi(\zeta)$, for various liquids and various substrates, and of its application to dynamical studies has been given by Teletzke, Davis, and Scriven (1984). In what follows we extract only some relatively simple limiting cases.

a. van der Waals forces

Here we have two regimes:

$$P(\zeta)=\begin{cases} A/(12\pi\zeta^2) & (\zeta<\lambda\!\!\!^{-};\ \text{nonretarded}) \quad (2.41)\\ B/(3\zeta^3) & (\zeta>\lambda\!\!\!^{-};\ \text{retarded})\,. \quad (2.42)\end{cases}$$

We shall restrict our attention to the case where the constants A and B are positive. This means that VW forces tend to thicken the liquid film (the "agonist" case in the nomenclature of de Gennes, 1983). The crossover length $\lambda\!\!\!^{-}=\lambda/2\pi$ is roughly related to a characteristic ultraviolet absorption wavelength λ of the medium, and is of order 800 Å.

b. Double-layer forces

If the liquid L is water, or more generally an ionic solution (of screening length κ_D^{-1}), the solid will usually create in the water a charge double layer, of thickness κ_D^{-1}. Since water has a high dielectric constant $\varepsilon\cong 80$, while the vapor phase has $\varepsilon\cong 1$, the electric field must (nearly) vanish at the liquid/gas interface. This means that the double layer is repelled by an electrostatic image. The asymptotic form of this repulsion at large ζ was computed long ago by Langmuir (1938) and Frumkin (1938). The "disjoining pressure" Π is exponential,

$$\Pi=C\exp(-2\kappa_D\zeta) \quad (\kappa_D\zeta>1)\,, \tag{2.43}$$

where (for monovalent ions, such as Na^+ and Cl^-) the prefactor is

$$C=64nk_BT\tanh(\psi_0 e/k_BT)\,. \tag{2.44}$$

Here ψ_0 is the potential at the solid surface, n is the number of ions/cm^3 in the water, and e is the unit charge. Comparisons between Eqs. (2.43), (2.44), and ellipsometric data on water films have been carried out by Callaghan and Baldry (1978). They find that Eq. (2.44) does not give a very good fit to their data, and use some more complicated models.

In the regime $\kappa_D\zeta<1$ the experimental data are somewhat surprising. Pashley (1980) concludes that $\Pi(\zeta)\sim\zeta^{-1}$ (for water on glass or silica) in the region $\zeta<400$ Å. There is no obvious theoretical explanation for this slow decrease.

On the other hand, the data reviewed by Israelashvili (1982), on double-layer forces between two closely spaced mica plates, do agree with the standard theory and thus with Eq. (2.44), in the large thickness limit. In any case, to obtain simple predictions on thick wetting films and contact lines, the form (2.44) is a natural starting point.

2. Final spreading equilibrium

It will be much simpler to discuss a one-dimensional problem where the liquid thickness ζ depends only on one coordinate x in the plane of the solid surfaces (Fig. 17). The ingredients in the free energy are listed below:

$$f=f_0+\int_{x_{\min}}^{x_{\max}} dx\left[-S+\frac{\gamma}{2}\left[\frac{d\zeta}{dx}\right]^2+P(\zeta)+G(\zeta)\right]. \tag{2.45}$$

Here $(x_{\min},x_{\max})$ represent the interval covered by liquid. S is defined by Eq. (2.3) for the "dry" case, and by a similar rule (replacing γ_{SO} by γ_{SV}) for the "moist" case. The

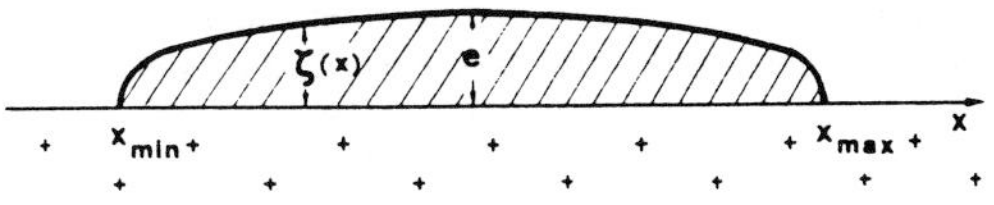

FIG. 17. The final "pancake" in complete dry spreading ($S>0$). Contrary to common belief, the equilibrium state is not a molecular film. Whenever $S \ll \gamma$ the thickness e is larger than the molecular size a.

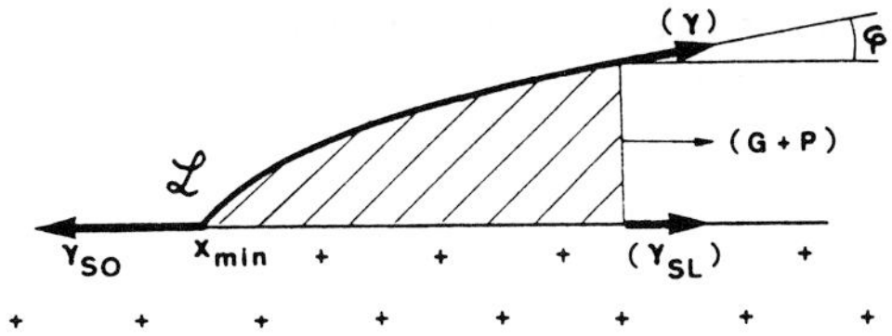

FIG. 18. The balance of forces on a liquid portion near the contact line $\mathscr{L}$.

γ term results from an expansion of the length element $ds^2=dx^2+d\zeta^2$, assuming $d\zeta/dx$ small (this will turn out to be the most interesting regime). The long-range forces show up in $P(\zeta)$. Note that $P(\zeta\rightarrow\infty)=0$. The interfacial energies S and γ appearing in Eq. (2.45) are the thermodynamic values, valid for thick films ($\zeta\rightarrow\infty$). They do incorporate contributions from the long-range forces. Finally $G(\zeta)$ describes gravitational and hydrostatic effects. These effects are often negligible for microscopic studies on contact lines and wetting films, but they show up in some special cases, and we keep them for this reason:

$$G(\zeta)=\rho g(\zeta^2/2+\zeta H) , \tag{2.46}$$

where ρ is the density difference between liquid and gas, g is the gravitational acceleration, and the H term has different meanings depending on the case being considered.

(a) For the "moist" case, the very existence of a liquid/vapor equilibrium requires that, in the experimental cell, a macroscopic reservoir of bulk liquid be present and in contact with the vapor. Then H is the difference in level between the solid plate and the liquid surface in the reservoir.

(b) For the "dry" case, the total volume of the spreading droplet, Ω, is fixed (no exchange), and we may think of the quantity

$$p_0=-\rho g H \tag{2.47}$$

as a Lagrange multiplier associated with this condition. After finding the droplet shape, imposing Ω fixes p_0.

The minimization of Eq. (2.45) with respect to $\zeta(x)$ leads to a standard equilibrium condition,

$$-\gamma\frac{d^2\zeta}{dx^2}+\frac{dP}{d\zeta}+\frac{dG}{d\zeta}=0 , \tag{2.48}$$

which has an important first integral,

$$\frac{\gamma}{2}\left[\frac{d\zeta}{dx}\right]^2=P(\zeta)+G(\zeta)-S . \tag{2.49}$$

The value of the integration constant is best understood from the balance of horizontal forces shown in Fig. 18. Let us consider, for instance, a fluid region extending up to a value of ζ where long-range forces are negligible $[P(\zeta)=0]$. The fluid region experiences on the right the forces $\gamma\cos\varphi+\gamma_{SL}$ (capillary) and $G(\zeta)$ (hydrostatic). To the left we have the force γ_{SO}. Noting that

$$\cos\varphi\cong 1-\tfrac{1}{2}(d\zeta/dx)^2 , \tag{2.50}$$

we recover (2.49). The calculation of all droplet shapes $\zeta(x)$ can now be performed by simple quadratures. We shall apply this scheme in the following section.

3. Partial wetting: microscopic structure of contact lines

a. Organic liquids: effects of van der Waals forces

Let us now consider the case of a nonzero contact angle θ_e ($\theta_e \ll 1$). In this case,

$$S=-\tfrac{1}{2}\gamma\theta_e^2 \tag{2.51}$$

is negative. We insert this value in Eq. (2.49) and discuss the core structure of the contact line $\mathscr{L}$.

In the vicinity of the contact line (Fig. 18) we may ignore gravitational forces as well as the macroscopic pressure difference p_0. Setting $G=0$ in Eq. (2.49), we then have

$$\left[\frac{d\zeta}{dx}\right]^2-\theta_e^2=2\gamma^{-1}P(\zeta) . \tag{2.52}$$

For most practical purposes here, the nonretarded form of $P(\zeta)$ [Eq. (2.44)] is adequate; we then define a molecular length a,

$$a^2=A/(6\pi\gamma) . \tag{2.53}$$

Solving Eq. (2.52) explicitly, we find a hyperbolic form,

$$\zeta^2=(\theta_e x)^2+(a/\theta_e)^2 . \tag{2.54}$$

Results equivalent to Eq. (2.54) for $\zeta \gg a$ were first obtained by Berry (1974). See also Joanny and de Gennes (1984b). Of course Eq. (2.54) is meaningful only in the regions where $d\zeta/dx \ll 1$, or equivalently $\zeta \gg a$. But, for small contact angles θ_e, the hyperbolic profile extends to thicknesses $\zeta\sim a\theta_e^{-1}$, which can be of order 100 Å, and is thus significant.

The limiting case $\theta_e=0$ (complete wetting) should also be mentioned at this point. Integrating (2.52) for this case, we reach a parabolic shape,

$$\zeta^2=2a(x-x_L) \quad (a<\zeta<\lambda\!\!\!^{-}) . \tag{2.55}$$

Beyond $\lambda\!\!\!^{-}$ we must switch to the retarded form for $P(\zeta)$, and we find a slightly different exponent,

$$\zeta^5 = B\gamma^{-1}[\tfrac{5}{3}(x - x_L)]^2 \quad (\zeta > \lambda) . \qquad (2.56)$$

Ultimately, when ζ gets large, the gravitational/hydrostatic terms $G(\zeta)$ come into play. One then returns to macroscopic physics (see Bouasse, 1924).

The profile for the opposite (and probably less frequent) case of *antagonist* VW forces ($A < 0$ and θ_e finite) has been discussed by Wayner (1982).

b. Water solutions: double-layer effects

If we insert Eq. (2.43) into Eq. (2.52), we find the modified wedge structure illustrated in Fig. 3(b): the local contact angle θ_1 near $\mathscr{L}$ tends to be larger than the thermodynamic angle θ_e. If we use Eq. (2.43) down to low values of the thickness ζ, we get

$$\theta_1^2 - \theta_e^2 = \frac{2P(0)}{\gamma} = \frac{2C}{\kappa_D \gamma} \quad (\theta_1 < 1) . \qquad (2.57)$$

Equation (2.57) is only indicative. In actual fact, at small ζ the exponential form (2.43) does not hold, and the VW terms may play a role as well. But, independently of these details, the perturbation around $\zeta = \kappa_D^{-1}$, indicated in Fig. 3(b), is meaningful.

Here again, the special case of complete wetting deserves a special mention. For $\theta_e = 0$, Eq. (2.52) gives a logarithmic profile:

$$\zeta = \kappa_D^{-1} \ln[1 + x\kappa_D^{3/2}(\gamma/C)^{1/2}] . \qquad (2.58)$$

Just as in the case of van der Waals forces, this holds only in the microscopic regime. Ultimately, at large ζ, the gravitational and hydrostatic terms become dominant, and we return to standard macroscopic forms.

4. "Complete" spreading: thickness of the wetting films

a. The "moist" case

A horizontal plate (at height H above reservoir level) is exposed to the vapor and completely covered by a wetting film. Here there does not exist a contact line giving us a balance of forces and specifying the integration constant in Eq. (2.49). But we obtain directly the equilibrium thickness e by minimization of van der Waals and gravitational energies:

$$\Pi(e) = \rho g (e + H) . \qquad (2.59)$$

This is a classical equation, discussed long ago by the Russian school (Deryagin, 1940; Dzyaloshinskii *et al.*, 1961), verified on Rollin films of He_4 (Brewer, 1978), on normal fluids (Deryagin *et al.*, 1978), and somewhat less clearly verified in experiments with consolute mixtures near the free surface (see the review by Moldover and Schmidt, 1983). Basically, when H is macroscopic (~ 1 cm) the thickness is small ($e \sim 300$ Å), and nonretarded interactions often prevail. On the other hand, the limit $H = 0$ does not seem to have been experimentally explored (it could be easily controlled by interference techniques). It would lead to very thick films ($e \sim 10$ μm) and to a rather direct determination of retarded VW forces:

$$e(H=0) = (B/\rho g)^{1/5} . \qquad (2.60)$$

b. Nonvolatile liquids (the "dry" case)

Here a spreading droplet ultimately becomes a flat "pancake," and we want to determine (i) the thickness of the pancake, and (ii) some information on the structure near the contact line (Fig. 17). For our one-dimensional problem the shape is ruled by Eq. (2.49), with $S > 0$. The maximum thickness corresponds to $d\zeta/dx = 0$, and this gives a condition

$$P(e) + G(e) = S . \qquad (2.61)$$

But we have two unknowns: the thickness e and the pressure p_0. To obtain a second condition, for a finite droplet, we should calculate the total volume Ω. Here, however, we shall restrict our attention to the limit of a very wide pancake ($\Omega \to \infty$). As usual, it is then convenient to think of Eq. (2.49) as the conservation of energy for a classical particle of position ζ at time x, with a mass γ, a potential energy $-(P+G)$, and a total energy $-S$.

For strongly negative values of the total energy $-S$, the particle oscillates in a wall near $\zeta = 0$, and the period is finite (i.e., the droplet size is finite). But if $-S$ is adjusted to coincide with the maximum of the potential, then the particle takes an infinite time to reach this maximum (infinitely wide pancake). The maximum then defines the position at long times (the macroscopic thickness of the pancake). This gives a hydrostatic condition identical in form to Eq. (2.59),

$$\Pi(e) = \rho g (H + e) , \qquad (2.62)$$

where $\Pi = -dP/de$ is always the disjoining pressure. Eliminating (H) between Eqs. (2.61) and (2.62), we obtain an explicit relation between the spreading coefficient S and the wetting film thickness:

$$P(e) + e\Pi(e) - \tfrac{1}{2}\rho g e^2 = S . \qquad (2.63)$$

To discuss Eq. (2.63), let us restrict our attention to van der Waals forces, with the simple limiting forms (2.42) or (2.43),

$$P(e) \sim e^{-m} , \qquad (2.64)$$

$$P(e) + e\Pi(e) = (m+1)P(e) . \qquad (2.65)$$

We start from $S \to 0$: here, the thickness e is large, and we must use the retarded form ($m = 3$). This gives

$$S = 4B/3e^3 - \rho g e^2/2 . \qquad (2.66)$$

The wetting film thickness at $S = 0$ is

$$e(S=0) = e^* = (8B/3\rho g)^{1/5} , \qquad (2.67)$$

differing only by a numerical coefficient from the thickness at $\theta_e=0$ in complete equilibrium [Eq. (2.60)]. Typically e^* equals several micrometers. If we now turn to small, positive values of the spreading coefficient S, we see from Eq. (2.66) that a useful dimensionless parameter is

$$u=S(e^*)^3/B\ . \qquad (2.68)$$

Numerically $B/(e^*)^3 \sim 10^{-7}$ ergs/cm^2, while $S\sim 0.1$ erg/cm^2. Thus u is immediately very large. Consequently, the gravitational term in Eq. (2.67) becomes completely negligible, and we may write (2.65) in the form

$$S=e\Pi(e)+P(e)\ . \qquad (2.69)$$

Equation (2.69) is the basic formula for dry spreading with finite S. It could be obtained more directly by ignoring the edge of the "pancake" and writing only the extensive part of the energy. If $\mathscr{A}$ is the pancake area, this is a sum of capillary and VW energies,

$$f=f_0-S\mathscr{A}+\mathscr{A}P(e)\ . \qquad (2.70)$$

This must be minimized with the constraint

$$e\mathscr{A}=\Omega=\text{const}\ , \qquad (2.71)$$

which brings one directly to Eq. (2.69).

For most practical purposes (finite S), the value of e deduced from Eq. (2.69) is of order 100 Å or less. The nonretarded form of VW interactions holds ($m=2$). This gives the final formula (Joanny and de Gennes, 1984b)

$$e=a\left[\frac{3\gamma}{2S}\right]^{1/2} \quad (a<e<\lambda)\ . \qquad (2.72)$$

Equation (2.72) provides the explanation for a very classical fact, observed long ago by Cooper and Nuttal (1915): liquids of large S spread more efficiently than liquids of small S. We do see in Eq. (2.72) that the final spreading film is thinner if S is larger, and this, in turn, implies that the equilibrium surface covered is larger.

Most of us believed that the explanation of the Cooper-Nuttal rule could be based on a different, dynamic effect, namely, more rapid spreading of drops of large S, but this is not true. As we shall see in Sec. IV, the rate of spreading of a droplet is essentially *independent of S*.

Let us close this discussion with two remarks. The first concerns the *edge* of the "pancake" realized after spreading. Returning to the minimization near $\zeta=0$, we can check that the edge is always parabolic, with the structure (2.55).

The second remark concerns the case where the dominant long-range force is due to a double layer. Then, neglecting gravitational effects and inserting (2.43) into (2.69) one arrives at

$$e=\frac{1}{2\kappa_D}\ln\left[\frac{C}{S\kappa_D}\right]\ . \qquad (2.73)$$

Equation (2.73) should hold for $S \ll C\kappa_D^{-1}$. At larger values of S, the discussion should include both VW and double-layer forces, and becomes more complex.

c. *Complete wetting: vertical wall*

The macroscopic analysis for a wetting liquid ($\theta_e=0$) near a *vertical* wall predicts a certain curved profile, with a contact line $\mathscr{L}$ at a level h_1 above the bulk liquid surface,

$$h_1=\sqrt{2}\kappa^{-1} \qquad (2.74)$$

(see Bouasse, 1924). Here κ^{-1} is a capillary length (not to be confused with the Debye screening length κ_D^{-1}) of order of magnitude 1 mm. We have

$$\kappa^2=g\rho\gamma\ . \qquad (2.75)$$

How is this modified when we switch on the VW forces? For $S=0$ exactly, the picture is essentially unaltered. For $S>0$, a film climbs up the wall. If x is now the altitude above the bulk liquid surface, and $\zeta(x)$ the film thickness, the equilibrium condition is (in the nonretarded regime)

$$\frac{d^2\zeta}{dx^2}+\frac{a^2}{\zeta^3}-\kappa^2x=0\ . \qquad (2.76)$$

The standard (loose) discussion of Eq. (2.76) separates out two regimes.

(i) *The film regime,* where the curvature term $d^2\zeta/dx^2$ is negligible, leaving an exact balance between VW and gravitational pressures:

$$\zeta=\zeta_1(x)=x^{-1/3}\kappa^{-2/3}a^{2/3}\ . \qquad (2.77)$$

[It has been realized recently (Joanny and de Gennes, 1984c) that the profile (2.77) does not hold up to arbitrarily large altitudes x. Returning to the full equation (2.76), one can show that the film is *truncated* at a certain $x=x_m$, such that $\zeta_1(x_m)=e(S)$, where $e(S)$ is the minimal thickness defined in Eq. (2.72).]

(ii) *The macroscopic regime,* where the VW forces are negligible.

The crossover between (i) and (ii) is nontrivial; it was discussed long ago by Deryagin (1940),[5] more recently by Renk, Wayner, and Homsy (1978), Adamson and Zebib (1980), and by Telo de Gama (unpublished), Legait and de Gennes (1984), and Evans and Marini (1985) in connection with experiments by Moldover and Gammon (1983) who were studying capillary rise between *two* plates. Moldover and Gammon had proposed that the effective interplate distance (to compute the capillary rise) be reduced by twice the thickness of a single film, $2\zeta_1(x_l)$, taken at the level of the nominal contact line $x=x_l$. Actually the presence of a second plate thickens the films, and the correction should be $3\zeta_1(x_l)$ rather than $2\zeta_1$ (in the

[5] I am indebted to Professor R. Evans for the Deryagin reference.

nonretarded regime). Similarly, for a single plate, the thickness of the film in the crossover region ($x_l \sim h_1$) is expected to be

$$\zeta \sim \zeta_1(h_1) \sim \kappa^{-1/3} a^{2/3} . \quad (2.78)$$

Finally let us consider the case of negligible van der Waals forces and dominant double-layer forces. Equilibrating gravitational pressures and double-layer pressures, we get a thickness

$$\zeta(x) = \frac{1}{2\kappa_D} \ln \left[\frac{B}{\rho g x} \right] , \quad (2.79)$$

and we find a film extending up to a finite height,

$$x_m = \frac{B}{\rho g} . \quad (2.80)$$

Equation (2.79) is limited by the restrictions mentioned after Eq. (2.58): above $x = x_m$ a residual film may be present, because of short-range disjoining pressures, but below $x = x_m$ the logarithmic profile should be visible.

III. THE WETTING TRANSITION

A. Experiments on related systems

1. Scope

A liquid/vapor interface L/V, in the vicinity of a solid S, may exhibit either a finite equilibrium contact angle θ_e (partial wetting) or a strictly vanishing contact angle (complete wetting). There may exist a particular temperature T_w at which we switch from one regime to the other. This is called the wetting transition temperature.

Unfortunately, the $L/V/S$ systems are usually not convenient for studies of this wetting transition. To change significantly the interfacial energies γ_{ij}, we would have to scan a rather broad temperature range. To maintain the L/V equilibrium then requires that one work at high pressures. Because of these practical difficulties, all current studies on T_w have been carried out with the other three-phase equilibria, where temperature variations (at atmospheric pressure) are feasible and have more spectacular effects on the interfacial energies. The two main examples are solid/liquid A/liquid B, free surface/liquid A/liquid B, where A and B are two coexisting phases of a binary mixture with a certain critical consolute temperature T_c. It turns out that, for such a case, the interfacial energies vary considerably when we consider the broad vicinity of T_c (typically a 30° interval). This makes the experimentation much simpler.

Thus, in this section, we shall broaden our subject and consider a variety of three-phase equilibria. The main emphasis, however, will be on solid/fluid/fluid systems. It is reasonable to assume that all these systems are rather similar (as far as the *static* properties are concerned). Thus the liquid/vapor wetting transitions, when observed, will probably follow the laws described here.

2. Wetting films

Numerous examples of complete wetting are found with solid/liquid/vapor systems. But the existence of wetting films, with $\theta_e = 0$, in other types of three-phase equilibria has been established only during the last decade.

An important, early experiment was carried out by Heady and Cahn (1972). Here the solid S is replaced by a vapor phase. The analogs of L and V are two liquids, made of a hydrocarbon (methyl cyclohexane) and of the fluorinated analog of this hydrocarbon. The fluorocarbon (hydrocarbon) system has a consolute point T_c. Below the critical temperature T_c we can have coexistence of two equilibrium phases; one of these is rich in fluorocarbon and will be called L; the other is rich in hydrocarbon and will be called V.

The original aim of the Heady-Cahn experiment was to study a quench into the two-phase region, followed by nucleation of L into V. They found, however, that, in the vicinity of the free surface, nucleation barriers could never be observed: droplets of L immediately began to drip from the surface. They concluded that, in the range of temperatures studied, a wetting film of L was always present near the surface S.

Another early observation of wetting films came from metallurgy (Zabel *et al.*, 1981). The L/V system here is a single crystal of niobium containing a significant fraction of dissolved hydrogen. The analog of S is again the free surface. In a certain temperature range the Nb/H system shows a two-phase equilibrium ($\alpha \leftrightarrow \alpha'$). Both phases are cubic, but they differ in their hydrogen content. The lattice spacing is swollen by the presence of H, and the nature of the phase present near the surface can thus be detected by x-ray reflections. The conclusion is that the α phase wets the interface, the thickness of the α film being of order one micron.

3. Wetting transitions with consolute pairs

A relatively simple measurement of θ_e was carried out by Pohl and Goldburg (1982) on a system of two liquid phases (A,B) (lutidine water mixtures) against common glass (S). The technique is based on capillary rise (Fig. 4) and allows for a plot of $\cos\theta_e$ as a function of temperature. Below a certain temperature $T = T_w$, $\cos\theta_e = 1$, while above $T = T_w$, $\cos\theta_e$ decreases (with a nearly constant slope).

Another important measurement was carried out by Moldover and Schmidt (1983) with S=free surface, A,B=alcohol + fluorocarbon mixtures. Macroscopic measurements of the contact angle indicated a wetting transition at $T_w = 311$ K, and complete wetting (by the fluororich phase) in the interval between T_w and $T_c = 363$ K. Ellipsometric measurements of the thickness e of the fluorocarbon film show a finite jump of $e(T)$ at $T = T_w$. This transition is clearly of first order.

To summarize, for these $S/A/B$ systems we find complete wetting in a finite temperature interval (T_w, T_c) near the critical point, and partial wetting far from T_c. There are more complicated cases. In particular, the pair cyclohexane-methanol, near the free surface S, shows a complex sequence of transitions, which occurs only in the presence of dilute contaminants (water, acetone). We shall come back to these impurity effects at the end of Sec. III.B. But the two examples (Pohl and Goldburg, 1982; Moldover and Schmidt, 1983) above are probably typical of a generic (impurity-free) situation.

A final remark: wetting transitions are also observed with solid films evaporated on a solid surface (for instance, organic solids on graphite, or oxygen on graphite). But the situation here is more complex for many reasons. For crystalline, epitaxial films, the discrete nature of the solid layers introduces a wealth of new transitions, and the elastic distortion fields induced by the substrate complicate the energy balance. For a recent discussion of these "solid on solid" problems see Gittes and Schick (1984).

B. Theory

1. The Cahn model

We follow first the simple and illuminating arguments of Cahn (1977) phrased in language adequate for a solid/liquid/vapor system.

(a) The first simplification is to describe the liquid/solid interface by a *continuum theory,* where the liquid number density $\rho(z)$ varies smoothly as a function of the distance z from the solid surface (see Fig. 21 below). This will be most adequate if we are dealing with temperatures T that are not too far from the critical point T_c. The hope is that most variations of $\rho(z)$ will take place over distances comparable to the correlation length ξ, and this ξ is larger than the intermolecular distance (a) in the liquid, when $T \sim T_c$.

(b) A second, important assumption is that the forces between solid and liquid are of *short range* ($\sim a$), and can, in fact, be described simply by adding a special energy $\gamma_c(\rho_s)$ at the solid surface. Here $\rho_s = \rho(z=0)$ is the liquid density "at the surface" and γ_c is a certain functional,

$$\gamma_c = \gamma_0 - \gamma_1 \rho_s + \tfrac{1}{2}\gamma_2 \rho_s^2 + \cdots . \qquad (3.1)$$

The γ_1 term (favoring large ρ_s) describes an attraction of the liquid by the solid. The γ_2 term represents a certain reduction of the liquid/liquid attractive interactions near the surface: a liquid molecule lying directly on the solid does not benefit from the same high number of liquid neighbors that it would have in the bulk. The parameters γ_1 and γ_2 describe the essential features at the interface. However, the Cahn approach is slightly more general than Eq. (3.1): any form of $\gamma_c(\rho_s)$ is acceptable. We may say that γ_c is the contribution to the solid/fluid interfacial energy which comes from direct contact. This is not all the interfacial energy. Another contribution γ_d will come from the distortions in the profile $\rho(z)$.

(c) A final (less important) approximation amounts to treating the fluid statistics by a mean-field theory. The specific form chosen for the free energy γ_d is a classical "gradient square" functional,

$$\gamma_d = \int dz \left[\tfrac{1}{2} L \left[\frac{d\rho}{dz} \right]^2 + W(\rho) \right] , \qquad (3.2)$$

$$W(\rho) = F(\rho) - \rho\mu - P . \qquad (3.3)$$

Here F is the free-energy density of the bulk liquid, μ its chemical potential, P its pressure. [For a complete justification of Eqs. (3.2) and (3.3) see Rowlinson and Widom, 1982.] We shall assume that μ and P correspond to the exact coexistence of liquid and vapor. Then $W(\rho)$ has two minima of equal height ($W=0$) for the two equilibrium densities $\rho=\rho_l$ (liquid) and $\rho=\rho_v$ (vapor). The general aspect of $W(\rho)$ is shown in Fig. 19. For simplicity we take L independent of ρ.

2. Determination of the surface density

To construct the density profile in the liquid $\rho(z)$ we optimize (3.2) and obtain

$$-L \frac{d^2\rho}{dz^2} + \frac{dW(\rho)}{d\rho} = 0 , \qquad (3.4)$$

from which we construct a first integral

$$\tfrac{1}{2} L \left[\frac{d\rho}{dz} \right]^2 = W(\rho) . \qquad (3.5)$$

There is no integration constant in Eq. (3.5). If we consider a point far in the bulk, where $\rho=\rho_b$ (ρ_b being either ρ_l or ρ_v), we must have $d\rho/dz=0$ and $W(\rho_b)=0$ as explained in Fig. 19. The simple form of Eq. (3.5) allows for a direct calculation of the distortion energy γ_d [Eq. (3.2)],

$$\gamma_d(\rho_s \rho_b) = \int_{\rho_b}^{\rho_s} L \frac{d\rho}{dz} d\rho = \int_{\rho_b}^{\rho_s} [2LW(\rho)]^{1/2} d\rho . \qquad (3.6)$$

The last step is to determine the surface density ρ_s by minimization of the total energy $\gamma_{\text{tot}} = \gamma_d + \gamma_c(\rho_s)$. The resulting condition is

$$-\gamma_c'(\rho_s) = [2LW(\rho_s)]^{1/2} \qquad (3.7)$$

[where $\gamma_c'(\rho) = d\gamma_c/d\rho$]. This leads to the graphical con-

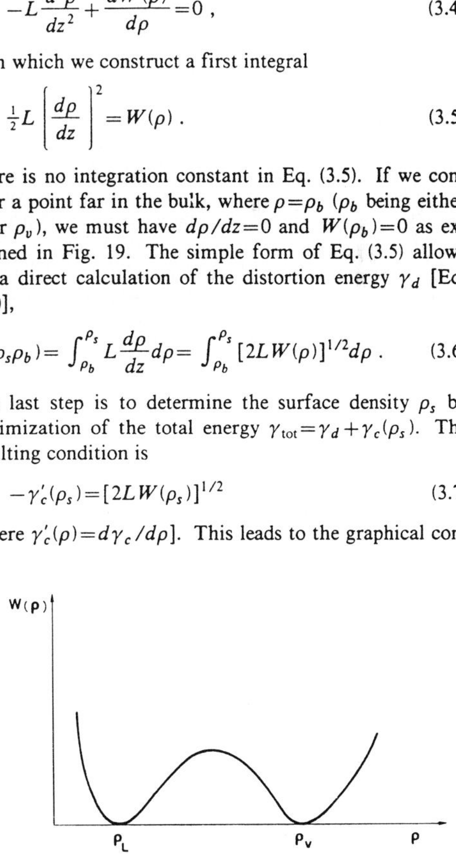

FIG. 19. The "effective free energy" $W(\rho)$ as a function of the density ρ.

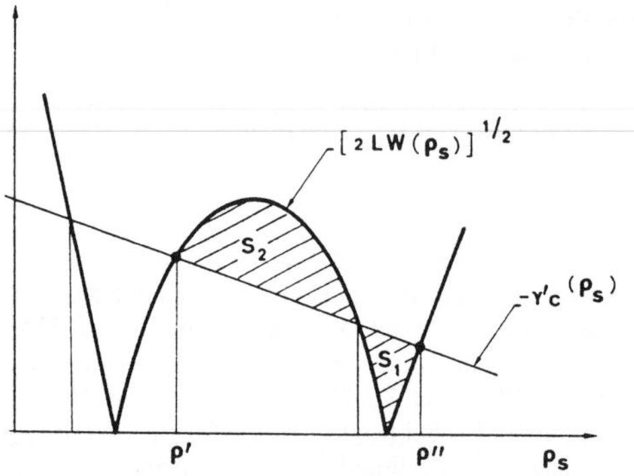

FIG. 20. The Cahn construction determining the density at the surface ρ_s. In the example displayed we have two locally stable roots (ρ',ρ''). The other two roots are unstable.

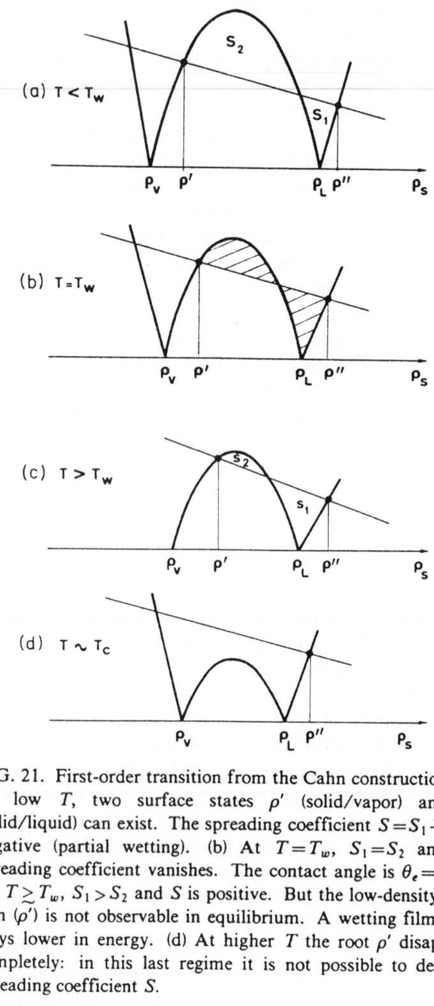

FIG. 21. First-order transition from the Cahn construction. (a) At low T, two surface states ρ' (solid/vapor) and ρ'' (solid/liquid) can exist. The spreading coefficient $S=S_1-S_2$ is negative (partial wetting). (b) At $T=T_w$, $S_1=S_2$ and the spreading coefficient vanishes. The contact angle is $\theta_e=0$. (c) At $T \gtrsim T_w$, $S_1>S_2$ and S is positive. But the low-density solution (ρ') is not observable in equilibrium. A wetting film is always lower in energy. (d) At higher T the root ρ' disappears completely: in this last regime it is not possible to define a spreading coefficient S.

struction of Fig. 20. Here, for simplicity, we have chosen the specific form of $\gamma_c(\rho_s)$ proposed in Eq. (3.1), and this gives a linear plot for $-\gamma_c'(\rho_s)=\gamma_1-\gamma_2\rho_s$.

3. Two types of wetting transitions

a. *First-order transitions*

If the slope (γ_2) is small, the condition (3.7) may give four roots for ρ_s. Two of these are locally stable, while the others correspond to a maximum of the free energy and are unstable. In this regime we find a competition between a state of low ρ_s $(\rho_s=\rho')$ describing a nearly "dry" solid in contact with the vapor $(\rho_b=\rho_v)$ and a state of high ρ_s $(\rho_s=\rho''>\rho_l)$ describing a wet solid in contact with the liquid $(\rho_b=\rho_l)$. The energies of these two states are (per cm^2 of solid surface)

$$\gamma_{sv}=\gamma_d(\rho_v,\rho')+\gamma_c(\rho') ,$$

$$\gamma_{sl}=\gamma_d(\rho_l,\rho'')+\gamma_c(\rho'') . \tag{3.8}$$

We must also remember that the liquid/vapor interfacial energy γ can be derived from the same analysis:

$$\gamma=\gamma_d(\rho_v,\rho_l) . \tag{3.9}$$

Let us discuss the "spreading coefficient" defined in Sec. II:

$$\gamma_{sv}-\gamma_{sl}-\gamma=S . \tag{3.10}$$

Using Eqs. (3.6), (3.8), and (3.9), one can check that S has a simple graphical interpretation in Fig. 20: $S=S_1-S_2$ is the difference of the two shaded areas.

Let us now vary the temperature, as indicated in Fig. 21.

(i) At $T \ll T_c$ the difference $\rho_l-\rho_v$ is large, and S_2 is larger than S_1. This gives $S<0$, i.e., $\cos\theta_e$ is finite (partial wetting).

(ii) If we raise T, the difference S_1-S_2 decreases and vanishes at a special temperature $T=T_w$. Here $S=0$ and $\theta_e=0$.

(iii) At temperatures $T>T_w$, $S_2<S_1$, and S is positive. As we know from Sec. I, this regime is unobservable in thermal equilibrium. Instead of building up a liquid/vapor interface with $\rho_s=\rho'$, the system prefers to achieve it in two steps, through a macroscopic film of L wetting the surface and giving a total surface energy $\gamma_{sl}+\gamma$. Thus, here, we keep $\theta_e=0$ (complete wetting). Ultimately, at high temperatures $(T\sim T_c)$, only one stable root is left, corresponding to a solid/liquid interface.

In this scenario the transition at T_w involves a jump from one energy minimum $(\rho_s=\rho')$ to a distinct minimum $(\rho_s=\rho'')$ and is clearly of *first order.* The plot of $\cos\theta_e$ versus temperature in the partial wetting regime has a finite slope, and intersects $\cos\theta_e=1$ at $T=T_w$.

b. Second-order transitions

If the slope (γ_2) of $-\gamma_c'(\rho_s)$ is large, at all temperatures T we find only *one root* ρ_s from the construction illustrated in Fig. 22.

(i) At low temperatures $\rho_s < \rho_l$, and we can construct two density profiles corresponding to two physical situations: one profile where $\rho(z)$ decreases from ρ_s to ρ_v (describing S/V) and one profile where $\rho(z)$ increases from ρ_s to ρ_l (describing S/L). Again a discussion of areas allows one to compare the surface energies. One finds a negative spreading coefficient, $S<0$ corresponding to partial wetting.

(ii) At high temperatures ($T > T_w$), the surface density ρ_s is higher than ρ_l; there is only one profile associated with ρ_s, where $\rho(z)$ decreases from ρ_s to ρ_l (S/L interface). The S/V interface must then involve a macroscopic film of L, and we have complete wetting. Clearly this scenario corresponds to a continuous ("second-order") transition. At $T=T_w$, $\rho_s=\rho_l$ exactly.

4. Special features of second-order transitions

Second-order wetting transitions are probably rather rare, for reasons to be explained below. But they have stimulated a considerable theoretical effort and deserve a few comments.

Let us consider a liquid film, of thickness ζ, covering the solid, and described by the density profile $\rho(z)$ shown in Fig. 23: most of the density drop takes place in a thickness ξ near the nominal interface (around $z=\zeta$). Far from this interface ($z \ll \zeta$) the difference between $\rho(z)$ and ρ_l is exponentially small. At the surface

$$\rho(z=0)=\rho_s=\rho_l-\varepsilon\ ,$$
$$\varepsilon \simeq (\rho_l-\rho_v)\exp-(\zeta/\xi)\ . \tag{3.11}$$

[Equation (3.11) can be derived from a complete integration of (3.5), taking for $W(\rho)$ the simplest polynomial form compatible with Fig. 19.] Near T_w (for T slightly lower than T_w) ε is small and positive, and the energy of the S/V interface may be expanded in powers of ε,

$$\gamma_{sv}=\gamma_{sv}(\varepsilon=0)-\alpha(T)\varepsilon+\tfrac{1}{2}\gamma_2\varepsilon^2+\cdots\ . \tag{3.12}$$

Here $\alpha(T)$ can be computed explicitly from Eq. (3.7) and is proportional to T_w-T ($\alpha>0$ for $T_w>T$). Equation (3.12) can be translated into a plot of film energy versus film thickness, since ε and ζ are related by (3.11):

$$\gamma_{sv}=\gamma_{sv}(T=T_w)-\alpha(\rho_l-\rho_v)\exp(-\zeta/\xi)$$
$$+\tfrac{1}{2}\gamma_2(\rho_l-\rho_v)^2\exp(-2\zeta/\xi)+\cdots\ . \tag{3.13}$$

For $T<T_w$ we have a weak attractive tail at large ζ and a finite repulsive tail at smaller ζ (Fig. 24). There is an optimal film thickness which corresponds to $\varepsilon=\alpha/\gamma_2$ or

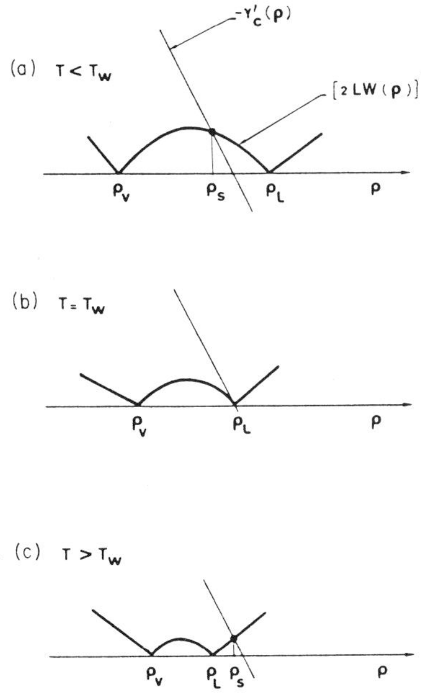

FIG. 22. Second-order transitions from the Cahn construction.

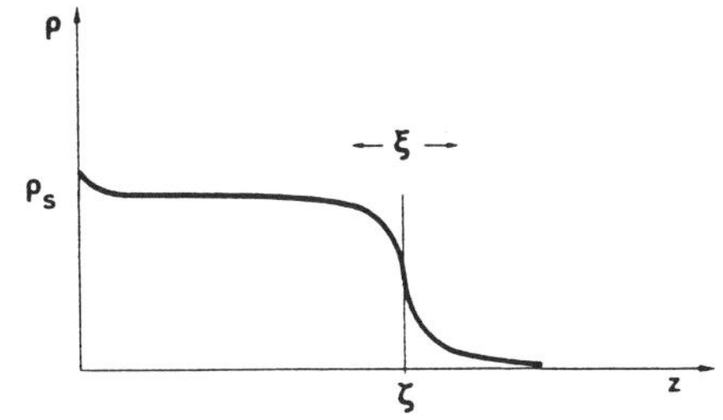

FIG. 23. Density profile for a thick film (thickness ζ larger than the interface width ξ). Depending on the coefficients in Eq. (3.1), the value at the surface (ρ_s) may be either larger or smaller than the bulk value in the liquid (ρ_L).

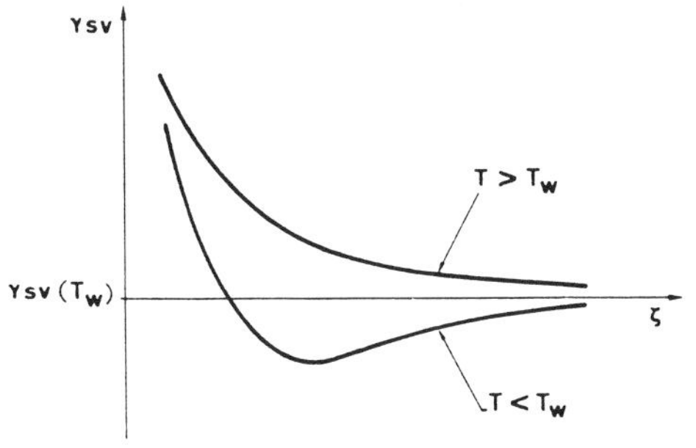

FIG. 24. Energy vs thickness for a film, in the vicinity of a second-order transition (at $T=T_w$). This type of plot (with exponential decay at large distance) would occur in the absence of long-range forces.

$$\zeta=\zeta^*=\xi\ln\frac{\gamma_2}{\alpha}\ . \tag{3.14}$$

This approach gives a very pictorial description of the second-order wetting transition (in a mean-field picture): there is a wetting film even for $T<T_w$, but its thickness diverges logarithmically at $T=T_w$. For $T>T_w$ the energy (3.13) becomes repulsive at all distances, and the film is macroscopic.

As usual, this mean-field picture is complicated by *fluctuation effects*. The L/V interface may undulate

$$\zeta(xy)=\zeta^*+u(xy)\ .$$

To order u^2 the fluctuation energy is

$$f=\int dx\,dy\left\{\gamma_{sv}(\zeta^*)+\tfrac{1}{2}\gamma''_{sv}(\zeta^*)u^2+\tfrac{1}{2}\gamma\left[\left[\frac{\partial u}{\partial x}\right]^2+\left[\frac{\partial u}{\partial y}\right]^2\right]\right\}, \tag{3.15}$$

where $\gamma''(\zeta^*)$ can be derived from Eq. (3.13) and turns out to be proportional to α^2. The form (3.15) leads to a correlation length ξ_u for the fluctuations:

$$\xi_u=\left[\frac{\gamma}{\gamma''(\zeta^*)}\right]^{1/2}\sim(T_w-T)^{-1}\ . \tag{3.16}$$

However, the expansion (3.15) of the energy to order u^2 is inadequate. If one computes from Eqs. (3.15) and (3.16) a mean-square average $\langle u^2\rangle$, one finds $u\sim\zeta^*$. Thus the special events where the fluctuating L/V interface touches the solid surface must be taken into consideration. This has been carried out in detail (Brézin, Halperin, and Leibler, 1983a,1983b) and gives rise to a very unusual critical exponent, whose value depends continuously on the parameter $kT_w/\xi^2\gamma$.

We shall not describe these delicate fluctuation effects in any detail, because they may often be short-circuited by long-range forces, as explained below.

5. Role of long-range forces

a. Simple estimates of wetting film energies

Let us assume that the short-range forces [described by $\gamma_c(\rho_s)$] lead by themselves to a second-order transition T_{w0} associated with the film energy (3.13). Then let us switch on an agonist van der Waals interaction, unretarded [Eq. (2.45)], with a positive Hamaker constant A. Then the plot of film energy versus thickness is still monotonous and repulsive for $T>T_{w0}$, and we have complete wetting in this temperature region. But for $T<T_{w0}$ the plot is deeply modified (Fig. 25). At large thicknesses ζ, the VW term dominates over the exponentials, and the energy is repulsive. At shorter distances the attractive exponential $-\alpha\exp(-\zeta/\xi)$ may create a trough when α is large enough. Finally, when α reaches a certain value α_1

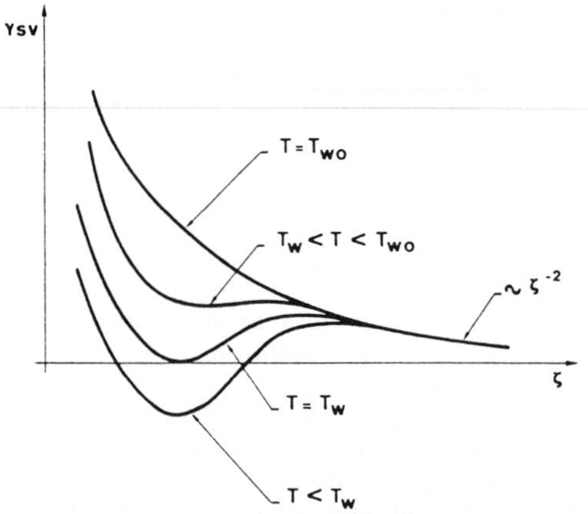

FIG. 25. Energy vs thickness with incorporation of long-range van der Waals forces. These forces are assumed to be "agonist" (i.e., they tend to thicken the film). They dominate at long distances.

(temperature T_1), the trough may give an energy equal to that of a macroscopic film $\gamma_{sv}(\zeta=\infty)$. At this point there is a certain optimal thickness ζ_1, and the system jumps by a first-order transition from $\zeta=\infty$ to $\zeta=\zeta_1$.

A similar description can be given for water solutions against an ionizable surface. Whenever the Debye screening length κ_D^{-1} is larger than the L/V interface thickness ξ, the Langmuir repulsive term [obtained from Eq. (2.57)] dominates the behavior of the film energy at large distances, and the energy plot is qualitatively similar to the preceding one. The transition point is shifted from T_{w0} to a lower value T_1, and the transition becomes first order.

Finally let us mention briefly some other situations.

(i) An "antagonist" VW force, corresponding to a negative Hamaker constant A. Here the long-range energy tends to shrink the wetting film, and the L/V interface is thus stuck near the solid wall (partial wetting). A transition can, however, occur (at $T<T_{w0}$) between two different minima, corresponding, respectively, to a thick film and a thin film.[6]

(ii) When the density of the V phase varies significantly with temperature, it may happen that we switch from the agonist to the antagonist case at one particular temperature $T=T_S$ (Lipowsky and Kroll, 1984; Dietrich and Schick, 1984). If, in the absence of long-range forces, there were a second-order wetting transition at $T=T_{w0}$, then, upon switching on VW forces, one would still expect a second-order transition, shifted to $T=T_S$ (provided that $T_S>T_{w0}$). This prediction can be understood by means of diagrams similar to Fig. 25.

Using naive VW calculations, one expects the Hamaker

[6]This was pointed out to me by M. Schick.

constant A to be a quadratic function of the polarizabilities α_i,

$$A \sim (\alpha_S - \alpha_L)(\alpha_L - \alpha_V) ,$$

and the switching temperature T_S corresponds to

$$\alpha_L(T_S) = \alpha_S .$$

Thus there will exist a second-order wetting transition at $T = T_S$ provided that a number of conditions are satisfied:

(1) The "bare" transition T_{w0} is of second order.

(2) There exists a T_S: the polarizability of the dense liquid must be higher than α_S, while the liquid polarizability at the critical point must be smaller than α_S.

(3) T_S is above T_{w0}.

(4) Although the preceding discussion involved only the leading term in the long-range potentials [$V(\zeta) \sim \zeta^{-2}$], we must make sure that at the temperature T_S, where this leading term vanishes, the next term ($\sim \zeta^{-3}$) does not upset the order of the transition. This has been analyzed (Dietrich and Schick, 1984; Ebner, Saam, and Sen, 1984; Kroll and Meister, 1984), and imposes another condition.

From an experimental point of view, the conditions (1)–(4) are not easy to satisfy simultaneously. But there remains some hope of finding a second-order transition by an intelligent tuning of parameters.

b. Limitations and improvements

The discussion on the effects of long-range forces based on Fig. 25 has the merit of being simple, and has been used by a number of authors (Pandit *et al.*, 1982; Hauge and Schick, 1983; Tarazona and Evans, 1983; Tarazona, Telo da Gama, and Evans, 1983; de Gennes, 1983; Teletzke *et al.*, 1983; Privman, 1984). The discussion, however, has certain limitations which should be kept in mind. (See the review by Sullivan and Telo da Gama, 1985.)

(1) It holds only when the L/V interface thickness ξ is not too large, so that the film has a well defined thickness ($\zeta \gg \xi$). Thus the close vicinity of the L/V critical point would require a special study.

(2) Fluctuations of the L/V interface could modify the effective film energy in a profound way. This possibility was mentioned in particular by Pandit *et al.* (1982), and was studied in detail by Nightingale *et al.* (1983), using a "solid on solid" model, but in a temperature regime (above roughening) that made it adequate for the L/V interface. They did not find any dramatic effects of the fluctuations. We can understand this as follows:[7] the fluctuating interface is restricted in its motions by the presence of the solid wall. This entropy reduction creates an effective potential $V_f(\zeta)$ (where f stands for fluctuation), and $V_f(\zeta)$ is a rapidly decreasing function of ζ. In the absence of long-range forces, $V_f(\zeta)$ is an essential component of the theory for second-order transitions. But in the presence of long-range forces, which decay much more slowly with distance than $V_f(\zeta)$, fluctuation effects become negligible. Thus, even if one finds a second-order transition at some $T = T_S$, as explained above, this transition is expected to be of the mean-field type.

(3) It is incorrect to assume exponential tails for the L/V interface [Eq. (3.11)] when the interactions between molecules in the fluid become long range. The $(\nabla\rho)^2$ expansion of Eq. (3.2) then becomes invalid. This was noticed long ago by Christiansen (see Plesner and Platz, 1968) and more recently by de Gennes (1981) and Rowlinson, Barker, and Henderson (1981). What happens when the $(\nabla\rho)^2$ expansion breaks down? A first attempt to answer this question is due to Sullivan (1979,1981). He chose a pair interaction,

$$U(r_{ij}) = -\frac{\alpha}{4\pi r_{ij}} e^{-\gamma_1 r_{ij}} , \tag{3.17}$$

and a solid/molecule interaction with the *same* range γ_1^{-1},

$$V(z_i) = -\varepsilon_w e^{-\gamma_1 z_i} . \tag{3.18}$$

Sullivan showed that this special choice allows for a simple solution of the complete (integral) mean-field equations for $\rho(z)$, without assuming a $(\nabla\rho)^2$ expansion. He reached some interesting physicochemical conclusions. High ε_w led to plots where $\theta_e(T)$ was decreasing (as it is in the Cahn theory), while low ε_w led to θ_e increasing with T. His wetting transitions were of second order, but later work of Teletzke *et al.* (1983) and Benner *et al.* (1984), with the same model, ultimately led to first-order transitions. (The more complex situation with different γ_1 parameters in U and V was studied by Hauge and Schick, 1983.) In any case, the exponential interactions (3.17) and (3.18) are not adequate to study the order of the transition. They lead to a film energy $p(z)$ which is also exponential, and thus more abruptly decreasing than the expected VW term ($\sim \zeta^{-2}$).

It is more instructive to keep realistic VW interactions; this was done in numerical calculations by Tarazona and Evans (1983). Using the standard 6-12 potential, they found wetting transitions that were constantly of first order. This, in our view, does not mean that second-order transitions are entirely ruled out. But the conditions for a second-order transition in the presence of long-range forces, as deduced from the analysis of Dietrich and Schick (1984), are so stringent that they were not met in any of the cases considered in recent numerical calculations.

Tarazona *et al.* (1983) also extended these calculations and considered the equilibrium thickness of wetting films. Their results agree with the macroscopic predictions of Fig. 25.

An interesting extension of these ideas has been worked out very recently by Evans and Tarazona (1984). Instead of a single plate, exposed to a liquid or a gas, they consid-

---

[7] I am indebted to M. Schick for this presentation.

er *two parallel plates,* separated by a gap of width H_g. Using the Sullivan trick, they compute the effective interfacial energies $2\gamma_{SL}(T,H_g)$ and $2\gamma_{SV}(T,H_g)$ as a function of H_g. For a given H_g there may exist a transition temperature $T_c(H_g)$ such that, at this temperature, $\gamma_{SL}=\gamma_{SV}$. For large H_g this T_c coincides with the wetting transition point T_w. For small H_g, they find that the transition line $T_c(H_g)$ ends up at a critical point (this, however, occurs when H_g is comparable to the range of the forces, and may thus be rather sensitive to detailed local effects).

c. Prewetting transitions

Our discussion in this section was restricted to cases of macroscopic coexistence between liquid and vapor. If we impose a vapor pressure p_v *below* the value $p_{eq}(T)$ for liquid/vapor equilibrium, we cannot maintain a macroscopic L phase, but we may still have a film of L near the solid. The equilibrium thickness of these films can be analyzed in terms of a graph very similar to Fig. 24: we need only add a term linear in ζ (with a slope proportional to the difference in chemical potential between vapor and liquid). In many cases one still finds competition between two minima, and a possible first-order transition between them—describing a switch from a thick film to a thin film. For more details on this "prewetting transition" the reader is referred to Ebner and Saam (1977), Tarazona and Evans (1983), and Nakanishi and Fisher (1982).

6. Impurity effects: facts and conjectures

a. Facts

We have already mentioned in Sec. III.A that chemical contaminants may have dramatic effects on the wetting behavior of a liquid/liquid system. This was recognized from direct observations of contact angles (Moldover and Cahn, 1980) and substantiated by various measurements on the cyclohexane/methanol system near the free surface S. The contaminant was either water (Beaglehole, 1983; Tverkrem and Jacobs, 1983) or acetone (Cohn and Jacobs, 1983)—but in the latter case water may also have been present as a second contaminant. The Beaglehole experiments on water effects were based on ellipsometry. They are summarized below in terms of three different regimes, for a sample with 0.3% water.

(i) Very close to the critical point ($T_{\rm on}<T<T_c$) no significant wetting film is observed.

(ii) In a certain temperature range ($T_{\rm off}<T<T_{\rm on}$) a wetting film of the β phase (methanol rich, heavy) is observed. The unfavorable gravitational potential of the β phase is compensated by the attractive force between S and methanol. Beaglehole also mentions *fluctuations* of the thickness (near $T_{\rm on}$) plus a long time *drift* of the thickness (in most of the temperature interval).

(iii) At lower temperatures $T<T_{\rm off}$ only a very thin residual film remains and methanol partially wets the free surface.

A very remarkable feature of these experiments is that, in the absence of water, the two transitions disappear: there is no detectable wetting film (at the free surface) at any $T<T_c$ for the pure system.

b. Tentative interpretations

(i) Surface transitions. One line of thought, advocated by Leibler (1984), is based on a possible similarity between surface transitions and the effect of He_3 impurities in superfluid He_4. Near a solid surface He_3 is depleted (less attracted than He_4) and superfluidity is favored. The result is the existence, in a small temperature range, of a two-dimensional superfluid phase near the solid, while the bulk He_4+He_3 system is still normal. Similar surface transitions—induced by water impurities—could occur in the cyclohexane/methanol system, and be responsible for anomalous behavior in a small region near T_{c0}. But it is not easy to interpret in these terms the lack of any observable film in region (i).

(ii) Macroscopic balance of interfacial tensions. The idea here is to describe the competition of water and methanol for the free surface, making use only of the macroscopic surface tensions γ_{ij} between the various partners (S, α, β, and $W\equiv$water). Let us assume the following structure for the γ_{ij}:

$$\begin{aligned}\gamma_{sw} &\sim \text{independent of } T\,,\\ \gamma_{s\alpha} &= \gamma_{sl}-m\gamma_s'\,,\\ \gamma_{s\beta} &= \gamma_{sl}+m\gamma_s'\,.\end{aligned} \tag{3.19}$$

Here γ_{sl} is the surface tension of the critical mixture. Following Cahn (1977) it is assumed that $\gamma_{s\alpha}-\gamma_{s\beta}$ is proportional to the difference in concentration between α and β, the latter being itself proportional to $m=(\Delta T/T_c)^{1/3}$. The coefficient γ_s' is postulated to be *positive*. This means that, in the absence of water, the free surface prefers the α phase. Since this is also the lighter phase, it will occupy a macroscopic region below the free surface, and no wetting film is expected (in agreement with the observations on water-free systems).

Let us now list the other interfacial tensions:

$$\gamma_{\alpha\beta}\equiv\gamma\sim\frac{kT}{\xi^2}\sim\gamma_1 m^4 \tag{3.20}$$

(where we have used a classical scaling ansatz and the approximation $\beta=\frac{1}{3},\nu=\frac{2}{3}$ for the standard critical exponents)

$$\begin{aligned}\gamma_{w\alpha} &= \gamma_{wl}+\gamma_w' m\,,\\ \gamma_{w\beta} &= \gamma_{wl}-\gamma_w' m\,.\end{aligned} \tag{3.21}$$

We expect γ_w' to be strongly positive (the water/methanol interfacial tension being much smaller than the water/hydrocarbon tension).

Having defined the surface tensions, we may now construct the spreading coefficients S, defined as in Sec. II. Here we shall call S_{ijk} the spreading coefficient for a film of (j) being spread between phase (i) and phase (k),

$$S_{ijk}=\gamma_{ik}-(\gamma_{ij}+\gamma_{jk}) . \tag{3.22}$$

As we know from Sec. II, a positive S means spontaneous spreading and buildup of a wetting film. The first spreading coefficient of interest is

$$\begin{aligned} S_{sw\beta}(T)&=\gamma_{sl}-\gamma_{sw}-\gamma_{wl}+m(\gamma_s'+\gamma_w') \\ &=S_{sw\beta}(T_c)+m(\gamma_s'+\gamma_w') . \end{aligned} \tag{3.23}$$

Let us assume that $S_{sw\beta}(T_c)<0$, so that, at the critical point, water does not wet the $S\beta$ interface. However, when we decrease T (increase m), because the coefficient $(\gamma_s'+\gamma_w')$ is expected to be strongly positive, we may rapidly reach a temperature $T_{\rm on}$ such that

$$S_{sw\beta}(T_{\rm on})=0 . \tag{3.24}$$

Below this temperature we do have a water film, and we must now consider the possible spreading of the β phase between water and α. Making use of the definition (3.22) and of the listed γ_{ij} values, we arrive at

$$S_{w\beta\alpha}=2m\gamma_w'-\gamma_1 m^4 , \tag{3.25}$$

and when m increases (T decreases) we may switch from positive $S_{w\beta\alpha}$ (wetting film of β) to negative $S_{w\beta\alpha}$. Thus we expect a second transition at $T=T_{\rm off}$, where

$$S_{w\beta\alpha}(T_{\rm off})=0 . \tag{3.26}$$

At this point, the system becomes unable to sustain a sequence gas$|$water$|\beta|\alpha$. It may then achieve one of the following conformations:

$$\text{gas} \mid \text{water} \mid \alpha \quad \text{if} \quad S_{sw\alpha}>0 ,$$

$$\text{gas} \mid \alpha \quad \text{if} \quad S_{sw\alpha}<0 .$$

The formula for $S_{sw\alpha}$ is obtained from (3.23) by interchanging m and $-m$,

$$S_{sw\alpha}=S_{sw\beta}(T_c)-m(\gamma_s'+\gamma_w') . \tag{3.27}$$

With our assumption, all terms in (3.27) are negative. We thus expect the water film to redissolve when we cool down below $T_{\rm off}$, and the α phase to be in direct contact with the surface, leaving only a very thin layer (ξ) of excess cyclohexane. This scenario thus appears compatible with the data. Clearly this model is tentative and very rough. The water film may be much too thin to justify these purely macroscopic arguments, and it may be that there is no sharp W/α interface (if water is entirely soluble in α).[8] But the trend is interesting. The slow drift, and the fluctuations, seen in the interval $T_{\rm off}<T<T_{\rm on}$, could be due to the slow buildup of water content at the surface by diffusion and random convection.

It may be worthwhile to mention at this point another strange system, studied by Ross and Kornbrekke (1984). This is cyclohexane/anilin against glass. Here, at low temperatures, $\cos\theta$ is an increasing function of T, and reaches unity at a special temperature T_1. But above T_1, $\cos\theta$ *decreases* smoothly. It may be that this reflects a very fundamental multicritical behavior. But it may also be an impurity effect. When we reach $T=T_1$, an anilin film spreads over the capillary wall and may trap an impurity (water?) from distant sources. Then, at $T>T_1$, we would be dealing with a different surface (glass wet by impurity).

Clearly, all the binary mixtures under discussion (hydrocarbon/polar liquid) have chemical difficulties, such as trapping of water or spontaneous decomposition, and these difficulties are enhanced when the surface S is glass, which is also amenable to ion exchange, etc. Full control of impurity effects will require long, patient experiments.

(iii) Effects of long-range forces. The above interpretation assumed that long-range forces do not play a major role in the *existence* of the films, although they clearly control the *thickness* of a film.

An opposite viewpoint has been taken by Law (1984). Further, using a certain (assumed) form for the VW energies of films when the thickness (ζ) is not very large compared to the width of the α/β interface, he attempts to discuss the near vicinity of the bulk critical point (domain of large ξ). He finds three roots for the equilibrium film thickness, and claims that, in a certain range, all three roots are locally stable. The competition between three states then gives interesting possibilities, but it is hard to see how an energy function could have three distinct minima without having two intermediate maxima, also showing up as roots; thus this calculation is open to some doubt.

Another line of thought has been suggested by Israelashvili (1984). He pointed out that in certain layer systems (with dipoles present) the retarded/nonretarded VW energies may be of opposite signs (in a certain temperature range). This could lead to new energy minima, or maxima, in Fig. 6, and lead to our phase transitions.

IV. THE DYNAMICS OF SPREADING

A. Macroscopic measurements

Many practical problems involve a liquid spreading on a solid and displacing a gas. However, long, patient efforts have been required to obtain quantitative data on these problems. First, one must eliminate hysteresis effects. As we have seen in Sec. II, this is not a dream. With a reasonably (but not perfectly) smooth and homogeneous surface, no hysteresis should be left. Second, one should choose a simple flow geometry—eliminating, for instance, gravitational effects (which are often important in practice, but not very fundamental). A loose, but useful, condition for getting rid of gravitation is that the

[8] I am indebted to G. Teletzke for this remark.

linear dimensions of the drop (or of the meniscus) studied be small when compared to the capillary length $\kappa^{-1}=(\rho g/\gamma)^{1/2}$. Similarly, with common viscous fluids, one wishes to eliminate all inertial effects. We shall restrict our attention to slow, viscous flows, except in one section, which will be devoted to the opposite case of superfluid helium 4, where inertia is dominant.

A general review on the significance of the experiments has been given by Dussan (1979). Two arrangements, which satisfy the above requirements, have been used in detailed experiments: forced flow in a capillary and spontaneous spreading of a drop on a horizontal solid.

The choice of materials has not been very systematic. The liquids, in particular, are often selected because they have a viscosity that falls in a convenient range, one typical example being silicone oils. In fact, these oils are polymer melts, which may have very anomalous flow fields in the vicinity of the solid; fortunately, however, as we shall see in the theoretical sections, there are reasons to believe that the macroscopic laws are weakly (logarithmically) sensitive to these special properties. Thus the experiments described below are probably meaningful, even when they make use of these oils.

Most experiments have been performed with fluids that wet the solid completely, but there remains a fundamental ambiguity, because this may correspond either to $S=0$ or to $S>0$. The case $S=0$ is expected to be exceptional for dry solids, but possible for moist solids ($\theta_e=0$). However, we know, from the static discussion of Sec. II, that in the moist case the thickness of the preexisting liquid film is not fixed, but depends on a control parameter. In complete L/V equilibrium, this is the height of the plate above the reservoir providing the equilibrium.

At present, we do not seem to have any data from experiments on spreading, in the moist case, with prescribed values of this control parameter. Most existing experiments refer to the dry case, and probably correspond to $S>0$. But clearly the role of initial conditions should be considered more precisely in future work.

1. Forced flow in a capillary

The geometry chosen by Hoffman (1975) is shown in Fig. 26. He used a glass capillary (with a diameter ~ 2 mm), inside which a fluid was forced with velocities varying over five decades. The dimensionless parameter

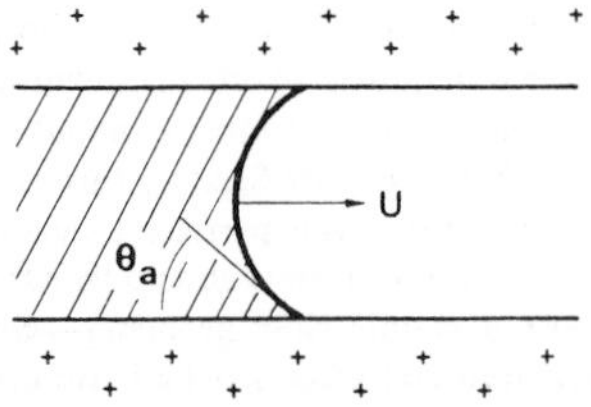

FIG. 26. Principle of the Hoffman experiments.

$$w=\frac{U\eta}{\gamma}=\frac{U}{V^*} \quad \text{(capillary number)} \tag{4.1}$$

(where η is the fluid viscosity) ranged between 10^{-4} and 10. Hoffman measured an *apparent contact angle* θ_a by a photographic technique. In a first series of experiments with silicone oils, he obtained conditions of complete wetting ($S\geq 0$) and found a rather universal relation between θ_a and w (Fig. 27):

$$w=F(\theta_a)\ . \tag{4.2}$$

Of particular interest is the low velocity limit ($w\rightarrow 0$), where the Hoffman data can be represented in the form

$$w=\text{const}\times\theta_a^m \tag{4.3}$$

with $m=3\pm 0.5$. Thus $\theta_a(w)$ first increases like $w^{1/3}$, and ultimately saturates at $w=1$, $\theta_a\rightarrow\pi$.

In a second series of experiments, Hoffman worked with other oils and industrial products, giving partial wetting ($\theta_e\neq 0$). He then found that his data could still be expressed in terms of the same F function, by writing

$$w=F(\theta_a)-F(\theta_e)\ . \tag{4.4}$$

However, the result (4.4) is much less documented than the data for complete wetting, and is open to some doubt.

These experiments clearly suffer from some defects. The materials were chosen mainly for their industrial interest, and many of them could carry contaminants. Ten years later, we are still lacking systematic data on pure (nonpolymeric) liquids, as well as on polymers of controlled molecular weight. It would be also be of interest to find out whether the apparent contact angle, as measured in photographs, at a given U, is independent of the capillary diameter. [Recent studies by Dussan and Ngan (1982) indicate that there is in fact some dependence.]

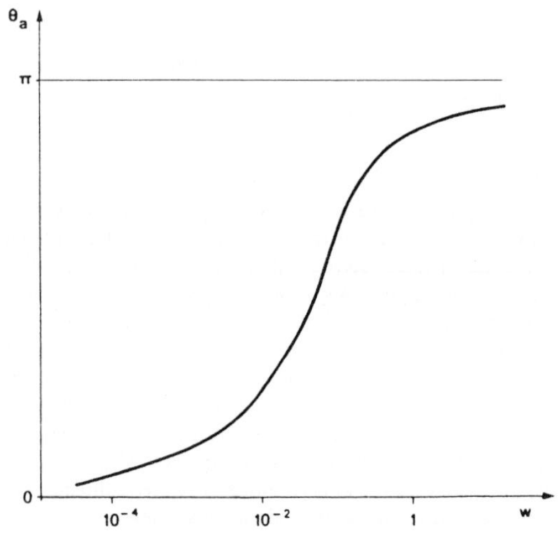

FIG. 27. Relation between reduced velocity $w=U\eta/\gamma$ and apparent contact angle θ_a (silicone oils on glass), after Hoffman (1975). At low w the relation is close to $\theta\sim w^{1/3}$.

However, in spite of these limitations, it is clear that the Hoffman experiments represent an important advance. The universal law (4.2) is very important. In particular, it is amazing to see that it holds without alteration for liquids of different S ($S>0$): *the magnitude of the spreading parameter has no influence.* We shall explain this in detail later.

2. Spreading of a droplet

Here most experiments have been performed for complete wetting ($S \geq 0$). The principle is shown in Fig. 28. In the classical approach, the expanding radius $R(t)$ is measured from photographs. In other studies, the nearly flat droplet is used as a lens. From the focal length of this lens one can go back to the apparent contact angle $\theta_a(t)$.

In the regimes where gravitation is negligible, the macroscopic shape of the droplet is found to be rather close to a spherical cap. Then the apparent contact angle, the drop thickness $h(t)$, and the radius $R(t)$ are related by

$$h=\tfrac{1}{2}R\theta_a , \tag{4.5}$$

$$\frac{\pi}{2}hR^2=\Omega , \tag{4.6}$$

valid for thin droplets ($\theta_a \ll 1$), this being the most important regime for complete spreading. Here Ω is the droplet volume and is assumed to be constant (weak evaporation).

Experimentally it is found that $R(t)$ increases rapidly at the early stages, and then very slowly. The data can often be represented (in terms of the wetted area πR^2) by a power law:

$$\pi R^2(t) \cong t^n \Omega^p . \tag{4.7}$$

The values of n obtained by different groups on different systems have been compared in a review by Marmur (1983). For water on glass, the early data of Hyppia (1948) suggest $n=0.22$, but Lelah and Marmur (1981) have found a strong temperature effect, with n ranging from 0.16 at 29 °C to 0.32 at 15 °C. One of the most accurate studies was performed by Tanner (1979) on silicone oils, and gave $n=0.21$. The exponent p has not been studied with the same detail, but is of order $\frac{2}{3}$ in the experiments of Lelah and Marmur (1981).

Some general remarks on the exponents n and p are in order at this point. Let us assume that the relation between apparent contact angle and velocity is correctly described by Eq. (4.3) and that the relevant velocity is the velocity of the contact line,

$$U=\frac{dR}{dt} \cong \frac{\gamma}{\eta}\theta_a^m \cong V^*\theta_a^m . \tag{4.8}$$

If we make use of the spherical cap approximation [Eqs. (4.5) and (4.6)] to eliminate θ_a, we arrive at

$$\frac{dR}{dt} \cong V^* \left[\frac{\Omega}{R^3}\right]^m , \tag{4.9}$$

and the spreading law becomes

$$R^{3m+1} \cong V^* t \Omega^m . \tag{4.10}$$

Thus we expect

$$n=\frac{2}{3m+1} , \tag{4.11}$$

$$p=\frac{2m}{3m+1} . \tag{4.12}$$

Taking the most probable value $m=3$ from the Hoffman data, we are then led to $n=0.20$ and $p=0.60$. We shall see below that $m=3$ is indeed expected theoretically for all cases of dry spreading ($S>0$).

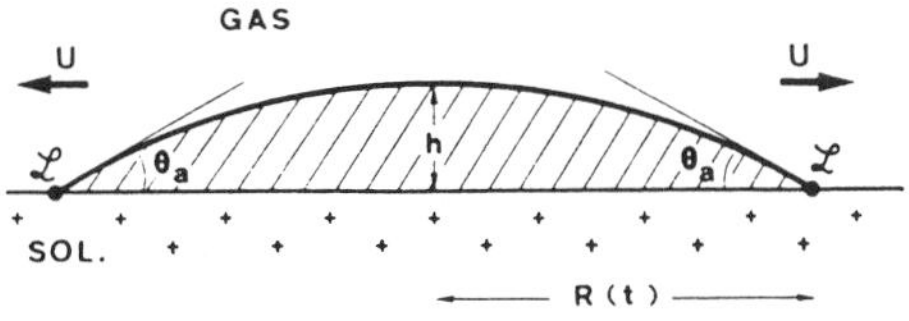

FIG. 28. A nearly flat droplet spreading on a solid: the macroscopic picture.

B. The precursor film

In the course of his pioneering work on wettability, Hardy (1919) recognized that a spreading droplet is announced by a precursor film (of submicrometer thickness), which shows up ahead of the nominal contact line. In particular, for droplets spreading on a solid surface, the film was revealed through its lubricating effects: a small test particle can slip more easily on the solid when the precursor is present. Hardy believed that these films occurred only with volatile liquids, which could condense ahead of the advancing droplet. This process may well exist, but more recent examples suggest that the film is present even in the absence of any vapor fraction (Bangham and Saweris, 1938; Chang *et al.*, 1982).

Detailed optical studies were carried out on the film at the Naval Research Laboratory (Bascom *et al.*, 1964), using nonpolar liquids which give complete spreading on steel. They found a precursor film, visible in ellipsometry at the late stages of spreading, with thicknesses of a few hundred angstroms. They also found (with impure liquids) a thicker, secondary film, probably due to Marangoni effects—a volatile contaminant being eliminated near the front and creating gradients of γ. We return to this impurity effect in Sec. IV.F.2, but omit it from the present discussion.

In one case, with a molten, viscous glass (which can be quenched and examined later), the film obtained by spreading on a metal was seen by electron microscopy (Radigan *et al.*, 1974). But the most brilliant detection of the film was based on electrical resistance measurements (Ghiradella *et al.*, 1975). The setup is slightly more complex [Fig. 29(a)], with a vertical plate and a conducting

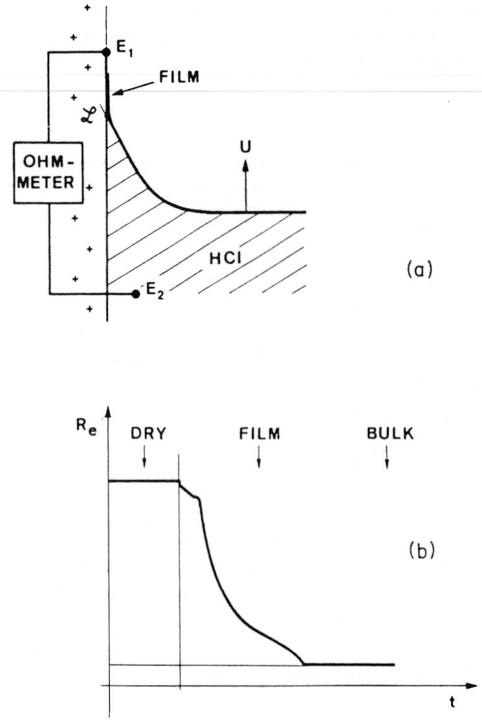

FIG. 29. Electrical detection of the precursor film (Ghiradella *et al.*, 1975). (a) Principle: the resistance $R_e(t)$ between two electrodes E_1E_2 is measured. (b) Typical record of $R_e(t)$: as soon as the film reaches (E_1), $R_e(t)$ drops.

liquid HCl that is set in vertical motion (velocity U). One measures the conductance between the bulk of the liquid (E_2) and an electrode (E_1) attached to the solid, when the nominal contact line $\mathscr{L}$ has not yet reached the electrode E_1. One finds a finite conductance occurring well in advance (e.g., when E_1 is one millimeter ahead of $\mathscr{L}$). A typical decrease of resistance with time is shown in Fig. 29(b). In principle, this could be translated into a profile $\zeta(x)$.

In practice, however, all the experiments we have described suffer from serious limitations: impurity effects resulting in gradients of the surface tension γ; choice of liquids, the polymer systems being in fact quite special, as explained in Secs. IV.B and IV.D; use of transient regimes (in the electrical experiments, where the velocity is imposed suddenly); noise and instability effects.

For all these reasons, we do not yet have a quantitative experimental law for the simplest (steady-state) film profile. But the situation should improve soon.

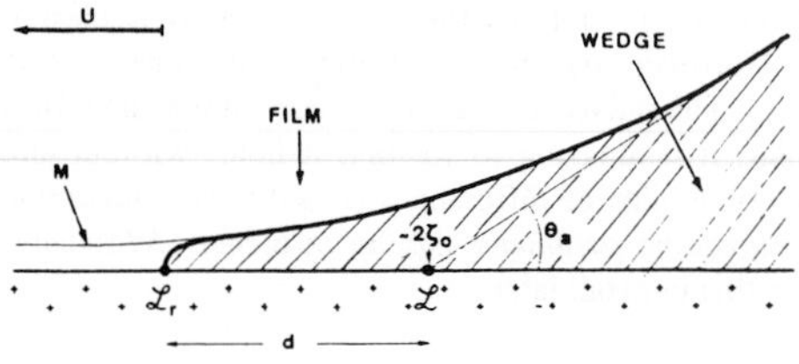

FIG. 30. A nearly flat droplet spreading on a solid: the microscopic picture.

C. Interpretation

1. Three types of dissipation

Let us restrict our attention to dry spreading ($S \geq 0$) of a pure, nonpolar liquid, attracted towards the solid by long-range van der Waals forces. Then the structure of the advancing front (with velocity U) corresponds to Fig. 30.

All macroscopic observations show the existence of a certain apparent contact angle θ_a. At distances $x \sim 100$ μm from the nominal contact line the fluid profile is very close to a *simple wedge* advancing along the solid. The motion of the liquid in this region has been probed in a clever experiment by Dussan and Davis (1974), marking the upper surface of the wedge with small spots of a dye (Fig. 31), and watching their motion (the liquid being highly viscous and the motion slow). They found a very characteristic rolling motion, reminiscent of a caterpillar

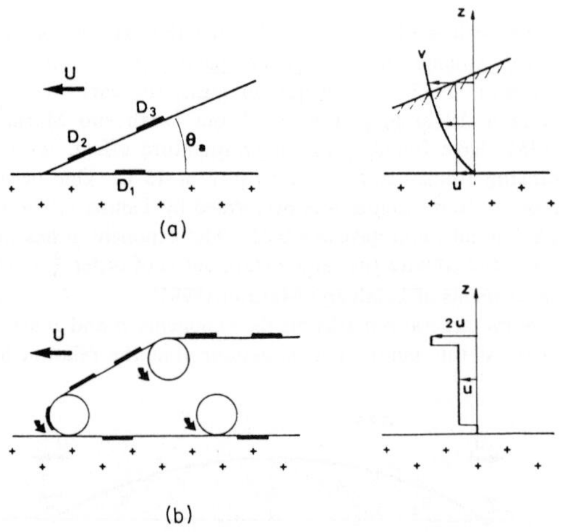

FIG. 31. (a) The Dussan-Davis experiment: spots of a dye $D_1D_2D_3$ are laid on the free surface of an advancing wedge. The spots slide down the wedge, and then get stuck to the solid. (b) The motion of a caterpillar vehicle is somewhat similar to the motion in the wedge [compare the velocity profiles $v(z)$].

vehicle. This rolling motion gives rise to viscous friction. We shall call the corresponding dissipation (per unit length of the contact line $\mathscr{L}$) $T\mathring{\Sigma}_w$, where the index w stands for "wedge."

Ahead of the wedge we have a *precursor* film (of typical thickness 100 Å) extending over a finite distance d. It has recently been realized (Hervet and de Gennes, 1984) that the viscous dissipation in this film is *very strong*. We shall call it $T\mathring{\Sigma}_f$.

The precursor film ends up by a *real contact line* $\mathscr{L}_r$ (shifted by a distance d from the nominal contact line $\mathscr{L}$). In the close vicinity of $\mathscr{L}_r$ we may have special losses, due to the attachment of liquid molecules to the solid. Some of the available free energy S may be transformed directly into heat. This third contribution is largely unknown; we shall call it $T\mathring{\Sigma}_l$ (where l stands for "local").

We can relate the total dissipation to the *unbalanced Young force F*, obtained by macroscopic considerations similar to Fig. 2:

$$F=\gamma_{SO}-\gamma_{SL}-\gamma\cos\theta_a \tag{4.13}$$

$$\cong S+\tfrac{1}{2}\gamma\theta_a^2 \quad (\theta_a \ll 1) . \tag{4.14}$$

Then the total dissipation is

$$FU=T(\mathring{\Sigma}_w+\mathring{\Sigma}_f+\mathring{\Sigma}_l) . \tag{4.15}$$

The common trend of the literature has been (i) to ignore $\mathring{\Sigma}_l$: we shall see that this may be correct in some favorable cases, and (ii) to ignore $\mathring{\Sigma}_f$: we shall see that this is grossly incorrect, and that, in fact,

$$T\mathring{\Sigma}_f=SU . \tag{4.16}$$

The free energy S is entirely burned in the film region.

2. Viscous losses in rolling motion

The flow patterns in a simple "wedge," advancing with constant velocity and angle θ_a [Fig. 31(a)], have been analyzed in a fundamental paper by Huh and Scriven (1971). They considered a very general case (arbitrary θ_a, nonzero viscosity in the gas phase). Here we shall present only a simplified view, holding for small θ_a and for negligible friction in the gas phase. This allows us to use the celebrated "lubrication approximation" of fluid mechanics. The idea is to treat the wedge as a nearly flat film, with a velocity profile

$$u_x(z)=u(z)$$

of the Poiseuille type [Fig. 31(a)]. On the solid side, u vanishes and on the gas side dw/dz vanishes (no tangential stress). The velocity U of the contact line is the z average of this profile,

$$U=\zeta^{-1}\int_0^{\zeta} dz\, u(z) . \tag{4.17}$$

(This may be checked by going to a frame moving with the line, where the solid slips at velocity $-U$, and where we find a steady state with 0 horizontal current.) One can then write

$$u(z)=\frac{3U}{2\zeta^2}(-z^2+2\zeta z) . \tag{4.18}$$

The viscous dissipation integrated over the film depth is

$$\int_0^{\zeta} dz\,\eta\left[\frac{du}{dz}\right]^2=\frac{3\eta U^2}{\zeta} \tag{4.19}$$

(where η is the fluid viscosity), and the total dissipation in the wedge is

$$T\mathring{\Sigma}_w=\int_{x_{\min}}^{x_{\max}}\frac{3\eta U^2}{\zeta}d\,|x\,|=\frac{3\eta U^2}{\theta_a}\ln\left|\frac{x_{\max}}{x_{\min}}\right| . \tag{4.20}$$

We expect $x_{\max}$ to be a cutoff related to the macroscopic size of the droplet $x_{\max}\sim R$. The cutoff $x_{\min}$ is more difficult to specify and will be discussed below. But without any cutoff ($x_{\min}\rightarrow 0$) the dissipation would diverge; as explained in Hellenic terms by Huh and Scriven (1971), "not even Herakles could sink a solid!"

Various physical processes may remove the singularity, and the choice of process depends very much on the choice of systems.

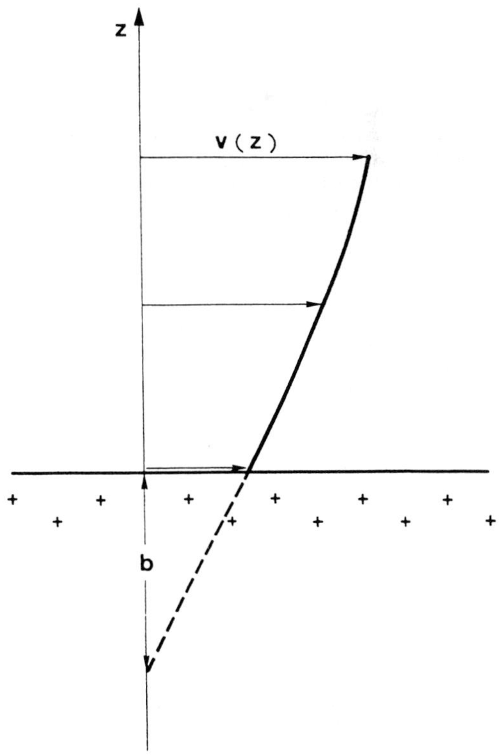

FIG. 32. The extrapolation length b characterizes the amount of slippage allowed in viscous flow near a solid surface.

a. Finite slippage at the solid surface

Instead of imposing zero velocity on the fluid (in the solid's frame) at the solid surface, it is natural to allow for a small amount of slippage, described by an extrapolation length b (Fig. 32). This may occur in various systems.

(i) A porous solid, saturated with the same fluid, will allow for an exponential tail in the velocity field. This could be found in nature—for instance, static contact angles have been studied on a "solid" made of a swollen silica gel (see Michaels and Dean, 1962)—but has not been studied systematically.

(ii) A rough solid surface may possibly be described along similar lines (Hocking, 1976). This, however, is not a very attractive situation, because roughness implies all the complications of hysteresis—discussed in Sec. II.C.

(iii) A normal liquid flowing over a smooth solid will display a length b comparable to the molecular size a. As we shall see, for this most important case, slippage is usually negligible, and the cutoff is provided by another effect (long-range VW forces).

(iv) Special physicochemical processes near the contact line have been proposed by Ruckenstein and Dunn (1976). They also lead to $b \sim a$.

(v) A polymer melt, flowing on a smooth surface (without any chemical attachment between polymer chains and the wall), is expected to show anomalously large values of b (de Gennes, 1979a). This generates special "foot structures" near the contact line, which are analyzed in Sec. IV.D.

A complete mechanical theory of macroscopic droplets spreading with finite slippage has been constructed by Huh and Mason (1977), and in more detail by Hocking (1977,1981) and Hocking and Rivers (1982). This theory is characterized by a matched asymptotic expansion between three regions—a "foot" where slippage dominates, a "wedge" with nearly constant slope, and a "central region" where the droplet is close to a spherical cap. This last feature deserves comment. In the thicker regions of the droplet, the flows are easy, and the mechanical pressure p_L equilibrates. The difference $p_L - p_V$ between the pressures in and out of the drop is constant, and this, in turn, through the Laplace-Young equation, means that the curvature of the L/V surface is constant, hence the spherical cap. The work of Hocking provides a precise proof of the above statement.

Quantitatively, as we shall see in Sec. IV.D, the result of slippage on the logarithmic cutoff $x_{\min}$ is simple: we expect

$$x_{\min} \cong b/\theta_a . \tag{4.21}$$

b. van der Waals forces

These lead to precursor films, and the film provides a cutoff for the dissipation $T\mathring{\Sigma}_w$. In cases of dry spreading with $S>0$, we shall see later that the cutoff is given by

$$x_{\min} \cong a/\theta_a^2 . \tag{4.22}$$

This dominates over slippage effects whenever the extrapolation length for slippage is comparable to the molecular size a [as can be seen by comparison of Eqs. (4.21) and (4.22)].

3. Structure and dissipation in the precursor film

The idea that a film can move because of a gradient of the disjoining pressure Π is not new (Deryagin, 1955; Lopez *et al.*, 1976; Starov, 1983). However, for the spreading of liquids on solids, progress, has been slow. A number of hydrodynamic flows have been solved numerically by Teletzke *et al.* (1983). The difficulty stems from the variety of experimental situations, revealed in the preceding paragraph. It is only by a patient separation of different physical regimes that one may hope to reach general laws (de Gennes, 1984a,1984b,1984c; Hervet and de Gennes, 1984).

Let us concentrate, then, on a "pure" case: a nonpolar liquid, giving complete wetting, and attracted towards the solid by VW forces. Transfer through the vapor is assumed to be negligible. There are no solute impurities, and we also exclude the case of polymers (the extrapolation length b for slippage is then negligible). Gravity effects are omitted. Finally, we restrict our attention to a *steady-state regime,* where the nominal contact line $\mathcal{L}$ moves with constant velocity $(-U)$ with respect to the solid (Fig. 31). We shall, in fact, work in the frame of the line, where the solid moves with a velocity $+U$. The choices of steady state, rather than transients, simplifies the equations enormously.

a. Hydrodynamic equations

Our starting point will be the pressure distribution in the film $p(x,z)$, which has the following structure:

$$p = p_G - \gamma \frac{d^2\zeta}{dx^2} + W(\zeta) - W(z) , \tag{4.23}$$

where p_G is the gas pressure; the second term represents the Young-Laplace capillary term. $W(\zeta)$ is the VW energy (per unit volume of liquid) between liquid and solid. In the nonretarded regime, which will be our main concern here, we have from Eq. (2.41)

$$W(z) = -\Pi(z) = -\frac{A}{6\pi z^3} . \tag{4.24}$$

Locally Eq. (4.23) describes a hydrostatic equilibrium. The vertical force acting on any volume element vanishes,

$$-\frac{\partial p}{\partial z} - \frac{\partial W}{\partial z} = 0 . \tag{4.25}$$

The term $W(\zeta)$ in (4.23) ensures that $p(\zeta)$ reduces to the Laplace-Young value.

In the lubrication approximation, the horizontal current J_S (in the frame of the solid) is given by a Poiseuille formula:

$$J_S=\frac{\zeta^3}{3\eta}\left[-\frac{\partial p}{\partial x}\right]. \tag{4.26}$$

In the frame of the line $\mathscr{L}$ this current becomes

$$J=U\zeta+J_S . \tag{4.27}$$

For steady-state solutions, J must be independent of x and t. In fact, for our problem (where there remains no film far ahead of the line $\mathscr{L}$), J must vanish exactly. This leads to the basic equation

$$\frac{3\eta U}{\zeta^2}=\frac{d}{dx}\left[-\gamma\frac{d^2\zeta}{dx^2}+W(\zeta)\right]. \tag{4.28}$$

Equation (4.28) describes not one, but many types of precursors, depending on the value of the spreading coefficient S. However, before discussing all these possibilities, it is instructive to return first to the macroscopic limit—i.e., to values of ζ that are large, so that $W(\zeta)$ becomes negligible. Then (4.28) reduces to an equation studied by Tanner (1979). We are particularly interested in solutions that have zero curvature for large ζ (i.e., that tend to behave nearly like a simple wedge in the macroscopic limit). They have the asymptotic form

$$\zeta(x)\to x\left[\frac{qU}{V^*}\ln\left[\frac{x}{x_1}\right]\right]^{1/3}, \tag{4.29}$$

and the slope varies very slowly with x. The constant x_1 will be determined later by matching Eq. (4.29) with appropriate solutions (at $x<0$) which describe a precursor film.

b. The "maximal" film

Let us consider first the profile marked M in Fig. 33. With this profile, we have a film that is present over *all* the solid surface. It will turn out that this "maximal" situation is relevant when the spreading coefficient S is positive and not too small ($S \gg \gamma\theta_a^2$). For the "maximal" case, there is no contact line $\mathscr{L}_r$ at the microscopic level, and the precursor is a nearly flat film. It is then permissible to drop the γ term in Eq. (4.28) over the whole film region ($x<0$).

Using (4.24) for $W(\zeta)$, and the definition (2.50) of the molecular length a, we can then reduce Eq. (4.28) to the simple form

$$\frac{a^2}{\zeta^2}\frac{d\zeta}{dx}=\frac{U}{V^*}\equiv w . \tag{4.30}$$

This can be integrated immediately to give the maximal film profile:

$$\zeta(x)=\frac{a^2}{w(x_2-x)}, \tag{4.31}$$

where x_2 is another integration constant, which we shall take to be equal to zero in what follows. We shall see that, in nearly all practical cases, Eq. (4.31) describes an important portion of the film, where the thickness is decreasing only slowly ($\zeta \sim |x|^{-1}$) with distance.

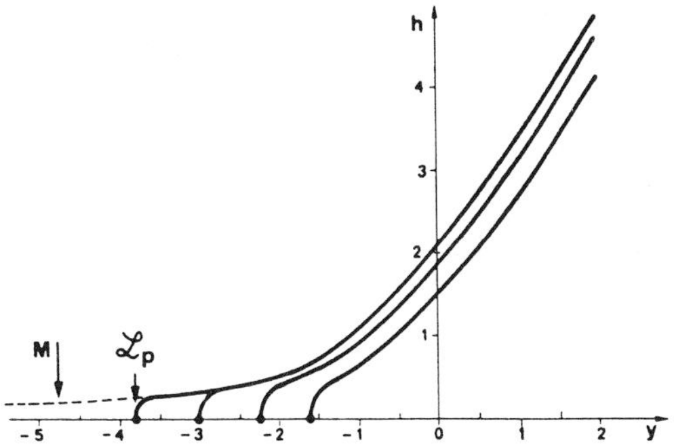

FIG. 33. A few numerical solutions of the film equation (4.33) for various values of the spreading parameter S (after Hervet and de Gennes, 1984). The larger S values correspond to the larger films.

c. Crossover between the maximal film and the macroscopic droplet

Let us first rescale (4.28) into an adimensional form,

$$x=x_0y ,$$
$$\zeta=\zeta_0h(y) ,$$
$$x_0=3^{-1/6}aw^{-2/3} , \tag{4.32}$$
$$\zeta_0=3^{1/6}aw^{-1/3} ,$$

$$h'h^{-2}-h'''h^2=1 , \tag{4.33}$$

where $h'\equiv dh/dy \cdots$. The "maximal" solution corresponds to $h(y\to-\infty)\to 0$. There exists a one-parameter family of solutions of Eq. (4.33) which satisfies this condition, with the asymptotic form

$$h_\alpha(y\to-\infty)=-\frac{1}{y}+\alpha\exp(y^3/3) . \tag{4.34}$$

The $\alpha=0$ solution corresponds to Eq. (4.31). The other solutions ($\alpha\neq 0$) are obtained by searching for small deviations from Eq. (4.31) and solving the corresponding linear equation in a WKBJ approximation. Starting from Eq. (4.34), one can extend the solutions $h_\alpha(y)$ towards positive y by numerical integration of Eq. (4.33) (Hervet and de Gennes, 1984).

One of these solutions ($\alpha=\bar{\alpha}\sim 0.38$) has the limiting property

$$h''_{\bar{\alpha}}(y\to+\infty)=0 . \tag{4.35}$$

This is the solution of interest (because the curvatures on the macroscopic side are always very weak on the scale of the film). At large, positive y the solution $h_{\bar{\alpha}}(y)$ does reach the asymptotic form announced in Eq. (4.29),

$$h_{\bar{a}}(y) \rightarrow y f^{1/3}(y), \quad f(y) \equiv 3\ln(0.4y) . \tag{4.36}$$

Thus the cutoff length x_1 defined in Eq. (4.29) is

$$x_1 = 2.5 x_0 ,$$

where x_0 is defined in Eq. (4.32)

The most important observable is the *apparent contact angle* θ_a, defined by a measurement at a macroscopic distance x from the contact line $\mathcal{L}$ $[x_1 \ll x \ll R(t)]$. Using (4.34), we find

$$\theta_a \equiv \frac{d\zeta}{dx} = \frac{\zeta_0}{x_0} f^{1/3}(1+f^{-1}) \cong \frac{\zeta_0}{x_0} f^{1/3} , \tag{4.37}$$

and returning to Eq. (4.32) this gives the fundamental formula

$$\theta_a^3 = 3 f w . \tag{4.38}$$

The main conclusions are the following.

(i) f is nearly constant, and $\theta_a \sim w^{1/3}$, as observed in the low-velocity experiments of Hoffman (1975) and Tanner (1979). An equation of this type was first derived theoretically by Fritz (1965), but for a slightly different problem (liquid spreading on a wet surface). It is probably fair to call Eq. (4.38) the *Tanner law*, because Tanner was the first to define it and to obtain it from experiments on a *dry* surface. He also interpreted his own experiments by similar ideas (although he did not take the precursor film into consideration). Of course, the exponent 3 in Eq. (4.38) is only approximate, because f is logarithmically dependent on w, but this is a minor refinement.

(ii) The width of the crossover region between film and droplet is of order $x_0 = a w^{-2/3}$ [Eq. (4.32)]. After making use of Eq. (4.38), this becomes

$$r_0 \cong a/\theta_a^2 . \tag{4.39}$$

(iii) The thickness of the film in the crossover region is

$$\zeta_0 \cong a/\theta_a . \tag{4.40}$$

Thus the precursor film has reality ($\zeta_0 \gg a$) only for situations of small θ_a. Typically $\theta_a = 10^{-2}$ and $a = 1$ Å, giving $\zeta_0 = 100$ Å (falling well into the range of nonretarded VW interactions).

d. Truncated films

We shall now show that the maximal film described above (covering the whole solid surface) corresponds to a certain limiting case of "dry" spreading, described by the inequality[9]

$$S \gg \tfrac{1}{2}\gamma\theta_a^2 . \tag{4.41}$$

This can be easily understood if we return to the static discussion of Sec. II.D. We saw there that a VW fluid with positive S does not spread on a dry solid down to a molecular layer, but in fact stops at a certain equilibrium thickness $e(S)$ [Eq. (2.72)],

$$e(S) \cong \left[\frac{\gamma}{S}\right]^{1/2} a . \tag{4.42}$$

It is natural to expect that the precursor film never thins down below this value. The maximal film solution is applicable only in the interval

$$e(S) < \zeta < \zeta_0 . \tag{4.43}$$

Setting $\zeta = e(S)$ in Eq. (4.31), we have a formula for the film width d:

$$d \cong \frac{a^2}{w e(S)} \cong \frac{a}{w}\left[\frac{S}{\gamma}\right]^{1/2} . \tag{4.44}$$

These predictions are entirely confirmed by detailed numerical solutions of Eq. (4.33) (Hervet and de Gennes, 1984). The truncated solutions are shown in Fig. 33. They start at certain contact line $\mathcal{L}_r$ of position $x_r = -d$. The initial rise of the profile near x_r is parabolic and identical to the static solution [Eq. (2.55)]. At higher x the solution merges with the maximal profile, provided that $d \gg x_0$. The latter condition also shows up in Eq. (4.43). To have a film, we must satisfy

$$e(S) < \zeta_0 , \tag{4.45}$$

and this requirement is equivalent both to $d > x_0$ and to Eq. (4.41). In most practical cases we expect to have Eq. (4.41) well satisfied, in which case a film will indeed be present.

e. Dissipation in the film

A remarkable feature of the Tanner law (4.38) is that the value of the spreading coefficient S does not play any role (even in the log term). The only requirement is that the macroscopic profile cross over into a maximal film, and this, as we just saw, is satisfied whenever $S \gg \gamma\theta_a^2$. The Tanner law expresses a certain relation between the flux U and the force F in the thermodynamic formulation of Eq. (4.15). The spreading coefficient S is the dominant term in the force, so there is an apparent paradox. How does S drop out of the energy balance?

The viscous dissipation computed from the hydrodynamic equation (4.28) does agree with Eq. (4.15). The dissipation in an interval $-\infty < x < x_{max}$ (where x_{max} is in the macroscopic region $x_{max} \gg x_0$) is exactly

$$T(\mathring{\Sigma}_w + \mathring{\Sigma}_f) = \int_{-\infty}^{x_{max}} \frac{3\eta U^2}{\zeta} dx . \tag{4.46}$$

Using the truncated solutions plus suitable integrations by parts (Hervet and de Gennes, 1984), we can transform this exactly into Eq. (4.15) (with the assumption $T\mathring{\Sigma}_l = 0$). Thus we have not forgotten anything in the energy balance. But where is the free energy S dissipated? We shall find the answer through a qualitative estimate of the

[9]This condition was not entirely appreciated in the original work (de Gennes, 1984a) on the maximal film.

losses in the film only,

$$T\mathring{\Sigma}_f=\int_{-d}^{0}\frac{3\eta U^2}{\zeta}dx\ , \qquad (4.47)$$

where we use an abrupt truncation at $x=-d$ and also arbitrarily decide that the film stops exactly at $x=0$. We use the maximal film solution in this interval, in the form (4.30), and transform (4.47) into

$$T\mathring{\Sigma}_f=3\eta U^2a^2w^{-1}\int_{e(S)}^{\zeta_0}\frac{d\zeta}{\zeta^3}\ . \qquad (4.48)$$

The leading term comes from the lower limit and is

$$T\mathring{\Sigma}_f=\frac{3\gamma a^2U}{2e^2(S)}=SU\ . \qquad (4.49)$$

We now have the answer: all the excess free energy S is burned in the film. The remaining contribution to F, namely, $\frac{1}{2}\gamma\theta_a^2$, is used up in the wedge, and imposes the relation (4.38) between reduced velocity (w) and wedge angle θ_a, where S does not appear.

4. Spreading over a wet surface (Bretherton, 1961; Fritz, 1965; Tanner, 1979)

Let us return now to macroscopic droplets (or capillaries) and consider a solid that was initially covered with a liquid film of constant thickness (e_0). On top of this we add, for instance, a droplet, and watch its spreading.

a. Macroscopic regime

If e_0 is large ($e_0>1000$ Å) we may omit all effects of long-range forces, and we are dealing only with capillary energies that are dissipated by viscous flows. This family of problems was discussed long ago by Landau and Levich (1942; see Chap. 12 of the book by Levich, 1962). They were more concerned with a plate being pulled out of a liquid, while our present problem is the analog of a wet plate being pushed in, but the basic equations are the same. We may summarize the results by saying that the logarithmic cutoff discussed in Sec. IV.B.2 is now provided by the original film,

$$x_{\min}\cong e_0/\theta_a\ . \qquad (4.50)$$

Inserting this into the wedge dissipation formula (4.20), we see that large $x_{\min}$ values correspond to a smaller logarithmic factor l, and should thus lead to larger velocities.

An interesting feature is that the profile is not monotonous: we expect small *capillary oscillations* ahead of the nominal contact line $\mathcal{L}$. Such an expectation arises from the linearized ($\zeta\to e_0$) steady-state equation, which is of third order and has exponential solutions with a complex decay length—leading to damped oscillations. These oscillations are, in fact, quite visible when moving a liquid in a prewetted test tube,[10] and are shown in photographs by Fritz (1965).

b. Microscopic regime

If e_0 is smaller than the precursor characteristic thickness ζ_0 [Eq. (4.32)], the main effects should occur in the precursor region. Let us discuss them briefly, starting from the current in the frame of the line [Eq. (4.27)]

$$J=U\zeta-\frac{A}{6\pi\eta\zeta}\frac{d\zeta}{dx}\ , \qquad (4.51)$$

where we have used Eqs. (4.23) and (4.24). In steady state J is constant, but the constant is now different from zero. In the far precursor region, the thickness $\zeta\to e_0$, and we must have

$$J=Ue_0\ . \qquad (4.52)$$

This leads to a profile

$$\zeta(x)=\frac{e_0}{1-\exp(q\,|x-x_r\,|)}\ , \qquad (4.53)$$

where x_2 is an integration constant, and

$$q=\frac{6\pi\eta e_0U}{A}\ . \qquad (4.54)$$

The result (4.53) describes a crossover between the maximal precursor [Eq. (4.31)] found in the region $\zeta>e_0$ (or $q\,|x_2-x\,|\ll 1$) and an exponential tail,

$$\zeta(x)-e_0=e_0\exp(-q\,|x-x_2\,|)\ , \qquad (4.55)$$

in the forward region ($\zeta-e_0\ll e_0$). We see that in the limit $e_0<\zeta_0$, the matching of the macroscopic solutions will be imposed by the conventional precursor, and the preexistant film of thickness e_0 should have only weak effects on macroscopic flows. More detailed numerical discussions of the various regimes have been carried out by Teletzke *et al.* (1983).

c. Spreading with obstacles

The ideas sketched in Secs. IV.A.4.a and IV.A.4.b above could be tested by relatively simple experiments using uniformly wet surfaces. But their main impact may be different. Some macroscopic experiments show that a moving contact line has a velocity sensitive to perturbations on the solid surface which lie *ahead of* $\mathcal{L}$.

(i) Bangham and Saweris (1938) noticed that a drop of methyl alcohol spreading on mica was slowed down when reaching the vicinity of a drop of butyl alcohol.

(ii) Lelah and Marmur (1981) showed that a water droplet spreading on a glass slide had its line $\mathcal{L}$ attracted towards the edge of the slide. This attraction was felt at distances ~ 1 mm.

All these situations are complex; one would like to know first what happens when two drops of the *same* liquid are spreading on a flat solid surface and on the verge of coalescing. Experimental studies on the macroscopic shapes would be easy to perform and (possibly) not too hard for theoretical analysis. Indeed, Teletzke, Davis, and Scriven (1984) have analyzed the Lelah-Marmur ex-

[10] I am indebted to Y. Pomeau for one of these observations.

periment (ii). They can explain the sign of the effect if they assume for the disjoining pressure $\Pi(\zeta)$ of water on glass the slow decrease ($\Pi \sim \zeta^{-1}$) extracted from empirical data by Pashley (1980).

D. The special case of polymer melts

1. Observations

We are concerned here with flexible polymer chains, which can exist in a liquid form and give complete spreading ($S > 0$). To study these polymer drops is not very easy; one encounters at least two difficulties: high viscosities (if the chains are long) and impurity content (many practical polymer systems contain additives, or catalysts, from the fabrication process). This led to serious complications in the early experiments by Bascom *et al.* (1964).

However, the spreading of polymers such as silicone oils is important for many industries (paints, adhesives, protective coatings). Also, from a more fundamental point of view, we shall see that it raises a very special problem. Three main experiments indicate an anomalous behavior.

(a) Early work by Schonhorn *et al.* (1966) showed that a certain characteristic length (independent of the original size R_0 of the droplet) came into the spreading laws. Later work by the same group (Radigan *et al.*, 1974), using electron microscopy, displayed a "protruding foot" near the contact line $\mathscr{L}$.

(b) Ogarev *et al.* (1974) studied polydimethylsiloxane chains of high molecular weight ($M \sim 10^6$) spreading on mica. (They had only one value of M, but they could vary the viscosity significantly by changing the temperature.) Their main conclusion (from our point of view) was that the macroscopic shape of the drops is not a spherical cap, as it is with nonpolymeric, pure liquids. The deviation from sphericity is most significant for small droplet volumes. There is indeed a spherical cap region in the center, but the cap is surrounded by a protruding "foot" (Fig. 34). The foot is macroscopic and has nothing to do with the precursor of submicron thickness which was discussed in Sec. IV.C.

(c) Sawicki (1978) carried out a series of systematic spreading with silicone drops of different chain lengths. His results on the apparent contact angle $\theta_a(t)$ are not far from the Tanner law ($\theta_a \sim t^{-0.3}$). However, when he compared the bulk viscosity to the apparent viscosity η_a required to fit a wedge model, he found η_a values that were too low. His interpretation was that the macromolecules were elongated in the film region, and that this led to a reduction of viscosities. But this non-Newtonian behavior is open to doubt at the very low shear stresses achieved in spreading. An alternate explanation will be presented below.

2. Interpretation

The "protruding foot" has been a source of confusion in the past. To keep things straight, we must (i) carefully

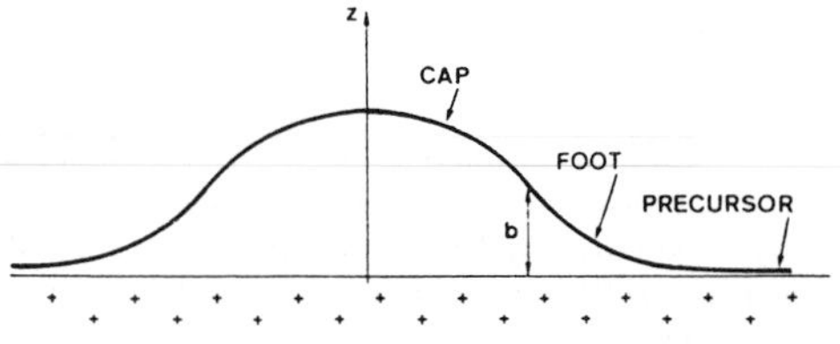

FIG. 34. Qualitative shape of a polymer melt during spreading.

discriminate between the foot and the precursor film (as explained above), and (ii) put in a different class the special profiles observed by Bascom *et al.* (1964), which are associated with volatile impurities.

a. Strong slippage

The basic idea, proposed independently of all spreading phenomena (de Gennes, 1979a), is that entangled polymers, flowing near a smooth, passive surface, should show a highly anomalous slippage, or, equivalently, a very large extrapolation length b (Fig. 32). Physically we may say that it is less expensive to concentrate the shear at the polymer/solid interface than to spread it over all the liquid (where entanglements oppose the shear very strongly). The resulting formula for b is

$$b = a\eta/\eta_0 , \qquad (4.56)$$

where a is a molecular size, η_0 is the viscosity of a liquid of monomers (with the same interactions, but no entanglements), and η is the melt viscosity (enormously enhanced by the entanglements).

In the "reptation" model for molten polymers, one expects (de Gennes, 1979b)

$$\eta = \eta_0 \frac{N^3}{N_e^2} , \qquad (4.57)$$

where N is the number of monomers per chain, while N_e is a characteristic number (the "number of monomers per entanglement"), of order 100. Experimental exponents are slightly higher than the value (3) predicted by (4.57), but the trend is clear: with high values of $N(10^4)$, one can expect to find b values up to one millimeter.

Early observations on the flow of polyethylene melts inside transparent capillaries give some support to these ideas (Galt and Maxwell, 1964). In these experiments, the velocity field was probed via tracer particles. It was found that the velocity at the well did *not* vanish in a significant fraction of the runs. Experiments by Kraynik and Schowalter (1981) detected the slip by hot film anemometry. Burton *et al.* (1983) used a Weissenberg rheogoniometer equipped with closely spaced parallel plates: for molten polystyrene ($M \sim 10^5$) the data indicate $b \sim 60$ μm. They are taken at relatively low shear rates: slippage does exist for Newtonian flows.

b. Droplet shapes
(Brochard and de Gennes, 1984)

Let us focus our attention on the most important case, where the droplet thickness at the center (h) is much larger than b, but b itself is macroscopic ($b > 1\ \mu m$). One then expects to find three regions.

A spherical cap, where $\zeta \gg b$. Here normal viscous flow takes place, slippage is negligible, and the apparent contact angle θ_a should follow the Tanner law [Eq. (4.3)].

A macroscopic foot, where $b > \zeta > \zeta_0$. In this region we expect to find a "plug flow" of polymer driven by capillary forces.

A precursor film, where $\zeta_0 > \zeta$, which differs from the normal precursor of Sec. IV.C because the flow is again of the "plug" type.

Let us start with the macroscopic domain. For finite b, the equation for the horizontal current J_S, which replaces (4.26), is

$$J_S = \frac{\zeta^2}{\eta}(b + \zeta/3)\left[-\frac{\partial p}{\partial x}\right] . \qquad (4.58)$$

If we work in a macroscopic regime, the only effect contributing to $\partial p/\partial x$ is capillarity,

$$-\frac{\partial p}{\partial x} \to \gamma \zeta''' , \qquad (4.59)$$

and the steady-state equation becomes

$$w \equiv \frac{U}{V^*} = -(b + \zeta/3)\zeta\zeta''' . \qquad (4.60)$$

At $\zeta \gg b$ we recover Tanner's equation, and the profile is a spherical cap. At $\zeta \ll b$, we can construct a special solution that describes the foot:

$$\zeta^2 = 8|x|^3/3\lambda , \qquad (4.61)$$

$$\lambda = b/w . \qquad (4.62)$$

Here $|x|$ is the horizontal distance measured from the nominal contact line $\mathcal{L}$. In principle, the crossover between the foot [Eq. (4.61)] and the spherical cap ($\zeta > b$) can be extracted from the work of Hocking (1977) and Hocking and Rivers (1982). For qualitative purposes, it is enough to note that the width f_1 of the foot must be such that Eq. (4.61), taken at $|x| = f_1$, gives $\zeta \sim b$. Thus

$$f_1 \cong (\lambda b^2)^{1/3} \cong b w^{-1/3} \cong b\theta_a^{-1} . \qquad (4.63)$$

The logarithmic cutoff discussed in Sec. IV.B is now expected to be $x_{\min} \sim f_1$, as announced in Eq. (4.21).

c. The polymer precursor

If we now go to a microscopic regime, where VW forces become dominant, the equation for the maximal film is obtained by ignoring capillarity and setting

$$-\frac{\partial p}{\partial x} \to \frac{A}{2\pi\zeta^4}\frac{\partial \zeta}{\partial x} \qquad (4.64)$$

(in the nonretarded regime). Inserting this into Eq. (4.58), choosing $\zeta \ll b$, and solving, one obtains

$$\zeta^2 = \frac{Ab}{4\pi\eta U}\frac{1}{|x - x_2|} , \qquad (4.65)$$

where x_2 is an integration constant. Thus the thickness of the maximal precursor should decrease even more slowly than in conventional fluids ($\zeta \sim |x|^{-1/2}$). Note that the prefactor $\eta/b = \eta_0/a$ should be *independent of molecular weight* (for entangled systems). The scaling form of the crossover length x_2 is

$$x_2 \cong (ab)^{1/2}\theta_a^{-3/2} . \qquad (4.66)$$

A truncation of the maximal precursor at finite S should occur as in Sec. IV.C.3, but has not yet been explored.

3. Perspectives

The existence of a macroscopic foot on spreading polymer droplets appears as a natural consequence of plug flows. It must be emphasized that the extrapolation length b, describing the plug flows, is very sensitive to the surface treatment (de Gennes, 1979a): if the surface is slightly rough, or if the surface can bind chemically to some of the chains (or to chain ends) b could be drastically reduced from the estimate (4.56). This means that a complete study of the spreading of polymer melts will require a delicate, coordinated effort between polymer science and surface science.

On the theoretical side, many aspects of polymer melts remain completely unexplored. The regimes at $h < b$ should not be ruled by the Tanner law, and may correspond to the observations of Schonhorn *et al.* (1966). Moreover, the discussion on surface friction, leading to Eq. (4.56) for b, has been carried out only in the limit of strongly entangled chains ($N \gg N_e$). With silicone oils of low N, which are of some practical interest, one needs an estimate of b in the opposite limit ($N < N_e$). We still expect an enhancement of b, and Eq. (4.56) may even remain qualitatively valid, but to prove this will require a delicate study of chain flow near a passive wall, which (fortunately) was not required in the entangled limit.

E. Spreading laws for superfluid He_4

A long experimental effort has been devoted to the statics of the Rollin film, and also to the small-amplitude oscillations of the film surface ("third sound"). Some studies have also been carried out on transport (from one reservoir to another, at a different level) via the film (see the review by Brewer, 1978). But we do not know of any observation on the horizontal spreading of droplets.

On the theoretical side, a recent study (Joanny, 1985) analyzes the motion in a simple inertial regime, at very low temperatures (no normal fluid, and no evaporation), without any vortex nucleation. All the work is restricted to the special case $S = 0$. We shall concentrate here on the *macroscopic regime.*

A force balance argument predicts that the dynamic contact angle *vanishes.* This property is in fact similar to Eq. (4.61) for polymer flow ($d\zeta/dx=0$ at the nominal line position $\mathscr{L}$), and the physical origin is the same: in both cases we have a plug flow and no singularity [in the dissipation, or in the horizontal hydrostatic forces $\zeta(-\partial p/\partial x)$] near the line $\mathscr{L}$. Thus the balance of forces at $\mathscr{L}$ is of the Young type, with no added terms, and $\theta_a=\theta_e=0$.

Let us, then, consider a droplet of original volume $\Omega\cong R_0^3$, spreading to a distance $R(t)\gg R_0$, under the sole action of capillary forces. The initial capillary energy was of order γR_0^2. When this is transformed into kinetic energy, with rms velocity U, we have

$$\Omega\tfrac{1}{2}\rho_L U^2\cong\gamma R_0^2\ , \tag{4.67}$$

$$U\cong\left[\frac{\gamma}{\rho_L R_0}\right]^{1/2} . \tag{4.68}$$

The final velocity of the line $\mathscr{L}$ should be constant and given by Eq. (4.68). The detailed shape of the droplet is not simple, but (within the lubrication approximation) Joanny was able to construct exact, self-similar solutions with $\theta_a=0$ and with the velocity (4.68). It would be of interest for the future to investigate the *stability* of these self-similar solutions.

The (more realistic) case where S is *positive* remains completely unexplored. It may be that the energy S is spent in the form of vortex lines nucleated at the solid wall.

F. Unsolved problems

1. Pure fluids

We have seen that a consistent picture may be constructed for the *dry spreading* of a simple fluid on a solid. On the other hand, we noticed that the situation of *moist spreading* requires the specification of a control parameter [H in the notation of Eq. (2.46)] and is thus not unique.

There remains, however, a long list of open questions connected with the spreading of pure fluids, some of which are specified below.

(a) Long-range forces *other than van der Waals* may come into play (especially double-layer effects, if the liquid is water).

(b) All our discussion assumed that the *local dissipation* $T\mathring{\Sigma}_{\text{local}}$ near the moving contact line was negligible. This need not be true. Let us list again the different types of dissipation (Sec. IV.C.1):

$$T\mathring{\Sigma}_f=SU\ , \tag{4.69}$$

$$T\mathring{\Sigma}_w=\tfrac{1}{2}\gamma\theta_a^2 U\sim\frac{\eta U^2}{\theta_a}\ , \tag{4.70}$$

$$T\mathring{\Sigma}_{\text{local}}=\tfrac{1}{2}\eta_l U^2\ . \tag{4.71}$$

In Eq. (4.71) we tentatively assumed that the local term could be described in terms of a simple friction coefficient η_l (which turns out to have the physical dimensions of a viscosity). If this assumption is correct, we usually expect that, in the limit $\theta_a\to 0$, the hierarchy is

$$T\mathring{\Sigma}_f>T\mathring{\Sigma}_w>T\mathring{\Sigma}_{\text{local}} \tag{4.72}$$

provided that

$$\theta_a<\eta/\eta_l\ . \tag{4.73}$$

But this statement may be useless if the local friction cannot be described in the form (4.71), or if the coefficient η_l is much larger than η [so that the conditions (4.73) never hold in practice]. A complete understanding of the local process will probably require calculations of molecular dynamics on specific examples.

(c) *Flow instabilities* may occur, even with viscous fluids. Optical observations by Williams (1977) on spreading droplets of various fluids show a wiggly contact line. This may be interpreted in (at least) three different ways: inhomogeneities in the solid surface (see the discussion on a *static* contact line in Sec. II); solute impurity effects (see Sec. IV.F.2 below); fundamental instabilities in precursor flow: it could be that the films described in Sec. IV.C.3 are intrinsically unstable, even for a pure liquid.

(d) *Transfer via the vapor* may be important even in dry spreading. Although the solid well ahead of the contact line is dry, it may receive a few molecules evaporated from the liquid interface. Even for mildly volatile liquids, this process may renormalize the effective value of the spreading parameter S. Our analysis suggests that this has not much effect on the Tanner law, but that it will change the width d of the precursor film.

(e) All our discussion ignored mechanical losses in the *gas* phase. As early as 1971 Huh and Scriven pointed out that even a gas may become important in the limit of $U>V^*$, where $\theta_a\to\pi$, and where we are actually dealing with a thin film of gas squeezed between liquid and solid. These effects become even more spectacular if we are dealing, not with liquid/gas, but rather with a liquid/liquid system. Pismen and Nir (1982) observed that a simple wedge solution was not acceptable in the macroscopic regime, and that self-similar solutions led to strange "spiral" configurations. Pumir and Pomeau (1984) propose a set of traveling waves following the contact line. This should be checked by experiments on consolute mixtures.

2. Effects of additives in the liquid

a. *Volatile impurities*

Their role in spreading was noticed early in the optical studies of Bascom *et al.* (1964): certain precursor structures occurred only with impure liquids. The explanation of Bascom *et al.* of these results is based on a local gradient of the surface tension, induced by evaporation near the tip. This interpretation is very plausible. It may be, however, that in some cases the renormalization of S by vapor signals—described in Sec. IV.F.1.c above—plays a role.

b. Surfactants

When they are insoluble in the bulk liquid, surfactants appear only in films at the various interfaces (S/L,S/V,L/V). If we know the surface concentrations Γ_{ij} in these various films, the conservation law gives a condition on the hydrodynamic velocities. Consider, for instance, the very simple case where $\Gamma_{LV} \equiv \Gamma$ is finite, while Γ_{SL} and Γ_{SV} vanish. Then the velocity at the free surface of the liquid must be just equal to the line velocity U, so that the surfactant "never catches up" with the line. This, in turn, imposes a certain modification of the velocity profile in the liquid film and a change of the numerical coefficients in the Tanner law (4.20). Considerations of this type were already present in the work of Huh and Scriven (1971).

For many practical purposes, the effects of surfactants are much more spectacular. For instance Lelah and Marmur (1981) find that a small amount of surfactant (above the critical micelle concentration) gives rise to a hydrodynamic instability in spreading. A well-known class of experiment makes use of a surfactant that attaches slowly to the S/L interface and makes it hydrophobic: a droplet spreads and then retracts. All these effects will require good models of adsorption/desorption kinetics (involving single-surfactant molecules or involving micelles).

c. Polymers

In solution, polymers also can be adsorbed on the S/L interface (and/or on the L/V interface). All the effects mentioned above may appear. Moreover, the hydrodynamic extrapolation length b of Fig. 32 may be drastically reduced by polymer adsorption.

ACKNOWLEDGMENTS

The first version of this review was based on a course given in Paris during 1983–1984, and I wish to thank the participants for their active criticism. Discussions with J. F. Joanny, F. Brochard, and H. Hervet were essential for the construction of a common viewpoint. I have also greatly benefited from written exchanges and/or oral disputes with D. Beaglehole, D. Chan, T. Davis, E. Dussan V., R. Evans, H. Frisch, S. Garoff, B. Hughes, J. Israelashvili, C. Knobler, B. Legait, A. Marmur, B. Maxwell, M. Moldover, M. Nightingale, B. Ninham, Y. Pomeau, F. Rondelez, L. Saraga, M. Schick, L. Schwartz, C. Taupin, G. Teletzke, and B. Widom. A critical reading of the manuscript by M. Adam and by D. Huse has been of considerable help.

REFERENCES

Adamson, A., and A. Zebib, 1980, J. Phys. Chem. **84**, 2619.

Bangham, D., and S. Saweris, 1938, Trans. Faraday Soc. **34**, 554.

Bascom, W., R. Cottington, and C. Singleterry, 1964, in *Contact Angle, Wettability and Adhesion,* edited by F. M. Fowkes, Advances in Chemistry Series, No. 43 (American Chemical Society, Washington, D.C.), p. 355.

Beaglehole, D., 1983, J. Phys. Chem. **87**, 4749.

Benner, R. E., G. F. Teletzke, L. E. Scriven, and H. T. Davis, 1984, J. Chem. Phys. **80**, 589.

Berry, M., 1974, J. Phys. A **7**, 231.

Bouasse, H., 1924, *Capillarité et phénomènes superficiels* (Delagrave, Paris).

Bretherton, F. P., 1961, J. Fluid Mech. **10**, 166.

Brewer, D. F., 1978, in *The Physics of Liquid and Solid Helium,* edited by K. Benneman and J. Ketterson (Wiley, New York), Part II, p. 573.

Brézin, E., B. I. Halperin, and S. Leibler, 1983a, J. Phys. (Paris) **44**, 775.

Brézin, E., B. I. Halperin, and S. Leibler, 1983b, Phys. Rev. Lett. **50**, 1387.

Brochard, F., and P. G. de Gennes, 1984, J. Phys. (Paris) Lett. **45**, L597.

Brockway, L., and R. Jones, 1964, in *Contact Angle, Wettability and Adhesion,* edited by F. M. Fowkes, Advances in Chemistry Series, No. 43 (American Chemical Society, Washington, D.C.), p. 275.

Burton, R. H., M. J. Folkes, K. A. Narh, and A. Keller, 1983, J. Mater. Sci. **18**, 315.

Cahn, J. W., 1977, J. Chem. Phys. **66**, 3667.

Callaghan, I., D. Everett, and J. P. Fletcher, 1983, J. Chem. Soc. Faraday Trans. **79**, 2723.

Callaghan, I. C., and K. W. Baldry, 1978, in *Wetting, Spreading and Adhesion,* edited by J. F. Padday (Academic, New York).

Chang, W. V., Y. M. Chang, L. J. Wang, and Z. G. Wang, 1982, *Organic Coatings and Applied Polymer Science Proceedings* (American Chemical Society, Washington, D.C.), Vol. 47.

Chappuis, J., 1984, in *Multiphase Science and Technology,* edited by G. F. Hewitt, J. Delhaye, and N. Zuber (Hemisphere, New York), p. 387.

Cohn, R., and D. Jacobs, 1983, J. Chem. Phys. **80**, 856.

Cooper, W., and W. Nuttal, 1915, J. Agr. Sci. **7**, 219.

Cox, R. G., 1983, J. Fluid Mech. **131**, 1.

de Gennes, P. G., 1979a, C. R. Acad. Sci. Ser. B **288**, 219.

de Gennes, P. G., 1979b, *Scaling Concepts in Polymer Physics* (Cornell University, Ithaca).

de Gennes, P. G., 1981, J. Phys. (Paris) Lett. **42**, 377.

de Gennes, P. G., 1983, C. R. Acad. Sci. **297 II**, 9.

de Gennes, P. G., 1984a, C. R. Acad. Sci. **298 II**, 111.

de Gennes, P. G., 1984b, C. R. Acad. Sci. **298 II**, 439.

de Gennes, P. G., 1984c, C. R. Acad. Sci. **298 II**, 475.

de Gennes, P. G., 1984d, C. R. Acad. Sci. **300II**, 129.

Deryagin, B., 1940, Zh. Fiz. Khim. **14**, 137.

Deryagin, B., 1955, Kolloidn. Zh. **17**, 191. This review describes the pioneering work of the Russian School.

Deryagin, B., and N. Churaev, 1976, Kolloidn. Zh. **38**, 438.

Deryagin B., Z. Zorin, N. Churaev, and V. Shishin, 1978, in *Wetting, Spreading and Adhesion,* edited by J. F. Padday (Academic, New York).

Dettre, R., and R. Johnson, 1964, in *Contact Angle, Wettability and Adhesion,* edited by F. M. Fowkes, Advances in Chemistry Series, No. 43 (American Chemical Society, Washington, D.C.), p. 136.

Dietrich, S., and M. Schick, 1984, Phys. Rev. B (in press).

Dussan, V. E., 1979, Annu. Rev. Fluid Mech. **11**, 371.

Dussan, V. E., and S. Davis, 1974, J. Fluid Mech. **65**, 71.

Dussan, V. E., and C. Ngan, 1982, J. Fluid Mech. **118**, 27.

Dzyaloshinskii, I. E., E. M. Lifshitz, and L. P. Pitaevskii, 1961, Adv. Phys. **10**, 165.
Ebner, C., and W. F. Saam, 1977, Phys. Rev. Lett. **38**, 1486.
Ebner, C., W. F. Saam, and A. K. Sen, 1984 (to be published).
Evans, R., and U. Marini, 1985, Chem. Phys. Lett. **114**, 415.
Evans, R., and P. Tarazona, 1984, Phys. Rev. Lett. **52**, 557.
Fowkes, F. M., 1962, J. Phys. Chem. **66**, 382.
Fowkes, F. M., 1964, Ed., *Contact Angle, Wettability and Adhesion*, Advances in Chemistry Series, No. 43 (American Chemical Society, Washington, D.C.).
Fox, H., and W. Zisman, 1950, J. Colloid Sci. **5**, 514.
Fritz, G., 1965, Z. Ang. Phys. **19**, 374.
Frumkin, A., 1938, J. Phys. Chem. USSR **12**, 337.
Galt, J., and B. Maxwell, 1964, Mod. Plastics, No. 12.
Garoff, S., and L. Schwartz, 1984, private communication.
Ghiradella, H., W. Radigan, and H. L. Frisch, 1975, J. Colloid Interface Sci. **51**, 522.
Girifalco, L. A., and R. J. Good, 1957, J. Phys. Chem. **61**, 904.
Gittes, F., and M. Schick, 1984, Phys. Rev. B **30**, 209.
Good, R. J., 1964, *Contact Angle, Wettability and Adhesion*, Advances in Chemistry Series, No. 43 (American Chemical Society, Washington, D.C.), p. 74.
Grinstein, G., and S. K. Ma, 1983, Phys. Rev. B **28**, 2588.
Guyon, E., J. Prost, C. Betrencourt, C. Boulet, and B. Volochine, 1982, Eur. J. Phys. **3**, 159.
Hardy, W., 1919, Philos. Mag. **38**, 49.
Hauge, E., and M. Schick, 1983, Phys. Rev. B **27**, 4288.
Heady, R., and J. Cahn, 1972, J. Chem. Phys. **58**, 896.
Hervet, H., and P. G. de Gennes, 1984, C. R. Acad. Sci. **299 II**, 499.
Hocking, L. M., 1976, J. Fluid Mech. **76**, 801.
Hocking, L. M., 1977, J. Fluid Mech. **79**, 209.
Hocking, L. M., 1981, Q. J. Mech. App. Math. **34**, 37.
Hocking, L. M., and A. Rivers, 1982, J. Fluid Mech. **121**, 425.
Hoffman, R., 1975, J. Colloid Interface Sci. **50**, 228.
Huh, C., and S. G. Mason, 1977, J. Colloid Interface Sci. **60**, 11.
Huh, C., and L. E. Scriven, 1971, J. Colloid Interface Sci. **35**, 85.
Huse, D., 1984, private communication.
Hyppia, J., 1948, Anal. Chem. **20**, 1039.
Imry, J., and S. K. Ma, 1975, Phys. Rev. Lett. **35**, 1399.
Israelashvili, J. N., 1982, Adv. Colloid Interface Sci. **16**, 31.
Israelashvili, J. N., 1984, private communication.
Joanny, J. F., 1985, Thèse, Université Paris VI.
Joanny, J. F., and P. G. de Gennes, 1984a, J. Chem. Phys. **81**, 552.
Joanny, J. F., and P. G. de Gennes, 1984b, C. R. Acad. Sci. **299 II**, 279.
Joanny, J. F., and P. G. De Gennes, 1984c, C. R. Acad. Sci. **299II**, 605.
Johnson, R., and R. Dettre, 1964, in *Contact Angle, Wettability and Adhesion*, edited by F. M. Fowkes, Advances in Chemistry Series, No. 43 (American Chemical Society, Washington, D.C.), p. 112.
Kravnik, A. M., and W. R. Schowalter, 1981, J. Rheology **25**, 95.
Kroll, D. M., and T. R. Meister, 1984, Phys. Rev. B (in press).
Landau, L. D., and V. G. Levich, 1942, Acta Physicochim. URSS **17**, 42.
Langmuir, I., 1938, J. Chem. Phys. **6**, 893.
Law, B., 1984 (unpublished).
Legait, B., and P. G. de Gennes, 1984, J. Phys. (Paris) Lett. **45**, 647.
Leibler, S., 1984, Thèse, Orsay.
Lelah, M., and A. Marmur, 1981, J. Colloid Interface Sci. **82**, 518.
Levich, V., 1962, *Physiochemical Hydrodynamics*, 2nd ed. (Prentice Hall, Englewood Cliffs, New Jersey).
Lipowsky, R., and D. M. Kroll, 1984, Phys. Rev. Lett. **52**, 2303.
Lopez, J., C. Miller, and E. Ruckenstein, 1976, J. Colloid Interface Sci. **56**, 460.
Lyklema, J., 1967, in *Study Week on Molecular Forces*, edited by Pontifical Academy of Science (North-Holland, Amsterdam and Wiley-Interscience, New York), pp. 181 and 221.
Marmur, A., 1983, Adv. Colloid Interface Sci. **19**, 75.
Mason, S. G., 1978, in *Wetting, Spreading and Adhesion*, edited by J. F. Padday (Adademic, New York), p. 321.
Michaels, A., and S. Dean, 1962, J. Phys. Chem. **66**, 1790.
Moldover, M., and J. Cahn, 1980, Science **207**, 1073.
Moldover, M., and R. Gammon, 1983, J. Chem. Phys. **80**, 528.
Moldover, M., and J. Schmidt, 1983, J. Chem. Phys. **79**, 379.
Nakanishi, H., and M. Fisher, 1982, Phys. Rev. Lett. **49**, 1565.
Nightingale, M. P., W. F. Saam, and M. Schick, 1983, Phys. Rev. Lett. **51**, 1275.
Ogarev, V., T. Timonina, V. Arslanov, and A. Trapeznikov, 1974, J. Adhesion **6**, 337.
Oliver, J., C. Huh, and S. Mason, 1977, J. Adhesion **8**, 223.
Overbeek, J. T. G., and A. Van Silfhout, 1967, in *Study Week on Molecular Forces*, edited by Pontifical Academy of Sciences (North-Holland, Amsterdam and Wiley-Interscience, New York), p. 143.
Owens, N. F., P. Richmond, and J. Mingins, 1978, in *Wetting, Spreading and Adhesion*, edited by J. F. Padday (Academic, New York), p. 127.
Padday, J. F., 1978, Ed., *Wetting, Spreading and Adhesion* (Academic, New York).
Pandit, R., M. Schick, and M. Wortis, 1982, Phys. Rev. B **26**, 5112.
Pashley, R. M., 1980, J. Colloid Interface Sci. **78**, 246.
Pismen, L. M., and A. Nir, 1982, Phys. Fluids **25**, 3.
Platikhanov, D., M. Nedyalkov, and A. Schelduko, 1980a, J. Colloid Interface Sci. **75**, 612.
Platikhanov, D., M. Nedyalkov, and A. Schelduko, 1980b, J. Colloid Interface Sci. **75**, 620.
Plesner, I., and O. Platz, 1968, J. Chem. Phys. **48**, 5361, footnote II.
Pohl, D., and W. Goldburg, 1982, Phys. Rev. Lett. **48**, 1111.
Pomeau, Y., 1983, C. R. Acad. Sci. **298 II**, 29.
Pomeau, Y., and J. Vannimenus, 1984, J. Colloid Interface Sci. (to be published).
Privman, V., 1984, J. Chem. Phys. **81**, 2463.
Pumir, A., and Y. Pomeau, 1984, C. R. Acad. Sci. (to be published).
Radigan, W., H. Ghiradella, H. L. Frisch, H. Schonhorn, and T. K. Kwei, 1974, J. Colloid Interface Sci. **49**, 241.
Renk, F., P. C. Wayner, Jr., and G. M. Homsy, 1978, J. Colloid Interface Sci. **67**, 408.
Ross, S., and R. Kornbrekke, 1984, J. Colloid Interface Sci. **99**, 446.
Rowlinson, J. S., J. A. Barker, and D. Henderson, 1981, private communication.
Rowlinson, J. S., and B. Widom, 1982, *Molecular Theory of Capillarity* (Oxford University, New York/London).
Ruckenstein, E., and M. Dunn, 1976, J. Colloid Interface Sci. **56**, 460; **59**, 137.
Sawicki, G., 1978, in *Wetting, Spreading and Adhesion*, edited by J. F. Padday (Academic, New York), p. 36.
Schonhorn, H., H. L. Frisch, and T. K. Kwei, 1966, J. Applied

Phys. **37**, 4967.
Snook, I. K., and W. Van Megen, 1979, J. Chem. Phys. **70**, 3099.
Snook, I. K., and W. Van Megen, 1980, J. Chem. Phys. **72**, 2907.
Starov, V. M., 1983, Kolloidnyi J. **45**, 1154.
Sullivan, D. E., 1979, Phys. Rev. B **20**, 3991.
Sullivan, D. E., 1982, Phys. Rev. A **25**, 1669.
Sullivan, D. E., and M. Telo da Gama, 1985, in *Fluctuation Effects on Surfaces*, edited by C. A. Croxston (Wiley, New York).
Tanner, L., 1979, J. Phys. D **12**, 1473.
Tarazona, P., and R. Evans, 1983, Mol. Phys. **48**, 799.
Tarazona, P., M. Telo da Gama, and R. Evans, 1983, Mol. Phys. **49**, 283.
Teletzke, G. F., H. T. Davis, and L. E. Scriven, 1984, J. Colloid Interface Sci. (to be published).
Teletzke, G. F., L. E. Scriven, and H. T. Davis, 1982, J. Chem. Phys. **77**, 5794.
Teletzke, G. F., L. E. Scriven, and H. T. Davis, 1983, J. Chem. Phys. **78**, 1431.
Telo da Gama, M., 1984, Mol. Phys. **52**, 585.
Trillat, J. and R. Fritz, 1937, J. Chim. Phys. **35**, 45.
Tverkrem, J., and D. Jacobs, 1983, Phys. Rev. A **27**, 2773.
Wayner, P. C., 1982, J. Colloid Interface Sci. **88**, 294.
Williams, R., 1977, Nature **266**, 153.
Young, T., 1805, Philos. Trans. R. Soc. London **95**, 65.
Zabel, H., B. Schönfeld, and S. Moss, 1981, J. Phys. Chem. Solids **42**, 897.
Zisman, W., 1964, in *Contact Angle, Wettability and Adhesion*, edited by F. M. Fowkes, Advances in Chemistry Series, No. 43 (American Chemical Society, Washington, D.C.), p. 1.

Afterthoughts: Wetting: statics and dynamics

The emphasis in this review is mainly on the dynamics of *total* wetting, with a non zero spreading coefficient S . The initial question was: where is the free energy S "burnt" when the liquid advances. The answer, explained in the text, is: it is "burned" in a precursor film.

During the last seven years, several problems connected with *partial wetting* have been solved:

a) The dynamics of dewetting — where a non-wettable surface grows a dry patch — has been understood and measured: C. Redon *et al., Phys. Rev. Lett.* **66**, 715 (1991).

b)The wetting of fibers (where the curvature of the fiber imposes a new component of film pressure) has been clarified: D. Queré *et al., Science* **249**, 1256 (1990).

c) The collective modes of one contact line have been studied: T. Ondarçuhu and M. Veyssié, *Nature* **352**, 418 (1991). On (a,b,c,), see paper 37 in the present book.

At the molecular scale, the most spectacular discovery was that of *terraced spreading*, observed by ellipsometry: F. Heslot *et al., Phys. Rev. Lett.* **62**, 1286 (1989). A theoretical discussion of the terraces is given in paper 34.

DYNAMICS OF DRYING AND FILM-THINNING

P.G. de Gennes

Collège de France 75231 Paris Cedex 05

(Abstract)

The spreading of non volatile liquids on ideal, wettable, solid surfaces is relatively well understood[1]. Here we transpose the same theoretical ideas to two problems where a metastable liquid film tends to thin out.

a) a thick soap film may nucleate a patch of much thinner "black" film;

b) a liquid film lying on a non wettable solid may nucleate a dry patch.

We consider here the dynamics of growth of the patch. They are basically the same for (a) and (b). The liquid which is expelled from the patch builds up a rim, which acts as a strong perturbation. The contact angle θ between patch and rim remains constantly much smaller than the equilibrium value θ_e ($\theta \to 0.7\ \theta_e$). The patch expands with a constant radial velocity, but the rim width b(t) (at time t) is expected to grow only like $t^{\frac{1}{2}}$.

I. PARTIAL / COMPLETE WETTING

Wetting and drying processes are important for many practical processes, involving paints, textiles, oil recovery, or the reinforcement of concrete by plastics... The simplest experiment amounts to put a droplet on a flat solid, and watch it's progressive spreading. But this requires a lot of care; impurities and surface irregularities can complicate matters enormously. Thus it is only recently that general laws have been clearly exhibited by experiments, and that the main dynamical classes have been identified.[1]

Let us start with a brief reminder of the equilibrium shape for a small liquid droplet on a horizontal solid plate (gravitational effects negligible).

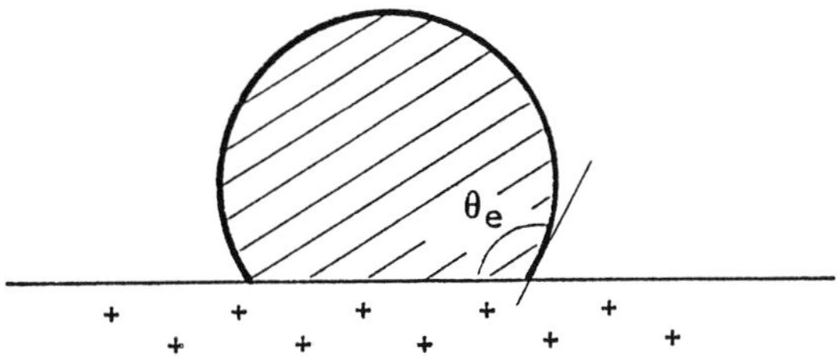

Fig.1 PARTIAL SPREADING

In situations of partial wetting the optimal state is a spherical cap, with a certain contact angle θ_e (fig.1). This angle is related to the interfacial energies γ_{SV} (solid/vapor), γ_{SL} (solid liquid) and $\gamma_{LV} = \gamma$ (liquid vapor) by the Young condition[2]

$$\gamma_{SL} + \gamma\cos\theta_e = \gamma_{SV} \tag{1}$$

When $\gamma_{SV} - \gamma_{SL} > \gamma$ this gives a finite angle θ_e. On the other hand, when $\gamma_{SV} - \gamma_{SL} < \gamma$, no real angle θ_e exists, and we enter a completely different regime, called complete wetting. To understand the physics chemical factors underlying this distinction, see the review by Zisman[2]. Complete wetting may occur in two distinct regimes.

a) moist spreading (when we have complete equilibrium with the vapor). Here, if the solid vacuum interfacial energy γ_{SO} is large, the real solid vapor interface will display an adsorbed liquid film, and the corresponding (renormalised) energy is

$$\gamma_{SV} = \gamma_{SL} + \gamma \tag{2}$$

This implies that $\theta_e = 0$ exactly, and is known as the Antonov rule.

b) in dry spreading (no vapor) this renormalisation does not take place, eq.(2) does not hold, and the deviations from it are measured by the so-called spreading parameter

$$S = \gamma_{SO} - \gamma_{SL} - \gamma \tag{3}$$

This S is positive for complete dry spreading.

What is the final state of a spreading droplet for complete wetting?

a) For moist spreading this is not a well posed question: to maintain the liquid/vapor equilibrium in the experimental cell we need a reservoir of liquid, and an essential parameter is the difference H in altitude between the spreading plate and the reservoir. The final film thickness depends on H.[1]

b) For dry spreading, there is a well defined answer, provided that S is known[3]. The final state is a thin film or "pancake" of area $\mathcal{A}$, thickness e, (fig.2).

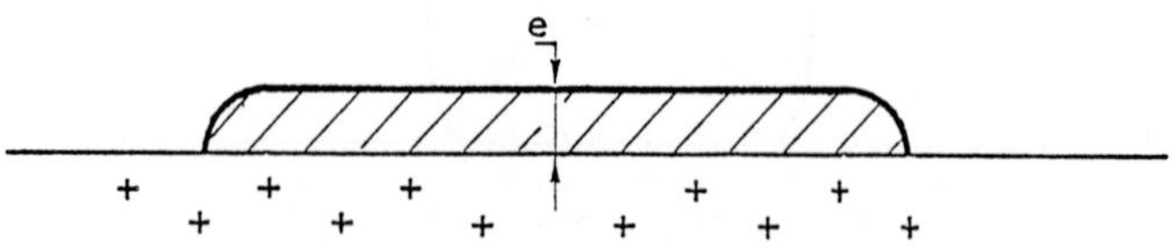

Fig. 2 "PANCAKE"

The free energy of the pancake is:

$$f = f_0 - S\mathfrak{a} + \mathfrak{a}P(e) \qquad (4)$$

where P(e) represents the long range effects of Van der Waals forces (which tend to thicken the film). The simplest form for P(e) is

$$P(e) = \frac{A}{12\pi e^2} \qquad (5)$$

where A is called the Hamaker constant[4]. Minimising (4) at constant volume (e) one finds

$$e = a\left(\frac{3\gamma}{2S}\right)^{\frac{1}{2}} \qquad (6)$$

where a is a molecular length defined by

$$a^2 = \frac{A}{6\pi\gamma} \qquad (7)$$

For small S/γ, this leads to sizeable thicknesses, which could be probed by ellipsometry: since He_4 spreads fast on solids, it may be a good candidate for these experiments. From general arguments[2] one should choose a solid of low polarisability (possibly H_2 solid) to reduce the S.

These considerations can be extended to films on vertical walls. The classical description of these films, constructed by the Russian school[5], and well verified on He_4 by thickness measurements[6] is based on a competition between Van der Waals and gravitational energies. However this film should be truncated at some finite altitude, where the thickness reduces to the value of eq.(6): again an experimental check on He_4 should be extremely interesting, but the altitudes involved are large (meters).

II. DYNAMICS OF DRY SPREADING

Macroscopic observations on a spreading droplet indicate that the shape near the (nominal) contact line is not far from a simple wedge, with a certain "apparent contact angle θ_a (fig.2). Experiments on viscous liquids pushed into capillaries or on spreading droplets[7,8,9] show that, in many cases of dry spreading, the angle θ_a is related to the line velocity U by

$$\theta_a^3 \sim U\eta/\gamma \qquad (U \to 0) \qquad (8)$$

where η = viscosity. The law (8) is very surprising at first sight. In the language of irreversible thermodynamics we may write the entropy source $T\Sigma$ (per unit length of line) in terms of a flux (U), and of a conjugate force F, which is the unbalanced Young force:

$$T\overset{\circ}{\Sigma} = FU = [S + \gamma(1-\cos\theta_a)]U \simeq [S + \tfrac{1}{2}\gamma\theta_a^2]U \quad (9)$$

How is it that the major component in the force (S) does not show up in the restult (8) ? The answer starts from an old observation by Hardy (1919) - summarised in Marmurs' review[9]: there is a precursor film moving ahead of the nominal contact line, with typical thicknesses 100 Å. The origin of this film is relatively obvious: it results from a balance between viscous forces and long range Van der Waals forces. For the simplest situation (when eq.5 holds) the thickness $\zeta(x)$ of the film at a distance x from the contact line decreases like $\zeta \sim 1/x$.[1] For more complex structures of the VW forces, numerical calculations of the film have been performed[10].

Two essential features emerge:

a) the precursor film is truncated: it stops abruptly when the local thickness becomes comparable to e(S) as given in eq.(6);

b) the dissipation in the film is huge. In fact, all the available free energy described by S is burned out in the film region. It is only the weak force $\frac{1}{2}\gamma\theta_a^2$ which operates in the macroscopic wedge, and this explains why S does not show up in the macroscopic law (8).

III. THINNING OF FILMS AND DRYING OF WET SURFACES

1) Situations of metastability

In many practical situations we deal with fluid films which are metastable:

a) a thick soap film (fig.3) floating on an or in a passive solvent, tends to reach a much smaller equilibrium thickness ("black" film) corresponding to a minimum in it's long range

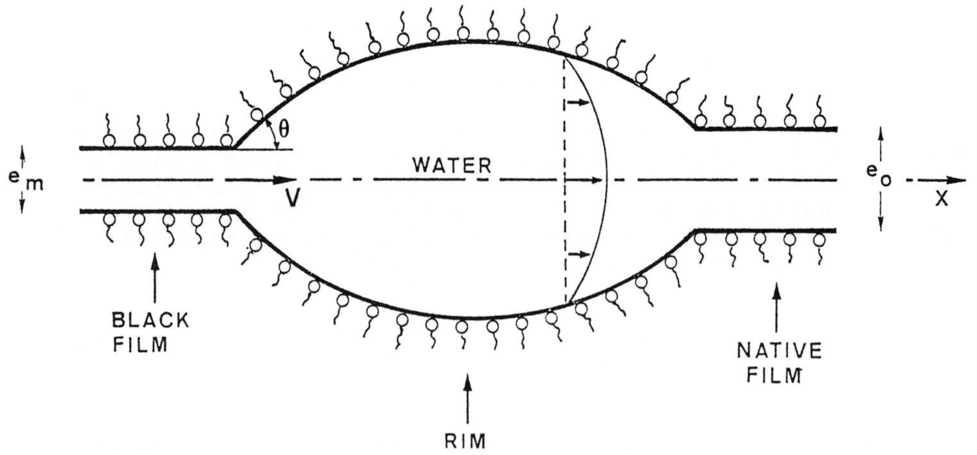

Fig.3 THINNING OF A SOAP FILM

energy P(e)[11]. Here the Van der Waals contribution to the energy P(e) is negative (the film is in a symmetric environment) and the minimum (e_m) results from a competition: Van der Waals attraction against short rang repulsions between the two sides (fig.4). If a black film has nucleated at some point, it will expand later as shown on fig.3.

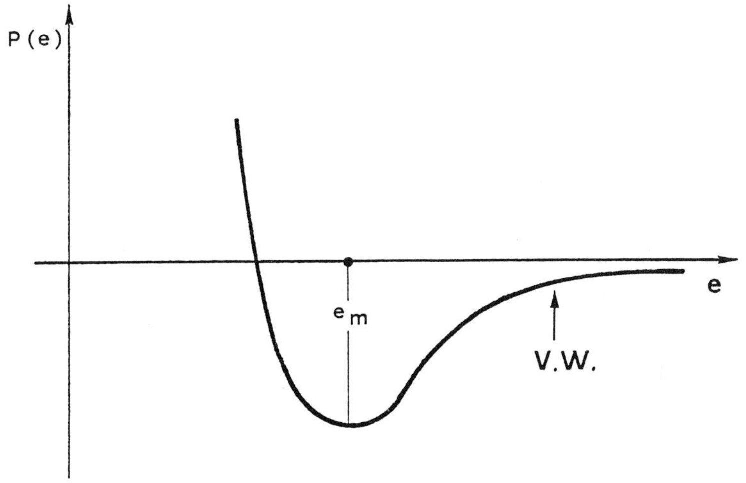

Fig.4 ENERGY/THICKNESS DIAGRAM FOR

b) a liquid film deposited on a non wettable solid (S < 0) is metastable: if a dry patch has nucleated (usually because of defects, or of mechanical noise) at some point on the solid, it will also expand (fig.5).

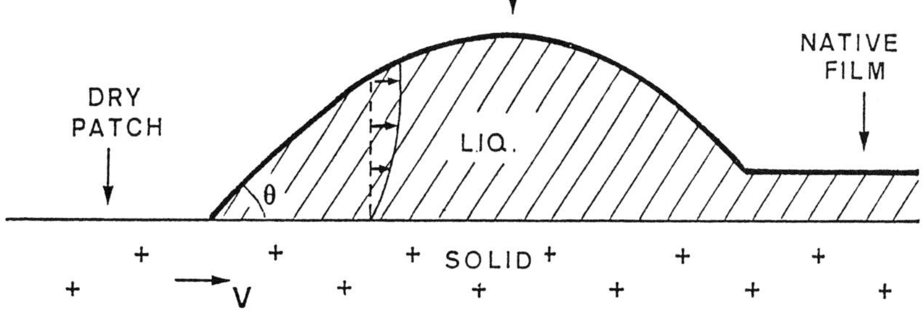

Fig.5 A SOAP FILM
DRYING OF A NON WETTABLE SOLID

2) Contact angles and Poiseuille flows

In the present text, we are not concerned by the nucleation process, but only by the growth of the new region (black film or dry patch). F. Brochard was the first to notice that both systems are ruled by the same equations.[12] To understand this, we note first that the central patch can exist in static equilibrium with the fluid only when a definite contact angle θ_e is reached. For the soap film problem, this contact angle is related to the (small) energy $P(e_m)$ and is itself small. If γ is the surface tension

$$2\gamma(1-\cos\theta_e) \sim \gamma\theta_e^2 = -P(e_m) \qquad (10)$$

For the drying problem θ_e is related to S, $(S < 0)$, and need not to be small: however, for simplicity, we shall still assume $\theta_e \ll 1$. In the soap film, water flows between two fixed walls of surfactant. In the drying problem, the liquid flows between a solid wall and a free surface. For both cases, we have simple Poiseuille flows, and the discussions of section II can be transposed. As before, we restrict our attention to one dimensional flows (along x).

3) Dynamics of the rim

A rim must appear ahead of the patch, and store all the liquid which has been expelled from the patch. As before, we expect the pressure in the rim to uniformize fast: then the rim cross section is a portion of a circle[13], making an angle θ with the flat patch, and an angle ϕ with the unperturbed film (fig.6). For clarity we shall now

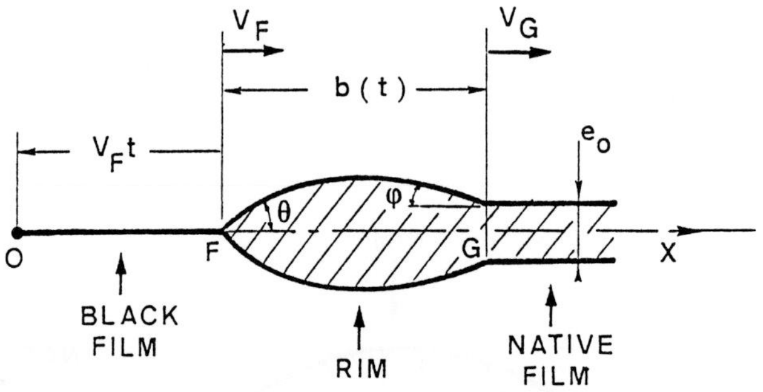

Fig.6 GEOMETRY OF PATCH AND RIM

discuss the film case which was analysed first[14]. The velocity V_F of the patch border has the form

$$V_F = V^*(\theta_e^2 - \theta^2)\theta \qquad (V^* = \text{const.}\gamma/\eta) \qquad (11)$$

Eq.(11) is derived from eq.(9), applied here to a condition of partial wetting $2\gamma(\cos\theta - \cos\theta_e)$ is the uncompensated Young force. In situations of partial wetting, there is no precursor film, and all the dissipation takes place in the wedge of angle θ.

A similar argument gives the velocity V_G of the outer end of the rim

$$V_G = V^*\phi^3 \tag{12}$$

because, at point G, the uncompensated Young force is simply $2\gamma(1 - \cos\phi)$.

We now assume, for simplicity, that the thickness of the patch (e_m) is very small ($e_m << e_o$), and set $e_m = 0$. The arc FG being a portion of a circle, we find

$$FG = b = \frac{e_o}{\theta - \phi} \tag{13}$$

and a cross sectional area for the rim

$$A = \frac{e_o^2(2\theta - \phi)}{3(\theta - \phi)^2} \tag{14}$$

The conservation of the volume inside the soap film imposes

$$A = (V_F t + b)e_o \tag{15}$$

where t is the time counted from the instant where F was at x = 0.

3) Fully developed growth

Eq.(15) shows that, when the time t increases, the cross sectional area A of the rim becomes large. Thus, returning to eq.(14), we see that the angles θ and ϕ must become equal $\theta(\infty) = \phi(\infty) = \theta_\infty$.

We shall check later the fact that the rim width b grows more slowly than the size of the dry region (Vt). If we accept this assumption, we must have (at large t) $V_F = V_G = V$. Eqs.(11) and (12) then give

$$\theta_\infty = \theta_e/\sqrt{2} \tag{16}$$

$$V = 2^{-3/2}\, V^*\theta_e^3 \tag{17}$$

For a typical soap film, $V^* = 3.10^3$ cm/sec and $\theta_e = 10^{-2}$, giving velocities V of order 10 microns/sec. Knowing V, we can now return to a discussion of the rim width b(t). Eq.(15) is practically equivalent to $A = Vte_o$, and this fixes the difference in angles through eq.(14)

$$\varepsilon \equiv \theta - \phi \sim \left(\frac{\theta_\infty e_o}{3Vt}\right)^{\frac{1}{2}} \tag{18}$$

$$b = e_o\varepsilon^{-1} = (3e_oVt/\theta_\infty)^{\frac{1}{2}} \tag{19}$$

All this holds when $\varepsilon << \theta_e$, or equivalently when

$$Vt >> e_0/\theta_e \tag{20}$$

and we can then check that $b/Vt = \varepsilon/\theta_e$ is indeed small as announced. When (20) is satisfied, we say that we have a fully developed growth.

IV. CONCLUSIONS

The thinning and drying processes should be ruled by rather simple laws, with a constant growth velocity V and a rim growing more slowly ($b \sim \sqrt{t}$). The major surprise is that, even at large times, the contact angle θ does not reach the equilibrium value: the growing rim creates a strong perturbation.

The velocity V depends critically on θ_e: for the soap films, with very small θ_e values, V is a slow velocity. But for the drying problem, where θ_e may be large, the velocity V may be of order 1 meter/sec.

Clearly we need systematic studies on both systems, with a "patch" which is nucleated by some external means (e.g. a focused laser pulse).

a) free soap films suspended in air are difficult to manipulate, and are very sensitive to parasitic effects of gravity (even if the film is nominally horizontal). One should preferably use a water film embedded in an organic solvent (comparable to the films present in a polygonal emulsion O/W with high oil content): here, by a suitable choice of density, the gravity problem can be efficiently suppressed.

b) for the drying problem, the problem is to work with a clean solid surface - avoiding any pinning of the contact line[15]. But drying is of such importance in applications that a systematic program would really deserve to be launched. Another geometry of interest is obtained with thin fibers, for which a theoretical discussion has recently been produced[12].

We should emphasize that our discussion of drying was limited to a one component liquid: more complex cases, with added surfactants or polymers, will require a separate analysis.

Acknowledgements: I have benefited from discussions with F. Brochard and H. Princen.

REFERENCES

1. P.G. de Gennes: In Rev. Mod. Phys. 57, 827, (1985).

2. W. Zisman: Advances in Chemistry 43, p.1, (ACS 1964).

3. J.-F. Joanny, P.G. de Gennes: C.R. Acad. Sci. (Paris) 299 II, 279, (1984).

4. See J. Israelashvili: Intermolecular and Surface Forces, Academic Press, NY, (1985).

5. L. Landau, I. Lifshitz: In Statistical Physics, Pergamon Press.

6. D.F. Brewer: In Physics of Solid and Liquid Helium (Benneman Ketterson eds.), Wiley, 1978.

7. R. Hoffmann: In J. Colloid Interface Sci. 50, 228, (1975).

8. L. Tanner: In J. Phys. D 12, 1473, (1979).

9. A. Marmur: In Adv. Colloid Interface Sci. 19, 75, (1983).

10. G. Teletzke, PhD Minneapolis, 1983.

11. K. Mysels, K. Shinoda, S. Frankel: In Soap Films, Pergamon, London, (1959).

12. F. Brochard, J.M. di Meglio, D. Quere: C.R. Acad. Sci. (Paris) 304 II, 553, (1987).

13. The pressure p in the rim is constant; outside we have the atmospheric pressure p_a. The difference $(p_a - p) = \gamma/R$ where R is the radius of curvature of the profile. Hence R is constant.

14. P.G. de Gennes: In C.R. Acad. Sci. (Paris), II 303, 1275, (1986).

15. J.-F. Joanny, P.G. de Gennes: In J. Chem. Phys. 81, 552, (1984).

Afterthought: Dynamics of drying and film-thinning

Film thinning is extremely hard to observe in neat conditions — even when the film is horizontal, the thick regions tend to bend it downwards. On the other hand, the growth of dry patches on a non -wettable surface (initially covered by a film) can be studied precisely, using very clean silicon wafers as the basic surface: C. Redon *et al.*, *Phys. Rev. Lett.* **66**, 715 (1991).

C. R. Acad. Sci. Paris, t. 307, Série II, p. 1841-1844, 1988 1841

Physique des surfaces et des interfaces/*Surface and Interphase Physics*

Tension superficielle des polymères fondus

Pierre-Gilles de GENNES

Résumé — La tension superficielle γ_N d'un polymère linéaire de N monomères est souvent un peu inférieure à sa limite pour N élevé : $\gamma_N = \gamma_\infty - \gamma_1 N^{-x}$ avec $x \sim 0,6$. Nous proposons une interprétation de ce résultat dans laquelle les groupes terminaux sont attirés par la surface. Si l'énergie attractive Δ dépasse un certain seuil, la fraction de sites superficiels occupés par des groupes terminaux est $\varphi_s \sim N^{-1/2}$ et $x = 1/2$. Mais, très souvent, on s'attend à ce que Δ soit en dessous du seuil, auquel cas le comportement doit être intermédiaire entre $x = 0,5$ et $x = 1$. Nous construisons une formule approchée pour ce régime intermédiaire.

Surface tension of molten polymers

Abstract — *The surface tension γ_N of a linear polymer (with N monomers per chain) is often slightly smaller than the limit value γ_∞ reached at high N: $\gamma_N = \gamma_\infty - \gamma_1$ with $x \sim 0.6$. We interpret this result in terms of an* attraction *between the terminal groups and the free surface. If the attractive energy Δ is larger than a certain crossover value, the fraction of surface sites occupied by terminal groups is expected to scale like $N^{-1/2}$ and this gives $x = 0.5$. However, in many practical situations, we expect smaller values of Δ, and the apparent exponent x should lie between 0.5 and 1. We propose a rough interpolation formula for this regime.*

INTRODUCTION. — Pour beaucoup de polymères fondus, la tension superficielle γ_N croît avec la masse moléculaire et tend finalement vers une limite γ_∞. Ceci suggère que les groupes terminaux ont un (faible) rôle tensioactif; un fluide formé totalement de groupes terminaux aurait une tension $\gamma_\infty - \Delta$ avec $\Delta > 0$. Les résultats de Gaines et Legrand ([1], [2]) se mettent assez bien sous la forme

$$\gamma_N = \gamma_\infty - \gamma_1 N^{-x} \tag{1}$$

avec $x \cong 2/3$. Ceci est surprenant à première vue : dans l'hypothèse où Δ est faible, on attendrait que la fraction de surface φ_s occupé par des groupes terminaux ait sa valeur non perturbée, soit $\varphi_s = \varphi_0 \equiv 2/N$ d'où $\gamma_1 = 2\Delta$ et $x = 1$. Une autre forme analytique

$$\gamma_N^{1/4} = \gamma_\infty^{1/4} - k N^{-1} \tag{2}$$

a été proposée par S. Wu et décrit assez bien les mesures [3]. Asymptotiquement pour $N \to \infty$, la loi (2) a encore la forme (1) avec $x = 1$. Mais l'origine de l'exposant 1/4 dans (2) est obscure.

Nous voulons proposer ici une interprétation un peu plus fondamentale de (1) : l'idée est que la fraction d'extrémités en surface φ_s est *augmentée* par l'attraction Δ. Ceci conduit à plusieurs régimes possibles, qui sont schématisés sur la figure.

II. LES TROIS RÉGIMES. — Le paramètre central est $u = \Delta a^2/kT$ où a^2 est une surface par monomère [4], et kT l'énergie thermique.

(a) Pour $u \ll 1$, on attend un interface non perturbé avec des chaînes idéales. Les chaînes qui touchent la surface occupent une épaisseur d'influence de l'ordre du rayon $R_0 = N^{1/2} a$. La concentration de groupes terminaux en surface est $\varphi_0 = 2/N$ et la tension superficielle vaut $\gamma_\infty - \Delta\varphi_0$ comme décrit plus haut.

(b) Pour $u \sim 1$, φ_s est augmenté : avec des distorsions élastiques faibles, on parvient à fixer sur la surface tous les groupes terminaux de la zone d'influence. Si tous ceux qui étaient dans l'épaisseur R_0 sont capturés, on tend vers $\varphi_s = (R_0/a)\,\varphi_0 \sim N^{-1/2}$.

Note présentée par Pierre-Gilles de GENNES.

1842 C. R. Acad. Sci. Paris, t. 307, Série II, p. 1841-1844, 1988

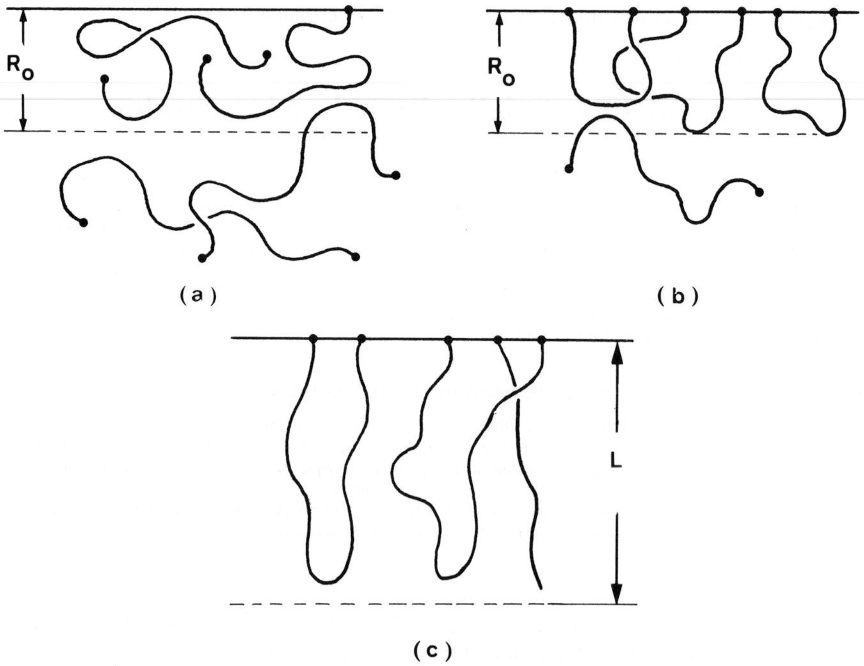

Trois régimes possibles pour la surface libre : (*a*) attraction très faible entre groupes terminaux et surface : seule une fraction $\varphi_0 = 2/N$ des sites de surface est occupé par des groupes terminaux; (*b*) attraction moyenne : tous les groupes terminaux initialement présents dans une épaisseur $R_0 = N^{1/2} a$ sont capturés par la surface, mais les chaînes sont peu déformées; (*c*) attraction forte : les chaînes sont étirées.

Three types of behavior at the free surface of a molten polymer: (a) weak attraction between end-groups and surface: a small fraction $\varphi_0 = 2/N$ of the surface sites is occupied by end-groups; (b) intermediate attractive energies: the end groups which were originally lying in a thickness $\sim R_0 = N^{1/2} a$ are trapped at the surface; (c) strong attractions: the chains are stretched.

(*c*) Pour $u > 1$, φ_s augmente et les chaînes doivent s'étirer pour arriver à mettre de nombreux groupes terminaux en surface [5] : l'épaisseur L de la zone d'influence devient supérieure à R_0 et on attend une énergie élastique par chaîne :

(3) $$f_{el} \sim k\,T\,\frac{L^2}{R_0^2} \qquad (L > R_0)$$

Dans toutes les formules telles que (3), nous ignorons les coefficients numériques.

L'épaisseur L est minimale, si toutes les chaînes de la région d'influence sont étirées, ce qui correspond à

(4) $$\Sigma\,L = N\,a^3$$

où $\Sigma = 2\,a^2\,\varphi_s^{-1}$ est la surface exposée par chaîne. Donc

(5) $$f_{el} \cong k\,TN\,\varphi_s^2$$

Finalement, l'énergie de surface prend la forme

(6) $$\gamma(\varphi_s) = \gamma_\infty + \frac{k\,T}{a^2}\,N\,\varphi_s^3 - \Delta\varphi_s$$

et l'optimum correspond à

(7) $$\varphi_s \cong (u/N)^{1/2}$$

(8) $$\gamma_N = \gamma_\infty - (\text{Cte})\,\Delta\,(u/N)^{1/2}$$

C. R. Acad. Sci. Paris, t. 307, Série II, p. 1841-1844, 1988

Donc, dans ce régime de « surfactant fort », on a bien une loi de type (1) avec $x=1/2$. Notons que, avec $u \sim 1$, on retrouve notre prédiction pour le régime (*b*).

III. Discussion. — (A) *Régimes de surfactant faible* ($u \gtrsim 1$). — Le régime de « surfactant fort » (*c*) serait le plus simple à comprendre : mais il n'est sûrement pas réalisé en pratique lorsque la structure chimique du groupe terminal est peu différente de celles du monomère. Par exemple, pour des interactions Van der Waals, associées à des polarisabilités α_t (groupe terminal) et α_m (monomère), on attend

$$\frac{\Delta}{\gamma_\infty} \sim \frac{\alpha_m^2-\alpha_t^2}{\alpha_m^2} = \varepsilon < 1 \qquad (9)$$

et ceci implique (en phase fondue) $u \lesssim 1$ (car $\gamma_\infty a^2 \sim k\mathrm{T}$). On doit donc souvent réaliser une situation intermédiaire, où l'exposant *apparent* de la tension superficielle x est comprise entre 0,5 et 1.

On peut essayer d'interpoler de façon un peu plus fine entre le régime (*a*) et le régime (*b*). Considérons d'abord (parce que l'écriture est un peu plus simple) le cas où une seule extrémité de chaque chaîne est tensioactive, alors que l'autre est sans effet. Alors on peut écrire la concentration de surface sous la forme

$$\varphi_s(\Delta) = \frac{1}{\mathrm{N}^{1/2}} \frac{a/\mathrm{R}_0\, e^u}{1+a/\mathrm{R}_0\, e^u} \qquad (10)$$

L'équation (10) exprime que chaque extrémité active peut être dans deux situations : soit repartie dans l'épaisseur d'influence (poids statistique 1), soit localisée à la surface [poids statistique $(a/\mathrm{R}_0)\,e^u$]. Notons que, pour $u \to 0$, l'équation (10) redonne $\varphi_s = \mathrm{N}^{-1}$. Mais pour $u \gg (1/2)\ln \mathrm{N}$, on a $\varphi_s \sim \mathrm{N}^{-1/2}$.

Si les deux extrémités de la chaîne sont tensioactives, on aurait, par un décompte analogue :

$$\varphi_s(\Delta) = \mathrm{N}^{-1/2} \frac{2(\mathrm{N}^{-1/2} e^u + \mathrm{N}^{-1} e^{2u})}{1+2\mathrm{N}^{-1/2} e^u + \mathrm{N}^{-1} e^{2u}} \qquad (10')$$

Dans tous les cas la tension superficielle s'obtient à partir de $\varphi_s(\Delta)$ par la formule

$$\gamma_\infty - \gamma = \int_0^\Delta d\Delta'\, \varphi_s(\Delta') \qquad (11)$$

$$= 2k\mathrm{T}\,a^{-2}\mathrm{N}^{-1/2}\ln(1+\mathrm{N}^{-1/2}e^u) \qquad (12)$$

L'équation (12) est acceptable entre $u=0$ et $u \cong 1$. Par choix convenable de u, elle permet de reproduire assez bien les données de la littérature ([1], [2], [3]). Mais, comme d'habitude, cet accord n'est pas en lui-même très probant : plusieurs formes analytiques très différentes donnent satisfaction.

(B) *Régimes de surfactant fort.* — Si le groupe terminal est très différent du monomère, plusieurs cas sont possibles :

(1) les groupes terminaux se groupent en paires, et suscitent par exemple certaines liaisons hydrogènes. Ce cas paraît avoir été observé [5] : les chaînes « effectives » sont beaucoup plus longues que N, et $\gamma \to \gamma_\infty$ pour tout N;

(2) les groupes terminaux ne s'associent pas, et on arrive au régime de surfactant fort [équation (8)];

1844 C. R. Acad. Sci. Paris, t. 307, Série II, p. 1841-1844, 1988

(3) les groupes terminaux tendent à s'ériger pour former une microphase (micellaire, lamellaire, ou autre...). Le coefficient χ de Flory qui donnait cette tendance à la ségrégation serait *grosso modo* de la forme

$$\left\{\begin{array}{l}\chi=\chi_1 2/N \\ \chi_1 \sim(\alpha_m-\alpha_t)^2\end{array}\right. \tag{10}$$

Le facteur N^{-1} dans (10) exprime la faible fraction de groupes terminaux. La démixtion survient en gros lorsque $\chi N \sim 1$ ou $\chi_1 \sim 1$. Elle ne doit pas se produire pour $\varepsilon<1$ (noter que $\chi_1 \sim \varepsilon^2$) mais elle peut se produire pour $\Delta/\gamma_{\infty} \sim 1$.

(C) *En conclusion*, il semble que le régime usuel soit celui de surfactant faible, pour lequel la formule (12) devrait permettre de systématiser les données, et de déterminer pour chaque système la constante de couplage u. Il serait possible de modifier u systématiquement en étudiant non pas la tension superficielle, mais la tension interfaciale entre le polymère et un non-solvent, qui reste un peu compatible avec les groupes terminaux.

Notons enfin que la présence de nombreux groupes terminaux en surface doit modifier profondément la cinétique de la *soudure polymère-polymère*.

Note remise le 27 septembre 1988, acceptée le 4 novembre 1988.

RÉFÉRENCES BIBLIOGRAPHIQUES

[1] D. LE GRAND et G. GAINES, *J. Colloid Interf. Sci.*, 31, 1969, p. 162-170.

[2] D. LE GRAND et G. GAINES, *J. Colloid Interf. Sci.*, 42, 1973, p. 181-190.

[3] SOUHENG WU, *Polymer Interfaces and Adhesion*, Marcel Dekker, 1982, chap. 3.

[4] Nous nous référons constamment à un modèle dans lequel les chaînes sont inscrites sur un réseau de maille *a*. Les groupes terminaux, eux aussi, occupent un site et un seul de ce réseau.

[5] Pour une discussion générale de polymères fixés à une extrémité sur un interface, *voir* par exemple P.-G. de GENNES, *Macromolécules*, 13, 1980, p. 1069-1075.

Laboratoire de Physique de la Matière condensée,
Collège de France, 75231 *Paris Cedex* 05.

Afterthought: Tension superficielle des polymères fondus

What is the origin of the coupling (u)?

1) There are obvious enthalpic contributions to Δ or u: if the specific polarisability α_e of the chain ends is smaller than the bulk α, one expects the ends to prefer the interface.

2) There is also an entropic term, first recognised by the IBM group, which can be understood as follows: there is an entropic restriction whenever a chain is "reflected" at the free surface. But there is no such restriction for a chain end. This tends to give an entropy shift of order 1 unit per chain end at the surface, or equivalently $u \sim 1$. Thus, if enthalpic effects are small, we are automatically in the intermediate coupling regime. These observations may have some relevance for polymer/polymer welding: see paper 38.

C. R. Acad. Sci. Paris, t. 310, Série II, p. 1601-1606, 1990 1601

Physique des surfaces et des interfaces/*Surface and Interface Physics*

Étalement d'une goutte stratifiée incompressible

Pierre-Gilles de GENNES et Anne-Marie CAZABAT

Résumé – L'étalement de films ultraminces de liquides simples sur une surface lisse horizontale conduit souvent à des *gouttes à gradins* d'une épaisseur moléculaire ([1]-[3]). Nous construisons ici un modèle pour la dynamique de l'étalement, en supposant que : (*a*) chaque gradin est un liquide bidimensionnel incompressible; (*b*) les molécules de la *n*-ième couche sont soumises à un potentiel (W_n) qui exprime leurs interactions à longue portée avec le solide; (*c*) on a un écoulement parallèle de cisaillement, et un écoulement normal de perméation.

La perméation intervient seulement dans un certain « ruban » de largeur ξ sur le bord de chaque gradin. Dans le cas usuel, où le rayon du *n*-ième gradin R_n est trs supérieur à ξ, la dissipation est dominée par le cisaillement, et on arrive à des lois simples pour la vitesse de dilatation (ou de contraction) $\dot{R}_n$.

Spreating of a stratified incompressible droplet

Abstract – *Very thin films of simple liquids often spread with well defined steps of molecular thickness* ([1]-[3]). *We construct a model for the dynamics of spreading assuming that:* (*a*) *each layer is an incompressible, 2 dimensional fluid;* (*b*) *the molecules in the (nth) layer experience a long range potential* (*e.g. Van der Waals*) *from the solid;* (*c*) *two types of flow occur: shear between layers and permeation normal to the layers.*

We find that permeation is important only in an annulus of small size ξ *near each step. Between steps, the viscous effects in simple shear dominate, and this leads to simple laws for the dilation* (*or contraction*) *of the various layers.*

I. OBJECTIFS. – Certains liquides simples, étalés en films très minces sur un support solide lisse, montrent des *gradins moléculaires* visibles en ellipsométrie ([1]-[3]). Ce phénomène est apparenté aux observations d'Israelashvili et coll. ([4]-[6]) sur les forces oscillantes existant entre deux plaques de mica séparées par du liquide. Dans une précédente Note, l'un de nous a analysé comment la présence de couches était associée à des plateaux du potentiel chimique μ en fonction de la quantité de molécules adsorbées [7]. Ici, nous construisons un modèle dynamique plus spécifique lorsque le système est fortement quantifié en couches, d'épaisseur $a=v/\Sigma$ (v = volume par molécule, Σ = surface par molécule).

1. Nous supposons que chaque couche se comporte comme un *liquide incompressible* ($\Sigma = \text{Cte}$). Ceci exclut certains cas : par exemple une monocouche de squalane finit par s'étaler aux temps très longs, avec un profil de concentration gaussien, ce qui suggère un gaz bidimensionnel plutôt qu'un liquide.

2. Nous discutons ici des liquides formés de **petites molécules**. Certaines observations montrent des gradins même avec des longues chaînes de polydimethyl-siloxane [8]. Mais la dynamique de chaînes (enchevêtrées) serait plus complexe.

3. L'étalement est entraîné par les forces à longue portée entre le support et le liquide. La goutte a N gradins de rayons $R_1 > R_2 > R_N$ (*fig.*). L'énergie libre requise pour former la goutte à partir d'un réservoir est :

$$f = \sum_{n=1}^{n=N} W_n \frac{\pi R_n^2}{\Sigma} \tag{1}$$

Note présentée par Pierre-Gilles de GENNES.

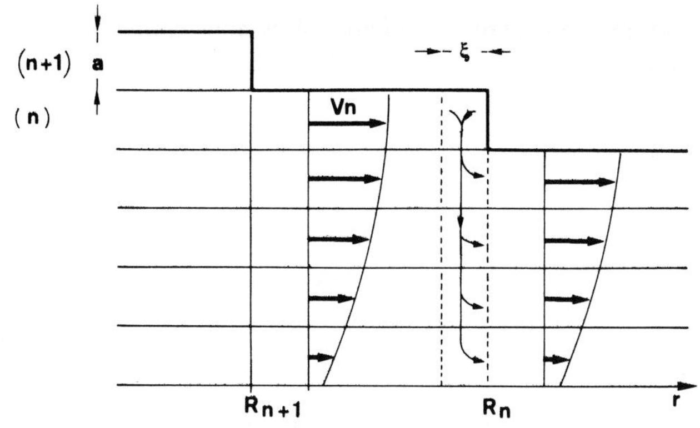

Écoulements dans une goutte à gradins.

Flow patterns in a terraced droplet.

Ici, (pour $n>1$), W_n représente l'interaction d'une molécule de la n-ième couche avec le substrat solide. Par exemple, dans un modèle simple de Van der Waals:

(2) $$W_n = -\frac{A\Sigma}{6\pi a^2 n^3} \qquad (n>1)$$

A étant une constante de Hamaker [9]. Le terme W_1 est spécial, car il contient les énergies capillaires dues aux forces à courte portée :

(3) $$W_1 = -S_0 \Sigma$$

où S_0 est le paramètre d'étalement [10] d'une monocouche. Il faut bien distinguer S_0 du paramètre d'étalement S d'un film épais :

(4) $$S = -\frac{1}{\Sigma}\sum_{n=1}^{\infty} W_n$$

II. Lois d'écoulement. — Nous attribuons à chaque couche (n) un champ de vitesses horizontal $\mathbf{V}_n(xy) = \mathbf{V}_n(\mathbf{r})$ (*fig.*). Mais nous permettons aussi à une couche de se vider dans la suivante, par un processus de *perméation* analogue à celui imaginé par Helfrich pour les smectiques [11]. Le courant J_z (volume/cm^2/s) normal aux couches, entre les couches n et $n+1$ vaut :

(5) $$J_{zn} = C_n(\mu_n - \mu_{n+1})$$

où μ_n est le potentiel chimique local

(6) $$\mu_n = \mu_0 + vp + W_n$$

μ_0 étant un potentiel de référence (qui sera omis dans la suite) et p la pression. L'équation de conservation s'écrit :

(7) $$a \operatorname{div} \mathbf{V}_n = J_{z(n-1)} - J_{zn} \qquad (1<n<N)$$

sauf pour les couches extrêmes ($n=1$ et $n=N$) où l'un des termes de (7) disparaît.

Dans le plan des couches, nous écrirons :

(8) $$-\nabla \mu_n = \zeta_n(\mathbf{V}_n - \mathbf{V}_{n-1}) + \zeta_{n+1}(\mathbf{V}_n - \mathbf{V}_{n+1})$$

où ζ_n est un coefficient de friction entre les couches $(n-1)$ et (n). Nous prendrons :

$$\zeta_n = \zeta = \eta\,\Sigma/a \qquad (n>1)$$

C. R. Acad. Sci. Paris, t. 310, Série II, p. 1601-1606, 1990

où η est la viscosité macroscopique du liquide. Mais nous garderons une valeur particulière à ζ_1, qui décrit la friction sur le solide. Les conditions aux limites sont $p_n=0$ (ou $\mu_n=W_n$) au bord de chaque gradin ($r=R_n$).

III. Exemple a deux couches. — 1. *Le ruban de perméation.* — Les équations hydrodynamiques, dans la région ($r<R_2$), se réduisent à :

$$\left\{\begin{array}{l} -\nabla\mu_1=\zeta_1 V_1+\zeta(V_1-V_2) \\ -\nabla\mu_2=\zeta(V_2-V_1) \end{array}\right. \tag{9}$$

$$\operatorname{div} \mathbf{V}_2=-\operatorname{div} \mathbf{V}_1=k(\mu_1-\mu_2) \tag{10}$$

(avec $k=C_1/a$). On en tire :

$$\xi^2\nabla^2(\mu_1-\mu_2)=\mu_1-\mu_2 \tag{11}$$

où ξ est une « largeur de perméation » microscopique, définie par :

$$\xi^{-2}=k(\zeta_1+4\zeta) \tag{12}$$

Dans toute la suite, nous supposerons $R_2 \gg \xi$. Alors la différence de potentiel $\mu_2-\mu_1$ n'est importante que dans un « ruban » ($R_2-\xi<r<R_2$) près du gradin :

$$\mu_2-\mu_1=\Delta \exp[(r-R_2)/\xi] \qquad (r<R_2) \tag{13}$$

Pour $r=R_2$, on a $\mu_2=W_2$, donc :

$$\mu_1(R_2)=W_2-\Delta \tag{14}$$

Tout le liquide qui disparaît de la deuxième couche descend par perméation dans la première :

$$2\pi R_2 a\dot{R}_2=\int_0^{R_2} J_{z_1}(r)\,2\pi r\,dr \cong -2\pi R_2 C_1\xi\Delta \tag{15}$$

2. *Équilibration des potentiels chimiques.* — Pour ($R_2<r<R_1$), on a simplement $-\nabla\mu_1=\zeta_1 V_1$ et div $V_1=0$, d'où $\nabla^2\mu_1=0$, et finalement :

$$\mu_1=(W_2-W_1-\Delta)\frac{\ln(R_1/r)}{\ln(R_1/R_2)}+W_1 \tag{16}$$

L'équation (16) satisfait à la condition (14) et à la condition aux limites $\mu_1(R_1)=W_1$. On peut alors calculer le flux :

$$2\pi r a V_1(r)=a\,2\pi R_1\dot{R}_1=-2\pi R_2\dot{R}_2 a \tag{17}$$

$$=2\pi(W_2-W_1-\Delta)\frac{a}{\zeta_1\ln(R_1/R_2)} \tag{18}$$

Les équations (15) et (18) permettent de trouver Δ

$$\Delta(k\xi\zeta_1 R_2\ln(R_1/R_2)+1)=W_2-W_1 \tag{19}$$

Il est important de noter que :

$$k\xi\zeta_1 R_2 \sim \frac{R_2}{\xi} \gg 1$$

On en déduit $\Delta/(W_2-W_1)\sim\xi/R_2\ll 1$. La perméation égalise rapidement les potentiels chimiques ($\mu_1=\mu_2$) au voisinage du gradin ($r=R_2$).

1604 C. R. Acad. Sci. Paris, t. 310, Série II, p. 1601-1606, 1990

3. *Évolution macroscopique.* — Puisque $\Delta \ll W_2 - W_1$, nous pouvons revenir à l'équation (18) et y écrire $\Delta \to 0$, ce qui donne :

$$R_1 \dot{R}_1 = -R_2 \dot{R}_2 = \frac{W_2 - W_1}{\zeta_1 \ln(R_2/R_1)} \tag{20}$$

IV. Système multicouches. — Pour un nombre N de couches supérieur à 2 (mais fini), on peut vérifier directement, à partir des équations (7) et (8), la propriété analogue de l'équation (11) : toutes les différences $\mu_n - \mu_{n-1}$ relaxent sur des longueurs ξ finies près du gradin. [Les ξ^2 sont ici les valeurs propres d'une matrice hermitique $(N-1)\times(N-1)$, et sont réelles et positives.]

Dans la limite $\xi \ll R$, on aboutit alors à une description hydrodynamique simplifiée :

(*a*) la perméation a pour effet d'uniformiser les potentiels chimiques pour $r = R_n$

$$\mu_p(r = R_n) = W_n, \qquad 1 \leqq p \leqq n \tag{21}$$

(*b*) la distribution de μ dans l'intervalle $(R_{n+1} < r < R_n)$ satisfait à $\nabla^2 \mu = 0$, d'où [comme dans l'équation (16)] :

$$\mu_p = (W_{n+1} - W_n)\frac{\ln(R_n/r)}{\ln(R_n/R_{n+1})} + W_n \tag{22}$$

(*c*) le champ de vitesse $\mathbf{V}_p(r)$ est parabolique

$$V_p(r) = A_0 p^2 + B_0 p + C_o \tag{23}$$

Les coefficients A_0, B_0, C_0 sont déterminés par comparaison avec les équations (8)

$$\begin{cases} A_0 = \dfrac{1}{2}\zeta^{-1}\nabla\mu \\ B_0 = -\left(n + \dfrac{1}{2}\right)\zeta^{-1}\nabla\mu \\ C_0 = -n(\zeta_1^{-1} - \zeta^{-1})\nabla\mu \end{cases} \tag{24}$$

Désignons par Q_n le flux radial dans l'intervalle $(R_{n+1} < r < R_n)$

$$Q_n = 2\pi r a \sum_1^n V_p = -\zeta^{-1}\nabla\mu F_n 2\pi r a \tag{25}$$

$$F_n = \frac{n(n+1)(2n+1)}{6} + n^2\left(\frac{\zeta}{\zeta_1} - 1\right) \tag{26}$$

Le gradient $\nabla\mu$ est connu par l'équation (22), d'où :

$$Q_n = 2\pi a \zeta^{-1}(W_{n+1} - W_n)\frac{F_n}{\ln(R_n/R_{n+1})} \tag{27}$$

Nous pouvons maintenant trouver les vitesses $\dot{R}_n$ des gradins par la relation de conservation :

$$2\pi a R_n \dot{R}_n = Q_n - Q_{n-1} \tag{28}$$

avec les cas spéciaux :

$$\begin{cases} 2\pi a R_N \dot{R}_N = -Q_{N-1} \\ 2\pi a R_1 \dot{R}_1 = +Q_1 \end{cases} \tag{29}$$

Il est important de noter que dans la région centrale $(r \gtrsim R_N - \xi)$, le potentiel chimique est constant $(\mu_N = W_N)$, et il n'y a aucun écoulement.

C. R. Acad. Sci. Paris, t. 310, Série II, p. 1601-1606, 1990 1605

V. Discussion. — 1. *Dilatation ou contraction.* — Les vitesses de gradin $\dot{R}_n$ définies par (28) et (29) sont positives pour $n=1$, négatives pour $n=N$, et satisfont à $\sum_n R_n \dot{R}_n = 0$.

Pour examiner le signe de $\dot{R}_n (1 < n < N)$, le cas limite où n est assez grand ($\gg 1$) est intéressant : on peut écrire (28) dans la limite continue, sous la forme :

$$\zeta \dot{R}_n R_n = \frac{\partial}{\partial n}\left[\frac{n^3}{3}\frac{\partial W}{\partial n}\frac{(-) R_n \partial n}{\partial R_n}\right] \tag{30}$$

Si par exemple W est contrôlé par des interactions de Van der Waals [équation (2)], on a :

$$\dot{R}_n R_n \to \frac{-A\Sigma}{6\pi a^2 \zeta}\frac{\partial}{\partial n}\left[\frac{\partial \ln(n)}{\partial \ln(R_n)}\right] \tag{31}$$

et suivant l'aspect des diagrammes ($\ln(n)/\ln(R_n)$) à l'instant étudié, le membre de droite peut être *positif ou négatif*. Par exemple, si nous partons d'une goutte en forme de calotte sphérique :

$$R_n^2\big|_{t=0} = R_1^2 \frac{N+1-n}{N} \tag{32}$$

nous trouverons pour les vitesses initiales, dans la limite continue :

$$R_n \dot{R}_n\big|_{t=0} = -\frac{A\Sigma}{6\pi a^2 \zeta}\frac{2(N+1)}{n^2} \tag{33}$$

c'est-à-dire que tous les gradins (sauf $n=1$) commencent par se rétracter.

2. *Autres situations dynamiques.* — Considérons un système de gradins parallèles (le long de y) qui avance (dans la direction $+x$), avec une vitesse constante U par rapport au support. Les gradins ont des positions $x_1 \ldots x_n \ldots$: dans un référentiel (K) qui avance avec la vitesse U, ces positions sont stationnaires. Si les vitesses dans le référentiel du solide sont $V_1(x), \ldots, V_n(x)$, la condition

$$-nU + \sum_1^n V_p(x) = 0 \tag{34}$$

assure un flux nul pour chaque section verticale dans le référentiel K. Dans chaque tranche ($x_n < x < x_{n+1}$), le gradient de potentiel chimique est constant :

$$\frac{d\mu}{dx} = \frac{W_n - W_{n+1}}{x_n - x_{n+1}} \tag{35}$$

Les profils $V_p(x)$ sont toujours paraboliques [équations (23), (24)] et :

$$U = \frac{1}{n}\sum_1^n V_p = \frac{F_n}{n\zeta}\left(-\frac{d\mu}{dx}\right) \tag{36}$$

où F_n est donné par (26). On en tire :

$$x_n - x_{n-1} = \frac{F_n}{3Un}(-W_n + W_{n+1}) \tag{37}$$

L'équation (34) détermine le profil des gradins. Il est facile de vérifier que dans la limite épaisse ($n \gg 1$), elle redonne l'équation hydrodynamique habituelle pour un écoulement de Poiseuille entraîné par le gradient de la pression de disjonction [12].

3. *Extensions possibles.* — Dans le modèle incompressible qui a été discuté ici, une monocouche reste immobile et compacte. En fait, avec certains liquides moléculaires comme le squalane, la monocouche « s'évapore » à deux dimensions : la cohésion de la phase liquide est forte pour les systèmes à plusieurs couches, mais faible pour un système strictement bidimensionnel. Il peut donc être nécessaire de compliquer le modèle pour la première monocouche ($r > R_2$) en abandonnant l'hypothèse d'incompressibilité. Les principes sont décrits dans la référence [7]. On peut garder la relation (9) : $\mathbf{V}_1 = -(1/\zeta_1)\nabla\mu$, mais le courant correspondant est $\psi\mathbf{V}$, où ψ est le taux de remplissage local de la couche. On aboutit à une équation de diffusion non linéaire :

$$\frac{\partial\psi}{\partial t} = \nabla\left[\frac{\psi}{\zeta_1}\frac{\partial\mu}{\partial\psi}\nabla\psi\right] \tag{38}$$

qui fait intervenir *deux* fonctions mal connues $\zeta_1(\psi)$ et $\mu(\psi)$, dépendant à la fois du substrat solide et du liquide.

Nous avons bénéficié de discussions constantes avec F. Brochard et F. Heslot.

Note remise le 9 avril 1990, acceptée le 20 avril 1990.

RÉFÉRENCES BIBLIOGRAPHIQUES

[1] F. HESLOT, N. FRAYSSE et A.-M. CAZABAT, *Nature*, 338, 1989, p. 1289-1290.

[2] F. HESLOT, A.-M. CAZABAT et P. LEVINSON, *Phys. Rev. Lett.*, 62, 1989, p. 1289-1292.

[3] F. HESLOT, A.-M. CAZABAT et N. FRAYSSE, *J. of Phys.* (*Condensed Matter, Liquids*), 1, 1989, p. 5793-5802.

[4] R. HORN et J. ISRAELASHVILI, *J. Chem. Phys.*, 75, 1981, p. 1400-1411.

[5] J. ISRAELASHVILI, *Intermolecular and surface forces*, Academic Press, N.Y., 1985.

[6] Pour une discussion récente de forces oscillantes calculables, *voir* P.-G. de GENNES, *Langmuir* (à paraître).

[7] A.-M. CAZABAT, *C.R. Acad. Sci. Paris*, 310, série II, 1990, p. 107-111.

[8] F. HESLOT, A.-M. CAZABAT, P. LEVINSON et N. FRAYSSE, soumis à *Physical Review Letters*.

[9] *Molecular forces*, Académie Pontificale des Sciences, North Holland, 1967.

[10] *Voir* par exemple W. ZISMAN, in *Contact angle, wettability and adhesion*, F. FOWKES éd.; *Adv. in Chemistry Sciences* (*A.C.S.*), n° 43, 1964, p. 1-55.

[11] W. HELFRICH, *Phys. Rev. Lett.*, 23, 1969, p. 372-375.

[12] P.-G. de GENNES, *Rev. Modern Physics*, 57, 1985, p. 827-863.

Collège de France, 11, *place Marcelin-Berthelot*, 75231 *Paris Cedex* 05.

Afterthought: Etalement d'une goutte stratifiée incompressible

Some checks on these theoretical ideas have been obtained quite recently by N. Fraysse (PhD, Paris, 1991) — although the logarithmic factor in Eq. (20) is not easy to display.

Dynamics of partial wetting

F. Brochard-Wyart [*] and P.G. de Gennes [**]

Abstract : Two broad classes of models have been used to describe the motion of a contact line when the contact angle θ deviates from the equilibrium value θ_e : a) an Eyring approach, emphasizing the microscopic jump of a single molecule at the tip. b) a hydrodynamic approach, concentrating on the viscous losses inside the liquid wedge of angle θ. In the present review, we compare the predictions from both models, for two critical experiments :

1) The pull out of a vertical plate from a fluid at rest -showing (for finite θ_e) a critical velocity V_c above which the plate is completely wet.

2) The velocity of growth of a dry patch for a non wettable surface covered by a flat liquid film -which turns out to vary like θ_e^3 at small θ_e.

The net conclusion is that, at small θ_e and for low velocities V, the hydrodynamic losses dominate, while at large θ_e and large V, the molecular features are probably important.

I. General aims

A small liquid drop, deposited on a low energy surface, forms a spherical cap, with a well defined equilibrium contact angle θ_e, related to the interfacial tensions by the celebrated Young equation [(1)]

$$\gamma_{so} = \gamma_{sL} + \gamma \cos \theta_e \tag{1}$$

(γ, γ_{so}, and γ_{sL} refer respectively to the liquid/air, solid/air, solid/liquid interfaces).

(*) SRI Université Paris 6
11, rue P. et M. Curie 75230 Paris Cedex 05 France

(**) Laboratoire de Physique de la Matière Condensée
Collège de France 11, place M. Berthelot 75231 Paris Cedex 05 France

We focus our attention here to situations where a) θ_e is finite (partial wetting) b) the surface is smooth and chemically homogeneous (negligible hysteresis) : this in our days may indeed be obtained with a few model systems such as mica, quartz, certain types of glasses, and (last but not least) silicon wafers -either bare or chemically modified to adjust the wettability.

With systems of this type, if we impose an angle θ different from θ_e to the contact line, the line will move at a certain velocity $U(\theta)$ (fig. 1). From the point of view of irreversible processes [2], the dissipation (per unit length of line) is :

$$T\dot{S} = \gamma (\cos \theta_e - \cos \theta)\, U \tag{2}$$

and is the product of a flux (U) by the conjugate force : the "non compensated Young Force".

$$F = \gamma_{so} - \gamma_{sL} - \gamma \cos \theta \tag{3}$$

Our aim is to construct a relation between flux and force. Two main lines of approach have been used for that purpose : molecular theories [3-5] and hydrodynamics theories [6-8]. We give a brief summary of both theories in section II. Then, in sections III and IV we discuss some recent experiments, for which their predictions are dramatically different. We compare the two in section V.

II. Theoretical background

1) The Eyring approach

In 1949, Yarnold and Mason [9] suggested that the law $U(\theta)$ could be controled by adsorption / desorption processes very near the contact line. Later, Blake and coworkers [3-5] transformed this idea into a quantitative theory. This focuses on the molecules which are in direct contact with the solid (fig. 2). The last such molecule will tend to hop from one adsorption site to the next, but may possibly return : this is described by two rates K_+ and K_- :

$$K_{\pm} = k \exp -\left(\frac{W \mp \frac{1}{2} F \lambda^2}{k_B T} \right) \tag{4}$$

Here W is an activation energy for hopping, λ is the distance between hopping sites, F is the force per unit length of eq. (3), F λ is the force per site i and F λ^2 is the energy shift due to a jump of length λ.

Ultimately, the velocity U is :

$$U(\theta) = \lambda (K_+ - K_-) \quad (5)$$

The general aspect of the resulting $U(\theta)$ is shown on fig. 3.

a) Of course, $U = 0$ for $\theta = \theta_e$. Near this point, the relation between flux and force is linear and $U \sim \theta - \theta_e$.

b) For $\theta = 0$, the velocity U does not vanish. This, as we shall see, is a major difference with the hydrodynamic picture.

Eq. (5) does give a good fit to a number of data collected by Blake, in particular in regimes of rather high velocity (which are of primary interest for many coating operations). An example is shown on fig. 4. [At low velocities, and with solids which are not ideal, the data becomes often very noisy].

An interesting observation, which is due to Gribanova (Toronto meeting on contact angles, to be published) concerns the activation energy W of éq. (4) : it turns out that W is sometimes very close to the heat of vaporization of the liquid : thus the jump idealised on fig. 2 may in fact be a jump through air, and the length λ may be somewhat larger than in the strict Blake picture.

2) The hydrodynamic approach [6-8]

The basic assumption here is exactly opposit to Blake's : namely that the molecular dissipation at the tip are negligible, and that the dominant losses are due to hydrodynamic shear flows in the liquid wedge of angle θ. Viscous flows in wedges have been studied long ago by Huh and Scriven [10]. For our purpose here, it is simpler to focus on the regimes $\theta << 1$, where a "lubrication approximation" can be used : the local flow velocities $v_x(z)$ are simply parabolic (fig. 5)

$$v_x(z) = \frac{3U}{2\zeta^2} (2\zeta z - z^2) \quad (6)$$

where U appears as the average of v_x over the liquid thickness $\zeta = \theta_x$.

$$\int_o^{\zeta} v_x(z)\,dz = \zeta U \tag{7}$$

The friction force F_v from the liquid onto the solid is :

$$F_v = \int_{x_{min}}^{x_{max}} \eta \left.\frac{\partial v_x}{\partial z}\right|_{z=o} dx \tag{8}$$

where η is the fluid viscosity, x_{min} is a molecular cut off below which the continuum theory breaks down, and x_{max} is another cut off, proportionnal to the horizontal size of the droplet. Eq. 8 gives :

$$F_v = \eta \frac{U}{\theta} \left(3 \ln \frac{x_{max}}{x_{min}}\right) \tag{9}$$

The logarithmic singularity in éq. (9) has caused a lot of agitation among hydrodynamicists -but is of minor scientific interest. The cut off x_{min} can be comparable to a molecular size (at large θ) or slightly larger (at small θ) when long range Van der Waals forces become important [11]. Also the assumption of an exact wedge ($\zeta = \theta_x$) is not strictly corect at finite velocities, where the pressure distribution associated with the Poiseuille flow (6) induces a slight curvature of the free surface : all these effects can be shown to be small at small velocities V [11]. The latter has been displayed by careful optical experiments near the contact line [12].

Combining éqs. (3) and (9) one arrives at :

$$U(\theta) = \frac{V^*}{6l} \theta (\theta^2 - \theta_e^2) \tag{10}$$

where $V^* = \gamma / \eta$ and $l = \ln (x_{max} / x_{min})$.

Eq. (10) is very different from éq. (5) at small θ, since it predicts $U \rightarrow 0$. Physically, this comes from the very high dissipations occurring in a thin wedge ($\theta \rightarrow 0$). The qualitative differences are shown on fig. 3.

Note that the velocity V has an extremum at $\theta = \theta_e / \sqrt{3}$

$$U_m = - \frac{V^*}{q\sqrt{3}\ l} \theta_e^3 \qquad (\theta_e << 1) \qquad (11)$$

III. Coating of non wettable surfaces

1) A basic experiment, performed by Petrov and Sedev [13] is idealized in fig. 6. A solid cylinder is pulled out from the liquid, at a constant velocity V. When V is smaller than a certain threshold value V_c, the contact line remains at a fixed altitude, and the cylinder is not coated. When $V > V_c$, the steady state corresponds to a wetting film, with a finite thickness e.

Petrov and Sedev have also investigated the transient growth of the film, when upward motion is started at t = 0. They find a film height L(t) increasing linearly with time :

$$L(t) \sim t\,(V - V_d) \qquad (12)$$

with V_d close to V_c.

2) Another elegant experiment is due to D. Queré (fig. 7). Here a liquid volume is pushed (by air pressure) along a very long capillary, and the thickness of the trailing film e is measured from the loss of the liquid volume in the droplet. Again a critical velocity V_c is found, and two important properties are found.

a) For fixed chemical properties but variable viscosity η (using different silicone oils), V_c is inversely proportionnal to η : This does suggest that the dissipation is indeed of hydrodynamic origin. (In the molecular theories, the activation energies need not coincide with the activation energy of $\eta^{(T)}$).

b) Above threshold, the thickness e is proportionnal to $V^{2/3}$, in agreement with the classical analysis of Landau and Levich [14].

3) Similar situations occur in the preparation of Langmuir Blodgett layers, where a plate is pulled out from a Langmuir trough.The analysis is slightly modified by the presence of surfactant films, but the physics is the same [6] : to obtain deposition, the surface monolayer must reach the wall, i.e. we need $V < V_c$.

IV. Dewetting experiments [15]

Consider a low energy surface (obtained by silanization of a silicon wafer) covered by a film of an organic liquid (alcanes, silicone oils,...). The film thickness e is chosen small enough so that gravity effects can be omitted : (e << 1 mm). The liquids are such that contact angle θ_e is small but not vanishing (partial wetting) : thus the film is metastable.

At time t = 0, a hole is opened at the center of the film (either by aspiration via a thin duct in the wafer, or by blowing air from above through a thin tube). A dry patch grows, with a circular shape and a radius R(t). The radius R(t) grows linearly with time : the corresponding velocity $\dot{R}$ ranges from cms/sec to microns/sec depending on the liquid viscosity. The experiments show that :

a) The velocity is inversely proportionnal to the viscosity η (again indicating a hydrodynamic effect).

b) The velocity is proportionnal to θ_e^3 (at fixed η).

These results can be interpreted in terms of a simple hydrodynamic picture [8,7], displayed on fig. 8. Ahead of the dry patch, the liquid forms a rim extending from A to B. Most of the dissipation is near the edges A and B, and nearly all the rim is at constant pressure : the rim profile is then (from Laplace's law) a portion of a circle. This implies that the dynamic angles at point A and at point B are equal $\theta_A = \theta_B = \theta$. Using éq. (10), we may write, at small angles ($\theta_e << 1$) :

$$V_A = \frac{V^*}{6l}\,\theta\,(\theta_e^2 - \theta^2) \tag{13}$$

$$V_B = \frac{V^*}{6l}\,\theta\,(\theta^2 - 0) \tag{14}$$

since at point B the equilibrium angle (between wedge and film is $\theta = 0$).

Let us now assume (and check later) that the size of the rim increases only slowly with time. Then the velocities at A and B must be equal $V_A = V_B$, and this imposes :

$$\theta = \theta_e / \sqrt{2} \tag{15}$$

$$V_A = V_B = \dot{R} = \frac{V^*}{12\sqrt{2}\, l}\, \theta_e^3 \qquad (\theta_e << 1) \tag{16}$$

Eq. (16) does agree with the main observations listed above. Our reader can also estimate that size L and the thickness $h = \frac{L\theta}{2}$ of the rim, writing (for h >> e) that all the liquid in the rim comes from a film of radius h :

$$\pi R^2(t)\, e = 2\pi R.\left(hL\, \tfrac{2}{3}\right)$$

$$= \frac{2\pi}{3}\, R\, \theta\, L^2 \tag{17}$$

This gives $L \sim t^{1/2}$ and $\frac{dL}{dt} \to 0$ at large times, as announced.

V. Discussion

Let us first focus our attention on the regimes of small θ_e and low velocities. The hydrodynamic model gives a velocity for dewetting (section IV).

$$\dot{R} = 0.059\, \frac{\theta_e^3}{l}\, V^* \tag{18}$$

and a critical velocity for pull out (section III, éq. (11)).

$$V_c = 0.064\, \frac{\theta_e^3}{l}\, V^* \tag{19}$$

We see that $\dot{R}$ and V_c are <u>very close to each other</u> : this does explain the observations on transients by Petrov and Sedev [13].

$$L(t) = t\,(V - \dot{R}) \cong t\,(V - V_c) \tag{20}$$

These experiments were performed before the set up of a hydrodynamic theory, and thus the authors tended to interpret their results in terms of the Eyring picture. But the relation $\dot{R} \sim \theta_e^3$ found in ref. (15) shows clearly that, <u>at low velocities</u>, the hydrodynamic dissipation is dominant. This can be understood from a simple physical argument, writing the dissipation as a sum of a hydrodynamic term and a molecular term :

$$T\dot{S} = \frac{31\,\eta}{\theta}\,U^2 + C\,U^2 \tag{21}$$

where from éq. (5) (linearised at small forces)

$$C = \exp\left(\frac{W}{k_B T}\right)\,(k\,\lambda^3)^{-1} \tag{22}$$

If the activation energy showing up in the viscosity η is comparable to W_1, the two terms in $T\dot{S}$ are similar, except for the factor θ^{-1}, which enlarges the hydrodynamic contribution at small angles.

To summarize : a complete discussion of the dynamics would in principle require both terms in éq. (21). We expect molecular features to be important mainly at high velocities and large angles -a frequent practical situation. On the other hand, for low angles, and when the displaced phase is more viscous than the displacing phase, hydrodynamics is essential, and this is well confirmed by experiments.

Acknowledgments : We have benefited from various discussions with T.D. Blake, J.M. di Meglio, D. Queré, C. Redon and F. Rondelez.

References

(1) For a general introduction to contact angles, see W. Zisman in "Contact angle, wettability and adhesion" (F.M. Fowkes Ed.) ; Adv. in Chemistry series n° 43 ; ACS Washington (1964) ; p. 1.

(2) P.G. de Gennes ; Rev. Mod. phys. ; 57 ; 827 ; (1985).

(3) T.D. Blake, J.M. Haynes ; J. Colloïd Interface Sci. ; 30 ; 421 ; (1969).

(4) T.D. Blake, K.J. Ruschak ; Nature 282 ; 489 ; (1979).

(5) T.D. Blake ; Aiche International Symposium on the mechanics of thin film coating ; New Orleans ; (1988). Paper 1a, "Wetting kinetics -how do wetting lines move ?"

(6) P.G. de Gennes ; Kolloïd Polymere Science 264 ; 463 ; (1986).

(7) P.G. de Gennes in "Physics of amphiphilic layers" (J. Meunier, D. Langevin, N. Boccara eds) ; Springer proc. in Physics n° 21 ; (1987) ; p. 64.

(8) F. Brochart-Wyart, J.M. di Meglio, D. Queré ; C.R. Acad. Sci. (Paris) ; 304 II ; 553 ; (1987).

(9) G. Yarnold, B. Mason ; Proc. Phys. Soc. (London) ; B 62 ; 121, 125 ; (1949).

(10) C. Huh, L. Scriven ; J. Colloïd Interface Sci. ; 35 ; 85 ; (1971).

(11) P.G. de Gennes, X. Hua, P. Levinson ; J. Fluid Mech. ; 212 ; 55 ; (1990).

(12) C. Ngan, E. Dussan ; J. Fluid Mech. ; 118 ; 27 ; (1982).

(13) J.G. Petrov, R.V. Sedev ; Colloïds and Surfaces ; 13 ; 313 ; (1942).

(14) L. Landau, V. Levich ; Acta Physicochemica USSR ; 17 ; 42 ; (1942).

(15) C. Redon, F. Brochard, F. Rondelez ; Phys. Rev. Lett.; 66 ; 715 ; (1991).

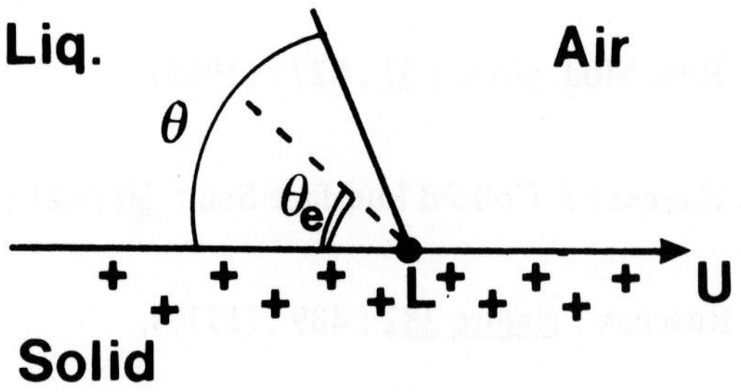

Fig. 1 : A triple line (solid, liquid, air) with a contact angle θ and a velocity U(θ). (In all our discussions, the liquid is assumed non volatile).

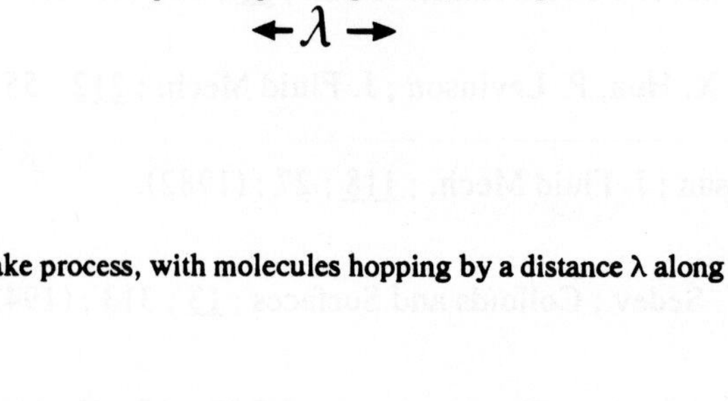

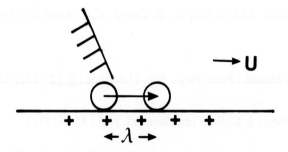

Fig. 2 : The Blake process, with molecules hopping by a distance λ along the solid surface.

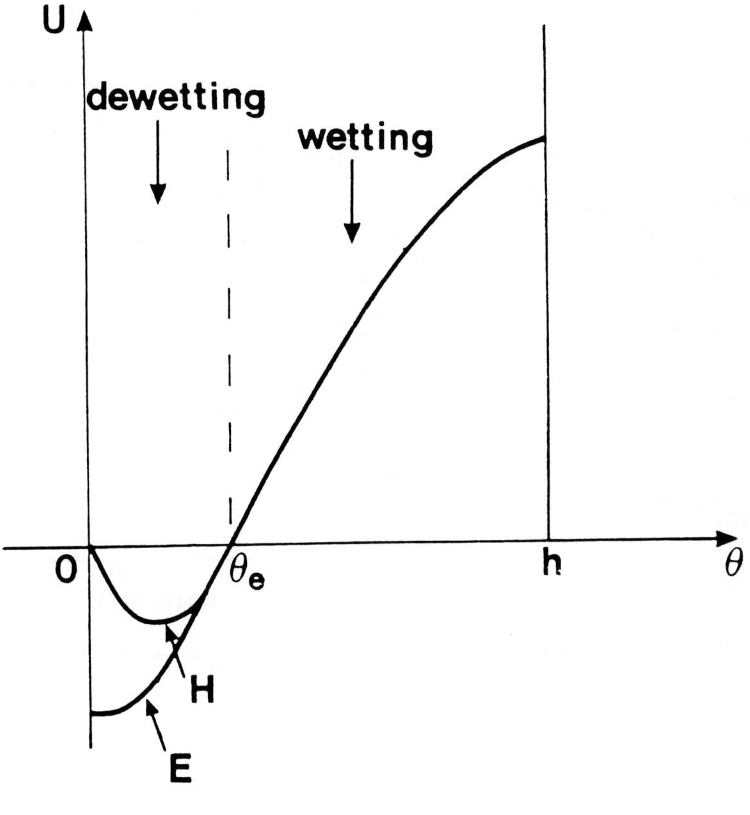

Fig. 3 : Qualitative plots of velocity versus contact angle (for an ideal surface) in the Eyring theory (E) and in the hydrodynamic theory (H).

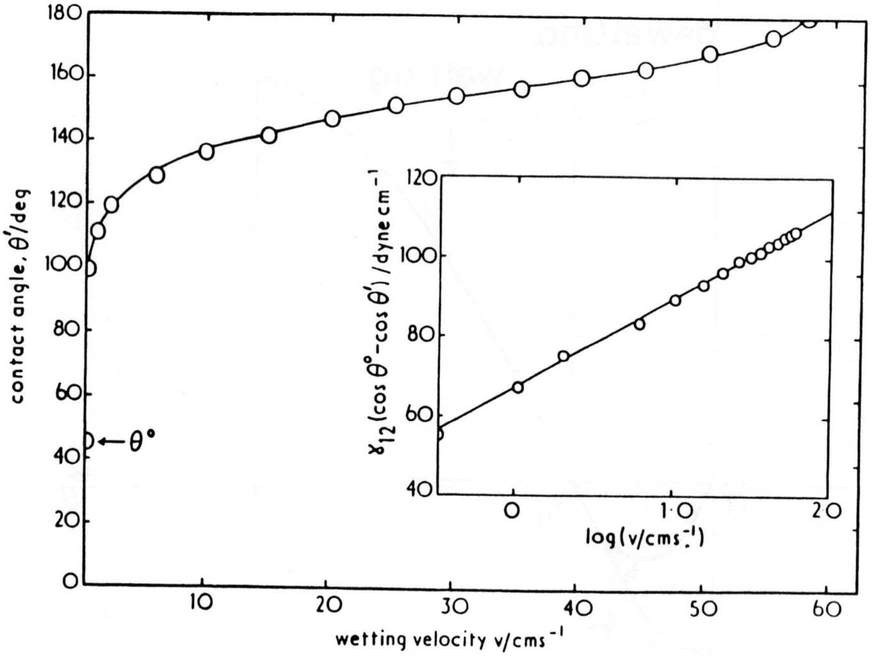

Fig. 4 : Apparent contact angle θ versus velocity V for 70 % glycerol/water solutions (after R. Burley and B. Kennedy in Wetting, Spreading and Adhesion, J.F. Padday ed., Academic Press 1978, p. 327). Insert : comparison with éq. (5) as shown in réf. 5.

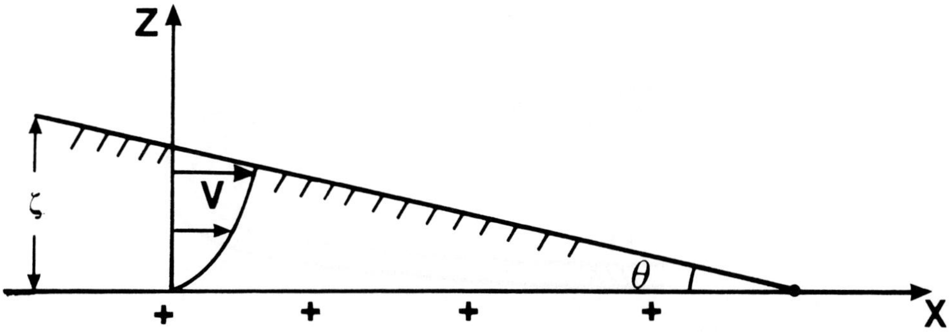

Fig. 5 : Poiseuille flow near a moving contact line (θ << 1).

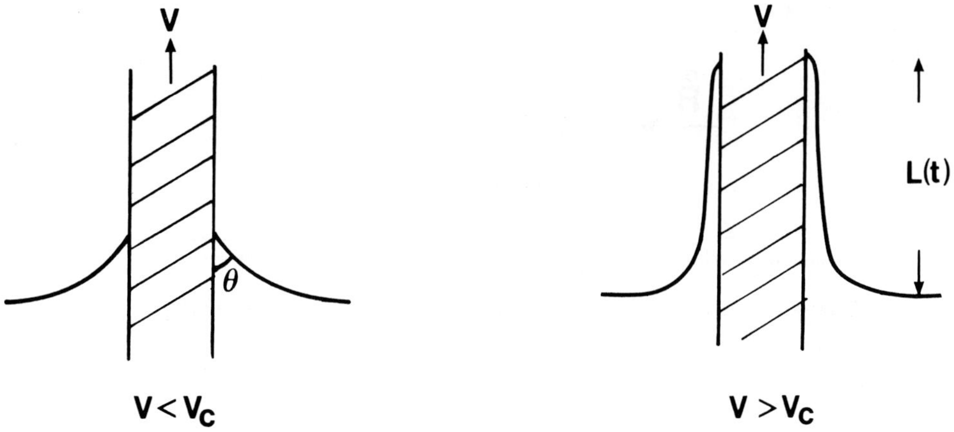

Fig. 6 : Pull out of a plate (or of a macroscopic cylinder) from a liquid environment.

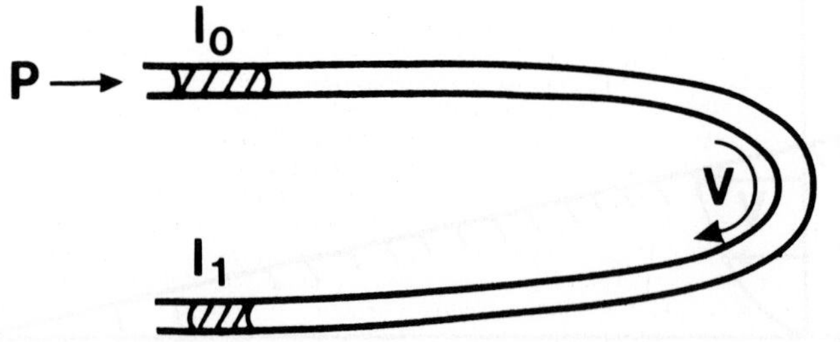

Fig. 7 : D. Queré's set up to measure the thickness of the remnant liquid film, when a droplet is pushed along a capillary at velocities $V > V_c$ (when V_c is defined on fig. 6). From the change in length of the droplet $(l_0 - l_1)$, one knows the film volume and ultimately the film thickness.

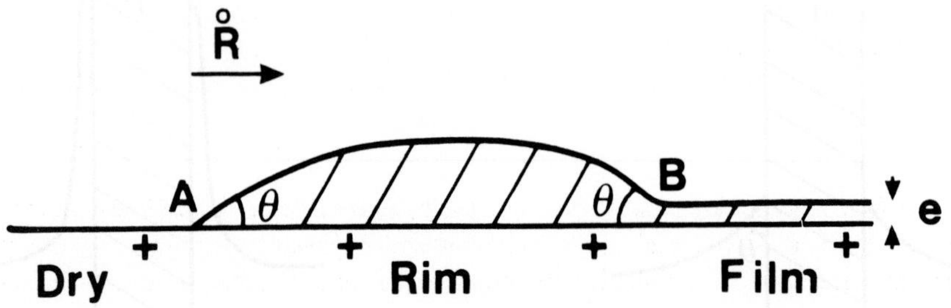

Fig. 8 : Growth of a dry patch by formation of a rim (AB). Inside the rim, the pressure is nearly constant ; thus the rim has a circular shape.

C. R. Acad. Sci. Paris, t. 307, Série II, p. 1949-1953, 1988 1949

Mécanique des solides/*Mechanics of Solids*

Fracture d'un adhésif faiblement réticulé

Pierre-Gilles de Gennes

Résumé — On analyse la répartition spatiale de la dissipation dans la fracture d'un polymère viscoélastique de temps de relaxation τ. Dans une large gamme de vitesse V, on doit distinguer *quatre* régions : (*a*) une zone de fissure, dans laquelle les chaînes sont fortement étirées; (*b*) une zone « dure » où le polymère se comporte comme un milieu rigide; (*c*) une zone fluide, où se produit la dissipation dominante; (*d*) une zone statique. On parvient ainsi à des formules pour l'énergie de fracture G fonction de la vitesse V, qui montrent un maximum, où G est plus grand que la dissipation de fissure $(G_0): G_{max} = G_0\,\mu_\infty/\mu_0$, où μ_∞ et μ_0 sont les modèles élastiques à haute et basse fréquence. Au-delà du maximum, G(V) retombe lorsque l'épaisseur de la région fluide devient comparable à celle du spécimen, et l'on retrouve finalement à $G = G_0$. Un 2e processus, analogue, doit intervenir lors de la transition vitreuse, où le module élastique augmente encore. Toute notre analyse est qualitative, mais elle donne une image physique plausible pour les expériences de Gent et Petrich sur les élastomères butadiène-styrène.

Fracture of a weakly crosslinked adhesive

Abstract — *We analyse the dissipative processes during fracture of a viscoelastic polymer with a relaxation time* τ. *In a large interval of velocities* V, *we must distinguish* four *spatial regions:* (*a*) *the crazing zone, where chains are strongly extended;* (*b*) *a "hard" region, where the polymer behaves like a solid of high modulus* μ_∞*;* (*c*) *a liquid zone, where a large viscous dissipation occurs;* (*d*) *a static zone where the polymer behaves like a solid of modulus* $\mu_\infty \ll \mu_\infty$. *The fracture energy* G(V) *starts from a relatively low value* G_0 *due to the irreversible crazing process, and rises up to a maximum* $G_{max} \sim G_0\,\mu_\infty/\mu_0$. *At higher velocities* G(V) *decreases as soon as the size of the liquid region becomes comparable to the thickness of the sample, and* G *ultimately falls back to* $G \sim G_0$. *A second process, with similar laws, is expected to take place at velocities which correspond to the rubber-glass transition. The whole analysis is qualitative but it provides a plausible picture for the experiments on Gent and Petrich of butadiene styrene rubbers.*

I. Introduction. — L'énergie de fracture d'un polymère G est très supérieure à l'énergie interfaciable de séparation *w*, et dépend fortement de la vitesse de propagation ([1]-[3]). Elle comprend notamment un terme de frottement statique G_0 associé à un étirement irréversible des chaînes. Nous avons suggéré jadis [4] une forme possible de G_0 pour un polymère non réticulé, dans le domaine où un mécanisme de succion des chaînes est dominant :

$$G_0 = \sigma_p\,\delta \tag{1}$$

où δ est la longueur des chaînes étirées, et σ_p une contrainte seuil de déformation physique. Mais à vitesse finie, une contribution importante à G résulte des pertes viscoélastiques : Gent et Schultz [5] ont montré que, pour divers rubans viscoélastiques, cette contribution a la forme

$$G = w\,[1 + \varphi\,(a_T\,V)] \tag{2}$$

où a_T est un facteur de Williams Landel Ferry qui permet de ramener sur une même courbe des points mesurés à des températures différentes. Il y a certaine similarité entre $\varphi(a_T V)$ et la partie dissipative du module élastique $\mu''(\omega)$ (où ω est une fréquence) [6]. Divers auteurs ont résolu des équations de mouvements viscoélastiques, soit sous une forme formelle [7], soit dans des versions simplifiées [8]. La référence [8], par exemple, suppose que la dissipation dominante se produit sur une petite longueur *l* (indépendante de V) et que la fréquence importante est donc $\omega = V\,l^{-1}$. Dans le présent texte, nous suivrons principalement une analyse récente de Langer et Barber [9], mais en la transposant au cas élastomère faiblement réticulé pour lesquels le module élastique statique μ_0

Note présentée par Pierre-Gilles de Gennes.

1950 C. R. Acad. Sci. Paris, t. 307, Série II, p. 1949-1953, 1988

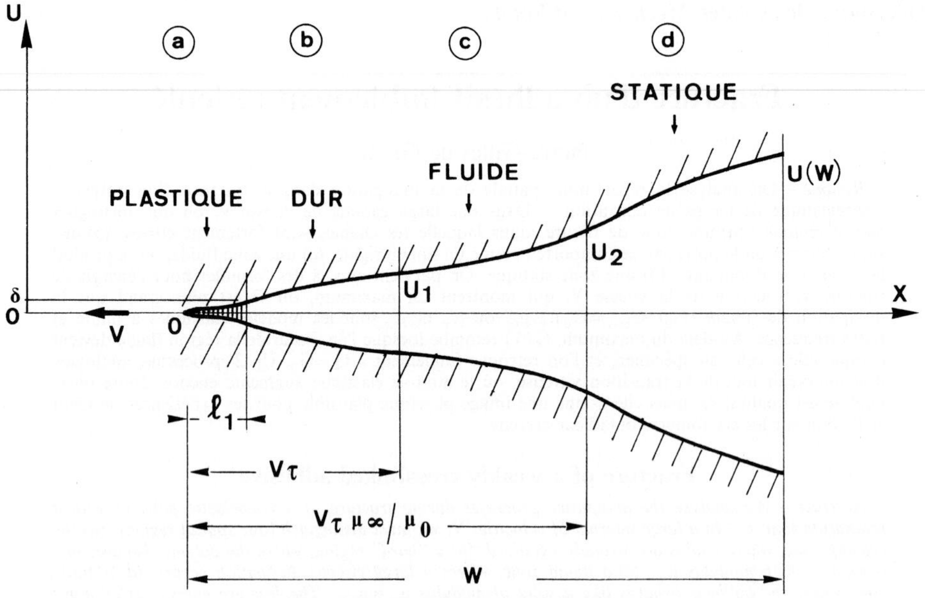

Fig. 1. — La « trompette » : fracture dans le régime $V > l_1/\tau$.
Les deux régions dissipatives sont (*a*) et (*c*).

Fig. 1. — The "trumpet": fracture shape in the regime $V > l_1/\tau$.
There are two regions of strong dissipation: (*a*) *and* (*c*).

est très inférieur au module de haute fréquence μ_∞

(3) $$\mu(\omega) = \mu_0 + (\mu_\infty - \mu_0)\frac{i\,\omega\tau}{1 + i\,\omega\tau}.$$

[Nous poserons $\eta = (\mu_\infty - \mu_0)\,\tau \cong \mu_\infty\,\tau$ = viscosité en régime fluide.] Notre analyse est qualitative, mais elle permet de dégager certains phénomènes importants : notamment le fait que, dans une gamme importante de vitesses, la dissipation viscoélastique dominante n'a *pas* lieu près de la tête de fissure, mais dans une région plus externe.

II. Le régime « trompette ». — Soit l_0 la longueur de la région plastiquement déformée dans la limite $V = 0$. Elle est définie par l'égalité des tensions élastiques et plastiques à l'échelle l_0

(4) $$\mu_0\,\delta/l_0 = \sigma_p.$$

Ici, $2\,\delta$ désigne l'ouverture en fin de fissure, et n'est pas nécessairement lié à la longueur des chaînes.

Aux vitesses $V \ll l_0/\tau$, le comportement reste statique partout et G est donné par l'équation (1). Aux vitesses intermédiaires $V \sim l_0/\tau$, la dissipation monte rapidement, et, comme le montre Langer, la longueur l augmente (ce qui exclut une correspondance exacte entre V/l_0 et ω). Dans tout ce qui suit, nous portons notre attention sur le régime $V\,t \gg l$ qui correspond encore à des vitesses assez faibles (par exemple $l_0 = 1$ μm et $t = 1$ s) mais à une situation de G fort.

Les différentes régions autour de la tête de fissure 0 sont représentées sur la figure 1 (chacune de ces régions correspond, en fait, à un anneau circulaire centré sur 0, mais

C. R. Acad. Sci. Paris, t. 307, Série II, p. 1949-1953, 1988

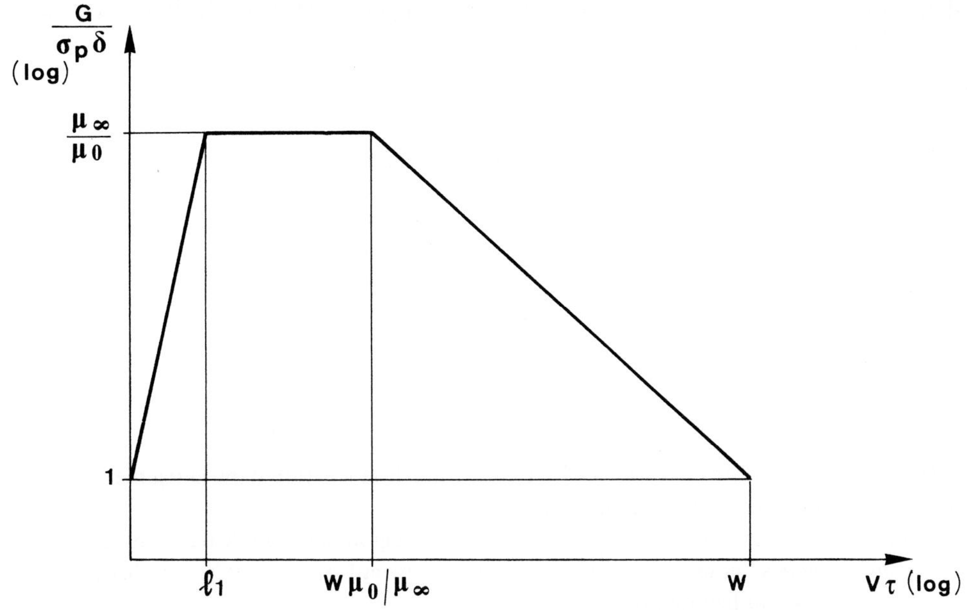

Fig. 2. — Énergie de fracture G en fonction de la vitesse V (aspect qualitatif). La figure 1 correspond au cas où $l_1 < V\tau < W\,\mu_0/\mu_\infty$. La partie décroissante est instable.

Fig. 2. — Fracture energy G versus velocity V (qualitative features). Figure 1 corresponds to the plateau regime $l_1 < V\tau < W\,\mu_0/\mu_\infty$. The decreasing portion is unstable.

nous représentons seulement l'intersection de ces anneaux avec le demi-plan de fracture) :

(*a*) La fissure à déformation plastique a ici une longueur l_1 telle que $\mu_\infty d/l_1 = \sigma_p$. Nous supposons $V\tau > l_1$.

(*b*) Il y a au-delà une région « dure » ($\mu = \mu_\infty$) dans laquelle la dissipation est faible, parce que les contraintes élastiques σ varient trop vite ($\omega > 1/\tau$). La limite de cette région correspond à une distance $x_1 = V\tau$.

(*c*) Plus loin on trouve une région « fluide » qui, dans notre cas s'étend très au-delà de x_1. En effet, sur l'équation (3) on voit que $\mu(\omega)$ devient imaginaire dès que $\omega\tau > \mu_0/\mu_\infty$. Spatialement, ceci correspond à une limite $x_2 = V\tau\,\mu_\infty/\mu_0 \gg x_1$.

(*d*) Finalement, on rentre dans une région « statique » avec le module élastique μ_0 (faible), qui s'étend de x_2 à W, où W est la largeur du spécimen.

Dans les régions (*b, c, d*) il est naturel de postuler que les contraintes σ ont la *même loi d'échelle* $\sigma \approx x^{-1/2}$, car les équations du mouvement (avec effets inertiels négligés) ont toujours la même forme $\nabla\sigma = 0$, les mêmes conditions de compatibilité sur les composantes σ_{ij} (en régime visqueux), et la même condition aux limites ($\sigma_n = 0$ sur la lèvre ouverte). Dans les régions (*b*) et (*d*), où le comportement est quasi statique, on a, en outre, $u(x)\,\sigma(x) = \text{Cte}$, ce qui implique

$$G = u(W)\,\sigma_\infty = u_2\,\sigma_2 \tag{5}$$

$$u_1\,\sigma_1 = \delta\,\sigma_p. \tag{6}$$

Ici, $u(x)$ mesure l'ouverture de la fracture ($u(l) = \delta$. fixé). Par contre, dans la région (*c*), le produit $u\sigma$ n'est pas constant. La loi d'échelle qui relie u et σ est :

$$\sigma = \eta \frac{d}{dx}\left(\frac{du}{dt}\right) = \eta V \frac{d^2 u}{dx^2} \tag{7}$$

donc, $u \sim x^{3/2}$ d'où la forme en « trompette » de la figure 1. Dans la région (*c*), $u\sigma \sim x$, donc, d'après (5) et (6) :

$$(8) \qquad \frac{G}{\delta\sigma_p} = \frac{u_2\sigma_2}{u_1\sigma_1} = \frac{x_2}{x_1} = \frac{\mu_\infty}{\mu_0}.$$

La dissipation se produit principalement dans la région (*c*) :

$$(9) \qquad T\dot{S}_v = (\mathrm{Cte}) \int_{x_1}^{x_2} \eta \left(\frac{Vu}{x^2}\right)^2 x\,dx = u_2\sigma_2 - u_1\sigma_1 = G - \sigma_p\delta.$$

On peut vérifier qu'il y a bien continuité de σ aux frontières x_1 et x_2 :

$$(10) \qquad \left\{ \begin{array}{l} \mu_0 \dfrac{u_2}{x_2} = \eta \dfrac{V u_2}{x_2^2} \\ \mu_\infty \dfrac{u_1}{x_1} = \eta \dfrac{V u_1}{x_1^2} \end{array} \right.$$

III. Régime des fortes vitesses. — La figure 1 décrit la fracture dans le régime $V\tau > l_1$ mais elle suppose aussi une autre inégalité $W > x_2$.

Si l'on renverse cette inégalité ($V\tau > W\mu_0/\mu_\infty$) la région statique disparaît de la zone active de la fracture, et est rejetée dans un domaine inactif où $u(x) = \mathrm{Cte}$. La dissipation viscoélastique est, au lieu de (9)

$$(11) \qquad T\dot{S}_v = (\mathrm{Cte}) \int_{x_1}^{W} \eta \left(\frac{Vu}{x^2}\right)^2 x\,dx$$

ce qui correspond à

$$(12) \qquad G = \sigma_w u_w = \sigma_p \delta \frac{W}{V\tau}$$

et $G(V)$ décroît (ce qui correspond à un régime instable). Finalement, lorsque x_1 devient comparable à W, la zone fluide disparaît complètement, et l'on retourne à $G = \sigma_p\delta$. L'ensemble de ces comportements est représenté qualitativement sur la figure 2.

IV. Discussion. — 1. Toute notre approche, basée uniquement sur des lois d'échelle, demanderait a être étayée par des calculs élastiques détaillés. Elle souffre, par ailleurs de simplifications extrêmes (σ_p et δ indépendants de V, un seul temps τ, etc.). Mais l'équation (8) montre bien l'intérêt des élastomères faiblement réticulés ($\mu_0 \ll \mu_\infty$) pour l'adhésion à V fini.

2. Pour une épaisseur de spécimen W très grande (fracture en volume), notre modèle prévoit un plateau, mais pas de maximum dans $G(V)$: il n'y a pas de ressemblance fondamentale entre $G(V)$ et $\mu''(\varpi = V l^{-1})$.

3. Par contre pour W faible, on attend un maximum $G = G_0 \mu_\infty/\mu_0$ suivi d'une chute de $G(V)$ dès que $V < W\tau^{-1}\mu_0/\mu_\alpha$ (par exemple, avec $W = 0{,}1$ mm, $\tau = 1$ s, $\mu_0/\mu_\infty = 0{,}1$, ceci correspond à un seuil de $V = 10$ µm/s). La présence de ce maximum et l'instabilité intrinsèque de la branche $G(V)$ décroissante, permettent de comprendre qualitativement le premier pic observé par Gent et Petrich [1] sur un élastomère butadiène-styrène. Ce pic disparaît dans un système fortement réticulé ($\mu_\infty \sim \mu_0$). D'une façon plus générale, l'activité de fortes largeurs W pour les *adhésifs en régime de choc* (V fini) est reconnue empiriquement depuis longtemps : elle s'explique bien par la figure 1.

4. A des vitesses plus élevées, un deuxième pic est observé dans $G(V)$; il doit être dû à une deuxième anomalie viscoélastique, liée à la traversée de la transition vitreuse,

C. R. Acad. Sci. Paris, t. 307, Série II, p. 1949-1953, 1988

comme il est suggéré dans la référence [1]. Le deuxième pic peut aussi être compris, à condition que le module élastique de la phase verre soit nettement plus élevé que le module de plateau du caoutchouc.

Nous avons bénéficié de discussions précieuses avec J. Langer et D. Maugis.

Note remise le 7 novembre 1988, acceptée le 8 novembre 1988.

RÉFÉRENCES BIBLIOGRAPHIQUES

[1] A. N. GENT et R. P. PETRICH, *Proc. Roy. Soc.*, A310, 1969, p. 433-450, *voir* aussi S. WU, *Polymer interface and adhesion*, Marcel Dekker, 1982.

[2] E. H. ANDREWS et A. J. KINLOCH, *Proc. Roy. Soc.*, A 332, 1973, p. 385-399.

[3] D. MAUGIS et M. BARQUINS, *J. Phys.*, D 11, 1978, p. 1989-2023, D. MAUGIS, *J. Materials Proc. Sci.*, 20, 1985, p. 3041-3073.

[4] P. G. de GENNES dans *Microscopic aspects of adhesion and lubrication*, J. M. GEORGES éd., 1982, p. 355-368, Elsevier.

[5] A. N. GENT et J. SCHULTZ, *J. adhesion*, 3, 1972, p. 281-294.

[6] L. MULLINS, *Trans. Inst. rubber*, 35, 1959, p. 213-222; J. GREENWOOD et K. L. JOHNSON, *Phil. Mag.*, A43, 1981, p. 697-711.

[7] *Voir* par ex. M. DAHAN et E. ZNATY, *C. R. Acad. Sci. Paris*, 292, série II, 1981, p. 481-484.

[8] R. M. CHRISTENSEN et E. M. WU, *Engineering Fracture mecanics*, 14, 1981, p. 215-225.

[9] J. LANGER, *Cours à l'école d'été de Physique théorique*, Les Houches, juin 1988 (à paraître).

Laboratoire de Physique de la Matière condensée, Collège de France,
11, *place Marcelin-Berthelot*, 75231 *Paris Cedex* 05.

Afterthought: Fracture d'un adhésif faiblement réticulé

This scaling estimate has been the starting point of a refined calculation by C.Y. Hui, D.B. Xu and E.J. Kramer (*J. Appl. Phys.*, to be published). This does confirm Eq. (8).

I have tried later to use the same concepts to provide a description of "tack" between a solid and a molten polymer [P.G. de Gennes, *Comptes Rendus Acad. Sci.* (*Paris*) **312 II**, 1415 (1991)]. This leads to the following prediction for the work of separation G:

$$G = W \frac{\theta}{\tau}$$

where W is the thermodynamic work, τ the reptation time, and θ the separation time ($\theta > \tau$). Unfortunately, in the most precise experiments on tack [A. Gent and H. Kim, *Rubber Chemistry and Technology* **63**, 613 (1990)], it is the *preparation* time of the polymer/solid interface which is varied, rather than the *separation* time θ. Thus I do not know whether this tack model makes any sense.

Polymer - Polymer Welding and Sliding

P.G. de Gennes (unpublished, 1992)

Collège de France 75231 Paris Cedex 05 France

Abstract : We discuss the interdigitation of polymer chains at the contact plane between two entangled polymer melts. For A / A interfaces (with the same polymer A on the both sides), we give a tentative picture of the fracture properties in the glassy state after partial interdigitation. For A / B interfaces, we also discuss various properties related to tangential slip.

I. Introduction

Polymer / polymer interfaces play a dominant rôle in various mechanical features : coextrusion, adhesive properties, toughness of polymer blends, being typical examples. We begin to have precise experimental informations on interfacial structures using refined probes such as neutron reflectances. At the same moment, the toughness of bulk (glassy) polymers, which craze under tension, begins to be understood through an original idea of H. Brown [(1)]. It is thus tempting to extend the Brown ideas to various systems of "weak junctions". The junction may be (i) a partly healed contact between two identical polymer blocks A / A, as in the experiments of the Lausanne group [(2) (3)] (ii) a contact between two different polymers A and B. In all our discussion, we shall assume these junctions to be perfect -with full contact between the two partners, and no gaps : experimental arguments for the existence of these good contacts have been presented by Kausch and coworkers [(2)].

Our aim here is (a) to give a brief reminder on the theoretical description of the weak junctions (b) to show how some basic mechanical properties can be related to the structure. One of the major conclusions, for the A / A case, is that chain ends play a crucial rôle. Thus any attraction between a chain end and the free surface of one A block will react significantly on the A / A mechanical properties after welding.

This type of attraction was first suggested by systematic experiments on melts by D. Legrand and G. Gaines [4] [5], showing that the surface tension γ of oligomers was often lower than the surface tension γ_∞ of a high polymer, and that the correction

$$\gamma_\infty - \gamma(N) \sim N^{-x} \qquad (1)$$

where N is the degree of polymerisation, and x an exponent of order 2/3. The fact that $x < 1$ shows that we are not dealing with a simple uniform dilution of chains ends (which would give a correction $\sim N^{-1}$). The most natural way of understanding the Legrand-Gaines result amounts to assume that the chain ends are attracted to the surface. The IBM group [6] has argued that a typical monomer along the chain suffers an entropy loss of order unity when it is located near the free surface, because the chain is "reflected" here, while the chain ends do not have this loss : thus one expects, on purely entropic grounds, a gain of free energy $\sim$ kT for each chain end brought to the surface. There are also enthalpic effects, which may increase or decrease the surface attraction. But, if, on the whole, the attraction is of order kT per chain end, we reach a simple regime [7], where all chains within one radius of gyration $R_o = N^{1/2} a$ of the surface, put their ends on the surface - and the deeper chains are unperturbed. This leads to a surface fraction of chain ends ϕ_s of order $\frac{2}{N} \cdot \frac{R_o}{a}$ (where a is a monomer size), and thus :

$$\phi_s = \sim N^{-1/2}$$

(or x = 1/2) in this regime. We shall call this the normal attractive regime. The value x = 2/3 observed by Legrand and Gaines, may be the result of a cross over between 0 attraction and normal attraction.

This interpretation of the Gaines results is still controversial : Dee and Sauer [8] interpret γ (N) not by an effect of chain ends, but from the empirical N dependence of the (P, V, T) equation of state oligomers (the mean feature here being the change of the equilibrium density ρ (N)), plus a standard mean field analysis of the interfacial energy [9], as related to the equation of state and to the range of the intermolecular forces (the latter being assumed independent of N). Dee and Sauer get remarkable fits to the Gaines data -without involving any special localisation of chain ends ! Here, however, we shall keep in mind constantly the possiblility of chain end segregation near the free surface : indeed, we shall see that some of the neutron data on partial healing of A / A interfaces are more easily understood in the normal attractive regime than in 0 attraction.

II. Healing of an A / A interface

A) Healing near the glass transition. The basic healing experiment is idealized on fig. (1). We start with two blocks of the same polymer, which we call H and D [For certain experiments (D) may be a deuterated polymer, while (H) is the usual -proton carrying-species]. The two blocks are put into close contact under a mild pressure, at a temperature close to the glass point T_g, during a time t. The polymer chains from H and D begin to intertwine, and build up a diffuse profile for the D concentration ϕ_D (fig. 2). We are interested here primarily in this interdigitation process, at times t smaller than the reptation time of the chains T_{rep}. This corresponds to spatial widths of the profile e(t) which are smaller than the coil radius R_o.

Most experiments have been performed with polystyrene, and with H and D chains of comparable length : $N_{H} \cong N_D = N$. The choice of N is non trivial :

a) We want $N >> N_e$ (the distance between entanglements).

b) We want $\chi_{HD} < 1$, where χ_{HD} is the (small) Flory parameter describing a weak trend for segregation between the H and D species.

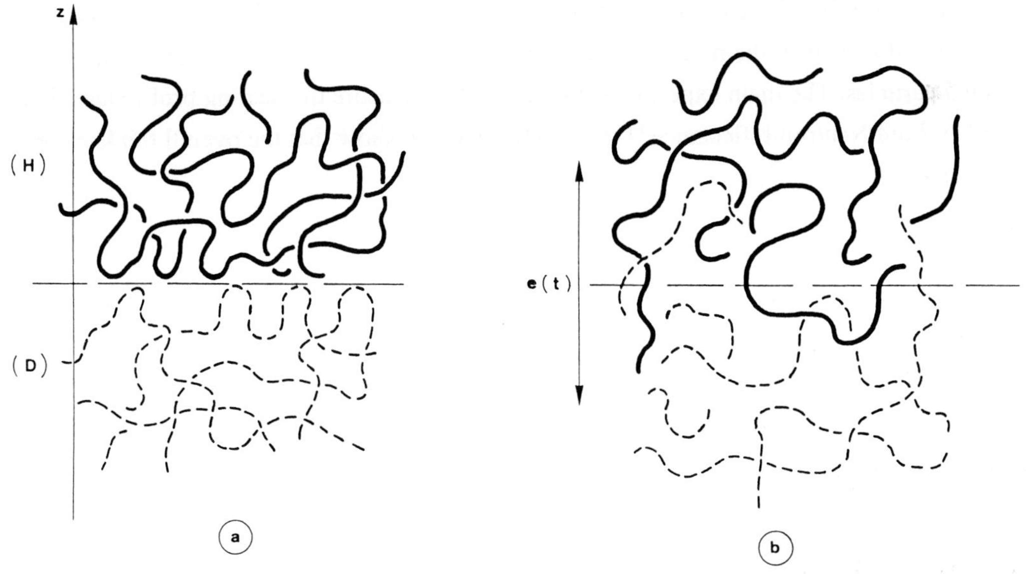

fig. 1. Schematic representation of an interface between two chemically identical polymers H and D, a) before healing all chains are "reflected" on the contact plane, b) after partial healing (during a time t), the chains interdigitate. We are concerned here by the regime where the thickness e(t) of the mixing region is larger than a tube diameter (d), but smaller than the coil size (R_0).

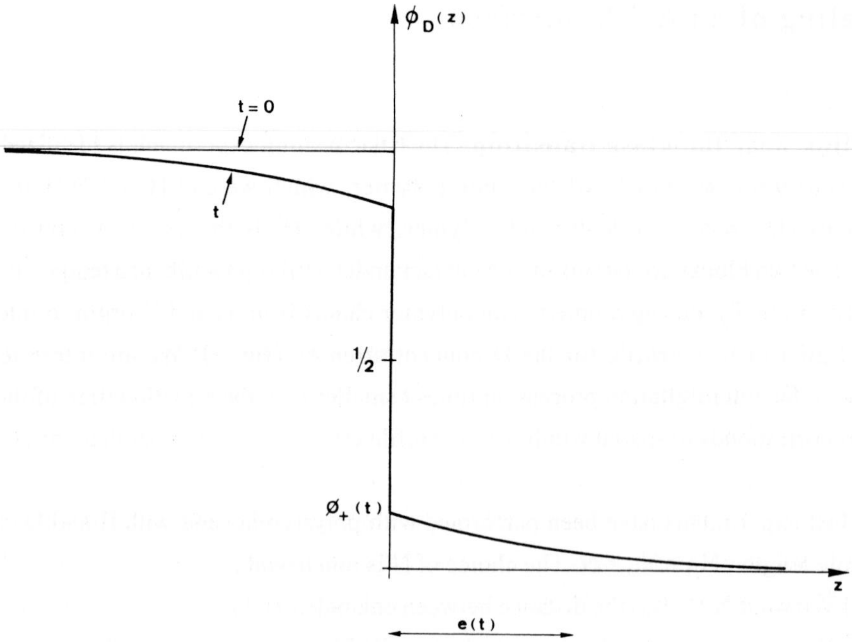

fig. 2. Concentration profile for the D species in the experiment of fig. 1. The tube diameter d is assumed to be much smaller than e(t) : then, the profile is discontinuous on the contact plane, because most chains are still reflected at this plane.

Typically N will be of order 2 000 – 6 000, while $N_e \sim 300$. The thickness e(t) of the partly healed zone is in the range of 100 Å - too small to be studied by forward scattering of charged particles. The main experimental tools used to measure the healing profile have been SIMS (10) and Neutron reflectance (11) (12). Most data do show that the overall thickness e(t) grows like $t^{1/4}$:

$$e(t) \sim R_o \left(\frac{t}{T_{rep}}\right)^{1/4} \qquad (t < T_{rep}) \tag{2}$$

This is the natural law for spatial motions of one labeled monomer in an entangled melt (13) : after a time t , the chain carrying this monomer has moved along its own tube by a curvilinear length :

$$s(t) = (D_t\, t)^{1/2} \tag{3}$$

where D_{tube} ($\sim N^{-1}$) is the tube diffusion coefficient. The corresponding distance as the crow flies is :

$$e(t) = (d\, s(t))^{1/2} \qquad (s > d) \qquad (4)$$

where $d = N_e^{1/2} a$ is the tube diameter, and (4) coincides with (2).

However, this simple agreement ignore an important fact, namely that near the contact surface, the chains were originally reflected, as is clear on fig. 1, and thus most of the tube motions do <u>not</u> give any intertwining. A first theoretical reflection on the problem was performed long ago by various authors (14) (15) (16), and will be summarized here.

a) For $N >> N_e$, it is reasonable to assume that the "hairpin" processes of fig. 3a are negligible, the entropy of a hairpin or a lattice model is one half of the entropy of a free chain more generally : hairpins are disfavored by a factor of order $\exp(-n / 2 N_e)$, where n is the contour length of the hairpin.

b) Then, at the times of interest (where $e(t) > d$), all the interdigitation is due to the motion of chain ends ; one of them will start from some initial position (within e(t) of the interface), and may cross (once or more) the interface. The number of monomers which it brings to the other side is a fraction of s(t). Thus the total number ν of monomers D going through the interface (per unit area) is of the form :

$$\nu \cong s(t) \int_{-e}^{0} \phi_e(z)\, dz \qquad (5)$$

where $\phi_e(z)$ is the initial distribution of chain ends. In the original discussions (14) (15) (16), it was assumed that $\phi_e(z)$ is uniform $\phi_e(z) = 2 / N$. But in our days, we know that chain ends may have been attracted to the original free surface of the block : as explained in the

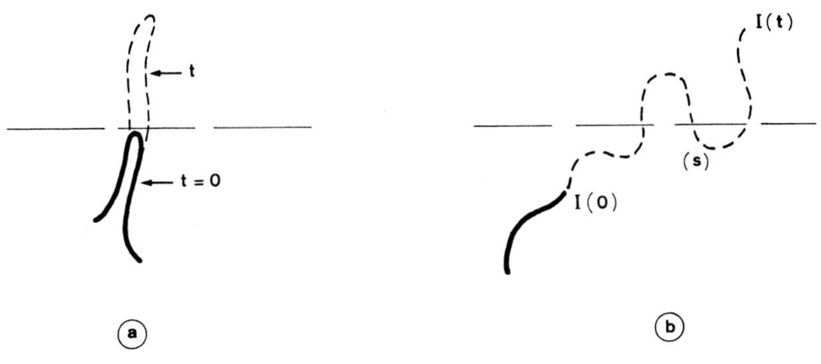

fig. 3. Two modes of interdigitation : (a) hairpins, (b) chain ends crossing the contact plane. For strongly entangled chains ($N >> N_e$), we expect process (b) to dominate.

introduction, in normal attractive conditions, this will bring another (dominant) contribution to the integral (5), proportionnal to $\phi_{so} = N^{-1/2}$. Thus we have two cases :

$$\left.\begin{array}{lll} v \sim s(t)\, e(t)\, N^{-1} & & \text{(no attraction)} \\ & & \\ v \sim s(t)\, a\, N^{-1/2} & & \text{(normal attraction)} \end{array}\right\} \qquad (6)$$

c)Because of the reflection of chains at the original interface, the profile is <u>discontinuous</u> (if we consider spatial scales larger than the tube diameter d). The general aspect is shown on fig. 2. Of major interest is the concentration ϕ_+ (t) of D monomers, on the H side, for $z \to 0$. We may write :

$$v \sim \phi_+ (t)\, e(t) \qquad (7)$$

Comparing (6) and (7), using the right normalisation factors, and inserting $\phi_{s0} \sim N^{-1/2}$, we then arrive at :

$$\left.\begin{array}{lll} \phi_+ \sim \left(\dfrac{t}{T_{rep}}\right)^{1/2} & & \text{(no attraction)} \\ & & \\ \phi_+ \sim \left(\dfrac{t}{T_{rep}}\right)^{1/4} & & \text{(normal attraction)} \end{array}\right\} \qquad (8)$$

Thus, when chain ends were originally numerous at the surface, ϕ_+ (t) rises more rapidly.

On the experimental side, the most recent data on the profile comes from the neutron reflectance experiments of Reiter and Steiner (12). They found that their profiles :

- could not be described by simple diffusion (giving an error function)
- could be described by the superposition of <u>two</u> errors functions E_{slow} and E_{fast}

$$\phi_D (z) = 2\phi_+ (t)\, E_{slow} \left(\frac{z}{e(t)}\right) + (1 - 2\phi_+)\, E_{fast} \left(\frac{z}{\sigma_c(t)}\right) \qquad (9)$$

where the errors functions E(z) are normalised by $E(0) = 1/2 \quad E(-\infty) = 1 \quad E(+\infty) = 0$.

For the "fast" component (describing what we called the discontinuity), they found $\sigma_c(t)$ very weakly dependent of time -increasing from ~ 20 Å to 30 Å in the time interval $0 < T_{rep}$. For the "slow" component, the result is $e(t) \sim t^{0.17}$ -not too far from eq. (2).

But their most interesting result is related to $\phi_+(t)$. They found $\phi_+(t) \sim t^{0.22}$, very close to the prediction of eq. (8) for normal attraction between chain ends and the free surface. What is nice is that they obtained this without being biased by any theoretical prediction !

Thus the Reiter-Steiner experiment does suggest that (in their conditions of sample separation) chain ends were originally attracted to the surface.

B) Mechanical toughness of partly healed A / A junctions. Kausch and coworkers (2) (3) have measured the fracture energy G_{1c} of partly healed A / A contacts : after healing over a time t, the sample is brought back to room temperature, where it is glassy, and then fractured along the junction. Experimentally, in most cases, the fracture energy G_{1c} increases with healing time : $G_{1c} \sim t^{1/2}$ $(t < T_{rep})$.

1) Bulk polymer toughness : the Brown model : To discuss this, let us first return to the case of bulk polymer fracture, in situations where crazing occurs, as shown on fig. (4). Over most of the crazed region, the stress σ (normal to the fracture plane), is nearly constant $\sigma = \sigma_y$, where σ_y is expected to describe plastic yield. However, near the crack tip, a stress concentration occurs, as emphasized by H. Brown (1). At distance x from the crack tip, smaller than the length h_f of the ultimate fibrils, we would expect :

$$\sigma(x) \sim \sigma_y \left(\frac{h_f}{x}\right)^{1/2} \qquad (x \leq h_f) \qquad (10)$$

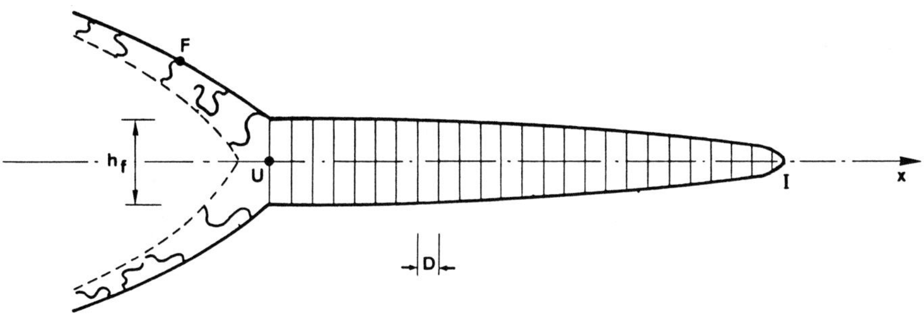

fig. 4. A craze (initiated at point I), terminating into a crack (at point U). The fibrils break at U, and the resulting half fibrils retract around point F.

Eq. (10) describes a square root singularity, as in a simple elastic medium. The coefficient is such that $\sigma \rightarrow \sigma_y$ for $x \sim h_f$ (beyond which we expect no stress concentration).

The singularity described above is cut off at the minimal value of x : $x \sim D$, where D is the interfibrillar distance. Thus the stress or the ultimate fiber is :

$$\sigma_1 \cong \sigma_y \left(\frac{h_f}{D}\right)^{1/2} \tag{11}$$

The length D is conditioned by capillary effects and cavitation instabilities at the point of birth I of the craze, and is expected to to be independent of molecular weight (17) (18). Typically $D \sim 200$ Å.

Let us now restrict our attention to cases where the ultimate fibril breaks by chemical scission of its polymer chains. (This will be correct if the chains are long enough : N above a certain limiting value N*). Then the rupture condition is :

$$\sigma_1 = \sigma_\chi \equiv \frac{f_\chi}{a^2} \tag{12}$$

where f_χ is the chemical force required to break a chain, a^2 is the chain cross section, and σ_χ the corresponding stress. f_χ is of the order U_{bond}/a, where U_{bond} is a covalent bonding energy, and is thus large (~ 1 nano Newton). The stress σ_χ may be described as a material parameter of the polymer : however, we should keep in mind that chain scission will be sensitive to all dopants present in the polymer : catalysts, antioxydants,...

Inserting (12) into (11), we arrive at the Brown formula for the ultimate fibril length :

$$h_f \cong \left(\frac{\sigma_X}{\sigma_1}\right)^2 D \tag{13}$$

Because $\sigma_\chi >> \sigma_1$, h_f can be of order several microns, as observed in materials like PS (18). We can now turn to an estimate of the fracture energy :

$$G \approx \int_0^{h_f} \sigma(h)\, dk \approx \sigma_1 h_y \approx \frac{\sigma_\chi^2}{\sigma_1} D \tag{14}$$

Eq. (14) reflects the fact that most of the fiber pulling took place under the stress σ_1. The essential feature of this Brown formula is that $G \sim \sigma_\chi^2$, and is thus very large : eq. (14) is the basic explanation for the toughness of glassy plastics.

In fact, H. Brown was led to this type of formula by a series of systematic experiments on the toughness of an AB interface with A = p. styrene B = p. methyl metacrylate [19] [20]. Here, the cohesion between the two is established via block copolymers AB lying at the surface (fig. 5). In a number of cases the blocks ruptured very near their junction point (as shown by SIMS). It was found that the energy G_{ic} was proportionnal to the square of the number n of bridging chains per unit area : this fits with eq. (14) since, in the present case, σ_χ is due to the copolymers only, and is thus proportional to n.

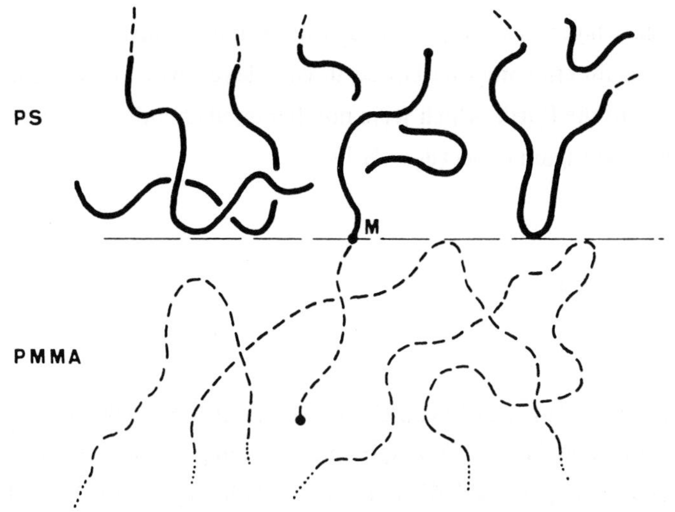

fig. 5. An interface between two incompatible polymers, decorated by block copolymer molecules M.

2) Transposition to partly healed interfaces : On fig. (3), the number of bridging chains per unit area is expected to be proportionnal to ϕ_+ (t) : any D monomer which has just crossed the border has a finite probability of being directly linked to the D side. Thus, to describe chemical rupture of the fibrils, we should perform the replacement :

$$\sigma_\chi \rightarrow 2\,\phi_+(t)\,\sigma_\chi \tag{15}$$

where the factor (2) is fixed by the condition that σ_χ returns to its bulk value at $t > T_{rep}$ ($\phi_+ \rightarrow 1/2$). Eq. (14) then gives :

$$G_{1c}(t) = G_{1c}\big|_{bulk} \cdot 4\,\phi_+^2(t) \tag{16}$$

If, and only if, the chain ends were originally numerous at the surface, we can then return to eq. (8) and write $\phi_+(t) \sim t^{1/4}$, giving the experimental form $G_{1c} \sim t^{1/2}$. Thus the Kausch law, combined with the Brown model, does suggest that a large number of chain ends were available at the interface when healing started.

A careful reader may be worried by the following point : in the Kausch experiments, the original blocks H and D were in fact obtained by rupture of one single sample. Could it be that chain ends were very numerous ($\phi_s > N^{-1/2}$, possibly $\phi_s \sim 1$) on the interface at t = 0 ? We do not believe this to be the case,as explained on fig. (4). The separation of the two blocks took place via fibril rupture, but after this, the half fibrils on both sides have probably retracted to build again a compact layer of polymer on each lip of the fracture (around point F) : in this retraction process, chain ends may be buried in each layer. Most of the chains in this layer belonged to portions of the fibrils which were not disrupted chemically. Thus, if the retraction led to an equilibrium, we again expect $\phi_s \sim N^{-1/2}$.

III. A / B interfaces

A) Interface structure for weakly incompatible pairs. The qualitative aspect of an A / B interface is shown on fig. (6). A simple understanding of the structure can be obtained, starting from an abrupt interface, and allowing one A chain to protrude in the B side (fig. 7). If m monomers are exposed in this process, the enthalpy required is :

$$\Delta H_m \sim m\,\chi\,kT \tag{17}$$

where χ is the Flory parameter [21] describing AB mixtures. The average value of m corresponds to $\Delta H_m \sim kT$, and is thus :

$$\overline{m} = \chi^{-1} \qquad (\chi < 1) \tag{18}$$

(We constantly assume that $\overline{m}$ is much smaller than the overall chain length N).

Since the protruding chain is a random walk, the width e of the interface is the size of this random walk :

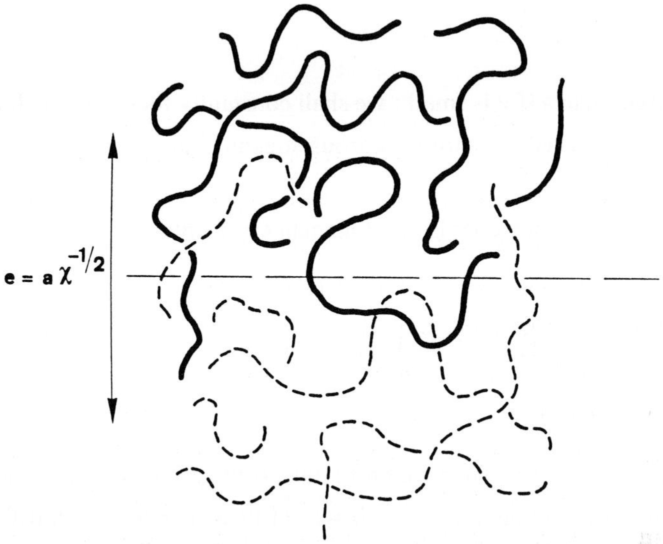

fig. 6. An interface between weakly incompatible polymers (χ << 1), with a width e.

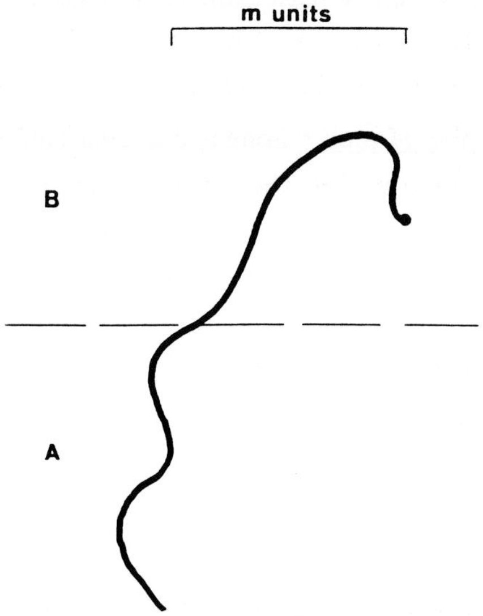

fig. 7. A simple approach to understand the structure of fig. 6 : the interface between the two polymers A and B is first assumed to be sharp. Then we allow one A chain to move partly into the B side, and discuss the resulting energy cost.

$$e \cong a\, \overline{m}^{1/2} = a\,\chi^{-1/2} \tag{19}$$

and e is much larger than a if χ is small : we shall constantly focus on this limit. Of course, the result (19) can be derived by more rigorous means, but the present approach is often illuminating.

The distribution of m values is the Boltzman exponential :

$$p_m = \frac{1}{\overline{m}} \exp(-\Delta H_m / kT) = \frac{1}{\overline{m}} \exp(-m/\overline{m}) \tag{20}$$

Of major interest for mechanical properties, is the probability that the protruding chain <u>entangles</u> with the surrounding matrix [2]. If we define an average chemical distance between entanglements N_e , we may write for the probability f of entanglements :

$$f = \sum_{N_e}^{\infty} p_m = \exp(-N_e\,\chi) \tag{21}$$

Of course, this formula is very approximate, because N_e needs not to be the same for the two partners A and B, and also not the same for the mixtures : a certain weighted average would then be required. But eq. (21) is still a reasonable starting point to discuss the mechanics of A / B contacts.

<u>B) Toughness of A / B interfaces :</u> Long ago a remarkable series of experiments was performed by Iyengar and Erickson [23]. They measured the adhesion energy G_{1c} of various polymers, on PET, by a 90° peeling test (at a fixed velocity 5 cm / sec.). The results were plotted as a function of the Hildebrand solubility parameter δ. They show a dramatic drop of G_{1c} as soon as the δ parameters differed by more than one unit.

Can we establish contact between these data and eq. (21) for the probability of entanglements ? Let us assume that :

a) G_{1c} associated to the post craze fracture of a glassy A / B junction.

b) The entangled A or B chains in the junction must break.

Then we may apply Brown's eq. (14), provided that the chemical rupture stress σ_χ is suitably reduced : only the entangled chains at the junction contribute. This could give :

$$\sigma_\chi \rightarrow \sigma_\chi\, f \tag{22}$$

and :

$$G_{1c} = G_o f^2 = G_o \exp(-2 N_e \chi) \qquad (23)$$

The result is an exponential drop in a $G_{1c}(\chi)$ plot : from the data, Iyengar and Erickson had proposed a different law : $G_{1c} \sim \exp[-k|\delta_A - \delta_B|]$ (Remember that, for simple Van der Waals interactions : $\chi \sim (\delta_A - \delta_B)^2$).

However, these differences are probably not very significative :

a) As already explained, N_e needs not be the same for all AB pairs : there need not be a universal plot $G_{1c}(\delta)$.

b) PET is always partly cristallised, and this complicates the picture.

The essential point is the rapid drop of G_{1c} when A and B become very different.

Of course, the best way of strengthening the AB interface, if A and B are strongly incompatible, amounts to bring an AB diblock copolymer at the interface [19] [20].

C) Tangential slip :

1) Long ago, we discussed the possibility of slip for a molten polymer against a solid surface [24]. One expects a significant slippage if : a) the polymer does not bind to the surface b) the polymer is not a glass (or a crystal) in the first few layers near the surface c) the surface is not too rough.

Experimentally, the usual (no-slip) boundary conditions at the wall are often found to hold [25]. Certain observations on transparent extruders [26], or in plane-plane rheometry [27], do suggest a significant slip, but it is not clear whether this holds even in the linear regime (small shear stresses at the wall) which was considered in ref. [24].

2) Let us now consider an interface A / B between two molten polymers, with $1 > \chi >> N^{-1}$. If χ is large, there will be no entanglements between A and B, and slippage may occur. This problem was first considered by Furukawa [28] -but without a full appreciation of the role of entanglements versus Rouse friction. A slightly improved (qualitative) discussion is given in ref. [29], and will be summarized here.

a) Consider first a case with no entanglements between A and B (fig. 8, 9). In the interfacial region, we have a steep velocity gradient $[V]/2e$, and a weak Rouse viscosity $\eta_R(e)$. This viscosity is itself scale dependent , because e is smaller than the coil size R_o. As argued by F. Brochard (unpublished), we expect :

$$\eta_R(e) = \eta_1 \frac{e^2}{a^2} \qquad (e < R_o) \qquad (24)$$

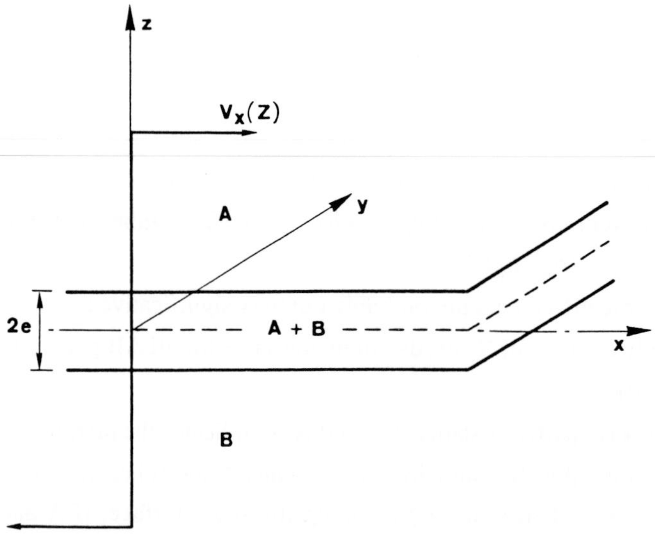

fig. 8. Geometry of flow lines when two molten polymers A and B slip on each other.

Note that eq. 24 gives the right form for very short scales : $\eta = \eta_1$, and also for long scales : $\eta \sim \eta_1 N$ for $e \gtrsim R_0$.

Outside of the interfacial region, we have the strong viscosity of an entangled melt (30) :

$$\eta = \eta_1 \frac{N^3}{N_e^2} \tag{25}$$

(where, for simplicity, we assume the same η_l and the same N for A and B). Writing that the stress $\sigma = \eta \frac{dv}{dz}$ is the same in both regions, we arrive at the extrapolation length b (as defined in fig. 9) :

$$b = e\left(\frac{\eta}{\eta_R(e)} - 1\right) \sim e\frac{\eta}{\eta_R(e)} \sim a\frac{N^3}{N_e^2}\chi^{1/2} \tag{26}$$

For instance, with $\chi = 0.1$, $N = 10^3$, $N_e = 10^2$ and a = 3 Å, we expect $b \sim 10$ microns.

b) Consider now a case where AB entanglement effects are dominant. Then a rough estimate of the inner viscosity is :

$$\eta_{in} = f\,\eta(e) \tag{27}$$

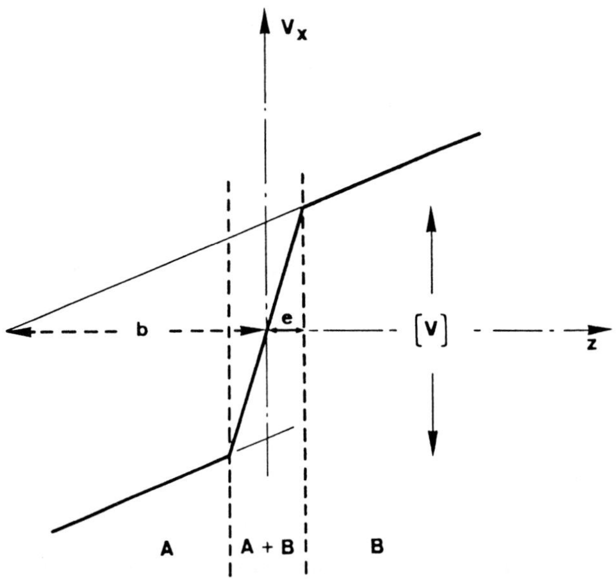

fig. 9. Velocity profile in an A / B slippage process.

where $\eta(e)$ is the scale dependent viscosity for an entangled melt. From the argument in ref.[31] we expect :

$$\eta(e) = \eta_R(e)\left(\frac{N}{N_e}\right)^2 \qquad (e < R_0) \tag{28}$$

We are thus led to an extrapolation length b, in the entangled regime, of the form :

$$b \sim \frac{\eta}{f\,\eta(e)} - 1 \sim \frac{\eta}{f\,\eta(e)} \tag{29}$$

c) in intermediate regimes, we should add up the two types of friction (each proportionnal to $\frac{1}{b}$), and we arrive at :

$$\frac{e}{b} = \frac{\eta_R(e) + f\,\eta(e)}{\eta} = \frac{\eta_R(e)}{\eta}\left[1 + f\left(\frac{N}{N_e}\right)^2\right] \tag{30}$$

The cross-over is obtained when :

$$f \equiv \exp(-N_e\chi) = \left(\frac{N_e}{N}\right)^2 \tag{31}$$

(an estimate slightly different from that of ref. [29], but the difference occurs only in a log.). Typically, with $N = 10^3$ and $N_e = 10^2$, we require $N_e \chi \lesssim 5$ to be entangled.

There are a number of semi quantitative data on coextrusion of layered polymer systems, which indicate rather large values of b. In particular, we should mention the experiments of Miroshnikov and Andreeva [32], where a "jelly roll" of alternating A / B layers was extruded, with controled layer thickness of order 50 microns. The apparent viscosity of this composite melt structure was $\eta_{app} \sim \eta / 20$ (where η is the average viscosity of A and B). This indicates that, for the particular system under study, the extrapolation length b was much larger than 50 microns.

D) Suppression of slippage by block copolymers [33]. Let us now assume that the number of mutual entanglements between A and B is exponentially small : if $N_e\chi >> 1$, the two sides are decoupled and easily slide over one another. We now add (per unit area of the interface) a number ν of AB copolymers with degrees of polymerization Z_A , $Z_B >> N_e$. Clearly the copolymers will connect the two blocks and increase the friction.

We assume that ν is small, i.e. $\nu Z_i a^2 < 1$. This places us in the "mushroom" regime, where two adjacents copolymers do not overlap. It will be seen that this is sufficient to block sliding.

1) Friction of a half block.

Consider now a polymer block (of Z_A monomers) which moves relative to the matrix at a velocity V which will be determined later. The semi-block cannot reptate because it is pinned at one end. In order to move, the matrix chains which traverse the "microgel" formed by Z_A must disentangle. There, number is $p_A = Z_A^{1/2}$. For the microgel to advance an entanglement spacing $d = N_e^{1/2} a$, each of the p_A chains must reptate along it's own tube, by a tube length $L_t = N_A N_e^{-1/2} a$. Thus the curvilinear velocity of the chain is [34]

$$V_c \cong V \frac{L_t}{d} \cong V \frac{N_A}{N_e} \tag{32}$$

The dissipation per chain is $\zeta_1 V_c^2 N_A$. The total dissipation is :

$$T\dot{S} \equiv \zeta_A V^2 = p_A \zeta_1 N_A \left(\frac{N_A}{N_e}\right)^2 \tag{33}$$

The friction coefficient ζ_A then has the Stokes form :

$$\zeta_A \cong \eta_A R(Z_A) \tag{34}$$

where $\eta_A = \zeta_1 a^{-1} N_A^3 / N_e^2$ is the matrix viscosity and $R(Z_A) = a Z_A^{1/2}$ is the mushroom size.

2) Interfacial friction .

Each copolymer moves at a velocity αV, such that the sum of the frictional forces acting upon it is zero :

$$\zeta_A \left(\frac{V}{2} - \alpha V\right) = \zeta_B \left(\frac{V}{2} + \alpha V\right) \tag{35}$$

This fixes α. The dissipation per unit interfacial area is then kV^2 , with

$$k = \nu \frac{\zeta_A \zeta_B}{\zeta_A + \zeta_B} \tag{36}$$

3) Discussion.

Consider the symmetric case ($N_A = N_B = N$, etc). Then:

$$k \cong \nu \zeta_1 N^3 Z^{1/2} N_e^{-2} \tag{37}$$

$$b \cong \frac{\eta}{k} \cong [\nu R(Z)]^{-1} \tag{38}$$

Equating (41) and (34), we estimate the minimum concentration ν^* in order to modify the sliding of the pure system :

$$\nu^* R^2 \cong N_e^2 N^{-3} Z^{1/2} << 1$$

In practice, trace amounts of copolymer is sufficient to prevent sliding.

The same effect is probably important to understand certain mechanical properties of the solid / polymer melt interface : We have seen that strong sliding is not generally observed. It is in fact sufficent for a few chains to be attached to special surface sites in order to supress sliding.

IV. Concluding remarks

The mechanical properties of polymer / polymer interfaces are clearly very sensitive to the detailed structure of the interface. We have seen here two major examples of this correlation a) the role of chain ends, and of their spatial distribution in A / A healing b) the role of entanglements in A / B fracture or in A / B slippage.

From a practical point of view, what can we do to modify the mechanical properties ? 1) For A / B systems, the most obvious additive is an A-B block copolymer. The main difficulty here is to bring the copolymer at the interface, since the kinetics of exchange between copolymer micelles and surfaces are very slow. 2) For A / A systems, we are facing an interesting chemical challenge : by suitable modifications of the chain ends, we may, or may not, encourage their segregation near the surface, and thus generate very different healing behaviors.

Acknowledgments : The author has greatly benefited from discussions and other exchanges with F. Brochard, H. Brown, A. Gent, H. Kausch, E. Kramer, G. Reiter, T. Russell, U. Steiner, and R. Wool.

REFERENCES

(1) H.R. Brown, *Macromolecules,* 1991, (to be published).

" *Annual Rev. Materials Sci.,* 1991, (to be published).

(2) H. Jud, H. Kausch, J. Williams, *J. Material Sci.,* 16, 204 (1981).

(3) H.H. Kausch, D. Petrovska Delacretaz, *Proc. IBM symposium on polymers,* Lech, 1990.

(4) D. Legrand, G. Gaines, *J. Colloid Interface Sci.,* 31, 162 (1969).

(5) " " " " 42, 181 (1973).

(6) A. Harihakan, S. Kumar, T. Russell, *Macromolecules,* 23, 3584 (1990).

(7) P.G. de Gennes, *C.R. Acad. Sci. (Paris),* 307, 1841 (1988).

(8) G. Dee, B. Sauer, *APS March meeting Cincinnati,* 1991, Abstract A 39-1.

(9) C. Poser, I. Sanchez, *J. Colloid Interface Sci.,* 69, 539 (1979).

(10) SIMS, S. G. Whitlow, R.P. Wool, *Macromolecules,* in print, 1991.

" " " paper presented at *Amer. Phys. Soc.,* 1989.

(11) T.P. Russel, A. Karim, A. Mansour, G. Felcher, *Macromolecules,* 21, 1890 (1988).

" " " " *Phys. Rev.,* B 42, 6846, (1990).

(12) G. Reiter, U. Steiner, *J. Phys. (Paris),* in print.

" " to be published in the proceedings of the Les Houches workshop on interfaces (D. Beysens and G. Forgacs editors, 1991).

(13) P.G. de Gennes, *J. Chem. Phys.,* 55, 572 (1971).

(14) " *C.R. Acad. Sci. (Paris),* B 291, 219 (1980).

(15) S. Prager, M. Tirrell, *J. Chem. Phys.,* 75, 5194 (1981).

(16) R.P. Wool, K. O'Connor, *J. Appl. Phys.,* 52, 5953 (1981).

R.P. Wool, B.L. Yuan, O. MC. Garel, *Polymer Engineering and Science,* 29, 1340 (1989).

(17) H.R. Brown, *Materials Sci Reports,* 2, 315 (1987).

(18) E. Kramer and L. Berger, *Adv. Polymer Sci.,* 91-92, p. 1 (1990).

(19) H.R. Brown, V. Deline, P. Green, *Nature,* 341, 221 (1989).

(20) K. Cho, H. Brown, D. Miller, *J. Pol. Sci. (Physics),* 28, 1699 (1990).

(21) P. Flory, *Principles of Polymer Chemistry,* Cornell U. Press.

(22) P.G. de Gennes, *C. R. Acad. Sci. (Paris),* 308 II, 1401 (1989).

(23) Y. Iyengar, D. Erickson, *J. Appl. Polymer Sci,* 11, 2311 (1967).

(24) P.G. de Gennes, *C.R. Acad. Sci. (Paris),* 288 B, 219 (1979).

(25) J. Meissner, *Ann. Rev. Fluid Mech,* 17, 45 (1985).

(26) J. Galt, B. Maxwell, *"Modern plastics",* (Mc Graw Hill ed.), Dec. 1964.

(27) R. Burton, M. Folkes, N. Karm, A. Keller, *J. Materials Sci.,* 18, 315 (1983).

(28) H. Furukawa, *Phys. Rev.,* A 40, 6403 (1989).

(29) F. Brochard-Wyart, P.G. de Gennes, S. Troian, *C. R. Acad. Sci. (Paris),* 310 II, 1169 (1990).

(30) M. Doi, S.F. Edwards, *"The theory of polymer dynamics"*, Clarendon Press, 1986.

(31) See also P.G. de Gennes, ***MRS Bulletin,*** 16, 20 (1991), for a simple interpretation of the linear viscosity.

(32) Miroshnikov, Andreeva, *Vysokomol Soedin. Ser. A (USSR),* 29, 579-82 (1987).

(33) F. Brochard, P.G. de Gennes, P. Pincus, *C.R. Acad. Sci. (Paris)*, 314II, 873 (1992).

(34) For a similar argument, see P.G. de Gennes, MRS Bulletin, January 1991, p. 20.

A model for contact angle hysteresis

J. F. Joanny and P. G. de Gennes
Collège de France 75231 Paris Cedex 05, France

(Received 10 February 1984; accepted 14 March 1984)

We discuss the behavior of a liquid partially wetting a solid surface, when the contact angle at equilibrium θ_0 is small, but finite. The solid is assumed to be either flat, but chemically heterogeneous (this in turn modulating the interfacial tensions), or rough. For weak heterogeneities, we expect no hysteresis, but the contact line becomes wiggly. For stronger heterogeneities, we first discuss the behavior of the contact line in the presence of a single, localized defect, and show that there may exist two stable positions for the line, obtained by a simple graphic construction. Hysteresis shows up when the strength of the defect is above a certain threshold. Extending this to a dilute system of defects, we obtain formulas for the "advancing" and "receding" contact angles θ_a, θ_r, in terms of the distribution of defect strength and defect sharpness. These formulas might be tested by controlled contamination of a solid surface.

I. INTRODUCTION

Wetting phenomena have been studied quantitatively during two centuries (at least).[1] The specific observable which we want to discuss here is the contact angle θ for the fluid–gas–solid system represented in Fig. 1.

If the solid surface is flat, smooth, and chemically homogeneous we are dealing with an ideal situation. Then, in equilibrium, the angle $\theta = \theta_0$ is related to the interfacial tensions defined in Fig. 1(a) through the Young–Dupré equation

$$\gamma_{SG} - \gamma_{LS} = \gamma \cos \theta_0. \tag{I.1}$$

This implies, for instance, that a vertical capillary containing a finite column of fluid, and displaying the angle θ_0 at both menisci (see Fig. 2) should always drain out by gravity forces (the Bertrand theorem).[1c] In practice we all know that, for thin capillaries, the theorem is not valid, the column does not move. An equilibrium is set between hydrostatic and capillary pressures, and this imposes that the angles at the rear and at the front be different. Equation (I.1) is then violated, and the natural interpretation of this violation is based on irregularities of the solid surface. There are two main types of irregularities:

(i) surface roughness[2–8] and (ii) inhomogeneous contamination of the surface.

Our first aim, in the present paper, is to focus on irregularity (ii), which is expected to be present often, and which is also amenable to a relatively simple theoretical treatment. The local inhomogeneity is described by a single function

$$-h(x, y) = \gamma_{SL}(x, y) - \gamma_{SG}(x, y) - [-\gamma_{SG} + \gamma_{SL}]_0, \tag{I.2}$$

where the brackets $[\ \ ]_0$ denotes the unperturbed value, and where the xy plane coincides with the solid surface.

In turns out, however, that most of the physics can be extended to incorporate a discussion of the effects of roughness. This is explained later in Sec. VI. Thus, for many practical situations our analysis covers both problems (i) and (ii).

All our work is based on macroscopic (19th century) concepts. The special effects of long range van der Waals forces, which are known to command the fine structure of the triple line,[3,4] are not incorporated here. Since the long range component is independent of surface contaminations, one may, in a first approximation, ignore it; this is in fact a safe assumption if, and only if, the size of the perturbed regions is larger than the range of the van der Waals forces (e.g., 300 Å).

In Sec. II we discuss the *elasticity of the triple line*, i.e., what sort of deformations does the line display when it is subjected to arbitrary external forces, and in particular to a very localized force. Then in Sec. III we consider the effects of weak inhomogeneities (small h values in Eq. I.2), they impose a certain wiggling of the line, but they do not create any hysteresis when h is small. To obtain hysteresis we need "strong" inhomogeneities. This regime is much more complex. To reach definite conclusions we focus our attention first on a single defect (see Sec. IV). Provided that the defect is small in size, we can construct the various allowed states, describing a line which is anchored to the defect, or free from it. Then we extend this to a distribution of defects on the surface (see Sec. V) and predict the macroscopic contact angles for an advancing line (θ_a) and for a receding line (θ_r) [see Fig. 1(b)].

The special case of "mesa defects" (see Fig. A1) is discussed at length in the Appendix. The word "mesa" means that the perturbation $h(x, y)$ has step-like singularities on the defect boundary. This, with a simple choice of defect shape (rectangle) leads to a relatively simple model, and is instruc-

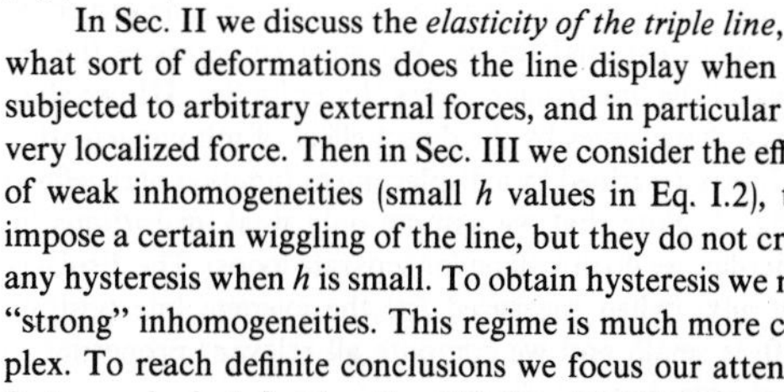

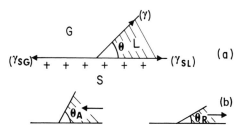

FIG. 1. Definition of the interfacial tensions (a) and of the receding and advancing contact angles (b).

552 J. Chem. Phys. 81 (1), 1 July 1984 0021-9606/84/130552-11$02.10

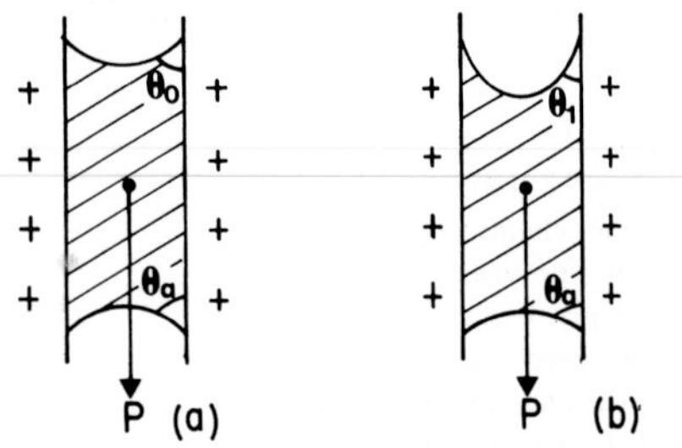

FIG. 2. A column of liquid in a capillary. In a, the rear and front contact angles are equal the column should drain out. In reality the rear and front contact angles are different and the column is pinned in the capillary b.

tive, But, on the other hand, mesa defects are misleading, because even for very small h, they always lead to some hysteresis. Finally in Sec. VI, the practical implications and limitations of our general model are reviewed.

II. A SLIGHTLY DEFORMED CONTACT LINE

We consider for the moment a perfectly smooth solid surface, but we assume that the contact line is deformed by some weak external forces. We want to compute the response of the line to such forces. Later (in Sec. III) we shall replace these forces by the direct effect of inhomogeneities. The line is described by a displacement function $\eta(x)$ defined in Fig. 3. Note that we assume only weak distortions (in particular, we do not allow for separate loops). The displacement $\eta(x)$ (when measured for the average line position $y=0$) is small, and we compute all relevant properties to lowest order in η.

Let us construct first the shape of the interface $z(x,y)$ associated with a given $\eta(x)$. This satisfies the Laplace condition

$$\frac{\partial^2 z}{\partial x^2}+\frac{\partial^2 z}{\partial y^2}=0$$

and can thus be written in the form

$$z(x,y)=\theta_0 y+\frac{1}{2\pi}\int_{-\infty}^{+\infty} dq\,\alpha_q e^{iqx}e^{-|q|y}. \tag{II.1}$$

The first term is the unperturbed profile. The second term describes corrections of wavelength $2\pi/|q|$, extending only up to a distance $|q|^{-1}$ from the average line position. We can then relate the amplitudes α_q to the imposed line shape $\eta(x)$

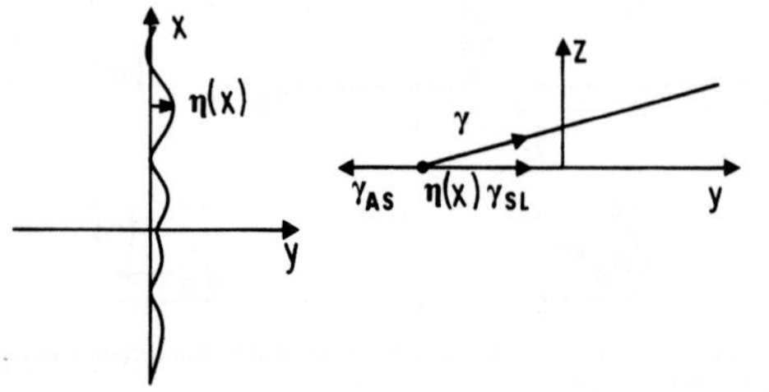

FIG. 3. Wiggly three phase contact line: $\eta(x)$ is the line displacement function in the y direction.

imposing that z vanishes on the line

$$z[x,\eta(x)]=0. \tag{II.2}$$

Introducing Fourier components through

$$\eta(x)=\frac{1}{2\pi}\int_{-\infty}^{+\infty}\tilde{\eta}(q)e^{iqx}\,dq$$

we have to first order in η:

$$\alpha_q=-\theta_0\tilde{\eta}(q). \tag{II.3}$$

Inserting this into Eq. (II.1) we arrive at an explicit specification for the profile. Inverting the Fourier transforms it can be written explicitly as

$$z(x,y)=\theta_0 y-\frac{\theta_0}{\pi}+\int_{-\infty}^{+\infty}\eta(x')\frac{y}{y^2+(x-x')^2}\,dx'. \tag{II.4}$$

Equation (II.4) may also be derived directly from a two-dimensional electrostatic analog, treating z as a potential and θ_0 as a charge density imposed on the contact line. Here we are mostly interested in the energy associated with a given line shape.

The correction to the capillary energy is

$$U_{\text{cap}}=\frac{1}{2}\int dx\,dy\gamma[(\nabla z)^2-\theta_0^2],$$

integrated on all the region $y>\eta(x)$:

$$U_{\text{cap}}=\frac{1}{q}\gamma\theta_0^2\int_{-\infty}^{+\infty}\alpha_q^2 q^2\frac{1}{|q|}\frac{dq}{2\pi}, \tag{II.5}$$

$$U_{\text{cap}}=\frac{1}{q}\gamma\theta_0^2\int_{-\infty}^{+\infty}|q||\tilde{\eta}(q)|^2\frac{dq}{2\pi}. \tag{II.6}$$

The reader may check that there is no linear term in $\tilde{\eta}(q)$ in this capillary energy, when one takes into account the Young equilibrium equation for the contact angle θ_0.

The unusual $|q|$ dependence comes from the integration of a q^2 energy over a thickness $|q|^{-1}$. Again we may Fourier transform this result and write formally

$$U_{\text{cap}}=\frac{\gamma\theta_0^2}{2\pi}\int dx\,dx'\frac{\eta(x)\eta(x')}{(x-x')^2}. \tag{II.7}$$

Of course cutoffs must be introduced in Eqs. (II.6) and (II.7) to prevent the singularities at large q (or $x\to x'$):the linearization leading to Eq. (II.3) assumes $q\tilde{\eta}\ll 1$. Keeping these restrictions in mind, we may write down the external force f which is required to create the distortion η:

$$F(x)=\frac{\delta U_{\text{cap}}}{\delta\eta(x)}=\frac{\gamma\theta_0^2}{\pi}\int dx'\frac{\eta(x')}{(x-x')^2}. \tag{II.8}$$

Physically the most interesting object is the response function $R(x-x')$ giving the deformation field $\eta(x')$ for a localized force $f(x')=f_1\delta(x-x')$. From Eq. (II.5) we find

$$\tilde{f}(q)=\gamma\theta_0^2|q|\tilde{\eta}(q). \tag{II.9}$$

This gives[10]

$$R(x)=\frac{1}{\pi\gamma\theta_0^2}\ln\frac{L}{|x|}, \tag{II.10}$$

where L is a large scale cutoff, provided by some macroscop-

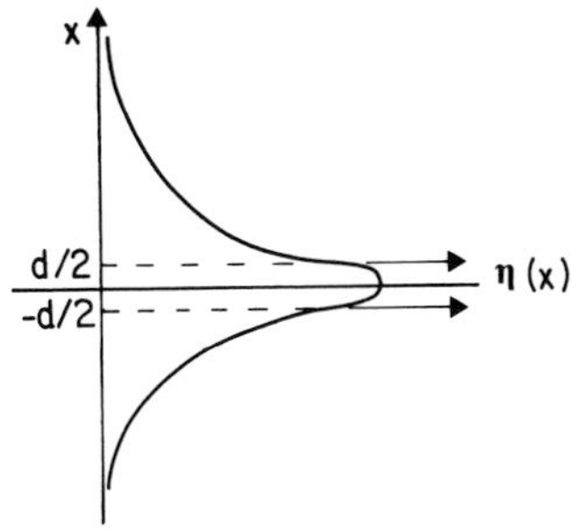

FIG. 4. Distortion of the contact line induced by a force f_1 (per unit length) acting over a distance d of the contact line.

ic sample size. The general aspect of the response is shown in Fig. 4, when the force f_1 is not strictly restricted to a point, but rather spread out over a small region of linear dimensions d, with a central value

$$\eta(d) = f_1 R\,(x \simeq d) = \frac{f_1}{\pi\gamma\theta_0^2}\ln\frac{L}{d} = \frac{f_1}{k}. \tag{II.11}$$

The associated energy is

$$U_{\rm cap} = \int f_1\, d\eta(d) = \frac{1}{2}k\eta^2(d), \tag{II.12}$$

k may be called the *spring constant* of the contact line for localized peturbations. It will be an essential parameter for our later discussions on hysteresis.

III. WEAK HETEROGENEITIES

Let us now consider a heterogeneous solid surface, described by a random fluctuation $h\,(x,y)$ of the work of adhesion, as explained in Sec. I. For the moment we take $h\,(x,y)$ to be very small, so that no hysteresis occurs. But, because of the heterogeneity, the contact line is distorted. We want to discuss the statistics of this wiggly line.

As explained in the introduction, $h\,(x,y)$ acts as a force pulling the contact line. To a first approximation, for weak distortions $\eta(x)$ we may simplify the force as follows:

$$f(x) = h\,[x,\eta(x)] \rightarrow h\,(x,0). \tag{III.1}$$

We can then insert this force [or its one dimensional Fourier transform $\tilde{f}(q)$ into Eq. (II.9)], solve for the deformation $\tilde{\eta}(q)$, and obtain

$$\begin{cases} \langle \tilde{\eta}(q)\rangle = 0 \\ \langle |\tilde{\eta}(q)|^2\rangle = \left(\dfrac{1}{\gamma\theta_0^2}\right)^2 q^{-2}|\tilde{f}(q)|^2. \end{cases} \tag{III.2}$$

Equation (III.2) has a certain similarity to the viscous motion of a particle [with position $\eta(x)$ at "time" x] under a random force $f(x)$. The instantaneous "velocity" $v = d\eta/dx$ is assumed proportional to the force

$$iq\tilde{\eta}(q) = \Lambda\tilde{f}(q) \tag{III.3}$$

and Eq. (III.2) is satisfied if we take $\Lambda = 1/\gamma\theta_0^2$ (because the factor i drops out when we take an amplitude squared). We expect that the long range behavior of this particle will be ruled by diffusion, with a diffusion coefficient D:

$$\langle[\eta(x) - \eta(0)]^2\rangle = 2D\,|x|, \tag{III.4}$$

where D is related to the correlation function of velocities by a classical formula

$$\begin{aligned} D &= \int_0^{+\infty} \langle v(0)v(x)\rangle dx \\ &= \Lambda^2 \int_0^{+\infty} \langle f(0)f(x)\rangle dx \\ &= \Lambda^2 \int_0^{+\infty} \langle h\,(0,0),\, h\,(x,0)\rangle dx. \end{aligned} \tag{III.5}$$

Equation (III.4) describes a very strong wiggly motion. For a macroscopic size of droplet L, the amplitudes of fluctuation behave like $\eta^2 \sim DL$, i.e., they increase like $L^{1/2}$. This result has also been obtained independently by Vannimenus and Pomeau (private communication).

As a specific example, let us choose a one-dimensional correlation function

$$\langle h\,(0,0)h\,(x,0)\rangle = h^2 e^{-|x|/\xi}, \tag{III.6}$$

where ξ is a correlation length for the surface inhomogeneities. Then we have

$$D = \left(\frac{h}{\gamma\theta_0^2}\right)^2 \xi. \tag{III.7}$$

In many practical cases we expect $h/\gamma\theta_0^2 \lesssim 1$ and thus $\eta^2 \sim L\xi$. For a droplet of millimetric size, with defects of correlation length $\xi \sim 10$ Å we would then expect fluctuations η of order one micron. Note that the opposite limit $h/\gamma\theta_0^2 > 1$ is not compatible with our assumptions. There is a natural limit corresponding to a local contact angle on the defect $\theta = 0$, which is

$$h = \tfrac{1}{2}\gamma\theta_0^2.$$

We end up this section with a brief discussion of the line energy (per unit length) or, more precisely, of the contributions to this energy originating from the heterogeneity. From the capillary term (III.6) and the coupling term $(-\tilde{\eta}_{(q)}\tilde{f}_{(q)})$ we find a correction per unit length of the form

$$\begin{aligned} \Delta F &= -\frac{1}{2}\int \frac{dq}{2\pi}\frac{|\tilde{f}(q)|^2}{\gamma\theta_0^2|q|} \\ &\sim -\frac{1}{2}\frac{\xi h^2}{\gamma\theta_0^2}\ln\frac{L}{\xi} \end{aligned} \tag{III.8}$$

[for the one-dimensional correlations see Eq. (III.6)].

It is not easy to see, however, how this energy could be detected in practice. Weak heterogeneities give small line energies and large heterogeneities are dominated by hysteretical effects, which we now begin to discuss.

IV. STRONG HETEROGENEITIES—THE SINGLE DEFECT PROBLEM

A. The defect force

We now consider specifically a distribution $h\,(x,y)$ with a peak (at $x = x_d$, $y = y_d$) and a certain small width $\Delta x = \Delta y = d$. A contact line anchored on such a "defect" is

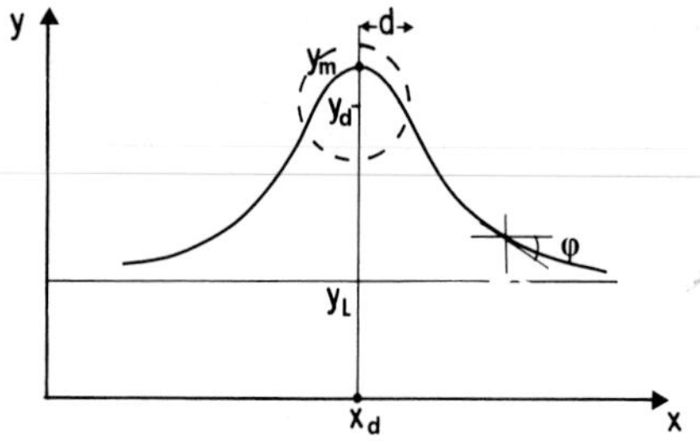

FIG. 5. A localized Gaussian defect of width d.

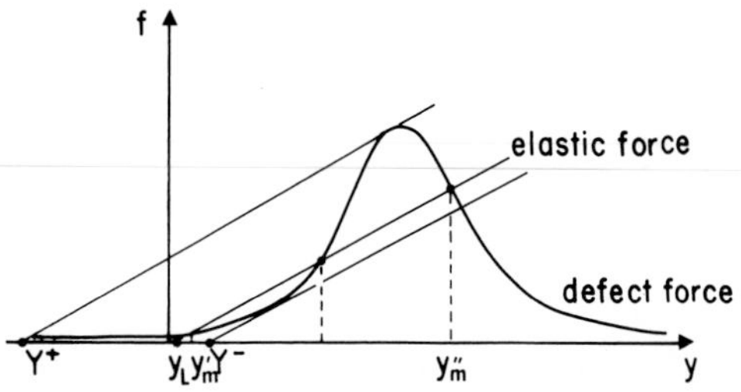

FIG. 6. Geometrical construction of the equilibrium positions of the contact line in the presence of a localized defect. The nominal line position is y_L. The maximum advancing position has two possible values y'_m and y''_m. In a receding experiment (retreating line) the anchoring by the defect ends for $y_L = Y_+$. An advancing line is captured by the defect when $y_L = Y_-$.

shown in Fig. 5. Far from the defect, the line returns to $y = y_L$.

For our purpose the important parameter is the total force exerted by the defect on the line

$$f_1 = \oint ds \cos\varphi\, h(x_s, y_s) = \int_{-\infty}^{+\infty} dx\, h[x, y_L + \eta(x)], \quad \text{(IV.1)}$$

where ds is the element of contact line (of position x_s, y_s). (The study is limited here to attractive defects $h > 0$. The case of repulsive defects is perfectly symmetric.)

φ is the angle between this element and the average line orientation (x) and $\eta = y_s - y_L$ is the distortion. The integral (IV.1) is dominated by the central region, where $\eta(x)$ is close to its maximum $\eta_m = y_m - y_L$. In what follows we approximate f_1 by the simpler form

$$f_1 \simeq \int_{-\infty}^{+\infty} dx\, h(x, y_m). \quad \text{(IV.2)}$$

The great merit of Eq. (IV.2) is to generate a force which does not involve the whole profile $\eta(x)$, but only its peak value (related to y_m). For a given functional form of the defect structure $h(x - x_d, y - y_d)$ we can then compute explicitly a force $f_1(y_m - y_d)$ by Eq. (IV.2). A simple example, to which we shall sometimes refer, is a Gaussian defect

$$h(x - x_d, y - y_d) = h_0 \exp - [(x - x_d)^2 + (y - y_d)^2]/2d^2. \quad \text{(IV.3)}$$

A nice feature of Eq. (IV.3) is that the simplified force (IV.2) is also Gaussian, and thus simple

$$f_1 = \sqrt{2\pi} h_0 d \exp - (y_m - y_d)^2/2d^2. \quad \text{(IV.4)}$$

Another merit of Eq. (II.3) is that, if the chemical contaminants which create the defect were spreading from an initial point source, simple diffusion would indeed generate a Gaussian form. However, in most of what follows, we can pursue our discussion without choosing a specific form for f_1, and this gives a much broader generality to our model.

B. Balance of forces

Far from the defect, the line has a fixed ordinate $y = y_L$. The line tip (at $y = y_m$) is then in equilibrium under the action of two forces. One is the force f_1 described above, and the other is a restoring force, tending to bring the line back at $y = y_L$. The spring constant for this second force has been derived in Eq. (II.11). Thus the balance of forces is simply

$$k(y_m - y_L) = f_1(y_m - y_d). \quad \text{(IV.5)}$$

This leads to a simple graphical construction (see Fig. 6). For a given distance between line and defect ($y_d - y_L$) we can find the equilibrium points y_m. For weak defects (f_1/k small) there is in general only one equilibrium point (no hysteresis). For stronger defects, we can have three equilibrium points. The smallest (y'_m) and largest (y''_m) of these are stable, while the intermediate one is unstable.

The onset of hysteresis corresponds to the particular case where the inflection point in $f_1(y)$ has a slope just equal to the spring constant k. For instance in the Gaussian model, this corresponds to

$$h_0 = h_e = \left(\frac{e}{2\pi}\right)^{1/2} k = \left(\frac{e}{2\pi}\right)^{1/2} \frac{\pi\gamma\theta_0^2}{\ln L/d}. \quad \text{(IV.6)}$$

Defects with $h_0 < h_c$ do not contribute to the hysteresis. Let us now concentrate on the interesting case $h_0 > h_c$, and assume that we gradually decrease the nominal position of the line y_L. Then we find anchoring with $y_m = y''_m$ up to the point where two roots y_m merge and disappear. At this moment the line position is $y_L = Y_+$, and suddenly the line snaps to a very weakly distorted profile $y_m = y'_m$. Similarly, in the reverse process (y_L is increasing) we get anchoring on $y = y'_m$ up to a certain position $y_L = Y_-$ at which the line jumps off to the other configuration $y_m = y''_m$.

C. Energy function

The energy associated to the defect is the sum of two contributions:

$$U_{\text{cap}} = \tfrac{1}{2}(y_m - y_L)^2 k \quad \text{(IV.7)}$$

an elastic energy discussed in Sec. II;

$$U_d = -\int_{-\infty}^{y_m} dy'_m\, f_1(y'_m - y_d) \quad \text{(IV.8)}$$

a defect energy.

Minimization of this energy with respect to y_m (at fixed nominal position of the contact line y_L) gives back the force balance (IV.5). The absolute minimum of the energy determines the stable line conformation. We may define a line position $y_L = Y_e$, where the two line conformations have exactly the same energy. This is defined by the Maxwell con-

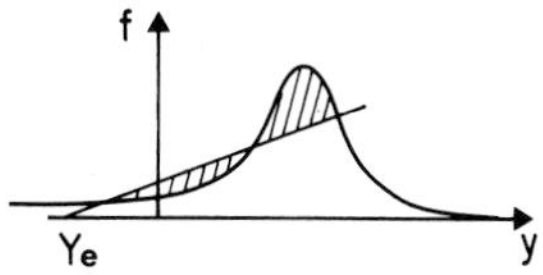

FIG. 7. Maxwell construction for the commuting position $y_L = Y_e$ at thermal equilibrium: the shaded areas are equal.

struction of Fig. 7. If we were able to shake the line gently (by thermal agitation for small defects or by mechanical vibrations for larger defects) the line would then commute reversibly from y'_m to y''_m at $y_L = Y_e$.

In real experiments such an agitation does not exist. In receding experiments, the line does not commute at $y_L = Y_e$, but remains in a metastable position up to $y_L = Y_+$. The jump of the line to its stable position is associated with an energy dissipation $E(Y_+)$ equal to the shaded area in Fig. 8. The corresponding dissipated energy for the advancing line $E(Y_-)$ is the shaded area of Fig. 8'.

Finally a simple expression can be obtained for the energy function by taking its derivative with respect to the nominal line position y_L:

$$U = U_{\text{cap}} + U_d = \text{cst} - \int_{Y_e}^{y_L} f_1(y_m - y_d)dy_L, \qquad \text{(IV.9)}$$

y_m is a function of y_L with two determinations y'_m and y''_m. According to the experiment studied one adequate determination is to be chosen.

V. MACROSCOPIC CONTACT ANGLES

A. A dilute system of defects

We now want to describe the effect of a distribution of surface defects on the macroscopic contact angles. Let us first assume that all defects are identical, and that they are spread at random on the solid surface, with a number of defects per cm^2 which we call n. We assume that the defects are well separated ($nd^2 \ll 1$). Then it is plausible to assume that the forces are simply additive. The macroscopic force per unit length is by the Young equation

$$\gamma(\cos\theta - \cos\theta_0),$$

and the sum of defect forces is

$$n \int dy_d\, f_1(y_m - y_d) = \gamma(\cos\theta - \cos\theta_0), \qquad \text{(V.1)}$$

where y_m and y_L are related by the balance of forces [Eq. (IV.5)] or equivalently by the construction of Fig. 6. The only delicate point in Eq. (V.1) is the choice of roots $y_m = y'_m$ or y''_m. For instance, when the line is pulled towards low y_L (receding experiment), we must use $y_m = y''_m$ whenever $y_L > Y_+$, and $y_m = y'_m$ wherever $y_L < Y_+$. Integration over the defect position y_d at a fixed line position y_L is equivalent to an integration over the line position y_L at a fixed defect position y_d. The result is expressed in terms of the energy function

$$\gamma(\cos\theta - \cos\theta_0) = m[-U(+\infty) + U(Y_+) + U(-\infty) - U(Y_-)], \qquad \text{(V.2)}$$

where Y_+ and Y_- are the positions of the two roots for y_m at the commutation of the contact line. In Eq. (V.2) the angle θ_0 corresponds to the absence of all defects.

At thermodynamic equilibrium the commutation between anchored line and free line occurs at $y_L = Y_e$ and the two positions have the same energy

$$\gamma(\cos\theta_E - \cos\theta_0) = n[U(-\infty) - U(+\infty)]. \qquad \text{(V.3)}$$

The difference $U(-\infty) - U(+\infty)$ depends only on U_d [Eq. (IV. 8)] because the line is undistorted when it is far from the defect ($U_{\text{cap}} = 0$). Then Eqs. (IV.8) and (IV.2) show that

$$U(-\infty) - U(+\infty) = \int dx\, dy\, h(x - x_d, y - y_d).$$

Thus, Eq. (V.3) describes simply the renormalization of interfacial tensions due to the average density of defects.

In the receding experiment the jump occurs at $y_L = Y_+$ and $U(Y_+) - U(Y_-)$ represents the dissipated energy $E(Y_+)$ for a single defect

$$\gamma(\cos\theta - \cos\theta_E) = n\,E(Y_+). \qquad \text{(V.4)}$$

In the advancing experiment the jump occurs at $y_L = Y_-$ and $U(Y_+) - U(Y_-)$ is the negative of the dissipated energy $E(Y_-)$ for a single defect

$$\gamma(\cos\theta_a - \cos\theta_E) = -nE(Y_-). \qquad \text{(V.5)}$$

In this limit of small defect density, we obtain a very general relationship between the advancing and receding angle and the total energy dissipated by one defect around a hysteresis cycle $W_d = E(Y_+) + E(Y_-)$:

$$\gamma(\cos\theta_r - \cos\theta_a) = nW_d. \qquad \text{(V.6)}$$

B. Renormalization of the spring constant

Our formula (II.11) for the spring constant described a line attracted by one defect and pinned at the sample edges ($x < L$). For the many defect problem we may guess that L must be replaced by the average distance between defects (as measured along the contact line) which we shall call b. This distance is given by

$$b = (nd)^{-1}. \qquad \text{(V.7)}$$

We shall now give two slightly more detailed justifications of this point; one based on a periodic system, the other based on the discussion of a pair of defects.

1. A periodic array

Let us consider a line of defects, located at the points $x = mb$ ($m =$ integer), $y = y_D$. Each defect gives a force f_1, and we want to analyze the shape of the pinned line.

The line shape can be studied by an expansion in a Four-

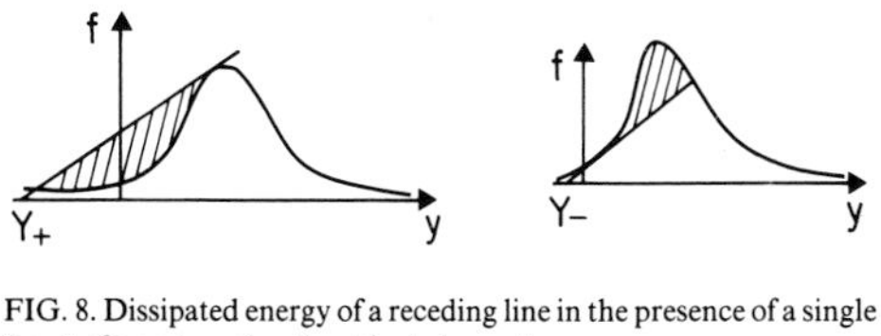

FIG. 8. Dissipated energy of a receding line in the presence of a single defect 8 and of an advancing line (shaded areas).

ier series of period b:

$$\eta = \sum_{p=-\infty}^{+\infty} \eta_p \exp\left(\frac{i2\pi x}{b}\right). \tag{V.8}$$

The constant term ($p = 0$) in this expansion represents the overall displacement of the line. If we focus on one defect, the other defects play the role of an effective medium which create this overall displacement. We are interested here in the perturbation to this average line.

The Fourier series of the periodic force is

$$f = \sum_p f_p \exp\left(\frac{i2\pi px}{b}\right),$$

$$f_p = \begin{cases} b^{-1} f_1, & p < bd^{-1} \\ 0, & p > bd^{-1}, \end{cases} \tag{V.9}$$

since the defect size (d) provides a cutoff.

The relation between force and displacement is still given by Eq. (II.9), for a wave vector $q = 2\pi pb^{-1}$:

$$f_p = \gamma\theta_0^2 |p| 2\pi b^{-1} \eta_p. \tag{V.8'}$$

The relative distortion of the contact line can be determined by using Eqs. (V.8) and (V.9):

$$\eta(x=0) = \frac{f_1}{2\pi\gamma\theta_0^2} \sum_{-b/d}^{b/d} \frac{1}{|p|} = \frac{f_1}{\pi\gamma\theta_0^2} \ln\left(\frac{b}{d}\right).$$

Thus this periodic problem leads to a renormalized elastic constant

$$\tilde{k} = \frac{\pi\gamma\theta_0^2}{\ln(n^{-1}d^{-2})} \tag{V.10}$$

in agreement with our qualitative prediction.

2. The two defect problem—a self-consistent argument

Let us consider two neighboring defects at a distance $x = b$ from each other, and at the same $y = y_D$. We want to look at the perturbation which they introduce on the *average* contact line. This perturbation can be determined from Eq. (II.2) for the response function

$$\eta(x) = \frac{f_1}{\pi\gamma\theta_0^2}\left(\ln\left|\frac{D}{k}\right| + \ln\left|\frac{D}{b-x}\right|\right). \tag{V.11}$$

D is not the macroscopic length L but a new integration constant chosen through a self-consistent argument. Between the two defects the contact line goes back to its average (perturbed) position. We find $D \sim b/2$.

The effective elastic constant is then

$$\tilde{k} = \frac{f_1}{\eta(d)} = \frac{\pi\gamma\theta_0^2}{\ln(b/d)}$$

and this leads us back to Eq. (V.10).

VI. CONCLUDING REMARKS

A. The role of defect size and shape

We have analyzed the effects of certain localized "defects" on the solid surface—chemical inhomogeneities, characterized by a "strength" h_0 (giving the local modification of $\gamma_{SL} - \gamma_{SG}$, and a geometrical size d). We find that this type of defect will effectively pin the contact line, if the strength is above a certain threshold

$$h_0 > h_c = \left(\frac{e}{2\pi}\right)^{1/2} \frac{\pi\gamma\theta_0^2}{\ln L/d} \tag{VI.1}$$

[where, to be specific, we have chosen a Gaussian structure for the defect, defined in Eq. (IV. 3)]. Equation (VI.1) gives a threshold h_c which depends only very weakly on the defect size d. This, however, results from a cancellation between two opposite effects, which we now discuss in slightly more detail. For this discussion, it is illuminating to consider an *anisotropic defect*, with two distinct sizes d_x along the direction of the (average) contact line, and d_y normal to it. The maximum pinning force f_1 is then of order $h_0 d_x$. But the onset of hysteresis (as explained in Fig. 6) occurs when the spring constant k of the line becomes equal to the maximum derivative $\partial f_1/\partial y$. Qualitatively

$$\left.\frac{\partial f_1}{\partial y}\right|_{\max} \simeq \frac{f_1}{dy} \simeq h_0 \frac{d_x}{d_y},$$

thus

$$h_c = \frac{d_y}{d_x} k. \tag{VI.2}$$

Equation (VI.2) agrees with Eq. (VI.1) when $d_y = d_x$, but gives much longer domain of hysteresis if $d_y < d_x$. More generally, if the defect *has sharp edges it will pin strongly*. We insist on this point in the Appendix, which is aimed at the so called mesa defect, where the perturbation $h(x, y)$ has step function singularities on the edges. In this mesa limit hysteresis is present even for very weak amplitudes h_0.

B. Defect statistics

Up to now we have assumed that all defects were identical. In practice they will be distributed in strength and in size. Let us, for instance, assume that the size d is fixed, but that we have a distribution $p(h_0)dh_0$ for the strengths, normalized by

$$\int_0^{+\infty} p(h_0)dh_0 = n, \tag{VI.3}$$

where n, is always the number of defects per unit area, and is assumed small ($nd^2 \ll 1$). For each defect, the crucial parameter, controlling the deviation from equilibrium contact angles, and defined in Eq. (V.6), is

$$W_d(h_0) = E(Y_+) + E(Y_-) \quad (h_0 \geqslant h_c). \tag{VI.4}$$

We have discussed the analytic structure of $W_d(h_0)$ in Sec. IV (see Fig. 9 also). The deviation from Young formula is then a weighted average on the W_d's

$$\gamma(\cos\theta_r - \cos\theta_a) = \int_{h_c}^{+\infty} p(h_0) W_d(h_0) dh_0. \tag{VI.5}$$

To illustrate this, consider the following example: $p(h_0)$ is a Gaussian

$$p(h_0) = \frac{n}{\sqrt{2\pi}} h^{-1} \exp(-h_0^2/2h)^2 \tag{VI.6}$$

and assume that the defects are rather weak ($h < h_c$).

Then, because $W_d(h_0) \sim (h_0 - h_c)^2$ near $h_0 \simeq h_c$ we find the following structure (this result is simply obtained by calculating the shaded area of Fig. 9 using a Taylor expansion

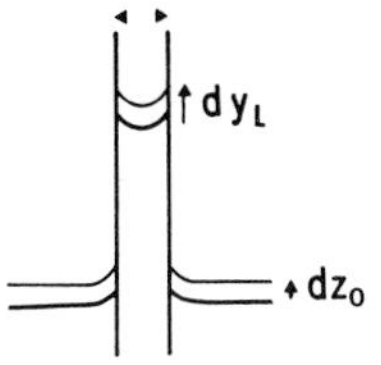

FIG. 9. A proposed experiment for the study of differential susceptibilities [Eq. (VI.11)] a slight change in the reference line (dz_0) induces a slight change in the contact angle ($d\theta$) and in the position of the contact line (dy_L). The displacements involved are of order of 1 μ: the angles θ and $\theta + d\theta$ are both inside the hysteresis interval (θ_r, θ_a).

force inflection point):

$$\gamma(\cos\theta_r - \cos\theta_a) \sim n \frac{k^2 d^2}{h_c} \exp - \frac{h_c^2}{2h^2}\left(\frac{h}{h_c}\right)^5. \quad \text{(VI.7)}$$

Thus, all hysteresis effects become exponentially small when $h \ll h_c$.

C. Limitations

The use of Equations such as Eq. (VI.7) must be conducted with care. On the one hand, it is true that for $h < h_c$ the number of pinning defects is small, and thus our assumption on dilution is easily satisfied. On the other hand, various complications may come in. The (few) remaining pinning defects are deforming the contact line very significantly. But our analysis assumed that the local inclination of the contact line (dz/dx in the notation of Sec. II) was small. Returning to Eq. (II.10) we see that this imposes $f_1 < \pi\gamma\theta_0^2|x|$, where $|x|$ is the distance from the observation point to the defect. The minimum meaningful value of $|x|$ is d, and thus the pinning force must be bounded by

$$f_1 \leqslant \pi\gamma\theta_0^2 d. \quad \text{(VI.8)}$$

But we also know from Eq. (VI.1) that hysteresis occurs only when

$$f_1 \geqslant dk = \frac{\pi\gamma\theta_0^2 d}{\ln L/d}. \quad \text{(VI.9)}$$

Writing $f_1 = h_0 d$ we see that the range of interest is

$$\frac{\pi\gamma\theta_0^2}{\ln L/d} < h_0 < \pi\gamma\theta_0^2. \quad \text{(VI.10)}$$

If h_0 becomes comparable to $\pi\gamma\theta_0^2$, we expect more exotic phenomena (a) strong distortions of the contact line, possibly leading to islands of unwetted regions in a wetted matrix (or reversely, to disconnected droplets of liquid on a dry solid), (b) for small defects ($d \lesssim 1\ \mu$) thermal agitation may allow hopping from one minimum ($y_m = y_m''$) to the other minimum ($y_m = y_m'$). We hope to return to these more delicate questions in future work.

D. Extension to surface roughness

All our discussion has been limited to heterogeneities coming from contaminants on the surface. However our model here is more general and can be extended to surface roughness.[8] Planar surfaces which are smooth but not perfect can also be described by an effective reduction of the work of adhesion $h(x, y)$.

Let us for example assume that the roughness of the surface is characterized at each point by a height $u(x, y)$ (Fig. 10). The contact line is parallel to the x axis; the Young

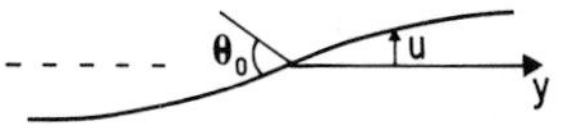

FIG. 10. A plane surface with smooth roughness.

contact angle θ_0 is the angle between the liquid gas interface and the actual surface. It is related to the contact angle with the average planar surface θ by

$$\theta_0 = \theta + \frac{du}{dy} \quad \left(\frac{\partial u}{\partial y} \ll 1\right). \quad \text{(VI.11)}$$

The Young equilibrium equation is then

$$\gamma \cos\left(\theta + \frac{\partial u}{\partial y}\right) = \gamma_{SG} - \gamma_{SL}$$

or

$$\gamma \cos\theta = \gamma_{SG} - \gamma_{SL} + h. \quad \text{(VI.12)}$$

The effective increase of work of adhesion is then

$$-h(x, y) = \cos\theta_0 \frac{\partial u}{\partial y}. \quad \text{(VI.13)}$$

Equation (VI.13) shows an equivalence for the two possible kinds of heterogeneities leading to hysteresis for smooth surfaces. The main practical difference is that a contaminated spot leads to an h function with one given sign and one peak (as in Fig. 5) while a rugosity bump leads [from Eq. (VI.13)] to an h function with a maximum and a minimum. But all our constructions remain valid for both cases.

E. Systematic experiments

The first experimental aim should be to produce *controlled defects*:

(i) On the 10 μ scale, they could be generated by the deposition of metallic (or other films) in the form of spots with well-defined sizes. This being a rather standard program in microelectronics. In most practical cases, these defects will have rather sharp edges, but with a residual smoothness often controlled by diffusion processes. The graphical construction of Fig. 6 should be essential to discuss such cases.

(ii) One could also generate the defects by chemical grafting (silanation,...) on surfaces which are accessible only at certain spots (using photoresist or photolabile film coverages and optical images to define these spots).

Having well specified the defects, one would also like to vary continously their relative strength (measured by $h_0/\pi\gamma\theta_0^2$). This is extremely delicate, but could possibly be achieved with suitable mixtures of two liquids.

Let us now list a few possible experiments: (i) the macroscopic contact angles θ_a, θ_r can be determined by relatively standard procedures. It is much more different to define an experiment leading to the thermodynamic value θ_e. Controlled mechanical agitation (e.g., sonication) could, lead to θ_e, but this point would require further study, (ii) the contact line, anchored on defects, may in principle display a *reversible susceptibility*

$$\chi = -\frac{d\bar{y}_L}{d(\gamma\cos\theta)} \quad \text{(VI.14)}$$

for *small* modulations of θ within the hysteresis limits θ_a, θ_r, and $\bar{y}_L$ being the average line position. For instance with the set up of Fig. 9 one could vary θ very slightly $\theta \to \theta + d\theta$ by shifting the reference z_0; then measure the changes in level of the contact line $(d\bar{y}_L)$ by some optical or electrical technique. However, it must be realized that the displacements d_2 involved are very small (smaller than $Y_+ - Y_-$ in Fig. 6): there experiments appear feasible only with rather large defects ($> 10\,\mu$).

Last but not least, one would like to observe directly the structure induced by the defects: (i) *statics*: the geometrical distortions of the contact near one defect, and the remarkable "Brownian path" $\eta(x)$ predicted at larger scales in Eq. (III.4), (ii) *dynamics*: when a contact line springs from one equilibrium position to the next, it must generate some form of *weak noise*, similar in origin to the Barkhausen noise of ferromagnets,[9] or to Haynes jumps for biphasic flows in porous media. It may be that these jumps can be detected optically or acoustically; again we hope to return to these questions later.

ACKNOWLEDGMENTS

We have greatly benefited from discussions with M. Dupeyrat, B. Legait, L. T. Minassian Saraga, Y. Pomeau, J. Vannimenus, and B. Widom, on various features of contact angles.

APPENDIX: MESA DEFECTS

In this Appendix, we study in great detail the contact angle hysteresis induced by mesa defects. These defects are small regions of space where the difference $\gamma_{SL} - \gamma_{SG}$ is reduced by a given quantity h. We focus here on attractive defects $h > 0$, the discussion would be identical for repulsive defects ($h < 0$).

1. Hysteresis of a single rectangular defect

The defect has the rectangular shape shown in Fig. A1. It sits at a distance 1 of the unperturbed contact line. We do not want to focus on the precise profile of the contact line but on its maximum advancing position $y = \eta(0)$. For a given value of y, we first derive the excess energy due to the defect $U(y,1)$. The equilibrium positions for y are the minima of this excess energy. Knowing these equilibrium positions and their stability, we build a hysteresis cycle for the advancing and receding experiments.

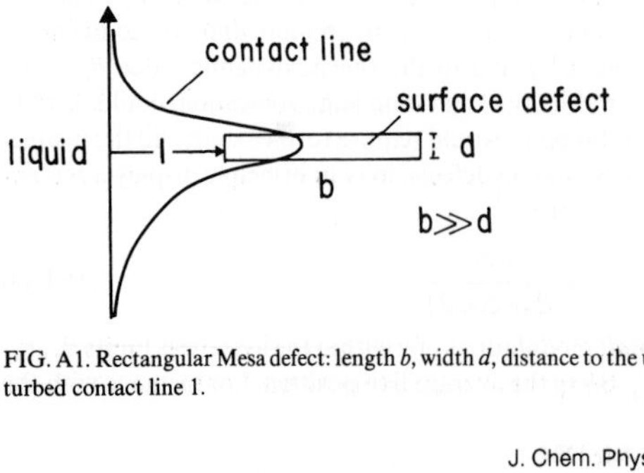

FIG. A1. Rectangular Mesa defect: length b, width d, distance to the unperturbed contact line 1.

a. Defect energy $U(y,1)$

The defect energy is the sum of two terms. (i) An elastic energy derived in Sec. II:

$$U_d = \frac{\pi\gamma\theta_0^2}{2\ln\frac{L}{d}}\,y^2; \tag{A1}$$

(ii) An adhesion energy corresponding to the defect surface covered by the liquid. This adhesion energy is proportional to the surface covered. To calculate this energy U_{adh}, we can in a good approximation ignore the small difference between $\eta(0)$ and $\eta(d/2)$ and consider that the defect surface covered by the drop is a rectangle.

At at given distance 1 between the defect and the contact line, the value of the total defect energy $U(y,1)$ depends on the value of y.

If $y < 1$, the liquid does not cover the defect $U_{\text{adh}} = 0$ and

$$U = U_d = \frac{1}{2}\frac{\pi\gamma\theta_0^2}{\ln\frac{L}{d}}\,y^2. \tag{A2}$$

If $1 < y < 1 + b$, the liquid covers the left part of the defect of area $d(y-x)$ and

$$U = \frac{1}{2}\frac{\pi\gamma\theta_0^2}{\ln\frac{L}{d}}\,y^2 - hd(y-x). \tag{A3}$$

If $1 + b > y$, the liquid covers the whole defect

$$U = \frac{1}{2}\frac{\pi\gamma\theta_0^2}{\ln\frac{L}{d}}\,y^2 - hdb. \tag{A4}$$

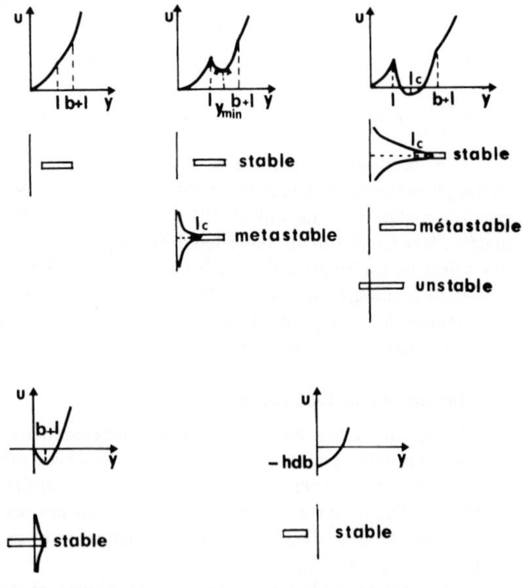

FIG. A2. Excess energy due to a rectangular Mesa defect for different positions of the defect as a function of the maximum distortion due to the defect y.

According to the different values of the distance between the contact line and the defect edge 1 we distinguish five different situations:

(i) $l > l_c = \dfrac{hd}{\pi\gamma\theta_0^2} \ln \dfrac{L}{d}$, free line.

The energy is sketched in Fig. A2a. There is one equilibrium value $y = 0$.

(ii) $l_e = \dfrac{l_e}{2} < l < l_c$, metastable capture.

The defect energy shows two minima. The undeformed profiles is stable but a metastable minimum $(U > 0)$ appears (Fig. A2b) corresponding to $y_{\min} = l_c$.

(iii) $l_c - b < l < l_e$, stable capture.

When $l = l_e$ the energy of the second minimum becomes negative and there is inversion of stability the stable profile corresponds to a capture of the contact line by the defect. The undeformed profile is metastable, it becomes unstable for $1 = 0$.

(iv) $-b < l < l_c - b$ pinning of the contact line on the right edge of the defect.

When the right edge of the defect is too close to the unperturbed contact line, the minimum of the defect energy remains pinned on that defect right edge. The undeformed profile is unstable.

(v) $1 < -b$ the liquid covers the whole defect.

b. Hysteresis of a rectangular defect

We can now describe a receding and an advancing experiment for the perfect plane with one rectangular defect. We study first the defect energy in the corresponding equilibrium state as a function of the contact line position R [Fig. (A3)]. This energy is the minimum value of $U(y,1)$ for the given relative position of the defect and the contact line.

1. Receding experiment. We start with a contact line at infinite distance R which completely covers the defect, the defect energy is $U(R) = -hdb$ [situation (v)]. When the position becomes equal to R_A the defect pins the contact line [situation (iv)].

The defect energy is $U(R) = -hdb + \frac{1}{2}\pi\gamma\theta_0^2 \times [(R - R_A)^2/\ln L/d]$. This pinning lasts up to point B where $R_A - R_B = l_c$. Decreasing the radius, we reach the stable capture region (iii) for which the defect energy is $U = -hdb + \frac{1}{2}\pi\gamma[(hd)^2/\pi\gamma\theta_0^2] \ln L/d - hd(R - R_A)$.

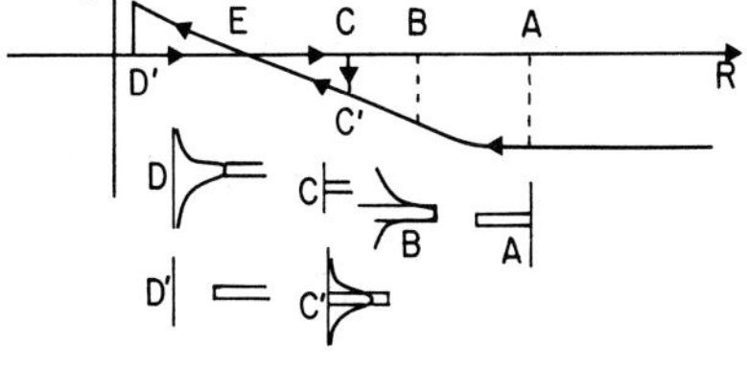

FIG. A3. Hysteresis cycle for a rectangular Mesa defect: energy–distance plot.

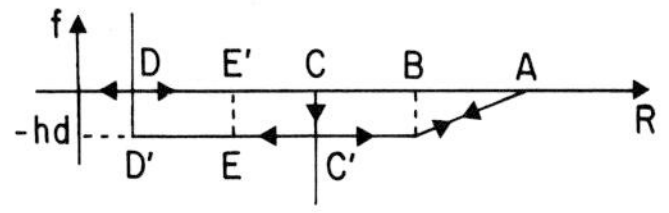

FIG. A4. Hysteresis cycle for a rectangular Mesa defect: force–distance plot.

For a thermodynamic equilibrium experiment, the capture ends at point E where this energy reaches zero $R_A - R_E = b - l_e = b - l_c/2$. In the receding experiment the equilibrium profile is the metastable one up to point D. $R_B - R_D = b$ where it becomes unstable. The defect does not perturb the contact line any longer.

2. Advancing experiment. At the beginning of the experiment the defect is completely uncovered, the liquid does not know of the existence of the defect. This unperturbed profile is stable up to point E, then metastable up to point C where the contact lines reaches the defect and jumps to its stable C' equilibrium conformation $R_B - R_C = b - l_c$. Increasing the radius we follow then the receding experiment.

This energy plot shows a clear hysteresis cycle $CC'EDD'EC$. Energy is dissipated at the two points where the contact line jumps C and D. The total dissipated energy when going around this hysteresis cycle is then

$$W_d = U_c - U_c' + U_D - U_D' = \frac{(hd^2)}{\pi\gamma\theta_0^2} \ln \frac{L}{d}. \quad \text{(A5)}$$

The hysteresis effects described in this section are very similar to the hysteresis of a first order magnetic transition. The role of the order parameter is played here by the variable y (maximum distortion of the contact line). Instead of the plot energy distance, a more familiar pair of variables to look at the hysteresis is the force $f = dU/dR$ distance plot. This is somehow analogous to the field magnetization plots of magnetic transitions (Fig. A4).

On this diagram, we can use the usual dissipation some rule: the total dissipated energfy is the area of the hysteresis cycle $DCC'D'$.

3. Square defect. The rectangular defects are easy to study but they are not very realistic because of their asymmetry. The same kind of study can be made with square defects. The major difference between the two is that the pinning regime is more important for square defects. The energy–radius plot is given in Fig. A5.

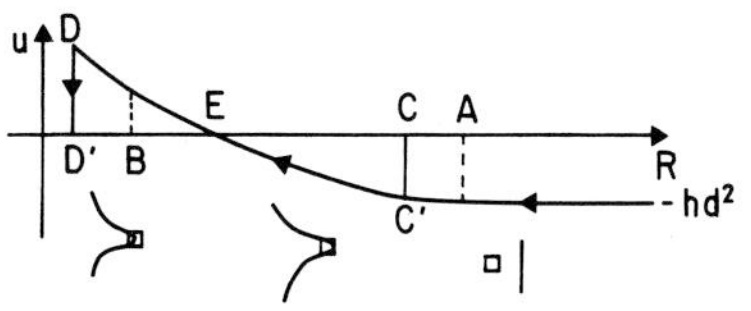

FIG. A5. Hysteresis cycle for a square Mesa defect: energy–distance plot.

$R > R_A \quad U = -hd^2$ defect covered,

$R_B < R < R_A \quad U = -hd^2 + \frac{1}{2}\pi\gamma\theta_0^2 \frac{(R-R_A)^2}{\ln L/d}$ pinning,

$R_D < R < R_B U = -hd^2 + \frac{(hd)^2}{2\pi\gamma\theta_0^2} \ln L/d$

$+ hd(R_B - R)$ metastable capture.

The position of the different points is given by

$$R_A - R_C = d, R_B - R_D = d,$$

$$R_C - R_E = l'_e = \left[\sqrt{\frac{2h}{\pi\gamma\theta_0^2} \ln L/d} - 1\right] d,$$

$$R_C - R_D = l_c.$$

The total dissipated energy is

$$W_d = U_c - U'_c + U_D - U'_D \simeq \frac{(hd)^2}{2\pi\gamma\theta_0^2} \ln \frac{L}{d}. \quad \text{(A6)}$$

2. Dilute system of defects—advancing and receding angles

a. Dilute system of defects

A single defect shows hysteresis but is not sufficient to explain the difference between receding and advancing angles. We need to introduce a certain distribution of defects on the surface. We will work here in the dilute limit where the surface fraction occupied by the defects is small. The centers of the defects are situated at random (but fixed) point $\mathbf{r}_i$ and the defect density is

$$n(\mathbf{r}) = \sum_i \delta(\mathbf{r} - \mathbf{r}_i). \quad \text{(A7)}$$

This defect density can be separated into two parts an average density n and a fluctuation $\delta n(\mathbf{r})$.

For a dilute system of defects, there are no interactions between the defects. The perturbation of the contact line due to the defects $\zeta(x)$ is the sum of the perturbations of all the defects.

$$\zeta(x) = \int\int dv\, dl\, n(x-v,l)\eta(v,l). \quad \text{(A8)}$$

The coordinate 1 (Fig. A6) is the distance between the defect and the contact line, $\eta(x,l)$ is the profile induced by a single defect at a distance l of the contact line. This profile depends on the type of experiment we are studying receding, advancing or thermodynamic equilibrium, through the domain of interation of the variable 1.

The experimentally measured contact angle is the slope of the liquid gas interface at a large distance from the perturbed contact line. In the small contact angle approximation

$$\theta = \lim_{y_{-2}+\alpha} \frac{\partial z}{\partial y}. \quad \text{(A9)}$$

The slope is given by Eq. (II.1) for a perturbation $\zeta(x)$ of the contact line

$$\theta(y) = \frac{\partial z}{\partial y} = \theta_0 + \frac{\theta_0}{2\pi} \int_{-\infty}^{\infty} dq \tilde{\zeta}(q)|q| e^{-|q|y} e^{iqx}. \quad \text{(A10)}$$

Inserting Eq. (A8), we get

$$\theta = \theta_0 \left[1 + \frac{1}{2\pi} \int dl \int_{-\infty}^{+\infty} dq \tilde{\eta}(q,l)\tilde{\eta}(q,l)|q| e^{-|q|y} e^{iqx}\right]. \quad \text{(A11)}$$

with

$$\tilde{\eta}(q,l)|q| = \int_{-\infty}^{+\infty} e^{-iqx} n(x,l) dx.$$

For a single defect the profile is related to the decrease in work of adhesion h through Eq. (II.9) which can be written in Fourier space

$$\tilde{\eta}(q,l) = -\frac{\pi}{|q|} \frac{1}{\pi\gamma\theta_0^2} \tilde{h}(q,l). \quad \text{(A12)}$$

The slope θ can be then be written

$$\theta(y) = \theta_0 \left[1 - \frac{1}{2\pi\gamma\theta_0^2} \int dl \int dq \tilde{h}(q) e^{-|q|y} e^{iqx} \tilde{\eta}(q,l)\right]. \quad \text{(A13)}$$

The defect density is separated into two parts. The Fourier transform of the average density has a Dirac function singularity at the origin and thus leads to a finite contribution to $\theta(y)$ even at infinite y. The Fourier transform of the fluctuation $\delta n(\mathbf{r})$ has no singularity at the origin it contributes to $\theta(y)$ at finite y but its contribution vanishes at infinity.

The contact angle is then given by

$$\theta = \theta_0 \left\{1 + \frac{n}{\gamma\theta_0^2} \int dl \int dx\, h[x,\eta(x),l]\right\}, \quad \text{(A14a)}$$

or equivalently

$$\gamma(\cos\theta - \cos\theta_0) = n \int dl \int dx\, h[x,\eta(x),l]. \quad \text{(A14b)}$$

We have here directly demonstrated for mesa defects the force balance of Eq. (V.1).

b. Thermodynamic, advancing, and receding angle

Before calculating the contact angle in the three different situations, we need take a further precaution. The equilibrium relationship between the profile and the surface tension (A12) is valid for stable and metastable capture but not in the so-called pinning regime. The results of Sec. IV suggest that in this regime the relevant reduction of the work of adhesion is not h but a smaller quantity $h_{\text{eff}} = \pi\gamma\theta_0^2/\ln L/dxl'$. l' being the distance between the contact line and the right edge of the defect. It is easily seen that using the bare h would produce too strong a force on the defect and the maximum advancing value $\eta(0)$ would be outside the defect area. Thus we need to use this effective value of h, which gives no force in the limit where the defect is completely covered $l' = 0$ and gives back h at the end of the pinning regime.

With this precaution it is simple to determine the advancing receding and thermodynamic angle for a surface with rectangular defects or square defects (see Fig. A4).

For rectangular defects

$$\begin{cases} \gamma(\cos\theta_E - \cos\theta_0) = nhdb & \text{(the variable } l \text{ is integrated between points } A \text{ and } E\text{)} \\ \gamma(\cos\theta_a - \cos\theta_0) = nhdb - \dfrac{n}{2}\dfrac{(hd)^2}{\pi\gamma\theta_0^2}\ln\dfrac{L}{d} & \text{(the variable } l \text{ is integrated between } C \text{ and } A\text{)} \\ \gamma(\cos\theta_r - \cos\theta_0) = nhdb + \dfrac{n}{2}\dfrac{(hd)^2}{\pi\gamma\theta_0^2}\ln\dfrac{L}{d} & \text{(the variable } l \text{ is integrated between } D \text{ and } A\text{).} \end{cases} \tag{A15}$$

For square defects (see Fig. A5)

$$\begin{cases} \gamma(\cos\theta_E - \cos\theta_0) = nhd^2 \\ \gamma(\cos\theta_a - \cos\theta_0) = n\dfrac{\pi\gamma\theta_0^2}{2\ln\dfrac{L}{d}}d^2 \\ \gamma(\cos\theta_r - \cos\theta_0) = \dfrac{n}{2}\dfrac{(hd)^2}{\pi\gamma\theta_0^2}\ln L/d + nhd^2. \end{cases} \tag{A16}$$

[1](a) An excellent introduction to capillarity problem is the book by Rowlinson and Widom, *Molecular Theory of Capillarity* (Oxford University, Oxford); (b) Contact angles are extensively discussed by A. W. Neuman in *Wetting Spreading and Adhesion*, edited by J. F. Padday (Academic, New York, 1978); (c) A description of the early knowledge (up to 1920) is the classic book: H. Bouasse, *Capillarité* (Delagrave, 1924).

[2](a) R. Johnson, R. Dettre and D. Brandeth, J. Colloid Interface Sci. **62**, 205 (1977); (b) R. Good and M. Koo, *ibid.* **71**, 283 (1979). (c) E. Wolfram and R. Faust in Ref. 1(b), p. 183. (d) Ottewill in Ref. 1(b), p. 183.

[3]S. G. Mason in Ref. 1(b), p. 321.

[4]C. Huh and S. G. Mason, J. Colloid Interface Sci. **60**, 11 (1977).

[5]A. W. Neuman, Adv. Colloid Interface Sci. **4**, 105 (1974).

[6]R. N. Wenzel, Ind. Eng. Chem. **28**, 988 (1936).

[7]R. Johnson and R. Dettre, *Surface Colloid Science*, edited by E. Matijevic (Interscience, New York, 1969), Vol. 2, p. 85.

[8]C. Cox, J. Fluid Mech. **131**, 1 (1983).

[9]Barkhausen, Phys. Z. **20**, 401 (1919).

[10]The Fourier transform of $\ln|x|/L$ is

$$\int_{-\infty}^{+\infty} e^{-iqx}\ln\frac{|x|}{L}\,dx = \frac{-\pi}{|q|}.$$

Dynamics of wetting with nonideal surfaces. The single defect problem

E. Raphaël and P. G. de Gennes
Collège de France, Physique de la Matière Condensée, 11, place Marcelin-Berthelot, 75231 Paris, Cedex 05, France

(Received 15 February 1989; accepted 10 March 1989)

Under static conditions, the macroscopic contact angle θ between a (partially wetting) liquid, a solid, and air, lies between two limiting values θ_r (receding) and θ_a (advancing). If we go beyond these limits (e.g., $\theta = \theta_a + \epsilon$, $\epsilon > 0$) the contact line moves with a certain macroscopic velocity $U(\epsilon)$. In the present paper, we discuss $U(\epsilon)$ (at small ϵ) for a very special situation where the contact line interacts only with one defect at a time. (This could be achieved inside a very thin capillary, of radius smaller than the average distance between defects.) Using earlier results on the elasticity and dynamics of the contact line in ideal conditions, we can describe the motions around "smooth" defects (where the local wettability does not change abruptly from point to point). For the single defect problem in a capillary, two nonequivalent experiments can be performed: (a) the *force F* is imposed (e.g., by the weight of the liquid column in the capillary). Here we define $\epsilon = (F - F_m)/F_m$, where F_m is the maximum pinning force which one defect can provide. We are led to a time averaged velocity $\overline{U} \sim \epsilon^{1/2}$. (b) The *velocity U* is imposed (e.g., by moving a horizontal column with a piston). Here the threshold force is not at $F = F_m$, but at a lower value $F = F_U$—obtained when the contact line, after moving through the defect, leaves it abruptly. Defining $\overline{\epsilon} = (\overline{F} - F_U)/F_U$, where $\overline{F}$ is the time average of the force, we find here $U \sim \overline{\epsilon}^{3/2}$. These conclusions are strictly restricted to the single defect problem (and to smooth defects). In practical situations, the contact line couples simultaneously to many defects: the resulting averages probably suppress the distinction between fixed force and fixed velocity.

I. INTRODUCTION

A. Motions on ideal surfaces

On a flat, homogeneous, surface, a partially wetting liquid reaches an equilibrium contact angle θ_0 defined by the Young condition.[1] If we impose a slightly different angle $\theta = \theta_0 + \epsilon$, the contact line moves with a velocity U. The relation between U and ϵ can be understood in rather simple terms[2]: the entropy source (the dissipation per unit length of line) is

$$T\overset{\circ}{S} = F_y U = \gamma(\cos\theta - \cos\theta_0)U, \tag{1.1}$$

where F_y is the unbalanced Young force (γ being the surface tension). In situations of partial wetting (where there is no precursor film[3]), the main dissipation is due to macroscopic flow. At small U, the fluid profile is nearly static and is a simple wedge of angle θ_0: for general θ_0, this dissipation has been calculated.[4] For small θ (which will turn out to be the most interesting case), the hydrodynamic dissipation has a simple form[5]:

$$T\overset{\circ}{S} = 3\frac{\eta U^2}{\theta} l, \tag{1.2}$$

where η is the fluid velocity, and l a logarithmic factor: $l = \ln(x_{\max}/x_{\min})$. Here $x_{\max}$ is the distance (from the contact line) at which the dynamic angle θ is measured, and $x_{\min}$ is a cut off at small distances. Most of the theoretical work performed in mechanics departments tends to describe $x_{\min}$ in terms of a slippage at the solid surface. This leads to $x_{\min} \sim a/\theta$ (a molecular size). On the other hand, long range Van der Waals forces modify the wedge profile over a much longer length a/θ^2 and suppress the singularity.[3,5] Thus Van der Waals forces dominate at small θ and $x_{\min} \sim a/\theta^2$. More generally, the dissipation by hydrodynamic losses is dominant only at θ small: we shall always assume $\theta \ll 1$ in what follows.

Equating the forms (1.1) and (1.2) for the dissipation, and specializing to small angles and low velocities, we get

$$U = \frac{\gamma}{3l\eta}\theta_0\frac{1}{2}(\theta^2 - \theta_0^2) \simeq \frac{\gamma\theta_0^2}{3l\eta}\epsilon \tag{1.3}$$

($\epsilon \ll \theta_0$).

Thus, for the ideal case, the relation between the force F (proportional to ϵ) and the velocity U is *linear.*

B. Anchoring defects

Practical surfaces have two types of defects: surface roughness and chemical contamination. The latter can be described in terms of a local modulation of the interfacial energies γ_{SL} (between solid and liquid) and γ_{SO} (between solid and air). The combination of interest is[6]

$$-h(x,y) = \gamma_{SL}(x,y) - \gamma_{SO}(x,y) - (\gamma_{SL} - \gamma_{SO})_0, \tag{1.4}$$

where x,y specify a point on the surface, and the bracket $(\ )_0$ denotes the unperturbed value. We shall be concerned with defects associated with a function h which is *localized* (a small spot of characteristic size d) and *smooth* (no discontinuities in h): a typical example being a Gaussian shape for $h(x,y)$. We assume that our defects are *dilute* (the distance between them is much larger than d). As shown in Ref. 6, the effects of surface roughness can also be described in terms of an h function.

 0021-9606/89/127577-08$02.10

The anchoring of a contact line on a single, smooth defect was analyzed in detail in Ref. 6. An essential ingredient is the *elasticity* of the contact line, which differs strikingly from the behavior of a violin string, as explained on Fig. 1. However, if the line is pinched at one point (say near a defect), and thus displaced from its average position y_L up to a local position y_m, there is a (nearly) hookean restoring force, of magnitude.

$$f_r = k(y_m - y_L)\,, \tag{1.5}$$

where

$$k = \frac{\pi\gamma\theta_0^2}{\ln L/d}\,. \tag{1.6}$$

Here L is the overall length of line involved, and d is the size of the anchoring defect. The force f_r must be balanced by the force f from the defect.

$$f(y_m) \simeq \int_{-\infty}^{+\infty} dx\, h(x,y_m)\,. \tag{1.7}$$

Equating Eqs. (1.5) and (1.7) one is led to the construction shown on Fig. 2. For a given position y_L of the line far from the defect, one can thus find one (or more) equilibrium positions for the anchoring point y_m.

When the amplitude of the h function is small, there is only one root y_m at any given y_L; we call this the weak defect regime. Reference 6 shows that in this regime (with many defects, but well separated and weak), we expect *no hysteresis*; if we move the line adiabatically (y_L increasing slowly with time) the position y_m is a smooth function of y_L and thus a smooth function of time: no special disssipation occurs when passing a defect. On the other hand, if we are in the strong defect regime, at some moment (when $y_L = Y_+$ on Fig. 2) the central point (y_m) jumps from an anchored position to a nearly unperturbed position ($y_m \sim y_L$): this jump causes an irreversible dissipation.

C. Relaxation of a pinched contact line

The relaxation modes of a contact line on an ideal surface have been analyzed in Ref. 7. This is done in two steps:

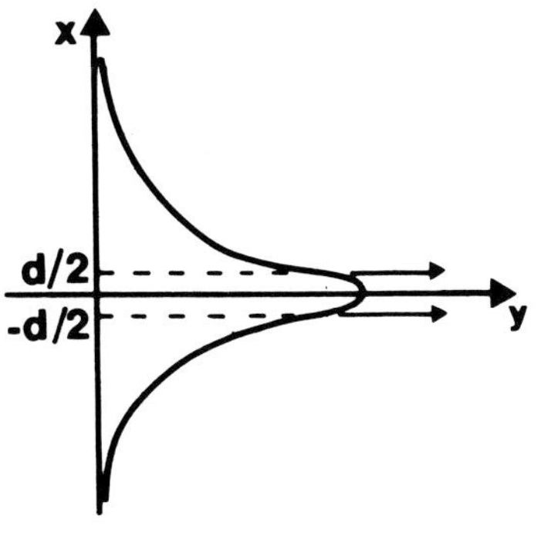

FIG. 1. Distortion of the contact line induced by a localized force acting over a distance d (under such a force, a violin string would be distorted into two straight segments).

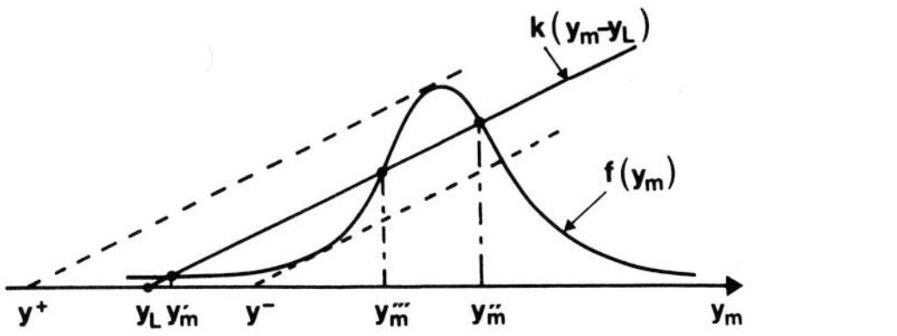

FIG. 2. Equilibrium positions of the anchoring point ($y = y_m$) of the line on the defect. For a given nominal line position (y_L), there may be three equilibrium positions, two of these (y'_m, y''_m) are locally stable. When $y_L = Y_+$, y_m jumps from an anchored position to a nearly unperturbed position. We call F_U the force exerted by the defect at the threshold $y_L = Y_+$.

(a) deriving a local restoring force for a distorted line, from the elastic theory of Ref. 6. (b) using an analog of Eq. (1.3) giving the local velocity in terms of the driving force. The result is a set of relaxation modes, for each wave vector q, with rates

$$\frac{1}{\tau_q} = \frac{\gamma\theta_0^3}{3l\eta}|q| \quad (\theta_0 < 1)\,. \tag{1.8}$$

An important application of these ideas is the *release of a pinched string*: at times $t < 0$, the contact line is pulled locally with a force $f\delta(x)$, and reaches the logarithmic equilibrium shape shown on Fig. 1. At $t = 0$ the force is suppressed, and Eq. (1.8) leads to a transient shape:

$$y(x,t) = -\frac{f}{2\pi\gamma\theta_0^2}\ln(x^2 + c^2t^2) + \text{const.}\,, \tag{1.9}$$

where the velocity c is defined by

$$c = \frac{\gamma\theta_0^3}{3l\eta} \quad (\theta_0 < 1) \tag{1.10}$$

and is very sensitive to the magnitude of the equilibrium angle θ_0. Equation (1.9) shows that a portion (of length 2 ct) of the pinched string, relaxes, while the outer parts ($|x| > ct$) are essentially unperturbed. These results will be useful when we describe a line separating from an anchoring defect, in the following sections.

II. SIMPLIFIED DYNAMICAL EQUATIONS

A. Fixed force

Our aim is now to describe motions of the contact line, not on an ideal surface, but in the presence of a defect. The methods follows Ref. 6: we do not describe the whole profile $y(x,t)$, but reduce it to two variables (Fig. 3) (a) the global position of the line $y_L(t)$, (b) the position of the tip $y_m(t)$.

Consider for instance a vertical capillary, of diameter R, containing a liquid column of height $H \gg R$. The driving force $F = \pi R^2 H\rho g$ is thus imposed (ρ = liquid density, g = gravitational acceleration).

If the total weight F is smaller than the maximum pinning force

$$F_m = f(y_d) \tag{2.a}$$

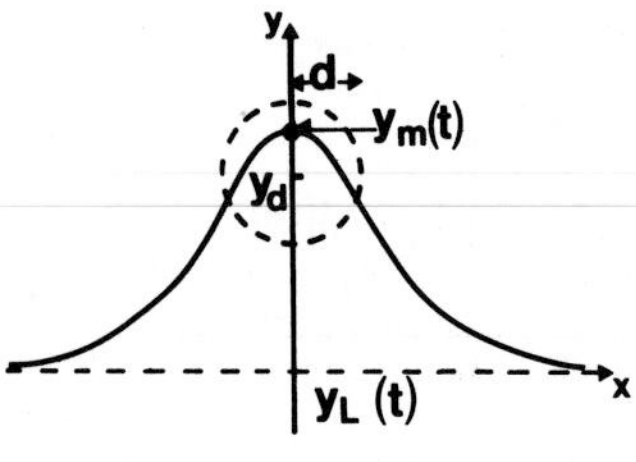

FIG. 3. A contact line anchored on a defect (of size d). A time t, the profile is characterized by the global position of the line [$y_L(t)$] and by the position of the tip [$y_m(t)$].

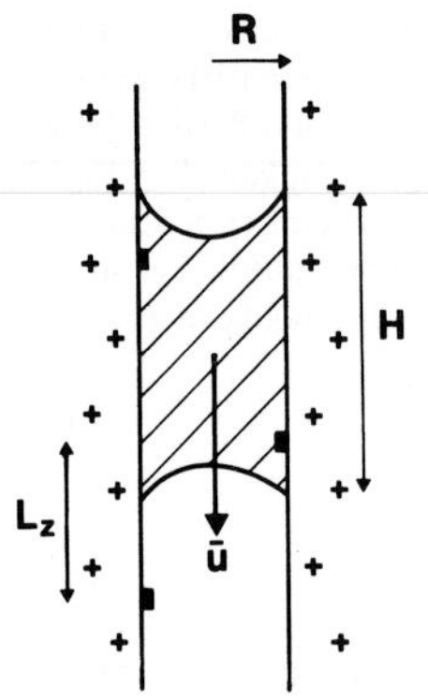

FIG. 5. A column of liquid, of height H, moving down a vertical capillary with a time averaged velocity $\bar{U}$. The black patches represent chemical contamination of the inner capillary surface. The average distance L_z between the defects is much greater than the capillary radius R.

which one defect can provide, the column is at rest, the gravity force being balanced by the force of the defect. The equilibrium values of y_L and y_m may be deduced from a simple graphical construction (see Fig. 4). For $F = F_m$, the equilibrium value of y_m coincides with y_d (y_d specifying the "center" of the defect).

We now assume that the driving force F is slightly larger than the threshold value F_m:

$$F = F_m(1+\epsilon) \quad (0<\epsilon \ll 1) . \tag{2.b}$$

The balance of forces is not possible any more and the fluid moves down the capillary (Fig. 5). The global velocity is

$$\frac{dy_L}{dt} = \mu_L [k(y_m - y_L) - \tilde{F}] . \tag{2.1}$$

Here μ_L is a mobility (ratio of velocity/total force) for the $q = 0$ mode, which we derive from Eqs. (1.1) and (1.2):

$$\mu_L = \frac{1}{2\pi R} \frac{\theta}{3\eta l} . \tag{2.2}$$

The force [the bracket in Eq. (2.1)] contains two terms: the first $k(y_m - y_L)$ is the spring force, with k defined in Eq. (1.6). The second term $\tilde{F}$ is *not* the total weight F for the following reason: as soon as the whole column starts moving down ($dy_L/dt \neq 0$) we have an important Poiseuille dissipation inside the whole column, and a resulting pressure drop. This leads to

$$\tilde{F} = \pi R^2 H \left(\rho g + 8 \frac{\eta}{R^2} \frac{dy_L}{dt} \right) . \tag{2.3}$$

Inserting Eq. (2.3) into (2.1) we arrive at

$$\frac{dy_L}{dt} = \mu_0 [k(y_m - y_L) - F] , \tag{2.4}$$

where μ_0 is a renormalized mobility.

$$\mu_0^{-1} = \mu_L^{-1} + 8\pi\eta H . \tag{2.5}$$

In the limit of interest ($H \gg R$) the Poiseuille term dominates:

$$\mu_0 \simeq \frac{1}{8\pi\eta H} \tag{2.6}$$

Equations (2.4) and (2.6) specify the global motion. Let us now discuss the motion of the tip. We write

$$\frac{dy_m}{dt} = \mu_m [k(y_L - y_m) + f(y_m - y_d)] . \tag{2.7}$$

Here μ_m is a mobility for modes of high wave vector q. Returning to Eq. (1.8), we see that $\mu_m \sim \text{const.}\ \theta q/\eta$. The q values of interest near the tip are of order d^{-1} (where d is the defect size). Thus we write

$$\mu_m = \text{const.} \frac{\theta}{\eta d} . \tag{2.8}$$

This estimate of μ_m is correct for a tip which is just being unhooked, and is thin ($\sim d$). After separation, the tip gets more diffuse [see Eq. (1.9)] and the mobility is expected to decrease. However, we shall see that the crucial step is at the onset of separation, and thus the choice (2.8) is reasonable.

Returning to Eq. (2.7), we find in the bracket two forces: first the spring force, and second the force f due to the defect [Eq. (1.7)]. This force depends on the distance between tip and defect $y_m - y_d$.

To summarize: the dynamics for fixed force is specified in terms of two equations (2.4) and (2.7). These equations are highly nonlinear via the defect force $f(y_m - y_d)$. Their solutions will be discussed in Sec. III.

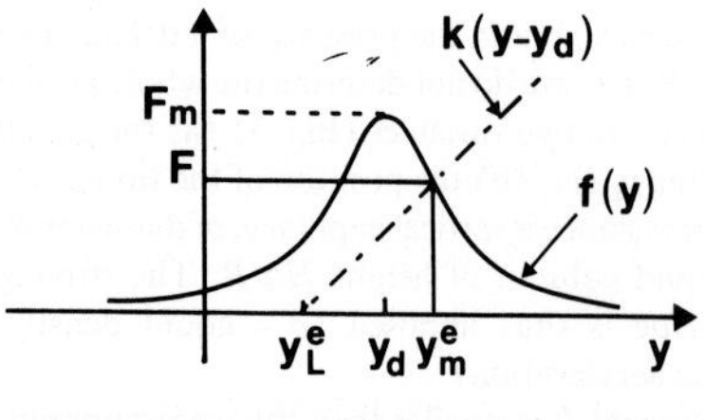

FIG. 4. Geometrical construction for the equilibrium position of the contact line (y_L^e) and of the tip (y_m^e). F is the total weight of the liquid column.

B. Fixed velocity

We now consider a different experiment (Fig. 6) where the liquid is horizontal (no gravity force) and pushed by a piston at constant speed U. We sit in the reference frame of

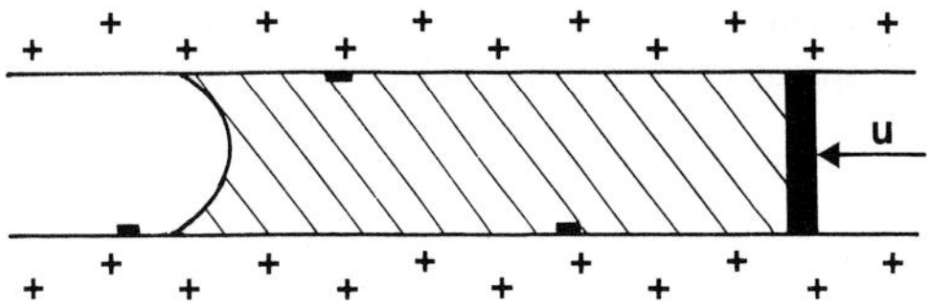

FIG. 6. A column of liquid in a horizontal capillary. The liquid is pushed by a piston at constant speed U. The black patches represent chemical contamination of the inner capillary surface.

the piston: thus $y_L(t) =$ constant (and we choose $y_L = 0$). We are left with a single variable of interest:

$$s = y_m - y_L . \tag{2.9}$$

The position of the defect is $y_d = Ut$, and the basic dynamical equation is

$$\frac{ds}{dt} = -\mu_m ks + \mu_m f(s - Ut) , \tag{2.10}$$

again highly nonlinear.

We shall discuss $s(t)$ in Sec. IV. Once $s(t)$ is determined, we know the instantaneous force due to the defect:

$$F(t) = ks(t) . \tag{2.11}$$

The total force experienced by the piston is the sum of $F(t)$ and of a Poiseuille term (linear in U) which is not interesting for us. Our main aim here is to compute the time average of $F(t)$ when the contact line passes a sequence of well separated defects (one at a time!).

III. DYNAMICS FOR FIXED FORCE

A. Adiabatic approximation

We now return to Eqs. (2.4) and (2.7) which describe the dynamics of the fixed force experiment. Introducing the new variables:

$$x = y_d - y_m \tag{3.1}$$

$$s = y_m - y_L \tag{3.2}$$

these two equations can be rearranged as

$$\frac{dx}{dt} = \mu_m [ks - f(x)] \tag{3.3}$$

$$\frac{ds}{dt} = -\frac{s}{t_1} + \mu_0 F + \mu_m f(x) , \tag{3.4}$$

where the time t_1 is defined by

$$\frac{1}{t_1} = k(\mu_0 + \mu_m) . \tag{3.5}$$

An approximate solution to Eq. (3.4) is obtained by setting the left-hand side of Eq. (3.4) equal to zero (*adiabatic approximation*). This leads to

$$s = t_1 [\mu_0 F + \mu_m f(x)] . \tag{3.6}$$

Inserting Eq. (3.6) into Eq. (3.3) we get

$$\frac{dx}{dt} = \tilde{\mu}[F - f(x)] , \tag{3.7}$$

where $\tilde{\mu}$ is a renormalized mobility:

$$\tilde{\mu} = \frac{\mu_0 \mu_m}{\mu_0 + \mu_m} . \tag{3.8}$$

Since $\mu_m \gg \mu_0$, $\tilde{\mu}$ differs only slightly from μ_0.

To estimate the range of validity of the adiabatic approximation, we have to calculate the ratio:

$$r = \frac{|ds/dt|}{s/t_1} \tag{3.9}$$

for the solution (3.6). Using Eqs. (3.6) and (3.7), we get

$$r = \frac{t_1 \mu_m \tilde{\mu} \left| \frac{df}{dx} \right| (F - f)}{\mu_0 F + \mu_m f} . \tag{3.10}$$

The above expression can be easily evaluated; in the Gaussian model, for instance, one finds

$$r \approx \frac{h_0}{\gamma \theta^2} \frac{d}{\theta H} . \tag{3.11}$$

Here h_0 characterizes the defect strength [see Eq. (1.5)]:

$$h(x,y) = h_0 \exp - (x^2 + y^2)/2d^2 \tag{3.12}$$

and is of the order of $\gamma\theta^2$. Thus, in the limit of interest ($H \gg d$), Eq. (3.11) leads to very small values of r ($r \ll 1$). Therefore, the adiabatic approximation is valid everywhere.

Thus the motion is described by Eqs. (3.7) and (3.6). As we shall see now, this leads to plots of $x(t)$ and $s(t)$ which are *smooth* functions of time.

B. Macroscopic velocity

We now want to compute the time averaged velocity $\overline{U}$ of the liquid column within the framework of the adiabatic approximation. The defects are assumed to be identical, and spread at random on the (inner) capillary surface (Fig. 5). If we call n the number of defects per cm^2, the average (vertical) distance L_z between defects is given by

$$2\pi R L_z n = 1 . \tag{3.13}$$

We assume that L_z is much greater than the capillary radius R ($L_z \gg R$) so that the contact line interacts only with one defect at a time.

The typical time $t(L_z)$ required by the tip to cover the distance L_z around one defect is given by [see Eq. (3.7)]

$$t(L_z) = \int_{-L_z/2}^{+L_z/2} dx \frac{1}{\tilde{\mu}[F - f(x)]} . \tag{3.14}$$

For a Gaussian defect [Eq. (3.12)], the force (1.7) exerted by the defect is also Gaussian:

$$f(x) = F_m \exp - (x^2/2d^2) \tag{3.15}$$

For that case, Eq. (3.14) leads to

$$t(L_z) \cong \frac{\sqrt{2}\pi d}{\tilde{\mu} F_m} \epsilon^{-1/2} . \tag{3.16}$$

The time averaged velocity $\overline{U}$ is simply defined by

$$\overline{U} = \frac{L_z}{t(L_z)} . \tag{3.17}$$

Inserting Eq. (3.16) into (3.17) we are led to

$$\bar{U} \sim \frac{1}{nRH}\left(\frac{h_0}{\eta}\right)\epsilon^{1/2} \quad [\epsilon \ll (d/L_z)^2] \tag{3.18}$$

For $\epsilon > (d/L_z)^2$, the contact line velocity is given by the Poiseville velocity: $F_m(1+\epsilon)/8\pi\eta H$.

IV. DYNAMICS FOR FIXED VELOCITY

A. The quasistatic distortion

We now consider the fixed velocity experiment of Fig. 6; the dynamics is specified by Eq. (2.10). In what follows, we shall always take the piston speed U to be small. The quasistatic distortion $s_a(t)$ is obtained by setting the left-hand side of Eq. (2.10) equal to zero. This leads to

$$s_a = t_2 \mu_m f(s_a - Ut)\,, \tag{4.1}$$

where the time t_2 is defined by

$$\frac{1}{t_2} = k\mu_m\,. \tag{4.2}$$

A schematic plot of s_a as a function of time is shown in Fig. 7. In the strong pinning regime s_a is a *multivalued* function of t; it jumps abruptly (from s_d to s^+) at $t = t_d$.

B. The perturbative expansion

Far away from the "disanchoring point" (t_d, s_d), (i.e., for $t \ll t_d$), we can solve Eq. (2.10) perturbatively by expanding $s - s_a$ in powers of U:

$$s(t) = s_a(t) + U\sigma_1(t) + U^2\sigma_2(t) + \cdots\,, \tag{4.3}$$

where the $\sigma_i(t)$ are of order 1. Substituting Eq. (4.3) into Eq. (2.10), we obtain for the leading correction to s_a:

$$\sigma_1(t) = t_2 \frac{k^{-1}f'(s_a - Ut)}{1 - k^{-1}f'(s_a - Ut)}\,. \tag{4.4}$$

Note that the sign of $\sigma_1(t)$ changes for $t = t_{\max}$ (see Fig. 7).

C. Behavior near the disanchoring point

We now want to study the behavior of $s(t)$ near the disanchoring point. In that region, the preceding expansion

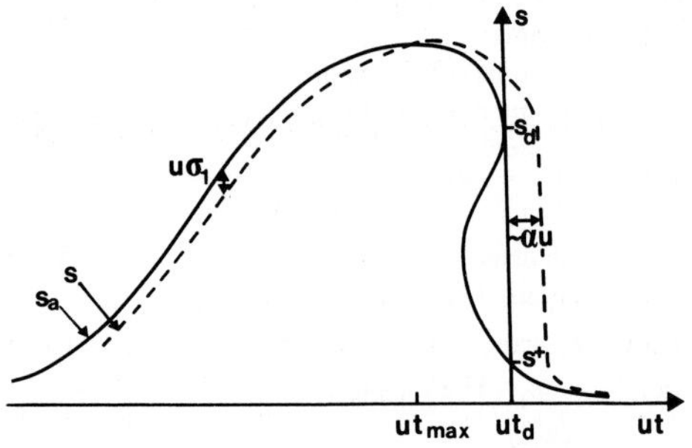

FIG. 7. Schematic plot of the quasistatic distortion s_a as a function of time (solid line). At $t = t_d$, s_a jumps abruptly from s_d to s^+. The dashed line represent the actual distortion $s(t)$ [Eq. (4.13)]. Note that the sign of the leading correction $U\sigma_1(t)$ to $s_a(t)$ changes at $t = t_{\max}$ where $s_a(t)$ is maximal.

(4.3) becomes invalid. However, in this region Eq. (2.10) can be approximated by the simpler form:

$$\frac{d\mu}{d\theta} = -\frac{U}{t_2}\theta - c\mu^2\,, \tag{4.5}$$

where we have introduced the new variables:

$$\theta = t - t_d \tag{4.6}$$

$$\mu = s - s_d\,. \tag{4.7}$$

Here c is a constant defined by

$$c = \frac{1}{2}\mu_m |f''(s_d - Ut_d)| \approx \frac{1}{dt_2}\,. \tag{4.8}$$

Equation (4.5) can be scaled by introducing

$$\theta = \alpha\tilde{\theta} \tag{4.9}$$

$$\mu = \beta\tilde{\mu} \tag{4.10}$$

with

$$\alpha = \left(\frac{t_2}{cU}\right)^{1/3} \sim t_2\left(\frac{d}{Ut_2}\right)^{1/3} \tag{4.11}$$

$$\beta = \frac{1}{c\alpha} \sim d\left(\frac{Ut_2}{d}\right)^{1/3}\,. \tag{4.12}$$

This leads to

$$\frac{d\tilde{\mu}}{d\tilde{\theta}} = -\tilde{\theta} - \tilde{\mu}^2\,. \tag{4.13}$$

Equation (4.13) is a Riccati equation which admits for general solution:

$$\tilde{\mu} = -\frac{A_i'(-\tilde{\theta}) + bB_i'(-\tilde{\theta})}{A_i(-\tilde{\theta}) + bB_i(-\tilde{\theta})}\,, \tag{4.14}$$

where A_i and B_i are the Airy functions,[8] A_i' and B_i' their derivatives; b is a constant of integration. To recover the quasi-static distortion:

$$\tilde{\mu}_a = +(-\tilde{\theta})^{1/2} \tag{4.15}$$

for large negative $\tilde{\theta}$, we must choose b equal to zero.[9] Whence

$$\tilde{\mu} = -\frac{A_i'(-\tilde{\theta})}{A_i(-\tilde{\theta})}\,. \tag{4.16}$$

The function $\tilde{\mu}$ is plotted in Fig. 8. At a value $\tilde{\theta}'' \simeq 2.339$ [corresponding to the first zero of the function $A_i(-\tilde{\theta})$], the function $\tilde{\mu}$ [Eq. (4.16)] diverges; the line separates from the defect. The simplified equation (4.5) ceases then to be valid.

To describe the late stages of the separation $(t > t'' = t_d + \alpha\tilde{\theta}'')$, we can approximate the term $f(s - Ut)$ of Eq. (2.10) by $f(s_a - Ut)$ where s_a now corresponds to the lower branch of Fig. 7. Neglecting the explicit time dependence of s_a this leads to

$$s(t) = \text{const.}\ \exp -\left(\frac{t}{t_2}\right) + s_a(t)\,. \tag{4.17}$$

Thus, the function s relaxes exponentially towards s_a in a time of order t_2. Note that this relaxation time is much smaller than α [Eq. (4.11)].

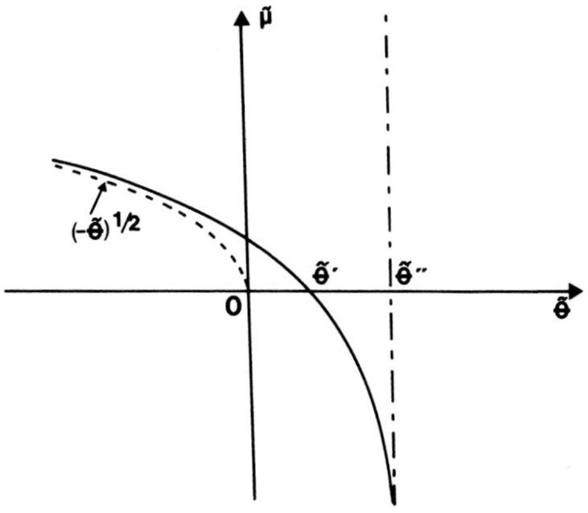

FIG. 8. Details of the disanchoring process $=\tilde{\mu}$ means the line position in reduced units [Eq. (4.10)] and $\tilde{\theta}$ is the reduced time [Eq. (4.9)]. At $\tilde{\theta}=\tilde{\theta}''$ the function $\tilde{\mu}$ diverges. The dashed line correspond to the quasi-staic solution $(-\tilde{\theta})^{1/2}$. For large negative $\tilde{\theta}$:$\tilde{\mu}\simeq(-\tilde{\theta})^{1/2}-(4\tilde{\theta})^{-1}$.

Let (t_r,s_r) be the coordinates of the crossover point between the two solutions (4.16) and (4.17). It is easy to see that t_r is contained between $t'=t_d+\alpha\tilde{\theta}'$ and $t''=t_d+\alpha\tilde{\theta}''$ [$\tilde{\theta}'$ corresponding to the first zero of the function $A_i'(-\tilde{\theta})$]. More precisely, it can be shown that

$$t''-t_r\simeq\sqrt{\frac{d}{s_d}}\,t_2\ll\alpha \tag{4.18}$$

and

$$\frac{s_d-s_r}{s_d}\simeq\sqrt{\frac{d}{s_d}}\ll 1\,. \tag{4.19}$$

A schematic representation of the behavior of $s(t)$ near $t=t_d$ is shown in Fig. 9. [Between $t=t_d$ and $t=t''$ the function $s_a(t)$ differs only slightly from the value $s^+\equiv s_a(t=t_d)$.]

The overall physical picture is the following:

- at times $t<t_d$ the contact line becomes progressively anchored — always remaining very close to its static conformation.

- at $t=t_d$ the line "springs off" and remains in the vicinity of the defect only for a time $\sim\alpha$.

- ultimately, the line relaxes to its unperturbed form in a (short) time t_2.

D. The time averaged force

We can now compute the time average of the instantaneous force (2.11):

$$\bar{F}=\frac{1}{T}\int_0^T ks(t)\,dt$$

$$=\frac{1}{T}\int_0^T ks_a(t)\,dt+\frac{1}{T}\int_0^T k\,[s(t)-s_a(t)]\,dt\,, \tag{4.20}$$

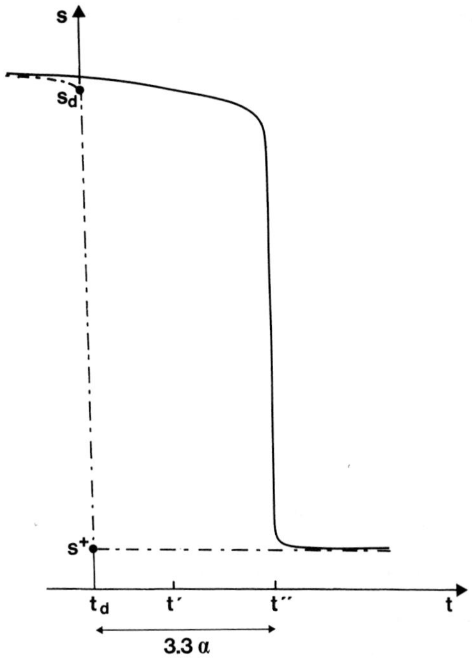

FIG. 9. Qualitative aspect of the line distortion $s(t)$ near the disanchoring point (t_d,s_d) (solid line). $s(t)$ differs markedly from the quasi-static distortion $s_a(t)$ (dashed line) only during the interval (t_d,t'') (of order 3.3 α). Between $t=t_d$ and $t=t''$ the function $s(t)$ is almost flat except for the immediate vicinity of $t=t''$ where it jumps abruptly in a time of order t_2.

where the period T is defined by

$$UT=L_z\,. \tag{4.21}$$

Here L_z is the average distance between defects [see Eq. (3.13)]. By using Eq. (4.1), the first integral in Eq. (4.20) can be rewritten as an integral over the defect position y_d:

$$\frac{1}{T}\int_0^T ks_a(t)\,dt=\frac{1}{L_z}\int_{-L_z/2}^{+L_z/2} f(s_a-y_d)\,dy_d\,. \tag{4.22}$$

We may approximate the integral (4.22) by the simpler form:

$$2\pi nR\int_{-\infty}^{+\infty} f(s_a-y_d)\,dy_d\,. \tag{4.23}$$

This corresponds exactly to the static force derived in Ref. 6. In the present experiment this static force corresponds to F_U, the force obtained when the contact line, after moving through the defect, leaves it abruptly. (Note that F_U is smaller than F_m.) Thus we are led to

$$\bar{F}-F_U\simeq\frac{1}{T}\int_0^T k\,[s(t)-s_a(t)]\,dt\,. \tag{4.24}$$

The above integral can be easily evaluated by using the results of Sec. IV C. The dominant contribution comes from the time (of order 3.3 α) from t_d to t'' where $s - s_a$ is of the order of $s_d - s^+$ (see Figs. 7 and 9):

$$\int_0^T k\,[s(t) - s_a(t)]\,dt \sim k\alpha(s_d - s^+) \qquad (4.25)$$

(where numerical factors have been ignored).
Inserting Eq. (4.25) into (4.24) we are led to

$$\overline{F} - F_U \sim F_m\left(1 - \frac{ks^+}{F_m}\right)(2\pi n dR)\left(\frac{Ut_2}{d}\right)^{2/3}. \qquad (4.26)$$

In the strong defect regime, F_U differs only slightly from F_m. We therefore obtain, for fixed velocity,

$$U \sim \bar{\epsilon}^{\,3/2}, \qquad (4.27)$$

where $\bar{\epsilon}$ is defined by

$$\bar{\epsilon} = \frac{\overline{F} - F_U}{F_U}. \qquad (4.28)$$

V. CONCLUDING REMARKS

(1) The single defect problem which is discussed here is important, because it separates the mechanical factors from the statistical features which show up when the line interacts simultaneously with many defects. Clearly, if we want informations on one particular type of defect, associated with a certain surface treatment, the single defect situation is much more informative.

It is not hopeless to achieve this situation, either with a thin capillary, or with a fiber. Very recent experiments by J. M. di Meglio (to be published) monitor the force experienced by a thin fiber pulled at constant U, and do see erratic oscillations which may be due to separate defects.

(2) The great surprise which came out of our discussion is the distinction between fixed force and fixed velocity, leading to different thresholds F_c and to different laws near the threshold. For simple mechanical systems, differences of this type are in fact often observed: the analog of the disanchoring situation is seen in force experiments between two plates, as measured by the Israelachvili technique,[10] where we have competition between a spring force and a nonlinear attraction. On the other hand, the new machine built by J. M. Georges[11] measures the force at fixed interplate velocity.

We hope to discuss the many defect situation in the future, and we expect a unique relation between force and velocity in this limit, where the contact line integrates simultaneous contributions from many defects, and averages them. However, this statistical problem is quite complex, as is already clear from its static counterpart[12] and many other surprises may occur.

(3) Returning to the single defect problem, we should repeat a severe limitation: we talked only about *smooth* defects. The opposite case of "mesa" defects seem to lead to very different structures, as observed recently by L. Léger and A. M. Guinet (unpublished). The disanchoring often proceeds here by coalescence of two line portions, as shown on Fig. 10. This process will be discussed separately.[13]

(4) Most of our figures, and of our numerical discussion, was based on "*chemical*" defects, where the h function is everywhere of the same sign. We should mention here the case of "*mechanical*" defects, where $h(x,y)$ is proportional to the local slope $\partial z/\partial y$ of the altitude profile $z(x,y)$ (see Ref. 6). The main difference is that a localized defect (a bump) gives an h function with the general aspect of Fig. 11, having now a maximum and a minimum of opposite signs. However, the distinction between weak and strong defects, the role of anchoring points, etc., remain the same, and we expect to retain the dynamical laws (3.18) and (4.27) when the defects are smooth bumps.

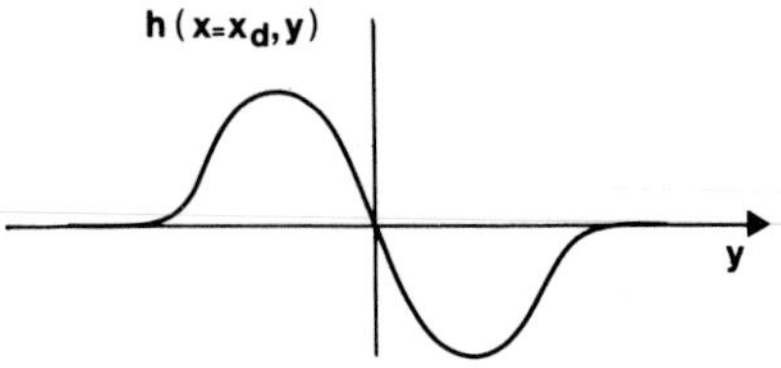

FIG. 11. General shape of the h function induced by a bump on the surface.

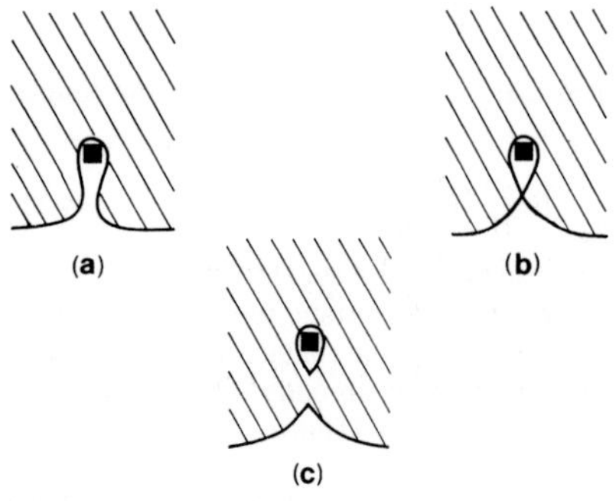

FIG. 10. A moving contact line anchored on a mesa defect (a). The disanchoring proceeds by the coalescence of two line portions (b). (Observations L. Léger and A. M. Guinet.)

ACKNOWLEDGMENTS

We have benefited from stimulating discussions with F. Brochard, L. Léger, J. M. Di Meglio, J. F. Joanny, and M. Robbins.

[1]W. Zisman, in *Contact Angle, Wettability and Adhesion*, edited by F. M. Fowkes, Advances in Chemistry Series No. 43 (ACS, Washington, D.C., 1964).
[2]P. G. de Gennes, Colloid Polym. Sci. **264**, 463 (1986).
[3]P. G. de Gennes, Rev. Mod. Phys. **57**, 827 (1985).
[4]C. Huh and L. E. Scriven, J. Colloid Interface Sci. **35**, 85 (1971).
[5]P. Levinson, H. Xua, and P. G. de Gennes (to be published).

[7]P. G. de Gennes, C. R. Acad. Sci. (Paris) II **302**, 731 (1986).
[8](a) M. Abramowitz and I. A. Stegun, *Handbook of Mathematical Functions*, National Bureau of Standards (U.S. GPO, Washington, D.C.). (b) A somewhat similar analysis concerning the pinning of charge density waves may be found in D. Fisher, Phys. Rev. B **31**, 1396 (1985).
[9]If $b \neq 0$, the leading behavior of $\tilde{\mu}$ [Eq. (4.14)] for large negative $\tilde{\theta}$ is given by $\tilde{\mu} \approx -(-\tilde{\theta})^{1/2}$.
[10]J. N. Israelachvili and P. M. McGuiggan, Science **241**, 795 (1988).
[11]A. Tonck, J. M. Georges, and J. L. Loubet, J. Colloid Interface Sci. **126**, 150 (1988).
[12]J. F. Joanny and P. G. de Gennes, J. Chem. Phys. **81**, 552 (1984); M. O. Robbins and J. F. Joanny, Europhys. Lett. **3**, 729 (1987); M. Cieplak and M. O. Robbins, Phys. Rev. Lett. **60**, 2042 (1988); R. Bruinsma, to appear in Conference Proceedings of the European Physical Society Meeting, Arcachon, France, 1988.
[13]E. Raphael (to be published).

C. R. Acad. Sci. Paris, t. 323, Série II *b*, p. 663-667, 1996
Surfaces, interfaces, films/*Surfaces, interfaces, films*

Éponges filantes

Pierre-Gilles de GENNES

Physique de la Matière Condensée, Collège de France,
11, place Marcelin-Berthelot, 75231 Paris CEDEX 05, France.

Résumé. Domingues Dos Santos et Ondarçuhu (1995) ont découvert les « gouttes filantes » contenant une solution de molécules réactives (chlorosilanes). Si une telle goutte est posée sur une surfae solide (verre, silice) comportant des groupes OH qui réagissent avec le silane, la goutte fuit les régions réagies de faible mouillabilité et se met en marche.

Nous discutons ici le remplacement de la goutte par une « éponge » : particule de latex, ou caoutchouc, gonflée par le solvant. Deux cas sont envisagés : *a*) glissement d'une pastille donnant des vitesses $\sim 1\ \mu$m/s ; *b*) roulement d'une sphère (donnant des vitesses comparables pour des particules de latex, mais plus élevées pour des sphères millimétriques de caoutchouc).

Spontaneous motion of reactive "sponges"

Abstract. *Domingues Dos Santos and Ondarçuhu have discovered that certain droplets, containing a reactive solute (chlorosilane) move spontaneously on a glass surface. We discuss here the replacement of the drop by a "sponge": swollen latex or rubber particule. Two cases are considered: a) sliding of a flat sponge leading to velocities of order 1 µ/sec.; b) rolling of a sphere, giving comparable velocities for a latex particle, and larger velocities for millimetric rubber spheres.*

I. Principes

1) Un effet mécano-chimique à une interface liquide/solide a été observé avec *a*) des gouttes de décane contenant un acide gras fluoré (Bain *et al.*, 1994) ; *b*) des gouttes d'hydrocarbone contenant une solution diluée de chlorosilane (Domingues Dos Santos et Ondarçuhu, 1995). Dans ce deuxième cas, par exemple, le silane réagit avec les groupes OH du support (verre) et le rend moins mouillable : la goutte fuit les régions réagies, et se met en route avec une vitesse V – qui dépend de sa taille, et de la concentration en réactif. Une théorie simpliste (à deux dimensions) de cet effet a été construite (Brochard-Wyart et de Gennes, 1995) et rend assez bien compte des observations.

Note présentée par Pierre-Gilles de Gennes.

1251-8069/96/03230663 \$ 2.00

P.-G. de Gennes

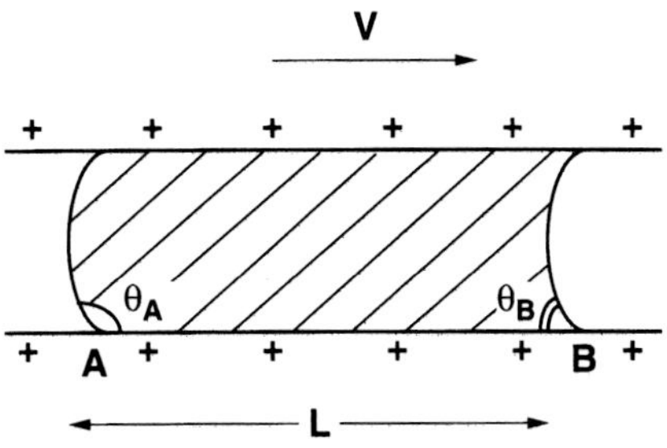

Fig. 1. – Une situation simple de « goutte filtante ». La goutte est enfermée dans un capillaire étroit, et contient un liquide réactif qui rend la surface du capillaire moins mouillante. Les angles de contact aux extémités (θ_A, θ_B) sont alors différents, et il en résulte une différence entre les pressions de Laplace aux deux extrémités.

Fig. 1. – A simple form of reactive wetting. A drop is confined into a thin capillary. It contains a liquid which reacts with the capillary wall and makes it less wettable. The contact angles (θ_A, θ_B) at both ends are then different: this creates a difference in Laplace pressure between the ends.

2) On peut envisager des processus analogues dans d'autres géométries : par exemple à l'intérieur d'un capillaire horizontal de diamètre D (*fig.* 1), une colonne liquide réactive de longueur $L \gg D$ va se déplacer à une vitesse V, régie par le bilan de pression :

$$p_A - p_B = \frac{-4}{D}(\cos\theta_A - \cos\theta_B) = 32\, L\, \eta\, \frac{V}{D^2}\,. \tag{1}$$

Dans l'équation (1) on peut (pour $L \gg D$) remplacer les angles de contact θ_A et θ_B par leurs valeurs d'équilibre local, déduites de la condition de Young :

$$\gamma \cos\theta_B = \gamma_{S0} - \gamma_{SL0} \tag{2}$$

$$\gamma \cos\theta_A = \gamma_S - \gamma_{SL} = \gamma_{S0} - \gamma_{SL0} - \gamma_1\,(1 - e^{-t/\tau}) \tag{3}$$

où γ_{S0} (γ_{SL0}) est l'énergie solide/air (solide/liquide) avant réaction, et γ_1 décrit l'effet de la réaction. t est le temps d'exposition d'un élément solide qui a vu passer la colonne $(V\,t = L)$. Le temps τ est le temps de réaction (cinétique du premier ordre) et est inversement proportionnel à la concentration en réactif dans le fluide. En reportant les équations (2), (3) dans l'équation (1), on arrive à une équation implicite pour $V\,(L)$:

$$\frac{D}{L}\,\frac{\gamma_1}{\eta}\left[1 - \exp\left(\frac{-L}{V\,\tau}\right)\right] = 32\,V. \tag{4}$$

Pour $L^2 \gg V^*\,\tau\,D$ (avec $V^* \equiv \gamma_1/\eta$) on a alors $V \sim V^*\,D/L$.

Pour $L^2 \gg V^*\,\tau\,D$ (ce qui est le cas usuel) :

$$V = V_0 \equiv \frac{1}{4\sqrt{2}}\left(V^*\,\frac{D}{\tau}\right)^{1/2}$$

Typiquement $V^* \sim 10$ m/s, $\tau \sim 10^3$ s, $D = 100\ \mu$m et $V_0 \cong 0{,}2$ mm/s.

3) Peut-on avoir un mouvement observable si on remplace la goutte par un gel gonflé (une « éponge ») ? Tout va dépendre de la friction entre le gel et le solide, qui est elle-même très sensible aux détails du système. Nous allons discuter ici deux cas :

a) pastille de gel glissant sur un solide (section II) ;

b) particule sphérique roulant sur la surface (section III).

II. Glissement d'une pastille rectangulaire

La géométrie est représentée sur la figure 2. On peut trouver la vitesse V par un bilan de dissipation :

$$T\dot{S} = k\,V^2\,L\,y = \gamma_1\,(1 - e^{-t/\tau})\,Dy\,V \tag{5}$$

où k est le coefficient de friction entre le gel gonflé et le solide. Il en résulte :

$$V = \frac{\gamma_1}{k\,L}\left[1 - \exp\left(\frac{-L}{V\,\tau}\right)\right]. \tag{6}$$

Ici encore on a deux régimes. le plus important en pratique correspond à $L^2 < \gamma_1\,\tau/k$ et donne une vitesse indépendante de L :

$$V^2 = \gamma_1/(k\,\tau). \tag{7}$$

Toute la difficulté réside dans l'estimation du coefficient k. Nous supposerons ici que le système solide/solvant/polymère est tel qu'il n'y a *pas adsorption* du polymère sur le solide. Si l'éponge est modérément gonflée (fraction en volume occupée par le polymère $\sim 50\,\%$), on peut peut-être écrire que les contacts polymère-solide sont lubrifiés par une couche de solvant d'épaisseur a comparable à la taille des monomères ($a \sim 1$ Å). Alors $k \sim \eta/a$ (η représentant la viscosité du solvant). Si l'on prend $\eta = 10^{-2}$ Po, $\gamma_1 = 10\,\mathrm{mJ/m^2}$, $\tau = 10^3$ s, on arrive alors à $V \sim 1$ μm.

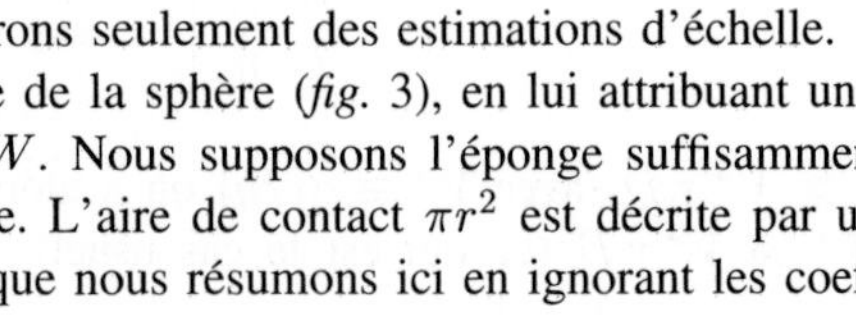

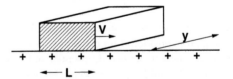

Fig. 2. – Un bloc de gel contenant un liquide réactif et glissant sur une surface solide.

Fig. 2. – A slice of gel containing a reactive liquid and gliding on a solid surface.

III. Frottement de roulement (éponge sphérique)

Pour ce problème, plus complexe, nous proposerons seulement des estimations d'échelle.

1) retournons d'abord à la conformation statique de la sphère (*fig.* 3), en lui attribuant un module élastique μ et une énergie d'adhésion de Dupré W. Nous supposons l'éponge suffisamment petite ($\varnothing < 1$ mm), pour que son poids soit négligeable. L'aire de contact πr^2 est décrite par un calcul classique de Johnson, Kendall et Roberts (1971), que nous résumons ici en ignorant les coefficients. L'énergie E est de la forme :

$$E \cong -W\,r^2 + \mu\left(\frac{u}{r}\right)^2 r^3 \tag{8}$$

P.-G. de Gennes

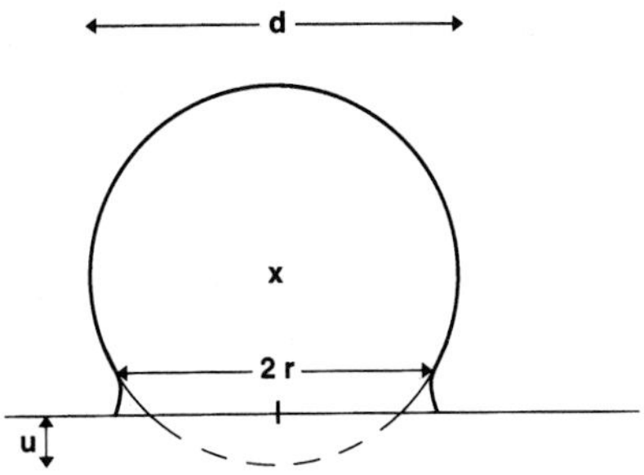

Fig. 3. – Une sphère de gel en roulement.

Fig. 3. – A rolling sphere of gel.

où le premier terme décrit l'adhésion, et le second l'effet de déformation élastiques : μ est le déplacement vertical, u/r la déformation, présente sur un volume $\sim r^3$. On attend $r^2 \sim ud$ (d=diamètre de la sphère). En optimisant l'expression (8), on arrive alors à :

$$r^3 \cong d^2\, h_0 \tag{9}$$

où $h_0 = W/\mu$ est une longueur caractéristique des effets capillaires en phase solide. Pour des solides durs, $h_0 \sim a$. Pour un gel gonflé, $h_0 \sim Na$, où N est le nombre de monomères entre points de branchement.

2) Considérons maintenant un roulement sans glissement à la vitesse V. Les vitesses verticales v développées dans la région de contact sont :

$$v = V\,\frac{\partial u}{\partial x} \sim V\,u/r.$$

Nous supposerons ici que la partie imaginaire du module élastique peut être décrite par une viscosité η, d'où une dissipation :

$$T\dot{S} \sim \eta\left(\frac{v}{r}\right)^2 r^3 = \eta\,V^2\,u^2/r = \eta\,V^2\,r^3/d^2 \tag{10}$$

où les gradients de vitesse v/r sont présents sur le volume r^3. En insérant dans l'équation (10) l'estimation statique pour r^3 [équation (9)], on arrive à :

$$T\dot{S} \cong \eta\,V^2\,h_0. \tag{11}$$

La longueur sur laquelle s'effectue la réaction chimique est de l'ordre de r, et le bilan d'énergie donne donc :

$$V\,r\,\gamma_1\,(1 - e^{-t/\tau}) \cong \eta\,V^2\,h_0 \tag{12}$$

soit :

$$V = \frac{\gamma_1}{\eta}\,\frac{r}{h_0}\left[1 - \exp\left(\frac{-r}{V\,\tau}\right)\right]. \tag{13}$$

Ici encore, le régime important est $r < V\,\tau$, et la vitesse correspondante est donnée par :

$$V^2 = \frac{\gamma_1}{\eta}\,\frac{d}{\tau}\left(\frac{d}{h_0}\right)^{1/3}. \tag{14}$$

La valeur de la viscosité η dépend beaucoup de la façon dont le gel a été réticulé, de la présence éventuelle de chaînes pendantes, ou d'oligomères libres, etc. Prenons à titre purement indicatif $\eta = 100\,\mathrm{Po}$, $\gamma_1 = 10\,\mathrm{multi}\,J/m^2$, $d = 1$ μm (pour une particule de latex), $\tau = 10^3$s et $h_0 = 5\mathrm{nm}$. On arrive alors à $V \sim 2$ μm/s. Si au contraire, nous utilisions une particule macroscopique ($d = 1$ mm), on attendrait $V \sim 0,2$ mm/s.

Mais, en fait, l'énergie (8) n'est correcte que pour des particules petites. Pour des particules millimétriques, il faut tenir compte du *poids* du grain. On arrive alors à l'estimation

$$V^2 = \frac{\gamma_1}{\eta}\,\frac{R}{\tau}\,x^{-1/3} \tag{15}$$

avec $x = \rho\, g\, R/\mu$ (ρ=accélération de la pesanteur). Typiquement, pour $R = 1$ mm, $x = 10^{-6}$ et $V \sim 1$ mm/s.

IV. Discussion

Le cas le plus intéressant est celui du roulement, qui devrait donner des frictions faibles, et des vitesses assez grandes pour des particules de caoutchouc macroscopiques. Il faut insister toutefois sur les diverses conditions physiques à réaliser :

1) pas d'adsorption du polymère (au moins sur des temps r/V) ;

2) structure bien sphérique (une particule facettée ne roulera pas) ;

3) effets visco-élastiques pas trop importants. Notre description de ces effets est très primaire : nous avons adopté pour le module complexe, fonction de la fréquence $\mu(\omega)$, une forme :

$$\mu(\omega) = \mu + i\,\omega\eta \tag{16}$$

a) souvent la forme empirique de $\mu(\omega)$ est bien plus complexe ;

b) même si l'équation (15) est acceptable, elle ne peut s'appliquer qu'à basse fréquence $\omega < 1/T$ où T est un temps de relaxation. Les fréquences impliquées dans le roulement sont $\omega \sim V/r$. En utilisant les équations (9) et (14), on arrive ainsi à :

$$\omega^2 \sim \frac{\gamma_1}{\eta\, h_0\, \tau} \sim \frac{\mu}{\eta\tau} \sim \frac{1}{T\,\tau} \tag{17}$$

et le régime basse fréquence suppose $\omega T < 1$, soit $\tau \gg T$.

Au total, vu la possibilité d'ajuster le temps de réaction τ (par la concentration en chlorosilane), il n'est peut-être pas impossible d'observer des « éponges filantes ».

Note remise le 2 septembre 1996, acceptée 7 octobre 1996.

Références bibliographiques

Bain C. D., Burnett-Hall G. D. et Montgomerie R. R., 1994. Rapid motion of liquid drops, *Nature*, 372, p. 414-415.

Brochard-Wyart F. et Gennes P.-G. de, 1995. Spontaneous motion of a reactive droplet, *C. R. Acad. Sci. Paris,* 321, série II, p. 285-288.

Domingues Dos Santos F. et Ondarçuhu T., 1995. Free running droplets, *Phys. Rev. Lett.*, 75, p. 2972-2976.

Johnson K., Kendall K. et Roberts A. 1971. Contact between a sphere and a plane, *Proc. Roy. Soc. (London),* A 324, p. 301.

Afterthoughts: **Éponges filantes**

1) Some colleagues (from mechanics departments) have taken bets with me on this idea: from their past experience, they are inclined to think that a "sponge" will be too sticky to roll. They may well be right, but I took stickiness into account in my discussion: I still wait for an experimental attempt.

2) In any case, a slightly different set up should work: this is based on a solid, porous, soaked with the reactive solution; (the sphere itself is not reactive with the liquid). The sphere lies on a horizontal, reactive plane, and a pendular droplet of reactive liquid forms an annulus around the contact point. This object should roll spontaneously with a sizeable velocity. Experiments of this type are under way at Institut Curie.

C. R. Acad. Sci. Paris, t. 324, Série II *b*, p. 257-260, 1997
Surfaces, interfaces, films/*Surfaces, interfaces, films*

Shocks in an inertial dewetting process

Françoise BROCHARD-WYART and Pierre-Gilles de GENNES

F. B.-W.: PSI, Institut Curie, 11, rue P.-&-M.-Curie, 75231 Paris cedex 05;
P.-G. de G.: Physique de la Matière Condensée, Collège de France, 11, place Marcelin-Berthelot, 75231 Paris cedex 05, France.

Abstract. When a film of millimetric thickness h_0 tends to dewet, we expect that a dry patch will grow at a constant velocity, and will be preceded by a 'flat rim' of thickness h^*, where h^* is exactly the equilibrium thickness, related to the spreading coefficient. The rim front should be identical to a 'shallow water shock'. For liquids of viscosities comparable to that of water, the Reynolds numbers are of order 100, and this inertial regime should be observable.

Chocs en démouillage inertiel

Résumé. *Lorsqu'on a initié le démouillage au centre d'un film liquide instable, d'épaisseur* h_0 *millimétrique, la région sèche doit grandir à vitesse constante, et être précédée d'un « bourrelet plat » d'épaisseur* h^**, où* h^* *est exactement l'épaisseur d'équilibre reliée au paramètre d'étalement* $S(<0)$*. Le bourrelet doit se terminer par un choc en eau peu profonde. Pour un liquide de viscosité comparable à celle de l'eau, les nombres de Reynolds sont* ~ 100, *et ce régime paraît observable.*

Version française abrégée

En régime visqueux, le démouillage de films d'eau sur une surface hydrophobe se fait avec une vitesse de croissance V des régions sèches :

$$V = (\text{cste})\ \gamma/\eta\theta_e^3$$

où γ est la tension superficielle, η la viscosité, et θ_e l'angle d'équilibre de Young (supposé < 1 dans cette formule) (C. Redon *et al.*, 1991 ; P. G. de Gennes, 1987 ; F. Brochard-Wyart, P. G. de Gennes, 1992). Dans ces situations, l'eau expulsée des régions sèches se rassemble en un fin bourrelet, dont le profil est un arc de cercle.

Quand on augmente l'épaisseur h_0 du film initial, on fait croître le nombre de Reynolds Re, et on observe des profils différents (C. Andrieu *et al.*, 1996 ; F. Brochard-Wyart *et al.*, 1995) relativement complexes pour $Re \sim 1$. Nous nous intéressons ici à la limite $Re >> 1$, qui est atteinte pour des films

Note présentée par Pierre-Gilles de GENNES.

millimétriques. Le profil attendu dans cette limite est représenté sur la *figure 1*. La force d'entraînement par unité de longueur est $-S > 0$ (S : paramètre d'étalement). En écrivant une forme généralisée de l'argument de Culik (Culik, 1960) pour les films de savon, on arrive à la condition (3) qui relie S à la vitesse V du bord, et à la vitesse V_S du choc. Ensuite, en utilisant les conditions d'Hugoniot en eau peu profonde, on arrive à déterminer tous les paramètres du profil. Notre analyse est unidimensionnelle, donc valable pour des largeurs de bourrelet $\ell < R$ (rayon de la zone sèche).

---

1. General aims

Our starting point is a relatively thick liquid film, of low viscosity — e.g. water — lying on a hydrophobic surface. The original thickness h_0 must be smaller than the threshold value h^* defined by:

$$\frac{1}{2}\rho g h^{*2} = -S \tag{1}$$

where ρ is the density, g the gravitational acceleration and S the spreading coefficient. Then the film is metastable. However, if a hole is opened (mechanically or optically) at the central point, a dry patch will grow, with a certain velocity V. Early studies of this dewetting process (de Gennes, 1987; Redon *et al.,* 1991; Brochard-Wyart and de Gennes, 1992) focused mainly on thin films ($h_0 \sim 10\ \mu$) where viscous effects are dominant, and where the rim profile is (nearly) a portion of a circle. At larger thicknesses, the profile becomes more complex (Brochard-Wyart *et al.,* 1995; Andrieu *et al.,* 1996) In the present note, we focus on thick films, where viscous effects can be omitted (except for a thin region near a shock front). We are led to predict a 'flat rim' represented in *figure 1.* Inside the rim (ignoring a thin boundary layer) we have a plug flow of velocity V. This terminates at a shock wave in shallow water advancing at velocity V_S. Our aim is to predict the velocities (V and V_S) and the rim dimensions (h_1 and ℓ) in terms of the driving force (eq. 1) and of the original thickness h_0. For simplicity, we stick to a one-dimensional geometry, assuming $\ell < R$.

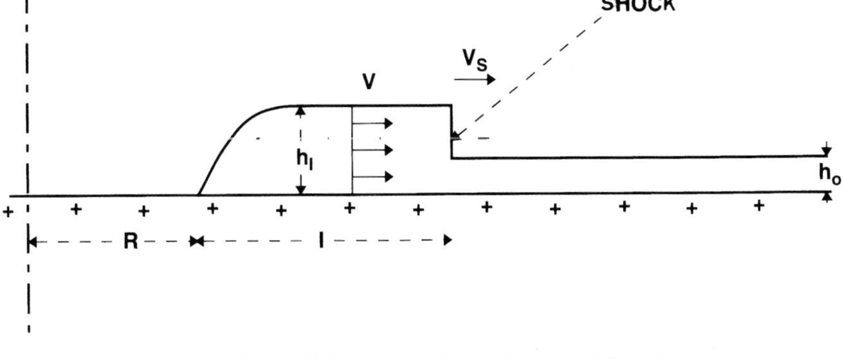

Fig. 1. – Proposed rim shape in inertial dewetting ($Re >> 1$).

Fig. 1. – Modèle de bourrelet en démouillage inertiel ($Re >> 1$).

2. Mass and momentum

2.1. *The driving force*

The rim is subjected to two forces (per unit length):

a) on the dry side: $-\gamma_{s0}$

b) on the wet side: $\gamma_{SL} + \gamma - \int_0^{h_0} p\mathrm{d}z$

where the γ_{ij}s are the interfacial tensions, and p is the pressure. Thus the overall force is:

$$-S - \frac{1}{2}\rho g h_0^2 = \frac{1}{2}\rho g(h^{*2} - h_0^2)$$

We shall now follow the spirit of a classical argument (Culik, 1960). Let M be the rim mass and P the rim momentum (both per unit length). We may write:

$$\frac{\mathrm{d}M}{\mathrm{d}t} = \rho h_0 V_S \tag{2}$$

This mass acquires a constant velocity V. Thus, Newton's law may be written in the form:

$$\frac{1}{2}\rho g(h^{*2} - h_0^2) = \frac{\mathrm{d}P}{\mathrm{d}t} = \frac{d}{\mathrm{d}t}(MV) = V\frac{\mathrm{d}M}{\mathrm{d}t} = \rho h_0 V V_S \tag{3}$$

2.2. *The Hugoniot conditions for the shock*

Shocks in shallow water are discussed, for instance, in the book of Landau and Lifshitz (1959). If we define v_0 and v_1 as the velocities (ahead and behind the shock) in a reference frame moving at the shock velocity, we have:

$$\left.\begin{aligned} v_0 &= -V_s \\ v_1 &= V - V_s \end{aligned}\right\} \tag{4}$$

The matter flux (for our incompressible fluid of density ρ) is:

$$\rho v_0 h_0 = \rho v_1 h_1 \tag{5}$$

and the momentum flux is:

$$\rho\left(v_0^2 + \frac{1}{2}gh_0\right) = \rho\left(v_1^2 + \frac{1}{2}gh_1\right) \tag{6}$$

Equations (5, 6) can be transformed into:

$$V = V_s\left(1 - \frac{h_0}{h_1}\right) \tag{7}$$

$$V_s^2 = \frac{1}{2}g\frac{h_1(h_0 + h_1)}{h_0} \tag{8}$$

Returning now to eq. (3), we find simply:

$$h_1 = h^* \tag{9}$$

3. Discussion

1) If the driving force $(-S)$ is relatively strong, we expect h^* (as defined in eq. [1]) to be of order 1 mm. The velocities are of order of $(gh^*)^{1/2}$, and the Reynolds numbers are:

$$Re = \frac{\rho V h_1}{\eta} \cong \frac{g^{1/2}(h^*)^{3/2}}{\nu} \cong 10^2 \tag{10}$$

where $\eta = \rho\nu$ is the viscosity.

Indeed, we are in an inertial regime.

2) Another question of importance concerns the width ℓ of the rim. To have a well defined flat rim, while remaining in a one-dimensional limit, we need:

$$h_1 << \ell << R$$

Note first that:

$$\frac{\ell}{R} = \frac{h_0}{h_1 - h_0} \tag{11}$$

Thus, the limit corresponds to $h_0 << h^*$. But the condition $\ell > h_1$ implies:

$$R > \frac{(h_1 - h_0)\, h_1}{h_0} \tag{12}$$

Thus, we cannot make h_0 too small, because the size of the experiment would become very large. Typically, we think of $h_0 = 0.1$ mm, $h_1 = 1$ mm, $R = 10$ cm, giving $\ell = 1$ cm. The difficulty of dealing with large R may be the reason why this inertial behaviour has not yet been fully observed.

Note remise le 15 novembre 1996, acceptée le 11 décembre 1996.

References

Andrieu C., Sykes C. and Brochard-Wyart F., 1996. Dynamics of fast dewetting on model solid substrate, *J. Adhesion*, 58, p. 15-24.

Brochard-Wyart F. and de Gennes P.-G., 1992. Dynamics of partial wetting, *Adv. Colloid Interface Sci.*, 39, p. 1-11.

Brochard-Wyart F., Raphael E. and Vovelle L., 1995. Démouillage en régime inertiel : apparitions d'ondes capillaires, *C. R. Acad. Sci. Paris,* 321, série II, p. 367-370.

Culik F.E.C., 1960. Bursting of soap films, *J. Appl. Phys.*, 30, p. 1128-1133.

De Gennes P.-G., 1987. Dynamics of drying and film twinning, In *Physics of Amphiphilic Layers, Springer Proc. in Phys.,* 21, p. 64-69, **Meunier J., Langevin D. and Boccara N.** ed., Springer.

Landau L.D. and Lifshitz E.M., 1959. *Fluid Mechanics*, Pergamon Press, London, 100.

Redon C., Brochard F. and Rondelez F., 1991. Dynamics of dewetting, *Phys. Rev. Lett.*, 66, p. 715.

J. Phys.: Condens. Matter **6** (1994) A9–A12. Printed in the UK

Dewetting of a water film between a solid and a rubber

F Brochard-Wyart† and P G de Gennes‡

† Physique des Surfaces et des Interfaces, Institut Curie, 11 Rue P et M Curie, 75005 Paris, France

‡ Collège de France, 11 Place M Berthelot, 75231 Paris Cédex 05, France

Received 4 October 1993, in final form 17 November 1993

Abstract. A water film (thickness e) between two hydrophobic solids is metastable. If contact is established (over an initial disc of radius $R(0) > R^*$, where R^* is a certain nucleation size), the dry path expands (up to a size $R(t)$ at time t), and the rejected water forms a rim (of width $l(t)$) around the patch. We predict $R(t) \sim t^{3/4}$ and $l(t) \sim t^{1/2}$. Typically, over 5×10^{-3} s we expect $R(t) \sim 1.6$ mm.

1. Aims

A water film, placed on a hydrophobic surface, nucleates dry patches, which grow with a constant velocity (Brochard-Wyart and de Gennes 1992, Redon *et al* 1991, Reiter 1992)

$$V = \text{constant} \times \gamma/\eta\theta_e^3 \qquad (\theta_e < 1) \tag{1}$$

where γ is the surface tension, η the viscosity, and θ_e the equilibrium contact angle of a water droplet on the solid. (Hysteresis is assumed to be negligible.) Ahead of the dry patch, the water builds up a rim; both the width and the height of the rim increase with time.

Our aim here is to extend this picture to the case shown in figure 1, where the water film is not at a solid/air interface, but rather at a solid/rubber interface. This is (remotely) related to the problems of driving on wet roads: here a given piece of the outer surface of the tyre is exposed to water during a contact time of order 5×10^{-3} s. The crucial question is whether, during this time, the water film has been expelled, thus restoring an acceptable level of adhesion.

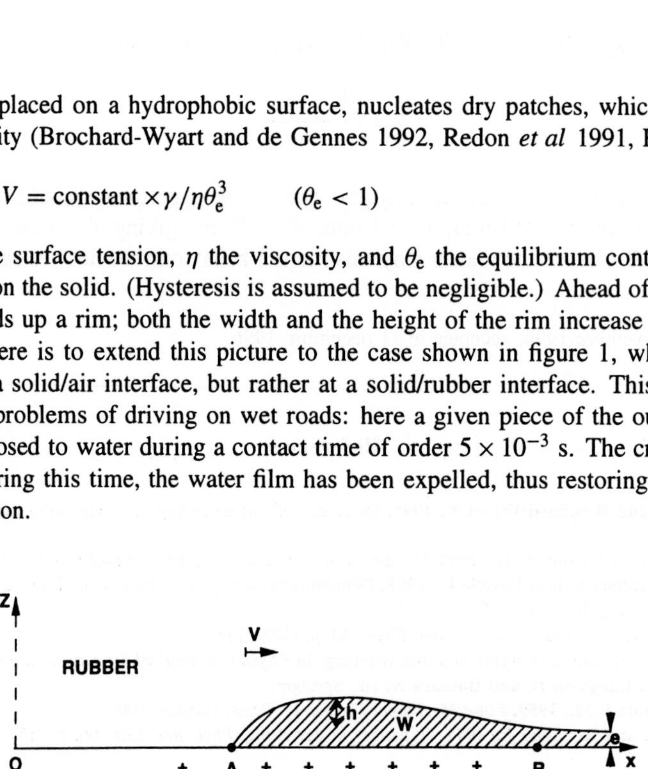

Figure 1. Proposed structure for a growing dry patch in a water film W between a solid S and a rubber. The dry patch was initiated near point O, and has a growing radius $R(t)$. The water rejected from the dry region builds up a rim of width $l(t)$.

In the present paper, we discuss this only for an idealized case, where both the solid and the rubber are completely flat. The solid is assumed to be perfectly rigid. The water and the rubber are taken as incompressible, but can of course be deformed. We also assume that the thickness e of the original film is relatively large ($\geqslant 1$ μm), so that the long-range effects of Van de Waals forces can be neglected.

A precise discussion of the rim shape requires the use of singular integral equations, relating the vertical stress at one point x on the rubber surface to the deformations of this surface at all other points. In the present paper, we try to avoid this complicated approach, using simple (but tentative) scaling arguments.

2. Two-parameter descriptions of the rim

(1) As shown in figure 1, we idealize the rim as a region of width l, and thickness $e+h \sim h$ ($h \gg e$). Our requirement of incompressibility then imposes

$$2\pi Rlh = \pi R^2 e \qquad (l \ll R). \tag{2}$$

(2) The driving force F for the drying process is a combination of interfacial energies γ_{ij}.

$$F = \gamma_{WR} + \gamma_{WS} - \gamma_{RS} \tag{3}$$

where W denotes water, R, rubber, and S, solid. We translate this into a pressure head p around point A:

$$p = F/h. \tag{4}$$

(3) This pressure head induces a Poiseuille flow in the rim, with an average velocity

$$V = -(\partial p/\partial x)(e+h)^2/12\eta \cong (\partial p/\partial x)h^2/12\eta. \tag{5}$$

Here, we estimate roughly

$$-\,\partial p/\partial x \cong p/l = F/hl. \tag{6}$$

(4) We also write that the deformation of the rubber around the rim is induced by the pressure p:

$$p \cong \mu h/l \tag{7}$$

where μ is the shear modulus of the rubber (4)

$$\mu = kT/Na^3 \tag{8}$$

N being an average number of monomer units between crosslinks, and a a monomer size.

3. Growth laws

Comparing (4) and (7), we obtain

$$h = (F/\mu)l/h = h_0 l/h \tag{9}$$

where $h_0 \equiv F/\mu$ is typically in the range 100–1000 Å.

The Poiseuille equations (5) and (6) can be rewritten as

$$V/V^* = h/l = h_0/h \tag{10}$$

where $V^* = F/\eta$.

Thus, using the two form of (10) plus (2)

$$(V/V^*)^2 = h_0/l \qquad (V/V^*)^3 \cong h_0^2/eR. \tag{11}$$

Writing $V \sim R/t$, this gives the scaling form of the growth law

$$R(t) \cong (V^* t)^{3/4} h_0^{1/2} e^{-1/4}. \tag{12}$$

We can then obtain the width $l(t)$ of the rim, using (10)

$$l = (Re)^{2/3} h_0^{-1/3}. \tag{13}$$

Thus whenever $R \gg e^2/h_0$, we expect $l \ll R$: the rim represents a small annulus around the dry patch. Finally, from (9), the increase in width of the rim should scale as

$$h = (l h_0)^{1/2} = (R e h_0)^{1/3}. \tag{14}$$

Again when $R \gg e^2/h_0$, we can check that $h > e$.

The related deformations

$$h/l = h_0/h = (h_0^2/Re)^{2/3} \tag{15}$$

are correspondingly small.

4. Discussion

(1) Consider for instance a hydrophobic solid with surface properties comparable to those of rubbers ($\gamma_{RS} \cong 0$), and assume $\gamma_{RW} = \gamma_{SW} = 35$ mJ m^{-2}. This leads to $V^* = 70$ m s^{-1}. Taking $e = 1$ μm, $h_0 = 100$ Å, and $t = 5 \times 10^{-3}$ s, we obtain from (12) a dry patch radius $R \sim 1.6$ mm. Of course, the coefficient in (12) is very uncertain, but we do see that spontaneous dewetting is not fully effective to restore ahesion between a tyre and a *flat* solid surface: we would require $R \sim 1$ cm, rather than 1 mm.

(2) Of course, in pratice, the tyre is facing a rough solid surface, with granularities ranging from millimetres down to micrometres. An interesting challenge for the future is to extend our primitive model towards random surfaces.

(3) Model experiments on ideal surfaces may still be useful, to check whether our simple picture has a real meaning.

(4) Clearly, near the tip of the wet region, we expect elastic stresses much larger than suggested by (7), but they should be localized in an area $\sim e^2$ (of the xz phase) near point A. They lead to a line energy T for the edge of the dry patch $T \sim \mu e^2$. Then the total energy becomes

$$E = 2\pi RT - \pi R^2 F \tag{16}$$

and the patch will grow only when R is larger than a certain *nucleation radius*

$$R^* = T/F \cong e^2/h_0. \tag{17}$$

It is worth noting that the condition $R \gg R^*$ automatically imposes $l \ll R$ and $h > e$, as discussed after (13).

(5) All our analysis was based on the low-frequency response of the rubber, described by a single modulus μ. Actually, the rim motion drives elastic perturbations at frequencies

$$\omega \sim V/l \sim R/lt. \tag{18}$$

$\omega/2\pi$ is expected to be in the 100 cycle range. If the rubber has a strong viscoelastic behaviour at these frequencies, the simple growth law (12) may be strongly modified.

(6) It may be worth noting that, at our level of description (incompressible materials), and overall pressure p_0 acting on the structure should have no effect.

Acknowledgments

This (very tentative) reflection was initiated by seminal discussions with Mr Bourret, Mr Raphaël, and Mr Vacherand.

References

Brochard-Wyart F and de Gennes P G 1992 *Adv. Colloid Interface Sci.* **39** 1
Redon C, Brochard-Wyart F and Rondelez R 1991 *Phys. Rev. Lett.* **66** 715
Reiter G 1992 *Phys. Rev. Lett.* **68** 75

Afterthought: Dewetting of a water film between a solid and a rubber

Since this paper was published, the problem was studied experimentally by P. Martin (PCC, Institut Curie, 11 rue Pierre et Marie Curie, 75005 Paris, Tel. 33 1 42 34 67 87). He did find that for current rubbers, dissipation inside the water film is the leading feature: the $R \sim t^{3/4}$ law does work. On the other hand, with very loosely crosslinked rubbers, dissipation in the rubber should alter the picture. Going even further, Martin replaced the rubber by a very viscous melt, with no crosslinks: here the laws are rather different.

PERGAMON

Chemical Engineering Science 56 (2001) 5449–5450

Chemical Engineering Science

www.elsevier.com/locate/ces

Some remarks on coalescence in emulsions or foams

Pierre-Gilles de Gennes *

Collège de France, 11 Place M. Berthelot, 75231 Paris Cedex 05, France

Abstract

Intrinsic coalescence (corresponding to the opening of a hole in a surfactant monolayer) is important only for poor surfactants. With good surfactants, and in usual industrial conditions, coalescence is *extrinsic*, and mainly due to dirt particles. In the present note, we first analyze the motions of *one* dirt particle inside an initially monodisperse O/W emulsion. The particle remains trapped at the surface of a growing oil droplet, of size $R(t) \sim t^{1/3}$. If we now go to a very dilute system of dirt particles, we expect that they generate first a collection of large drops. Ultimately, these large drops come into contact, and one may end up with an oil matrix containing droplets of the O/W emulsion. © 2001 Elsevier Science Ltd. All rights reserved.

Keywords: Emulsions; Foams; Coalescence; Soap films

1. Introduction

The classical view (Kaschiev & Exerowa, 1980) on *intrinsic* coalescence is sketched in Fig. 1: (a) we start with a black film, close to a simple bilayer, (b) a depleted region occurs in one of the monolayers, (c) if the width of this depleted region is large enough (comparable to the length L of the surfactant tails), we may arrive at a nascent hole.

What is the barrier energy B for process (b)? If the Langmuir pressure is Π and the depleted area is L^2, the work required for expelling the surfactant is

$$B \cong \Pi L^2. \tag{1}$$

A modified version of this argument considers the spontaneous fluctuations of (two dimensional) concentration Γ in the monolayer, and gives

$$B = \Gamma \frac{d\Pi}{d\Gamma} L^2. \tag{2}$$

In any case, B is rapidly an increasing function of L (because of the L^2 factor and also because Π increases with L).[1] With good surfactants (L large) B is $\sim$100 kT, and the opening of holes is utterly forbidden.

We may also consider process (c), where a natural estimate of the barrier C is based on the bending energy constant K:

$$C \sim \frac{K}{L^2} L^2 \sim K^* \quad \text{(court)}. \tag{3}$$

Again K is a rapidly increasing function of L. For good surfactants, intrinsic coalescence is a myth.

We must then think of extrinsic processes. They are well known from studies on antifoaming (Garrett, 1997). If we consider O/W emulsion with a small water faction, the most efficient agent is a slightly hydrophobic grain, of size larger than the water film. Another powerful agent is a droplet of silicone oil, which spreads (totally or partially) on the water film, and thins it hydrodynamically.

In practical (industrial) conditions, these extrinsic processes are dominant; dirt particles control the coalescence. Our aim, in the present note, is to discuss the trail of one dirt particle inside an emulsion.

* Tel.: +33-1-40-79-45-00; fax: +33-1-45-35-14-74.
E-mail address: pierre-gilles.degennes@espci.fr (P.-G. de Gennes).

[1] To get a rough idea about the dependence of Π on L, we can use a simple model where (a) the area per polar head is fixed and (b) the chains are normal to the interface. Then the entropy loss due to this orientation increases linearly with L, and so does Π.

PII: S0009-2509(01)00170-1

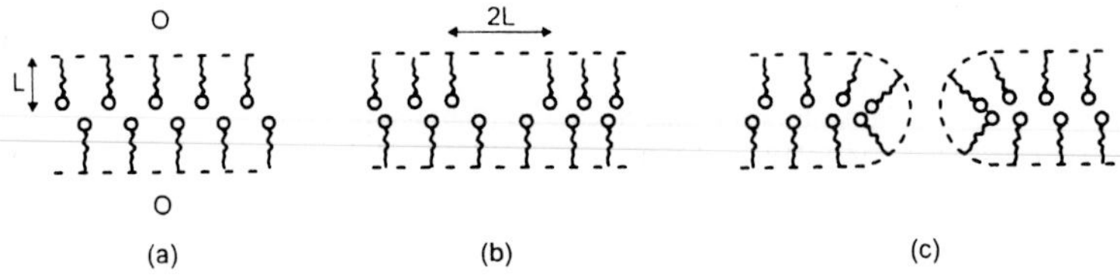

Fig. 1.

We choose a monodisperse O/W emulsion (drop diameter d) as our starting system. Emulsions of this type have been prepared by J. Bibette and coworkers, and their behavior in intrinsic coalescence has been studied Deminiere, Colin, Leal Calderon, and Bibette (1998). Intrinsic processes retain the monodispersity (for unclear reasons). As we shall see, the expected behavior is very different when dilute particles are acting.

2. Trail of a single particle

Let us assume that the particle, standing at the surface of one droplet, initiates a rupture after a certain (average) time τ. After the destruction of the local film, we postulate that the particle remains attached to a neighboring O/W interface, and that it is displaced by $\sim d$. We are led to think of a random walk, with a diffusion coefficient $D = d^2/\tau$.

But, in fact, we are *not* dealing with simple random walks, as explained in de Gennes (1998): it is generating a growing oil drop, of volume Ω increasing linearly with time:

$$\Omega \cong d^3 \frac{t}{\tau}. \tag{4}$$

This drop must be in capillary equilibrium with its neighbors, with a uniform Laplace pressure in the oil: thus it must be spherical, with a radius:

$$R \sim \Omega^{1/3} \sim d(t/\tau)^{1/3}. \tag{5}$$

The dirt particle constantly moves at the outer surface of the drop. It diffuses fast on this surface, with a characteristic time:

$$\tau_d \sim \frac{R^2}{D} \sim \tau \left(\frac{R}{d}\right)^2 \sim \left(\frac{t}{\tau}\right)^{2/3}. \tag{6}$$

Thus, $\tau_d \ll t$ when $t \gg \tau$.

3. Extension to a very dilute system of dirt particles

Let us call $\bar{\alpha}$ the average number of dirt particles per emulsion droplet in the initial state ($\bar{\alpha} \ll 1$). Here, we first expect a growth of drops (each drop being associated with one particle) with a slowly increasing radius given by Eq. (5). Then, at a certain moment ($t = t^* = \tau/\bar{\alpha}$) the drops should enter into contact.

At $t > t^*$, the original emulsion is not a majority component. The water films must thicken, because the interfacial area has considerably decreased. Then, the coalescence kinetics are expected to slow down. It is conceivable that, in certain cases, coalescence is blocked, leading to a mixed state with small droplets in the interstices between the large drops. Indeed, there are observations of this type.

On the other hand, if the blocking is not complete, we may arrive at a fusion between large drops, leading to a continuous oil phase, still containing (for a while) some islands of O/W emulsion.

Other complications could also occur: the progressive drop of the O/W area raises the surfactant concentration in the water phase. This may lead to the formations of bulk lipid–water phases—with possible reactions on the coalescence rates (Deminiere, Colin, Leal Calderon, & Bibette, 1998).

But the simple ideas sketched above may hopefully stimulate some experiments on the action of a controlled population of "dirt" particles on an emulsion.

Of course, from a practical standpoint, if we wish to minimize the role of "dirt", we should not purify merely our system; we should rather arrange by surface treatments so that the particles become (a) preferentially soluble in the film phase, as discussed in Garrett (1997) and (b) smooth.

References

Deminiere, B., Colin, A., Leal Calderon, F., & Bibette, J. (1998). In B. Binks (Ed.), *Modern aspects of emulsion science*. Cambridge, UK: The Royal Society of Chemistry.

de Gennes, P. G. (1998). *C. R. Academy of Science* (Paris), 326 II, 331.

Garrett, P. R. (1997). *Defoaming. Surfactant series*, no. 45, NY: Marcel Dekker.

Kaschiev, D., & Exerowa, D. (1980). *Journal of Colloid Interface Science*, *77*, 501.

ELSEVIER

Colloids and Surfaces
A: Physicochemical and Engineering Aspects 186 (2001) 7–10

COLLOIDS AND SURFACES A

www.elsevier.nl/locate/colsurfa

Two remarks on wetting and emulsions

P.G. de Gennes *

Collège de France, 11 place M. Berthelot, 75231 Paris Cedex 05, France

Abstract

This paper is extracted from an opening address given at the workshop 'Wetting: from microscopic origins to industrial applications' (Giens, May 6–12, 2000). It discusses two special points a) the nature of line energies for a contact line b) the aging of emulsions. © 2001 Elsevier Science B.V. All rights reserved.

Keywords: Emulsions; Wetting; Line energies

1. Introduction

These notes are dedicated to the memory of Karol Mysels. I met him only rather late (in the 1980s). But I learned a large number of important things from him. I also treasure a certain group of his original slides on soap films, which he gave to me, and which I have used in a number (≈ 200) of talks in high schools. Karol was both a gifted experimentalist and a deep thinker. It was one of my great prides to have Karol and Estrella as participants for the Stockholm ceremonies, in which I was involved in 1991.

The following notes are a (clumsy) attempt to contribute to some things in which Karol was deeply interested: the basics of wetting, and the stability of foams and emulsions. The notes are rough, and would have been much improved if he had still been with us... We shall not forget him.

* Tel.: + 33-1-40794500; fax: + 33-1-45351474.
E-mail address: pierre-gilles.degennes@espci.fr (P.G. de Gennes).

2. Line energies

If we pinch a violin string (Fig. 1), we obtain a static triangular form. This is associated with a standard line tension $\Im$. For small displacements u with Fourier transforms u_q the energy has the standard form:

$$E = \sum_q \frac{1}{2} \Im q^2 |u_q|^2 \qquad (1)$$

On the other hand, if we pinch a contact line (for instance between a fluid, a solid, and air) we find a very different form (Fig. 2). This originates from a different form of the energy:

$$E = \sum_q \frac{1}{2} \gamma f(\theta_e) |q| |u_q|^2 \qquad (2)$$

where θ_e is the equilibrium contact angle, γ is the surface tension, and $f(\theta_e)$ is a dimensionless function, discussed in refs[1] and [2]. The $|q|$ dependance is very singular: it comes from the integration of q^2 contributions over the width of the perturbed fluid region (q^{-1}).

PII: S0927-7757(01)00476-9

Eq. (2) corresponds to what I call "fringe elasticity". It holds for scales $q^{-1.}$ which are smaller than the Laplace length:

$$\kappa^{-1} = \left(\frac{\gamma}{\rho g}\right)^{1/2} \tag{3}$$

ρ: density, g: gravitational acceleration.

It is well known that we should add in Eq. (2) an intriasic line tension $\Im_0$, transforming Eq. (2) into:

$$E = \sum_q \left\{\frac{1}{2}\gamma f(\theta_e)|q| + \Im_0 q^2\right\} \tag{4}$$

Molecular models give values of $\Im_0$ (positive or negative), which are of order:

$$\Im_0 = \gamma a \tag{5}$$

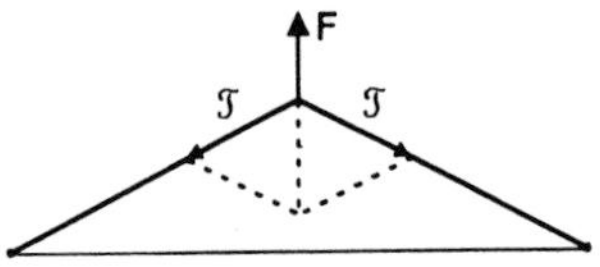

Fig. 1.

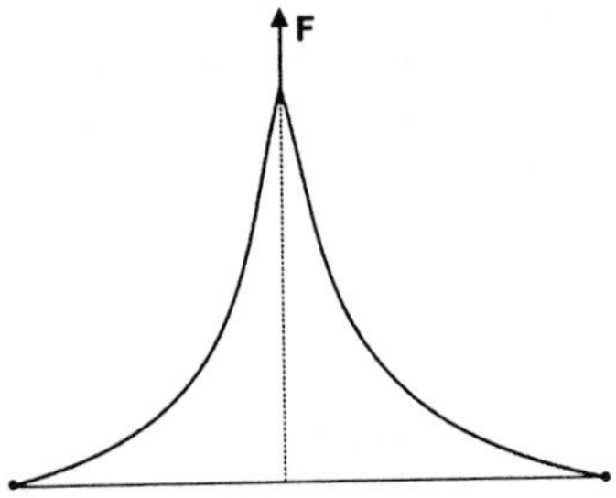

Fig. 2.

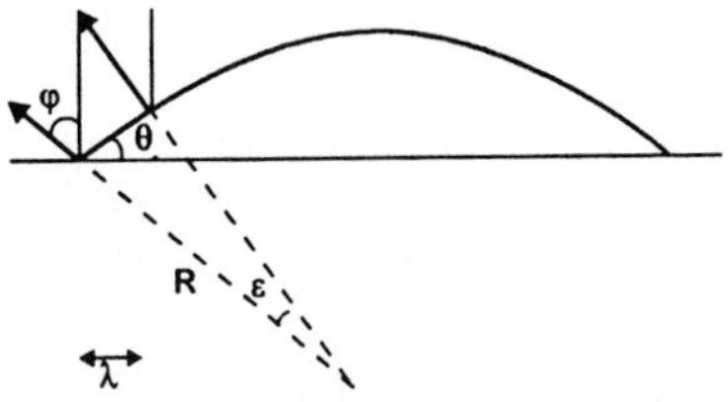

Fig. 3.

where a is a molecular size (a few angstroms).

At submillimeter scales ($a < q^{-1} < \kappa^{-1}$) this correction should be unobservable. However, a number of groups [3–7] have tried to measure $\Im_0$ by looking at the contact angle in small droplets. The result can be stated in the form:

$$\Im_0 = \gamma \ell \tag{6}$$

where $|\ell|$ is anomalously large (of order 1 micron).

My own view is that this is an artefact of optical methods. One example, based on ray optics, is illustrated on Fig. 3. In principle, we determine θ through the maximum deflection angle of a reflected ray $\varphi = 2\theta_e$. However, because of diffraction effects, the actual last measurement point has a weaker deflection $\varphi - \varepsilon$, where $\varepsilon \sim \lambda/R$ (R: radius of curvature; λ: optical wavelength). This correction has exactly the same structure than the inclusion of $\Im_0$, and gives an apparent $\Im_0$:

$$\Im_0/\text{app} \sim \gamma\lambda \tag{7}$$

This is the artefact. I believe that all optical methods have the same defect.

3. Aging of emulsions

Certain O/W emulsions, which are monodisperse, can coalesce slowly, with droplets growing in size, but remaining monodisperse[8]. This occurs in the absence of contaminants. In practice, many practical emulsions do contain some small particles, with a wettability such that the particle gets hooked at the O/W interface, although it is rather hydrophobic. These grains induce the rupture of oil films by the so called Garrett process [9]. If the grains are very dilute (so that each grain works individually), this can lead to a rough, polydisperse emulsion. Our aim here, is to discuss this, at the level of scaling laws, following the lines of ref. [10].

Start with a single grain: when trapped on a film, it induces rupture within a certain time τ. When the film is locally destroyed, we postulate that the particle binds to a neighboring interface—the size of the jump being comparable to the diameter of the initial droplets.

We are thus led to think first of a random walk with a diffusion coefficient:

$$D = d^2/\tau \tag{8}$$

3.1. Definition of a simple limit

This walk destroys some droplets: the overall amount of interface present decreases, and this could alter the control parameters: we want to avoid this to reach a simple discussion:

1. We assume that there is a plentiful reservoir of surfactant (concentration above *cmc*).
2. The thickness of the water films could increase, and this would reduce the coalescence rates. If we estimate the diffusion coefficient D_W of *water* in the structure, we find that it is related to the Poiseuille films in the Plateau borders (of diameter h) induced by gradients of the pressure:

$$p \cong p_0 - \gamma/h \tag{9}$$

The (qualitative) result is:

$$D_W \cong \frac{\gamma}{\eta_W} h \tag{10}$$

where η is the water viscosity. In all what follows, we assume $D_W \gg D$. Then water can diffuse fast from the "wounded" region to the bulk of the monodisperse emulsion. It is then plausible to assume that the water films remain at constant thickness.

3.2. Grooth laws in the simple limit

A major point is the following: the grain does *not* follow an arbitrary random walk in the emulsion. If it "digs a tunnel" in the network of droplets, the Laplace pressure at the bottom of the tunnel forces it to retract very soon.

This retraction would follow the Washburn law for fluid motion in a capillary, with a certain diffusion coefficient D_r (*r* stands for retraction). We assume (as in agreement with usual conditions) that $D_r > D_W > D$.

In this situation, the grain can move only at the interface between a coarse oil drop and the unperturbed fine emulsion, as shown on Fig. 4.

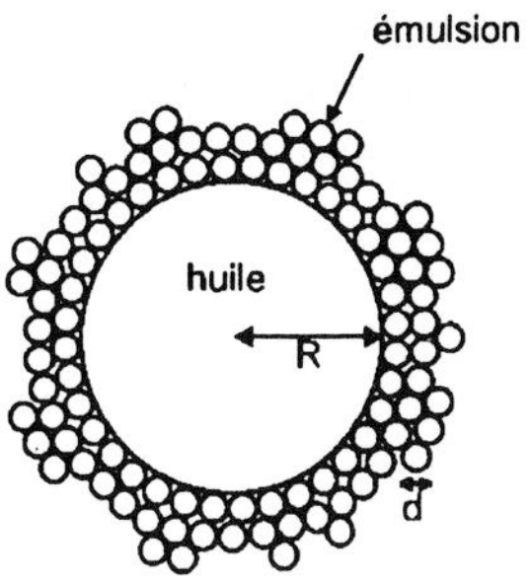

Fig. 4.

It builds up a coarse drop of radius $R(t)$ increasing with time t. The number of ruptured droplets is $t/\tau(>>1)$, and the oil volume of the coarse drop is (omitting coefficients):

$$R^3 = \frac{t}{\tau} d^3 = dDt \tag{11}$$

During time t, the grain has moved at random on the surface of the coarse drop, spanning a length s such that $s^2 = Dt$. We see from this that:

$$\frac{s^2}{R^2} = \frac{R}{d} >> 1 \tag{12}$$

This means that the particle explores fast the surface, and ensures that the coarse droplet is essentially spherical. (For more details on the roughness of the interface, see ref.[5].

Our conclusions are the following:

If we start, at $t = 0$, with a monodisperse emulsion, containing a very dilute suspension of active grains, we expect a first stage, where each grain generates one coarse droplet of size $R(t)$, growing like $t^{1/3}$.

At a certain moment ($t = t^*$), these coarse droplets enter into contact. At $t > t^*$, the original fine emulsion is not the main component, and the growth laws must become more complex. Even if D_W is fast, the water films will thicken, and coalescence should slow down, or even stop, as often observed. On the other hand, if τ has become long, one could possibly arrive at a fusion of the coarse droplets, leading to an inverted (W/O) emulsion.

On the whole, we are still very far from understanding the aging of emulsions. But we clearly have an intrinsic blow up of the walls (which here are water films) plus an extrinsic process due to grains (or other external objects). Monodisperse emulsions may allow us to separate neatly the two processes.

References

[1] J.F. Joanny, P.G. de Gennes, A model for contact angle hysteresis, J. Chem. Phys. 81 (1984) 552–562.
[2] F. Brochard-Wyart, D. Quéré, P.G. de Gennes, Gouttes, bulles et perles (2001) (in press).
[3] D. Li, A. Neuman, Contact angles, Colloids Surf. 43 (1990) 195.
[4] J. Drelich, J. Wilbur, J.D. Miller, G. Whitesides, Contact Angles for Liguid Drops at a Model Heterogenous Surface Consisting of Alternating and Parallel Hydrophobic/Hydrophobic Strips, Langmuir 12 (1996) 1913.
[5] J. Drelich, Contact angles and line tensions, Polymer J. 71 (1997) 525.
[6] A. Amirfazli, D. Kwok, J. Gaydos, A. Neuman, Line Tension Measurements through Drop Size Dependence of Contact Angle, J. Colloid Interface Sci. 205 (1998) 1.
[7] J.Y. Wang, S. Betelu, B.M. Law, Line Tension Effects near First-Order Wetting Transitions, Phys. Rev. Lett. 83 (1999) 3677.
[8] B. Deminière, A. Colin, F. Leal Calderon, J. Bibette, Cell Growth in a 3D Cellular System Undergoing Coalescence, Phys. Rev. Lett. 82 (1) (1999) 229–232.
[9] P.R. Garrett, Defoaming, in: Surfactant Science Series, vol. 45, Marcel Dekker, New York, NY, 1997.
[10] P.G. de Gennes, Progression d'un agent de coalescence dans une émulsion, C. R. Acad. Sci. (Paris) 328 (2) (1998) 331–335.

Adhesion induced by mobile binders: Dynamics

F. Brochard-Wyart[†] and P. G. de Gennes[‡§]

[†]PCC Institute Curie, 11 Rue P. et M. Curie, 75005 Paris, France; and [‡]College de France, 11 Place M. Berthelot, 75231 Paris Cedex 05, France

Contributed by P. G. de Gennes, April 12, 2002

We consider a vesicle bilayer loaded with molecules that can bind (upon contact) with a solid surface, following the classical model of Bell, Dembo, and Bongrand. We are interested in situations where the contact area varies with time: we assume that binders can then migrate via diffusion. The resulting dissipation and lag create a retarded force on the contact line, which could be significant in squeezing or rolling experiments. However, there are two cases where we expect the lag force to be ineffective: (*i*) *separation* by shrinking of an adhesive patch (where the Evans "tear out" process turns out to be less costly) and (*ii*) spontaneous growth of a patch from a point contact. In this last case, the lag force is weak, and we give detailed predictions for the growth laws.

Cell adhesion is based on a set of bridging molecules ("binders") that can attach to specific ligands on the opposite surface. The density Γ of binders per unit area is originally rather small. But, when facing a surface with enough ligands, the binders converge towards this surface and build up a more concentrated adhesive region (Fig. 1).

The equilibrium picture for this process has been described long ago by Bell, Dembo, and Bongrand (1). They showed that the effective work of adhesion G is equal to the difference between the two-dimensional (2D) osmotic pressures of the binder, inside and outside of the sticky region. (In practice, the osmotic pressure inside is largely dominant.)

However, the nature of this osmotic pressure is delicate: ref. 1 was mainly based on an ideal gas behavior of the binders in two dimensions. But there are strong proofs of cooperativity in binding, as discussed in ref. 2: a nominal adhesive zone is often fragmented into smaller patches of high binder density.

The explanation provided in ref. 2 is essentially the following: in the absence of binder, the protective glycocalix forces the two opposed surfaces to remain rather distant (say 3 nm away). When one binder molecule adheres, it forces the two sides to become locally closer; the resulting thinned region acts as an attractor for other binders.

The net result is a phase transition between a 2D "gas phase" of binders, extending over unbound regions, and a "liquid phase" (much more concentrated) occupying the sticky regions.

Our aim, in the present work, is to extend some of these ideas to the *dynamics*, i.e., to situations where the contact area is time-dependent.

If (for instance) we tend to decrease the contact area, we are faced with two possible types of behavior: (*i*) a *diffusion* response where the binders remain coupled to the surface but move and become more crowded inside the patch, and (*ii*) a *reaction* response, where some binders decouple from the surface.

The "reaction" dynamics has been analyzed on some typical cases by Evans and coworkers (3). They showed that the bonds appear strong if they are loaded fast, and weaker if they are loaded slowly.

In the present work, we analyze the diffusion mode, and the resulting lag force, opposing the motion. The final question for separation experiments will be to compare the lag force to the Evans force (we give some comments on this point in *Competition Between Tear Out and Diffusion*).

For simplicity, we begin by assuming that the contact is made of a *single patch*, rather than the structure of micropatches plus blisters, which is often observed in practice (2). We shall incorporate the multipatch systems in *Spontaneous Growth of a Patch*. From an experimental point of view, it may be possible to achieve a single patch by very slow expansion of a single contact, with relatively high numbers of binders/vesicle.

On the other hand, we can lump many features of the binder–binder interaction into the osmotic pressure, which is then higher than the ideal gas value.

Our starting point is shown on Fig. 1: a vesicle establishes contact with a solid wall via a certain number of "binder" molecules. In the contact area, each binder experiences an attractive potential U. In dilute conditions, the surface concentrations inside (Γ_i) and outside (Γ_0) of binders are related by

$$\Gamma_i = \exp(U/kT)\Gamma_0 \equiv f\Gamma_0, \qquad [1]$$

where f is large. To each concentration is associated an osmotic pressure $\Pi(\Gamma_i)$ and an osmotic "rigidity" $K \equiv \Gamma_i d\Pi/d\Gamma_i$.

We now modify the contact area either by squeezing the vesicle (Fig. 1*a*) or by rolling it (Fig. 1*b*). For instance, if we decrease the contact area during a time t, a number of binders move inward to stay in the attractive region. This means that the concentration Γ is increased in an annulus of size $(Dt)^{1/2}$ near the contact line. The result is an excess osmotic pressure Π, or a lag force F, opposing the motion.

The major process (for $\Gamma_i \gg \Gamma_0$) is therefore taking place at a fixed number of bonds; thus, we do not depend on the complex (multistep) bonding processes discussed in ref. 4.

At first sight, we might hope to describe the movement through a velocity-dependent separation energy $G(V)$, as is usually done in adhesion science (5–7). But this approach does not make sense here. For instance, if we increase persistently the contact area, the system has to bring in binders from very distant regions, and one cannot generate a steady-state solution for a moving line at constant velocity V.

We present here two (equivalent) ways of reformulating the problem. The first way is based on a very compact scaling argument, described in *Scaling Structure of the Lag Force*. The second way makes use of a complete analysis for small sinusoidal perturbations (*Small Oscillations*). In *The Rolling Problem*, we discuss an extension of these results to certain rolling motions. In *Spontaneous Growth of a Patch*, using the same ideas, we analyze the spontaneous growth of a single patch after contact. Here, we find that for this growth problem, in most realistic cases, the lag force is negligible, because the motions are very slow. In *Competition Between Tear Out and Diffusion*, we compare the tear out process and the diffusion process for separation experiments.

Scaling Structure of the Lag Force

As shown in Fig. 2*a*, we consider a single contact line and restrict our attention to time scales t, such that the diffusion length $\sqrt{Dt}$ is small when compared to the size of the contact zone. D is the diffusion coefficient (for realistic binders, it is expected to be very small, of order 10^{-11} to 10^{-9} cm^2/sec).

Abbreviation: 2D, two-dimensional.

§To whom reprint requests should be addressed. E-mail: pierre-gilles.degennes@espci.fr.

www.pnas.org/cgi/doi/10.1073/pnas.112221299

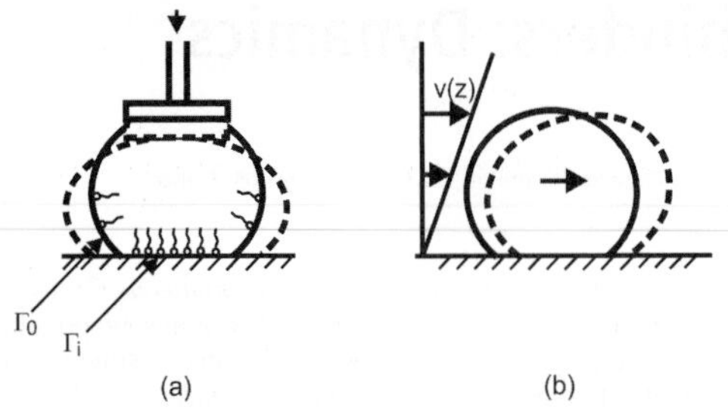

Fig. 1. Typical modifications of the contact between a vesicle and a binding flat surface. The vesicle wall is represented by a single line. The binder molecules are represented by a head + tail system, (*a*) squeezing and (*b*) rolling by a hydrodynamic flow *V*(*z*).

The line moves with a small velocity $V(t)$, and we want to compute the opposing force $F(t)$ to first order in V, as explained after Eq. **1**. This force will be proportional to the internal osmotic rigidity at equilibrium $K = \Gamma_i d\Pi/d\Gamma_i$. Then, its most general structure is

$$F(t) = K\int^{t} dt' V(t')\, R(t-t'), \qquad [2]$$

where R is a certain response function, which dimensionally must be an inverse length. The only available ingredients to define a length are the diffusion constant D and the time interval $t - t'$. Thus, we must have

$$R(t) = \alpha[D(t-t')]^{-1/2}, \qquad [3]$$

where α is a numerical constant.

We see, in Eqs. **2** and **3**, why the situation of constant velocity is not acceptable: at fixed V, the integral over t' diverges (physically it would be cut off by the finite size of our specimen; Eq. **2** holds only for sizes $\gg \sqrt{Dt}$).

It is also important to notice that the force F has a mixture of reactive and viscous behaviors. Indeed, if we Fourier transform Eq. **2**, we find that, at a given frequency ω, $F_\omega/v_\omega \sim (i\omega)^{-1/2}$ has both a real and an imaginary component. This will appear naturally in the next section.

Small Oscillations

We assume here that the speed of the line V is modulated sinusoidally. This situation is described in Fig. 2*b*:

$$V = V_i e^{i\omega t}. \qquad [4]$$

This creates, at point x, a small deviation $\delta\Gamma(xt)$ from the initial (local equilibrium) value. This is ruled by a diffusion equation:

$$\frac{\partial}{\partial t}\delta\Gamma \equiv i\omega\delta\Gamma \equiv D\frac{\partial^2\Gamma}{\partial x^2}. \qquad [5]$$

The perturbations then decay exponentially from the unperturbed line position:

$$\left.\begin{array}{l}\delta\Gamma(x) = \delta\Gamma_0 \exp(-\varkappa x)\ (x>0)\\ \delta\Gamma(x) = \delta\Gamma_i \exp(+\varkappa x)\ (x<0)\end{array}\right\}, \qquad [6]$$

where the characteristic length $\varkappa^{-1}$ is complex and is defined by

$$\varkappa^2 = i\omega/D. \qquad [7]$$

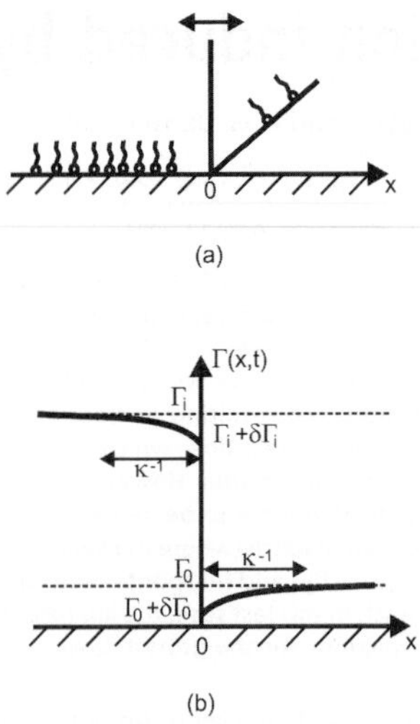

Fig. 2. Concentration profiles of the binder Γ(*x*) for a line oscillating back and forth (*a*) general aspect (*b*) profiles at an instant where the line moves to the right: more attractive sites are presented, and the binders diffuse towards this region. The main effect is a strong drop of the 2D osmotic pressure at the inner side (*x* = 0−) opposing the motion.

PHYSICS

(We choose to define $\varkappa$ as the root of Eq. **7**, with a positive real part.) We must now supplement Eq. **5** by two boundary conditions at the interface.

(*i*) There is a rapid equilibrium at the interface, implying

$$(\Gamma_i + \delta\Gamma_i)/(\Gamma_0 + \delta\Gamma_0) = e^{\beta V} = f, \qquad [8]$$

and giving, by comparison with Eq. **2**,

$$\delta\Gamma_i/\delta\Gamma_0 = f. \qquad [9]$$

(*ii*) We must match the currents at the boundary; in the reference frame moving with the contact line, at velocity V, the current is

$$J = -Dd\Gamma/dx - V\Gamma. \qquad [10]$$

It must be continuous at the line, and this gives

$$-D[d\Gamma/dx|_{0^-} - d\Gamma/dx|_{0^+}] = V[(\Gamma_i + \delta\Gamma|_{0^-}) - (\Gamma_0 + \delta\Gamma|_{0^+})]$$

$$-D[d\Gamma/dx|_{0^-} - d\Gamma/dx|_{0^+}] \equiv V[\Gamma_i - \Gamma_0] \qquad [11]$$

to first order in V. Making use of Eq. **6**, this gives

$$-D\varkappa[\delta\Gamma_i + \delta\Gamma_0] = V(\Gamma_i - \Gamma_0). \qquad [12]$$

Eqs. **12** and **8** give us the complete solution. But we may simplify things in our limit $f \gg 1$: then $\delta\Gamma_0 \ll \delta\Gamma_i$, and the contribution to the lag force due to the external region is negligible. We replace Eq. **12** by

$$\delta\Gamma_i/\Gamma_i = V/(D\varkappa). \qquad [13]$$

The main role of the lag force F is then the increase of osmotic pressure just inside the contact line:

$$F = K\delta\Gamma_i/\Gamma_i = KV/(D\varkappa) = K(i\omega D)^{-1/2}V. \qquad [14]$$

This has exactly the form required by Eqs. **2** and **3**, with $\alpha = \pi^{-1/2}$.

Starting from Eq. **3**, we choose to use time intervals of order 1 sec, *a* diffusion constant, $D = 10^{-9}$ cm^2/sec, and a velocity V of order 1 μm/sec. For the surface concentration Γ_i, we assume an area per binder in the contact zone of order 1000 Å^2, and we use the perfect gas law for an estimate of K. This gives forces of order 1 mJ/m^2. Thus, the lag force is not negligible.

How would we measure F? We might possibly use an Evans *et al.* (8) set up, where the pressure in a micropipette allows us to modulate the surface tension γ at some low frequency. We would then measure the modulation of the contact angle θ and of the contact radius (giving V). Then we would write a dynamic form of the Young equation:

$$G + F(V) = \bar{\gamma}(1 - \cos\theta). \qquad [15]$$

The Rolling Problem

In the last two sections, we considered a single contact line, moving over times t (or at frequencies $\omega \sim t^{-1}$), such that the diffusion length $\varkappa^{-1}$ is much smaller than the sample size. This is adequate for squeezing experiments at small amplitudes of drive.

However, it may be tempting to measure the lag force differently, through a rolling experiment, as shown in Fig. 1*b*.

How would we drive the rolling? The first, naive, idea is to impose a density difference between the vesicle and the surrounding water (via a passive solute) and to tilt the support plane: the vesicle should roll under the Archimedes force, as observed in ref. 9. But, in our case of strong adhesion, this force is much too small for our purposes. A better approach would be to impose a tangential flow on the vesicle, with a certain velocity V_0 at the level of the vesicle center. Then, we might write a rough balance of force (ignoring hydrodynamic wall effects and rotation effects):

$$6\pi\eta L_0(V_0 - V) = F_{\text{lag}}(V), \qquad [16]$$

where η is the viscosity, and L_0 is the vesicle diameter. We shall now derive F_{lag}, for a contact area of diameter L, in the limit $\varkappa L \ll 1$. The equations of the last section are not valid here.

Then, instead of having exponential decays in the concentration profile, we go to a constant concentration gradient inside. From the diffusion equation in steady state we get

$$d\Gamma/dx = V\Gamma_i/D. \qquad [17]$$

Integrating the pressures over the circular contact line, we arrive at a total lag force:

$$F_{\text{lag}} = \frac{\pi}{4} L^2 \frac{K}{D} V \, (V < D/L). \qquad [18]$$

A number of remarks are useful at this point:

(*i*) Note the difference in dimension between Eqs. **18** and **15**: Eq. **15** gives a force per unit length, while Eq. **18** is the total force.

(*ii*) Eq. **18** holds when $\varkappa^{-1} > L$, where the effective modulation frequency $\omega \sim V/L$. Thus, we must have $V < D/L$.

(*iii*) For the single line problem (with $\varkappa^{-1} < L$), we never reached a steady-state regime at constant V. But we reach it here when Eq. **17** holds.

Returning to Eq. **16**, we can now compare the drift velocity V to the applied velocity V_0. It turns out that V/V_0 is very small:

$$V/V_0 = 6\pi\eta D/(KL). \qquad [19]$$

Even with anomalously high values of D(10^{-6} cm^2/sec), we get $V/V_0 \sim 10^{-3}$. The difficulty is that, at these high contrasts, the vesicle is probably very strongly distorted by the flow; then the hydrodynamic friction is not properly estimated by Eq. **16**.

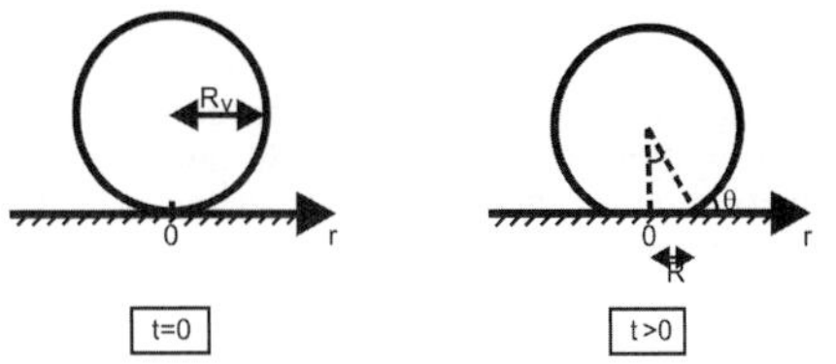

Fig. 3. Growth of an adhesive patch in idealized conditions. At $t = 0$, the vesicle enters into contact with the surface and is under 0 surface tension. At $t > 0$, an adhesive patch of radius r builds up by migration of binder molecules. The contact angle θ increases with time.

Spontaneous Growth of a Patch

The problem is described in Fig. 3. We start from a spherical vesicle under a small initial tension $\gamma_0 > 0$. (The case $\gamma_0 = 0$ would lead to large fluctuations and the possibility of more than one contact).

After a time t, we assume that a single patch has grown, with a radius $R(t) = R_v\theta(t)$, where R_v, is the vesicle radius, and θ (assumed small) is the external contact angle. Experiments of this type have been performed in particular in Paris (10, 11) and Munich (12). In the following sections, we present our (naive) theoretical views on this problem.

The Surface Tension γ. The contact has imposed an increase of area ΔA for the vesicle. The relative increase is

$$\Delta A/A = \theta^4/16. \qquad [20]$$

The classic formula for the surface tension γ superposes fluctuation effects (of small γ) and intrinsic elasticity (for large γ). It is

$$\Delta A/A = f(\gamma) - f(\gamma_0)$$

$$f(\gamma) = \frac{kT}{8\pi K_b}\ln\left(1 + \frac{R_v^2\gamma}{6K_b}\right) + \frac{\gamma}{E_2}. \qquad [21]$$

Here, K_b is the bending modulus of the bilayer ($K_b \sim 10kT$ in typical situations). The logarithmic term in Eq. **21** describes the smoothing-out of fluctuations by the tension γ. The last term corresponds to the intrinsic elasticity of the membrane, with a large elastic modulus E_2.

The fluctuation regime holds whenever

$$\gamma < E_2 kT/(8\pi K_b), \qquad [22]$$

and this is well satisfied for our purposes. We may also safely assume that $\gamma > K_b/6R_v^2$, and rewrite Eq. **21** in the following compact form:

$$\frac{\Delta A}{A} = \frac{kT}{8\pi K_b}\ln\frac{\gamma}{\gamma_0} = \frac{1}{16}\left(\frac{R}{R_v'}\right)^4, \qquad [23]$$

or equivalently:

$$\frac{\gamma}{\gamma_0} = \exp\left[\frac{\pi}{2}\frac{K_b}{kT}\left(\frac{R}{R_v}\right)^4\right]. \qquad [24]$$

Taking $K_b/kT = 10$ and $\theta = R/R_v = 0.1$, we see that the argument in the exponential is of order 10^{-3}. Thus, $\gamma = \gamma_0$. The surface tension should remain constant during the growth of the patch.

Establishment of a Nonspecific Contact. At early times, the binders cannot move. They maintain a concentration near Γ_0 at all points. Their contribution to the adhesion energy G (and to the lag force) is negligible. We can set G equal to G_0, a small value due to nonspecific interactions (e.g., van der Waals) between bilayer and wall. This assumption would not be valid for the experiments of ref. 12, where a peptide analog of the glycocalix is present and suppresses G_0. The corresponding contact angle at equilibrium is θ_0, defined by:

$$G_0 = \gamma_0(1 - \cos\theta_0) \cong \gamma_0\theta_0^2/2. \quad [25]$$

The only force opposing the growth of θ from $\theta = 0$ to $\theta = \theta_0$ is the classical force due to viscous flow in the wedge of angle $\theta[t]$. This force has already been used in this context by di Meglio and coworkers (10).

The basic balance between Young force and viscous force reads (13):

$$3\ell\eta\theta^{-1}dR/dt = \gamma_0(\theta_0^2 - \theta^2)/2, \quad [26]$$

where ℓ is a logarithmic factor of order 10, and η is the viscosity of water. This may be rewritten as

$$d\theta/dt = V^* R_\eta^{-1}\theta(\theta_0^2 - \theta^2), \quad [27]$$

with $V^* = \gamma_0/(6\ell\eta)$. Thus, the rise time for the nonspecific contact is

$$\tau = R_v/V^*\theta_0^3. \quad [28]$$

and is of order 1 min for $\theta_0 = 0.1$.

The diffusion length over the time τ is a fraction of microns, while the final radius $R_0 = \theta_0 R_v$ is of order 1 μm. Thus, indeed diffusion was weak during this first stage.

Accumulation of Binders: The Perfect Gas Regime. We now redefine the time t as starting at the end of the first step. At $t > 0$, the accumulation of binders becomes important. There is a nearly uniform concentration Γ_i in the patch. The binders come from a region of radius $\sqrt{Dt}$ outside. (We assume now that this region is large, $\sqrt{Dt} > R$.) Then the number conservation of binders imposes:

$$(\Gamma_i - \Gamma_0) = k\Gamma_0 \frac{Dt}{R^2}. \quad [29]$$

One can derive the factor k from the solution for steady-state diffusion in two dimensions, in quasi-static conditions ($Dt \gg R^2$).

The result is (for $f \gg 1$)

$$k = \frac{2}{\ln(R_v/R)} \sim 1, \quad [30]$$

and we shall set $k = 1$, for simplicity, in what follows.

Eq. **29** must be supplemented by a balance of forces at the contact line. Here again, we assume quasi-static conditions.

The lag force is negligible when diffusion is fast ($Dt > R^2$), and we may write

$$\frac{1}{2}\gamma_0\theta^2 = G = \frac{1}{2}\gamma_0\theta_0^2 + \Pi(\Gamma_i). \quad [31]$$

For the moment, let us assume a perfect gas law for the 2D gas of binders $\Pi(\Gamma_i) = kT\Gamma_i$. Comparing Eqs. **29** and **31**, we arrive at the growth law:

$$\theta^2(\theta^2 - \theta_0^2) = \varepsilon Dt/R_v^{-2}, \quad [32]$$

where

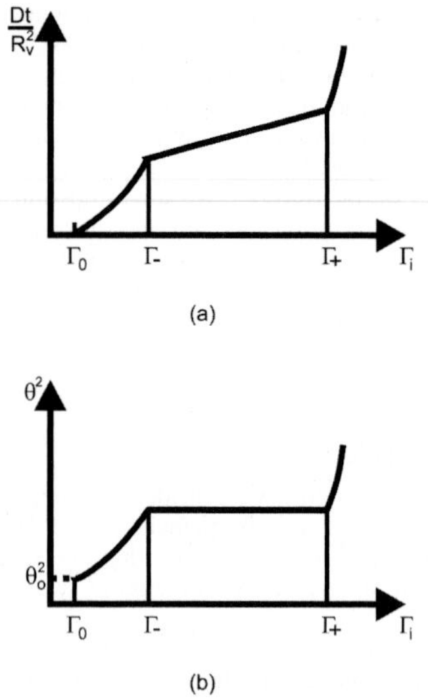

Fig. 4. Spontaneous growth laws for a patch, in the presence of a cooperative phase transition of the binders, giving a plateau in osmotic pressure within an interval (Γ_-, Γ_+) of concentrations. (*a*) Relation between time t and patch concentration $\Gamma_i(t)$ as derived from Eq. **36**. (*b*) Value of the contact angle as a function of the internal concentration Γ_i.

$$\varepsilon = 2kT\Gamma_0/\gamma_0 \quad [33]$$

is a small dimensionless parameter.

Eq. **32** is our final answer for perfect gas conditions. Note first that the quasi-static assumption ($R^2 < Dt$) makes sense. Indeed, from Eq. **29** we see that

$$y \equiv R^2/(Dt) = \Gamma_0/(\Gamma_i - \Gamma_0).$$

Thus, whenever we have reached interesting values of Γ_i (much larger than Γ_0), we do expect $y \ll 1$.

Ultimately, at $\theta \gg \theta_0$, Eq. **32** reduces to a simple power law:

$$R^4 = \varepsilon Dt R_v^2. \quad [34]$$

This is not far from the observations of refs. 10 and 11.

Modification Due to a Phase Transition of the Binders. Eqs. **32–34** assumed an ideal gas behavior for the binders. But, in many cases, the binders attract each other as explained in ref. 2. The osmotic pressure rises linearly at small Γ, and then reaches a 2D gas/2D liquid coexistence plateau in an interval $\Gamma_- < \Gamma < \Gamma_+$. The lower end Γ_- is conditioned by the Bruinsma interactions (2). The upper end Γ_+ is due to the finite number of receptor sites, available on the support surface. For $\Gamma > \Gamma_+$, the osmotic pressure rises very high. Note incidentally that the equilibrium condition (1) is modified and becomes

$$\int_{\Gamma_0}^{\Gamma_i} \Gamma_i^{-1} d\Pi = U \equiv kT\ln nf. \quad [35]$$

But this modification will not play a major role in what follows. If we return to Eqs. **31** and **29** (with $k = 1$), we arrive at an implicit equation for $\Gamma_i(t)$:

$$Dt/R_v^2 = 2\Pi(\Gamma_i)(\Gamma_i - \Gamma_0)/\gamma_0\Gamma_0. \quad [36]$$

The general aspect of this relation is shown on Fig. 4. At relatively low concentrations ($\Gamma_i < \Gamma_-$), we essentially retain the perfect gas behavior. For $\Gamma_- < \Gamma_i < \Gamma_+$, the contact region will contain islands of the dense phase, with a well defined osmotic pressure (the plateau value Π_p). In this region, the concentration $\Gamma_i(t)$ increases linearly with time and rather fast. Ultimately, we reach Γ_+, and beyond this point, the growth is very slow.

These effects also show up in the contact angle $\theta(t)$. In the dilute regime ($\Gamma < \Gamma_-$), Eq. **32** still holds. When we enter the plateau region, the contact angle should be locked by Eq. **31** at a constant value. Ultimately, at $\Gamma > \Gamma_+$, we expect a very slow growth of $\theta(t)$ and $R(t)$.

Competition Between Tear Out and Diffusion

Here we start from an adhesive patch at equilibrium and (by some external means) we tend to *decrease* the contact area. As mentioned in the introduction, we can think of two scenarios: *tear out*, where some bridges are broken, and diffusion, where the binders migrate, but the number of bridges is constant. Clearly, the diffusion scenario is limited in time: if the patch becomes very small, Γ_i reaches a saturation value Γ_{max}, where all binders are side by side. Beyond this point, tear out must prevail. We assume here $\Gamma_i < \Gamma_{max}$.

We want to compare the horizontal forces corresponding to both scenarios: F for the diffusion mode, and F_E (where E stands for Evans) for the tear out process. We consider a contact line moving at a prescribed velocity V and first construct a simple estimate for $F_E(V)$ based on the model of ref. 3.

At a microscopic scale, we consider one couple binder/receptor and assume that this couple begins to be separated by a vertical distance z. In the simplest case, with a single barrier of activation energy B, we expect a rate equation of the form:

$$\frac{dz}{dt} = V_0 \exp\left[-\frac{1}{kT}(B - \varphi a)\right]. \qquad [37]$$

Here $dz/dt = V\,dz/dx$, where V is the line velocity and x defines the horizontal location of the binder, while φ is the pull out force on one binder, a is a molecular length, and V_0 is a typical thermal velocity (of order 10 m/sec). Eq. **37** may be rewritten in the form:

$$a\varphi = B + kT\left[\ln\left(\frac{V}{V_0}\right) + \ln\left(\frac{dz}{dx}\right)\right]. \qquad [38]$$

We can now construct the entropy loss due to the motion as an integral over all sites near the line that are partially detached. We call this $T\dot{S}$ (per unit length of line in the y direction):

$$T\dot{S} = \Gamma_i \int dx \varphi dz/dt = V\Gamma_i z_m a^{-1}\left[B + kT\ln\left(\frac{V}{V_0}\right) - \ell\right], \qquad [39]$$

where z_m is the overall distance required for separation (~1 nm), and

$$\ell = -z_m^{-1}\int dx \ln(dz/dx) dz/dx \qquad [40]$$

is a constant of order unity.

We now derive from Eq. **39** the horizontal friction force F_E:

$$F_E = \frac{\partial}{\partial V} T\dot{S} = \Gamma_i \frac{z_m}{a}\left[B + kT\ln\left(\frac{V}{V_0}\right) - (\ell - 1)kt\right]. \qquad [41]$$

(In what follows, for our rough estimates, we shall set $\ell - 1 = 0$).

We can now compare this Evans force to the lag force F derived in *Scaling Structure of the Lag Force*: for a duration t (or a frequency t^{-1}), we replace Eq. **2** by the simplified form:

$$F = KV(D/t)^{-1/2}. \qquad [42]$$

From Eqs. **41** and **42**, we get the ratio:

$$r \equiv \frac{F}{F_E} = \frac{V}{V_1}\frac{1}{\ln(V/V^*)}, \qquad [43]$$

with

$$V_1 = \left(\frac{D}{t}\right)^{1/2}\frac{\Gamma_i kT}{K} \qquad [44]$$

$$V^* = V_0 \exp(-B/kT). \qquad [45]$$

The plot of $r(V)$ shows a minimum at $V = eV^* = 2.7V^*$ and $r = 2.7V^*/V_1$.

(*i*) If $V_1 < 2.7V^*$, the ratio r is always larger than unity: the reaction process demands less force and dominates the separation.

(*ii*) If $V_1 > 2.7V^*$, there is an interval (around $3V^*$) where $r < 1$, and, in this interval, the lag force may be dominant.

Thus, the crucial parameter is

$$y = eV^*/V_1. \qquad [46]$$

Let us make a rough estimate of y, using Eqs. **44** and **45**, taking $\Gamma_i kT/K \cong 1$, and assuming that the diffusion constant D is controlled by the *same barrier* B, which opposes separation. Hopping inside the adhesion patch demands a separation binder/receptor:

$$D = aV_0 \exp(-B/kT), \qquad [47]$$

where a is a molecular diameter.

We choose $V_0 = 10$ m/sec, $t = 100$ sec, and $a = 1$ nm. Then,

$$y = (V_0 t/a)^{1/2}\exp(-B/2kT) \cong 10^6\exp(-B/2kT), \qquad [48]$$

and the lag force plays a role only if

$$B/kT \gtrsim 28. \qquad [49]$$

The conclusion is that for most practical separation experiments ($B/kT \sim 15$), tear out should dominate over diffusion.

Discussion

Our calculations of the patch growth in the diffusion regime are crude for a number of reasons. (*i*) We treated the outer region as a large reservoir of binders. But, in reality, the total amount of binders available in our vesicle is fixed, and the growth of the patch may stop trivially, because all binders have been used. (*ii*) We ignored the complexity of the contact line: on the outer side of the line, the angle θ shows up only after a certain distance $\lambda = (K_b/\gamma)^{1/2}$. All our discussion assumes $R > \lambda$. (*iii*) The diffusion constants may be very different in the unbound/bound regions. The bound binders must break out from their receptor site to be able to move, and the diffusion constant D_i, inside the adhesive patch, should thus be small. On the other hand (and especially for vesicles without any cytoskeleton), the diffusion D_0 in the unbound region may be much faster.

There is, however, a certain rule of the thumb: in the squeezing and rolling problems of previous sections, it is the internal diffusion D_i that controls the force lag, and we can put $D = D_i$. On the other hand, in the growth problem of the last section, what limits the growth is the external diffusion D_0 towards the patch, and we should put $D = D_0$.

Summary

We expect the lag forces to be important in certain (not all) squeezing or rolling experiments. But their observation is delicate: the simplest procedure may be to use a modulated squeezing and to monitor simultaneously the modulations of the radius R and of the contact angle θ. (θ is an independent variable in this case: Eq. **20** does not hold?) The angle θ gives us the force $\gamma(1 - \cos\theta)$, and we could end up with an experimental relation $F(V)$. However, we may face a complex situation where the lag force acts upon squeezing, while the tear out process dominates in the other half period.

We have benefited from the helpful advice of R. Bruinsma.

1. Bell, G., Dembo, M. & Bongrand, P. (1985) *Biophys. J.* **45,** 1051–1083.
2. Bruinsma, R. & Sackman, E. (2001) *C. R. Acad. Sci. (Paris)* **2,** 803–810.
3. Merkel, R., Nassoy, P., Leung, A., Ritchie, K. & Evans, E. (1999) *Nature (London)* **397,** 50–52.
4. Springer, T. A. (1994) *Cell* **76,** 301–320.
5. Ahagon, A. & Gent, A. (1975) *J. Polym. Sci. Phys. Ed.* **13,** 1285–1297.
6. Wu, S. (1982) *Polymer Interface and Adhesion* (Dekker, New York).
7. Brown, H. (1991) *Annu. Rev. Mat. Sci.* **21,** 463–489.
8. Evans, E., Berk, D. & Leung, A. (1991) *Biophys. J.* **59,** 838–850.
9. Abkarian, M., Lartigue, C. & Viallat, A. (2001) *Phys. Rev. E* **63,** 041906–041915.
10. Bernard, A. L., Guédeau, M. A., Jullien, L. & di Meglio, J. M. (2000) *Langmuir* **16,** 6809–6815.
11. Bernard, A. L., Guédeau, M. A., Jullien, L. & di Meglio, J. M. (1999) *Europhys. Lett.* **46,** 101–106.
12. Nardi, J., Bruinsma, R. & Sackmann, E. (1998) *Phys. Rev. E* **58,** 6340–6359.
13. Boulbich, A., Guttenberg, Z. & Sackmann, E. (2001) *Biophys. J.* **81,** 2743–2760.
14. de Gennes, P. G., Brochard, F. & Quéré, D. (2002) *Gouttes, Bulles, Perles et Ondes* (Collection Echelles, Belin, Paris).

PHYSICS

Part VI. Chirality

C. R. Acad. Sc. Paris, t. 270 (6 avril 1970). Série B — 891

PHYSIQUE MOLÉCULAIRE. — *Sur l'impossibilité de certaines synthèses asymétriques.* Note (*) de M. **Pierre-Gilles de Gennes**, présentée par M. Alfred Kastler.

On montre que, même en présence de deux champs (électrique $\vec{E}$ et magnétique $\vec{H}$) parallèles, un système racémique de réactifs chimiques ne peut pas évoluer vers un état d'équilibre final optiquement actif. Par contre, si l'état final est hors d'équilibre, une dissymétrie reste possible.

P. Curie a établi dans un article célèbre (1) qu'un système de deux champs $\vec{E}$ et $\vec{H}$ parallèles n'a pas la symétrie droite-gauche. Il a conclu qu'il est « peut-être possible » de synthétiser un matériau optiquement actif, à partir de constituants inactifs, en présence de tels champs. Dans la présente Note nous montrons qu'une telle synthèse asymétrique est en fait impossible, si l'état final envisagé est un état d'équilibre : la preuve repose sur une autre considération de symétrie (renversement du sens du temps).

Soit $\mathcal{H}_{H,E}$ l'hamiltonien du milieu réactif, en présence des champs $\vec{H}$ et $\vec{E}$. Nous supposerons que les interactions élémentaires entre particules du milieu conservent la parité et sont invariantes par renversement du temps ce qui est correct pour toutes les forces présentes à l'échelle atomique. L'équilibre final après réaction sera décrit par une matrice densité :

$$(1) \qquad \left\{ \begin{array}{l} \rho = Z_{H,E}^{-1}[\exp - \beta \mathcal{H}_{H,E}], \\ \text{Trace}\, \rho = 1. \end{array} \right.$$

Pour définir la chiralité éventuelle des produits de réaction, nous mesurerons une observable pseudoscalaire X : par exemple X pourrait être le produit mixte $(\vec{a} \wedge \vec{b}).\vec{c}$ des vecteurs reliant un carbone tétraédrique à trois de ses quatre ligands (différents) a, b, c. La moyenne de X est

$$(2) \qquad \langle X \rangle = \text{Trace}[\rho X].$$

Pour discuter sa valeur, nous introduirons deux opérations de symétrie :

(A) une réflection par rapport à un plan contenant H. Soit U l'opérateur unitaire associé à cette transformation. Puisque X est pseudoscalaire

$$(3) \qquad XU = -UX.$$

Par ailleurs, conformément aux remarques de Curie :

$$(4) \qquad \mathcal{H}_{H,E} U = U \mathcal{H}_{-H,E};$$

892 — Série B C. R. Acad. Sc. Paris, t. 270 (6 avril 1970).

(B) le renversement du temps, décrit par l'opérateur antiunitaire θ, l'observable X est inchangée par renversement du temps :

$$X\theta = \theta X. \tag{5}$$

Dans l'hamiltonien, le renversement du temps est équivalent à un renversement de H :

$$\mathcal{H}_{H,E}\,\theta = \theta\,\mathcal{H}_{-H,E}. \tag{6}$$

Il en résulte en particulier que

$$Z_{-H,E} = \mathrm{Trace}[\theta^{-1}\exp(-\beta\mathcal{H}_{HE})\,\theta] = Z_{H,E} = Z.$$

Dans ces conditions :

$$\begin{aligned} \langle UXU^{-1}\rangle &= Z^{-1}\,\mathrm{Trace}[\exp(-\beta\mathcal{H}_{H,E})\,UXU^{-1}] \\ & Z^{-1}\,\mathrm{Trace}[U^{-1}\exp(-\beta\mathcal{H}_{H,E})\,UX] \\ & Z^{-1}\,\mathrm{Trace}[\exp(-\beta\mathcal{H}_{-H,E})\,X] \\ & Z^{-1}\,\mathrm{Trace}[\theta^{-1}\exp(-\beta\mathcal{H}_{H,E})\,\theta X] \\ & Z^{-1}\,\mathrm{Trace}[\exp(-\beta\mathcal{H}_{H,E})\,\theta X\theta^{-1}] \\ & Z^{-1}\,\mathrm{Trace}[\exp(-\beta\mathcal{H}_{H,E})\,X] = \langle X\rangle. \end{aligned} \tag{7}$$

En comparant (7) et (3) on conclut que $\langle X\rangle = 0$: dans tout état d'équilibre en présence de deux champs $\vec{H}$ et $\vec{E}$, on doit trouver autant de molécules droites que de molécules gauches.

Ce théorème exclut certaines possibilités de synthèse asymétrique qui avaient été envisagées dans le passé ([2]). Il implique par exemple que les températures de cristallisation de deux inverses optiques A_+ et A_- sont strictement égales, même en présence de $\vec{H}$ et $\vec{E}$.

Mais il faut souligner que le théorème s'applique seulement si la situation finale correspond réellement à un équilibre thermodynamique. A titre de contre-exemple, considérons l'expérience suivante : un faisceau lumineux issu d'une source S traverse un matériau M doué de dichroïsme circulaire magnétique, puis rencontre une cuve de réaction C. Un champ H est appliqué sur M, parallèlement à la direction SC du faisceau. Du point de vue de la symétrie spatiale, les vecteurs $\vec{H}$ et $\vec{SC}$ jouent le même rôle que $\vec{H}$ et $\vec{E}$ dans le problème de Curie. A la sortie de M, on peut, dans des conditions favorables, obtenir une lumière entièrement polarisée, par exemple à droite σ_+. La cellule C contient initialement un mélange racémique de deux inverses optiques A_+ et A_-. Les absorptions de A_+ et A_- pour une lumière σ_+ sont différentes : si, par exemple, A_+ absorbe seul, on pourra éventuellement observer une photolyse :

$$A_+ + h\nu_+ \rightarrow B_+.$$

Après un certain temps d'irradiation, le contenu de la cuve perd la symétrie droite-gauche et devient optiquement actif : cet effet, suggéré par Van

C. R. Acad. Sc. Paris, t. 270 (6 avril 1970). Série B — 893

t'Hoff et Cotton, a été effectivement observé (2). Mais il ne viole pas le théorème ci-dessus, car il correspond à une situation finale (telle que $A_- + B_+$) qui est hors d'équilibre : soit le produit B est métastable (et après avoir coupé l'irradiation l'équilibre vrai correspond à l'état initial $A_+ + A_-$); soit les réactifs (A_+ et A_-) étaient métastables, et l'équilibre vrai correspond à ($B_+ + B_-$) (3).

Un autre cas non justiciable de notre théorème serait celui d'un mélange racémique conducteur : ici, à cause de l'effet Joule, l'état de régime permanent n'est plus un état d'équilibre, et il n'est pas impossible que, en présence de H et E parallèles, une dissymétrie optique apparaisse.

MM. C. Sadron et G. Spach ont attiré notre attention sur ces problèmes.

(*) Séance du 16 mars 1970.

(1) P. Curie, *J. Phys.*, 3e série, 3, 1894, p. 393.

(2) Pour une revue récente sur les synthèses asymétriques, *voir* A. Amariglio, H. Amariglio et X. Duval, *Ann. Chim.*, 3, 1968, p. 5.

(3) Une discussion détaillée de ces régimes cinétiques a été faite par Kuhn : *voir* W. Kuhn, E. Knopf, *Natürwissenchaften* 18, 1930, p. 183, et références incluses. Nous sommes redevables au prof. J. P. Mathieu pour avoir souligné la concordance entre le travail de Kuhn et la présente note.

(*Laboratoire de Physique des Solides,*
Bât. 510, *Faculté des Sciences,*
91-*Orsay, Essonne.*)

Pierre Curie et le rôle de la symétrie dans les lois physiques

P. G. de Gennes

Ecole Supérieure de Physique et de Chimie, 10, rue de Vauquelin, 75005 Paris, France.

Pierre et Jacques Curie découvraient il y a cent ans le phénomène de *piézoélectricité* : comment un cristal, de suffisamment basse symétrie, développe une polarisation lorsqu'il est soumis à une contrainte mécanique. Pierre Curie n'a alors que 21 ans ! Et pourtant ses notes lapidaires aux Comptes Rendus datées de 1880 et 1881 [1] portent en elles tout l'essentiel du phénomène : charges superficielles proportionnelles à la pression, indépendantes de l'épaisseur du cristal, etc. Lorsque quelques mois plus tard, Lippmann prédit l'effet inverse (déformation du cristal sous champ électrique), les frères Curie imaginent tout de suite un dispositif admirablement simple pour détecter ces très faibles déformations. Ils couplent mécaniquement le cristal étudié avec un deuxième cristal piézoélectrique qui joue le rôle d'un détecteur ; ce dernier transforme le signal mécanique du premier cristal en un signal électrique mesuré à l'électromètre ! Les résultats confirment brillamment la prédiction de Lippmann. Et pourtant, les moyens engagés sont très modestes — les champs électriques, par exemple, sont calibrés par la longueur d'une décharge dans l'air...

Très peu de temps après, Pierre Curie est nommé chef de travaux à l'Ecole de Physique et Chimie. Cette école vient d'être créée par le Conseil de Paris ; elle est installée dans des locaux de fortune. Mais, remarquablement animée par Schutzenberger, elle met en place un enseignement pragmatique, où le travail de laboratoire joue un rôle central. Son jeune chef de travaux en physique (il n'a alors que 25 ans) va mener de front trois actions remarquables : enseignement, recherche instrumentale, et recherche théorique.

Pour apprécier le rôle de Curie enseignant, il n'est que de lire le témoignage émouvant de Paul Langevin [2]. En ce qui concerne l'instrumentation, Pierre Curie réalise à cette époque des progrès très importants. D'une part, avec un cristal piézoélectrique, il est à même d'engendrer des quantités de charge bien contrôlées, et donc d'avoir des électromètres étalonnés d'une précision remarquable. Ce point peut paraître mineur : mais il s'avèrera essentiel 15 ans plus tard, pour l'étude quantitative de la radioactivité α. D'autre part, Curie met au point une balance sensible, avec amortissement optimum, qui va permettre des pesées rapides : ici encore, sous un aspect modeste, il s'agit d'un instrument important : c'est avec cette balance que Curie établira ultérieurement les lois fondamentales du paramagnétisme et du ferromagnétisme.

Il est impressionnant de voir que, pendant ces mêmes années où il construit des appareils si bien adaptés au futur, Pierre Curie mène aussi une recherche théorique profonde. Sa bonne éducation cristallographique lui a fait comprendre très tôt l'importance des considérations de symétrie. Mais, dès les premières expériences sur la piézoélectricité, il est autre chose qu'un cristallographe, il fait agir des champs électriques ou (plus tard) magnétiques sur des cristaux. Tout ceci le conduit à une réflexion globale, décrite dans l'article de 1894, sur la *symétrie dans les phénomènes physiques* [1].

Il n'est pas exagéré de dire que cet article est l'une des contributions scientifiques les plus profondes du XIXe siècle. D'une part Curie y dégage les propriétés de symétrie associée à des *champs* (vecteurs, pseudovecteurs, scalaires, pseudoscalaires, ...). D'autre part, il y énonce un principe essentiel, reliant la symétrie des « effets » à la symétrie des « causes ». Ce qu'on entend ici par « cause » et par « effet » ne prête pas à de grands discours philosophiques, mais se sent bien sur des exemples concrets : par exemple si l'on applique un champ électrique **E** et un champ magnétique **H** croisés sur un métal liquide (conducteur isotrope) on peut observer un certain courant **J** qui n'est pas parallèle à **E**. Dans cet « effet Hall » on peut définir **E** et **H** comme les causes, et **J** comme l'effet. Curie observe que le seul élément de symétrie des causes est le plan perpendiculaire à **H** ; donc, si le courant **J** est fixé de façon *unique* par la donnée de **H** et **E** il doit être dans ce plan.

Un autre exemple, sur lequel je voudrais m'étendre plus longtemps, est celui des *synthèses asymétriques* [3]. On savait à l'époque, grâce à Pasteur, que certaines molécules ne sont pas superposables à leur image dans un miroir. De telles molécules « chirales » existent sous deux formes : droite et gauche. Les forces mises en jeu dans les réactions chimiques usuelles ne distinguent pas la droite et la gauche. Donc, si l'on fait la synthèse complète d'un stérol, par exemple, on trouve à la fin des opérations 50 % de stérol droit et 50 % de stérol gauche. Pour la pharmacologie, c'est en général une seule de ces deux espèces qui est utile. D'où l'intérêt des synthèses « asymétriques », où l'on essaye de privilégier l'une des espèces au moyen d'agents extérieurs qui distinguent la droite de la gauche. Pasteur avait déjà, dans ce but, appliqué un champ magnétique **H** — sans succès. Curie montre dans l'article de 1894 qu'un champ **H** ne peut pas, à lui seul, induire une dissymétrie droite gauche (puisqu'un tel champ admet un plan de symétrie perpendiculaire à son axe). Mais il observe en outre que si deux champs **E** et **H** colinéaires sont appliqués au bain de réaction, « il n'est pas impossible » qu'une certaine dissymétrie soit engendrée. On devine évidemment que cette dissymétrie sera faible, puisque les couplages entre une molécule et des champs **E** ou **H** réalisés au laboratoire sont en général très faibles devant les énergies mises en jeu par une réaction chimique (ou même par l'agitation thermique).

La question est donc restée en sommeil pendant très longtemps. Il y a une dizaine d'année, stimulé par des réflexions très intéressantes de G. Spach [4], j'ai essayé de situer les effets possibles. Une première conclusion, pessimiste, est liée à l'existence d'une autre symétrie (peu explorée au temps de Curie) : la symétrie par renversement du temps. Renverser le sens de t revient à changer le signe des courants, donc celui des champs **H** qu'ils engendrent. Le champ **E** par contre n'est pas affecté. On tire de cette remarque les conclusions suivantes [5, 6] :

a) si on fait, à partir d'un système initial achiral, une réaction chimique (conduisant à des produits droits ou gauches) sous deux champs **E** et **H** parallèles, après achèvement de la réaction, on doit garder encore un mélange racémique : pas d'effet à l'équilibre ;

b) par contre, les cinétiques peuvent différer : la vitesse de la réaction « droite » et celle de la réaction « gauche » ne sont pas exactement identiques. Mais si l'on écrit les constantes de vitesse sous la forme :

$$k = \tilde{k} \exp(-E/kT) \tag{1}$$

les énergies d'activation E sont les mêmes pour les deux réactions. C'est seulement le préfacteur $\tilde{k}$ qui est différent $\tilde{k}_D \neq \tilde{k}_G$.

Il reste à estimer l'ordre de grandeur de la dissymétrie sur $\tilde{k}$ que nous pouvons définir *via*

$$\varepsilon = \frac{\tilde{k}_D - \tilde{k}_G}{\tilde{k}_D + \tilde{k}_G}\,. \tag{2}$$

Une conjecture naturelle, qui a été citée dans la littérature [7], revient à comparer les énergies d'un dipôle électrique p ($\to pE$) et d'un dipôle magnétique μ ($\to \mu H$) à l'énergie thermique kT, soit :

$$\varepsilon \sim \frac{pE}{kT}\,\frac{\mu H}{kT} \qquad \left(\frac{pE}{kT} < 1, \frac{\mu H}{kT} < 1\right). \tag{3}$$

Les deux moments seraient les moments de transition associés à la réaction : en particulier μ serait comparable au magnéton de Bohr électronique μ_e si l'une des étapes impliquait une sélection dans un sous-niveau Zeeman orbital. Pour les réactifs usuels (diamagnétiques), je pense que la situation est un peu plus complexe, et que, en général, ce n'est pas le magnéton électronique qui intervient, mais plutôt le magnéton *nucléaire* μ_N qui est bien plus petit.

On peut essayer de comprendre qualitativement le phénomène, sur un exemple schématique (Fig. 1). On considère ici une macromolécule en bâtonnet, supposée porteuse d'un dipôle permanent p grand, et donc complètement alignée dans le champ E. Cette macromolécule a une structure en hélice soit droite, soit gauche, selon l'empilement des monomères successifs. Un monomère supplémentaire ($n+1$) a rejoint l'extrémité en croissance. Nous supposons qu'il tourne autour de l'axe de la molécule, et qu'il s'arrête dès qu'il atteint l'une des deux positions (α) ou (β) en contact avec le monomère précédent (n). Ces deux positions correspondent respectivement à une portion d'hélice droite ou gauche. Supposons que le monomère porte une charge élémentaire $-e$ sur l'axe de rotation et une charge $+e$ à l'extérieur ; la rotation implique alors un groupe tournant, de charge e, avec une certaine masse M. Dans le champ H un tel groupe a une fréquence de Larmor

$$\Omega_L = \frac{eH}{Mc}\,. \tag{4}$$

A cette fréquence Ω_L se superpose une rotation d'agitation thermique, dont les fréquences sont de l'ordre de Ω_t, où

$$\tfrac{1}{2} I\Omega_t^2 = \tfrac{1}{2} kT \tag{5}$$

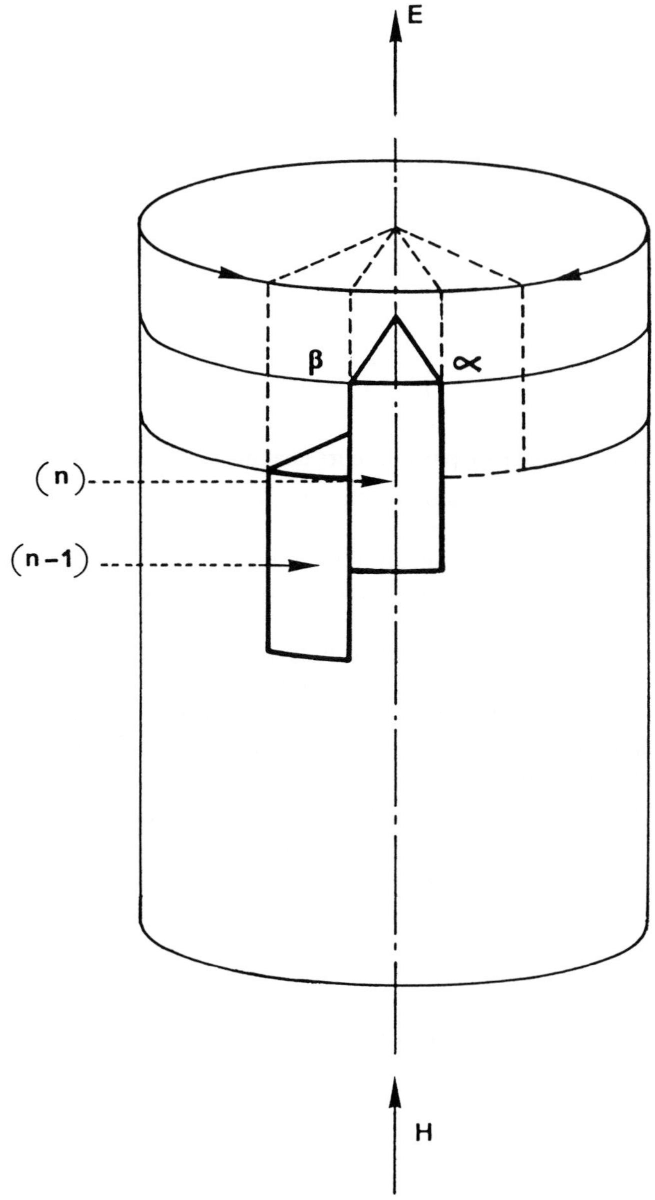

Fig. 1. — Modèle mécaniste schématique pour l'attachement d'un monomère $(n + 1)$ sur une macromolécule rigide alignée par un fort champ électrique vertical E. On suppose que le monomère $(n + 1)$ se fixe d'abord à la cote convenable, puis qu'il tourne dans le plan horizontal pour s'arrêter au contact du monomère précédent (n) en position (α) ou (β). L'existence d'une précession de Larmor autour du champ magnétique H (parallèle à E) introduit une dissymétrie entre (α) et (β).

(I = moment d'inertie du groupe tournant). La distribution des fréquences de rotation Ω est légèrement excentrée

$$p(\Omega) = (2\pi)^{-1/2}\,\Omega_t^{-1}\exp - \left[(\Omega - \Omega_L)^2/(2\,\Omega_t)^2\right]. \tag{6}$$

Dans le présent schéma, nous négligeons les collisions entre le monomère tournant et les molécules de solvant : alors tous les $\Omega > 0$ donnent une fixation en (α) et tous les $\Omega < 0$ donnent (β). Le taux de chiralité $\varepsilon = p_\alpha - p_\beta$ du produit obtenu est déduit de

$$p_\alpha = \int_0^\infty \mathrm{d}\Omega\, p(\Omega) = 1 - p_\beta \tag{7}$$

$$\varepsilon = p_\alpha - p_\beta = \left(\frac{2}{\pi}\right)^{1/2}\frac{\Omega_L}{\Omega_t}. \tag{8}$$

Ce résultat diffère de l'éq. (3) par plusieurs facteurs :

a) un facteur trivial : le champ électrique s'est éliminé parce que nous avons supposé un fort alignement ($pE > kT$) ;

b) un facteur numérique mineur : multipliant haut et bas par $\hbar$, le dénominateur de (8) est proportionnel à $\hbar\Omega_t$ plutôt que kT. A partir de (5) on peut évaluer la différence, qui n'est pas dramatique ;

c) un facteur essentiel : au numérateur nous avons $\hbar\Omega_L$, qui d'après (4) fait intervenir la fréquence de Larmor pour un groupe lourd. Dans un cas très optimiste, où la masse M se réduit à celle d'un proton, $\hbar\Omega_L$ correspond à un magnéton nucléaire. En pratique, dans un cas de ce genre, avec $H = 10^5$ G on attend $\varepsilon \sim 10^{-4}$ ce qui est trop faible pour être détecté de nos jours.

Il y a bien d'autres exemples possibles, dans lesquels on s'efforce de créer, ou de sélectionner une espèce chirale par l'application de champs macroscopiques. Aussi, si on dispose d'un mélange égal de cristaux d'une espèce droite D et de cristaux d'une espèce gauche G, on peut espérer les séparer en les plaçant dans un *fluide tournant* : idée qui remonte à Pasteur [8], mais qui a été reprise récemment [9]. J'en décrirai une variante, qui est instructive (Fig. 2). Un cylindre horizontal contient une suspension de cristaux D ou G, et tourne autour de son axe à une vitesse Ω. On s'arrange pour que les cristaux, eux, ne tournent pas avec le fluide : par exemple, s'ils sont allongés, et de constante diélectrique plus élevée que le solvant, on les place dans un champ électrique vertical, qui les alignera. Alors les cristaux D doivent se « visser dans le fluide », donc avancer vers une extrémité du cylindre avec une vitesse v, et les cristaux G doivent aller vers l'autre extrémité avec la vitesse $-v$. Du point de vue de Curie, les « causes » ont un plan de symétrie (vertical, normal au cylindre). Les effets aussi : les molécules D qui partent à droite sont l'image exacte dans le plan de symétrie des molécules G qui partent à gauche.

Du point de vue pratique on prévoit :

$$V = \delta R \Omega \tag{9}$$

où δ est un « coefficient d'asymétrie » sans dimension décrivant le caractère G ou D de la surface extérieure du cristal. R est la dimension du cristal. Avec $R = 1$ mm,

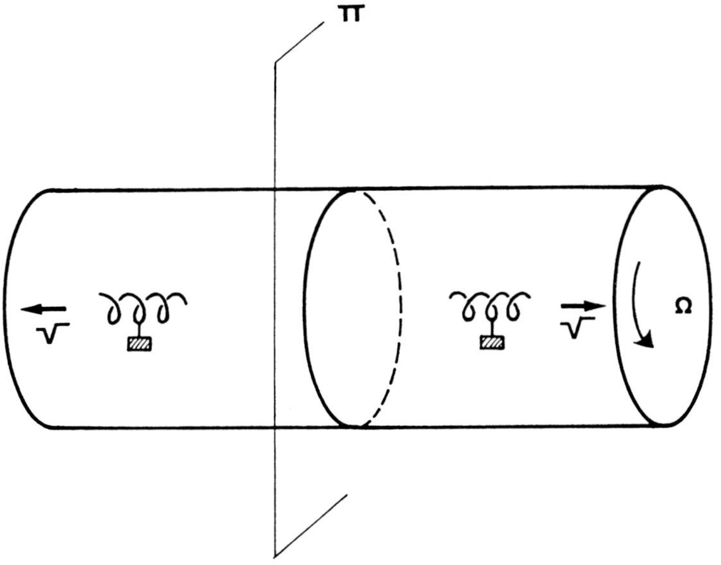

FIG. 2. — Dédoublement mécanique de cristaux racémiques. Ici, au lieu de cristaux, on montre le système modèle de A. Martinet [10] : des hélices lestées droites et gauches, placées dans un fluide en rotation. Le plan (Π) est un plan de symétrie des « causes ». Les effets correspondent à des hélices droites allant dans un sens, et des hélices gauches dans l'autre : (Π) reste un plan miroir des effets.

$\delta = 10^{-3}$ et $\Omega = 10\ \mathrm{s}^{-1}$ on atteindrait des vitesses v de 0,5 mm/min. L'expérience n'a pas encore été tentée sur des cristaux, mais elle a été faite par A. Martinet [10] sur des objets modèles (solides macroscopiques en forme d'hélice, lestés par une « quille »). Ici l'objet ne tourne pas avec le fluide à cause du couple de pesanteur sur la quille. L'expérience est délicate car les hélices prennent facilement des orientations erratiques : mais en éliminant ces complications, on arrive aux résultats (préliminaires) de la figure 3, qui montrent des vitesses v bien proportionnelles à Ω.

Terminons ces remarques sur le principe de Curie par une observation générale : quand on a défini des « causes », et que l'on veut étudier un « effet », il est essentiel de savoir si l'effet est uniquement déterminé par les causes. La figure 4 nous montre un contre-exemple classique : le *flambage d'une poutre*, soumise à des forces opposées aux extrémités. Les « causes » (les forces) ont ici un axe de symétrie (l'axe initial de la poutre). Donc, si la conformation finale de la poutre (effet) est unique, la poutre doit rester colinéaire à l'axe. Mais en réalité, au-dessus d'un certain seuil en force, la poutre « flambe » : elle prend la forme d'un arc, et choisit pour ce faire un certain plan de flexion repéré par un angle φ. Il y a une infinité d'états finaux différant par le choix de φ. Le principe de Curie nous dit ici que l'*ensemble* des états finaux a même symétrie que les causes : mais chacun de ces états a une symétrie plus basse.

Le flambage est un cas de *symétrie brisée* : notion qui a pris une importance extraordinairement grande dans toute la physique contemporaine, et que nous verrons constamment apparaître dans ce colloque.

Par son travail après 1894, P. Curie a été sans doute le premier expérimentateur à analyser quantitativement un phénomène de symétrie brisée au niveau des atomes :

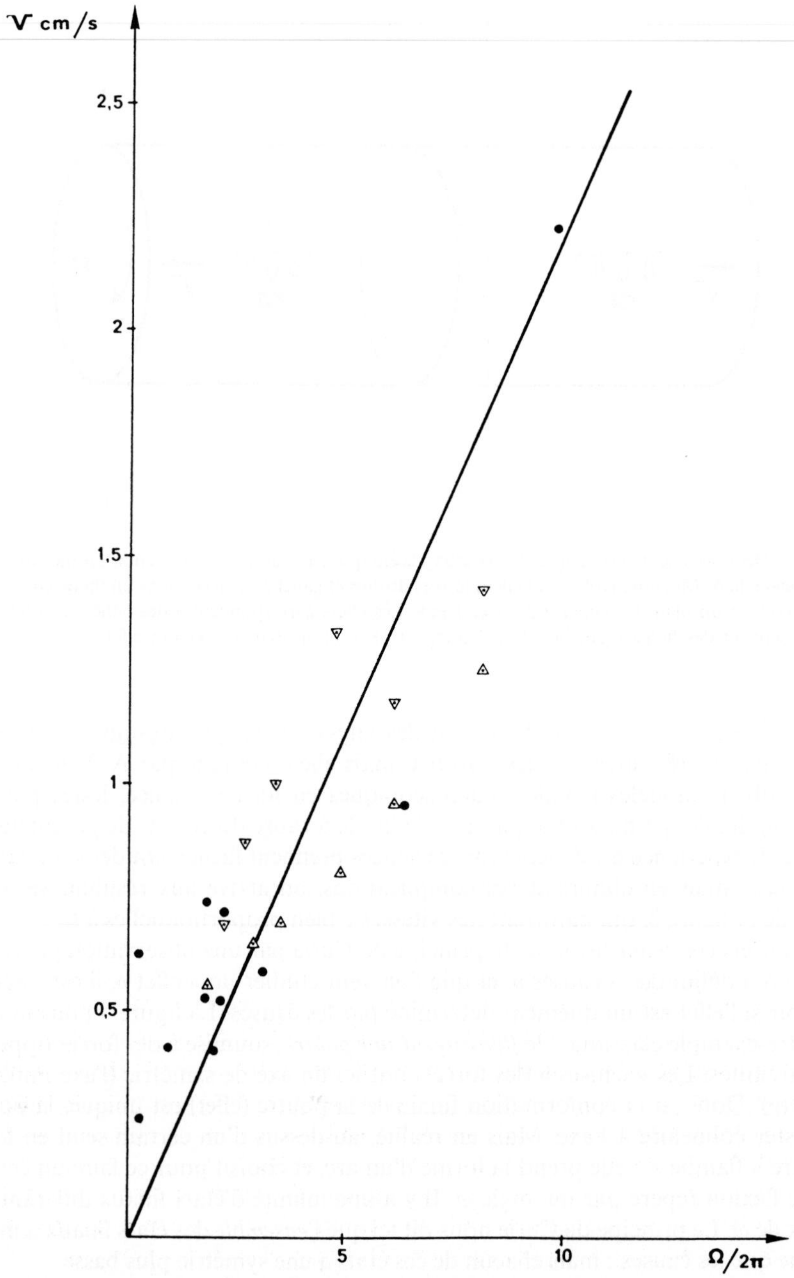

FIG. 3. — Mesures de A. Martinet sur la situation de la figure 2 avec des hélices lestées macroscopiques (matériau : bois. Pas : 45 mm). Le « rendement » qV/Ω (où q est le vecteur d'onde des hélices) est de l'ordre de 10 %. Les deux types de points correspondent aux deux sens de rotation.

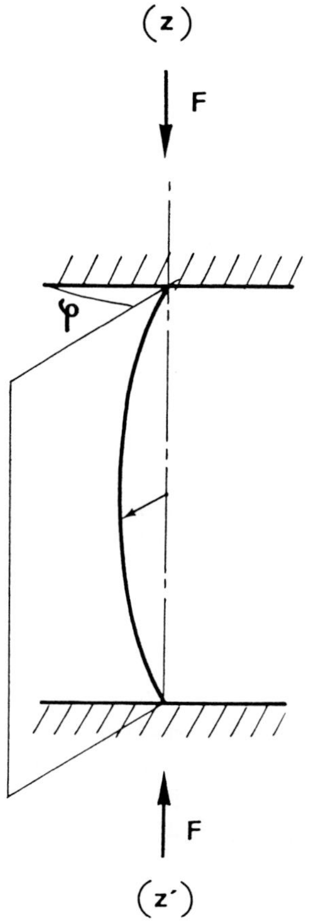

FIG. 4. — Flambage d'une poutre : l'axe ZZ' est un axe de rotation C_∞ des causes. Mais, en dessus du seuil de flambage, il y a une infinité d'états finaux (effets) correspondant à des choix différents de l'angle φ.

quand il a établi en détail les lois de l'aimantation spontanée du fer en dessous de la température critique (que nous appelons maintenant, le point de Curie T_c) : dans un cylindre de fer monodomaine, pour des températures $T < T_c$ il y a, en champ magnétique nul, *deux* états possibles d'aimantations opposées et non pas seulement un état d'aimantation nulle. Une grande partie du bagage culturel actuel sur les symétries brisées s'est formé à partir de cet exemple magnétique.

En 1894, donc, Curie a commencé ses études sur le magnétisme. A ce moment il vient d'être nommé professeur dans notre Ecole. Ceci lui donne des moyens de travail un peu plus développés, mais encore bien modestes (Voir la préface de [1]). C'est aussi en 1894 qu'il accueille une étudiante polonaise, qui doit travailler sur le magnétisme des alliages de fer. En 1895 il épouse cette étudiante, qui devient Marie Curie. En 1898 tous deux abordent l'étude de la radioactivité, En 1904 le prix Nobel leur est décerné. En 1906 Pierre Curie meurt par accident. Il n'a que quarante-sept ans.

De toute cette œuvre qu'il nous laisse, nous ne voulons extraire ici que deux thèmes particuliers : piézoélectricité et symétrie. Sur chacun de ces deux thèmes, le développement ultérieur a été considérable. La piézoélectricité, par exemple, a permis la transformation de signaux électriques en signaux acoustiques : le *sonar* est né, grâce à l'action énergique de Paul Langevin, dans les murs de cette maison. De nos jours encore, on y travaille à des dispositifs acousto-électriques sophistiqués, qui permettront peut-être de simplifier le dialogue entre un ordinateur et son utilisateur. Et il y aurait beaucoup d'autres exemples à citer ! Mais si nous prenons l'autre thème, celui des symétries, et de leur rôle dans les lois physiques, l'explosion des connaissances durant le dernier demi-siècle y est peut-être encore plus impressionnante. C'est de ces deux remarques qu'est née l'idée de ce colloque, qui célèbre le centenaire du premier montage piézoélectrique — idée due à E. Dieulesaint, et développée par une collaboration remarquable entre lui, N. Boccara et J. Lewiner. Je suis particulièrement heureux de pouvoir les remercier ici ; de remercier aussi la Ville de Paris — notre autorité de tutelle — et l'industrie française, qui les ont généreusement aidés.

Du point de vue de notre Ecole, le style de travail de Pierre Curie, avec sa combinaison expérience/théorie et aussi avec cette sélection de problèmes fondamentaux qui seront plus tard les pilotes d'applications extrêmement vastes — reste, après cent ans, le modèle que nous voulons suivre.

Remerciements

G. Spach m'a fait connaître jadis l'article de Curie, et toute la réflexion sur les synthèses asymétriques qui est décrite ici lui doit beaucoup. Je remercie également M. Kagan, J. Jacques, J. P. Mathieu, W. Rhodes, R. C. Dougherty pour des échanges extrêmement utiles dans ce domaine, et A. Martinet pour m'avoir permis de décrire ses expériences de 1977.

References

[1] P. Curie. *Oeuvres* (Gauthier-Villars, Paris) 1908.
[2] P. Langevin, Amicale de la Société des Anciens Ecoles de l'E.S.P.C.I. (1904).
[3] Voir par exemple la revue de A. Amariglio, *Annales de Chimie* **3** (1968) 5.
[4] G. Spach, communication privée.
[5] P. G. de Gennes, *C. R. Acad. Sci.* **B 270** (1970) 891.
[6] C. Mead, A. Moscowitz, H. Wynberg, F. Meuwese, *Tetrahedron lett.* (1973) 1063.
[7] W. Rhodes, R. C. Dougherty, *J. Am. Chem. Soc.* **100** (1978) 6247.
[8] Voir note (5) de la réf. [7].
[9] D. Howard, E. Lightfoot, J. Hirschfelder, *AI. Chem. Eng. J.* **22** (1976) 794.
[10] A. Martinet, expériences non publiées (1977).

C. R. Acad. Sci. Paris, t. 307, Série II, p. 233-237, 1988 233

Chimie physique/*Physical Chemistry*

Discrimination chirale dans une monocouche de Langmuir

David ANDELMAN et Pierre-Gilles de GENNES

Résumé — On envisage des tensioactifs « tripodes », qui posent à la surface de l'eau, trois groupes fonctionnels (i=A, B, C) liés à un carbone asymétrique, dont la 4e valence porte une chaîne aliphatique. Deux molécules au contact sont supposées former *deux* liaisons intermoléculaires (ij et $i'j'$). A chacune de ces liaisons (ij) est associée une énergie V_{ij} qui est négative (attractive) si (i) et (j) ont tendance à s'associer.

On forme par comptage les fonctions de partition Z_{++}, Z_{+-} pour une *paire* de molécules de même chiralité (Z_{++}) ou de chiralité opposée (Z_{+-}). Si $\Delta \equiv Z_{++} - Z_{+-} > 0$, on aura tendance à la ségrégation chirale dans une monocouche dense (cas homochiral HOC). Si $\Delta < 0$, on n'attend pas de ségrégation (cas hétérochiral HEC). Ce modèle conduit à certaines règles pratiques :

(α) si les seules liaisons possibles sont entre groupes identiques ($V_{ij} \rightarrow +\infty$ pour $i \neq j$) → HEC;

(β) si toutes les interactions sont du type dispersion (avec $V_{ij} = -M\,\alpha_i \alpha_j$, où α_i est la polarisabilité du groupe i) → HEC;

(γ) si A est un groupe apolaire, B un groupe chargé (+), et C un groupe chargé (−) → HOC. On a la même prédiction si les charges sont remplacées par des dipôles verticaux;

(δ) si A est un groupe aliphatique, B un groupe aromatique, et C un groupe chargé → HEC;

(ε) si l'une des liaisons entre groupes identiques (par ex. AA) est beaucoup plus forte que les autres (V_{AA} très négatif) → HEC;

(φ) si un des groupes (A) est « passif » (V_{Aj} indépendant de j) et si V_{BC} est fortement attractif → HOC.

Chiral discrimination in a Langmuir monolayer

Abstract — *We consider detergents with a "tripod" shape, containing three functional groups (i=A, B, C) bound to an asymmetric carbon atom, which lie at the water surface. The fourth valence of the carbon carries an aliphatic chain. Two neighboring detergent molecules are assumed to associate via* two *intermolecular bonds (ij and $i'j'$). To each of these bonds is associated an energy V_{ij} which is negative (attractive) if i and j do tend to associate.*

By direct counting, we write down the partition functions Z_{++}, Z_{+-} for a pair *of molecules with the same chirality (Z_{++}) or with opposite chiralities (Z_{+-}). If $\Delta \equiv Z_{++} - Z_{+-} > 0$, we expect a homochiral case (HOC) leading to segregation in dense phases. If $\Delta < 0$, we have the reverse, heterochiral case (HEC).*

This model leads to certain practical rules:

(α) *if the only allowed bonds are between indentical groups ($V_{ij} \rightarrow +\infty$ if $i \neq j$) → HEC;*

(β) *if all interactions are of the Van der Waals type (with $V_{ij} = -M\,\alpha_i \alpha_j$, where α_i is the polarisability of group i) → HEC;*

(γ) *if A is apolar, B is charged (+) and C is also charged but of opposite sign (−) → HOC. A similar conclusion holds if we replace the charges by dipoles normal to the water surface;*

(δ) *if A = aliphatic, B = aromatic, C = charged → HEC;*

(ε) *if one of the bonds between identical groups (e. g. AA) is much stronger than the others → HEC;*

(φ) *if one of the groups (A) is "passive" (V_{Aj} independent of j) and if V_{BC} is strongly attractive → HOC.*

I. PRINCIPES. — Si deux antipodes L et D sont peu compatibles en phase dense, on peut espérer les séparer par dédoublement spontané à partir d'un mélange racémique [1]. Nous sommes loin de pouvoir prédire ces situations « homochirales » à partir de la structure moléculaire [2]. Nous nous restreignons ici à des phases *bidimensionnelles* (monocouches de Langmuir) qui ont deux avantages importants : (*a*) les arrangements moléculaires sont plus simples à classer; (*b*) la pression de surface Π peut être variée dans un intervalle large. Les équations d'état (pression Π/surface A) d'un antipode et du mélange racémique correspondant ont été étudiées sur de l'eau ultrapure par l'école d'Arnett [3] : par exemple, avec un stéaramide chiral (sur une solution sulfurique), le racémique est moins compressible qu'un énantiomère pur, ce qui décrit une tendance homochirale (HOC). Un résultat de même signe a été vu par Bouloussa et Dupeyrat [4]

Note présentée par Pierre-Gilles de GENNES.

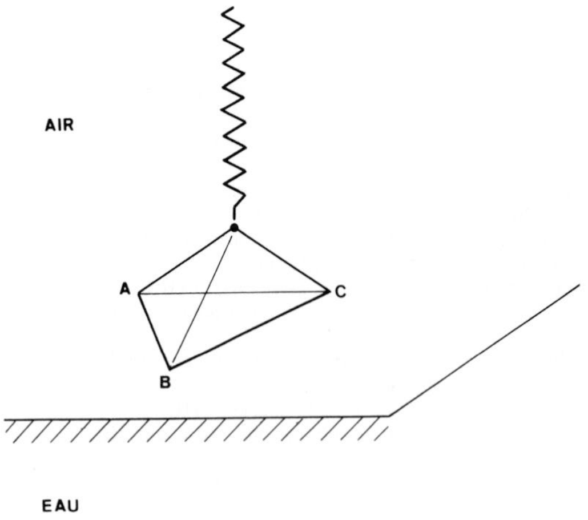

Fig. 1

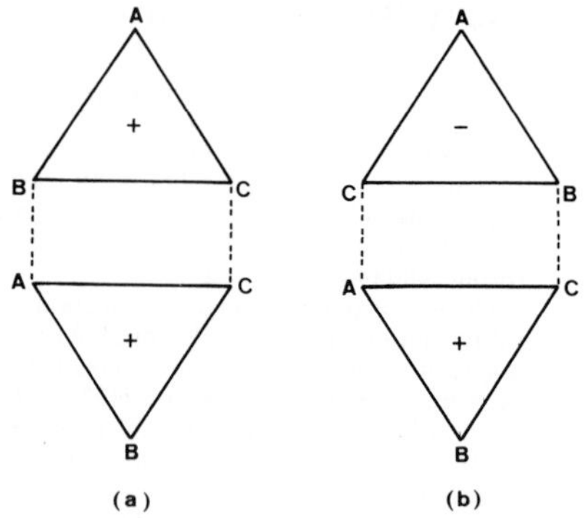

Fig. 2

Fig. 1. — Une molécule « tripode » à l'interface eau air. Pour l'énantiomère représenté (+), le cycle ABC est parcouru dans le sens trigonométrique pour un observateur regardant vers le bas.

Fig. 1. — A "tripod" molecule at the water air interface. For the enantiomer (+) which is shown, the cycle ABC, as seen from above, is counterclockwise.

Fig. 2. — Exemples d'associations entre tripodes permises par le modèle, pour deux molécules : (*a*) de même chiralité; (*b*) de chiralité opposée. Les interactions prises en compte sont définies par des traits pointillés.

Fig. 2. — Examples of associations between two tripods which are allowed in the model: (a) homochiral; (b) heterochiral. The interactions which are taken into account are marked by dotted lines.

sur une miristylalanine. Par contre, sur le système hexadécanol-thiophosphate 2 glycinol (qui comporte deux atomes chiraux) étudié récemment par Guedeau et Dvolaitsky [5], le signe est inverse (hétérochiral HEC). Enfin, dans certains cas, comme celui de la dipalmitoyl phosphatidyl choline [6], le carbone chiral est masqué et les isothermes de Langmuir sont les mêmes pour l'énantiomère et le racémique.

Notre objectif est de prédire la tendance (HOC ou HEC) à partir de considérations chimiques simples, pour une molécule « tripode » comportant un carbone chiral, lié à trois groupes fonctionnels ABC qui reposeront sur la surface de l'eau, et à une longue chaîne aliphatique. A, B, C, peuvent être constitués par exemple par un méthyle, un chlore, une fonction alcool ou amine, etc. Il y a deux énantiomères (+) et (−), représentés dans notre modèle par des triangles équilatéraux d'orientations opposées (*fig.* 1). L'existence de triades orientées ABC a déjà été utilisée dès 1961 par Amaya [7] pour décrire certaines associations chirales à trois dimensions.

Nous formulons ensuite deux hypothèses restrictives :

(*a*) nous discutons seulement l'association de *deux* molécules : si les paires (+ +) sont favorisées vis-à-vis des paires (+ −), la tendance est homochirale. Ceci est analogue en esprit à une équation d'état discutée à partir du 2e viriel. (Dans un article séparé [8], nous analyserons l'équation d'état de la monocouche dense et les transitions fluide solide);

C. R. Acad. Sci. Paris, t. 307, Série II, p. 233-237, 1988

(*b*) nous postulons que l'association en paires se fait par établissement de *deux*[1] liaisons entre groupes (*fig.* 2). (Les situations à une seule liaison ne contribueraient pas de toute façon aux effets chiraux.) Ces « liaisons » intermoléculaires peuvent être de type Van der Waals, hydrogène, interactions coulombiennes entre charges ou dipole dipole, etc.

II. MÉTHODE. — La liaison entre groupe (i) et groupe (j) est caractérisée par une énergie V_{ij}. Au total il y a donc dans notre modèle six constantes d'interaction V_{AA}, V_{AB}... Connaissant les V_{ij}, on calcule facilement la fonction de partition des paires. Par exemple, sur la figure 2 *a*, on lit une contribution à la fonction Z_{++} relative à une paire homochirale

(1) $$Z_{++}=\exp[-(V_{AB}+V_{CC})/kT]+\ldots$$

(T = température, k = constante de Boltzmann). On posera

(2) $$f_{ij}=\exp[-V_{ij}/kT]$$

Pour discuter le poids relatif des paires homo et héterochirales, on peut former le rapport Z_{++}/Z_{+-} (qui est la quantité utile pour une formulation thermodynamique détaillée) ou la différence

(3) $$\Delta=Z_{++}-Z_{+-}$$

qui se prête mieux à une discussion simple ($\Delta>0\rightarrow$ HOC).

Le comptage explicite donne

(4) $$\Delta=-[f_{AA}f_{BB}+f_{BB}f_{CC}+f_{CC}f_{AA}]+f_{AB}^2+f_{BC}^2+f_{CA}^2 -2[f_{AB}f_{BC}+f_{BC}f_{CA}+f_{CA}f_{AB}]+2[f_{AA}f_{BC}+f_{BB}f_{CA}+f_{CC}f_{AB}].$$

Nous allons voir que cette expression d'aspect austère se simplifie pour beaucoup de cas pratiques.

III. APPLICATIONS. — 1. Dans la limite de hautes températures ($V_{ij}/kT \ll 1$) $\Delta=0$ au 1er ordre en V. Pour des molécules en rotation libre, on sait que les forces à deux corps n'induisent pas d'effets chiraux [9].

2. Si les seules liaisons permises sont entre groupes identiques ($f_{ij}=0$ pour $i\neq j$) $\Delta<0$ ($\rightarrow$ HEC).

3. Si deux groupes identiques ne peuvent jamais venir en contact ($f_{ii}=0$), il reste trois paramètres $f_{AB}=x$; $f_{BC}=y$; $f_{CA}=z$ et

(5) $$\Delta=x^2+y^2+z^2-2(xy+yz+zx).$$

Dans le 1er quadrant (x, y, $z>0$), les régions HEC sont à l'intérieur d'un cône de révolution centré sur l'axe 111 et tangent aux plans $x=0$, $y=0$, $z=0$. L'extérieur est HOC.

4. Si un groupe (A) est « passif » ($f_{Ai}=w$ indépendant de i)

(6) $$\Delta=(f_{BC}-w)^2-(f_{BB}-w)(f_{CC}-w).$$

En particulier si l'interaction BC est fortement attractive (f_{BC} grand), on a $\Delta>0$ (HOC).

5. Considérons le cas d'un groupe A peu sélectif ($f_{AB}=f_{AC}=v$) et de groupes B et C chargés électriquement. Si les répulsions coulombennes au contact sont fortes, $f_{BB}=f_{CC}=0$, alors que $f_{BC}=w\gg 1$.

Alors, en posant $f_{AA}=u$, on trouve

(7) $$\Delta=w^2+2w(u-2v).$$

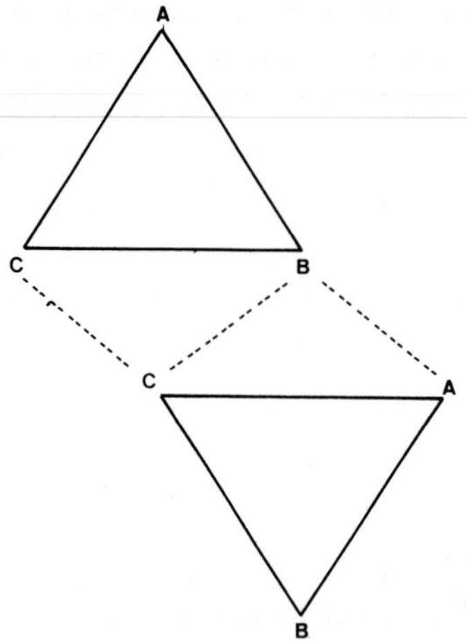

Fig. 3. — Exemple d'association qui n'est pas prise en compte par le modèle.

Fig. 3. — An association which is not incorporated in the model.

Dès que $w > 2(2v-u) \rightarrow$ HOC.

(Cette conclusion subsiste, si l'on remplace les charges sur B et C par des dipoles *verticaux* = deux dipoles opposés s'attirent.)

6. Si toutes les interactions sont purement Van der Waals

$$V_{ij} = -M\alpha_i\alpha_j \tag{8}$$

(où α_i est la polarisabilité du groupe i) Δ est $\leqq 0$ ($\rightarrow$ HEC).

Nous ne sommes pas parvenus à démontrer cette propriété de façon analytique, mais nous l'avons vérifiée par analyse numérique et par l'étude de divers cas particuliers. Par exemple dans la limite des températures élevées, on trouve

$$\Delta = -\frac{1}{2}\left(\frac{M}{kT}\right)^3 (\alpha_A-\alpha_B)^2(\alpha_B-\alpha_C)^2(\alpha_C-\alpha_A)^2 \leqq 0. \tag{9}$$

7. Supposons que les groupes A et B ont uniquement des interactions de type Van der Waals (par ex. A = méthyle, B = phényle), alors que le groupe C est chargé ($f_{CC}=0$) et interagit peu avec A et B ($f_{CA}=f_{CB}\sim 0$) parce que C préfère être hydraté. Alors :

$$\Delta = \exp\left(\frac{2M}{kT}\alpha_A\alpha_B\right)\left\{1-\exp\left[\frac{M}{kT}(\alpha_A-\alpha_B)^2\right]\right\} \tag{10}$$

et un argument de convexité montre que $\Delta < 0$ ($\rightarrow$ HEC).

IV. Discussion. — Les surfactants tripodes définissent probablement le meilleur système modèle pour voir des effets de chiralité dans une monocouche. Mais notre modèle ne leur rend pas totalement justice :

(*a*) il se peut que deux molécules adjacentes ne disposent pas leurs groupes (*ii'*) et (*jj'*) sous forme de deux liaisons (*ij*) (*i'j'*) mais réalisent par exemple une disposition où un groupe est à égale distance de deux autres (*fig.* 3). De telles associations donneraient un décompte très différent, qui reste à faire;

(*b*) il est clair que les associations de paires ne peuvent pas prédire avec rigueur les associations en phase solide. Ce problème est envisagé dans un article séparé, où nous discutons d'état fondamental des tripodes denses, au moins pour certains modèles simples sur réseau.

C. R. Acad. Sci. Paris, t. 307, Série II, p. 233-237, 1988

Mais, malgré ses fortes limitations, le présent modèle montre comment des considérations physicochimiques simples peuvent aider à prédire la ségrégation chirale.

([1]) C'est à ce niveau que notre problème s'écarte complètement de celui de la référence [7], où trois groupes d'une molécule s'associent à trois groupes de la deuxième.

Nous avons bénéficié de discussions étendues avec O. Bouloussa, M. Dvolaitsky et J. Jacques.

Note reçue le 22 avril 1988, acceptée le 25 avril 1988.

RÉFÉRENCES BIBLIOGRAPHIQUES

[1] J. JACQUES, A. COLLET et S. H. WILEN, *Enantiomers, Racemates and Resolutions*, J. Wiley and Sons, New York, 1981.

[2] P. E. SCHIPPER, *Aust. J. Chem.*, 35, 1513, 1982; 28, 1975, p. 1161; *Optical activity and chiral discrimination*, S. F. MASON éd., Reidel, Dordrecht, 1979.

[3] M. V. STEWARD et E. M. ARNETT dans *Topics in Stereochemistry*, N. L. ALLINGER, E. L. ELIEL et S. H. WILEN éd., J. Wiley and Sons, New York, 1982, p. 195-262; E. M. ARNETT, J. CHAS, B. J. KINZIG, M. V. STEWART, O. THOMSON et R. J. VERBIAR, *J. Am. Chem. Soc.*, 104, 1982, p. 389-400.

[4] O. BOULOUSSA et M. DUPEYRAT, *Biochim. Biophys. Acta* (à paraître).

[5] M. DVOLAITSKY et M. A. GUEDEAU-BONDEVILLE, dans *Proceedings of the International Symposium on New Trends in Physics and Physical Chemistry of Polymers*, 3nd. chemical Congress of North America, Toronto, June 1988.

[6] E. M. ARNETT et J. M. GOLD, *J. Am. Chem. Soc.*, 104, 1982, p. 636-639; D. A. WISHER, T. ROSARIO-JANSEN et M.-D. TSAI, *J. Am. Chem. Soc.*, 108, 1986, p. 8064-8068.

[7] K. AMAYA, *Bull. Chem. Soc. Jpn.*, 34, 1961, p. 1689-1693; *ibid.*, 1962, p. 1803-1806; *ibid.*, 53, 1980, p. 3510-3512.

[8] D. ANDELMAN et P. G. DE GENNES, *C.R. Acad. Sci. Paris* (à paraître).

[9] L. SALEM, X. CHAPUISAT, G. SEGAL, P. C. HIBERTY, C. MINOT, C. LEFORESTIER et P. SAUTET, *J. Am. Chem. Soc.*, 109, 1987, p. 2887-2894.

Laboratoire de Physique de la Matière condensée,
Collège de France, 75231 *Paris Cedex* 05.

Afterthoughts: Discrimination chirale dans une monocouche de Langmuir

This paper has been expanded by D. Andelman [*J.Am. Chem. Soc.* **111**, 6536 (1989)], and in the latter form, did attract some attention (the editor of *Nature* wrote a special comment about it...). In my mind, the model is a very crude starting point, for three reasons:

a) On the theoretical side, as pointed out in Fig. 3, we drop a very important class of conformations, which might completely upset the predictions.

b) The role of water in between the polar heads is ignored.

c) On the practical side, it turns out to be surprisingly difficult to build up the tripod molecules which we had in mind: usually one of the three "feet" is far too large.

But it remains true that (i) monolayers are a good system for fundamental studies on chirality, (ii) the present "quasi chemical" approach (with interactions between a small number of functional groups) is probably more instructive than sophisticated calculations trying to incorporate all atoms.

Part VII. Granular Matter

ELSEVIER

Physica A 261 (1998) 267–293

PHYSICA A

Minireview

Reflections on the mechanics of granular matter

P.G. de Gennes

Collège de France, Physique de la Matière Condensée, 11 place Marcelin-Berthelot, 75231 Paris Cedex 05, France

Received 16 September 1998

Abstract

During recent years, a rather basic conflict has emerged between departments of mechanics/physics concerning the description of granular media. Experts from mechanics measure stress/strain relations, using the so-called triaxial tests, and then use these data to predict the behavior of a sample under given boundary conditions. Some physicists have a different view: they have claimed that it is not possible to define a proper displacement field in a heap of sand, and that the notion of *strains* is thus ambiguous. In the present text, we conclude that the crucial features are the following: (a) a heap is usually formed from a flowing phase of sand, (b) there is (empirically) a sharp interface between the flowing phase and the frozen heap below, (c) we may define for each grain a displacement, which is measured from the moment when it froze. (This displacement is due, for instance, to a compaction of the heap under its own weight.) We also present some aspects of the dynamics, including dune motions and surface flows of grains. Bouchaud, Cates and coworkers have constructed a very compact description for thin flows. We discuss the practical consequences of this picture, emphasizing some possible extensions for thicker flows. This suggests a number of possible experiments on avalanches. © 1998 Elsevier Science B.V. All rights reserved.

1. Introduction

Granular matter refers to particle systems where the size (d) is larger than 1 μm. Below 1 μm, thermal agitation is important, and Brownian motion can be seen. Above 1 μm, thermal agitation is negligible. We are interested here in many particle systems, at zero temperature, occupying a large variety of metastable states: if we pour sand on a table, it is likely to go to a ground state, with a monolayer of grains giving the lowest gravitational energy. But in reality the sand remains as a heap; the shape of the heap and the stress distribution inside, depend critically on how the heap was made. Hence, arise many difficulties.

We describe sand: a desert like the Sahara provides us with a gigantic lab model. The grains are silica (rounded by collisions) of $\sim$100 μm in size. They form ripples and

0378-4371/98/$ – see front matter
PII: S0378-4371(98)00438-5

dunes. These deserts have fascinated a number of great men – Lawrence and Thesinger in Arabia, Monod in western Sahara, and Bagnold in the Lybian desert. Bagnold's book "Physics of blown sand and sand dunes", published in 1941, remains a basic reference 60 years later [1]. We shall give an "idealized summary" of his views in Section 3.

Of course, there are many other important granular systems in nature: snow is a glaring example; but snow is frightfully complex, because water can show up in all its natural states, and the resulting phase transitions imply deep macroscopic consequences. In the present text, we shall try to concentrate on *dry* systems. This may be sand, but it may also be mustard seed (the latter being very convenient for certain nuclear resonance studies).

Many industrial products are powders:

- "clinkers" (the starting point of cement) are complex mixtures of silicoaluminates, calcium silicates, etc.
- "builders" are an important part of a commercial detergent: they are based on inorganic particles such as calcium carbonate.
- most pharmaceutical products are derived from powders, obtained by precipitation, crystallization, or prilling (prilling is based on a molten thread of material, which breaks into droplets via the Rayleigh instability; the droplets then reach a cool region where they freeze, giving grains with a very well-defined size).

If we measure it by tons, the first material which man manipulates is water; the second is granular matter. But in our supposedly sophisticated 20th century, the manipulation of powders still involves some very clumsy and/or dangerous operations.

(1) *Milling* is slow, inefficient, and generates a very broad distribution of final sizes.

(2) The smaller size component of these distributions is often *toxic*.

(3) Many powders, when dispersed in air, achieve a composition which is ideal for strong *detonations*. Certain workshops or silos explode unexpectedly. One of the main reasons for this is electrostatic: many grains, when manipulated, hit each other or hit a wall, generating triboelectric charges, which ultimately end up in sparks. To understand this, a new type of mass spectrometry is now set up, where the particles are grains rather than molecules. They are studied after a sequence of wall collisions; here, the interest is more in the charge than in the weight.

(4) When feeding, for instance, a glass furnace with a mixture of oxides, one finds that the corresponding flow of oxide in the hoppers can lead to *segregation* – thus giving dangerous inhomogeneities in the final glass: the manipulation of mixtures is delicate.

Certain other operations are quite successful, although their basic principles are only partly understood: for instance, by injecting a gas at the bottom of a large column filled with catalytic particles, one can transform them into a *fluidized bed*. This is crucial for many processes – such as the production of polyethylene. But the dynamics of these beds is still not fully understood.

We see, at this level, the importance of fundamental research in granular matter: this has been appreciated very early in Mechanical and Chemical Engineering; physicists have joined in more recently. For them, granular matter is a new type of condensed matter; as fundamental as liquid, or solid; and showing in fact two states: one liquid-like, one solid-like. But, there is yet no consensus on the description of these two states. Granular matter, in 1998, is at the level of solid state physics in 1930.

There are some excellent reviews [2] but very few textbooks – apart from [1,3]. The most recent one is (at the moment) published only in French [4].

In the present short survey, we shall talk first about the *statics* of heaps and silo (Section 2). Then, in Section 3, we shall describe some dynamical aspects: fluidization, surface flows and avalanches, saltation and dune motion, etc.

2. Statics

2.1. Preparing a granular sample

"We fill a glass column with sand". This innocent statement hides many subtleties. Did we fill it from a jet of sand near the axis, or did we sprinkle the sand over the whole section? Did we shake the object after filling?

The first, obvious, problem is *compaction*. Bernal [5] and Scott [6] measured the average density of containers filled with ball bearings: they were in fact concerned with models for amorphous systems at the atomic level, but their results are of wider utility. Computer simulations [7] indicate that the maximum volume fraction achieved in a random packing of spheres is $\phi_{rp} = 0.64$ – significantly smaller than the face centered cubic (or hexagonal) compact packing $\phi_{\max} = 0.74$. Compaction is favored by the own weight of the grains. Immersing the grains in a fluid of matched density [8], one can study weaker compactions and more or less reach the connectivity limit, or, as it is called, the random loose stacked limit, which for spheres, is around $\phi_{\min} = 0.56$.

When a powder is gently shaken, it densifies. In fact, a useful method for characterization of a new granular material is based on tapping a vertical column [9]. Powders which compact fast, are expected to flow easily, while powders which compact slowly, more or less refuse to flow. Fundamental studies on the compaction of non-cohesive grains have been performed by the Chicago group [10,11]. The density plots ϕ_n (after n taps) depend on the amplitude of the taps. At small amplitudes, they follow a logarithmic law:

$$\phi_n = \frac{a}{\ell n(n) + b} . \tag{2.1}$$

Many frustrated, frozen, systems are expected to show similar forms of creep [12,13]. The simplest interpretation of Eq. (2.1) is based on free volume models [10,14] which are familiar from the physics of glasses. The case of strong tapping is more complex [11], but some relevant simulations and modelisations have been performed [15,78,79].

Even if we do not perform any tapping, we must specify how the grains were brought in: there is a critical moment, where the grains stop and adopt a *frozen conformation*. For instance, if we build a heap of sand from an axial jet falling on the center, we create avalanches from the center towards the edges; the freezing process takes place via grains which roll and stop.

The distinction between rolling and frozen grains is crucial. It is reminiscent of a phase transition. If we accept it, we may describe the later evolution of the frozen phrase by a *displacement field* $\underset{\sim}{u}(x,y,z,t)$. This is defined by the following "gedanke experiment". We focus our attention on one rolling grain, and watch when it stops, at a certain point x,y,z. This will define the origin of its displacements. Later, with other grains added and loading the system, our grain will move by an amount $\underset{\sim}{u}(x,y,z,t)$. Its position will thus depend on the whole history of loading. The resulting displacement field is continuous. Inside the frozen phase, we may define deformations ∇u. We may also define a (coarse grained average) stress field $\sigma_{\alpha\beta}$, and relate it by some empirical relation to the deformations.

This procedure is essentially what has been used in the mechanics departments: see, for instance, the review by Biarez and Gourves [16]. But the precise definition of $\underset{\sim}{u}$ is not always stated, and thus the very notion of a displacement field has been questioned by a number of physicists (for a recent summary, see Cates et al. [17,80]).

The present author's belief is that $\underset{\sim}{u}$ is well defined, provided that there is a sharp distinction between fluid particles and frozen particles. [1] But we shall come back to this discussion later in Section 2.

Another important point is the role of *boundary conditions*, on the frozen piece:

(a) At the free surface: a heap, for instance, shrinks under its own weight, and this renormalises the relation between deformations and displacements: a simple example is given in the appendix.

(b) At the interface between the grains and a solid wall, the normal displacements must, of course, be continuous. The delicate part is the description of friction, i.e., of tangential stresses σ_t at the surface. The natural scheme is as follows:

(i) If the tangential component of $\underset{\sim}{u}(\underset{\sim}{u_t})$ has grown monotonically and is large enough, the reaction σ_t from the wall is opposed to $\underset{\sim}{u_t}$. For a cohesionless interface, we may write the classical Amontons law (see for instance [18]):

$$\sigma_t = \mu_f \sigma_n ,$$

where σ_n is the normal stress, and μ_f is a friction coefficient. We call this regime "fully mobilized friction".

(ii) If the tangential displacement $|\underset{\sim}{u_t}|$ is smaller than a certain microscopic length Δ, the friction is only partly mobilized. We call Δ the "anchoring length" [19,81]. It is usually related to the size of microscopic asperities. For macroscopic solids in contact, Δ is of order of 1 μm.

An example of partial mobilization is presented in Section 2.2.

(iii) If we reverse the displacements (as may happen in experiments where weight and thermal expansions are in conflict) the friction force will reverse fully, only if we move backwards by more than 2Δ.

[1] This may exclude certain complex problems such as tapping.

Thus, the state of friction may be influenced by minute displacements of the grains (of order Δ) with respect to the container walls. In recent experiment on columns [20], the apparent weight, at the bottom, was found to vary cyclically between day and night: as pointed out by the authors, this is probably due to thermal expansion, inducing some (very small) relative displacements between the grains and the lateral walls, and changing drastically the mobilization of friction.

To summarize: the definition of an initial state, in an experiment on granular matter, requires great care. Many theories, and some experiments, suffer from a lack of precise definitions.

2.2. Macroscopic stress fields

2.2.1. The general problem

Over more than a hundred years, the static distribution of stresses in a granular sample has been analyzed in departments of Applied Mechanics, Geotechnical Engineering, and Chemical Engineering. What is usually done is determine the relations between stress and strain on model samples, using the so-called triaxial tests. Then, these data are integrated into the problem at hand, with the material divided into finite elements (see for instance [21]).

In a number of cases, the problem can be simplified, assuming that the sample has not experienced any dangerous stress since the moment, when the grains "froze" together: this leads to a *quasi-elastic description*, which is simple. I shall try to make these statements more concrete by choosing one example: a silo filled with grain.

2.2.2. The Janssen picture for a silo

The filled silo is shown in Fig. 1. The central observation is that stresses, measured with gauges at the bottom, are generally much smaller than the hydrostatic pressure $\rho g H$ which would be present in a liquid (ρ: density, g: gravitational acceleration, H: column height). A first modelization for this was given long ago by Janssen et al. [22] and Lord Rayleigh [23].

(a) Janssen assumes that the horizontal stresses in the granular medium $(\sigma_{xx}, \sigma_{yy})$ are proportional to the vertical stresses:

$$\sigma_{xx} = \sigma_{yy} = k_j \sigma_{zz} = -k_j p(z) , \tag{2.2}$$

where k_j is a phenomenological coefficient, and $p = -\sigma_{zz}$ is a pressure.

(b) An important item is the friction between the grains and the vertical walls. The walls endure a stress σ_{rz}. The equilibrium condition for a horizontal slice of grain (area πR^2, height dz) gives:

$$-\rho g + \frac{\partial p}{\partial z} = \frac{2}{R} \sigma_{rz}|_{r=R} \tag{2.3}$$

(where r is a radical coordinate, and z is measured positive towards the bottom).

Janssen *assumes* that, everywhere on the walls, the friction force has reached its maximum allowed value – given by the celebrated law of da Vinci and Amontons [18]:

$$\sigma_{rz} = -\mu_f \sigma_{rr} = -\mu_f k_j p\,, \tag{2.4}$$

where μ_f is the coefficient of friction between grains and wall.

Accepting Eqs. (2.2) and (2.4), and incorporating them into Eq. (2.3), Janssen arrives at

$$\frac{\partial p}{\partial z} + \frac{2\mu_f}{R} k_j p = \rho g\,. \tag{2.5}$$

This introduces a characteristic length:

$$\lambda = \frac{R}{2\mu_f k_j} \tag{2.6}$$

and leads to pressure profiles of the form

$$p(z) = p_\infty[1 - \exp(-z/\lambda)] \tag{2.7}$$

with $p_\infty = \rho g \lambda$. Near the free surface ($z < \lambda$) the pressure is hydrostatic ($p \sim \rho g z$). But at larger depths ($z > \lambda$) $p \to p_\infty$: all the weight is carried by the walls.

2.2.3. *Critique of the Janssen model*

This picture is simple, and does give the gross features of stress distributions in silos. But the two assumptions are open to some doubt.

(a) If we take a (excellent) book describing the problem as seen by the mechanics department [24], we find that relation (1) is criticized: a constitutive relation of this sort might be acceptable if x, y, z were the principal axes of the stress tensor – but in fact, in the Janssen model, we also need nonvanishing off diagonal components σ_{xz}, σ_{yz}.

(b) For the contact with the wall, it is entirely arbitrary to assume full mobilization of the friction, as in Eq. (2.4). In fact, any value σ_{rz}/σ_{rr} below threshold would be acceptable. Some tutorial examples of this condition and of its mechanical consequences are presented in Duran's book [4]. I discussed some related ambiguities in a recent note [18] emphasizing the role of the anchoring length.

2.2.4. *Quasi-elastic model*

When a granular sample is prepared, we start from grains in motion, and each grain freezes at a certain moment. This defines our reference state: (i) the origin of the grain displacements is the point of freezing (ii) the reference density (to define deformations is the density achieved immediately upon freezing (see the appendix for a detailed explanation of this point)).

If we fill a silo from the center, we have continuous avalanches running towards the walls, which stop and leave us with a certain slope.

As we shall see in Section 3, this final slope, in a "closed cell" geometry like the silo, should always be *below critical*: we do not expect to be close to an instability in shear, and the material is under compression everywhere. In situations like this, we may try to describe the granular medium as a *quasi-elastic medium*. The word "quasi" must be explained at this point.

When we have a granular system in a certain state of compaction, it will show a resistance to compression, measured by a macroscopic bulk modulus K. But the forces are mediated by small contact regions between two adjacent grains, and the contact areas increase with pressure. The result is that $K(p)$ increases with p. For spheroidal objects and purely Hertzian contacts, one would expect $K \sim p^{1/3}$, while most experiments are closer to $K \sim p^{1/2}$ [25]. Various interpretations of the $p^{1/2}$ law have been proposed [26,27].

Evesque and the present author [28] recently used the quasi-elastic picture to describe displacements and stresses in a silo. The displacements correspond to a slight collapse of the column under its own weight. They increase during filling: their description involves the whole sample history. (Also the displacements are slightly smaller near the walls than in the center: this creates the shear stresses which worried Nedermann.)

The result is a Janssen relation of form (2.2), with a value of k_j which depends only on the Poisson ratio σ_p of the material:

$$k_j = \frac{\sigma_p}{1-\sigma_p} . \tag{2.8}$$

Although the elastic moduli do depend on pressure, it may be that σ_p and k_j are pressure independent: then the Janssen pressure profile should hold, provided that mobilization of the wall friction is complete. For long columns ($H \gg \lambda$) the maximum displacement is achieved at mid-height, and is

$$|u|_{\max} = \frac{\lambda^2}{\lambda_c} , \tag{2.9}$$

where $\lambda_c = E/\rho g$ is what we call the compaction length (E= Young modulus; ρ= density). Mobilization is indeed complete if $|u|_{\max} \gg \Delta$ (the anchoring length), or equivalently $\lambda > H^*$, where

$$H^* = (\Delta\lambda_c)^{1/2} . \tag{2.10}$$

In this formula, Δ is very small, but λ_c is very large. Typical values of H^* depend on E, but these may be centimetric. Thus, if the quasi-elastic model makes sense, the Janssen picture should hold for silos ($\lambda \cong$ meters, $\lambda > H^*$) but not necessarily for laboratory columns ($\lambda \cong 1$ cm).

2.2.5. Stress distribution in a heap

Below a heap of sand, the distribution of normal pressures on the floor is not easy to guess. In some cases, the pressure is not a maximum at the center point. This has led to a vast number of physical conjectures, describing "arches" in the structure [29,30]. In their most recent form [31], what is assumed is that, in a heap, the principal axes of the stress are fixed by the deposition procedure. Near the free surface, following the pioneering work of Coulomb, it is usually assumed that (for a material of zero cohesion) the shear and normal components of the stress (τ and σ_n) are related by the condition:

$$\tau = \sigma_n \mu_i = \sigma_n \tan \theta_{\max} , \tag{2.11}$$

where μ_i is an interval friction coefficient and $\tan \theta_{\max}$ is the resulting slope. Eq. (2.25) should hold for a dry system with no cohesion between grains. In a two-dimensional geometry, this corresponds to a principal axes which is at an angle $2\theta_{\max}$ from the horizontal [24]. The assumption of Wittmer et al. [31], is that this orientation is retained in all the left-hand side of the heap (plus a mirror symmetry for the right-hand side). Once this is accepted, the equilibrium conditions incorporating gravity, naturally lead to a "channeling of forces" along the principal axes, and to a distribution of loads on the bottom which has two peaks. More generally, in the descriptions of Bouchaud, Cates et al. [30,31], the transmission of stresses is described by *hyperbolic* equations, leading to certain preferred directions. In the classical approach from continuum mechanics, the transmission is ruled by *elliptic* equations. In the first picture, all the heap is pictured as being in some sort of critical state. In the second picture, we are far from criticality, and the heap is not dramatically different from a conventional solid – although the sample history is important for a clear definition of deformations.

The "critical" view has been challenged by Savage [32], and by Goddard [33]. Savage gave a detailed review of the experimental and theoretical literature. He makes the following claims:

(a) For two-dimensional heaps ("wedges") with a rigid support plane, there is no dip in the experiments.

(b) If the support is (very slightly) deformable, the stress field changes deeply, and a dip occurs. This is another example of the role of minute displacements which was already emphasized in Section 2.5.

(c) For the 3d case ("cones") the results are extremely sensitive to the details of the deposition procedure.

The most recent data on cones are by Brockbank et al. [34]: they use an accurate optical measurement of the local load under a conical heap of steel balls. The balls, in the bottom layer, deform the support, which is made of a transparent rubber film (~2 mm in thickness) lying over a glass surface. They do find a dip with steel, and also with glass heads of diameter 0.18 mm. But, when going to larger glass beads (~0.6 mm) the dip disappears.

Savage also describes finite element calculations, where one imposes the Mohr Coulomb conditions (to which we come back in Section 3) at the free surface of a wedge. If we had assumed a quasi-elastic description inside, we would have found an inconsistency: there is a region, just below the surface, which becomes unstable towards shear and slippage. Thus Savage uses Mohr Coulomb in a finite sheet near the surface, plus elastic laws in the inner part: with a rigid support he finds no dip. But, with a deformable support, he gets a dip.

The Savage methodology is similar in spirit to the quasi-elastic method presented in Section 3.2, although the details of the boundary conditions could possibly be altered.

For instance, there may exist an extra simplification – which I already announced in connection with the silos. If we look at the formation of the heap (as we shall do in Section 3) we find that the slope angle upon deposition should be slightly *lower* than the critical angle $\theta_{\max}$. Thus, our system is prepared in non-critical conditions: all the samples may then be described as quasi-elastic. This, in fact, should not bring in very great differences from the results obtained by Savage.

But there is a certain doubt, formulated by Cates and others: if the grains were glued together by microscopic glue patches at the contact point, indeed we might define displacements, deformations, and use the Savage picture. But there is no glue! Certain grains might then be under tension (even if we are under a global compressive load): mechanical integrity is not granted.

In reply to this, the present author proposes three observations, which tend to support the classical view from mechanics.

(i) *Shear tests*: under compressive load (in conditions without fracture) the stress strain relations are clearly history dependent, but do not display (as far as we can tell) any singular power laws.

(ii) *Lack of criticality*: if we examine the local density in a horizontal bed of sand, or the volume fraction ϕ as a function of depth, we find that ϕ is nearly constant, and significantly larger than the critical value $\phi_{\min}$ mentioned in Section 2.1.[2] For these practical ϕ value (as we shall see in Section 2.3) the few indications available on correlation lengths ξ suggest that ξ is not large (at most of order 5–10 grain diameters). The singularities linked with arches, with tensile microcracks, should thus be confined to very small scales $\Delta x < \xi$.

(iii) *Texture*: One of the features which the physicists really wanted to incorporate, is the possible importance of an internal *texture*. If we look at the contacts $(1, 2, \ldots i, \ldots, p)$ of a grain in the structure, we can form two characteristic tensors: one is purely geometrical and defines preferred directions of contact. It is

$$Q_{\alpha\beta} = \overline{\sum_i x_\alpha^{(i)} x_\beta^{(i)}} \tag{2.12}$$

(where x_α are the distances measured from the center of gravity of the grain). $Q_{\alpha\beta}$ is also called the "fabric tensor" [34–36]. It is related to the "ellipsoid of contacts" introduced by Biarez and Wiendick [37]. The other tensor is the static stress:

$$\sigma_{\alpha\beta} = \frac{1}{2} \sum_i (x_\alpha^{(i)} F_\beta^{(i)} + x_\beta F_\alpha^{(i)}), \tag{2.13}$$

where $\underset{\sim}{F}^i$ is the force transmitted at contact (i). There is no reason for the axes of these two tensors to coincide. For instance, in an ideal hexagonal crystal, one major axis of the Q tensor is the hexagonal axis, while the stresses can have any set of principal axes. In the heap problem, I am personally inclined to believe that the deposition process

[2] Note that although ϕ is nearly constant, in a sand bed elastic moduli increase dramatically with depth: this is the basis of the "quasi-elastic" model.

freezes a certain structure for the Q tensor, but not for the stress tensor. However, this is still open to discussion. Recent arguments defending the opposite view point have been given by Cates et al. [17].

The presence of a non-trivial Q tensor (or "texture") can modify the quasi-elastic model: instead of using an isotropic medium, we may need an anisotropic medium. In its simplest version, we would assume that the coarse-grained average $Q_{\alpha\beta}$ has two degenerate eigenvalues, and a third eigenvalue, along a certain unit vector (the director) $\underset{\sim}{n}(\underset{\sim}{r})$. Thus, a complete discussion of static problems (in the absence of strong shear bands) would involve an extra field $\underset{\sim}{n}$ defined by the construction of the sample. This refinement may modify the load distribution under a heap. But, conceptually, it is, in my opinion, minor. Texture effects should not alter deeply the quasi-elastic picture.

2.2.6. *Strong deformations*

Sophisticated tools have been designed for measuring the yield stress τ_y of granular materials in simple shear ([38]; for a review, see for instance [3]). There is an elastic response at low shears, followed by yield at a certain value of the stress τ_y:

$$\tau_y = C + \mu p_n \,, \tag{2.14}$$

where p_n is the normal pressure. The constant C represents adhesive interactions between grains, and μ is a friction coefficient. An important feature of these strongly sheared systems – emphasized long ago by Reynolds is *dilatancy*: when the material was originally rather compact, and is forced to yield, it increases in volume [39]. This can be qualitatively understood by thinking of two compact layers of spheres sliding over each other.

In some cases, these strong deformations, with dilatancy, are present over large volumes. In other cases, they may be concentrated on *slip bands* (see for instance [40,41]). If we remove sand with a bulldozer, slip bands will start from the bottom edge of the moving plate. Sometimes, the size of these slip bands is large and depends on the imposed boundary conditions (on the sharpness of the plate edge). But there seems to be a minimal thickness for a slip band: for spheroidal grains, without cohesion, it may be of order 5–10 grain diameters. We shall come back to this thickness when discussing microscopic properties.

2.3. *Microscopic features*

2.3.1. *Correlation lengths*

We have talked about macroscopic stresses σ_{ij}: they must represent some coarse grained averages over a certain volume. The implicit assumption here is that, indeed, a granular medium can be considered as homogeneous at large scales. This is not obvious: if we were talking about non-compacted material, with a density close to the lower limit $\phi_{\min} = 0.56$, we might have a structure of weakly connected clusters (similar to percolation clusters). Exactly at threshold ($\phi = \phi_{\min}$) a structure like this would probably be self-similar, and not homogeneous at all. However, in real life, we

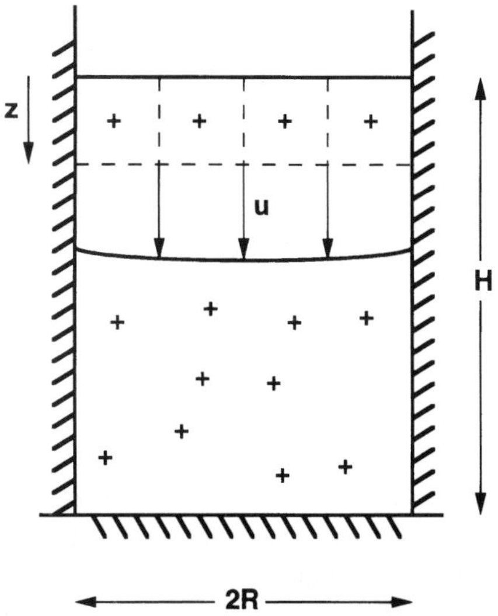

Fig. 1. A silo filled with granular material: the material falls slightly under its own weight, by an amount u (For a precise definition of u, see the appendix.) The width of the silo has been exaggerated to display the expected profile of u in a quasi-elastic model.

always operate on systems with $\phi > \phi_{\min}$, and we can expect that, at scales larger than a certain correlation length $\xi(\phi)$, our system may be treated as homogeneous.

Various experiments [42,82] and simulations [43–46] have investigated the local distribution of forces between grains. The central conclusion is that there are *force channels*, which build up a certain mesh with a characteristic size ξ. For spherical objects and ϕ values in the usual range, this ξ is somewhat larger than the grain diameter $d(\xi/d \sim 5\text{–}10)$.

This network is obviously sensitive to variations in size among the grains. This "polydispersity" is always present, and plays an important role in the actual value of ξ.

It may well be that the minimum thickness of a slip band (as introduced in Section 2.2) is equal (within coefficients) to the correlation length ξ. Thus, we have at least two empirical ways of estimating ξ for a given system.

2.3.2. *Fluctuations of the local load*

It is also of interest to probe the local distribution of forces on all grains in contact with a supporting wall, as shown in Fig. 1. This has been done in experiments by the Chicago group [42,47], (together with some simulations). Their trick is to lay the granular sample on a sequence carbon paper/white paper/solid. There is an empirical relation between the size of the dots printed by each grain on the white paper, and the force (w) with which it presses the ground. What Liu et al. found was a distribution of w, of the form

$$p(w) = \frac{w^2}{2\bar{w}^3} e^{-2w/\bar{w}} . \quad (2.15)$$

Liu et al. [42,82] constructed a simple model for this statistical behavior, ignoring the vector character of the forces. They stipulated that each grain receives a load (w) from the three neighbors above it:

$$w = q_1 w_1 + q_2 w_2 + q_3 w_3 , \quad (2.16)$$

where w_1, w_2, w_3 are the loads on the "parents", and q_1, q_2, q_3 are 3 coupling factors statistically distributed between 0 and 1, and independent. Conversely, each parent sends some of its weight, on three "children" with fractions q'_1, q'_2, q'_3, and these fractions satisfy the sum rule $\Sigma q'_1 = 1$. But apart from this constraint, all the q'_s are independent.

Law (2.15) can be understood as follows:

(a) for $w \gg \bar{w}$, we must have $q_1, q_2, q_3 \sim 1$, and we can then factorize $p(w) \sim p(w_1)p(w_2)p(w_3)$, with $w = w_1 + w_2 + w_3$. As pointed out by Witten, this condition is similar to the problem of a Bolzmann distribution of energies in thermal physics, and the solution is exponential $p(w) \sim \exp(-\alpha w)$.

(b) for $w \ll \bar{w}$, the weights carried by the three parents are much larger than w, and the probability $p(w)$ is essentially proportional to the phase space available in (q_1, q_2, q_3) where the q_s are linked by Eq. (2.16). This corresponds to a triangle of edges. $(0_1, 0, w/w_1)$, $(0_1, 0, w/w_2)$, $(0_1, 0, w/w_3)$ in the (q_1, q_2, q_3) space, with an area $\sim w^2$. (However, on the experimental side, the more recent data of Mueth et al. [47] give a different law.)

To summarize: (i) the fluctuations of w are comparable to the average ($\bar{w}$). (ii) the tail of the distribution at large w is exponential. The probabilities q_1, for very small loads ($w \to 0$), are still open to discussion.

A subsidiary question is: what are the correlations $\langle w(\underset{\sim}{x}) w(\underset{\sim}{y}) \rangle$ between grains at different locations (x, y) on the ground. The natural guess is that the range of these correlations is the correlation length ξ.

Of course, the model should be refined by introducing the vector character of the forces. The vectorial features are crucial when the *average* load is variable from point to point on the bottom plate. Consider, for instance, a horizontal slab of grains, with a thickness H, and a very large aspect ratio. Impose a weak localized force F, downwards, at the center of the upper surface ($x = y = 0$) (Fig. 2).

(a) The scalar model of Eq. (2.30) would give an average load profile on the bottom plate with a peak at the center and a width $\Delta x \sim \Delta y \sim \sqrt{dH}$ (where d is the grain diameter).

(b) With a tensorial stress field, and a quasi-elastic model, we expect $\Delta x \sim \Delta y \sim H$.

(c) With "singular" models which predict transmission of the weight only in special directions [30], the load would be concentrated on a ring, and disorder would make this ring slightly diffuse.

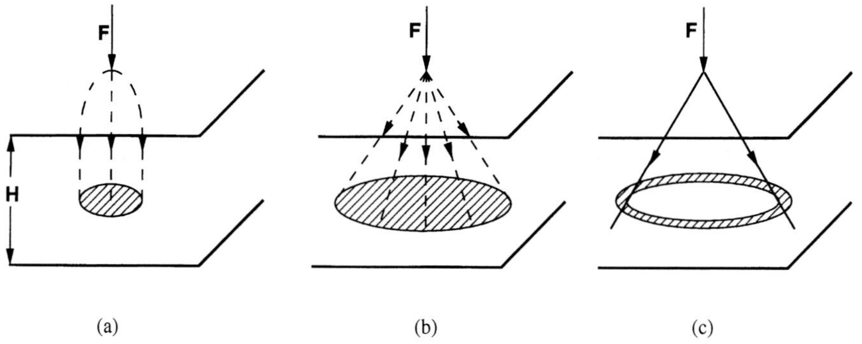

Fig. 2. A crucial experiment, which to the author's knowledge has not yet been performed in a completely conclusive way. A bed of sand is deposited uniformly on a large flat surface, and fills a height H. A small local force F is applied vertically at one point of the top surface. What are the resulting extra loads on the bottom plate? (a) In the "elliptic" models, used in soil mechanics, the load is spread over a region of size $\sim H$. (b) In the "hyperbolic" models of Bouchaud et al., the load is distributed over an annulus. (c) In some "scalar" models, the equations are parabolic.

3. Dynamics

3.1. The hour glass

For thousands of years, man has known that the amount of sand passing through an hourglass per unit time, is independent of the height of the sand column at the top. This is a natural consequence of the Janssen analysis for static pressure distributions: the pressure at the bottom of a long column is independent of the height of sand in the column, because most of the weight is transmitted to the lateral walls. And, whatever the exact shape of the column, the pressure in the bottom region scales like p_∞ in Eq. (2.7), namely $p_\infty \sim \rho g D$, where D is the diameter of the neck.

Many years ago, Hagen [48] studied the quantity of sand flowing per unit time (Q) as a function of the neck diameter D, for D much larger than the grain size. The result is $Q \sim D^{5/2}$. This can be, to some extent, understood by a heuristic argument: in the neck, inertial effects dominate, and this suggests a velocity V which is related to pressure by something like a Bernouilli formula, giving $p_\infty = \rho V^2$. Then, the current Q should scale as

$$Q = \rho D^2 V = \rho g^{1/2} D^{5/2} \,. \tag{3.1}$$

At smaller diameters ($D = D_{\min}$), the flow stops: the present author's conjecture is that $D_{\min} \cong \xi$, the static correlation length which was discussed in Section 2.3.

Detailed discussions of this complex flow (at $D \gg D_{\min}$) are summarized in Ref. [49].

3.2. Shear flows

Starting with Bagnold [50], many experiments have been performed on shear flows (in annular cells, to avoid end effects) [51–53]. Another approach is based on *surface flows* – on the slope of heap of sand. Savage [54] studied these flows (via an immersed optical fiber) for a system of glass beads, and found typical thicknesses for the flowing regions which were in the range of $20d$ – again possibly related to ξ. Observations of flows in a vertical Hell Shaw cell (i.e. between two transparent plates), have been performed by Drake [55]: he distinguishes five zones, out of which I tend to memorize three: pure solid, mixed phase with solid regions and fluid regions, and fluid.

A considerable number of simulations have also been performed on these flows, but they do not give a very clear message. On the theoretical side, one major line of approach was started by Bagnold [50]. He was concerned with rapid flows, which bring the density down to a collision-dominated regime. Thus he thought in terms of a gas of particles, with a relative velocity $v_{\mathrm{rel}} \sim d|\nabla v|$. The standard kinetic theory of gases predicts a viscosity η_{eff}, scaling like:

$$\eta_{eff} = \rho d v_{\mathrm{rel}} = \rho d^2 |\nabla v| , \tag{3.2}$$

where the average interparticle distance is taken to be of order d, although the volume fraction ϕ is relatively small. (This is a critical assumption because, near $\phi = \phi_{\mathrm{min}}$, some very large clusters may be present.) Eq. (3.2) leads to unusual profiles in shear flow, which are in reasonable agreement with data on annular flows.

More generally, the temptation of introducing a gas model with a granular temperature proportional to v_{rel}^2 is very natural. But inelastic collisions tend to make these "hot gases" instable, as shown by simulations: see for instance, Mac Namara and Young [56]. The gas tends to condense into small droplets: if a fluctuation has nucleated a droplet, an incoming particle hitting the droplet cools down fast, because of multiple collisions in the dense region. Thus, the incoming particle sticks to the droplet.

In a one-dimensional system, with inelastic particles floating between two walls, one hot, one cold, simulations show an oscillatory motion of aggregates which has nothing to do with the classical transport of heat in a gas [57].

3.3. Fluidization

It is often important, technically, to fluidize a granular bed. There are two lines of approach:

(a) by shaking a container vertically.

(b) by blowing a gas from the bottom, generating a "fluidised bed".

3.3.1. Shaking

A container of sand is vibrated vertically with an amplitude $A \cos \omega t$. (Typically $A = 1$ mn and $\omega / 2\pi = 20$ Hertz.) The maximum acceleration is $A\omega^2$, and the crucial parameter is $\Gamma = A\omega^2 / g$. We assume that there is *no air* around (a mild vacuum is enough). With air, gas pressures can build up, and generate special effects – some of which were already understood by Faraday [58].

When $\Gamma > 1$, the sand in the column may experience free fall during some part of the period. Let us start by one-dimensional picture. Then, at a certain instant, the material hits the ground, and a pressure wave moves up from this point, at a certain acoustic velocity. When this pressure wave hits the free surface, it liberates some particles, which fly up and then fall down again. A certain thickness e_f near the top is fluidized. The value of e_f depends on the inelasticity of the collisions. Let us define a restitution coefficient ε for a head on collision of two grains, in a frame linked to their center of gravity: the relative velocity after collision is reduced by a factor ε. Then a simple one-dimensional model [59] leads to:

$$e_f = \frac{\pi d}{1-\varepsilon} \tag{3.3}$$

and this agrees rather well with the experiments [59]. Qualitatively, if the inelasticity is very strong ($\varepsilon \to 0$), a fluidized particle, when it hits the solid part, does not recoil. It is then liberated by the next pressure wave from the bottom, and falls back to stick again. The fluidized thickness is reduced to the last layer. On the other hand, if ε is close to unity, we can have a train of fluidized particles, the pressure wave is damped only gradually inside this train, over a height $\sim e_f$. The pressure wave retains an ability to propel each particle upwards in the train.

This presentation is acceptable when the lateral walls of the container do not play a major role. In the opposite case, more complex things happen: the lateral walls tend to induce friction, and global convection can show up.

In three dimensions, fluidized layers may show remarkable instabilities, giving rise to parametric instabilities, to a regular array of peaks, or to more complex (chaotic) behaviors [60–62].

3.3.2. Blowing

The idea of performing a gas-phase reaction with a solid catalyst in granular form, which floats in a current of gas, emerged around 1937, in connection with the cracking of hydrocarbons (for a review, see Ref. [63]). It is now used in many large-scale industrial processes.

We blow the gas from the bottom, with a certain velocity V. At a certain critical velocity V_0, the lift from the air compensates the weight, and the system switches from a solid to a fluid. When V increases beyond V_0, a number of complex phenomena occur: instabilities and turbulence show up. In the most important regimes, collisions between grains are probably dominating the losses. A very important feature here, is the occurrence of *bubbles*, which are often a great nuisance for practical purposes. These bubbles look strangely similar to bubbles in a liquid. They have a sharp interface with the fluid of particles – although there is nothing like surface tension in this system.

My own (naive) interpretation of these sharp interfaces is again based on the clustering instabilities of granular gases (mentioned in Section 3.2). If we had a diffuse interface, particles coming from the center of the bubbles would hit the denser "fluid"

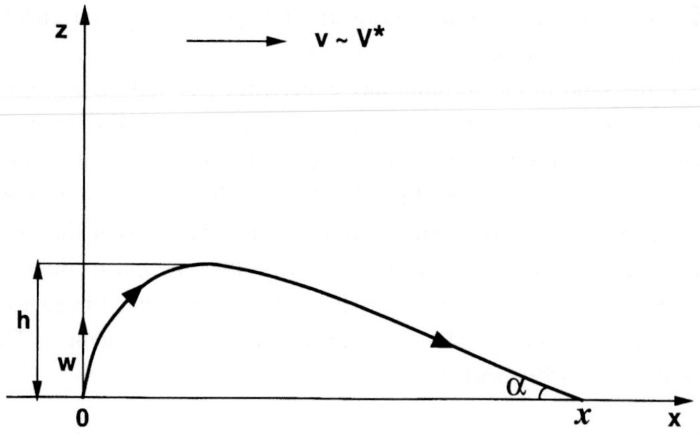

Fig. 3. An average trajectory for saltation of sand grains under a strong wind ($V^* > V_s^*$). The grain is kicked out by other grains at point 0, with a vertical velocity w. It progressively adjusts to the wind velocity, drifts, and ultimately lands with an angle $\alpha \sim h/x$. Typically, $h = 10\,\text{cm}$ and $x = 1\,\text{m}$.

and lose their energy by a cascade of inelastic collisions: thus, they would stick, and we would return to a sharp interface. The thickness of the interface may resemble the thickness of the fluidized layer in Eq. (3.3).

3.4. Wind over sand

3.4.1. Saltation

Our vision here is based on the pioneering work of Bagnold [1]. We shall present a very naive, simplified, form of his ideas. We start with a wind profile defined by the standard form for wall turbulence:

$$\frac{dv(z)}{dz} = \frac{V^*}{z}, \tag{3.4}$$

where $v(z)$ is the average horizontal component, and V^* is related to the shear stress on the ground:

$$\tau = \rho_a V^{*2} \tag{3.5}$$

(ρ_a being the density of air). When V^* increases, we reach a certain threshold V_s^* at which grains begin to jump from the ground. (Hence the word "saltation" comes from the latin word *saltare* = to jump.) An *average* trajectory is shown in Fig. 3: the grain starts with a vertical velocity w, reaches a height h, and drifts in the wind, progressively acquiring a horizontal velocity comparable to V^*. The "landing velocity" is of order V^*. Ultimately, the grain falls back after a time of flight t. The vertical motion is simply gravitational. Thus we expect:

$$\begin{aligned} gt &\sim w, \\ gh &\sim w^2, \end{aligned} \tag{3.6}$$

w is much smaller than the landing velocity V^*, and we may postulate an inelastic law of the form

$$w^2 = \eta V^{*2}\,, \tag{3.7}$$

where η is typically of order 10^{-2}.

This leads us to:

$$h = V^{*2}\eta/g \tag{3.8}$$

and to a horizontal distance of travel:

$$x = V^* t = \frac{V^{*2}\eta^{1/2}}{g}\,. \tag{3.9}$$

Thus, the angle of landing α (defined in Fig. 2) is $\alpha \sim \eta^{1/2} \sim 10^{-1}$. The whole picture holds if, and only if, the grain in flight has had enough time to equilibrate its velocity with the ambient wind. If we return to the simple problem of a grain moving with a relative velocity v, with respect to the air, and experiencing an air friction ($\rho_a v^2 d^2$),we arrive at an equilibration length:

$$\ell = d\frac{\rho}{\rho_a}\,. \tag{3.10}$$

For $d = 0.1$ mm (typical of the Sahara), ℓ is in the range of meters. It is much larger on Mars (ρ_a small), and smaller on Venus (ρ_a large).

Saltation can exist only if x, as given by Eq. (3.9) is larger than ℓ. This suggests a scaling structure for the threshold velocity:

$$V_s^* = \eta^{-1/4}(gd)^{1/2}\left(\frac{\rho}{\rho_a}\right)^{1/2}, \tag{3.11}$$

which is typically a few meters/s.

At velocities V^* larger than V_s^*, we do have sand rising above the ground, and there is a certain transport of mass Q downwind (per unit length and unit time). Most of this Q comes from the flying particles (a smaller component comes from "creep" on the ground due to the impacts). Comparing the momentum transport (i.e. the stress τ) and the mass transport (Q), Bagnold arrived at

$$Q = t\tau \cong \eta^{1/2}\rho a V^{*3}/g\,. \tag{3.12}$$

Eq. (3.12) wraps under the rug a certain number of complications: Q depends (weakly) on grain size, etc. But it gives an insight.

3.4.2. Dunes

The typical dune in a steady wind is a "barkane", with the crescent shape of Fig. 4. There is a gentle windward slope, of angle ψ. Clearly, from a hydrodynamic point of view, the velocity V^* tends to be increased on this side, increasing Q, while on the leeward side, Q is decreased (in fact nearly zero). This tends to accumulate sand at the crest, and induces the birth of the dune. But the process stops when $\psi \sim \alpha$ (the landing angle of Fig. 3). Whenever $\psi > \alpha$, the trajectory of Fig. 3 is significantly reduced in

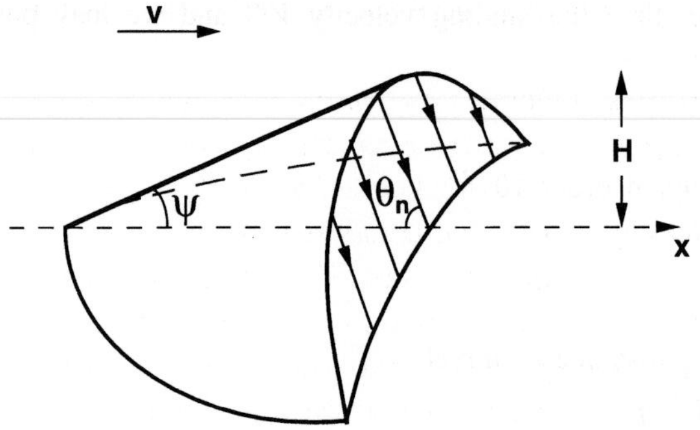

Fig. 4. A typical dune (barkane) as observed under winds of constant direction. The height H can range from 1 to 100 m. The angle ψ is small ($\sim 10^\circ$). The angle θ_m is related to avalanche processes, and is large ($\sim 30^\circ$).

length, and we arrive at $x < \ell$: i.e. the grain has not enough flight time to capture the energy from the wind. Thus $\psi \sim \alpha$.

The behavior on the leeward side is not dominated by the wind, but by avalanches. We discuss this aspect in the next paragraph.

From these scaling laws, we can also estimate the speed (c) of a dune. Since the leeward side is essentially in a situation of zero wind, we are just receiving, at the crest, a saltation flux Q (Eq. (3.12)). In a time interval dt, the mass transported (per unit of length) must be $cH\,dt = Q\,dt$ giving:

$$c = \frac{Q}{\rho^h} = \frac{\rho a}{H\rho}\eta^{1/2}V^{*3}/g\,. \tag{3.13}$$

In particular, if V^* is near threshold (Eq. (3.11)), c takes a very simple form:

$$c \to c_s = (\text{const.})V_c^* d/H\,. \tag{3.14}$$

The velocity is inversely proportional to height. Large dunes ($H \sim 100$ m) do not move. Small barkanes ($H \sim 3$ m) move with velocities in the range of 10 m/yr.

A final word of caution: (a) the ideas of Bagnold are oversimplified in this presentation (b) during the last 50 years, a considerable amount of experimentation in wind tunnels, and of simulations have been performed (see for instance [53]): the trajectories are more complex, they are modified by wind fluctuations, etc. The details of the *landing* process would greatly benefit from more experiments and simulations.

3.5. Avalanches

3.5.1. The Coulomb view

As already mentioned in Section 2, Coulomb (who was at the time a military engineer) noticed that a granular system, with a slope angle θ, larger than a critical value $\theta_{\max}$, would be unstable. He related the angle $\theta_{\max}$ to the frictional properties of the material. For granular materials, with negligible adhesive forces, this leads to $\tan\theta_{\max} = \mu_i$, where μ_i is an internal friction coefficient (for a precise review on these aspects, see [24]). The instability generates an avalanche. What we need is a detailed scenario for the avalanche.

We note first that the Coulomb argument is not quite complete: (a) it does not tell us at what angle $\theta_{\max} + \varepsilon$ the process will actually start (b) it does not tell us which gliding plane is preferred among all those of angle ($\theta_{\max}$).

My own (tentative) answer is that the thickness of the excess layer must be of order ξ; and the excess angle ε must be of order ξ/L, where L is the size of the free surface.

Thus, at the moment of onset, our picture is that a layer of thickness $\sim \xi$ starts to slip. It shall then undergo various processes: (i) the layer of grains involved is fluidized by collisions on the underlying heap (ii) it is amplified because the rolling grains destabilize some other grains below. The steady-state flow has been studied in detailed simulations [64]. It shows a sharp boundary between rolling grains and immobile grains: this observation is the starting point of most current theories.

The amplification process was considered in some detail by Bouchaud et al. in a classic paper [65] which we refer to as "BCRE".

3.5.2. BCRE equations and their modification

BCRE discuss surface flows on a slope of profile $h(x,t)$ and slope $tg\theta \cong \theta = \partial h/\partial x$, with a certain amount $R(x,t)$ of rolling species. The rate equation for the profile is written in the form

$$\frac{\partial h}{\partial t} = \gamma R(\theta_n - \theta)\,. \tag{3.15}$$

This gives erosion for $\theta > \theta_n$, and accretion for $\theta < \theta_n$.

We call θ_n the neutral angle. This notation differs from BCRE who called it θ_r (the angle of repose). Our point is that different experiments can lead to different angles or repose, not always equal to θ_n.

For the rolling species, BCRE write:

$$\frac{\partial R}{\partial t} = -\frac{\partial h}{\partial t} + v\frac{\partial R}{\partial x}(+\text{ diffusion terms})\,, \tag{3.16}$$

where γ is a characteristic frequency, and v a flow velocity, assumed to be nonvanishing (and approximately constant) for $\theta \sim \theta_n$. For simple grain shapes (spheroidal) and average levels of inelastic collisions, we expect $v \sim \gamma d \sim (gd)^{1/2}$, where d is the grain diameter and g the gravitational acceleration. Eq. (3.16) gives $\partial h/\partial t$ as linear in R: this should hold at small R, when the rolling grains act independently. But, when $R > d$, this is not acceptable.

This difficulty shows up in particular when we observe "uphill waves". As pointed out by BCRE these waves are a natural consequence of Eq. (3.15): assume that R is constant, Eq. (3.15), then shows that an accident in slope moves upward, with a velocity $v_{\text{up}} = \gamma R$. These uphill waves are often observed in everyday life.

However, it is not natural to assume that their velocity v_{up} can become very large for large R. This lead us [66] to propose a modified version of BCRE, valid for flows which involve large R values, and of the form

$$\frac{\partial h}{\partial t} = v_{\text{up}}(\theta_n - \theta) \quad (R > \xi), \tag{3.17}$$

where v_{up} is a constant, comparable to v.

Remark. *In the present problems, the diffusion terms in* Eq. (3.16) *turn out to be small, when compared to the convective terms* (*of order d/l, where L is the size of the sample*): *we omit them systematically. But they may be important for some other issues.*

3.5.3. *A simple case of thick avalanches*

A basic example, is a two-dimensional silo, fed from a point at the top, with a rate $2Q$, and extending over a horizontal span $2L$: the height profile moves upward with a constant velocity Q/L. The profiles have been analyzed within the BCRE equations. With the modified version, the R profile remains unaltered:

$$R = \frac{x}{L}\frac{Q}{v}, \tag{3.18}$$

but for thick avalanches ($Q > v\xi$), the angle is modified and *differs from the neutral angle*: setting $\partial h/\partial t = Q/L$, we arrive at

$$\theta_n - \theta = \frac{Q}{Lv_{\text{up}}} \quad (Q > v\xi). \tag{3.19}$$

Thus, we expect a slope which is now dependent on the rate of filling: this might be tested in experiments or in simulations. Current experiments, however, have been complicated by another feature: if the grains are fed in from a significant altitude z_f above the free surface, the rolling species arrives with a high energy and does not reach the steady-state velocity v: all results become dependent on z_f; the BCRE analysis is really suited only for the limit $z_f \to 0$.

3.5.4. *Downhill and uphill motions*

Our starting point now is a supercritical slope, extending over a horizontal span L with an angle $\theta = \theta_{\max} + \varepsilon$. Following the ideas of Section 1, the excess angle ε is taken to be small (of order ξ/L). It will turn out that the exact values of ε is not important: as soon as the avalanche starts, the population of rolling species grows rapidly and becomes independent of ε (for ε small): this means that our scenarios have a certain level of universality. The crucial feature is that grains roll down, but profiles move uphill.

It is convenient to introduce a reduced profile:

$$\tilde{h}(x,t) = h - \theta_n x. \tag{3.20}$$

Following BCRE, we constantly assume that the angles θ are not very large, and write $tg\theta \sim \theta$: this simplifies the notation. Ultimately, we may write Eqs. (3.16) and

(3.17) in the following compact form:

$$\frac{\partial R}{\partial t} = v_{\mathrm{up}} \frac{\partial \tilde{h}}{\partial x} + v \frac{\partial R}{\partial x} , \tag{3.21}$$

$$\frac{\partial \tilde{h}}{\partial t} = -v_{\mathrm{up}} \frac{\partial \tilde{h}}{\partial x} . \tag{3.22}$$

Another important condition is that we must have $R > 0$. If we reach $R = 0$ in a certain interval of x, this means that the system is locally frozen, and we must then impose:

$$\frac{\partial \tilde{h}}{\partial t} = 0 . \tag{3.23}$$

One central feature of the modified Eqs. (3.21), (3.22) is that, whenever $R > 0$, they are linear. The reduced profile $\tilde{h}$ is decoupled from R, and follows a very simple wave equation:

$$\tilde{h}(x,t) = w(x - v_{\mathrm{up}}t) , \tag{3.24}$$

where w is an arbitrary function describing uphill waves.

It is also possible to find a linear combination of $R(x,t)$ and $\tilde{h}(x,t)$ which moves downhill. Let us put:

$$R(x,t) + \lambda \tilde{h}(x,t) = u(x,t) , \tag{3.25}$$

where λ is an unknown constant. Inserting Eq. (2.12) into Eq. (2.7), we arrive at

$$\frac{\partial u}{\partial t} - v \frac{\partial u}{\partial x} \left[v_{\mathrm{up}} - \lambda (v_{\mathrm{up}} + v) \right] \frac{\partial \tilde{h}}{\partial x} . \tag{3.26}$$

Thus, if we choose:

$$\lambda = \frac{v_{\mathrm{up}}}{v + v_{\mathrm{up}}} , \tag{3.27}$$

we find that u is ruled by a simple wave equation, and we may set:

$$u(x,t) = u(x + vt) . \tag{3.28}$$

We can rewrite Eq. (3.25) in the form

$$R(x,t) = u(x + vt) - \lambda w(x - v_{\mathrm{up}}t) . \tag{3.29}$$

Eqs. (3.24) and (3.29) represent the formal solution in all regions where $R > 0$. This solution leads in fact to a great variety of avalanche regimes [66]. One of the major conclusions concerns the final state. In an "open cell" geometry, where the sand falls out from the bottom end, the final angle is the neutral angle θ_n. In a closed cell, the sand accumulates near the bottom, and the final angle is smaller than θ_n. It is predicted to be $\theta_f = 2\theta_n - \theta_{\mathrm{max}}$.

This conclusion for a closed cell is probably valid also for an avalanche terminating on a horizontal plane. This suggests that the classical notion of a unique angle of repose is oversimplified. There are indeed some experimental data on θ_f for closed cells which show a difference [67].

On the whole, the main merit of this description of surface flows is simplicity: it is the cheapest picture incorporating downhill and uphill waves. But we might need a more realistic relation between rolling velocity and erosion rate.

4. Mixing and demixing

With conventional fluids, shaking induces mixing. With granular systems, the opposite is usually true. This is a serious nuisance in many operations of chemical engineering, where the starting point is a mixture of powders with carefully proportioned ingredients.

4.1. Demixing in surface flows

A typical example is obtained with rotating cylinders (Oyama, 1939): a homogeneous mixture A + B, rotated slowly in a horizontal cylinder, segregates after a few minutes into slices of A and B, each slice being normal to the rotation axis. Explanations have been proposed [37,68]. The general idea is that, if the chemical nature, or the shape, of the particles A and B are different, their natural slope angle (e.g. the neutral angle θ_n) will be different. Assume for instance that $\theta_n(A) > \theta_n(B)$. This means that B rolls down more easily than A. Assume that a slice is rich in A, while the neighboring slice is rich in B. Then there will be a crest (on the high-level side) at the A rich slice, and a valley at the B rich slice. Particles will roll from the crest, and they will be mostly B, since B rolls better. This rolling in not along the crest line, but mainly towards the valleys, i.e. towards the adjacent B sheet: segregation increases.

This problem is complex, because of its three-dimensional nature. One can reach a slightly simpler geometry by setting up avalanche flows of mixtures in a two-dimensional Hele Shaw cell. This was studied first by Makse et al. [69,83]. They found that, with certain AB-pairs, successive strata of A and B can build up. Various theoretical schemes have been proposed [70–72]. Two features come into play:

(a) A difference in size, for grains which are otherwise identical (in shape and in composition).

(b) A difference in the neutral angle, as in the preceding discussion.

Depending on the relative importance of (a) and (b), many types of behaviors are expected.

Another spectacular example of demixing occurs with a sheet of grains flowing over a solid slope [73]. The front edge of this sheet is stable if the grains are monodisperse. But, if there is a distribution in sizes or shapes, segregation occurs and generates an instability at the front.

4.2. Segregation by shaking

The classical starting point here is the "Brazil nut effect". A truck carries a homogeneous mixture of large and small nuts. After some time of travel, the larger nuts concentrate at the top.

Many experiments have been performed with one larger grain in a sea of smaller grains; some of them are complicated by macroscopic convective flows which are induced by the shaking. Studies by Duran et al. [74], in Hele Shaw cells, with weak agitation, avoid the convective feature, and measure an upwards velocity as a function of the size ratio. A crude description of the effect is the following: the larger grain appears to act as the key stone of an arch, based on smaller grains on both sides. Below this arch is a low-density region, which tends to be filled later by smaller grains upon agitation. After filling, agitation rebuilds an arch, etc. At each cycle, some small grains have moved downwards, and the large grain has moved upwards.

5. Concluding remarks

The science of granular materials started with outstanding pioneers: Coulomb, Reynolds, Bagnold... . In recent years, it benefited from the impact of very novel techniques, e.g. nuclear imaging of grains at rest or in motion [75]. A strong stimulus also came from computer simulations which have not been adequately described in the present text, because of the author's inexperience. It is clear that virtual experiments with controlled, simplified interactions between grains can have a major impact. A review of the tools, and of certain difficulties, can be found in Duran [4]. Recent advances are described in the proceedings of the Cargèse Workshop [76].

However, in spite of these powerful means, and even for the simplest "dry" systems, the statistical physics of grains is still in its infancy. Some basic notions may emerge: (a) the sharp distinction between a fluid phase and a frozen phase, with the resulting possibility of defining a displacement field to describe the evolution of the frozen phase, (b) a displacement field containing a memory of all the sample history, (c) the possibility of describing surface flows with equations coupling the two phases, and reduced to a simplicity which is reminiscent of the Landau Ginsburg picture of phase transitions.

But we are still left with strong disputes, and large sectors of unraveled complexity.

Two fundamentally different pictures of the static behavior of heaps are facing each other: one represents the material as a deformable solid, the other assumes a completely singular state of matter, with stress fields transmitted along special directions, and with microscopic instabilities(earthquakes) occurring all the time (see for instance [77]).

We have to know more! Here are some examples:

(a) The problem already raised in Fig. 2: if we press *gently* at the free surface of a large, flat bed of sand, are the stresses below widely spread (as expected from a quasi elastic solid) or are they localized on a cone (as expected in "singular" models)? The word "gently" is important here: if we go to strong, local loads, we shall of course, generate shear bands.

(b) Acoustic propagation in a granular bed: it is mainly controlled by the (non linear) quasi elastic features plus mild effects of disorder. Or is it qualitatively different, because a sound wave, even at small amplitudes, starts some sort of earthquake?

(c) Decompaction: if we open the bottom of a vertical column, we see pieces of solid-like matter which separate from each other. Can we think of this as propagation of fractures in a quasi-elastic solid, or is it completely different?

(d) Similarly, when we perform a sequence of taps on a column, as mentioned in Section 2, should we visualise the grains during the tap as a solid with microcracks; or as a liquid (if the amplitudes are high enough).

We mentioned some current uncertainties for the solid phase. There are uncertainties of comparable magnitude for the fluid phases. Think, for instance, of fluidized beds: an intelligent literature (describing both transport and macroscopic instabilities) has been built up, but we are still looking for a unified vision. The link between mechanics, tribology, statistical physics, surface chemistry,... remains to be built.

Acknowledgements

My education on the physics of grains is due mainly to J.P. Bouchaud, J. Duran, P. Evesque, H. Herrmann and T. Witten. I am deeply thankful for their patient explanations. However, in many instances, the perspectives which are proposed here, do not coincide with their own views: they should not be held responsible for my (naive) attempts at unification.

I also greatly benefited from discussions with (and/or) messages from: R. Behringer, J. Biarez, D. Bideau, T. Boutreux, M. Cates, S. Coppersmith, S. Fauve, H. Jaeger, J. Kakalios, L. Limat, H. Makse, J. J. Moreau, S. Nagel, E. Raphael, J. Rajchenbach, P. Rognon, S. Roux.

Appendix: an example of boundary conditions at a free surface

To show the special character of the displacement field $\underset{\sim}{u}$ introduced in Section 2, we briefly discuss a simple example, where a bed of sand is fed by a uniform "mist" of particles, falling vertically and slowly (Fig. 5). The mass flux is J. We also assume, for simplicity, that the bed is close to a certain density ρ, and that near this density, we can use a fully elastic description, with a Young modulus E. The mass flux J imposes a certain growth velocity $v = J/\rho$.

At time t, each particle in the bed is at a height $z(t)$. It landed at an earlier time t_0 and at a certain height z. The displacement u is defined by

$$z(t) = z + u(z,t)\,. \tag{A.1}$$

At first sight, we would think that dilation is:

$$\theta = \frac{\partial z(t)}{\partial z} - 1 \qquad (?)\,. \tag{A.2}$$

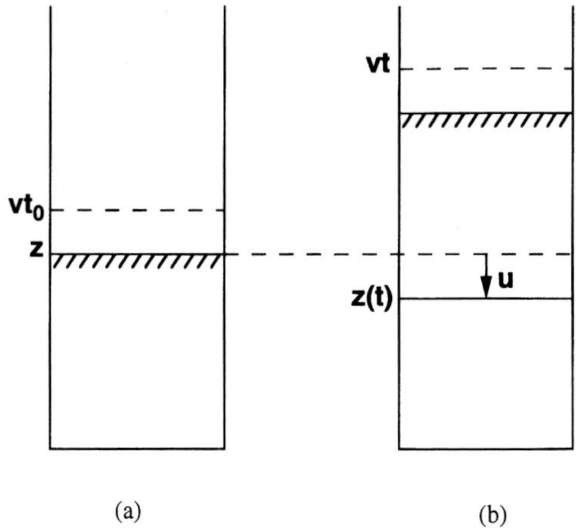

Fig. 5. Definition of a displacement for a sand bed deposited by a gentle vertical current $J=\rho v$ (ρ : density of bed near the surface). (a) For one particular grain, landing occurred at time t_0 and position z. This z is smaller than vt_0, because the column contracts under its own weight. (b) Later, the structure became more compact, and the grain reached a lower position $z(t)$ at time t. The displacement u is the difference $z(t)-z$.

However, Eq. (A.2) is wrong. The increment dz in Eq. (2) has to be examined more closely. When we vary $t_0 \to t_0 + dt_0$, we get

$$dz = dz_a + dz_c\,, \tag{A.3}$$

where dz_a describes accretion, $(dz_a = v dt_0)$ and dz_c describes compaction of the bed. A simple elastic calculation shows that

$$dz_c = -\frac{vt_0}{\lambda_c} d(vt_0)\,, \tag{A.4}$$

where $\lambda_c = E/\rho g$ is huge: all the calculation is carried out to first order in z/λ_c.

The correct form for the dilation is

$$\theta = \frac{\partial z(t)}{\partial z_a} - 1 = \frac{\partial u}{\partial z} - \frac{z}{\lambda_c}\,. \tag{A.5}$$

The weight equilibrium imposes a pressure p:

$$p = -E\theta = \rho g(vt - z) \tag{A.6}$$

giving:

$$u = -\frac{z}{\lambda_c}(vt - z)\,. \tag{A.7}$$

This shows, in particular, that the boundary conditions $u=0$ and $p=0$ at the surface, are compatible, because of the special form of Eq. (A.5).

References

[1] E.R. Bagnold, Chapman & Hall, London, Physics of Blown Sand and Sand Dunes, 1941.
[2] H. Jaeger, S. Nagel, R. Behringer, Rev. Mod. Phys. 68 (1996) 1259.

[3] R. Brown, J.C. Richards, Principles of Powder Mechanics, Pergamon, Oxford, 1970.
[4] J. Duran, Sables, Poudres et Grains, Eyrolles, Paris, 1997.
[5] J.D. Bernal, Proc. Roy. Soc. (London) A 280 (1964) 299.
[6] G.D. Scott, Nature 194 (1962) 956.
[7] J.L. Finney, Proc. Roy. Soc. A 319 (1970) 479.
[8] G. Onoda, E. Lininger, Phys. Rev. Lett. 64 (1990) 2727.
[9] E.T. Selig, R. Ladd, 1973, American Soc., for Testing of Materials, Publication no. 523, Baltimore.
[10] J.B. Knight, C. Fandrich, C. Lau, H. Jaeger, S. Nagel, Phys. Rev. E 51 (1995) 3957.
[11] E.R. Nowak, J.B. Knight, E. Ben Haim, H. Jaeger, S. Nagel, Phys. Rev. E, to be published.
[12] A. Coniglio, H. Hermann, Physica A 225 (1996) 1.
[13] M. Nicodemi, A. Coniglio, H. Herrmann, Phys. Rev. E 55 (1997) 3962.
[14] T. Boutreux, P.G. de Gennes, C.R. Acad. Sci. (Paris) 324 (1997) 85–89. T. Boutreux, P.G. de Gennes, Physica A 244 (1977) 59.
[15] G. Barker, A. Mehta, Phys. Rev. Lett. 67 (1991) 394.
[16] J. Biarez, M. Gourves, Powders and Grains, Balkena, Rotterdam, 1989.
[17] M. Cates, J.P. Wittmer, J.P. Bouchaud, P. Claudin, Jamming, J de Physique (1998).
[18] P. Bowden, D. Tabor, Friction: an introduction, Doubleday, New York, 1973.
[19] P.G. de Gennes, (1997) C.R. Acad. Sci. (Paris) 325 II 7–14.
[20] L. Vanel, E. Clément, J. Lanuza, J. Duran, in: H. Herrmann, M. Luding, N. Hovi (Eds.), Physics of Dry Granular Media, Kluwer, Dordrecht, 1998.
[21] A.N. Schofield, C.P. Wroth, Critical State of Soil Mechanics, McGraw-Hill, New York, 1968.
[22] H.A. Janssen, Z. Vereins, Deutsch Eng. 39(25) (1895) 1045.
[23] Rayleigh (Lord), Philos. Mag. 36 (1906) 11, 61, 129, 206.
[24] R. Nedermann, Statics and Kinematics of Granular Materials, Cambridge University Press, Cambridge, 1992.
[25] J. Duffy, R. Mindlin, J. App. Mech. (ASME) 24 (1957) 585.
[26] J. Goddard, Proc. Roy. Soc. (London) 430 (1990) 105.
[27] P.G. de Gennes, Europhys. Lett. 35 (1996) 145–149.
[28] P. Evesque, P.G. de Gennes, C.R. Acad. Sci. (Paris) (1998), submitted.
[29] S.F. Edwards, C.C. Mounfield, Physica A 226 (1996) 1, 12, 25.
[30] J.P. Bouchaud, M. Cates, P. Claudin, J. Physique II (France) (1995) 639.
[31] J.P. Wittmer, M. Cates, P. Claudin, J. Physique I (France) 7 (1997) 39–80.
[32] S.B. Savage, in: Behringer, Jenkins (Eds.), Powders and Grains, Balkena, Rotterdam, 1997, p.185.
[33] J. Goddard, in: H. Herrmann, J.P. Hovi, S. Luding (Eds.), Physics of granular media, Kluwer, Dordrecht, 1998.
[34] R. Brockbank, J.M. Huntlley, R.C. Ball, J. Phys. II (France) 7 (1997) 1521–1532.
[35] M. Oda, 1972, Soils and Foundations, vol. 2, p. 1.
[36] M. Oda, T. Sudoo, in: Biarez, Gourves (Eds.), Powders and Grains, Balkema, Rotterdam, 1989, pp. 156 and 161.
[37] J. Biarez, K. Wiendick, C. R. Acad. Sci. (Paris) 256 (1963) 1217.
[38] A.W. Jenike, Bull. 108, Utah Eng. Exp. Station, University of Utah, 1961.
[39] O. Reynolds, Philos. Mag. 5(50) (1885) 469.
[40] J. Desrues, in: Bideau, D., Dodds, J., (Eds.), Physics of Granular Media, Nova Sci. Pub., Les Houches Series, 1991, p. 127.
[41] H.J. Tillemans, H.J. Herrmann, Physica A 217 (1995) 261.
[42] C.H. Liu, S. Nagel, D. Shechter, S. Coppersmith, S. Majumdar, O. Narayan, T. Witten, Science 269 (1995) 513.
[43] J.J. Moreau, Eur. J. Mech. A Solids 13 (1994) 93.
[44] X. Zhuang, A. Didwanis, J. Goddard, J. Comp. Phys. 121 (1995) 331.
[45] S. Ouaguenouni, J.N. Roux, Europhys. Lett. 32 (1995) 449.
[46] F. Radjai, D. Wolf, S. Roux, M. Jean, J.J. Moreau, in: R. Behringer, J. Jenkins (Eds.), Powders and Grains, Balkema, Rotterdam, 1997, p. 211.
[47] D. Mueth, H. Jaeger, S. Nagel, Phys. Rev. E 57 (1998) 3164.
[48] G. Hagen, 1852, Berl. Monats. Akad. Wiss, p. 35.
[49] R. Nedermann, U. Tuzun, S. Savage, G. Houlsby, Chem. Eng. Sci. 37 (1982) 1597.
[50] E.R. Bagnold, Proc. Roy. Soc. A 255 (1954) 49.

[51] S.B. Savage, M. Sayed, J. Fluid. Mech. 142 (1984) 391.
[52] D. Hanes, D. Inman, Proc. Roy. Soc. 150 (1985) 357.
[53] A. Ahn, C. Brennen, R. Sabersky, in: M. Satake, J. Jenkins (Eds.), Micromechanics of Granular Materials, Elsevier, Amsterdam, 1988.
[54] S.B. Savage, J. Fluid. Mech. 92 (1974) 53.
[55] J.J. Drake, J. Fluid Mech. 121 (1991) 225.
[56] S. Mc Namara, W.R. Young, Phys. Rev. E 53 (1996) 5089.
[57] Y. Du, H. Li, L. Kadanoff, Phys. Rev. Lett. 74 (1995) 1268.
[58] M. Faraday, Phil. Trans. Roy. Soc. (London) 121 (1831) 299.
[59] B. Bernu, R. Mazichi, J. Phys. A 23 (1990) 5745.
[60] E. Clement, S. Luding, A. Blumen, J. Rajchenbach, J. Duran, Int. J. Mod. Phys. 1807 (1993) 9–10.
[61] F. Melo, P. Umbanhowar, H. Swinney, Phys. Rev. Lett. 75 (1995) 3838.
[62] P. Umbanhowar, F. Melo, H. Swinney, Nature 382 (1996) 793.
[63] F.A. Zenz, in: M. El Fayed, L. Otten (Eds.), Handbook of Powder Science and Technology, ch. 11, Chapman & Hall, London, 1997.
[64] P.A. Thompson, G.S. Grest, Phys. Rev. Lett. 67 (1991) 1751.
[65] J.P. Bouchaud, M. Cates, R. Prakash, S.F. Edwards, J. Phys. (France) 4 (1994) 1383.
[66] T. Boutreux, E. Raphaël, P.G. de Gennes, 1998 (to be published).
[67] P. Evesque, Phys. Rev. A 43 (1991) 2720.
[68] D. Levine, Looking inside granular materials, Physics World, vol. 30, 1997, p. 26–28.
[69] H. Makse, P. Cizeau, H.E. Stanley, Phys. Rev. Lett. 78 (1997) 3298.
[70] T. Boutreux, P.G. de Gennes, J. Phys. (France) 6 (1996) 1295.
[71] H. Makse, Phys. Rev. E 56 (1997) 7008.
[72] T. Boutreux 1998, Thèse, Université Paris 6, to be published.
[73] O. Pouliquen, J. Delour, S.B. Savage, Nature 386 (1997) 816.
[74] J. Duran, T. Mazozi, E. Clement, J. Rajchenbach, Phys. Rev. E 50 (1994) 5138.
[75] M. Nakagawa, S. Altobelli, A. Caprihan, E. Fukushima, E.K. Jeong, Exp. Fluids 16 (1993) 54.
[76] H. Herrmann, 1997, On the shape of a sand pile, Proc. the Cargèse Workshop, see ref [20].
[77] B. Miller, C. O'Hern, R.P. Behringer, Phys. Rev. Lett. 77 (1996) 3110.
[78] G. Barker, A. Mehta, Phys. Rev. A 45 (1992) 3435.
[79] G. Barker, A. Mehta, Phys. Rev. E 47 (1993) 184.
[80] M. Cates, J.P. Wittmer, J.P. Bouchaud, P. Claudin, 1998, Phil. Trans. Roy. Soc. (London), to be published.
[81] P.G. de Gennes, in: R. Behringer, J. Jenkins (Eds.), Powders and Grains, Balkema, Rotterdam, 1997, p. 3.
[82] C.H. Liu, Science 269 (1995) 513.
[83] H. Makse, S. Havlin, P. King, H.E. Stanley, Nature 386 (1997) 379.

Afterthought: Reflections on the mechanics of granular matter

A recent (careful) experiment by Guillaume Reydellet (thèse, Université Paris 6, 2002) has solved the basic question raised in Fig. 2: the experimental answer is of type (b), in agreement with the classical view from mechanics.

J. Phys.: Condens. Matter **12** (2000) A499–A505. Printed in the UK PII: S0953-8984(00)06910-1

Effect of topographic convergence on erosion processes

Pierre-Gilles de Gennes
Collège de France, 11 place M Berthelot, 75231 Paris Cedex 05, France

Received 12 August 1999

Abstract. We consider a basin made of an impermeable soil, eroded by water flow (not by weathering). Our approach is based on standard deterministic models, within a 'streamlet' picture where the flow is always directed downhill. The central assumption is that erosion occurs only when the surface shear stress τ (due to water flow) is above a certain threshold τ_c. Some features of the landscape are simplified by the existence of τ_c, and do not depend on detailed assumptions on the behaviour above τ_c. In particular, if the initial landscape is a U-shaped valley, we can construct the 'line of attack'—i.e. the border of the eroded regions—by a simple prescription.

1. Principles

The oral version of this paper covered both problems of dry sand and wet sand. For the written version, in view of the restrictions on size, we select only one topic related to wet systems: namely erosion processes.

A deterministic description of water basins implies a detailed knowledge of many erosion and sedimentation processes [1, 2]. Erosion may be due to weathering conditions (e.g. the impact of rain droplets) [3] or to tear-off by surface flow: here we focus on the latter case. Early mechanical models [4] postulated a local sediment flux F which would be uniquely determined by the local slope $|\nabla z| = \theta$ and the local water flux Q. This is somewhat oversimplified, since F is a sum from upstream contributions which occurred at different slopes and smaller fluxes. Another important feature, particularly emphasized in [5], is that erosion occurs only when the tangential stress due to flow,

$$\tau = \rho g h \theta \tag{1}$$

is above a certain threshold τ_c. In (1) ρ is the water density, g is the gravitational acceleration and h is the local water thickness.

Our aim in the present article is to emphasize two facts.

(1) The existence of a threshold stress is enough to fix the 'line of attack', i.e. the border of the eroded regions after the onset of rain on a given landscape. We show this in section 2, with two landscapes (an inclined plane and a tilted U-shaped valley) where topographic convergence is important. The concept of a line of attack is well known. It provides one answer to the classical question: '*where do channels begin*?' raised originally by Montgomery and Dietrich [6, 7].

(2) Consider now a mature basin, experiencing erosion, sedimentation (plus a slight uplift velocity U allowing for steady-state regimes). We show in section 3 that there is a good surprise: at small U, the flow velocity in the eroded regions remains always very close to the threshold value V_c (associated with τ_c). It is then possible to predict some universal

features of the profiles, which are not sensitive to the details of the erosion process beyond threshold.

Points (1) and (2) are simple illustrations of the ideas described in [5], but they do bring some simplification. Let us list here the main assumptions involved.

(a) The solid is impermeable: rain flows only at the surface, with a local thickness h, a local velocity $\boldsymbol{V}$ and the resulting local flux $\boldsymbol{Q} = \boldsymbol{V}h$.

(b) The material is not sensitive directly to droplet impacts, but is eroded only by surface flows, when $\tau > \tau_c$.

(c) The hydrodynamic flow is turbulent, and is locally in a steady state with the following relation between velocity and slope $|\nabla z| = \theta$:

$$V^2 = kgh\theta \tag{2}$$

where k is a numerical constant. Of course, (2) assumes that the slopes vary very smoothly, i.e. that curvature effects are negligible.

(d) The flow velocity points downhill. This neglects certain lateral exchanges between 'streamlets', which are weak (see the appendix). We do assume that there is indeed a well defined downhill direction at all points. Mathematically, this corresponds to $\theta \neq 0$ everywhere. Physically, this means that our landscapes do not have any lake-forming regions.

(e) The critical stress τ_c is dependent only on V, not on the slope. If we accept (1) and (2) the local stress is $\tau = k^{-1}\rho V^2$. Then the threshold stress τ_c corresponds to a fixed threshold velocity V_c:

$$\rho V_c^2 = k\tau_c. \tag{3}$$

For laboratory experiments, we would need rather small values of V_c (say 10 cm s^{-1}) and this implies very weak solids ($\tau_c \sim 10^{-4}$ atm $= 10^2$ dynes cm^{-2}).

2. Flows below threshold and the line of attack

The main feature of our systems, emphasized in [5], is the convergence of streamlets, as indicated in figure 1. Consider an element dl along an isolevel line. This collects the water from an upstream area dA and must have

$$Vh\,\mathrm{d}l = p\,\mathrm{d}A \tag{4}$$

where p is the rainfall per unit area and unit time.

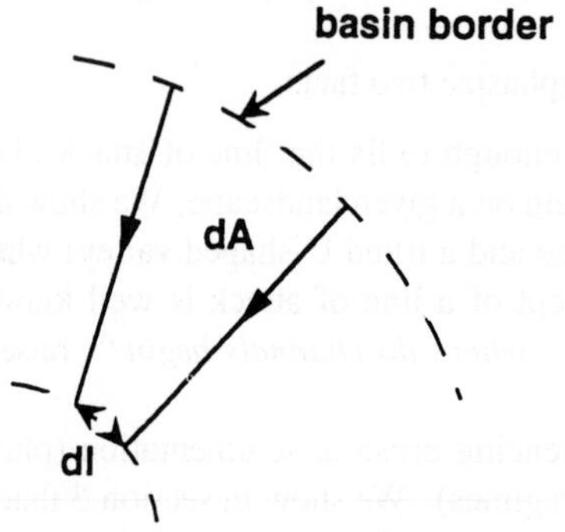

Figure 1. Definition of the collection factor dA/dl.

If we compare this to (2), we arrive at the basic relation

$$V^3 = kg\theta p \frac{\mathrm{d}A}{\mathrm{d}l}. \tag{5}$$

We call $\mathrm{d}A/\mathrm{d}l$ the collection factor. The role of the collection factor was emphasized long ago [6].

We shall now illustrate (5) with two examples.

2.1. *Inclined plane*

Here the slope is constant $\theta = \theta_0$ and the collection factor is simply equal to the distance x from the crest line. We may thus rewrite (5) in the form

$$\left(\frac{V}{V_c}\right)^3 = \frac{x}{L} \tag{6}$$

where L is a characteristic length

$$L = \frac{V_c^3}{kg\theta_0 p}. \tag{7}$$

The top part of the landscape ($0 < x < L$) is not eroded. The line of attack corresponds to $x = L$. The length L is inversely proportional to the rainfall p and to the slope θ_0. It is also proportional to $V_c^3 \sim (\tau_c)^{3/2}$.

For a laboratory experiment, with $V_c \sim 10$ cm s^{-1}, $\theta_0 \sim 1$, and $p \sim 10^{-3}$ cm s^{-1}, equation (7) leads to $L \sim 10$ m.

2.2. *U-shaped valley*

2.2.1. Geometry. We assume that the initial landscape is parabolic (figure 2(*a*)):

$$z = z_0 - \theta_0 x + \frac{y^2\theta_0}{2R}. \tag{8}$$

The lines of equal altitude correspond to

$$y^2 = 2R\left(x - \frac{z_0}{\theta_0}\right) \tag{9}$$

and the length R is their radius of curvature near the median line ($y = 0$). The local derivative $\mathrm{d}y/\mathrm{d}x$ along an isolevel line is given by

$$\frac{\mathrm{d}y}{\mathrm{d}x} = \frac{R}{y}. \tag{10}$$

The flow lines (normal to the isolevel lines) are ruled by

$$\frac{\mathrm{d}y}{\mathrm{d}x} = -\frac{y}{R} \tag{11}$$

$$y = a\exp\left(\frac{x_s - x}{R}\right) \tag{12}$$

where x_s corresponds to the starting point (at $y = a$).

The area A collected at level x inside a strip of small width y near the median is the hatched area of figure 2(*b*), and

$$A = a\left[x - R\ln\left(\frac{a}{y}\right) + R\right]. \tag{13}$$

Thus the collection factor (at small y) is

$$\frac{\mathrm{d}A}{\mathrm{d}l} \sim \frac{\mathrm{d}A}{\mathrm{d}y} = \frac{Ra}{y}. \tag{14}$$

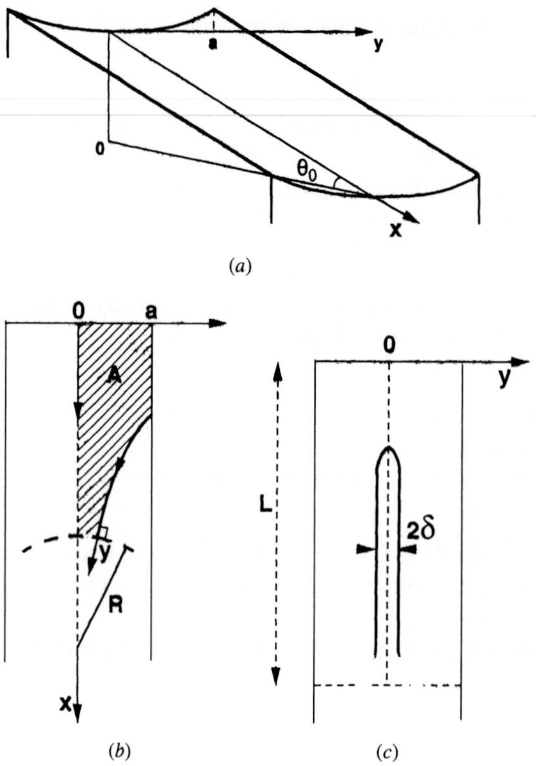

Figure 2. U-shaped valley: (*a*) the landscape, (*b*) isolevel lines (broken lines) and flow lines (arrows), and (*c*) aspect of the line of attack, separating uneroded regions (uphill) from eroded regions (downhill).

2.2.2. Line of attack. This is obtained by imposing $V = V_c$ in (5) and making use of (14). For small y, the slope is $\theta \cong \theta_0$. This leads to

$$y = \delta \equiv \frac{Ra}{L}. \tag{15}$$

Thus the line of attack is now *parallel to the median*. Equation (15) is the central result of this section. A few comments are useful here:

- the half width δ of the eroded region is proportional to the rainfall p and is a decreasing function of the threshold stress τ_c ($\delta \sim \tau_c^{-3/2}$) and
- the regimes of interest correspond to $y \ll a$ or $R \ll L$.

The distance x cannot be smaller than R: if $x < R$, the starting point is not on the lateral crest, but is at $x = 0$, and the collection factor is weaker. This ultimately leads to a shape for the line of attack which is qualitatively represented in figure 2(*c*).

3. Mature basins

Our aim now is to describe both erosion and sedimentation by an extension of the same ideas. Let us call $R(x, y, t)$ the local amount of moving solid (measured in terms of an equivalent height). To make things more palpable, we now write down a specific model for erosion and sedimentation:

$$\frac{\partial z}{\partial t} = -\frac{K}{V_s}(V^2 - V_e^2) + w\frac{R}{h} + U. \tag{16}$$

The first term describes erosion, and assumes that the rate is proportional to $\tau - \tau_c$. This linear law may be insufficient in practice, as pointed out in [5, ch 6]. It turns out, however that more general power laws lead to the same qualitative conclusions. Thus, for simplicity, we present our results with the linear model. The second term describes sedimentation, assumed to be proportional to the volume fraction of solid R/h in the flow. The last term is a very small uplift velocity. K is a material constant. It is dimensionless (and small), w has the dimensions of a velocity: it will in general depend on V, but for our purposes all these features are not essential.

Consider first the one-dimensional case where z depends only on the down slope coordinate x. The general aspect with uplift is shown in figure 3. There is a top region with no erosion, no sedimentation, and an uplift velocity U. This stops at a certain line of attack ($x = x_a(t)$). Below this point ($x > x_a$) we can have a steady-state solution with $\partial z/\partial t = 0$. We also have a book-keeping for the sediment, which in a steady state is produced at a rate U in all of the interval $x - x_a$

$$VR = U(x - x_a).$$

Comparing this with (4), we see that

$$\frac{R}{h} = \frac{x - x_a}{x}\frac{U}{p}$$

and (16), in a steady state, gives us

$$\frac{V^2 - V_c^2}{V_c} = K^{-1}U\left(1 + \frac{x - x_a}{x}\frac{W}{p}\right). \tag{17}$$

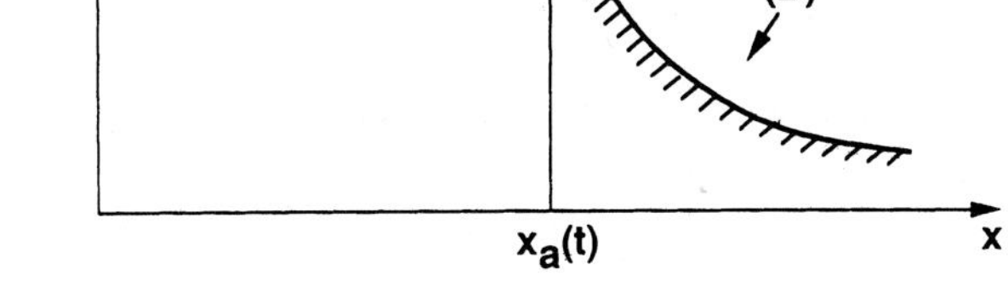

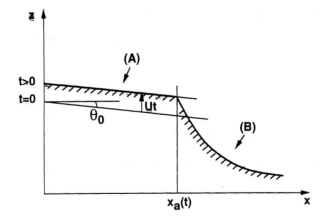

Figure 3. Predicted one-dimensional profiles for a mature basin with a slow uplift velocity U, (A) unperturbed region and (B) domain of erosion and sedimentation.

The crucial point is that U is geologically small: thus V is *necessarily close to* V_c. This remains true if we replace (16) by more realistic power laws.

This allows us to find immediately the steady-state profile: using (5) with $V = V_s$, we obtain

$$\theta \equiv \frac{-\mathrm{d}z}{\mathrm{d}x} = \theta_0\frac{L}{x}. \tag{18}$$

Thus the profile is expected to be logarithmic:

$$z(x) = z_1 - \theta_0 L \ln \frac{x}{L} \tag{19}$$

where z is a constant, which depends on the boundary conditions downstream. It is important to notice that the line of attack ($x = x_a$) moves slightly upward during time. On the line, we have

$$z = Ut + z_0 - \theta_0 x_a = z_1 - \theta_0 L \ln \frac{x_a}{L} \tag{20}$$

and this is an implicit equation for $x_a(t)$.

These considerations can be extended to more general basin shapes: x is replaced by $\mathrm{d}A/\mathrm{d}l$ and $x - x_a$ is replaced by $\mathrm{d}\tilde{A}/\mathrm{d}l$, where $\tilde{A}$ is the area collected between the line of attack and the level of observation.

4. Concluding remarks

(1) The main predictions from the model are: (a) the original position of the line of attack (equation (7) or (15)); (b) the steady-state profile, in the one-dimensional case, with a slope inversely proportional to the distance from the crest (equation (20)). The latter differs significantly from field observations (see [5, section 1.2.10]). This may mean that the weakly cohesive systems which we have in mind are very different from natural basins, but hopefully these laws could be compared to laboratory experiments. The solid material must be weak (low V_c). However, erosion should not be dominated by the direct impact of rain droplets. This is feasible is the droplet diameter is small; for $d = 1\ \mu$m, the fall velocity of the droplets is of order 1 cm s^{-1}, and this is much smaller than V_c ($\sim$10 cm s^{-1}). Thus erosion by flow may be the leading process.

(2) An obvious question is related to the stability of the flows that we have described. Starting from an inclined plane as in section 2, can we have valleys forming spontaneously? The answer can be obtained from our discussion of U-type valleys: ultimately, *no linear instability is expected* in the region $x < L$, because to start erosion we need a concentration factor $\mathrm{d}A/\mathrm{d}l$, which is significantly larger than x, and this would impose a valley system of finite amplitude.

(3) Another open question, with the U-shaped valleys, concerns the evolution of the 'gully' (of initial width 2δ) which appears near the median line. We postpone the discussion for a later study.

(4) The mechanism of erosion initiation is definitely not the unique process responsible for channel-like patterns. However, in some simple systems, the deterministic models (initiated by Smith and Bretherton [4], Rodriguez-Iturbe and Rinaldo [5] and others) do lead to scaling predictions which are universal—i.e. independent of the detailed laws for erosion and sedimentation. The only crucial feature is the existence of a velocity threshold.

Acknowledgment

This text has greatly benefited from comments and criticisms by A Rinaldo.

Appendix. Exchanges between streamlets

The limits of the streamlet approximation appear clearly if we think of a flow on an inclined plane, with constant flux $V_0 h_0$ and constant velocity. Assume that in a small region near the starting point ($z = z_0$) the height h is increased $h = h_0 + h_1(y)$. Then the corresponding 'streamlet' will spread out. Ignoring capillary effects (gravity regime) we expect a transverse flow velocity:

$$V_y \sim \frac{-V_0}{\theta_0}\frac{\partial h_1}{\partial y} \tag{A1}$$

leading to a diffusion equation

$$\frac{\partial h_1}{\partial t} = -V_0\frac{\partial h_1}{\partial x} + \frac{-V_0 h_0}{\theta_0}\frac{\partial^2 h_1}{\partial y^2}. \tag{A2}$$

If your streamlet was infinitely narrow at the start, after a distance x downhill, it then reaches a width y_d of order

$$y_d \sim \left(\frac{h_0 x}{\theta_0}\right)^{1/2}. \tag{A3}$$

All our discussion of the convergence in a U-valley (section 2) assumed that the corresponding y_d is smaller than δ. For the U-valley problem, the uphill length of a flow line is of order R (rather than x) and thus $y_d \sim (h_0 R/\theta_0)^{1/2}$.

References

[1] Eagleson P S 1970 *Dynamic Hydrology* (New York: McGraw-Hill)
[2] Scheidegger A 1970 *Theoretical Geomorphology* (Berlin: Springer)
[3] Howard A D 1994 *Water Resour. Res.* **30** 2107–17
[4] Smith T and Bretherton F 1972 *Water Resour. Res.* **8** 1506–29
[5] Rodriguez-Iturbe I and Rinaldo A 1997 *Fractal River Basins* (Cambridge, UK: Cambridge University Press) (and references therein)
[6] Montgomery D and Dietrich W 1988 *Nature* **336** 232–4
[7] Montgomery D and Dietrich W 1992 *Science* **255** 826–9